PRODUCTION
FORMULATION
MEDICATION

THE HANDBOOK OF
FEEDSTUFFS

PRODUCTION
FORMULATION
MEDICATION

AN ENCYCLOPEDIC PRESENTATION

of Economic Plants and other Feedstuffs
Plant Diseases and Insect Enemies
Herbicides, Insecticides, Drugs

with explanation of Agricultural, Botanical, Chemical,
Nutritional and other Scientific Terms

THE HANDBOOK OF
FEEDSTUFFS

PRODUCTION
FORMULATION
MEDICATION

By **RUDOLPH SEIDEN, Ch. E., D. Sc.**
> *Consultant on Veterinary Pharmaceuticals and Agricultural Chemicals*

In assocation with **W. H. Pfander, Ph. D.**
> *Professor of Animal Husbandry, University of Missouri*

Foreword by Grant Cannon
> *Managing Editor, The Farm Quarterly*

SPRINGER SCIENCE+BUSINESS MEDIA, LLC

Copyright, 1957
Springer Science+Business Media New York
Originally published by Springer Publishing Company, Inc. in 1957
Softcover reprint of the hardcover 1st edition 1957

ISBN 978-3-662-39293-5   ISBN 978-3-662-40326-6 (eBook)
DOI 10.1007/978-3-662-40326-6

Library of Congress Catalog Card Number: 57-9182

To the memory
of my father
BERNARD SEIDEN
(1880-1955)
who handled many of the innumerable
details of putting this book together.
I can never thank him for all he did,
for he died suddenly while at work
on the manuscript.

FOREWORD

I have been waiting a long time for this book—ever since Dr. Seiden mentioned his plan for it some four years ago—and many times I have wished that it were available. For I have felt a need for it and I am sure that it will be of very real value to farmers, feed suppliers, agricultural students and those who advise farmers and farm youth everywhere.

Now that I have seen this handbook, I am glad that Dr. Seiden and Dr. Pfander took time in gathering the material, for thoroughness and completeness is certainly, along with accuracy, the major value of this book. Checking the book against the most exhaustive indexes on animal feeding as well as on the production of forages and other feedstuffs has convinced me of its amazingly thorough coverage; comparison of its text with that of other authoritative works shows the accuracy of the work done by Dr. Seiden and his associate.

One difficulty that any writer has when he works on an agricultural story is trying to hold the material within bounds. How can you talk about feeding hogs, for example, without discussing sanitation and health and housing and temporary fencing and crop rotations and a host of other related subjects? If you let yourself go, you have a long and rambling book, and its value for the reader diminishes as it becomes more difficult to locate needed facts. I like the arrangement of this book because any specific fact can be found instantly, while the related material is called to your attention by cross-references and is there if you have the time to read it. The arrangement should be of unusual value to the busy farmer and feed manufacturer or dealer as well as to the serious agricultural student. For the agricultural agent, vocational agricultural leader, and other advisers who at any time are expected to come up with the complete and correct recommendation for every farm problem, this book is a god-send.

Having read Dr. Seiden's other books, having edited articles he has written for The Farm Quarterly and having followed this book from the time it was just an idea; as well as having known and written about the work Dr. Pfander has done in animal nutrition, it is a pleasure to see their joint effort in print. It is an honest book, carefully done, that fills a need felt by farmers and the rest of us in agriculture.

GRANT C. CANNON
Managing Editor, THE FARM QUARTERLY

PREFACE

This is the first encyclopedic compilation of facts and figures dealing with feedstuffs. Editor and publisher friends encouraged me to write this book, perhaps because two other books of mine, Livestock Health Encyclopedia and Poultry Handbook, have been well received and have brought me some repute as a lexicographer.

In general, I have limited the source material of this book to officially approved statements and recommendations by experts in the Federal Government and by scientists at Agricultural Experiment Stations, Agricultural Extension Services, and State Universities or Agricultural Colleges. Thousands of publications were reviewed and hundreds were selected for condensation to form the basic text of this book. Feedstuffs of all kinds are listed together with information on (1) their production on ranches and farms or in factories and (2) the utilization of the feedstuffs by the various species of farm animals, including poultry. Around this basic core are gathered the essential facts that relate to feed and feeding problems.

Professor Pfander, an outstanding animal nutritionist, has contributed the articles and the many tables dealing with rations. My own special field, for the past twenty years, has been veterinary pharmaceuticals.

As one important feature of this book, the reader will here find complete and much-needed information on medicated feeds—explanations, warnings, and practical advice not available in any other single publication. The use of drugs in feedstuffs and drinking water is increasing; at present, about 25 percent of feeds manufactured in the U.S. contain drugs. Their actions must be understood; they can be dangerous. Indications and doses for the different farm animals must be known to those who use drugs in feeds.

The nutritional aspects and the economics of feeding are considered —how to balance the diet for all species of animals, for all age groups, for all purposes; and how to combine ingredients for highest profit. In the livestock rations generally used, there are about 300 specific ingredients, most of them valuable for one purpose or another but some of them not worth the money they cost.

Descriptive articles, many of them illustrated, are devoted to practically all plants used in the U.S. as feedstuffs: grasses and cereals, legumes, root and leaf crops. The aim of this handbook is not only to be encyclopedic (it describes more pasture and range crops than does any other feed-book) but also to be clear in detail as well as overall.

Inconsistencies and mix-ups in nomenclature were cleared up and, with the approval of experts, some plants were newly classified and grouped, as for instance, the sorghums.

Typically, an article dealing with a forage plant will give its Latin (scientific) name and synonymous common names of the plant; its botanical characteristics; its preference for climates, types of soils, and fertilizers; rotation programs that have proven successful; utilization as pasture, silage, hay, etc., and as officially recognized feed product (meal, bran, grit, cake, and so on); the dangers; and the valuable varieties of the plant.

Each plant species is described in a separate article. The varieties are listed in their proper alphabetical spot, with reference to the species; genera and families will also be found. The same method of description and reference is used also for disease-causing micro-organisms, such as fungi and bacteria.

Other major subjects are dangers to animals, such as bloat; poisonous plants, plant diseases, insect pests, weeds (points that are often of greatest economical importance); the pesticides officially recommended for the control of plant diseases and insects, with data for correct application and with necessary warnings; mineral feedstuffs (again, more complete than in other feed-books); trace elements; growth stimulants, such as certain antibiotics; sulfas; arsenicals; vitamins; hormones; urea; amino acids; and other chemicals of importance to stockmen and poultrymen.

Also described are production of special feed products and animal and plant by-products of value as feedstuffs; feedstuff composition (carbohydrate, fat protein, fiber, etc.); feed control regulations; pasture and range management; and equipment and implements used for cultivation, harvesting, and storage of crops.

Depending on the importance of the subject, the articles vary in length from several pages to concise dictionary-type entries, many of the latter defining botanical, agricultural, nutritional, chemical, and other scientific terms.

Acknowledgment and thanks are due to the many from whose published works I used text or illustrations; generally, the texts were abstracted, often rearranged, revised, and combined with other information. Similarly, most of the illustrations were "edited" to fit purpose and format of this book. The reader will find abbreviated references at the end of most articles and complete bibliographic data of all sources at the end of the book.

Many experts contributed to the book by advising me, whenever I asked their help, on matters relating to their special fields. I wish to thank all of them—listed here alphabetically and without their official and academic rank:

Bernard A. App, Entomologist
Reed W. Bailey, Forester
Sidney F. Blake, Botanist
Victor R. Boswell, Horticulturist
Franklin A. Coffman, Agronomist
Reynold G. Dahms, Entomologist
William A. Dayton, Dendrologist
Angus A. Hanson, Agronomist
Mason A. Hein, Agronomist
Paul R. Henson, Agronomist
F. J. Hermann, Botanist

R. A. Hollis, Information Officer
Eugen A. Hollowell, Agronomist
Herbert W. Johnson, Agronomist
Kermith W. Kreitlow, Pathologist
Robert W. Leukel, Pathologist
Arthur W. Lindquist, Entomologist
John H. Martin, Agronomist
P. W. Oman, Entomologist
Fred W. Poos, Entomologist
T. Ray Stanton, Agronomist
Arlo M. Vance, Entomologist

ALL IN THE SERVICES OF THE U. S. DEPARTMENT OF AGRICULTURE;

R. B. Becker, Dairy Husbandman
Arthur A. Case, Professor of Veterinary Medicine
Larue C. Chapman, Professor of Agronomy
Charles E. Denman, Professor of Agronomy
Franklin P. Ferguson, Editor
Jack R. Harlan, Professor of Forage Crops
B. B. Higgins, Botanist
L. D. Kintner, Professor of Veterinary Pathology
Nicholas W. Kramer, Agronomist

J. J. Norris, Professor of Animal Husbandry
Stuart M. Pady, Professor of Botany and Plant Pathology
Reginald H. Painter, Professor of Entomology
John M. Poehlman, Professor of Field Crops
Henry H. Rampton, Agronomist
John B. Sieglinger, Professor of Agronomy (Sorghum Research)
Ben R. Spears, Agronomist
Erdman West, Botanist and Mycologist
C. W. Wingo, Professor of Entomology

ALL ASSOCIATED WITH AGRICULTURAL EXPERIMENT STATIONS AND/OR STATE COLLEGES;

Mary D. Alexander, Editor

Leslie E. Bopst, Executive Secretary, Association of American Feed Control Officials

Alice Gray, Scientific Assistant of the American Museum of Natural History

J. R. McNeill, Seed Breeder, Spur, Texas.

Thanks also to my wife, Juliette. She patiently typed and repeatedly retyped the pages of the manuscript and helped in its preparation. For her encouragement through the years, I am grateful beyond words.

Kansas City 10, Mo.
March 1957.

RUDOLPH SEIDEN

Explanations for the Reader

Several kinds of type are used to make the use of the book easier and more efficient.

The alphabetical key words are in **BOLD CAPITAL** letters. **Bold** print is used for such headings as Danger, Caution, Control, etc.

In the text, words in SMALL CAPITAL letters are the key words of other articles that should be read for additional information. The equal sign (=) shows that two key words are synonyms and that the subject is discussed under the term printed in small capital letters. Small capital letters are also used for major subheadings. "NOTE" indicates a comment or explanation by the author of this Handbook.

Words in *italic* letters indicate scientific or popular synonyms; italics are also used for some of the subheadings and in various ways for emphasis.

References to the literature are given in the form of simple keys (from A.1. to W.36.) at the end of many articles. The Bibliography, arranged according to these keys, is printed at the end of the Handbook, pages 581-590. The Authors of Publications referred to in the text will be found on pages 573-580.

Illustrations and Tables are placed in nearly all instances within the article to which they belong. Occasionally they are placed on a page facing the article or otherwise near it.

The following abbreviations are used in the text:

ad lib.—*ad libitum (at pleasure)*	I.U.—*International unit(s)*
a.m.—*ante meridiem (before noon)*	lb.—*pound(s)*
approx.—*approximately*	mcg.—*microgram(s)*
av.—*average*	mg.—*milligram(s)*
B.—*Bacillus*	mm.—*millimeter(s)*
bu.—*bushel(s)*	No.—*Number*
C.—*Centigrade*	oz.—*ounce(s)*
cc.—*cubic centimeter(s)*	p.m.—*post meridiem (afternoon)*
cm.—*centimeter*	ppm.—*parts per million*
cu.—*cubic*	pt.—*pint(s)*
cwt.—*hundred weight*	qt.—*quart(s)*
e.g.—*exempli gratia (for example)*	sing.—*singular*
etc.—*et cetera*	spp.—*species*
f.—*forma* (in Latin names of species)	sq.—*square*
F.—*Fahrenheit*	tablesp.—*tablespoonful(s)*
F.D.A.—*Food and Drug Administration*	teasp.—*teaspoonful(s)*
fl. oz.—*fluid ounce(s)*	U.S.—*United States*
ft.—*foot, feet*	U.S.P. unit(s)—*U.S. Pharmacopoeia unit(s)*
gal.—*gallon(s)*	
gm.—*gram(s)*	
gr.—*grain(s)*	var.—*variety* (in Latin names of species)
i.e.—*id est (that is)*	w/w—*weight by weight*
in.—*inch(es)*	yd.—*yard(s)*

Abbreviations used in "Authors of Publications" and "Bibliography" are explained on page 591.

A

0-9-27 and other three-numbered combinations refer to FERTILIZERS.

38-11 designates one of the inbred lines of CORN hybrids.

7078. *See* COMBINE-7078.

A is an outstanding inbred of CORN hybrids.

A.A.F.C.O. = ASSOCIATION OF AMERICAN FEED CONTROL OFFICIALS.

ABOMASUM is the *fourth* STOMACH of a ruminant.

ABOVE-GROUND SILOS. *See* SILOS.

ABRUZZES. *See* RYE (variety).

ABSORPTION is the holding (by capillary action) of a liquid in the pores of a solid; as water is held by the soil. In nutrition, however, this definition does not fit the events that take place in the gut. Here absorption is rather a process for transfer from the gastrointestinal tract to the blood; it is *active transport* instead of capillary action.—*See also* OSMOSIS.

Simple sugars, like *glucose*, can be absorbed directly from the STOMACH. Most of the other NUTRIENTS must be digested in the small INTESTINE before they can be absorbed into the blood stream through the mucous lining of the intestinal wall. In the small intestine, the CARBOHYDRATES are split into simple sugars, PROTEINS into *amino acids*, and FATS into *glycerin* and *fatty acids* and then absorbed. The latter, however, are sometimes neutralized to *soaps* before absorption. Fats are also emulsified and absorbed into the lymph system. Fat-soluble vitamins (A, D, E) are absorbed with fat and with B-COMPLEX vitamins and VITAMIN C from the small intestine. MINERALS are absorbed as such from the small intestine and WATER is absorbed unchanged all along the digestive tract.—*See also* DIGESTION.

ABUNDANCE. *See* COMMON OAT (variety).

ACADIAN. *See* SOYBEANS (variety).

ACCEPTED ANTIOXIDANTS.
See ANTIOXIDANTS.

ACCEPTED MOLD INHIBITORS.
See MOLD INHIBITORS.

ACERATAGALLIA. *A. sanguinolenta* = CLOVER LEAFHOPPER.

ACETIC ACID is a colorless liquid with a pungent odor; it forms when green forage changes into SILAGE and is one of the major acids formed in the rumen. In the body it is used to form CHOLESTEROL, steroid HORMONES (which are closely related to the STEROLS) and other substances.

Vinegar is weak (5 percent) acetic acid. —*See also* BACTERIA.

ACHENBACH is a southern type of SMOOTH BROME.

ACHENE (pronounced ay-*keen*), or *akene*, is a small, dry, one-seeded, and one-celled fruit; e.g., the fruit of SEDGES.

ACHILLEA. The *Achillea* spp. are known as YARROWS. *A. lanulosa* = *western yarrow*; *A. millefollium* = *common yarrow*.

ACID is a chemical compound usually characterized by a sour taste; it forms SALTS when neutralized (i.e., when united with ALKALIES). There are two types of acids—the INORGANIC ACIDS and the ORGANIC ACIDS.

Many substances, including soil, react *acid.*—*See also* pH; AMINO ACIDS; SILAGE CROPS.

ACID-BASE BALANCE is a term used to explain the necessity of balancing the intake of ACIDS and ALKALIES (or bases) contained in the food or feed. The animal system is capable of balancing most such differences, thus keeping the blood (very close to) neutral, i.e., at pH 7.35.

ACIDIFY means: to make ACID.

1

ACIDITY is the state of being ACID in reaction, which is evident to the taste as *sourness*. The degree of acidity of a solution or soil may be expressed as pH.

ACID PHOSPHATE = SUPERPHOSPHATE.

ACME = *Mediterranean wheat*.
See COMMON WHEAT.

ACORNS are the fruits (nuts) of OAK species. They are used as feed for swine and other livestock. When consumed in large amounts, acorns may cause POISONING.

Acorn calf is a name derived from the wrong assumption that this congenital deformity of a calf results from its dam's eating too many acorns during gestation. It is true, however, that eating too many acorns may prevent the utilization of some essential food element and thereby aid in producing the deformities. More acorn calves are found in the Oak Belt of the Californian Sierra Nevada foothills than in areas where no acorns are consumed.—*See also* POISONOUS PLANTS.

ACRE, a land measure, equals 43,560 sq. ft., 4,840 sq. yd., or 160 sq. rods. A square, the sides of which are 210 ft. long, represents approx. 1 acre.

ACTINEA. The genus *Actinea* belongs to the COMPOSITE family. Among its many species are those belonging to the RUBBERWEEDS. At least two of them are poisonous to livestock, namely the *bitter rubberweed* (*A. adorata*) and *Colorado rubberweed* (*A. richardsoni*), also known as *pingue*.

ACTIVE DRY YEAST. See YEAST.

ACUMEN is a sharp, tapering point; as, the acumen of a leaf.

ACUTE means: terminating abruptly and sharply (in an angle of less than 90°); in disease, it means: severe symptoms and short course.

ADAMS. See SOYBEAN (variety).

ADAPTABILITY is the ability to become adjusted to the environment, especially to climate and/or soil.

ADENOSTOMA. *A. fasciculatum*
= CHAMISE.

ADSORPTION is the ability of a substance to hold on its surface (due to adhesion) gases, liquids, or substances dissolved in a liquid.—*See also* BENTONITE; CHARCOAL.

ADVANCE. See RYE (variety).

AECIOSPORES. See RUSTS.

AERATION is exposure to the action of *air*, as in HAY making.

AERIAL ROOTS are those arising from the stem above the ground.

AEROBIC BACTERIA are micro-organisms which act in the presence of *oxygen* (air).—*See also* SILAGE.

AEROSOL is a suspension of minute liquid or solid particles in a gas which is kept under pressure; when released, it distributes the suspended particles.—*See also* INSECTICIDES; PARATHION.

AFRICAN ALFALFA belongs to the nonhardy ALFALFAS.

AFRICAN LOVEGRASS is a name applied to the several LOVEGRASSES which originated in Africa; erroneously, this name is sometimes used for WEEPING LOVEGRASS.

AFRICAN MILLET = *sourless sorgo;* it is one of the forage SORGHUMS.

AFRICAN RUNNER is a runner-type PEANUT.

AFRICAN SORGO. The White African sorgo belongs to the forage SORGHUMS.

AFTERMATH is the second (shorter) growth of meadow plants in the same season, after cutting a hay or seed crop. *Aftermath pastures* are supplemental PASTURES.

AGNES PEA. See FIELD PEA (variety).

AGRICULTURAL EXPERIMENT STATIONS. Unless otherwise indicated, the station should be addressed: Agricultural Experiment Station—at the post office listed.

ALABAMA: Auburn.

ALASKA: Palmer.

ARIZONA: Tucson.

ARKANSAS: Fayetteville.

CALIFORNIA: Berkeley 4.

COLORADO: Fort Collins.

CONNECTICUT: Connecticut Agricultural Experiment Station, New Haven 4.
Storrs Agricultural Experiment Station, Storrs.

DELAWARE: Newark.

FLORIDA: Gainesville.

GEORGIA: Georgia Agricultural Experiment Station, Experiment.
  Coastal Plain Experiment Station, Tifton.
HAWAII: Honolulu 14.
IDAHO: Moscow.
ILLINOIS: Urbana.
INDIANA: Lafayette.
IOWA: Ames.
KANSAS: Manhattan.
KENTUCKY: Lexington 29.
LOUISIANA: University Station, Baton Rouge 3.
MAINE: Orono.
MARYLAND: College Park.
MASSACHUSETTS: Amherst.
MICHIGAN: East Lansing.
MINNESOTA: University Farm, St. Paul 1.
MISSISSIPPI: State College.
MISSOURI: Columbia.
MONTANA: Bozeman.
NEBRASKA: Lincoln 1.
NEVADA: Reno.
NEW HAMPSHIRE: Durham.
NEW JERSEY: New Brunswick.
NEW MEXICO: State College.
NEW YORK: State Agricultural Experiment Station, Geneva.
  Agricultural Experiment Station at Cornell University, Ithaca.
NORTH CAROLINA: State College Station, Raleigh.
NORTH DAKOTA: State College Station, Fargo.
OHIO: Wooster.
OKLAHOMA: Stillwater.
OREGON: Corvallis.
PENNSYLVANIA: State College.
PUERTO RICO: Rio Piedras.
RHODE ISLAND: Kingston.
SOUTH CAROLINA: Clemson.
SOUTH DAKOTA: Brookings.
TENNESSEE: Knoxville 16.
TEXAS: College Station.
UTAH: Logan.
VERMONT: Burlington.
VIRGINIA: Virginia Agricultural Experiment Station, Blacksburg.
  Truck Experiment Station, Norfolk.
WASHINGTON: Pullman.
WEST VIRGINIA: Morgantown.
WISCONSIN: Madison 6.
WYOMING: Laramie.

AGRICULTURAL VARIETY
  = *agronomic variety. See* VARIETY.
AGRONOMIC VARIETY. *See* VARIETY.
AGRONOMY is the science of crop production and soil management.
AGROPYRON. The *Agropyron* spp., or WHEATGRASSES, include the following grasses: *A. intermedium* = INTERMEDIATE WHEATGRASS; *A. cristatum* = FAIRWAY CRESTED WHEATGRASS; *A. desertorum* (mixed with *A. cristatum*) = STANDARD CRESTED WHEATGRASS; *A. smithii* = WESTERN WHEATGRASS; *A. trachycaulum* = SLENDER WHEATGRASS; *A. spicatum* = BLUEBUNCH WHEATGRASS; *A. repens* = QUACKGRASS; *A. elongatum* = TALL WHEATGRASS; and *A. trichophorum* = PUBESCENT WHEATGRASS.—*See also* REE WHEATGRASS.
AGROSTEMMA. *A. githago*
                                     = CORNCOCKLE.
AGROSTIS. To the genus *Agrostis* belong the BENTGRASSES (e.g., *A. tenuis* = COLONIAL BENT) and *A. alba* = REDTOP.
AGROTIS. *A. orthogonia*
                        = PALE WESTERN CUTWORM.
AHMEDNAGER No. 1.
                      See SAFFLOWER (variety).
AIR is a gas mixture, the *atmosphere*, which consists chiefly of nitrogen (78 percent by volume) and oxygen (not quite 21 percent) and traces of many other gases. No animal can live without oxygen.
  Poorly *ventilated* livestock housing causes discomfort to animals and may endanger their health.—*See also* AERATION.
AJAX OAT. *See* COMMON OAT (variety).
AJAX SORGHUM is one of the intermediate-type grain SORGHUMS.—*See also* IMPERIAL KAFIR.
AKENE = ACHENE.
ALABAMA RUNNER is a runner-type PEANUT.
ALABAMA VELVETBEAN.
          See DEERING VELVETBEAN (variety).
ALAND. *See* RYE (variety).
ALANINE is one of the dispensable AMINO ACIDS.
ALASKA PEA. *See* FIELD PEA (variety).
ALASKA WHEAT.
              See POULARD WHEAT (variety).
ALBERTA RED = *Turkey.*
              See COMMON WHEAT.
ALBIT. *See* CLUB WHEAT (variety).

ALBUMINS are PROTEINS which are soluble in water and dilute salt solutions. They occur in eggs, body fluids, and in the PROTOPLASM of plants.—*See also* MILK PRODUCTS.

ALCOHOL is a term usually reserved for *ethyl alcohol*. (There exist many other alcohols.) It is a liquid obtained as a result of the fermentation of SUGARS by yeast or other micro-organisms. The sugars occur in plant juices or can be produced by the breaking down of starch with the help of ENZYMES. Alcohols are easily oxidized to ORGANIC ACIDS.—*See also* SILAGE; STEROLS.

ALDRIN is one of the CHLORINATED HYDROCARBON insecticides. It forms crystals which are stable and unsoluble in water, but very soluble in many organic solvents; it is compatible with most fertilizers, HERBICIDES, FUNGICIDES, and INSECTICIDES. Aldrin is available in form of emulsifiable concentrates, oil solutions, wettable powders, and dusts. It has a high insecticidal toxicity, but its residual action is short.

NOTE: The chemical aldrin of 85 percent strength contains 77.9 percent of the pure compound.

A 5-percent dust of aldrin applied at the rate of 40 lb. per acre controls the SOUTHERN CORN ROOTWORM, WIREWORMS, and many other insect pests.

Caution: Aldrin is readily absorbed through the skin. It is toxic, causing symptoms of poisoning in one to four hours and death within twenty-four hours. Avoid skin contact, inhalation, and food (feed) contamination.

ALEURONE is a PROTEIN in granular form which occurs in the outer layer of the ENDOSPERM of a ripe seed or grain.

ALFALFA. *See* ALFALFAS.

ALFALFA CATERPILLAR (*Colias philodice eurytheme*). If ALFALFA is grown for hay in the Southwest, one can expect the alfalfa caterpillar to infest the crop. Occasionally widespread outbreaks cause serious losses. The insect may also be present in alfalfa grown for seed or pasture and may occur in other parts of the country.

The alfalfa caterpillar is the larva of a small butterfly, yellow with a black border on its wings. Yellow spots in this border distinguish the female from the male.

These butterflies appear in the spring; as early as February in southern Arizona, in April in the San Joaquin Valley of California.

The females lay their eggs singly on the upper surface of the leaves. These eggs are white at first, but become reddish

Alfalfa caterpillar and its butterfly. (U.S.D.A.)

brown before they hatch. A female may lay as many as 500 eggs in her life of about two weeks.

After three to seven days the eggs hatch into small caterpillars which feed on the leaves and soon take on a green color. It is difficult to see them. Their feeding at first gives the leaves a shot-hole appearance, but eventually the caterpillars may strip the leaves and sometimes even consume the stems.

When full-grown, twelve to fifteen days after hatching, the caterpillars attach themselves to alfalfa stalks with the head-end up and change to pale-green pupae suspended by slender silken threads. After one to three months they emerge as adult butterflies and start a new generation.

The time required for the life cycle

varies with the temperature. In southern Arizona it ranges from about twenty-six days in the summer to sixty-five days in the early spring, and there are six to seven generations a year. Farther north the insect develops more slowly and fewer generations are completed annually.

Control: There are several natural enemies of the alfalfa caterpillar.

A VIRUS DISEASE attacks both larvae and pupae, and is present probably wherever alfalfa is grown in this country.

Insect parasites are important also. One of the most effective is a small WASP known as *Apanteles medicaginis*. Its bright-yellow or white cocoons can often be seen on the leaves of alfalfa. The adult wasps that emerge from these cocoons lay eggs in the young alfalfa caterpillar, and the larvae hatching from these eggs feed inside the caterpillar and destroy it. Another parasite is a FLY called *Phorocera claripennis*, which lays its eggs in the half-grown alfalfa caterpillar and develops similarly to the wasp. Other parasites lay their eggs in the pupa or egg of the caterpillar.

Several insects kill alfalfa caterpillars by preying on them. The CORN EARWORM and LADY BEETLE are helpful in this respect. Caterpillars containing parasites can be distinguished by their distorted shape.

The time when alfalfa is cut has a marked effect on the life of the stand and on the yield and feeding value of the hay. However, if large numbers of full-grown caterpillars can be seen, the alfalfa should be cut before it becomes severely damaged. By cutting alfalfa close to the ground and clean, food and shelter for the caterpillars and butterflies will be removed. If possible, the closely mowed field should be flooded after removal of the hay. In the Southwest this flooding together with the heat of the midday sun will kill most of the caterpillars that are left on the bare soil or short stubble.

METHOXYCHLOR is the only insecticide now recommended for use on alfalfa grown for hay.

Reference: U.2.

ALFALFA DODDER. See SMALL-SEEDED ALFALFA DODDER; LARGE-SEEDED ALFALFA DODDER; DODDERS.

ALFALFA DWARF = DWARF DISEASE.

ALFALFA FACTOR = *grass factor*.

See U.G.F.

ALFALFA HAY. See ALFALFAS; HAY GRADING; HAY MEASURING.

ALFALFA-HAY STANDARDS are divided into grade and class. The *grade* refers to the quality of the HAY; the *class* to the kind or mixture of the ALFALFA plants.

Three factors are used to determine grade: leafiness, green color, and foreign material. *Leafiness* is considered to be most important, because it reflects the protein content of the hay. The percentage of *green color* in a general way indicates the CAROTENE (or vitamin A) content. *Foreign material* becomes important only when alfalfa contains weeds or other foreign material in excess of the quantities permitted for several grades.

Mixtures of alfalfa with other hay crops are listed in the standards by classes in which the class name describes the kind of mixture. The class name "Alfalfa" describes hay that is practically pure alfalfa; "Alfalfa Light Grass Mixed" describes a mixture which contains 5 to 20 percent of GRASS; "Alfalfa Heavy Grass Mixed" describes hay with 21 to 60 percent of grass.

Those who purchase alfalfa hay have found certain specific grades satisfactory for certain types of livestock: U.S. No. 1 Leafy Alfalfa for poultry, swine and rabbits; U.S. No. 1 or U.S. No. 2 Leafy Alfalfa for dairy cattle; U.S. No. 2 Alfalfa for horses; U.S. No. 1, U.S. No. 2 Leafy, or U.S. No. 3 Leafy Alfalfa or Alfalfa Light Grass Mixed for sheep; and low-grade hay, such as U.S. No. 3 and U.S. No. 3 Green Alfalfa for stocker and feeder cattle.

Reference: W.19.

ALFALFA HOPPER.

See THREE-CORNERED ALFALFA HOPPER.

ALFALFA LEAF MEAL. See ALFALFAS.

ALFALFA LOOPER (*Autographa californica*) is an insect pest which occurs particularly in the West Coast region. The color of the larvae varies from cream to dark green. Fully grown larvae are about 1 in. long and crawl in a looping fashion. They feed on various weeds, vegetables, and crop plants, including ALFALFA, but rarely become numerous enough to cause serious injury.

Control: Infestations normally develop in the first crop of alfalfa a short time before the date of cutting. The most practicable means of control is to cut the first crop as soon as damage starts to become severe. Cure and remove the hay from the field as soon as possible.

Reference: M.27.

ALFALFA MEAL. *See* ALFALFAS.

ALFALFA-MOLASSES FEEDS are combinations of ALFALFA meal with usually from 20 to 40 percent molasses. They are fed to cattle and sheep, but must be considered as carbonaceous roughage rather than protein supplement.

ALFALFA PASTURE. *See* ALFALFAS.

ALFALFA PELLETS are made from high-grade ALFALFA *hay* after dehydration by processing in a pelleting machine. They are cleaner and easier to feed in adverse weather since they will not blow in the wind and will not ball up or become moldy in rain or snow. Since these pellets are alfalfa, they are usually fed as a roughage, but some producers mix them with grain and feed them as a concentrate. For dairy cows fed on grass hay, the addition of ½ lb. alfalfa pellets per 100 lb. live weight of cow has given good returns in some tests.

Pellet alfalfa is more digestible than the finely ground dehydrated alfalfa. It has the further advantage over alfalfa hay of possessing a higher CAROTENE content. Baled alfalfa when received at the farm may contain less than one-half of the carotene values of the pellets, and the carotene values of the baled hay decline at greater rate.

Reference: B.11.

ALFALFA PRODUCTS. *See* ALFALFAS.

ALFALFA RUST (*Uromyces striatus*) is a minor disease of ALFALFA. In years when there has been moderate to heavy rainfall during the summer and early fall, alfalfa sometimes is infected heavily during the latter part of the growing season. The rust generally does the greatest damage to the crop that is being grown for seed. Occasionally, newly seeded alfalfa is infected.

Alfalfa rust which is closely related to LEAF RUST, is characterized by cinnamon-brown pustules on the lower surface of the leaves. The color is due to the mass of powdery spores which has broken open to the surface of the leaves.

Control: To date there has been no practical control of this disease.

Reference: G.7.

ALFALFAS are *Medicago* spp.; they are divided into five groups. Of these, three groups belong to the species *M. satina*, namely *common alfalfas, Turkestan alfalfas,* and *nonhardy alfalfas:* they will be discussed in this section. The other two groups are the VARIEGATED ALFALFAS of the species *M. media* and the commercially unimportant YELLOW-FLOWERED ALFALFAS of the species *M. falcata*. All alfalfas belong to the LEGUMES.

Alfalfa is one of the most palatable and nutritious crops grown for forage in the United States. The hay is rich in proteins, minerals, and vitamins; better grades are low in fiber.

The alfalfa plant is a herbaceous perennial; its flowers are borne in loose bunches, or RACEMES, and are mostly of a purplish color. The pods in which the seed is produced are twisted spirally in one or two turns; each pod contains several small kidney-shaped seeds.

The stems, which are usually not more than ⅛ in. in diameter, are erect and commonly reach their maximum height of several feet at the time of blossoming. The semiwoody base of the plant, known as the crown, gives rise to new stems every four to six weeks throughout the active growing season. This makes it possible to take several cuttings of hay during the year. The root system is characterized by a distinct taproot, which may extend 30 ft. or more in depth. The taproot has few to many branch roots. The leaves are in threes, and are arranged alternately on the stem.

The most important alfalfa-producing states are Wisconsin, Minnesota, Nebraska, Iowa, Michigan, Kansas, and California.

Average annual acre yield of alfalfa hay in the United States is about 2¼ tons. In general, the highest average production per acre is in those states where most of the crop is grown under irrigation; California has averaged more than 4 tons per acre annually for several years, with individual farms in the southernmost part of the state producing more than 8 tons per acre.

Purple alfalfa, one of the common alfalfas.
(U.S.D.A.)

Alfalfa possesses a remarkable adaptability to a wide range of climatic and soil conditions. Although the crop requires considerable moisture to produce profitable yields of hay and pasturage, it does best in a relatively dry atmosphere where irrigation water is supplied.

In the United States alfalfa succeeds at altitudes ranging from below sea level to 8,000 ft. It withstands hot weather well, but is seriously affected by the cold weather of winter and early spring.

Deep loams with open, porous subsoils are undoubtedly best for alfalfa, but where other conditions are favorable the plant adapts to many kinds of soils. Very sandy or very compact soils should be avoided. Because of its root system, alfalfa does not thrive on a soil that has impervious (not penetrable) subsoil, hardpan, or bedrock near the surface.

On strongly saline and alkali soils, alfalfa makes little or no growth.

Good surface drainage and underdrainage are necessary if alfalfa is to thrive. During the growing season complete submergence for one or two days may do considerable injury, but when the plants are dormant they may remain under water several days with no damage. Where good drainage does not exist naturally, it must be supplied by artificial means before alfalfa can be expected to succeed. Tile drains placed 3 ft. below the surface will ordinarily lower the water table sufficiently to insure the satisfactory growth of alfalfa.

Alfalfa may be grown successfully after almost any crop if the soil is properly prepared following removal of the preceding crop. It is best to precede alfalfa for one to two years with some cultivated crop, such as CORN or POTATOES in the North, and corn, TOBACCO, or COTTON in the South.

Except in the extreme North, SMALL-GRAIN stubble may usually be worked up in time for late summer sowing, provided the land has previously been treated to destroy weeds. Crops such as COWPEAS and SOYBEANS that smother weeds may be used advantageously in some areas to precede late-summer- or fall-sown alfalfa. In some areas, particularly in the Southeast, one of the recommended practices is to plant a legume cover crop, such as LESPEDEZA (with small grains) or soybeans, in the spring and disk it in thoroughly four to six weeks before planting time.

The tender nature of young alfalfa plants requires that the soil be in excellent tilth at sowing time.

Alfalfa stands out among legumes as a heavy consumer of lime. The soils of the Pacific slope in the northwestern United States have a high lime requirement. All sandy soils, wherever they are located, should be checked for needs of lime. Ordinarily not less than 2 tons of ground limestone per acre are needed, and in many cases much larger quantities. One application usually lasts several years.

On most soils east of the 95th meridian, alfalfa responds to phosphate and potash. Good barnyard manure furnishes not only organic matter but also various plant-food elements required by alfalfa. If manure is not available, the organic matter may be supplied by plowing under some green manure crop. In the North such crops as RYE, VETCHES, and CANADA FIELD PEAS may be used; farther south BUR-CLOVERS, CRIMSON CLOVER, soybeans, and lespedeza.

Alfalfa has shown excellent response to commercial fertilizers, particularly in the eastern states. As much as 700 to 1,000 lb. 2-12-12 fertilizer per acre is recommended at seeding, with possibly smaller quantities in the midwestern and northern regions. Thereafter, the fields should be topdressed annually with phosphate and potash; in many cases an annual application of 400 to 600 lb. 0-14-14, 0-10-20, 0-9-27, per acre is required. Light applications of 20 to 35 lb. borax per acre at time of seeding have been beneficial in many areas in the eastern states.

West of the 97th meridian, many of the soils do not require any special fertilizer treatment for alfalfa. On some of the lands of the Pacific Northwest, particularly in Oregon, applications of 50 to 100 lb. flowers of sulfur per acre have given greatly increased yields. On the same soils gypsum has increased the yields of alfalfa.

Most of the soils in the Great Plains and western states, with the exception of those in the Pacific Northwest coastal regions, are naturally supplied with proper bacteria for inoculating alfalfa. In the eastern part of the country, however, it is nearly always advisable to inoculate at the time of the sowing. Fields that within the past few years have successfully grown alfalfa, SWEETCLOVER, bur-clover, or BLACK MEDIC will not ordinarily need further INOCULATION for alfalfa.

To inoculate, use one of the artificial cultures put out by commercial firms, or scatter soil from a successful alfalfa field or from the roots of sweetclover or bur-clover plants. The bacteria will live in soil for many years, provided the soil is kept well supplied with organic matter and lime.

Alfalfa flowers must be "tripped" by pollinating insects to set seed. Pollinating insects include the solitary bees, bumblebees, honeybees, and others.

In the northern half of the United States seed is often sown early in spring with a nurse crop. When sown alone, late-spring or early-summer sowing is generally best. In the Southeast the most favorable time varies from the middle of August in the latitude of Washington, D.C., to late October or early November along the Gulf Coast. February and March sowings are sometimes successful in the extreme South. In the northern part of the dry-farming area of the Great Plains it is necessary to sow seed as early in the spring as the land can be put into shape. In the southern part of this area, good stands are often obtained from late-summer and early-fall sowings. In the irrigated districts of the Southwest, October is the best month for sowing alfalfa, although good stands are obtained by sowing almost any time between October 1 and April 15.

If the crop is to be used primarily as *pasture*, alfalfa is usually sown in mixture with some adapted grass. North of the Kansas-Oklahoma line BROME GRASS is most generally used for this purpose, although some CRESTED WHEAT GRASS and SLENDER WHEATGRASS are used. In the Northwest, where alfalfa is grown under irrigation and at high altitudes, TIMOTHY or brome grass, or both, are often sown with alfalfa, thus prolonging the usefulness of the meadow; CLOVER—e.g., RED CLOVER or LADINO CLOVER—is sometimes added to the mixture. In the southern Great Plains and the Southeast alfalfa is usually sown alone. In the irrigated valleys of the Southwest, BARLEY is frequently sown in the fall in established stands of alfalfa, such fields being grazed during the winter and early spring. In thin stands of alfalfa, SUDANGRASS is sometimes sown in the spring or early summer to increase the yields. In the Corn Belt it is a rather common practice to sow timothy or brome grass with alfalfa. In Ohio, a mixture of timothy, clover, and alfalfa is preferable to alfalfa alone. In the Northeast a similar mixture in which adapted ORCHARDGRASS replaces timothy, is being used to a considerable extent.

If a nurse crop—e.g., flax, WINTER WHEAT, rye, or canning peas—can be used without decreasing the stand of alfalfa, it is desirable, as it gives some return from the land while alfalfa is becoming established. The chief objections to a nurse crop are: it draws rather heavily on the moisture supply of the soil; and being harvested at a hot time of the year, the sudden change from the shade afforded by the nurse crop is likely to injure the alfalfa seedlings.

Even where weeds are rank, it is better to delay clipping until the young alfalfa plants are in bloom.

Alfalfa should not be pastured until after the first year and should never be pastured closely, as the grazing down of the crowns often kills the plants. Rotation grazing is recommended to maintain stands and for high production. Cattle should never be allowed on a field when the ground is wet or frozen.

Most of the alfalfa grown in this country is cut for *hay*. In a few areas relatively large acreages are harvested for dehydration and production of alfalfa meal. The number of cuttings obtained annually varies from two, and sometimes only one in the North and semiarid sections, to eight or more in the irrigated areas in the Southwest. From thirty to forty days are required to produce a HAY crop.

Highest yields usually are obtained when growth is permitted to reach the full-bloom stage before cutting, but this lowers the nutritive value of the forage. Hay of the highest feeding value results from early cutting, when the number of the leaves is greater; they contain twice as much protein as the stems. CAROTENE is also at a high level in forage that is cut before it reaches the blossoming stage. In general, stands survive in best condition, and acre yields of protein tend to be highest when harvesting is delayed until the plants are at stages between one-fourth and one-half of full bloom. This early removal of top growth, however, limits the amount of food re-stored in the roots. Repeated and continuous cutting at early or premature stages gradually lowers the reserves during the season and from year to year. Coincident with this, there is a weakening and loss of plants resulting in poorer hay yields.

It is important that fall growth is sufficient to permit the manufacture and storage of large quantities of food in the crown and roots so that winter losses are kept small. This may be accomplished by cutting the last hay of the season thirty days or more before top growth of the plants is killed by the frost. In regions where the winters are characterized by intermittent freeze-free periods of two to three weeks duration, alfalfa may start growth several times during the winter. In order that plants thus weakened may recuperate it is especially important that harvest of the first cutting be delayed until between the ¼- and ½-bloom stage.

Although presence of flowers on alfalfa is commonly used to determine when the crop is ready to be cut, this criterion sometimes has questionable value, particularly in the East where alfalfa frequently blooms sparingly. In such cases a suitable guide to determine time for harvest is the color of the field; when it begins to get yellowish, the alfalfa should be cut. Young shoots developing from the crown indicate also that the alfalfa should be cut.

ALFALFA HAY, to be classified as high grade, must have high purity, a high percentage of clinging leaves, and a green color. Of the various vitamins found in alfalfa hay, carotene (VITAMIN-A value) VITAMIN C, VITAMIN D, and VITAMIN B₂ (riboflavin) are the ones most important. Vitamin A is the one the livestock feeder associates with quality.

In making hay, the shorter the time consumed in the curing process, the better the chances are for getting high quality; it is necessary to handle hay when partially dry or slightly tough. Alfalfa hay with the leaves dry and stems slightly tough contains about 25 percent moisture and can be stacked or stored loose in the barn safely. Hay can be baled from the windrow with a similar amount of moisture if the bales are piled to allow for ventilation. New-mown hay contains from 70 to 80 percent moisture, depending on atmospheric conditions when cut. Well air-dried hay contains about 12 percent moisture.

In western areas or where there is little rain, *overdrying* (or *"leaf shatter"*) is the most common cause of loss of quality; in humid areas, *rain damage*—leaching of nutrients, mold, and physical loss—is more important. For this reason, large amounts of alfalfa from the first cutting are made into SILAGE. The best method of curing is to allow the hay to wilt in the swath and finish curing in the windrow.

Barn-cured alfalfa hay is coming into prominence particularly where field curing is difficult because of weather conditions. Curing in the barn is accomplished with the hay drier. Hay that is to be barn-cured usually is cut in the morning, allowed to wilt in the swath from four to five hours, and brought to the barn to finish curing. Hay handled in this manner will be reduced in the field to approximately 45 percent moisture, and dried by the drier in the barn to a satisfactory storage content of 20 percent. The main advantage of a hay drier is that it saves leaves and color, the chief sources of carotene and protein.

Alfalfa silage, properly made, is a satisfactory method of harvesting alfalfa to conserve the feeding value of the plant. Dehydration, however, is best, and good hay will save as many nutrients as silage.

Alfalfa straw consists of the stems and leaves after the crop is threshed and the seeds are collected. Most of this straw is fed to cattle, and in some cases cattle are carried through the winter on it with no supplemental feed. This is not recommended, for the straw, even when it is of good quality and eaten in large quantities, is barely a maintenance ration. Calves and old cows have difficulties in chewing it properly. It is not considered very satisfactory for dairy cows. On the other hand, work horses and mules do very well on alfalfa straw if a liberal supplemental grain ration is fed. Sheep seem to do well on alfalfa straw provided their teeth are in good condition.

Unless care is taken in feeding the straw, its fine material may be lost. To avoid such waste, the most satisfactory method is feeding the straw in light-bottom mangers, and giving only enough for one feed at a time.

The chief value of the alfalfa straw is that it furnishes some roughage, and the best results are secured when it is fed with a grain ration. To avoid gastrointestinal disturbances the animals must not be forced to eat large quantities of the straw and they should have access to salt and water at all times.

Alfalfa products are important nutritional ingredients of many mixed feeds.

> NOTE: It is permissible to include on the label of alfalfa products a guaranty of their crude carotene content expressed in international units per gram or pound, or in milligrams per pound, and accompanied by an expiration date.

Officially recognized are the following alfalfa products:

Chopped alfalfa, or *cut alfalfa*, which is reasonably free of other crop plants, weeds, and mold and is not ground finely enough to become meal.

The chopping of alfalfa is done by running the field-cured hay through an ensilage cutter. It then is stacked in the field. The chief advantage of chopping hay is that the stock will consume both leaves and stems. Most of the chopped hay will heat sufficiently to be browned.

Alfalfa meal is obtained by grinding alfalfa hay which is reasonably free of other crop plants, weeds, and mold. It must contain not more than 33 percent crude fiber. Alfalfa meal is a product of cured alfalfa, usually sun-cured, stacked, and ground later on the farm or by a commercial mill. The quality of the meal depends entirely on the hay used.

The following guaranties are officially recommended for the various grades of alfalfa meal:

For 13 percent *protein alfalfa meal*, crude fiber not more than 33 percent.

For 15 percent protein alfalfa meal, crude fiber not more than 30 percent.

For 17 percent protein alfalfa meal, crude fiber not more than 27 percent.

For 20 percent protein alfalfa meal, crude fiber not more than 22 percent.

Brand names, such as "Doe's .....% alfalfa meal with animal fat" shall be used to show that the product is a mixture and not simply alfalfa meal. The chemical name of the ANTIOXIDANT(s) shall be listed on the ingredient statement.

Alfalfa leaf meal is the ground product consisting chiefly of alfalfa leaves. It must be reasonably free of other crop plants and weeds and shall contain not less than 20 percent crude protein and not more than 18 percent crude fiber.

NOTE: Recently, extra high quality alfalfa leaf meals containing 25 percent crude protein have been placed on the market.

Leaf meal, if made from high quality hay, is a high quality protein feed used largely in the making of mixed feeds.

Alfalfa stem meal is the ground product remaining after the separation of the leafy material from alfalfa hay or alfalfa meal. It must be reasonably free from other crop plants and weeds.

Extracted alfalfa meal is the dried residue remaining after extraction of alfalfa meal with organic solvents to remove chlorophyll, carotene, and other fat-soluble materials.

Dehydrated alfalfa is available in different grades. The term "dehydrated" may precede the name of any alfalfa product, provided that the freshly cut alfalfa, having a moisture content of not less than 50 percent, has been artificially dried at a temperature of at least 212° F.; that the drying process covers a period of not more than forty minutes; and that there is no admixture of sun-cured alfalfa.

Dehydrating alfalfa saves the food value of the crop. It is used largely by the mixed feed manufacturers. Dehydrating prevents the loss of leaves and insures high protein and carotene content. The carotene content of dehydrated alfalfa is much higher than that of sun-cured hay. To hold a high carotene content it is necessary to store the dehydrated product in a cool place or in oxygen-free atmosphere. Some dehydrators use closed tanks with nitrogen gas. The loss of carotene during storage will average 3 percent per month when the temperatures are 45° F. or less, and 18 percent when temperatures range 66° F. or higher. This indicates that the loss of carotene can be large, and emphasizes the necessity of buying dehydrated meal on a guaranteed carotene basis.

Dangers: There is always danger from BLOATING when pasturing alfalfa with cattle or sheep; this danger will be reduced greatly if the livestock is well fed before being turned onto the alfalfa pasture and if a good supply of salt and water is accessible at all times.

Weeds constitute one of the most serious problems in the alfalfa production. Troublesome weeds in the northeastern states include BLUEGRASS species, QUACKGRASS, and CHICKWEED; in the southeastern and southern regions, CRABGRASS, FOXTAIL grasses, JOHNSONGRASS, BERMUDA-GRASS, CHEATGRASS, certain annual RYEGRASS species (other than Italian ryegrass), chickweed, KNOTWEED, PIGWEED, and LAMBSQUARTERS; in the North Central region, the foxtail grasses, cheatweed, chickweed, quackgrass, and such perennials as CANADA-THISTLE and PLANTAIN; in the irrigated section of the West and the Pacific Coast states, WILD BARLEY, foxtail grasses, pigweed, lambsquarters, SHEPHERD'S-PURSE, and YELLOW STAR-THISTLE.

DODDERS are very objectionable in seed-producing areas, but seldom give much trouble in fields that are devoted to the production of hay.

Alfalfa is susceptible to more than 75

diseases caused by fungi, bacteria, viruses, and nematodes; among the most important diseases attacking alfalfa are BACTERIAL WILT; FUSARIUM WILT OF ALFALFA; ROOT ROTS, (e.g., VIOLET ROOT ROT), CROWN ROTS and CROWN WART; *stem diseases*, such as SOUTHERN ANTHRACNOSE, BLACKSTEM OF CLOVER, and CHARCOAL ROT; *leaf diseases*, e.g., COMMON LEAF SPOT, YELLOW LEAF BLOTCH, DOWNY MILDEW OF ALFALFA, SUMMER BLACKSTEM, ALFALFA RUST, PLEOSPORA LEAF SPOT, PEPPER SPOT, BLACKPATCH, BACTERIAL LEAF AND STEM SPOT, etc.; the *virus diseases* DWARF DISEASE (or alfalfa dwarf) and "WITCHES'-BROOM"; *nematodes*, especially the STEM NEMATODE and the ROOT-KNOT NEMATODE.

Several insects infest alfalfa and may cause damage to the hay or seed crop if not controlled; this is particularly true of GRASSHOPPERS, the ALFALFA WEEVIL, LYGUS BUGS, WEBWORMS, the VARIEGATED CUTWORM and other species of CUTWORMS, the ARMYWORM, the FALL ARMYWORM, the ARMY CUTWORM, the CORN EARWORM, the POTATO LEAFHOPPER and other LEAFHOPPERS, the PEA APHID, the YELLOW CLOVER APHID, the ALFALFA CATERPILLAR, the WESTERN HARVESTER ANT, the RED HARVESTER ANT, the MEADOW SPITTLEBUG, the CLOVER-SEED CHALCID, the GARDEN WEBWORM, the SOUTHERN CORN ROOTWORM, CLOVER ROOT CURCULIOS, the VELVETBEAN CATERPILLAR, the CLOVER LEAF WEEVIL, the ALFALFA LOOPER, the MORMON CRICKET, the THREE-CORNERED ALFALFA HOPPER, various species of BEETLES, and STINK BUGS.

The most troublesome animal pests encountered in growing a crop of alfalfa are rodents, e.g., POCKET GOPHERS, ground squirrels, MOLES, prairie dogs, and FIELD MICE.

Classifications: In addition to the *variegated alfalfa* group and the *yellow-flowered alfalfa* group, which belong to different *Medicago* spp. (and therefore are discussed elsewhere) the alfalfas of the species *M. sativa* are subdivided in these three groups.

(1) COMMON ALFALFAS

The common alfalfas are also called *purple alfalfas*, or (in Europe) *lucerne*; they are by far the most important alfalfa species. Having been grown many generations under different climatic conditions, the original "common alfalfa" has differentiated into regional varieties which are generally adapted in the latitudes where they have been grown for a long period of time. They are distinguished by the name of the states where grown; e.g., *Kansas Common (alfalfa)*, one of the widely grown alfalfas; *Montana Common (alfalfa)*, *Dakota Common (alfalfa)*, and other *Northern commons; Arizona Common (alfalfa)*, *Oklahoma Common (alfalfa)* and other *Southern commons*, etc. In purchasing common alfalfa seed an effort should be made to ascertain its origin and to obtain it from a source where the winters approximate in severity those of the region where the seed is to be sown. Most common alfalfas are susceptible to BACTERIAL WILT.

Among the newer varieties of common alfalfas, the following are promising:

Buffalo (alfalfa), obtained from Kansas Common, is important because of its resistance to BACTERIAL WILT, and its seed and forage productivity. Buffalo has a flower ranging from a light blue to a reddish-purple. It is upright in type of growth, has a medium-sized stem, and makes a medium to leafy quality of hay. When grown by itself, Buffalo cannot be distinguished from ordinary Kansas Common. Its growth in spring and fall is a little more upright than that of Kansas Common and it makes a slightly more rapid recovery after cutting.

Buffalo has a higher stand survival in northern alfalfa areas of the United States than Kansas Common.

California Common 49 was developed from California Common for tolerance to the DWARF DISEASE. It is susceptible to BACTERIAL WILT.

Caliverde is highly resistant to BACTERIAL WILT, COMMON LEAF SPOT, and MILDEW. It was developed by crossing California Common with a Turkestan selection and backcrossing to California Common. Caliverde is identical to California Common in growth habit, recovery after cutting, and limited hardiness. It is adapted to areas in the Southwest where winterkilling is not severe.

Williamsburg (alfalfa) is a selection from Kansas Common. It maintains its stand under eastern Virginia conditions because of its high resistance to STEM ROT; but it is not WILT-resistant. Williamsburg competes better with summer weeds than most varieties.

(2) TURKESTAN ALFALFAS

These varieties originated from Turkestan (also spelled Turkistan). They are purple-flowered and differ slightly from the common alfalfas in their habits of growth. This group includes the following:

(Commercial) Turkestan alfalfa, grown from seed imported from the most important seed-producing districts in Turkestan, is generally equal or superior to Grimm alfalfa in hardiness and is also resistant to BACTERIAL WILT. In the United States, Turkestan alfalfa has usually been less productive than some of our domestic alfalfas. It has given fairly good results in the central and northern Great Plains where wilt is a problem. In the East and South it has been unsatisfactory; it has a tendency to become dormant early in fall and is susceptible to yellowing and to certain leaf diseases that cause the shedding of many leaves before the plants have reached the best stage for harvesting.

Hardistan (alfalfa) is resistant to WILT and cold. It is very similar to the commercial importations from Turkestan. The variety is promising where wilt is prevalent in the upper Mississippi and Missouri Valleys, but is not more satisfactory in the East and South than most of the commercial lots from Turkestan.

Nemastan (alfalfa) is resistant to the STEM NEMATODE. It is usually planted only in Utah and Nevada where this pest is prevalent. Nemastan is highly susceptible to the leaf spot diseases.

Orestan (alfalfa) is a WILT-resistant variety grown in Oregon. It is now being replaced by improved disease-resistant types.

(3) NONHARDY ALFALFAS

This group of alfalfas is grown only in the extreme southern regions of the United States. It is characterized by the lack of cold-resistance, upright habits of growth, quick recovering after cutting, and long periods of growth.

Most of the nonhardy alfalfas are importations from South America, India, and Africa. Of some local importance are the following:

African (alfalfa), selected from an Egyptian introduction, is a rapid grower that recovers quickly after cutting. It is suited to the deep Southwest.

Chilean (alfalfa) is used in the Southwest. It is very similar to smooth Peruvian.

Indian (alfalfa), now being used in the Southwest, was introduced from India.

Peruvian (alfalfa) seldom survives winters where the temperature falls below 10° F. and is suited only for the South and Southwest.

Strains: Two strains of Peruvian alfalfa are recognized, the *smooth-leaved Peruvian* and the *hairy Peruvian.*

"Poor man's alfalfa" = SERICEA.

References: G.6; M.3; B.7; G.7; W.19; F.6.

See also MEDICAGO; ALFALFA PELLETS; ALFALFA-HAY STANDARDS; LEGUME FEED-PRODUCTS; SILAGE CROPS; RANGE PLANTS; PASTURES; PASTURE PLANTS; BUCKWHEATS; OATS; CORN; TALL FESCUE; SMOOTH BROME; MOUNTAIN BROME; ORCHARDGRASS; RUSSIAN WILD-RYE; WEEPING LOVEGRASS; BIG BLUEGRASS; ALSIKE CLOVER; BLUE PANICGRASS; INTERMEDIATE WHEATGRASS; MESQUITE; HAIRY VETCH; FIELD BEANS; FIELD PEA; MUNGBEAN; TREFOILS; BIG TREFOIL; KUDZU; BROCCOLI; CHILEAN DODDER; LARGE-SEEDED ALFALFA DODDER; SMALL-SEEDED ALFALFA DODDER; FIELD DODDER; CLOVER DODDER; ALMOND HULLS; LEGUME BACTERIA; STRAW MEAL; PRUSSIC ACID POISONING; INSECTICIDES; WHITE GRUBS; WHEATS.

ALFALFA SEED SCREENINGS are not very palatable to livestock and should be finely ground to increase their digestibility.

ALFALFA SILAGE. *See* SILAGE CROPS.

ALFALFA-STEM MEAL. *See* ALFALFAS.

ALFALFA STRAW. *See* ALFALFAS.

ALFALFA WEEVIL (*Hypera postica*) is a pest which has spread from Utah into Arizona, California, Colorado, Idaho, Montana, Nebraska, Nevada, Oregon, South Dakota, and Wyoming. The larvae feed on the growing tips, leaves, and buds of ALFALFA and may destroy most of the feed value of a hay crop or may prevent the profitable production of seed. The weevil is essentially a pest of first-growth alfalfa. When the first growth is cut for hay, weevil larvae feed upon the basal shoots and retard the second growth for a few days to several weeks. This is especially serious in dry-land farming or second-crop seed production.

Alfalfa weevil. (U.13.)

The insects winter chiefly as adults, mostly in the fields. Soon after the snow melts, the females lay their first eggs in pieces of dead stems on the ground. After the spring growth of alfalfa is about 6 in. high, the weevils gradually shift their egg laying to the growing plant stems. The number of eggs per female averages about 400. Hatching begins in April, but larvae do not become numerous enough to cause economic crop damage until late May or early June, about the time the first growth of alfalfa produces buds. Meanwhile, all of the early larvae and many of the later ones have become parasitized by a tiny wasp, commonly called the "WEEVIL PARASITE" (*Bathyplectes curculionis*). Starting in the middle of May, the weevil larvae complete their growth, drop to the ground, and spin cocoons, usually attaching them to fallen leaves. Parasitized larvae die after they

spin their cocoons; healthy larvae pupate inside their cocoons in from seven to ten days. Weevil adults then leave their cocoons; since they remain sexually immature until fall or spring, there is only one generation of weevils a year.

Control: Maintain a dense, vigorously growing stand of alfalfa. Cut first and second crops when most plants are in the bud stage. Mow the field clean and remove the hay as soon as it is cured. Do not irrigate the fields for seven to ten days after cutting.

To kill the adult alfalfa weevil, apply DIELDRIN or CHLORDANE as a spray when the spring growth of alfalfa is 1 to 2 in. tall.

To kill the larvae, dust or spray the crop as soon as plants become noticeably riddled, but before many have become gray, with CALCIUM ARSENATE, METHOXYCHLOR, or PARATHION. Do not cut hay treated with calcium arsenate or methoxychlor for seven to ten days after treatment. Leave parathion-treated hay uncut for at least two weeks.

To protect alfalfa for seed, apply the dieldrin or chlordane treatment as is done for hay crops, to kill adults in the early spring. When the plants reach the bud stage of development, treat with DDT as a dust or as a spray. This treatment is prescribed to control LYGUS BUGS and several other pests of seed alfalfa as well as alfalfa weevil.

Reference: U.3.

ALFILERIA (*Erodium circutarium*), also known as *filaree*, *pinclover*, and *stocksbill*, thrives well in the desert ranges of Arizona and in the foothills of California. It spreads very rapidly and furnishes choice spring forage for all classes of livestock and game animals. Its nutritive value is high.

Reference: U.6.

See also RANGE PLANTS.

ALGAE (sing. *alga*) represent the lowest section of plant life. They live submerged beneath the water; e.g., seaweeds and pond scum. They contain CHLOROPHYLL and some CAROTENE. Algae are valuable as a source of IODINE. They are closely related to the FUNGI.

Algal means: pertaining to algae.

ALGERIA KAFIR = *Bishop kafir;* belongs to the grain SORGHUMS.

ALICEL. *See* CLUB WHEAT (variety).

ALIMENTARY TRACT
= DIGESTIVE TRACT.

ALISMA. *A. plantago-aquatica*
= WATERPLANTAIN.

ALKALI is any inorganic substance with *alkaline,* or *basic,* properties. Alkalies are able to neutralize ACIDS, thus forming SALTS. Many soils react alkaline.—*See also* pH.

ALKALI DISEASE.
See SELENIUM POISONING.

ALKALI LOVEGRASS (*Eragrostris obtusiflora*) grows in alkaline soils of Arizona, New Mexico, and Mexico. It rarely flowers and is a rigid GRASS which furnishes a large part of the forage in the extreme alkaline areas of Sulphur Springs Valley (Arizona).—*See also* LOVEGRASSES.

ALKALINE. *See* ALKALI.

ALKALI SACATON (*Sporobolus airoides*), also known as *bigplume bunchgrass, finetop saltgrass,* and *hairgrass dropseed,* is a robust perennial and an important forage plant in the Southwest. This DROP-SEED species endures much alkali; it grows on the alkaline flats, rocky sites, bottom lands, and along drainages in the desert areas. Chemical analyses of alkali sacaton show very high mineral content.

An abundance of herbage is produced by this species; it is eaten freely by cattle and horses and is often utilized closely. It should be grazed during the growing season, because the foliage becomes tough and unpalatable as it matures and does not provide good winter forage. In some parts of the Southwest, where moisture is adequate to produce a good cover, patches of alkali sacaton are fenced for pasture and, if kept closely cropped, afford good grazing.

Alkali sacaton has deep, coarse roots. The stems are smooth, stout, leafy, 1 to 3 ft. high, and grow in dense bunches commonly from about 8 to 12 in. diameter. On favorable sites, the grass sometimes forms a uniform cover approaching a sod. The numerous basal leaves are up to 18 in. long and about ⅛ in. wide at the base and taper to long, slender, inrolled points.

The leaves are smooth beneath but rough above, with the sheaths sparsely hairy at their throatlike portion.

Alkali sacaton. Plant, glumes, and floret. (H.26.)

Alkali sacaton produces an abundant supply of exceptionally long-lived seeds which enable this species to extend its stands rather vigorously on favorable areas.

References: U.6; H.1.

See also RANGE PLANTS; GRASSES.

ALKALOIDS are organic, nitrogen-containing substances, basic (alkaline) in re-action, and often found in POISONOUS PLANTS; e.g., NICOTINE and STRYCHNINE.

Alkaloid-containing LUPINES, which are poisonous, are called *bitter*, nonalkaloid lupines, *sweet.—See. also* CROTALARIAS; DUTCHMAN'S BREECHES; WATERHEMLOCKS; LARKSPURS; DEATHCAMASS; RAGWORT; GROUNDSELS; HARMEL PEGANUM; SABA-DILLA POWDER.

ALLEGHANY BARBERRY.
See BARBERRY BUSHES.

ALLENROLFEA. *A. occidentalis*
= PICKLEWEED.

ALLIUM. *A. canadense* = WILD ONION.

ALLUVIAL SOIL of plains and valleys has been deposited by running water. It con-sists of finely divided material washed down from high grounds. In general, allu-vial soils are desirable areas for agriculture.

Alluvial plains are subject to periodic overflow, except the higher levels which form river terraces.

ALMOND HULL MEAL
= GROUND ALMOND HULLS.

ALMOND HULLS of the soft-shell *almond* variety are of better feed value than hulls of the hard-shell variety.

Almond hulls are a by-product of the almond industry. After the almonds are harvested, the *nuts* with attached hull are processed through a machine which re-moves the hull. The hulls contain 10 to 30 percent moisture, 10 to 17 percent fiber, 4 to 7 percent ash, 1 to 4 percent fat, 2 to 5 percent protein, and 50 to 60 percent nitrogen-free extract; the latter includes total sugars ranging from 18.3 to 30.56 percent. The nitrogen-free extract portion of the hulls is very digestible.

Almond hulls, properly dried, are very satisfactory for ruminants when fed in combination with such feeds as BARLEY and ALFALFA hay which compensate for the lack of protein in the hulls. The thick, fleshy type of hull is preferable for live-stock feeding. Almond hulls can also be *ensiled.*

Reference: W.33.

See also PLANT BY-PRODUCTS.

ALMOND HULLS SILAGE.
See SILAGE CROPS.

ALOPECURUS. *A. pratensis* = MEADOW

FOXTAIL; *A. arundinaceus* = REED FOX-TAIL.—*See also* FOXTAIL; WEEDS.

ALPHA BARLEY.
See TWO-ROWED BARLEY (variety).

ALPHA TOCOPHEROL. *See* VITAMIN E.

ALPINE = *Mountain rye. See* RYE.

ALSIKE CLOVER (*Trifolium hybridum*) is known also as *Swedish clover* or *hybrid clover.*

This perennial LEGUME is usually treated agriculturally as a biennial. Many smooth stalks come from the crown and bear smooth leaves, each with three leaflets. The heads of flowers—the oldest below, the youngest at the top of the stem—are mixed pink and white or all white or all pink. Under favorable conditions the stalks grow 3 to 5 ft. long; on dry soil they may be only 18 in. in height.

Alsike clover grows best in a cool climate. It withstands severe winters better than RED CLOVER. It is very gen-erally grown north of the Ohio and Poto-mac Rivers and as far west as the Dakota-Minnesota boundary. It also is grown in Idaho and on the Pacific Coast of Washing-ton and Oregon and, to a limited extent, in Virginia, Kentucky, Tennessee, and Missouri.

Alsike clover prefers a rather heavy silt or clay soil with plenty of moisture. It thrives on good loams. Usually it does not do well on dry, sandy, or gravelly soils. It responds to an application of lime, but is not as sensitive to acidity as red clover, and can be successfully grown on many wet, cold, and "sour" soils.

In the northern states the seed is usually sown in early spring, though in many places in late summer, for instance in New Jersey; in Indiana and Ohio seed is some-times sown in CORN at the last working. Seeding in July with BUCKWHEAT is success-fully practiced in southern Michigan. In the South this clover is seeded in the fall.

Unless a seed crop is wanted, alsike clover is usually seeded with TIMOTHY, red clover, or both. A common practice is to sow timothy in the fall and to sow either alsike alone or alsike and red clover in the spring (a common formula being alsike clover, 1 part; red clover, 1 to 2 parts; timothy, 2 to 3 parts). Nodule-forming bacteria are present on most clover land. On new land they may be wanting, especially on burned-over land and newly drained marsh land. In such cases the seed should be inoculated. This may be done by scattering soil from a field on which white, red, alsike, or CRIMSON CLOVER has made a good growth or by applying a pure culture of the proper bacteria to the seed.

Alsike clover makes a good quality of *hay;* its composition is similar to that of red clover, but its protein is more digestible than red clover, though where conditions are favorable for red clover it will yield more hay than alsike clover. Because alsike clover is finer stemmed, there is less waste than in red clover, as stock clean it up better. However, because of its slender stems, alsike clover is usually sown with timothy or red clover; their more upright stems support those of the alsike clover, thus adding materially to the weight of hay cut. Alsike hay is greener than red clover hay; unless the stand is extremely thick it cures more readily. The best time to cut alsike clover for hay is when it is in full bloom. The presence of much ripe seed is said to be one cause of the slobbering of horses fed on the hay. The fine stems of alsike clover do not get hard as quickly as the stems of red clover. When seeded with red clover and timothy, the hay will be at its best if cut when the red clover is in full bloom and timothy is in early bloom.

Besides the common mixture of alsike clover, red clover, and timothy, various other mixtures are recommended for different situations and sections; e.g., SWEET-CLOVER, alsike or red clover, and timothy; alsike clover, timothy, and REDTOP (used on rather "sour" soils on which the timothy fails to last long); redtop and alsike (on heavy, wet land). In many places alsike clover and ALFALFA are seeded together, in others the clover with BROME GRASS. Alsike clover is also mixed with redtop, timothy, and MEADOW FESCUE (in Idaho); with RYEGRASS or with red clover, timothy, ORCHARDGRASS, and ryegrass (on the Pacific Coast); with orchardgrass, TALL OATGRASS, redtop, and white clover (in Virginia); with alfalfa, meadow fescue, orchardgrass and timothy (in Missouri);etc.

Alsike clover is especially suited for use in *pasture* mixtures. Its vigorous growth enables it to hold its own against the competition of other plants, and its persistence in seeding provides for its continuance. For permanent pasture on low or "sour" land alsike clover is especially valuable

Alsike clover. (U.S.D.A.)

and may be used alone or, better, with redtop.

In the course of a rotation the fields are usually plowed after three to five years, but alsike clover has been pastured for twelve and as many as thirty years. When grazed or cut early, it sends out new branches at once, and these produce flowers and seed when not more than a few inches long; in this way the ground is constantly full of seed and the pasture keeps in good shape. Alsike clover in pasture is more hardy than red clover; it will withstand more trampling than red clover and does not heave out so readily. Because it endures close grazing and trampling it is valuable also for hog pasture.

When alsike clover is fed to animals, the manure carries a large quantity of seed, and enough clover may volunteer in fields where the manure is spread to make a good seeding.

On many farms sloughs and swales are covered with marsh grasses and sedges. When such places are not actually swampy, alsike clover will improve the wild hay or may even run the wild grasses out and occupy the ground. It will endure a certain amount of overflow without damage; a 1-year-old stand may remain under water for several weeks, without being killed.

Good *silage* is made from alsike clover either alone or with some grass crop or corn. It must be thoroughly packed. The crop is put into the silo fresh and moist. If it gets rather dry between the time of cutting and putting it into the silo, a stream of water is run on it. Air pockets must be avoided or spoiling will result.

In most cases alsike clover does not make a second growth worth cutting; however, on river-bottom lands and on moist, rich lands in the North two crops may be secured, especially if the first crop is cut rather early, when just coming into bloom. A few days' delay here may result in the loss of the second crop. The clover should be allowed to make at least 4 in. growth before winter.

Alsike clover is recommended as a soil improver on land where red clover will not catch.

Dangers: Alsike clover is not seriously,

if at all, affected with diseases such as trouble red clover. (Of local importance are SOOTY BLOTCH and BLACKPATCH.) The same may be said in regard to insects. Alsike clover is little troubled by the CLOVER ROOT-BORER. The CLOVER APHIDS give some trouble, but these pests may be destroyed by close spring grazing of infested fields.

Like other clovers, alsike clover will cause BLOAT if stock are allowed to eat freely of the fresh plants, especially early in the season. Another injurious effect that seems confined to alsike clover, is the disease TRIFOLIOSIS which affects horses and mules.

Reference: P.2.

See also TRUE CLOVERS; INOCULATION; WHITE CLOVER; RICE; PASTURES; PASTURE PLANTS; GRAZING; SILAGE CROPS; HAY.

ALSTROUM. *See* SPELT (variety).

ALTA FESCUE is a strain of TALL FESCUE.—*See also* SUBCLOVER; BIRDSFOOT TREFOIL; RICE; GRAZING.

ALTERNARIA. The *Alternaria* spp. are FUNGI which cause ALTERNARIA LEAF SPOT.

ALTERNARIA LEAF SPOT, a FUNGUS disease, is caused by *Alternaria* spp. The spots are dark brown, rather large, and show concentric rings. This disease does not, in general, cause serious damage to SOYBEANS.

Reference: C.9.

ALTERNATE LEAVES are placed on a stem singly, first one and then another— not oppositely or in pairs.

ALTRIPLEX. *A. canescens* = FOURWING SALTBUSH; *A. gardneri* = GARDNER SALTBUSH; *A. confertifolia* = SHADSCALE.

ALUMINUM SILICATE. *See* BENTONITE.

ALYCECLOVER (*Alysicarpus vaginalis*) is a summer annual; it does not belong to the true clovers. In thin stands it tends to spread and be moderately branched, but in thick stands it is ascending and little branched. It attains a height of about 3 ft. on moderately fertile soil. The stems are rather coarse but fairly leafy. The leaves are broadly oval and borne the entire length of the stems on short leafstalks. The seed, which is quite small, is borne in jointed pods.

Alyceclover is adapted to the area adjacent to the Gulf of Mexico and is being grown in many places throughout that region, most extensively in Florida and Mississippi. Its principal uses are for hay and soil improvement. It also makes good pasturage.

The fertilizer requirements for alyceclover are about the same as for most other LEGUMES of the South. It does not tolerate wet lands and makes poor growth on soils of low fertility. Inoculation has not been needed.

Alyceclover should be seeded early in May. When the plants are allowed to mature and shatter seeds, a volunteer crop is assured for the following year; when used for green manure, the crop is volunteered for several years. Yields may reach 10 tons of green forage or 1 to 2 tons of hay per acre.

The crop can be handled with ordinary farm machinery. The cut hay dries readily and in good weather can be stacked or baled the day after it is cut.

Danger: Alyceclover is very susceptible to the ROOT-KNOT NEMATODE.

References: M.3; M.11.

See also TRUE CLOVERS.

ALYSICARPUS. *A. vaginalis*
= ALYCECLOVER.

AMARANTHUS. Among the *Amaranthus* spp., or PIGWEEDS, is *A. retroflexus* = ROUGH PIGWEED.

AMBER SORGOS belong to the forage SORGHUMS. Among the Amber varieties are (1) *Red Amber* and (2) *Black Amber* (formerly called *Early Amber*), including the locally important *Minnesota Amber*, *Waconia Amber*, *Dakota Amber*, and *Dakota Amber 39-30.—See also* COWPEA.

AMBROSIA. *A. artemisiaefolia*
= RAGWEED.

AMERICAN DAIRY SCIENCE ASSOCIATION has set up the standards of SILAGE.

AMERICAN-EGYPTIAN COTTON.
See GOSSYPIUM.

AMERICAN SINGLE-CUT CLOVER.
See RED CLOVER (variety).

AMINO ACIDS are the basic building blocks of PROTEINS. Since living organisms build their tissues from amino acids, it is important that the ration provide both the proper kind and amount. As many as 45 amino acids have been found in nature; most familiar products contain at least 22. Before the proteins can be absorbed from the small intestine into the blood system, they must be digested, i.e., split in amino acids. Each amino acid contains one or more *amino* (NH_2) groups.

Essential, or *indispensible*, amino acids are those required by the animal for tissue growth. They are not formed by the body at a rate needed for maximum growth and must therefore be in the feed. Among them are the following: *arginine*, HISTIDINE, ISOLEUCINE, *leucine*, LYSINE, METHIONINE, PHENYLALANINE, THREONINE, TRYPTOPHAN, and VALINE; the chick requires GLYCINE in addition. Among the dispensable amino acids are ALANINE, CYSTINE, GLUTAMIC ACID, and TYROSINE. However, cystine can replace part of the methionine, and tyrosine replaces some phenylalanine.

Herbivorous animals, and particularly ruminants, are less definite in their requirements for amino acids than are chickens, pigs, and man because the bacteria in the former's gastrointestinal tracts convert many compounds to bacterial protein which the animal can use. Different protein sources contain different types and percentages of amino acids, as shown in one of the tables; e.g., SOYBEANS contain most of the amino acids needed by growing chicks, but are slightly deficient in the sulfur-containing amino acids methionine and cystine. The cereal grains are all deficient in lysine. Tankage is deficient in tryptophan.

NOTE: *Methionine hydroxy analogue calcium*, which contains a minimum of 95 percent *dl-2-hydroxy-4-methylthiobutyric acid calcium*, is tentatively recognized as feedingstuff; 1.2 lb. of it are molecularly equivalent to 1.0 lb. (of 100 percent) dl-*methionine*, an essential amino acid which the Association of American Feed Control Officials lists among the VITAMINS.

THE AMINO ACID CONTENT OF SOME FEEDS AND TISSUES IN PERCENT

	Arginine	Histidine	Isoleucine	Leucine	Lysine	Methionine	Cystine	Phenylalanine	Tyrosine	Threonine	Tryptophan	Valine	Glycine
Alfalfa meal, 17%	0.8	0.3	0.8	1.2	0.8	0.3	0.3	0.7	0.8	0.6	0.3	0.8	
Barley, average	0.5	0.2	0.4	0.7	0.3	0.2	0.2	0.5	0.2	0.3	0.1	0.5	0.3
Blood meal	3.2	3.8	1.0	9.9	6.9	1.1	1.5	5.3	3.0	4.4	1.1	6.2	
Buttermilk, dried	1.1	0.8	2.5	3.3	2.3	0.7	0.4	1.5	1.2	1.4	0.4	2.6	0.7
Casein	3.4	2.6	5.5	8.5	6.7	2.8	0.3	4.7	5.6	3.5	1.1	5.9	1.8
Cereal pasture	1.5	0.4	1.3	1.5	1.9	0.4	0.4	1.1	1.2	1.1	0.2	1.3	
Corn	0.4	0.2	0.4	1.5	0.2	0.3	0.1	0.4	0.5	0.3	0.1	0.5	0.4
Corn gluten meal	1.3	0.8	2.1	8.0	0.7	1.0	0.6	2.5	2.6	1.4	0.3	2.3	1.7
Cottonseed meal, 41%	3.5	1.0	1.5	2.2	1.5	0.7	0.9	2.2	1.2	1.3	0.5	1.9	2.3
Distillers' dark grains	0.9	0.4	1.7	2.0	0.1	0.5		1.7	0.5	1.0	0.2	1.6	
Dist. solubles, av.	1.0	0.7	1.6	1.7	1.0	0.5	0.2	1.7	0.6	1.0	0.2	1.5	0.3
Eggs, contents	0.9	0.3	1.0	1.2	0.9	0.5	0.3	0.8	0.6	0.6	0.2	1.0	3.8
Fish meal	3.7	1.4	3.0	5.1	5.0	1.8	1.1	2.6	1.8	2.6	0.7	3.6	2.2
Fish solubles	1.4	1.2	0.5	0.7	1.4	0.5	0.4	0.6	0.2	0.6	0.3	0.8	
Linseed meal, 37%	2.7	0.7	1.7	2.2	1.0	0.7	0.6	1.8	1.8	1.4	0.5	2.1	2.4
Liver meal	3.9	1.6	3.1	5.4	4.1	1.6	0.9	3.1	2.3	2.5	0.8	3.7	
Meat scraps, 50%	3.8	1.2	2.3	4.0	3.5	1.0	0.6	2.2	1.9	2.0	0.4	2.8	2.4
Meat and bone scraps	3.0	0.9	1.4	3.1	2.7	0.7	0.6	1.7	1.0	1.8	0.4	2.3	0.2
Milk, skimmed, dried	1.2	0.9	2.1	3.4	2.6	0.9	0.4	1.6	1.3	1.4	0.4	2.5	0.5
Milo	0.4	0.2	0.5	1.3	0.3	0.1	0.2	0.5	0.2	0.3	0.1	0.5	
Muscle, animal	1.5	0.7	1.2	1.6	2.0	0.6	0.2	1.0	0.8	1.0	0.3	1.1	1.0
Oats	0.6	0.3	0.5	0.8	0.4	0.2	0.2	0.5	0.4	0.4	0.2	0.7	0.3
Oats, rolled	0.9	0.3	0.8	1.2	0.5	0.3	0.3	0.9	0.7	0.5	0.2	1.0	
Peanut meal, 44%	4.4	0.9	1.6	2.6	1.3	0.5	0.7	2.3	1.9	0.9	0.5	2.3	2.5
Potato meal, white	0.4	0.2	0.3	0.5	0.5	0.1		0.3		0.4	0.2	0.4	
Rice bran	0.6	0.2	0.4	0.6	0.5		0.1	0.4		0.3	0.1	0.6	
Rye	0.5	0.2	0.5	0.7	0.4	0.2		0.6	0.2	0.4	0.1	0.6	
Soybean meal, 44%	3.0	1.1	2.2	3.2	2.8	0.7	0.7	2.2	1.5	1.8	0.6	2.4	
Tankage, 60%	3.6	1.6	1.7	4.8	4.1	0.6		2.9	1.7	1.9	0.4	3.3	7.3
Wheat, average	0.5	0.2	0.5	0.8	0.3	0.2	0.2	0.7	0.5	0.4	0.2	0.6	0.9
Wheat mids or shorts	0.8	0.4	0.8	0.9	0.4	0.2	0.2	0.7	0.2	0.4	0.2	0.8	0.4
Whey, dried	0.3	0.2	0.7	1.0	0.7	0.3	0.1	0.4	0.1	0.6	0.2	0.7	0.2
Yeast, dried brewers'	2.0	1.1	2.6	3.3	3.3	0.8	0.5	1.8	1.5	2.4	0.5	2.5	3.4

AMINO ACID REQUIREMENTS, IN PERCENT OF DIET

Amino Acid	Poults	Chicks	Laying Hens	Pigs
Arginine	1.6	1.2	1.6	0.2
Histidine	...	0.15	...	0.40
Leucine	...	1.4	1.4	0.80
Isoleucine	0.72	0.6	...	0.7
Lysine	1.5	0.9	0.5	1.0
Methionine [1]	0.5–0.8	0.5–0.9	0.28–0.43	0.3–0.6
Phenylalanine [2]	...	0.9–1.6	...	0.46
Threonine	0.6	0.6	0.15	0.4
Tryptophan	0.26	0.25	...	0.2
Valine	...	0.8	...	0.4
Glycine	0.9	1.5	...	...
Protein	24	20	15	...

[1] Lower value is absolute requirement; remainder can be either methionine or cystine.
[2] Lower value is absolute requirement; the remainder can be either phenylalanine or tyrosine.

The tables show (1) the amino acid content of some feedstuffs and (2) the requirements of several animal species for amino acids.

See also ABSORPTION; BROWNING REACTION; U.G.F.

AMINOBENZENEARSONIC ACID.
See ARSANILIC ACID.

AMINOBENZOIC ACID.
Para (or *p*)-*aminobenzoic acid* = PABA.

2-AMINO-5-NITROTHIAZOLE, or *enheptin*, is used in controlling *blackhead*, an infectious poultry disease which has caused great losses particularly in turkeys. The drug forms a yellow powder which is only slightly soluble in water. For continuous medication, it is mixed in feed at the rate of 0.05 percent.

Caution: Medication should be discontinued two weeks before the birds are slaughtered.
Reference: S.11.

AMMANNIA. *A. coccinea* = REDSTEM.

"AMMATE." *See* AMMONIUM SULFAMATE.

AMMONIA—consisting of nitrogen and hydrogen: NH_3—is a colorless, irritating gas with an intense, pungent odor. It is very soluble in water, forming *ammonium hydroxide*, an alkaline solution, which is commonly called *ammonia water*. Ammonia is used in the manufacture of various fertilizers and feedstuffs.

Inhalation of ammonia gas (obtained as fumes from the official preparation *Stronger ammonia water*) may supplement the treatment of animals affected with PRUSSIC ACID POISONING.

AMMONIATED BEET PULP.
See AMMONIATED PLANT PRODUCT.

AMMONIATED MOLASSES.
See MOLASSES.

AMMONIATED PLANT PRODUCT—a tentatively accepted feedstuff—results from a combination of AMMONIA with plant products in such a manner that a material of stable composition results. The source of the material shall be part of the official name of the product, as, for example, *ammoniated beet pulp*, or *ammoniated molasses*. It shall be recognized as ingredient in feeds for ruminants only. On the label of this product or any mixture containing it, the crude protein guaranty shall be followed by a guaranty of the maximum equivalent crude protein from nonprotein nitrogen.

Reference: F.6.

See also MISCELLANEOUS PRODUCTS; MOLASSES.

AMMONIA WATER. *See* AMMONIA.

AMMONIUM BICARBONATE.
See AMMONIUM CARBONATE.

AMMONIUM CARBONATE is available as a white powder or cubes and has a strong odor of AMMONIA; it must be kept tightly closed in a cool place. The commercial grade is not a true carbonate, but a mixture which contains *ammonium bicarbonate*.

Micro-organisms in the rumen can convert ammonium carbonate to protein.—
See also NONPROTEIN NITROGEN.

AMMONIUM HYDROXIDE.
See AMMONIA.

AMMONIUM NITRATE, available in form of white granules, decomposes under explosion on heating. It is very soluble in water. The high nitrogen content (33.5 percent in the commercial grade) makes it a valuable fertilizer.

AMMONIUM PHOSPHATES are commercially available in two water-soluble forms: (1) *monobasic* and (2) *dibasic*. Both are used as fertilizers—the first one being richer in phosphorus, the second richer in nitrogen. They are used also in feedstuffs; taken internally, they have a diuretic effect and acidify the urine.—*See also* NONPROTEIN NITROGEN.

AMMONIUM SULFAMATE, also known as "*Ammate*," is a water-soluble, nonpoisonous powder used to kill BRUSHES. For instance, to kill BARBERRY BUSHES, their canes are cut off; then the HERBICIDE is applied in dry form to all freshly cut surfaces; 1 lb. suffices for killing several medium-sized bushes.

AMMONIUM SULFATE occurs in white granules or crystals which are very soluble in water. It contains (in pure form) 21.2 percent nitrogen and is widely used as fertilizer.

AMSINCKIA. *A. intermedia* = TARWEED.

AMYLASE is an ENZYME.

AMYLOPECTIN is a phosphorus-containing STARCH (amylum). It is less soluble than AMYLOSE, the other fraction found in all starch granules, except in waxy CORN which consists of amylopectin only.

AMYLOSE is a fraction of STARCH. It is free from phosphorus and thus distinguished from AMYLOPECTIN. The latter forms the outer layer of starch granules (and all of the waxy CORN); amylose, which is more soluble, forms the inner substance.

ANABRUS. *A. simplex*
= MORMON CRICKET.

ANDREW OAT.
See COMMON OAT (variety).

ANDROPOGON. The *Andropogon* spp. are known as BLUESTEMS; they include the following GRASSES: *A. gerardi* = BIG BLUESTEM; *A. scoparius* = LITTLE BLUESTEM; *A. virginicus* = BROOMSEDGE; *A. ischaemum* = TURKESTAN BLUESTEM (with its varieties *Elkan bluestem* and *King Ranch bluestem*); and *A. caucasian* = CAUCASIAN BLUESTEM.

ANEMIA. *See* MACROCYTIC ANEMIA; IRON; VITAMIN B_{12}.

ANGULAR LEAF SPOT, caused by the FUNGUS *Mycosphaerella puericola*, may affect KUDZU. The lesions are confined to the leaves and are dark brown and often have a yellowish margin. When lesions coalesce (unite) over large areas of the leaf, defoliation results. Leaves having many lesions gradually die and fall off.

Reference: W.15.

ANIMAL BY-PRODUCTS.
See ANIMAL PRODUCTS.

ANIMAL FAT.
See ANIMAL PRODUCTS; FATS.

ANIMAL LIVER AND GLANDULAR MEAL. See ANIMAL PRODUCTS.

ANIMAL LIVER MEAL.
See ANIMAL PRODUCTS.

ANIMAL PRODUCTS include by-products from *poultry dressing plants* and *fisheries*, as well as *rendering plant* products, and *packing house* products and by-products. They contain proteins that are good sources of lysine and other amino acids in which grain proteins are deficient; some of these animal products, however, are low in tryptophan. They have long been used to supplement grain rations. Some of them are also good sources of calcium, phosphorus, NIACIN, VITAMIN B_1, and VITAMIN B_{12}. Many animal products are relatively rich in FAT. This should be stabilized or the products become rancid. The following animal products are officially recognized feedstuffs:

Blood meal, which is dried, ground blood, is prepared by heating blood until coagulated, pressing out the excess moisture, and drying and grinding the solid residue. It is higher in protein than other animal by-products, but its protein is generally considered to be of poor quality—though relatively rich in some of the essential amino acids—maybe because some of it is damaged by heat during processing. It is not a good source of vitamins.

Blood flour is dried blood, prepared by

special processes and reduced to a fine powder.

Meat is the clean, wholesome flesh derived from slaughtered mammals; the term is limited to striate muscle that is skeletal or found in the tongue, diaphragm, heart, or esophagus (but it does not include the lips, snout, or ears), with or without the accompanying and overlying fat and portions of skin, sinew, nerve, and blood vessels.

The term "meat" when applied to animals other than cattle, swine, sheep, and goats shall be used in qualified form, as, for example, "horse meat," "reindeer meat," "whale meat," etc.

Meat by-products consist of any non-rendered, clean, wholesome parts of the carcass of slaughtered mammals other than meat, such as lungs, spleens, kidneys, brains, stomach, and intestines free from their contents, but meat by-products do not include skin, horns, teeth, hoofs, and bones.

When applied to animals other than cattle, swine, sheep, and goats the term shall be used in qualified form, as for example, "horse meat by-products," "reindeer meat by-products," etc.

Meat meal, or *meat scrap*, is the finely ground, dry-rendered residue from animal tissues exclusive of hair, hoof, horn, hide trimmings, blood meal, manure, and stomach contents, except in such traces as might occur unavoidably in good factory practice. When these products contain more than 4.4 percent phosphorus, they shall be designated either *meat and bonemeal* or *meat and bone scrap*. The products must be sold according to their protein content.

> NOTE: Meat scrap or meat and bone scrap must not be used as a designation of food for dogs and fur-bearing animals.

In modern *dry-rendering*, the fat of animal meat scraps, trimmings, offal, or whole carcasses is removed by dry cooking in an open steam-jacket vessel. After the water of the raw material is evaporated and the excess fat drained off, the residue is pressed and ground.

Meat meal and meat and bone meal are widely used in poultry feeds; in swine feeding meat meal is used like tankage as a main supplement to corn.

Digester tankage, meat meal tankage, or *feeding tankage*, is the finely ground, dried residue from animal tissues exclusive of hair, hoof, horn, manure, and stomach contents, except in such traces as might occur unavoidably in good factory practice, especially prepared for feeding purposes by tanking under live steam or dry-rendering, or a mixture of the products. When these products contain more than 4.4 percent phosphorus, they must be designated *digester tankage with bone, meat and bone meal digester tankage, meat and bone meal tankage*, or *feeding tankage with bone*. The product must be sold according to its protein content.

The process of cooking rejected carcasses, etc. in closed digester tanks under pressure—so as to skim off most of the fat—is called *tanking under live steam*. The watery portion is then removed and concentrated to a gluey mass, or *stick;* the latter is returned to the (pressed) solid portion in the tank, mixed well, dried, and ground.

> NOTE: The dry-rendering process gives a lighter and less odoriferous product than digester-tanking.

Tankage is used primarily as swine feed to supplement corn. Like other animal-protein supplements, tankages (and meat meal) remain an integral part of the more dependable rations for growing pigs and for the breeding herd.

Whale meal is prepared from the clean, dried, undecomposed flesh of the whale, after part of the oil has been extracted. If it contains more than 3 percent salt (NaCl), the amount of salt must constitute a part of the brand name; in no case shall the salt content of whale meal exceed 7 percent.

Animal liver meal is the product obtained by drying and grinding liver from slaughtered mammals. It is valued in poultry feeds for its VITAMIN B_2, VITAMIN B_{12}, and possibly other factors now unknown.

Animal liver and glandular meal is obtained by drying and grinding liver and other glandular tissue from slaughtered mammals. At least 50 percent of the dry weight of the product must be derived from liver. It may contain one-third less VITAMIN B$_2$ than liver meal but otherwise is similar in composition and use.

Extracted animal liver meal is obtained by drying and grinding the residue of animal liver tissue from which a large portion of the vitamins and/or minerals has been removed.

Poultry by-product meal consists of the ground, dry-rendered, clean, wholesome parts of the carcass of slaughtered poultry, such as head, feet, underdeveloped eggs, gizzard and intestines exclusive of feathers, and gizzard and intestinal contents except in such trace amounts as might occur unavoidably in good factory practice.

NOTE: *Animal fat*—a tentatively accepted animal product—is a mixture of FATS of a quality suitable for feeding, obtained from animal tissues in the commercial processes of rendering.

References: F.6; E.12.

See also BY-PRODUCT FEEDSTUFFS; SULFUR DIOXIDE; DEHYDRATED TANKAGE.

ANIMAL PROTEIN. *See* PROTEINS.

ANIMAL PROTEIN FACTOR = APF.

ANIMAL STARCH = GLYCOGEN.

ANISE SEED, or *anise*, the dried ripe fruit of *Pimpinella anisum*, is a stomachic STIMULANT, CARMINATIVE, and flavoring agent sometimes used as a FEED INGREDIENT.

ANNUAL is a plant which completes its entire life cycle from germinating seedling to seed production and death within a year. In colder climates typical annuals do not survive the winter, but the so-called *winter-annuals* germinate in late fall or spring, are dormant through the winter, and complete their life history the following spring.

Reference: D.9.

See also BIENNIAL.

ANNUAL BLUEGRASS (*Poa annua*), as the name implies, an annual, normally begins growth in late summer or early fall from seed produced earlier the same year. It is more dwarfed than KENTUCKY BLUEGRASS and lacks creeping rootstocks (RHIZOMES). It also has shorter, broader leaves than the other BLUEGRASSES; the leaves are shiny on the dorsal side. The color, a light green, is distinctive. The grass is of little economic importance as pasturage.

Reference: H.1.

See also GRASSES.

ANNUAL LESPEDEZAS include two important species: (1) COMMON LESPEDEZA with the varieties *Kobe* and *Tennessee 76*, and (2) KOREAN LESPEDEZA (with the varieties *Harbin, Climax, Late Korean*, and *Early Korean*). They are leafy LEGUMES, less than 2 ft. tall. Korean has more prominent STIPULES and a wider leaf than common lespedeza. Seed is borne along the stem on the latter; on Korean it is borne in clusters at the tips of branches arising from the leaf axils. Hairs on stem of Korean turn upward; those of common lespedeza turn downward.

Both species are used as reseeding annuals in permanent pastures in combination with introduced GRASSES and with other legumes; in overgrazed and depleted native PASTURES; as temporary summer pasture in pure stands or in combination with CEREALS; and as seed and hay crop.

Livestock gains per acre and per head are greater in July, August, and September from pastures containing an annual lespedeza as the predominant species than from similar pastures without lespedeza. The palatability declines rapidly after the plant reaches full-bloom stage.

Dangers: BACTERIAL WILT is the most serious disease of annual lespedezas.

Reference: E.7.

See also LESPEDEZAS.

ANNUAL PASTURES.

See PASTURES; PASTURE MANAGEMENT.

ANOREXIA means: loss of appetite.—

See also VITAMINS.

ANT. *See* ANTS.

ANTENNA, or *feeler*, is one of the two movable organs of sensation projecting from the head of insects.

ANTHER is a usually two-celled pollen sack. It is the pollen-bearing part of the male floral organ, better known as STAMEN.—*See also* GRASSES.

ANTHOXANTHUM. *A. odoratum*
= SWEET VERNALGRASS.

ANTHRACNOSE. *See* ANTHRACNOSES.

ANTHRACNOSE OF GRASS is caused

by *Colletotrichum graminicola* which occurs commonly on the leaves of forage and grain SORGHUMS, JOHNSONGRASS, RYE, WHEAT, ORCHARDGRASS, and other GRASSES grown in the humid areas of the South. Broomcorn is especially susceptible to this LEAF DISEASE. The fungus, which also causes COLLETOTRICHUM STALK ROT, may be carried on the seed or may live on dead or on decaying plant refuse or in the soil. Infection often causes spots to develop on the leaves when the plants are still in the seedling stage; later the disease may spread to other leaves as they appear. Usually the leaves are not affected severely until about the middle of the growing season.

The spots on the leaves first are $\frac{1}{16}$ to $\frac{1}{8}$ in. in diameter, tan to reddish-purple in color, depending on the variety. Later they enlarge and may unite to cover large areas. The leaf midrib is often strikingly discolored. The centers of the spots fade to a grayish-tan color, and examination with a hand lens will reveal numerous pinpoint black specks with short, stiff hairs. These are the fruiting bodies of the fungus which, under moist conditions, produce pinkish spore masses. The spores are spread by rain and wind to other leaves.

Defoliation due to anthracnose reduces the value of the plants for forage and may reduce the sugar content of the sorghum stalks.

Control: Clean culture and rotation should reduce the losses due to anthracnose. The principal means of control, however, is the growing of resistant varieties.

The Atlas and Planter sorghums are highly resistant. Most grain sorghums are less susceptible to anthracnose than forage sorghums and broomcorn; hegari, Western Blackhull kafir, and Martin are resistant.

Reference: L.1.

See also ANTHRACNOSES; FUNGUS LEAF DISEASES.

ANTHRACNOSES are various FUNGUS DISEASES usually characterized by ulcerlike areas on the host. The fungi involved are *Colletotrichum* spp., *Glomerella* spp., and *Kabatiella* spp. They cause ANTHRACNOSE OF GRASS; LESPEDEZA ANTHRACNOSE, LUPINE ANTHRACNOSE, CROTALARIA

ANTHRACNOSE, PEA ANTHRACNOSE, NORTHERN ANTHRACNOSE, SOUTHERN ANTHRACNOSE, VETCH ANTHRACNOSE, and FALSE ANTHRACNOSE OF VETCH (all of these affecting LEGUMES); and various anthracnoses affecting other plants.

ANTIBIOTIC FEED SUPPLEMENT.
See VITAMINS; ANTIBIOTICS.

ANTIBIOTICS are chemical agents (produced either by micro-organisms or synthetically) that have the ability to inhibit the growth of micro-organisms or even to destroy them. They have become important for their therapeutic effects in many diseases and as admixtures to feedstuffs because they are capable of increasing the *rate of growth* of the young of various animal species. If more than 50 gm. antibiotics is contained in 1 ton of finished feed, it is considered a *medicated feed* and the LABEL must indicate disease conditions, dosage, etc.

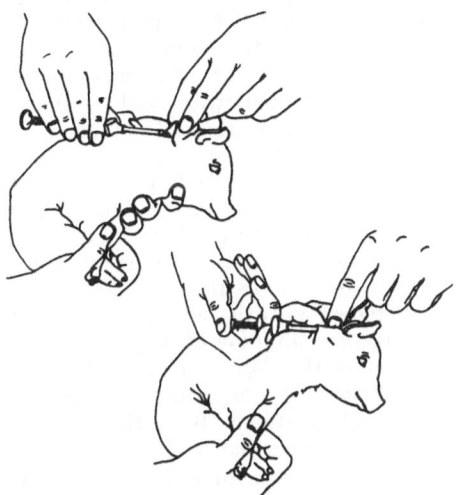

Implantation of bacitracin pellet. *Left*, with the thumb and forefinger of the left hand, grasp the loose skin behind the ear of the baby pig and place the point of the implanter on the skin. *Right*, force the point through the skin and advance it about 1 in. beyond the skin opening. Push the pellet through the needle with the plunger.

Antibiotics are also capable of controlling some *plant diseases*, especially those caused by FUNGI and BACTERIA.

ANTIBIOTICS AND THEIR COMPARATIVE VALUES
IN THE RATIONS OF FARM ANIMALS

	Swine		Poultry		Dairy calves	
	Gain	Feed eff.	Gain	Feed eff.	Gain	Feed eff.
Basal ration †............	100	100	100	100	100	100
Aureomycin.............	115*	96	115	98	124	90
Gramicidin..............	...	...	98	...	...	...
Bacitracin..............	111	97	116	...	...	...
Chloromycetin...........	98	96	93	...	...	...
Neomycin...............	93	88	105	...	...	...
Penicillin...............	110	95	118	94	...	...
Polymyxin..............	96	100	...	...	...	...
Streptomycin...........	109	97	123	...	...	...
Subtilin................	89	130	...	...	...	...
Terramycin.............	117*	97	...	...	...	...

* When compared directly these two antibiotics—aureomycin and terramycin—were equal.
† The gain and feed efficiency of animals fed the basal ration (without antibiotics) is assigned the value of 100. Various antibiotics were added to the basal ration and the comparative values are related to the basal ration as follows:
If it took 350 lb. basal ration or 336 lb. basal + aureomycin to product 100 lb. gain the feed efficiency of the aureomycin-containing feed is 350 : 336 = 100 : x, or 96.

Antibiotic feed supplements are feedstuffs containing one or more antibiotics having growth-promoting properties. It is assumed that the antibiotics in the animal's digestive tract eliminate microorganisms which otherwise would compete for the available nutrients from the consumed feed, or eliminate micro-organisms secreting toxins which could retard animal growth; the antibiotics may also stimulate intestinal bacteria to produce B-COMPLEX VITAMIN fractions (especially FOLIC ACID, NIACIN, and vitamins B_1, $_2$, and $_{12}$) and/or yet unknown growth factors; they may encourage consumption of larger amounts of feed, and thus speed up growth; they may enhance the absorption of nutrients into the blood-stream from the mucous lining of the intestinal wall; finally, they may provide help in the utilization of protein and vitamins, thus permitting savings on these nutrients in otherwise well-balanced rations.

Among the antibiotics used in feedstuffs are AUREOMYCIN (now a tradename for *chlortetracycline*), BACITRACIN, CHLOROMYCETIN, PENICILLIN, STREPTOMYCIN, and TERRAMYCIN (now a trade name for *oxytetracycline*); of minor importance are GRAMICIDIN, NEOMYCIN, POLYMYXIN, and SUBTILIN. The relative values of anti-biotics in the rations of livestock and poultry are shown in the table.

Swine: The effects of different antibiotics on performance of growing-fattening pigs fed a corn-soybean meal ration in dry lot are not uniform. On the basis of average daily gains made by unthrifty pigs, *chlortetracycline* (aureomycin) and *oxytetracycline* (terramycin) give the greatest response (27 percent of basal ration), followed by procaine penicillin (10 percent), chloromycetin (7 percent), and streptomycin (5 percent). The response of thrifty pigs to antibiotics is much smaller.

The greatest effect can be observed during the first part of the feeding period, or at the lighter weights, with a decline in the effects as the pigs gain weight. Unthrifty pigs usually do best on a high protein ration (18 to 20 percent). However, the response to antibiotics is greatest on lower protein levels (10 to 12 percent).

Suckling pigs, given a single implantation of *bacitracin* subcutaneously, in form of a 1,000 unit pellet, showed a growth response averaging 11.3 percent at weaning time (56 days old) in at least one test, but later work failed to confirm this observation.

Ruminants: An increase in growth of calves will result from feeding *chlortetracycline* (aureomycin), *oxytetracycline* (terramycin) and possibly other antibiotics up to about six months of age; the greatest response, however, usually occurs at eight to twelve weeks of age. The economic value depends on two factors: (1) whether the extra weight is to be marketed during the six-month period and (2) whether the extra gains pay for the antibiotics used. In the case of herd replacement, no particular advantage is gained from the increased body weight because the differences do not persist until the age of first calving. The antibiotics are also useful in reducing the incidence of scours.

NOTE: No consistent advantages in weight gains of antibiotic-fed *lambs* have been obtained.

Poultry: Best results with antibiotics are obtained in poultry with penicillin, bacitracin, aureomycin, and terramycin.

The growth response of chicks to antibiotics when fed in chick-starting rations depends on the quality of the rations. The better their quality with respect to protein and vitamins, the less the growth response obtained by supplementing the ration with antibiotics. The greatest percentage-growth response with antibiotics is obtained on the poorer all-plant protein rations. The greatest weight is obtained on the better quality rations (plant protein plus fish meal) supplemented with antibiotics; in this ration, penicillin and bacitracin maintain an increased percentage-growth for a prolonged period of time.

The addition of aureomycin or terramycin to different types of basal laying diets has not been shown to affect the performance of laying birds with respect to rate of lay, gain in body-weight, feed consumption, mortality, egg weight, egg quality, and hatchability of fertile eggs.

GENERAL CONCLUSIONS

Response to antibiotics is usually greater with plant protein rations; greatest improvement over basal has been with peanut meal-corn rations in Florida; some rations containing fish meal are not improved by antibiotics.

Response is greater with young or unthrifty animals on infected ground, or chicks in "old" buildings. Addition of antibiotics will often help runts. Response on pasture is less than in dry lot.

RECOMMENDED DOSAGES

Poultry: Penicillin 2 mg./lb. feed; other antibiotics 5 to 10 mg./lb. feed.

Swine: Growing, fattening: 5 to 10 mg./lb. feed; creep feeds: 20 mg./lb.; protein supplements 25 mg./lb.

Calves: 50 mg./calf a day until twelve weeks of age. Heavier levels have sometimes been successful in reducing scours.

Precautions: Remove antibiotics from calf rations at twelve to sixteen weeks. If antibiotics are removed from the rations of swine, there may be a set-back with decreased rates of gain and some scouring. At the present time, antibiotics are being tried in the rations of adult ruminants. Results have been variable. Some favorable responses with pelleted sheep rations and high roughage rations for cattle have been reported.

References: L.13; N.5; M.46; B.26; H.46.

See also VITAMINS; FERMENTATION PRODUCTS; BLOOD.

ANTICARSIA. *A. gemmatilis*
= VELVETBEAN CATERPILLAR.

ANTIDOTES are agents which are used to counteract the effects of poisons; e.g., METHYLENE BLUE; SODIUM NITRITE, and SODIUM THIOSULFATE.

ANTIENZYMES are substances which inhibit the action of ENZYMES. Among the many substances which contain antienzymes are legume proteins (e.g., soybean proteins).—*See also* OILSEED MEALS.

ANTIHEMORRHAGIC FACTOR
= VITAMIN K.

ANTI-INFECTIVE VITAMIN
= VITAMIN A.

ANTINEUROTIC VITAMIN
= VITAMIN B₁.

ANTIOXIDANTS are chemicals which retard the oxydation of such substances as fats. They conserve the nutritive value of feeds which contain fats.

Officially *accepted antioxidants* for feedstuffs are: RESIN GUAIAC; LECITHIN and NORDIHYDROGUAIARETIC ACID; CITRIC ACID; PHOSPHORIC ACID; PROPYL GALLATE; THIODIPROPIONIC ACID; DILAURYL THIODIPROPIONATE; DISTEARYL THIODIPROPIONATE;

BUTYLATED HYDROXYANISOLE; BUTYLATED HYDROXYTOLUENE; ETHOXY - TRIMETHYL DIHYDROQUINOLINE (for poultry feed only); VITAMIN C; VITAMIN E.

ANTIPELLAGRA VITAMIN = NIACIN.

ANTIRACHITIC VITAMIN
= VITAMIN D_3.—*See also* VITAMIN D.

ANTISTERILITY VITAMIN.
See VITAMIN E.

ANTIXEROPHTHALMIA VITAMIN
= VITAMIN A.

ANTS are insects which live in nests under the ground or in tree trunks. The nests consist of numerous galleries. Adult ants may be males, females, or sexless workers. The eggs develop to larvae and then to nymphs from which the mature ants emerge. There are many species of ants. Before attempting to control ants it is necessary to locate their nests.

Control: When nests are located, the ant colonies can be completely eradicated by applying CHLORDANE.

Reference: M.27.

See also RED HARVESTER ANT; WESTERN HARVESTER ANT; CORNFIELD ANT; FIRE ANT.

ANURAPHIS. *A. maidi-radicis* = CORN ROOT APHID; *A. bakeri* = CLOVER APHID.

ANUS. *See* DIGESTIVE TRACT.

A.O.A.C. stands for *Association of Official Agricultural Chemists*, an organization which was formed for the purpose of developing dependable methods for analyzing feeds, feed supplements, drugs, fertilizers, etc.

A.O.A.C. chick units is a term used to express VITAMIN D potency. 1 U.S.P. unit = 1 International unit = 1.33 A.O.A.C. chick units VITAMIN D_3.

APANTELES. *A. medicaginis*, a tiny WASP, is a natural enemy of the ALFALFA CATERPILLAR.

A. militaris, another very small insect, is an important enemy of the ARMYWORM. It pierces the caterpillar with its sting, laying its eggs inside the armyworm's body where they quickly hatch; the maggots, having eaten their fill, bore their way outward and spin little silken cocoons in a mass together, somewhat resembling grains of rice entangled in cotton.

APEX is the tip of anything in botany; it means the end farthest from the point of attachment of an organ.

APF, or *animal protein factor*, is a term which referred to an undetermined vitamin in animal products. Antibiotic fermentation residues also were said to have APF activity. The term APF must not be used on labels or in advertisements. This term has been replaced with two other terms: ANTIBIOTIC SUPPLEMENT and VITAMIN B_{12} SUPPLEMENT.—*See also* VITAMINS.

APHELENCHOIDES. *A. oryzae*, a nematode, causes WHITE TIP on rice.

APHIDS, or *plant lice*, are minute, soft-bodied insects of different genera and of green, gray, red, or black color. They feed by sucking sap through their beaks, which they insert into tissues of practically all kinds of plants. They serve as carriers of plant VIRUSES. Most species produce winged and wingless individuals. The former are chiefly responsible for the spread of virus diseases.

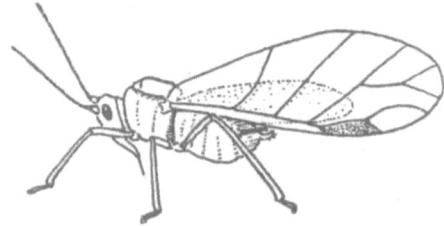

Aphid. (U.13.)

The aphids present on clover and other legumes cause injury, but the numbers observed in many regions do not indicate that special control measures need be taken. If an insecticide is to be used (under observation of the proper precautions), DDT or ROTENONE is recommended.

Among the aphids are PEA APHID, CLOVER APHID, YELLOW CLOVER APHID, CORN LEAF APHID and CORN ROOT APHID.

Reference: C.10.

See also SUBCLOVER; VETCHES; MEADOW FOXTAIL; SUDANGRASS; BUCKWHEATS; RED LEAF; BACTERIAL STRIPE OF OATS; HALO BLIGHT.

APHIS. *A. maidis* = CORN LEAF APHID.

APLOPAPPUS. *A. heterophyllus*
= RAYLESS GOLDENROD.

APPLE PECTIN.
                    See DRIED APPLE PECTIN PULP.
APPLE POMACE is a PLANT BY-PRODUCT obtained after expressing juice from apples. It may be fed to cattle fresh, as silage, or dry. DRIED APPLE POMACE is an official feedstuff.

Reference: E.12.
See also SILAGE CROPS.

APPLER. *See* RED OAT (variety).

APPLES. *See* FRUITS; APPLE POMACE; DRIED APPLE PECTIN PULP.

APPRESSED means: lying flat or closely against another organ or part—not spreading; for example, leaves lying against the stem.

APUS. *Apus* spp. = TADPOLE SHRIMPS.

AQUATIC means: of, or pertaining to, water; growing in water; e.g., an aquatic GRASS.

AQUEOUS means: watery. In an aqueous solution, water is used as the solvent.

ARACHIDONIC ACID, which occurs in animal fats, is an important unsaturated FATTY ACID.—*See also* VITAMIN F.

ARACHIS. *A. hypogaea* = PEANUT.

ARASAN is a fungicide containing 50 percent THIRAM or *bis(dimethylthiocarbamyl) disulfide;* it has been found excellent for SEED TREATMENT of CORN and SORGHUM. It controls the KERNEL SMUTS and greatly improves emergence and stand when soil conditions after planting are unfavorable for germination and growth. Apply it at the rate of 2 oz. per bushel in the manner and with the precautions described for COPPER CARBONATE. It does not injure sorghum seed, even if the seed is treated several months before planting.

NOTE: Arasan in the form of *Arasan SF*, containing 75 percent thiram, may be applied by the slurry method; 1 gal. water containing 1½ lb. of this material will treat 30 bu. sorghum seed. Used at the rate of 2 percent of the weight of shelled seed, Arasan is an effective seed treatment for PEANUTS.

For the control of GRASS SMUTS, 4 oz. Arasan per bushel seed is recommended; the seed must be planted not sooner than two days and not later than two months after treatment.

If used in slurry form at the rate of 1 oz. per 1 bu. rice seed, Arasan controls BROWN SPOT OF RICE.

Caution: Arasan is relatively non-poisonous to farm animals and human beings, but may irritate the skin of anyone allergic to sulfur. As it is irritating to the nose and throat, an effective dust mask should be worn when applying it.

NOTE: When fed to hens, 10 to 50 ppm. Arasan may reduce calcification of eggs; and 100 ppm. may prevent the development of hard-shelled eggs. In controlled experiments, egg production dropped from 70 to 100 percent when half of the corn fed to hens was replaced with Arasan-treated corn.

Reference: L.1.
See also TERSAN.

ARCTIC BARLEY
          =*O.A.C.21.* See SIX-ROWED BARLEY.

ARGENTINE BAHIAGRASS.
                    See BAHIAGRASS (variety).

ARGENTINE WHEAT
          = *Turkey wheat.* See COMMON WHEAT.

ARGININE is one of the essential AMINO ACIDS.

ARGON is a gas which occurs in the atmosphere.—*See also* OXYGEN.

ARID is a dry climate with an annual precipitation of (usually) less than 10 in. and requiring irrigation for crop production.—*See also* SEMIARID.

ARISOY. *See* SOYBEAN (variety).

ARISTASTOMA. The FUNGUS *A. oeconomicum* causes WHITE SPOT.

ARISTIDIA. *A. fendleriana*
          = FENDLER THREE-AWN.

ARIVAT BARLEY.
                    See SIX-ROWED BARLEY.

ARIZONA BROWN STINK BUG
                    = BROWN COTTON BUG.

ARIZONA COMMON ALFALFA is one of the common ALFALFAS.

ARKAN. *See* SOYBEAN (variety).

ARKSOY and **ARKSOY 29's** are *soybean* varieties.

ARLINGTON OAT.
                    See COMMON OAT (variety).

ARLINGTON VELVETBEAN.
          See DEERING VELVETBEAN (variety).

ARMREDO. *See* SOYBEAN (variety).

ARMSBY FEEDING STANDARDS are expressed in terms of (1) *net energy* values in calories and (2) *digestible true* (not crude) *protein*. However, no recommendations are made for dry matter. These standards are rarely used.

ARMY CUTWORM (*Chorizagrotis auxiliaris*) is one of the CUTWORM species. It is injurious to ALFALFA in the early spring; it likes also SMALL GRAINS, GRASSES, and CORN.

Reference: G.7.

See also ARMYWORM.

ARMYWORM (*Cirphis unipuncta*), or *true armyworm*, is a pest which occurs throughout the United States east of the Rocky Mountains; it has also been found in New Mexico, Arizona, and California.

The armyworm is often confused with the FALL ARMYWORM. The ARMY CUTWORM is occasionally mistaken for the armyworm. The VARIEGATED CUTWORM often occurs in large numbers along with the armyworm where LEGUMES are present, especially in ALFALFA fields containing much GRASS.

The loss in money to the farmer caused by the armyworm has been great in the past.

Armyworm larva. (B.16.)

By preference the armyworm feeds on grasses, both wild and cultivated, particularly TIMOTHY and BLUEGRASS; next, on the grasslike grains, such as the several varieties of MILLET. WHEAT, CORN, OATS, BARLEY, RYE, SERICEA, GRAIN SORGHUMS, and SOYBEANS, are also among its favorites, as is alfalfa, particularly in the southwestern states.

The armyworm usually appears in the fields very suddenly. It seems certain that the moths (its parents) at times fly in great numbers for many miles with the prevailing wind and congregate to deposit their eggs on favorable food plants.

Generally, outbreaks of the armyworm are more common following cold, backward springs; they should be looked for first in neglected portions of fields in which rank growths of wild grasses or lodged and fallen unripe grains are found. These should be examined frequently and closely, especially from late April to the early part of July, to see whether any small greenish caterpillars are feeding near the ground under overhanging leaf blades.

The MOTH, or *miller*, measures about 1½ in. across the expanded wings. It is brownish gray, with a single small white spot near the center of the front pair of wings, the hind wings being somewhat darker along the hind edges. The parents of the armyworm fly only at night.

The eggs laid by the moths resemble white beads, considerably smaller than the head of a pin; they are deposited in masses or rows. In from eight to ten days very small greenish "worms" hatch from the eggs; high temperatures may shorten this period and low temperatures lengthen it.

As the caterpillars grow and feed, their skins become too tight for them; they grow a flexible one underneath, shed the old one, and continue feeding greedily until the new skin has stretched to its limit. This occurs several times until the caterpillar becomes full-grown in three to four weeks. It is then nearly naked, smooth, striped, and about 1½ in. long. The stripes, one along each side and a broad one down the center of the back, are dark and often nearly black. The color of the body between the dark stripes ranges from greenish to reddish brown. The head is pale greenish brown, finely mottled with darker brown.

When an army of these worms is at work in a field, a rustling sound can be heard as they devour every blade in sight. After consuming the food supply, the caterpillars crawl away in search of other food.

When the full-grown caterpillar ceases feeding, it usually burrows under litter on the ground, under clods, or into the soil to a depth of 2 or 3 in. and by twisting and turning forms a cavity, or cell, where the *pupa* is formed.

The pupa, at first reddish or chestnut brown, turns almost black as the time for emergence of the moth approaches. It wriggles its tail vigorously on being disturbed.

When the moth crawls from the pupal case it has not yet developed its wings; after about an hour they are completely developed, and the insect is capable of flying. If undisturbed, however, the moths will usually remain at rest for several hours before they fly away to mate and lay their eggs. The male moth is usually considerably smaller than the female. It takes from seven to eight weeks for the development from egg to moth.

The armyworm moth is strongly attracted to light at night, and frequently large numbers of the moths are seen about outdoor lights at the time of an outbreak.

Control: The armyworm has many natural enemies. A common and most effective one is a medium-sized, gray fly, *Winthemia quadripustulata*. This parasite often becomes numerous enough to control the armyworm completely in a given locality. Another important foe is a very small wasplike insect, *Apanteles militaris*. The maggots of both insects eat the inside of the caterpillar. There are several other insect enemies, e.g. *Calosoma calidum*, *Enicospilus purgatus*, and a *Sphex* sp.

More than 40 species of native wild birds eat the armyworm in its various stages. Among them are the redwinged blackbird, purple grackle, robin, crow, yellow-headed blackbird, grasshopper sparrow, song sparrow, and crowbird. Domestic fowls of all kinds, if allowed to roam over infested fields, will devour the caterpillars and pupae. Skunks and toads also eat thousands of the caterpillars and pupae.

The importance of watchfulness on the part of the farmer in combating the armyworm cannot be too greatly emphasized. If the invaded area is small, the caterpillars can be destroyed by mowing off the grass or grain, scattering straw over the spot, and burning it.

Poisoned baits have long been used to destroy the armyworm and many different species of cutworms. An effective bait is POISONED BRAN MASH, containing Paris green, arsenic trioxide, toxaphene, sodium fluosilicate, or calcium arsenate. The bait may be used in alfalfa fields and cornfields where it is desired to save the crop for forage, although some temporary injury

from burning may result if the bait is scattered too thickly on the plants. For best results the bait should be spread late in the afternoon. When the worms are migrating, a swath of the bait can be spread in front of them with good effect.

Excellent control of the armyworm has been obtained with a spray of TOXAPHENE applied by aircraft. A spray containing DDT has also given satisfactory control when applied by aircraft or with ground equipment. Dusts containing toxaphene or DDT have provided good control.

If the worms are not discovered until they have begun to travel in a mass, many of them can often be destroyed by running a ditch completely around the infested area. The worms will fall into it and can easily be crushed by dragging a log back and forth in the ditch. Sometimes the ditches are not dragged but made so that the far side is as steep as possible, and post holes about 18 in. deep are sunk in the bottom of the ditch about every 20 ft. The worms will crawl along the bottom of the ditch and fall into the holes, where they may be crushed or sprinkled with a little KEROSENE or other light oil. Sometimes, in extremely hot weather, they die in these holes without any treatment.

Reference: W.12.

See also LYGUS BUGS.

ARNAUTKA.
　　　See DURUM WHEAT (white variety).

AROMATICS are substances having an agreeable taste and/or odor from the presence of an essential oil. They have sometimes medicinal value by stimulating appetite, but are more often used to flavor feeds; e.g., FENUGREEK SEED.—*See also* FEED INGREDIENTS.

ARRHENATHERUM. *A. elatus*
　　　　　　　　= TALL OATGRASS.

ARROWGRASS = ARROW PODGRASS.

ARROWHEAD (*Sagittaria latifolia*), also called *waterlily*, is a perennial which occurs as a serious WEED pest in RICE fields. Its leaves are shaped like arrows.

ARROW PODGRASS (*Triglochin maritima*) is commonly called *arrowgrass*. In ruminants it will cause difficult breathing, spasms, coma, and other symptoms of illness of short duration when the animals eat, within a few minutes, about 1 lb. of green leaves and stems per 100 lb. body weight. This *cyanogenetic* plant is toxic only under drought conditions.—*See also* POISONOUS PLANTS; CYANOGETIC GLUCOSIDES.

Arrow podgrass. (D.19.)

ARROYO is a steep-sided gully where water once flowed.

ARSANILIC ACID, or *para-aminobenzenearsonic acid*, forms colorless crystals. Its sodium salt, SODIUM ARSANILATE, is water-soluble.—*See also* ARSONIC ACIDS.

ARSENIC is a CHEMICAL ELEMENT found in nature in form of minerals. It is very poisonous.

White arsenic = ARSENIC TRIOXIDE.—*See also* ARSENICALS.

ARSENICALS are ARSENIC-containing chemical compounds used as insecticides, fungicides, herbicides, or drugs; e.g., CALCIUM ARSENATE, PARIS GREEN, ARSONIC ACIDS, SODIUM ARSANILATE, and SODIUM ARSENITE.

When used to kill BRUSHES and trees, the arsenical should be applied as a liquid in gashes cut completely around the plant, preferably in the fall months.

Caution: Livestock must be kept out of the area until the next grazing season or until rains have leached the poison into the soil.—*See also* TONICS.

ARSENIC TONIC. See TONICS; ARSONIC ACIDS; ARSENICALS.

ARSENIC TRIOXIDE—commonly called *white arsenic* or "arsenic"—forms a white, crystalline powder or amorphous lumps and is soluble in boiling water and solutions of alkalies. It is very poisonous. It is sometimes used in place of PARIS GREEN in POISONED BRAN MASH for the control of GRASSHOPPERS. When this grasshopper bait falls on leaves, its arsenic may cause spots.—*See also* NONPARASITIC SORGHUM-LEAF DISCOLORATIONS; SODIUM ARSENITE.

ARSONIC ACIDS are organic ARSENIC compounds. Among their derivatives, known as *arsonic compounds*, are ARSANILIC ACID, 3-NITRO-4-HYDROXYPHENYLARSONIC ACID, and SODIUM ARSANILATE (the sodium salt of ARSANILIC ACID). These compounds are used at the rate of 0.01 percent in hog and poultry feeds and sometimes also in drinking water because of a variety of actions they stimulate (in addition to being *arsenic tonics*); e.g., they improve feathering and pigmentation in poultry, control enteritis in hogs, and promote growth in the young of both species.

Caution: These ARSENICALS are poisonous when used in larger than the recommended amounts. Feeding of medicated rations must be discontinued a few days before slaughter.

ARTEMISIA. The genus *Artemisia* includes the following sagebrush species: *A. spinescens* = BUD SAGEBRUSH; *A. nova* = BLACK SAGEBRUSH; *A. tridentata* = BIG SAGEBRUSH; *A. cana* = SILVER SAGEBRUSH; and *A. tripartita* = THREETIP SAGEBRUSH; *A. frigida* = FRINGED SAGEBRUSH.

ARTICHOKES are plants which belong to three genera; the most important of them is the JERUSALEM ARTICHOKE.—*See also* GRAZING; INULIN.

ARTIFICIAL HAY DRYING.
                              See HAY DRYING.

ASAHI. *See* RICE (variety).

ASCENDING means: upcurved; growing or directed upward.

ASCLEPIAS. The *Asclepias* spp., or MILK-WEEDS, include *A. eriocarpa* = BROADLEAF MILKWEED; *A. galioides* = WHORLED MILKWEED; *A. mexicana* = MEXICAN WHORLED MILKWEED; *A. labriformis*, and *A. latifolia*.

ASCOCHYTA. The FUNGUS *Ascochyta sorghina* causes the leaf disease ROUGH SPOT; *A. imperfecta* is the cause of BLACK STEM OF CLOVER; *A. gossypii* causes ASCOCHYTA BLIGHT OF COTTON and ASCO-CHYTA STEM CANKER; *A. pisi* causes LEAF AND POD SPOT; *A. pinodella* is the cause of ASCOCHYTA FOOT ROT and its sexual form (called *Mycosphaerella pinodes*) causes ASCOCHYTA BLIGHT OF PEA.—*See also* ASCOCHYTA BLIGHT COMPLEX.

ASCOCHYTA BLIGHT COMPLEX, which affects FIELD PEAS, is divided into two phases: (1) ASCOCHYTA FOOT ROT and (2) ASCOCHYTA BLIGHT OF PEA. Closely related to the Ascochyta blight complex and often considered as its third phase is the LEAF AND POD SPOT.

Control: The various phases are best controlled by avoiding, eliminating, or reducing the causal organisms—i.e., by the use of disease-free seed, crop rotation, and sanitation. Do not plant seed grown in the humid sections of the East and Middle West.

References: S.13; M.24.

ASCOCHYTA BLIGHT OF COTTON is a cotton disease caused by the FUNGUS *Ascochyta gossypii*. On LUPINES this fungus causes ASCOCHYTA STEM CANKER.

ASCOCHYTA BLIGHT OF PEA is caused by the FUNGUS *Mycosphaerella pinodes* (the sexual stage of *Ascochyta pinodella*). This disease is one phase of the ASCOCHYTA BLIGHT COMPLEX which often affects FIELD PEAS. First, small purplish specks are found on the leaves. Then the spots enlarge into round, target-like spots and, if numerous, join to make irregular brownish-purple blotches. The fungus may also blight the blossoms and young pods and cause withering, distortion, and eventual dropping.

Reference: S.13.

ASCOCHYTA FOOT ROT, a phase of the ASCOCHYTA BLIGHT COMPLEX which affects FIELD PEAS, is caused chiefly by the FUNGUS *Ascochyta pinodella;* however, its sexual stage, *Micospaerella pinodes* and the related *A. pisi*, may sometimes attack the stem and root at the soil line and produce a bluish-black foot rot. It is most severe when it is caused by *A. pinodella* and mildest when caused by *A. pisi*.

Reference: S.13.

ASCOCHYTA STEM CANKER affects LUPINES, especially the BLUE LUPINES. It is caused by the *Ascochyta gossypii*, the same FUNGUS that causes ASCOCHYTA BLIGHT OF COTTON. Thus far the disease has been found only in fields where COTTON had been grown during the preceding summer and in which the old cotton stalks were not properly turned under.

Infected lupine seedlings may be girdled by a dark-brown to black canker around the stem that resembles the canker caused by the ANTHRACNOSE fungus. Plants of any age may be killed. Plants approaching maturity may wilt suddenly and die.

Control: Ascochyta stem canker usually can be controlled by turning the cotton stalks under completely before planting lupine, or lupine should follow a nonsusceptible crop.

Reference: W.8.

ASCORBIC ACID = VITAMIN C.

ASEXUAL means: without sex. A grass which propagates by ROOTSTOCKS exhibits *asexual reproduction* (VEGETATIVE REPRODUCTION).

ASH, or *mineral matter*, is the nonvolatile mineral residue obtained after complete burning of organic substances, such as plants. The animal body contains about 4 percent ash and most feedstuffs from 1 to 10 percent.—*See also* WOOD ASHES; NITROGEN-FREE EXTRACT; U.G.F.

ASIATIC BLUESTEMS. *See* TURKESTAN BLUESTEM; CAUCASIAN BLUESTEM.

ASIATIC COTTONS. *See* GOSSYPIUM.

ASPARAGUS. *Wild asparagus* = SKELETONWEED.

ASPARAGUS BEAN. *See* COWPEA (variety).

ASPERGILLUS. The *Aspergillus* spp. are

among the FUNGI causing SEED ROT of sorghum and CONCEALED DAMAGE IN SEED.

ASSASSIN FLIES = ROBBER FLIES.

ASSOCIATED PLANTS form a unit of vegetation, or an *association.—See also* SAGEBRUSHES.

ASSOCIATION OF AMERICAN FEED CONTROL OFFICIALS, at times abbreviated *A.A.F.C.O.*, is an organization which establishes the official definitions of feed ingredients and the LABEL requirements for the various feed formulas.—*See also* MINERAL FEEDS.

ASSOCIATION OF OFFICIAL AGRI-CULTURAL CHEMISTS = A.O.A.C.

ASTER is a name which refers to two related genera of plants, the true or *hardy asters* (which are *Aster* spp.), and the *China aster* (or *Callistephus* spp.); both have daisy-like flowers.

The PARRY ASTER (*Aster parryi*) is a woody species which grows on ranges; it is a POISONOUS PLANT.—*See also* COMPOSITE.

ASTRALGUS. To the *Astralgus* spp. belong the LOCOWEEDS. *A. drummondii*, or *red locoweed*, and *A. tenellus* are nontoxic; other species, e.g., *A. tetrapterus*, are poisonous, but do not produce true locoism.

The POISONVETCHES also belong to this species; they include the following: *A. bisulcatus* = *two-grooved poisonvetch; A. tetrapterus* = *four-winged poisonvetch; A. convallarius* = *timber poisonvetch;* and *A. sabulosus* = *straight-stem poisonvetch.* —*See also* POISONOUS PLANTS.

ASTROCARYUM. The *Astrocaryum* spp. are known as TUCUM PALMS.

ATLANTIC ALFALFA is one of the VARIEGATED ALFALFAS.

ATLAS BARLEY.

See SIX-ROWED BARLEY (variety).

ATLAS SORGO is one of the forage SORGHUMS.

ATMOSPHERE. See AIR.

ATOM is the smallest unit of a CHEMICAL ELEMENT that enters into the composition of MOLECULES.

ATRIPLEX. *A. confertifolia* = SHADSCALE SALTBUSH; *A. gardneri* = GARDNER SALT-BUSH; *A. canescens* = FOURWING SALTBUSH.

ATROPINE is an ALKALOID which acts as circulatory stimulant. Veterinarians sometimes use it in form of the water-soluble *atropine sulfate* (an intense poison) to supplement the injection of SODIUM NITRITE and SODIUM THIOSULFATE in the treatment of PRUSSIC ACID POISONING.

AUGUSTINEGRASS.

See ST. AUGUSTINEGRASS.

AUREOMYCIN is now a trade name for *chlortetracycline;* this ANTIBIOTIC is obtained from a soil organism, the MOLD *Streptomyces aureofaciens*. It is available as *aureomycin hydrochloride*, a crystalline, golden-yellow powder which is only slightly soluble in water. Aureomycin is widely used as a therapeutic agent and in feedstuffs as growth stimulant for poultry, swine, and calves.

AUSTIN. *See* COMMON WHEAT (variety).

AUSTRIAN WINTER FIELD PEA is one of the FIELD PEA varieties.—*See also* ROUGH PEA; OATS.

AUTOGRAPHA. *A. california*
= ALFALFA LOOPER.

AVAILABLE ENERGY
= *metabolizable energy. See* ENERGY.

AVENA. The genus *Avena* includes (1) the cultivated OAT species, *A. sativa* = COMMON OAT and *A. byzantina* = RED OAT, and (2) the wild growing species, *A. sterilis* = WILD RED OAT; *A. fatua* = COMMON WILD OAT; *A. brevis* = SHORT OAT; *A. barbata* = SLENDER OAT; and *A. strigosa* = SAND OAT.

AVIDIN is a protein isolated from raw egg-white. It is capable of binding BIOTIN.

AVOYELLES. *See* SOYBEAN (variety).

AWN is a slender bristle at the end or on the back or edge of an organ; e.g., the "beards" of wheat, oats, barley, or rye.

AWNED means: provided with AWNS; bearded.

AWNLESS means: without AWN or beard.

AXIL is the angle formed between plant stem and any leaf, branch, or organ arising from it.

AXILLARY means: of, or petaining to, an axil.

AXIS is the main stem of a flower or PANICLE.

AXONOPUS. *A. affinio* = CARPETGRASS.

AXTELL is one of the forage SORGHUMS.

AZALEA. *A. occidentalis*
= WESTERN AZALEA.

AZOR. *See* RYE (variety).

B

BAART and **BAART 38** are varieties of COMMON WHEAT.

BABASSU (*Orbignya speciosa*) is a South American palm. Its hard-shelled seed is used in the manufacture of oil and its by-product, BABASSU OIL MEAL.

BABASSU OIL MEAL is the ground residue obtained after the extraction of part of the oil from BABASSU kernels by crushing, cooking, and mechanical pressure. It must be designated and sold according to its protein content.

This officially recognized OILSEED MEAL resembles COCONUT OIL MEAL in color, odor and chemical composition. It is palatable to livestock; too large a proportion of it in the ration causes scouring.

References: F.6; F.8.

BACCHARIS. Some of the *Baccharis* spp. are poisonous; e.g., *B. ramulosa*, or *Yerba-de-pasmo*. Its leaves, when eaten by range cattle because of scarcity of feed in fall and early winter, soon cause extreme prostration and severe inflammation of the stomach.—*See also* POISONOUS PLANTS.

BACILLUS is a rodlike BACTERIUM; e.g., *B. subtilis*, a strain of which is used for the production of BACITRACIN.

BACITRACIN, an ANTIBIOTIC obtained from a *Bacillus subtilis* strain, forms a grayish-white, water-soluble powder. It is used as drug and also in feedstuffs as GROWTH factor for hogs and poultry.

NOTE: 1 Bacitracin unit is equivalent to 23.8 mcg.

BACOPA. *B. rotundifolia*
= WATER HYSSOP.

BACTERIA. See BACTERIUM.

BACTERIAL BLIGHT OF PEA, caused by *Pseudonomas pisi*, attacks not only the leaves, but all above-ground parts of the FIELD PEA plant, producing olive-green to olive-brown, water-soaked areas. The blight may kill the entire infected plant.

Control: No satisfactory measures are known; if practical, only clean seed should be sown and rotation should be practiced.

Reference: M.24.

BACTERIAL BLIGHT OF SOYBEAN, caused by *Pseudomonas glycinea*, is one of the most widespread SOYBEAN diseases. It is usually one of the first leaf spot diseases to appear on young plants. On infected plants, small, angular spots varying from yellow to brown develop on the leaves. The brown, central area of these spots is usually surrounded by a water-soaked margin. These spots later become dry and sunken; they are frequently surrounded by a narrow, yellow border which is more noticeable on the top side of the leaf. Under some conditions, the infection travels especially along the veins.

In severe infection, the leaves—which often have a torn, ragged appearance—drop off the plants.

NOTE: Beating rain may also cause a ragged appearance of the leaves, which should not be confused with disease symptoms.

Although bacterial blight develops more extensively in cool, rainy weather, it appears to a limited extent throughout the summer. In most years the disease is probably checked appreciably by the onset of hot weather. While the disease is most commonly found on the leaves, it can also infect the stems, leaf stalks (PETIOLES), and pods. It is seed-borne.

Susceptibility to bacterial blight varies considerably among soybean varieties. *Hawkeye* is resistant.

Control: Wherever possible seed of resistant varieties should be used.

Reference: C.9.

See also BACTERIUM; BACTERIAL PUSTULE; BACTERIAL LEAF BLIGHT.

BACTERIAL DISEASES include plant diseases caused by a great variety of BACTERIA; among them are BACTERIAL BLIGHT OF PEA; BACTERIAL BLIGHT OF SOYBEAN; BACTERIAL LEAF BLIGHT; BACTERIAL LEAF AND STEM SPOT; BACTERIAL LEAF DISEASES of sorghum; BACTERIAL PUSTULE; BACTERIAL SPOT; BACTERIAL STREAK; BACTERIAL STRIPE OF OATS; BACTERIAL STRIPE OF SORGHUM; BACTERIAL WILT; STEWART'S WILT; WILDFIRE; and HALO BLIGHT.—*See also* WEAK NECK; ANTIBIOTICS; LEAF DISEASES.

BACTERIAL LEAF AND STEM SPOT, caused by *Pseudomonas alfalfae*, is one of the minor diseases of ALFALFA.

BACTERIAL LEAF BLIGHT of CORN,

caused by *Pseudomonas alboprecipitans*, is a comparatively new disease. It has been severe only in localized areas in some of the southern states and in Nebraska, Kansas, and Indiana.

Long, narrow streaks on the leaves are a characteristic symptom. At first they are water-soaked, olive-green or oily in appearance. Later they turn light brown or tan and have a reddish-brown margin. The streaks may run almost the entire length of the leaf and are often most abundant near the mid rib. Badly diseased leaves shred readily, especially when exposed to wind and driving rain. The organism may also attack the stalk, causing a brown ROT and a shredding of the pith. Infection of the stalks usually takes place at or just above the point where ears are attached. Most plants showing this STALK ROT produce multiple ears which are sterile and may become rotted.

Apparently the bacteria are able to enter the host through stomata (breathing pores) and probably through wounds.

In addition to corn, the disease may attack other GRASSES, including some of the SMALL GRAINS. Nothing is known regarding the inheritance of resistance to the disease.

Control: The disease is of minor economic importance. Thus far no control measures have been developed.

Reference: U.7.

BACTERIAL LEAF DISEASES of sorghum—BACTERIAL STRIPE OF SORGHUM, BACTERIAL STREAK, and BACTERIAL SPOT—

formerly called *red spot* or *sorghum blight*, are likely to be found in the United States wherever SORGHUMS are grown. The organisms that cause these diseases are believed to be carried over from one season to another on the seed, on infected plant material in or on the soil, and occasionally on plants that overwinter. They may be spread from one leaf or plant to another by wind and splashing rain and also by insects.

The bacterial diseases usually do not cause serious losses, because they generally do not develop fully until the plants have reached their full size. During warm (75° to 85° F.), moist seasons, however, they may spread rapidly from the lower to the upper leaves until half to two-thirds of the leaf surface is destroyed. This materially reduces the forage value of the crop and may also interfere with the proper filling of the kernels.

Control: Disposing of old infected plant litter and infected plants that overwinter, along with crop rotation, will reduce the quantity of infective material present in the fields the next season. SEED TREATMENT before planting will keep the disease from being carried over on the seed. Leoti sorgo, Cody, shallu, Tift Sudangrass, Piper Sudangrass, and Sweet Sudangrass, and certain crosses with these varieties are somewhat resistant to all three bacterial diseases.

Reference: L.1.

See also BACTERIAL LEAF BLIGHT; STEWART'S WILT; BACTERIAL BLIGHT OF PEA; WILDFIRE; BACTERIAL BLIGHT OF SOYBEAN; BACTERIAL PUSTULE; HALO BLIGHT; BACTERIAL SPOT; BACTERIAL WILTS; WEAK NECK; BACTERIAL STRIPE OF OATS; SUDANGRASS.

BACTERIAL PUSTULE, caused by *Xanthomonas phaseoli* var. *sojensis*, is a

disease of SOYBEANS, primarily of the leaves, though it sometimes infects the pods. The first symptoms are small, yellow-green spots with reddish-brown centers, most conspicuous on the upper surface of the leaves. The central portion of the individual spot appears slightly raised. It develops into a small pustule, especially on the underside of the leaf.

Many infections on the same leaf produce a large, yellow-to-brown area, dotted with small, darker brown spots. The dead areas of older leaves frequently break up; the leaf then has a ragged look. In later stages the pustules rupture and dry. When this occurs, the disease may become hard to tell from BACTERIAL BLIGHT. Bacterial blight, however, in the early stages develops a narrow, water-soaked area around the center of the spot, whereas bacterial pustule does not. Severe infection often causes plants to loose their leaves.

The disease is apparently carried over from year to year in infected leaves and may be borne on the seeds. It is a warm-weather disease. Many southern varieties of soybean are resistant to it.

No control is known.

Reference: C.9.

See also WILDFIRE.

BACTERIAL SPOT, caused by *Pseudomonas syringae*, attacks the leaves of forage and grain SORGHUMS, broomcorn, JOHNSONGRASS, PEARLMILLET, FOXTAIL MILLET, and CORN. On sorghums the spots appear first on the lower leaves, and gradually spread to the upper leaves as the plants approach maturity. The spots may occur on any part of the leaf and usually are circular to irregularly elliptical and from $\frac{1}{25}$ to $\frac{1}{8}$ in. in diameter. At first they look dark green and water-soaked, but in a few hours become dry and light-colored in the center, which usually is surrounded by a red border. The smaller lesions turn often red throughout, with tiny, somewhat sunken centers. The color varies somewhat in different varieties. Frequently the spots are so numerous that they unite into large diseased areas and cause the death of the whole leaf.

Control: Bacterial spot is controlled like the other BACTERIAL LEAF DISEASES through sanitation, SEED TREATMENT, and selection of resistant varieties, such as Leoti sorgo, Codi, shallu, Tift Sudangrass, Piper Sudangrass, and Sweet Sudangrass.

Reference: L.1.

BACTERIAL STREAK, caused by *Xanthomonas holcicola*, may be seen on the leaves of grain SORGHUMS, SUDANGRASS, and JOHNSONGRASS as narrow, almost transparent streaks about $\frac{1}{8}$ in. wide and 1 to 6 in. long. This BACTERIAL LEAF DISEASE may occur on plants from the seedling stage to near maturity. At first, light-yellow beadlike drops of exudate stand out on the young streaks. Later, narrow red-brown margins or blotches of color appear in the streaks which become red throughout after a few days. Parts of the streaks may broaden into elongated spots with tan centers and narrow red margins. When very numerous, the streaks may join to form long, irregular areas covering a considerable part of the leaf blade, and there may be dead tissue between the reddish-brown streaks. At this stage the bacterial exudate has dried to thin white or cream-colored scales.

The kafirs are relatively resistant to bacterial streak.

Control: Bacterial streak is controlled like all other BACTERIAL LEAF DISEASES through sanitation, SEED TREATMENT, and selection of resistant sorghum varieties, such as Leoti sorgo, Cody, shallu, Tift Sudangrass, Piper Sudangrass, and Sweet Sudangrass.

Reference: L.1.

BACTERIAL STRIPE OF OATS is a minor disease of OATS caused by *Pseudomonas striafaciens*. Like the related HALO BLIGHT, it is a cool-weather disease which may occur sporadically during winter, but disappears quickly as warm weather returns. Both diseases are spread by insects, so their distribution during peaks of infection is associated with APHID and LEAFHOPPER outbreaks. The long water-soaked stripes appear mostly on the leaves.

Control: SEED TREATMENT and crop rotation reduce the primary phase of this disease.

Reference: M.30.

See also CERESAN M.

BACTERIAL STRIPE OF SORGHUM, caused by *Pseudomonas andropogoni*, is the most abundant of the BACTERIAL LEAF DISEASES. It attacks forage, grain, and sweet SORGHUMS, broomcorn, and SUDANGRASS; the forage sorghums seem to be more susceptible than the other grasses of this genus.

The disease is characterized by narrow, irregular stripes, which usually are red and first seen on the lower leaves. The stripes are $\frac{1}{4}$ to 9 in. or more long and tend to be confined between the leaf veins but may join together so as to cover a large part of the leaf surface. Abundant bacterial exudate occurs on the stripes. Unless washed off by rains, this dries and forms red crusts or thin scales, especially on the lower side of the leaves. The color of the stripes varies somewhat on different varieties; e.g., they are light brick red on Red Amber sorgo, dark purplish red on common Sudangrass, and brownish red on kafir.

Control: Bacterial stripe is controlled like the other bacterial leaf diseases, through sanitation, SEED TREATMENT, and selection of somewhat resistant sorghum varieties, such as Leoti sorgo, Cody, shallu, Tift Sudangrass, Piper Sudangrass, and Sweet Sudangrass.

Reference: L.1.

BACTERIAL WILT affects many plants. *Bacterium solonacearum* attacks PEANUTS in form of a seedling disease of minor importance.

Most important is the bacterial wilt of ALFALFA, caused by the bacterium *Corynebacterium insidiosum*. The disease is known in most of our alfalfa-growing regions; it is most serious in central and northern areas of abundant rainfall or irrigation and frequent winter injury. It diminishes in importance with low rainfall and with long growing seasons in the southern states. Because the disease does not usually become destructive until the third crop year, it is of small importance where alfalfa is grown in short rotations.

Bacterial wilt of alfalfa is caused by bacteria that have the ability to grow between the living cells of the plants where these may be exposed, as in a wound, and to enter the water-carrying vessels. Here they increase and are distributed extensively through the plant by moving water until the vessels are obstructed. If the bacteria advance so rapidly that many young cells are killed, the plant wilts and dies quickly. Usually the bacteria grow so slowly that the infected plant is able to survive for a time with a habit of growth so conspicuously different from the usual one that the disease is readily recognized. The first obvious symptom of wilt is a dwarfing and yellowing of the entire plant. Stems are short, leaves are small and pale, many yellow at the edges, which curl upward, and the growth is slow. Such wilted plants become most conspicuous after cutting, when the subsequent crop is about half grown.

By the time the disease can be detected by the foliage, it has already done much damage to the root. In fact, bacterial wilt is most readily distinguished from other WILTS by the appearance of the taproot. Because the bacteria develop in the newest growth and discolor it, a yellow ring is found just beneath the bark. This yellowing may extend far down the root by the time the stems and leaves are conspicuously dwarfed.

Three other alfalfa diseases may be easily confused with bacterial wilt. They are WITCHES'-BROOM, DWARF DISEASE, and a FUSARIUM WILT.

Control: Diseased plants should be thoroughly turned under and allowed to decay before a new seeding is made. Late fall cutting or any cutting treatment that reduces the vigor of the stand should be avoided. Cultivating fields and pasturing when the ground is wet may lead to a rapid spread of the wilt. Under excessive irrigation the disease is increased, expecially when fields are not allowed to dry out to some extent during part of the summer.

Four wilt-resistant varieties of alfalfa have been produced—*Ranger*, *Buffalo*, *Caliverde*, and *Vernal*. Wherever bacterial wilt has been established, the use of a wilt-resistant variety adapted to the region is recommended.

References: J.2; J.3.

See also STEWART'S WILT; ANNUAL LESPEDEZA; VARIEGATED ALFALFAS; SEEDLING DISEASES.

BACTERIAL WILT OF CORN
= STEWART'S WILT.

BACTERICIDE is any substance which destroys BACTERIA.—*See also* SEED TREATMENT; BACTERIOSTAT.

BACTERIOSTAT is an agent which arrests the growth of BACTERIA.

Bacteriostatic means: arresting the growth of bacteria, but not killing them. —*See also* BACTERICIDE.

BACTERIUM (plural: *bacteria*) is any one-celled vegetable micro-organism which multiplies by fission (splitting into parts) or through SPORES. There exist various families of bacteria, each consisting of a number of genera with many species; thus, a total of over a thousand species result— some useful, others harmless, and many dangerous.

The bacteria are widely distributed in air, water, soil, bodies of living plants and animals, and dead organic matter. Lacking CHLOROPHYLL, they cannot produce their own food and must get it already prepared from other sources. Many of the bacteria living in the bodies of plants cause *plant diseases*. They enter the plant through a natural opening, such as a water pore, or through a wound. Once inside they multiply by dividing in two, and migrate among the cells of the plant. Bacteria, however, exert also many beneficial effects in agriculture (e.g., atmospheric NITROGEN FIXATION) and industries (e.g., fermentation).

Some bacteria are necessary for fiber digestion in stomach or intestine. They attack the cellulose and pentosans of the cell walls of plant feeds (roughage), breaking them into acetic, propionic, butyric, and other organic acids, gases, and heat which can be used to maintain normal body temperatures of the animal. If too much gas is formed—for instance after intake of large quantities of fermentable feed such as legumes—BLOAT may result.

Rod-shaped bacteria are called BACILLI; e.g., *Bacillus subtilis*. Bacteria of round or ovoid form are known as *cocci*; e.g., *Staphylococcus aureus*.

Among the more important bacterium species are the following: *Bacterium solonaceum* and *B. stewartii; Corynebacterium insidiosum; Phytomonas* spp.; *Pseudomonas* spp.—e.g., *P. alboprecipitans, P. alfalfae, P. andropogoni, P. coronafaciens, P. glycinea, P. phaseoli, P. pisi, P. striafaciens, P. syringae,* and *P. tabaci; Rhyzobium* spp.; and the *Xanthomonas* spp., e.g., *X. phaseoli* var. *sojensis.*

See also AEROBIC BACTERIA; LACTIC BACTERIA; LEGUME BACTERIA; BACTERIAL DISEASES; SYMBIOSIS; LEGUMES; CHARCOAL ROT; STALK ROT.

BAGASSE (pronounced ba-*gahs*) is a term applied to the mill residues from the sugar industry. Bagasse consists of the crushed stalks from which the juice has been pressed. The term is used for residues not only of SUGARCANE, but also of other plants, such as SORGHUMS, or SUGAR BEETS.

Bagasse pith is used for the production of DRIED MOLASSES.—*See also* MOLASSES.

BAHAM GRASS = BERMUDA-GRASS.

BAHIAGRASS (*Paspalum notatum*) is also called *Bahia* or *common Bahiagrass*.

It is a low-growing perennial which spreads by short, heavy runners. It forms a dense, tough sod even on droughty, sandy soils. It survives the winter in the southern Coastal Plains when temperatures are not too severe.

Bahia grass is used primarily as a pasture grass.

Dangers: Bahiagrass, particularly Argentine Bahiagrass, is at times quite seriously affected by the fungi ERGOT and FUSARIUM species.

Varieties: The use of three new varieties of Bahiagrass is now being increased in the United States: *Paraguay Bahiagrass, Pensacola Bahiagrass,* and *Argentine Bahiagrass.* They are more winter-hardy and have narrower and more hairy leaves than the common Bahiagrass. Argentine Bahia gives higher yields of forage than other Bahiagrasses.

References: H.1; K.1.

See also PAPSALUM GRASSES; GRASSES; PASTURE PLANTS.

BAIT. *Poisoned baits*, such as POISONED BRAN MASH, are used to control insect pests or rodents.—*See also* DRY BAIT; INSECTICIDES; METALDEHYDE; STRYCHNINE-POISONED GRAIN.

BAKERY WASTE, such as *stale bread, cakes,* or *crackers,* may be used in place of part of the grain in livestock rations; its best use is to feed it to swine. When dried, it is similar to grain in chemical composition, but has poor quality protein and is low in vitamins.

Bakery waste has a feeding value of about 75 percent of farm grains. When fed in large amounts, it must be supplemented with vitamin A, calcium, and protein supplements of extra good quality.

BAKING SODA = SODIUM BICARBONATE.

BALANCED RATION is the daily feed allowance which contains the right proportions of NUTRIENTS for the proper nourishment of an animal.—*See also* RATION.

BALBO. *See* RYE (variety).

BALDWIN = *Goens.*

          See COMMON WHEAT (variety).

BALED HAY. *See* HAY.

BALL CLOVER (*Trifolium nigrescens*) is one of the TRUE CLOVERS of some local importance.

BALTIC ALFALFA belongs to the VARIEGATED ALFALFAS.

BANNER, or *standard*, is the topmost PETAL of a LEGUME flower.

BANNER OAT.

See COMMON OAT (variety).

BANNOCK. See COMMON OAT (variety).

BARB is a sharp, downwardly or backwardly projecting twin point terminating a BRISTLE.

Barbed means: beset with barbs.

BARBERRY BUSHES (*Berberis* spp.) are often a source of rust menace. Three kinds are of importance in spreading stem rust in the United States.

The *European barberry* is an erect bush, often more than 6 ft. high when mature. The outer bark of the stems is gray, the inner bark yellow and the edges of the dark green leaves are saw-toothed. Usually there are three spines under each group of leaves. Red berries hang in bunches as currants do in the fall. The European barberry, scattered over most of the northern grain-growing areas, is the most important of the rust-spreading bushes.

Barberry bush. (U.12.)

The *Alleghany barberry*, limited to Virginia and West Virginia, and the *Colorado barberry* of southwestern Colorado also spread stem rust. These native plants are similar to the European barberry, but are low shrubs that grow in patches.

Control: Put common salt (SODIUM CHLORIDE) about the crowns or on the ground around the canes of the barberry bushes; apply AMMONIUM SULFAMATE to the stubs; or treat the stubs with 2,4-D.

Reference: U.10.

See also WHEAT STEM RUST.

BARCHET. See SOYBEAN (variety).

BARD VETCH (*Vicia monantha*) is similar to HAIRY VETCH in general habit of growth. It succeeds well in the irrigated areas of the Yuma and Imperial Valleys of the Southwest. Farther north in the West, it cannot compete with the other VETCHES, and in the Cotton Belt east of the Mississippi River it has never succeeded.

Reference: M.18.

See also LEGUMES; INOCULATION.

BARE LAND. See DUST STORMS.

BARILLA = HALOGETON.

BARLEY. See BARLEYS.

BARLEY BLACK = *Congo cowpea.*

See COWPEA (variety).

BARLEY BRAN is a product which consists mainly of *barley hulls* with only a small percentage of BRAN. The term, therefore, is a misnomer and should not be used.

BARLEY BY-PRODUCTS are *malt sprouts, brewers' dried grains,* and other feedstuffs classified as BREWERS' PRODUCTS.

BARLEY FEED. See BARLEYS.

BARLEY HAY. See BARLEYS.

BARLEY HULLS. See BARLEYS.

BARLEY MIXED FEED. See BARLEYS.

BARLEY PASTURE. See BARLEYS.

BARLEYS (*Hordeum* spp.) belong to the GRASSES; they are among the most important cereal crops of northern countries and are extensively used for food and livestock feed where other small grains are not sufficiently hardy.

The length of the stem varies with variety and environment from 1 to 4 ft. The diameter varies from 2 to 6 mm. A single leaf arises at each node of the stem; the leaves are borne alternately on opposite sides of the stem. The head at the top of the stem consists of the flowers of the plant arranged in SPIKELETS.

The lateral spread of barley roots varies from 6 to 12 in.; they penetrate 3.0 to 6.5

ft. deep. The grains are 8 to 12 mm. long,
3 to 4 mm. wide, and 2 to 3 mm. thick; the
kernels are hulled or they are hulless
(naked).

The barley varieties can be divided according to their long, short, or missing
beards, or awns, as follows: (1) bearded
with barbed beards—*rough-awned;* (2)
bearded with smooth beards—*smooth-
awned;* (3) *semismooth-awned;* and (4)
beardless—either *hooded* (with a small,
three-forked hood replacing the awn) or
awnless. In general the yields of the rough-
awned types are higher than those of the
(preferred) smooth-awned types; both
types have higher yields and better
quality than the beardless types, but the
latter can be handled and threshed with
less discomfort.

The barleys grow well in a cold, humid
climate. They will stand more heat under
semiarid conditions than under humid
conditions. Early seeding is advantageous.
When grown in warmer climate the barleys
are often winter-sown.

The best barley soils are well-drained
loams. Barleys cannot stand "wet feet,"
and poorly drained soils do not produce a
good crop in regions of frequent rains; the
grain is likely to become coarse, and the
plants are more subject to diseases. On
the other hand, on light, sandy soils
growth is not maintained at a uniform
rate, and ripening is often hastened by
drought.

Barley is not a poor-land crop, and
commercial fertilizers give profitable returns when applied to soil with low fertility.
Standard recommendations for small grain
fertilization apply for barleys; usually 400
to 500 lb. per acre of a complete fertilizer
is sufficient. Barley of high quality is
seldom produced on land that is too fertile.
On rich lands, fertilizer is best applied to
the preceding crop, as the residual effect is
sufficient for the barley plants. Where
SCAB is serious, the barleys should follow
some crop other than CORN or small grain.

When the barleys are grown on disked
corn land, the soil should be prepared so
that the seed will be uniformly covered in
drilling. The seed should be treated with
a disinfectant before planting.

Feed barley grown solely for feed is
usually or relatively high quality. When it
is the result of an unsuccessful attempt to
grow *malting barley* (for cash revenue), it is
usually second-rate feeding material.

Barleys. Plant with group of spikelets and
floret; also, spike of beardless barley. (H.26.)

Spring barley varieties are not winter-hardy; they are best adapted to the northern and western states, generally outside the Corn Belt. They are of great value as a grain feed that can be substituted for corn. In balanced rations barley serves principally as a source of carbohydrates. Although it is higher in protein than corn, it must be supplemented with a protein concentrate or with LEGUME hay to obtain best results. Barley has no vitamin-A activity.

Winter barley acreage is increasing in the southeastern states where the crop is often pastured or grown as a cover crop, in addition to being used as a feed grain.

While the winter-hardy varieties are grown on isolated farms in nearly every state, the main areas of winter barley lie south and east of a curved line running from New York City through Kansas City and western Texas. Seeded early, the winter barleys produce abundant *pasture*, particularly in the South where growth is not stopped by cold weather. When the crop is pastured in spring, prolonged grazing will reduce the grain yield. Seeded in fall, winter barley serves as a *cover crop* and on hilly land does much to control soil erosion. Since the barleys are among the earliest crops harvested in the South, they often serve as a "filler feed crop" when the supply of corn runs out.

The barley varieties are well suited as a companion (nurse) crop for the small-seeded legumes and grasses. Since the barley plant usually ripens earlier than wheat or OATS, has shorter straw, and is less leafy, it robs the legume or grass to a lesser extent of moisture, plant nutrients, and sunlight. Barley fits well into a 1-year rotation with LESPEDEZA or SOYBEANS in areas where these legumes are adapted.

In many sections along the colder limit of the winter barley area, the plants are cut somewhat green and fed as *barley hay* or made into *silage*.

On favorable soils, barley yields more pounds per acre than any other small grain. Almost the equal of corn as a feed, it can be fed to all classes of livestock and is constantly increasing in popularity with dairy farmers. Like corn, it is prized as a feed for hogs, because it produces firm pork.

The value of clean and *sound barley* for feeding hogs depends largely upon its bushel weight. Barley weighing 48 lb. a bushel has almost the same value as corn. For other livestock the high weight is not such an important factor. In fattening cattle, *scabby barley*, which has a higher fiber content, greater bulk and, therefore, a considerably lighter weight than sound barley, has practically the same value, pound for pound, as sound barley. Ordinarily, hogs will not eat scabby barley.

There is little difference in the feeding value of the barley varieties. The farmer growing barley for feed ordinarily should use the highest yielding variety. The percentage of hull is higher on small-kerneled than on large-kerneled varieties. Feeding tests, however, have shown little difference in the value of the different kinds if they are plump and well-grown.

The *barley kernel* is too hard to use satisfactorily as feed without preparation. It is preferably used as *ground barley* or as *rolled barley*. Grinding is the most common treatment in the Central States and in the eastern areas. The grain should not be ground fine, because it then becomes a pasty mass when chewed by animals, and results in reduced consumption. When barley is ground, the machine should be set to merely crack the grain rather than pulverize it. In the West, much of the barley is rolled. This is an ideal method of preparation. A jet of steam softens and moistens the kernel so that little loose material results. At the same time the kernel is flattened to a soft, easily crushed disk. *Soaking* in water is sometimes practiced when feeding whole barley to swine. There is no advantage in soaking ground barley.

When barley is to be mixed with chicken feed the *hulls* are sometimes removed. This separation need not be complete. The removal of most of the hull increases the palatability for chickens and greatly improves the quality of the feed by reducing the quantity of roughage.

Barley straw is not equal to oat straw but is usually superior to wheat straw in feeding value. There is some objection to feeding the straw from bearded varieties because of the discomfort the beards may cause to animals. Occasionally irritation of the mouth develops in feeding barley straw, but usually this has not been a serious factor.

Silage is sometimes prepared from barley when its kernels are in the milk stage or early dough stage.

The following barley products are officially recognized:

Barley hulls—consisting of the outer covering of the barley.

Barley feed—the entire by-product resulting from the manufacture of PEARL BARLEY from clean barley. In total digestible nutrients, barley feed is equal if not superior to barley; its supply is rather limited, however.

Barley mixed feed—the entire offal from the milling of barley flour from clean barley; it is composed of barley hulls and barley middlings.

Ground barley—the entire product obtained by grinding clean barley, containing not less than 90 percent pure barley and not more than 10 percent weed seeds and other foreign materials, and not more than 6 percent crude fiber (not intentionally added).

Rolled barley, or *crimped barley*, is heavy barley that has been rolled to desired thickness. It shall contain not less than 90 percent pure barley and not more than 10 percent weed seeds or other foreign materials, and not more than 6 percent crude fiber.

Dangers: The most important diseases of barley are WHEAT SCAB, WHEAT STEM RUST, BARLEY STRIPE, COVERED SMUT OF BARLEY, POWDERY MILDEW, TAKE-ALL, ERGOT, and LOOSE SMUT OF BARLEY.

CHINCH BUGS are a potential hazard to barley production in some areas. Insect injury caused by GRASSHOPPERS, STINK BUGS, the HESSIAN FLY, the ARMY CUTWORM, and the EUROPEAN CORN BORER also occurs on barley.

FIELD BINDWEED, too, is a danger in barley fields.

Species: More than 5,000 varieties of barley have been tested in the United States of which only a few are under cultivation on farms. They belong to two distinct species; namely, SIX-ROWED BARLEY and TWO-ROWED BARLEY. A number of uncultivated species of barley are aggressive weeds.

References: H.27; H.26; G.9; F.5; U.6; W.20; W.21; A.6; E.12.

See also CEREAL GRAIN BY-PRODUCTS; BY-PRODUCT FEEDSTUFFS; WHEAT JOINTWORM; WHEAT STEM SAWFLY; STEM NEMATODE; ALFALFAS; WHEATS; BUCKWHEATS; VETCHES; RICE; RYE; SPELT; FIELD PEA; BREWERS' PRODUCTS; SILAGE CROPS; WILD BARLEY; FOXTAIL BARLEY; SILAGE.

BARLEY SILAGE. *See* BARLEYS.

BARLEY STRIPE is a fungus disease incited by *Helminthosporium gramineum* which occurs on BARLEY only. Seedling infection results in fungus invasion of all plant parts, and spores are produced during the period of barley flowering. Seed infection occurs from wind-borne spores.

All leaves of a diseased plant generally are affected. Pale-green stripes appear on the young leaves as they unfold. The stripes turn brown and spread to the leaf sheath when the leaves reach full development. All SPIKES of affected plants are blighted. Kernels affected from wind-borne spores do not show the disease in the dormant seed.

Control: Resistance to the known races of the fungus has been obtained and incorporated into a number of commercial varieties of barley.

Reference: D.11.

BARN-CURED ALFALFA HAY.

See ALFALFAS.

BARN CURING. *See* HAY.

BARN-STORED HAY. *See* HAY.

BARNYARDGRASS (*Echinochloa crusgalli*) is a GRASS which occurs throughout the United States in moist, open places. Though often a troublesome weed in RICE, COTTON, and fields of beans, it is of considerable value as forage. The seeds are eaten by birds.

Variety: *Japanese millet* grows in the cooler regions of the United States. It can be used for green feed, silage, or hay.

Reference: W.16.

See also STEM ROT OF RICE.

BARNYARD MANURE. *See* MANURE.

BARRIERS are used for the control of migrating CHINCH BUGS.—*See also* DUST BARRIER and CHEMICAL BARRIERS.

BASE = ALKALI.

Basic means: having the properties of a base; alkaline.—*See also* ACID-BASE BALANCE.

BASIC COPPER CARBONATE.
See COPPER CARBONATE.

BASIC COPPER SULFATE
= COPPER SUBSULFATE.

BASIC CUPRIC CARBONATE.
See COPPER CARBONATE.

BASIC CUPRIC SULFATE
= COPPER SUBSULFATE.

BASIC SLAG, or *Thomas phosphate,* a by-product of the steel manufacture, is often used as a fertilizer instead of SUPER-PHOSPHATE.—*See also* SLAG POISONING.

BASIDIOSPORES. *See* RUSTS.

BASSIA (*B. hyssopifolia*) is one of the (fair) forbs found on RANGES.—*See also* RANGE PLANTS.

BAST is a term used for the strong, woody fibers in the bark of trees, flax, and other plants from which ropes and cords are often manufactured.

BATHYPLECTES. *B. curculionis.*
See WEEVIL PARASITE.

BAVENDER SPECIAL.
See SOYBEAN variety.

BAY BREWING = *California Coast barley. See* SIX-ROWED BARLEY.

B-COMPLEX VITAMINS are the various factors which form the *vitamin-B complex;* among them are VITAMIN B$_1$ (thiamine hydrochloride); VITAMIN B$_2$ (riboflavin); VITAMIN B$_6$ (pyridoxine hydrochloride); VITAMIN B$_{12}$; NIACIN; PANTO-THENIC ACID; BIOTIN; CHOLINE; FOLIC ACID (now called pteroylglutamic acid); and PABA (para-aminobenzoic acid).

They belong to the so-called *water-soluble* vitamins. Liver and yeast are good sources, but the pure compounds or their salts (e.g., thiamine hydrochloride, calcium pantothenate) are usually used to fortify natural rations.—*See also* YEAST.

BEACHWOOD = *Red May.*
See COMMON WHEAT (variety).

BEAN. *See* BEANS.
Yard-long bean = ASPARAGUS BEAN.

BEAN BETTLE.
See MEXICAN BEAN BETTLE.

BEAN PASTURE. *See* FIELD BEANS.

BEANS is a term applied to (1) fleshy seed PODS, (2) SEEDS (e.g., soybean seed), (3) many LEGUMES, e.g., FIELD BEANS (including COMMON BEAN, TEPARY BEAN, LIMA BEAN, and MUNGBEAN), CAROB BEAN, and VELVET BEAN. Among the most important *bean enemies* are GRASSHOPPERS, the POTATO LEAFHOPPER, STINK BUGS, and the MEXICAN BEAN BETTLE.—*See also* LEGUME BACTERIA; CHARCOAL ROT; JACK-BEAN.

BEAN STRAW. *See* FIELD BEANS.

"BEARD." *See* AWN.

BEARDED BLUEBUNCH WHEAT-GRASS = BLUEBUNCH WHEATGRASS.—*See also* RANGE PLANTS.

BEARDED BLUESTEM WHEAT
= *Fulcaster. See* COMMON WHEAT (variety).

BEARDGRASS. *Texas yellow beardgrass,* commonly known as *King Ranch bluegrass,* is a variety of TURKESTAN BLUESTEM.

BEARDLESS BLUEBUNCH WHEAT-GRASS (*Agropyron inerme*) is closely related to BLUEBUNCH WHEATGRASS and differs only in that it lacks AWNS. Many stockmen prefer it because lack of awns makes the plants more palatable, especially during the late stages of growth.

Reference: H.1.

See also WHEATGRASSES; GRASSES.

BEAUVERIA. *B. globuliferia,* a white FUNGUS, is the most destructive natural enemy of the CHINCH BUG.

BEAVER, also called *Beaver milo,* is not a true milo, but one of the kafir-milo derivatives which belong to the grain SORGHUMS.

BEECHER.
See SIX-ROWED BARLEY (variety).

BEEF CATTLE RATIONS. The RATIONS for beef cattle, as given here, are adequate for *commercial* herds. *Purebred* breeders may wish to use more liberal rations to keep their animals in top condition for show and advertising purposes.

FEEDING PREGNANT COWS

In ordinary beef cattle production, calves are weaned in the fall and the pregnant cow is wintered on low quality hays, straws, stover, or dried grasses on the range. The cow's main nutrient need is for energy. If the roughage dry matter contains 50 percent T.D.N., mature cows will usually winter well. The low quality forage should be supplemented with some digestible protein, carotene, and MINERALS, especially PHOSPHORUS and COBALT.

Usually, 5 lb. of good quality hay plus all of the low quality ROUGHAGE, such as straw, stover, or cobs that the cow can eat, will be satisfactory. If no good quality LEGUME HAY is available, a protein, VITAMIN, and mineral supplement that supplies ½ to 1 lb. oil meal or its equivalent and some good source of VITAMIN A, TRACE ELEMENTS, and phosphorus should be fed. SILAGES are excellent feed and if available may be full fed. It is more economical to limit silage to about 40 lb. per day.

During the last month of pregnancy the cow should get a slight increase in feed by improving the quality of hay that is offered or adding small amounts of silage or 1 to 2 lb. grain to the ration. If calves are born in January, February, or March in northern sections of the country, it will be necessary to provide up to 4 lb. grain per day for the cow that is nursing her calf. Depending on the use to be made of the calf, it may be advantageous to offer it a small amount of creep feed to supplement the cow's milk. Usually the calf can do very well on a mixture of 1 part of oil meal and 10 parts of CORN or other grain. During the summer, beef cows on good pasture will not need supplemental feeds other than salt. Sample rations are shown in the table.

RATIONS FOR WINTERING PREGNANT BEEF COWS *

1. *Rations ordinarily recommended which use maximum amounts of roughage and minimum amounts of concentrates*

Ration:	1	2	3	4	5	6
Legume hay, good, lb.	5–7	5–7	5–7	...	...	...
Grass hay, good, lb.	12–18	...	...	10–15	20	...
Corn silage, lb.	...	25–45	...	...	...	40–50
Legume silage, lb.	...	...	...	20–30	...	...
Corn or sorghum stover, lb.	...	...	15–20	...	...	...
Oil meal, lb.	...	...	...	...	1	1

2. *Rations that use low grade roughages in maximum quantities*

Ration:	1	2	3	4	5	6
Corn cobs or cotton seed hulls, lb.	10–12	10–14	10–14	12–15	...	...
Corn stalks or straw, lb.	...	...	...	...	10–15	15–18
Alfalfa leaf meal, lb.	...	1–2	...	...	...	...
Legume hay, good, lb.	5–7	...	2–3	...	5–10	...
Oil meal, lb.	0–½	1–1½	1	1½–2	...	1½
Molasses, lb.	...	...	1	1–2	...	1–2
Vitamin A, I.U.	...	...	15,000	30,000	...	30,000

Use corn and/or oats (1-4 lb.) to keep the animals in desired condition.

3. *Emergency rations that utilize minimum amounts of roughage and enough drought relief concentrates to balance the ration*

Ration:	1	2	3	4	5	6
Corn or sorghum stover, lb.	5	...	...	...	...	...
Low grade hay, lb.	...	5	...	...	...	...
Stalks or straw, lb.	...	...	5	...	...	...
Ground oats, lb.	5	7	6	...	...	...
Ground corn, lb. †	1–4	1–4	1–4	...	...	...
Oil meal, lb.	1	1–1¼	1–1½	...	...	...
Vitamin A, I.U.	30,000	30,000	30,000	...	...	...

* Cattle on all rations should have free access to iodized salt that has had 1 oz. cobalt chloride or cobalt sulfate mixed with 100 lb. salt and a mineral mixture of limestone, bone meal and salt (1:1:1).

† Amount of corn will vary with condition of cattle.

WINTERING RATIONS FOR STEERS AND HEIFERS

Many beef production systems defer full feeding of animals that are to be marketed as fat stock until they are one to two years old. Replacement heifers can be handled in the same way. Steers are wintered during their first and sometimes their second year on low energy rations that restrict gains to 1 to 1½ lb. per day. At times, low grade cattle are purchased that may be two to three years old and fed wintering type rations. Rations permitting gains of 1 lb. per day can be based on a full feed of hay with additional grain or protein if needed. Silages are also entirely satisfactory for wintering stockers or feeders and may be supplemented as needed. A full feed of corn silage supplemented with 1 lb. protein will produce 1 to 1½ lb. gain per day. Other silages contain less energy than corn silage and need supplementing with about 1 lb. grain per 100 lb. of body weight to produce the gain obtained with corn silage and protein supplement. See table for examples.

FATTENING OR FINISHING RATIONS

After a beef animal has considerable growth, it is usual to switch it to a fattening or finishing ration. Light weight beeves and baby beeves are sometimes kept on fattening rations from birth. Fattening rations must be energy-rich to meet the maintenance needs and provide excess energy for fat production. Limited amounts of hay or fibrous feeds are used so that the animal's capacity can be used for grain consumption. Corn is the best grain for cattle fattening rations. The rate of grain feeding will be about 1½ to 2 lb. per 100 lb. of body weight. Oil meals can be used to supply protein if needed. With a full feed of grain and 3 to 4 lb. alfalfa hay, only small amounts of other feeds will be required. If roughage other than good legume hay is used, improved performance can often be obtained by adding trace minerals or molasses to the ration.

Steroid-like HORMONES, e.g., STILBESTROL or HEXESTROL, may be added to fattening rations at the rate of 10 mg. per animal per day.

Sample rations are shown in the tables.

RATIONS FOR GROWING OR WINTERING BEEF CATTLE

1. Calves, 450 lb. initial weight

Ration:	1	2	3	4	5
Legume hay, lb.	12–15	3–5	...	2	...
Grass hay, lb.	...	...	10–15	...	...
Straw, cobs, hulls, or stover, lb.	...	...	...	8–10	...
Corn or sorghum silage, lb.	...	20–25	...	...	...
Small grain, pulp or grass silage, lb.	...	...	...	...	25–30
Grain, lb.	...	...	...	...	1–5
Molasses, lb.	...	...	1	1	...
Oil meal, lb.	...	...	1	1½	1
Minerals, free choice	+	+	+	+	+
Vitamin-A supplement, 15,000 I.U.	...	...	?	+	...

2. Yearlings, 700 lb. initial weight

Ration:	1	2	3	4	5
Legume hay, lb.	5	...	5	...	...
Grass hay, lb.	12	...	...	...	4
Straw, cobs, hulls, or stover, lb.	...	...	10	10	...
Corn or sorghum silage, lb.	...	...	...	20	...
Beet or citrus pulp, lb.	...	...	...	...	50
Molasses, lb.	...	...	1	...	...
Grain, lb.	...	...	...	...	...
Oil meal, lb.	...	1	1–0	1–2	1
Minerals, free choice	+	+	+	+	+
Standover pasture or cured grass, free choice	...	+	...	...	...

RATIONS FOR FATTENING BEEF CATTLE

1. *Calves, 450 lb. initial weight*

Ration:	1	2	3	4	5
Legume hay, lb..............	3	3	2	3	...
Grass hay, lb.................	...	...	2	3	3
Corn or sorghum silage, lb.....	10–15	...	...	...	...
Corn, sorghum, or barley, lb. ..	6–10	10–15	10–15	5–8	8–10
Citrus or beet pulp, lb.........	...	...	...	5–8	...
Molasses, lb..................	...	...	...	...	1
Oil meal equivalent, lb........	1½	1	1½	1	1½
Salt and minerals, free choice..	+	+	+	+	

2. *Yearlings, 800 lb. initial weight*

Ration:	1	2	3	4	5	6
Legume hay, lb..............	3	5	...	...	5	...
Grass hay, lb.................	...	...	5	3	...	3–5
Corn or sorghum silage, lb.....	50	...	...	...	...	...
Legume silage, lb.............	...	...	...	20–25	...	...
Grain, lb....................	3–6	11	8	10–12	16	16
Molasses, lb..................	...	...	...	...	...	1½
Molasses dried pulp, lb........	...	5	8	...	...	...
Oil meal equivalent, lb........	2	...	1	½	1	1½
Salt and minerals, free choice..	+	+	+	+	+	+

BEES are insects which live in colonies. *Solitary bees*—especially most of the *wild bees*—show no tendency toward neighborliness and live in small groups. These and the *social* bees (e.g., the *honeybees*), as well as *bumblebees*, are important as pollinating insects. Each colony comprises a queen (the only female), drones (males), and workers (undeveloped females) which collect the pollen and nectar for the colony, form the wax (in glands) and use it to build combs. Workers live in summer six to seven weeks, but in winter as long as six months. There are up to 100,000 of them in a colony.

The drone dies during the act of mating; the young queen then lays up to 5,000 eggs within twenty-four hours.—*See also* POLLINATION.

BEET. *See* BEETS.

BEETLES form the largest insect group; most bettle species have hardened skins and hard wing covers. The worm-like grubs, or larvae, change to pupae which are not enclosed in cocoons and are able to wiggle. In most cases, larvae as well as adult bettles possess chewing mouthparts and are therefore injurious.

Among the more important beetles are the FLEA BEETLE, CORN FLEA BEETLE, JAPANESE BEETLE, MEXICAN BEAN BEETLE, SPOTTED CUCUMBER BEETLE, WESTERN SPOTTED CUCUMBER BEETLE (and its subspecies, the SOUTHERN CORN ROOTWORM); also the WEEVILS (e.g., the PEA WEEVIL), the WHITE-FRINGED BEETLES, the CLOVER ROOT CURCULIOS, and the CLICK BEETLES. —*See also* ALFALFAS; CORN; GROUND BEETLES; GRAIN BEETLES; LADY BEETLES.

BEET MOLASSES
= *feeding beet molasses. See* MOLASSES.

BEET PULP, the residue after extraction of the sugar-containing juice, is a byproduct from the manufacture of sugar and MOLASSES from SUGAR BEETS. It is low in protein and fat and high in fiber, but the fiber is well digested by ruminants. Beet pulp is fed either wet or dry, principally to cattle.

It is also used as feedstuff in form of *dried molasses-beet pulp* and *ammoniated beet pulp*, and SILAGE.

DRIED BEET PULP is an officially recognized feedstuff.

Reference: E.12.

See also PLANT BY-PRODUCTS; SILAGE CROPS; AMMONIATED PLANT PRODUCTS.

BEET PULP SILAGE. *See* SILAGE CROPS.

BEETS (*Beta* spp.) is a name applied to many root plants, among them the economically most important SUGAR BEET.

BEET SUGAR. *See* SUCROSE.

BEET TOPS are PLANT BY-PRODUCTS obtained from the tops of SUGAR BEETS. These tops consist of the crowns and leaves of the root plant and are of no value for the sugar production, but they are of value as feedstuffs. They are fed, after curing, to livestock, especially in feeding lots; or they are used as one of the SILAGE CROPS. Beet tops, if fed in large quantities, may act as laxative.

BEET TOP SILAGE. *See* SILAGE CROPS.

BEET WEBWORM (*Loxostege sticticalis*) is one of the WEBWORMS. It occurs in the Great Plains and Intermountain area and feeds on plants that belong to 33 families. In the East the pests will not invade indefinitely an area with more than about 25 in. precipitation annually; the western edge of distribution is correlated with an annual precipitation of a little more than 10 in. To the south the distribution seems to relate to an average annual temperature of about 55° F.; thus, the insect does not extend beyond southern Kansas, northern Oklahoma and New Mexico.

Reference: M.16.

BEGGARWEED.

   See FLORIDA BEGGARWEED.

BELDI-GIANT.

   See SIX-ROWED BARLEY (variety).

BELONOLAIMUS. *B. gracilis*

     = STING NEMATODE.

BELOTURKA = *Kubanka.*

     See DURUM WHEAT.

BELOW-GROUND SILO. *See* SILOS.

BENT = *bentgrass. See* BENTGRASSES.

BENTGRASSES belong to the genus *Agrostis* which includes the *bent* (or *bentgrass*) species and the species REDTOP; they have a habit of creeping growth.

Most of the bentgrasses, particularly the so-called *fine bentgrasses*, are well adapted for putting greens and, in mixture with other GRASSES, for lawns; one of them, the COLONIAL BENT, is also used as a pasture grass.

Reference: H.1.

See also GRAZING; PASTURE PLANTS.

BENTON. *See* COMMON OAT (variety).

BENTONITE, like other clays, is a (colloidal) hydrated aluminum silicate. It is of volcanic origin, swells greatly upon wetting, and is characterized by its dispersing and adsorbing qualities. Bentonite is used as filler in wet marshes and as a binding agent in feed pellets.

BENZENE HEXACHLORIDE, or *BHC,* one of the CHLORINATED HYDROCARBONS used for their insecticidal properties, occurs as mixture of isomeres; the most effective of them is the *gamma isomere,* also called LINDANE. Benzene hexachloride is a stomach and contact poison for many insect pests. It has a characteristic musty odor, but lindane is nearly odorless. For this reason, lindane-containing insecticides are often preferred.

Caution: Avoid inhaling dust or mist from sprays; avoid contact with skin; avoid contamination of feedstuffs.

BERBERIS. The *Berberis* spp. are known as BARBERRY BUSHES.

BERMUDA-GRASS (*Cynodon dactylon*) is also called *wiregrass, dog's tooth grass, joint grass, Bahama grass, devil grass,* and (in Australia) *couch grass.* It is common in all tropical and subtropical parts of the world, including the southern half of the United States from Virginia to Florida and westward to Arizona and California.

Bermuda-grass is a long-lived perennial with a habit of spreading growth. The erect, flowering branches are 6 to 12 in. high (and may reach 18 in.), depending on the fertility and moisture of the soil. The leaves are short, flat, bluish-green, and usually 1 to 4 in. long (and may reach 6 to 8 in.). At the base of each leaf is a fringe of white hairs; the leaf sheath is compressed and slightly hairy. The flowers are in slender SPIKES, three to six in a cluster.

This grass grows well on almost any soil that is fertile and not too wet, but prefers heavy soils. It thrives in warm or hot weather. It usually does not survive heavy freezes, although it has lived through temperatures of 10° F.

Bermuda-grass may be propagated by seed, runners, or underground rootstocks:

1. Because the *seeds* are small and light, a well-prepared seedbed is desirable. Spring

seedings are generally best. The seed should be covered with a cultipacker or a light harrow.

2. Many methods are used in planting *runners*, which vary from a few inches to 3 to 4 ft. in length and under favorable conditions may grow 15 to 20 ft. in a season. The common practice is to plow furrows 4 to 6 ft. apart, drop runners 2 or 3 ft. apart in the furrow and cover them by plowing or with the foot. Deep planting is important if the runners are not watered when set. Rolling or cultipacking the soil after plantings is also desirable.

3. The *rootstocks* (which may become RUNNERS or STOLENS on hard soils), are thick and white. A widely used planting method consists in dropping the rootstocks into shallow furrows behind the turning plow and covering them with 2 to 3 in. soil.

A complete fertilizer, such as 4-8-4 or 6-6-6, should be applied at 400 to 600 lb. per acre just ahead of planting or seeding. For rapid establishment, nitrogen fertilizer should be applied in midsummer at rates of 100 to 200 lb. SODIUM NITRATE or its equivalent. It is characteristic for Bermuda-grass that the addition of nitrogen to the soil will greatly increase the protein content of the forage.

The principal use of Bermuda-grass is for *pasture* and lawns, but it is also used for *hay*. It is very useful for small paddocks that must withstand heavy tramping and use. Bermuda-grass is palatable and nutritious even after frost in the fall.

Bermuda-grass responds to cultivation; only old, undisturbed pasture sods become *weedy* and unproductive. The best methods to maintain a productive PASTURE is by shallow plowing every three to five years, fertilizing, and maintaining the stand of LEGUMES in it. If good legumes—such as annual lespedeza or crimson clover—are grown with Bermuda-grass, the animal returns may be several times those obtained from native grasses.

Dangers: Insects, especially HARVESTER ANTS and BILLBUGS, cause some injury. In Florida, MOLE CRICKETS have killed out large areas. Diseases that attack Bermuda-grass are due to the ROOT-KNOT NEMATODE and to the fungus species *Rhizoctonia*,

Helminthosporium, and *Sclerotium;* in livestock, fungus-infected Bermuda-grass may cause a nervous disorder which is sometimes called the "shakes." Control is usually obtained by merely cutting off the seed heads of the affected plants with a mower.

Bermuda-grass. Plant, spikelet, and floret. (H.26.)

Varieties: Improved varieties are more vigorous in growth than common Bermuda-grass and more disease-resistant. *Suwannee Bermuda* shows promise in Florida. The hybrid *Coastal Bermuda* makes more vegetative growth, grows later in the fall, tolerates more frost, and is more resistant to diseases than common Bermuda-grass; but Coastal Bermuda will not persist as well when it is overgrazed and will not compete successfully with common Bermuda in a closely grazed pasture sod.

When properly managed, Coastal Bermuda produces high yields of good quality hay. It should be cut for hay four to five times a year—whenever the growth is 12 to 15 in. high. The hay is easily and quickly cured. With heavy nitrogen fertilization, yields up to 10 tons per acre have been obtained.

Midland Bermuda-grass, a new variety, is a hybrid between Coastal Bermuda and a hardy Bermuda variety from Indiana. On fertile soils, it is two to four times as productive as common Bermuda-grass; where fertility is low, it is no more productive than common Bermuda-grass. Midland has somewhat less desirable forage than Coastal, but is much more winterhardy. It is palatable and has good disease resistance.

References: H.1; E.2; B.2; H.24.; W.36.

See also GRASSES; LAPPA CLOVER; BURCLOVERS; LESPEDEZAS; ALFALFAS; LUPINES; CROTALARIAS; RHIZOCTONIA; HELMINTHOSPORIUM; SCLEROTIUM; SILAGE CROPS; PASTURE MANAGEMENT; PASTURE PLANTS; GRAZING.

BERMUDA HAY. *See* BERMUDA-GRASS.

BERSEEM (*Trifolium alexandrinum*), or *Egyptian clover*, is adapted to hot climates; it is killed at 20°F. and is not suitable for acid soils. It is not very popular in the United States.

BETA. The *Beta* spp. include the three important *B. vulgaris* varieties: *crassa* = SUGAR BEET, *cicla* = SWISS CHARD, and *macrorhiza* = MANGEL.

BETA-CAROTENE. *See* VITAMIN A.

BETAINE HYDROCHLORIDE forms water-soluble crystals which contain 76.25 percent *betaine* (which occurs in sugar beet MOLASSES, wheat germ, COTTONSEED, etc.). Betaine hydrochloride can replace part of the choline in rations.—*See also* VITAMINS.

BETA-TOCOPHEROL. *See* VITAMIN E.

BHA = BUTYLATED HYDROXYANISOLE.

BHC = BENZENE HEXACHLORIDE.

BI- is a prefix signifying two, twice, or double.

BICOLOR (*Lespedeza bicolor*), a LESPEDEZA species also called *common bicolor*, is a

permanent perennial shrub that grows 5 to 10 ft. high. It is one of the LEGUMES and supplies its own nitrogen after the first year. Its seeds are eaten readily by bobwhite quails; the bark and leaves are eaten by rabbits; and the flowers are attractive to honeybees. The use of bicolor has greatly advanced the conservation of farm game and the management of farmlands suitable for wildlife. This lespedeza species produces seed from northern Florida to Kentucky and Virginia.

Bicolor will not grow in wet, poorly drained soils. It will not stand grazing, because cattle and horses like it too well. Where deer populations are high, the deer will badly damage the bicolor. Bicolor should be grown as a game-food strip in an open place in the woods. It is useful on highway fills, dikes, and spoil banks.

Fertilizer is important; potash and phosphates produce vigorous growth and greatly increase seed yields.

Bicolor can be established by planting seeds or transplanting nursery-grown one-year-old seedlings.

Variety: *Natob* (*lespedeza*), early maturing and hardy, is in general appearance similar to common bicolor except that the leaves are lighter green. The seeds of Natob are 40 to 50 percent larger than those of common bicolor.

The outstanding quality of Natob is that its seed matures at an early date, usually early in October. The adapted area includes most of the eastern United States north of Virginia, Tennessee, and Arkansas.

No insects or diseases have been observed to seriously affect Natob except the JAPANESE BEETLE, which attacks both the leaves and flower parts and seems to prefer the tender flower buds.

References: D.7; C.8.

BIENNIAL means: enduring for two years.

Biennials are herbs which germinate in the spring of one year and flower, fruit, and die in the fall or winter of the succeeding year. Thus, *winter-annuals* that germinate in the fall of one year and die the following spring are not true biennials.

BIG BLUEGRASS (*Poa ampla*) is a robust, perennial, native bunchgrass that grows in all parts of the West. Plants are tufted and vigorous, about 2 to 4 ft. tall, and have numerous basal leaves. Leaf blades are pale green, flat, and about ⅜ in. wide and 8 to 16 in. long; PANICLES are erect, dense, and from 4 to 10 in. long. The fibrous roots penetrate deeply.

Although seldom found in dense stands, big bluegrass is an important range species because of its heavy forage production, palatability, and tendency to begin growth in early spring and continue into the fall. For regrassing abandoned farm land and depleted range land it is one of the most useful of the BLUEGRASSES.

Heavy, continuous grazing and severe trampling are injurious to native stands, but if big bluegrass is grazed in moderation it responds satisfactorily.

Mixed field plantings of big bluegrass and LEGUMES have produced high forage yields of good to excellent quality. This is particularly true of grass-ALFALFA mixtures adapted to areas receiving limited rainfall.

Dangers: Big bluegrass may be attacked by plant diseases such as GRASS SMUTS.

Variety: Plant-selection work has resulted in development of an improved variety of big bluegrass which is named *Sherman big bluegrass.*

Reference: H.1.

See also GRASSES.

BIG BLUESTEM (*Andropogon gerardi*) is a vigorous, coarse, perennial bunchgrass that occurs widely over most of the United States. Its major distribution is in the region of the tall-grass prairie in the Central States and along the eastern edge of the Great Plains where it is often found in close association with LITTLE BLUESTEM.

Plants grow 6 ft. tall under favorable conditions of soil and moisture. Leaf blades are about 12 in. long and from ¼ to ½ in. wide. The leaves may be hairy near the base, and the sheaths are usually hairy. The flowering stalks are stout and solid. The extensive root system penetrates deeply. The grass grows well on most soil types; it abounds on moist, well-drained loams of high fertility.

Growth starts in late spring and continues throughout the summer. The leafy

forage is palatable to all classes of livestock. It makes good-quality hay if mowed before the stemmy seed heads have formed. Much of the native hay marketed in the Midwest consists of big bluestem.

The species may be seeded alone or in mixture with other grasses. It has been

Big bluestem. Plant and spikelet. (H.26.)

planted chiefly to retire cropped land for permanent meadow or pasture use. Seedlings should be planted on a well-prepared, firm seedbed free from weeds, and protected during the period of establishment.

Reference: H.1.

See also BLUESTEMS; GRASSES; INDIAN GRASS.

BIG CLUB 43. *See* CLUB WHEAT (variety).

BIGHEAD. *See* HORSEBRUSHES.

BIGPLUME BUNCHGRASS

= ALKALI SACATON.

BIG SAGEBRUSH (*Artemisia tridentata*), often called simply *sage* or *sagebrush*, is the most common of the SAGEBRUSHES. It is a large shrub with silvery-green leaves and grows on a great variety of soils in western North America. Despite low palatability, this shrub is important to western stockmen. It has high fat content and often furnishes considerable feed for sheep and goats, especially on winter ranges. Cattle browse the plants only lightly and horses seldom eat more than a few flower heads.

Reference: U.6.

See also RANGE PLANTS.

BIG TREFOIL (*Lotus uliginosus*), a leafy LEGUME which resembles ALFALFA, is one of the two agriculturally important TREFOILS. It produces fine stems up to 36 in. long and has taproots with fibrous branching (laterals) as well as underground rootstocks (RHYZOMES). Each leaf has five leaflets, three borne at the apex of a short PETIOLE and two opposite, at its base. Its seed is much smaller than that of BIRDSFOOT TREFOIL.

Big trefoil is a long-lived perennial but succeeds only in regions with comparatively mild climate, especially in western Oregon, northern California, Georgia, and Florida.

In the grazing stage (2 to 5 in.) the dry matter contains about 28 percent protein; older growth suitable for hay contains about 21 percent protein. It is also high in vitamins, especially VITAMIN B_1, VITAMIN B_2, and NIACIN.

Its adaption to low, wet, and brackish land is unique among pasture legumes. A good mixture for such areas is REED FOXTAIL 5 lb. and big trefoil 2 lb. per acre.

Seeding vigor of big trefoil is only fair and it benefits from application of phosphate fertilizers. A specific inoculant is required. The new stand should be grazed very carefully, if at all, during the first season so that the plants can become well established.

Dangers: Big trefoil is subject to several diseases, particularly RHYZOCTONIA, STALK ROT, and SOUTHERN ANTHRACNOSE. BLACKPATCH has been observed on big trefoil in South Georgia. If the growth becomes very rank during the summer these diseases cause the plants to defoliate. The defoliation at first is concentrated in small areas of the planting, but the disease will spread over the entire planting, especially in moist areas. In cooler weather, however, the defoliated plants will initiate new growth and produce good grazing. Keeping the growth from becoming too rank by a good grazing management system will reduce the effect of the disease.

Varieties: Two distinct types of big trefoil are recognized: *hairy-leaved* and *smooth-leaved*. Good varieties of the two types do not differ in production or amount of ground cover, but the hairy type produces more seed.

References: M.3; H.22; W.13.

See also INOCULATION; LEGUME BACTERIA.

Illustration: See BIRDSFOOT TREFOIL.

BILE. *See* DIGESTIVE TRACT.

BILLBUGS (*Calendra* spp.) destroy or injure CORN, WHEAT, and other small grain crops, TIMOTHY, BLUEGRASSES, BERMUDAGRASS, JOHNSONGRASS, RICE, SUGARCANE, and PEANUTS. The most conspicuous damage by the adult billbugs is done to young corn plants. The most costly damage is undoubtedly by the grubs in cutting the underground portions of plants, especially those grown for hay and pasture.

Billbugs have only one generation yearly and are generally dependent on grass sods or wild sedges and rushes. Corn, sugarcane, and timothy probably are the only crops in which they can perpetuate themselves within the plant tissues. The other host plants admit of inside feeding only during the early part of the grub stage, after which feeding is completed among the fibrous roots.

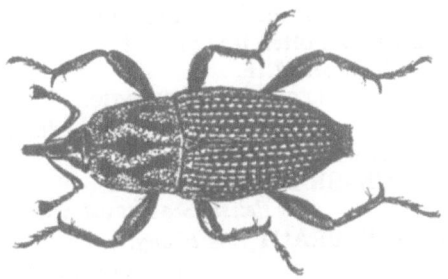

Timothy billbug. (S.31.)

Control: Efficient means of preventing crop losses from billbugs include clean cultivation, especially the complete elimination of wild sedges and rushes; suitable crop rotations; summer or early fall breaking of cultivated or infested wild sods; early planting of crops menaced by billbugs; and the protection of birds, especially ground feeders like the bobwhite and the shore birds. Hand picking has occasionally resulted in effectual control of billbug outbreaks in cornfields and on turf.

Community adoption of a 2-year maximum period for timothy; early fall plowing, either of cultivated sod or of plain swamp sod; or the planting of flax or cotton for the first crop will so reduce billbug injury as to render it of little account.

Reference: S.31.

BILOXI SOYBEAN.
      See SOYBEAN (variety).

BINDER is a machine which cuts the crop and binds it into bundles with twine. The *grain binder* is used for drilled or broadcast crops, the *corn binder* for corn and other crops grown in rows.

BINDWEED. *See* FIELD BINDWEED.

Knot bindweed = WILD BUCKWHEAT.

BIOCHEMICAL means pertaining to the chemical changes in the living organism. Biochemistry is the chemistry of the living cells.—*See also* METABOLISM.

BIOLOGY is the science of living matter, its occurrence, functions, evolutions, etc. It includes BOTANY and zoology.—*See also* PHYSIOLOGY.

BIOTIN, one of the B-COMPLEX VITAMINS, is widely distributed in foods and feeds. Good sources of biotin are yeast, milk products, cane molasses, soybean meal, dried greens, and grain sorghums (which contain more biotin than corn). Lack of biotin in poultry feed causes certain types of skin inflammation.

Biotin deficiencies seldom occur on natural rations. Raw egg-white contains a protein, *avidin*, that binds biotin in the intestine and prevents its absorption.—*See also* VITAMINS.

BIRDSFOOT TREFOIL (*Lotus corniculatus*) can be distinguished from BIG TREFOIL by the absence of underground rootstocks; it has, however, taproots and produces many fibrous roots. Its flower stalks carry fewer seed pods than does the big trefoil; the spreading pods of this TREFOIL species resemble a bird's foot.

Birdsfoot trefoil. (M.3.)

Birdsfoot trefoil, which is more winter-hardy than big trefoil, is grown in eastern New York, in western Oregon, and in northern California. This LEGUME is often used in mixtures with GRASSES on bottom land and foothill soils for pastures, silage, or hay. A good pasture mixture is REED FOXTAIL 5 lb. and birdsfoot trefoil 4 lb.

per acre; a good hay or silage mixture is ORCHARDGRASS 4 lb. and birdsfoot trefoil 4 lb. per acre. ALTA TALL FESCUE may be added to this mixture. When birdsfoot trefoil is planted, the seed must be inoculated with especially prepared cultures.

Varieties: The value of birdsfoot trefoil depends on the variety used. The several varieties are of two general types: *narrow-leaved* and prostrate; *broad-leaved* and erect, which is more productive. Varieties also differ in seedling vigor, coarseness, date of maturity, and production.

References: M.3; H.22.

See also INOCULATION; LEGUME BACTERIA; DODDERS; PASTURES.

BIRD'S PEPPER = PEPPERGRASS.

BIRD VETCH (*Vicia cracca*) is a wild VETCH species. It is a very winter-hardy perennial which occurs in considerable abundance in some places of the northern part of the United States.

Reference: M.18.

See also LEGUMES; INOCULATION.

BIS(DIMETHYLTHIOCARBAMYL) - DISULFIDE = THIRAM.

See also ARASAN; TERSAN.

BISHOP, also called *Bishop kafir* and *Algeria*, is not a true kafir, but one of the kafir-milo derivatives which belong to the grain SORGHUMS.

BITTER. *See* BITTERS.

BITTER CLOVER = SOURCLOVER.

BITTER LUPINES are the ALKALOID-containing LUPINES; they are poisonous. The nonalkaloid lupines are usually called *sweet lupines.*—*See also* BLUE LUPINE; WHITE LUPINE; YELLOW LUPINE.

BITTER RUBBERWEED. *See* RUBBER-WEEDS; ACTINEA; POISONOUS PLANTS.

BITTERS, or *bitter tonics*, are bitter-tasting drugs; e.g., GENTIAN. They are used to increase the appetite and to stimulate the activity of the stomach.—*See also* FEED INGREDIENTS; STIMULANTS; TONICS.

BITTERVETCH (*Vicia ervilia*) is more nearly upright in growth than other VETCHES. The seeds are small and conical or pyramidal in shape. In the western part of the United States this vetch has made good growth and produced good crops of seed. The seed is used for stock feed, especially for sheep. In the Cotton

Belt it has made comparatively little growth and often is winterkilled.

Reference: M.18.

See also LEGUMES; INOCULATION.

BLACK AMBERS are AMBER-type forage SORGHUMS.

BLACK-BEARDED DURUM
= *Peliss. See* DURUM WHEAT.

BLACK BEAUTY = *Ebony.*
See SOYBEAN.

BLACK CHAFF = *Blackhull wheat.*
See COMMON WHEAT.

BLACK CHOKECHERRY.
See WILD CHERRY; POISONOUS PLANTS.

"BLACK" COWPEAS is a group name applied to the black-seeded varieties of COWPEA.

BLACK GRAMA (*Bouteloua eriopoda*), also know as *woollyfoot* or *crowfoot grama*, is by far the most important forage grass on millions of acres of semidesert grasslands in Arizona, New Mexico, and southwestern Texas; it thrives also in other parts of the Southwest.

As a prime indicator of range value, black grama ranks second only to BLUE GRAMA over the entire Southwest. The

Black grama. (C.2.)

two sometimes occur together in mixed stands, but black grama prefers the better-drained soils in the short-grass country and the warmer and lower semidesert grasslands. It may be easily distinguished from other GRAMA GRASSES by widely creeping runners (stolons) which root at the joints and send up new shoots that later become separate plants.

The grass is nutritious at all times of the year. Although fully 90 percent of the growth is produced during the summer rainy season (usually July, August, and September), it ordinarily cures well on the stalk, and the stems remain green several inches up from the ground. Except under occasional conditions of extreme dryness, black grama retains its nutritive value through the dry spring when most other range vegetation is parched and harsh. It is a dependable forage relished all year long by cattle and horses; it is usually eaten more sparingly by goats and sheep. When black and blue gramas occur together the latter is generally preferred.

Natural revegetation is the most practical method of maintaining and restoring black grama on the range. It spreads or reproduces naturally in three ways:

1. By means of *seed:* This is negligible; although flower stalks and flowers are usually abundant, few seeds mature.

2. By *runners:* Successful spreading by runners requires two successive favorable growing seasons, the first for the new plants to get started, and the second for them to become rooted and firmly established as individuals. A black grama tuft may produce from one to nine runners, but trampling by livestock breaks off many before they become rooted.

3. By *tillering:* Tillering and the breaking up of black grama tufts into new plants is generally the most effective means of revegetation. Under conservative *grazing,* spread during a given growing season depends chiefly on the vigor resulting from climatic conditions of the preceding year. The stockman, therefore, has almost a year's advance notice of changes in the stand of black grama: above-average rainfall produces an increased stand the following year, provided grazing has been moderate; below-average rainfall causes a decline in stand the next year.

On firm loam or gravelly soils not subject to severe wind erosion, proper grazing utilizes about 50 percent of the total growth of black grama on a weight basis by the end of the grazing season, which usually means that from 70 to 85 percent of the height growth is taken on most plants. On sandy soils, subject to rapid blowing, grazing use should not exceed 40 percent of the total growth. On ranges in good condition the grazed stubble should not be shorter than 2 or 3 in. above the ground in June, under both year-long and seasonal winter use. In addition, 20 percent of the flower stalks and most of the runners should be ungrazed. Uniform grazing to within 1 to 2 in. of the ground constitutes severe overutilization.

Proper use is roughly proportional to the length of the season remaining; thus, if 50 percent of the grama should be taken on year-long ranges by the end of the growth year in June, only 25 percent should be taken by December or January.

To preserve the natural soil-protective qualities of black grama, slopes steeper than 50 percent (50 feet rise to 100 feet distance) should not be grazed.

Stocking which aims to utilize 65 percent of forage production in the average year safeguards the rancher in most years, and in a drought year necessary reductions can be handled without undue sacrifice in the breeding herd. The management goal should be to graze lightly in good years and to avoid grazing heavier than proper in the bad ones.

Reference: C.2.

See also GRASSES.

BLACK GREASEWOOD

= GREASEWOOD.

BLACKHAWK. See SOYBEAN (variety).

BLACK HEAD = LOOSE SMUT OF WHEAT.

BLACKHEAD is a poultry disease which causes great losses, particularly among turkeys. To control this condition, AMINO-NITROTHIAZOLE mixed in feed may be fed to the flock continuously at a concentration of 0.05 percent, until two weeks before the birds are slaughtered (or at higher rates for a two-week period of time).

Blacklaurel.(U.6.)

BLACKHULL KAFIRS—*Standard Black-hull, Western Blackhull,* and *Texas Black-hull*—are grain SORGHUMS.

BLACKHULL WHEAT.

See COMMON WHEAT (winter variety).

BLACK JAP=*Black Spanish broomcorn.*
See SORGHUMS.

BLACK KERNEL is a RICE disease, caused by the FUNGUS *Curvularia lunata,* that often blackens the entire kernel inside and out. Small, slightly raised, seedlike fungus bodies dot the surface of the kernels. In other cases the affected kernels show only surface discoloration. Rice kernels that are severely infected with kernel spot fungi usually germinate poorly and rarely produce good plants.

Reference: T.2.

BLACKLAUREL (*Leucothoe davisiae*) is a POISONOUS PLANT. When sheep eat about 3 oz. leaves in a day, they develop such symptoms of poisoning as salivation, vomiting, and weakness.

BLACK LOOSE SMUT, caused by the FUNGUS *Ustilago avenae,* is one of the OAT smuts. It transforms the kernels and the entire oat head into a mass of black spores.

Control: SEED TREATMENT with CERE-SAN M controls black loose smut easily. In addition, disease-resistant oat varieties are available.

Reference: M.30.

BLACK MEDIC (*Medicago lupulina*), a close relative of the ALFALFAS, has been frequently introduced, but has found no distinctive place in our agriculture. In the southern United States this LEGUME grows well with some of the better pasture grasses. In the warmer regions, such as Argentina and Hawaii, it is relatively more important.

Black medic may be attacked by BLACK-PATCH and other legume diseases.

Apparently there are annual and perennial strains of black medic.

Reference: M.3.

See also INOCULATION; PASTURE PLANTS.

BLACK MEDITERRANEAN
= *Rudy. See* COMMON WHEAT.

BLACK NIGHTSHADE (*Solanum nigrum*) is a SAPONIN-containing POISONOUS PLANT. It affects ruminants and poultry when they eat the green fruits and leaves of the plant. Symptoms of the poisoning are thirst, diarrhea, loss of appetite, weakness, and lack of co-ordination.

BLACKPATCH is a disease which attacks the stems, leaves, and blossom heads of LEGUMES. The causal FUNGUS, not as yet identified, grows over the surface of the leaves, penetrating here and there to produce large, black lesions. The disease may be carried into new fields in the seed since no spores are known. Blackpatch has been observed on RED CLOVER, ALSIKE CLOVER, WHITE CLOVER, ALFALFA, BLACK MEDIC, KOREAN LESPEDEZA, and BIG TREFOIL.

Periods of heavy rainfall and high humidity are ideally suited for the growth and spread of the blackpatch fungus which is one of the major factors that reduce seed yield during wet seasons.

Control: Tests indicate that FERMATE is a possible agent for controlling the blackpatch fungus .

References: E.5; H.21.
See also FUNGUS DISEASES.

Black nightshade. (H.48.)

BLACK POD of PEANUT is quite common in the varieties having large pods and seeds. Frequently many pods are found discolored when removed from the soil. The discolored pods usually contain dead, shriveled seed, or collapsed seedcoats.

The primary cause of the condition is deficiency of CALCIUM in the soil which causes weak cell walls in the rapidly growing pods and seeds. Drought, occurring during the early stage of seed development, causes collapse of the seed and of the pod tissues. Various soil fungi may then attack the dead pod tissues, causing discoloration. If the drought conditions occur when the pods approach maturity, their tissues do not collapse or discolor. On maturity the seeds may show one or more faded spots, indicating points where the seedcoat tissues have collapsed. A gradation between the two types of injury may be found in a single lot of peanuts.

Control: An application of GYPSUM six to eight weeks after planting will usually prevent this type of damage.

Reference: B.10.

BLACK SAGEBRUSH (*Artemisia nova*) is an excellent browse on winter range, but on spring, summer, and fall ranges this shrub may not be considered valuable.— *See also* RANGE PLANTS.

Illustration: See SAGEBRUSHES.

BLACK SHEATH ROT is a disease caused by the FUNGUS *Ophiobolus oryzinus*. It is found on RICE in Arkansas, Louisiana, and Texas. In the earlier stages the disease is difficult to distinguish from STEM ROT OF RICE because blackening of the sheath occurs in both. The disease usually appears in late July and in August. The tissues of the sheaths attacked by the fungus soon begin to rot, so that in severe cases the leaf sheaths are entirely rotted away at the water line; the stems are rarely attacked as they are in stem rot. The fungus overwinters on rice stubble and also on the sheaths of the CATTAIL.

Reference: T.2.

See also REDDISH-BROWN SHEATH ROT.

BLACK SMUT = LOOSE SMUT OF WHEAT.

BLACK SPANISH is a *broomcorn* variety.
See SORGHUMS.

BLACKSTEM OF CLOVER, caused by the FUNGUS species *Ascochyta imperfecta* and *Phoma trifolii*, is a disease that has similar symptoms on different species of CLOVER, particularly SWEETCLOVERS and TRUE CLOVERS.

Blackstem of clover is characterized by dark-brown or black spots on the stems and leaves. Severely infected stems turn dark and die. Immature spore-bearing bodies develop in the older stems in the late fall and winter, and by the following spring the stems often are covered with them. The leaf lesions vary in size and shape. Heavily diseased leaves turn yellow and wither before they drop. Defoliation commences with the lower leaves and progresses upwards. PETIOLE (leafstalk) lesions occur as small, black or brown, elongated spots and cause the leaves to wither and drop off.

Various phases of the disease appear any time during the growing season, and forage losses result primarily from falling leaves or partial or complete destruction of leaves and stems. Seed yields may be reduced considerably when floral parts become infected.

A. imperfecta attacks also ALFALFA.

Control is accomplished by proper management and use of adapted clover varieties. There are, however, no commercial alfalfa varieties resistant to blackstem. Crop rotation is practical since the fungus survives in crop residue and consequently can infect a new planting on the same field. Removal of old stems from fields early in the spring before the crop starts to grow will aid in reducing the disease. This may be accomplished by burning, but that method is not generally recommended.

References: T.1; G.7.

See also SUMMER BLACKSTEM.

BLACKSTEM OF OATS, or *speckled blotch*, caused by the FUNGUS *Leptosphaeria avenaria*, is a disease that is becoming more prevalent. The *blackstem stage* is generally noticed at the point where the oat leaf joins the stem. Frequently a fluffy white mass of fungus material can be seen within the hollow portion of infected stems. The *speckled-blotch stage* on the leaves is often masked by other more noticeable leaf diseases, such as CROWN RUST or OAT LEAF SPOT.

Control: Resistant varieties are being bred for states in which this disease has become of importance.

Reference: M.30.

BLACKSTRAP MOLASSES = *feeding cane molasses. See* MOLASSES; SILAGE.
BLACK VORONEZH.

See PROSO (variety).
BLACKWELL = *Blackwell switchgrass.*

See SWITCHGRASS (variety).
BLACK WINTER EMMER is a WHEAT variety which belongs to the EMMER species.
BLADE, or *lamina*, is the portion of a leaf above the sheath.
BLAST, or *blasting*, is a term applied to (1) the injury of grains or seeds or (2) the sudden death of young buds or INFLORESCENCES: e.g., blast of the SORGHUMS is caused by the SORGHUM MIDGE, but BLAST OF RICE is a FUNGUS disease.—*See also* BUCKWHEATS; BLIGHT.
BLAST OF RICE is a disease that is caused by the FUNGUS *Piricularia oryzae*. It has at times been serious in most of the RICE-growing states, with the exception of California. The disease is characterized by the breaking-over of the "necks" and branches of the heads, by the blasting of entire heads, and by dark-brown spots on the leaves and hulls. The spots tend to be long and narrow on young leaves and more or less round on old ones, similar in appearance to BROWN SPOT OF RICE and, in some cases, almost indistinguishable from it except by microscopic examination. The joints (nodes) of the stem may also be attacked.

Rice plants may be affected by the blast fungus while they are young, prior to irrigation, or after the water has been drained from the field. Under such attacks the leaves of infected plants wither and dry, and infected areas may occur at the junction of the leaf blade and leaf sheath. This symptom has not been observed in the brown spot disease.

A severe infection will reduce yields and lower the market value of the rice. Severe outbreaks of blast on newly cleared or newly broken land unquestionably result from infected grasses that were growing on the land before it was sown to rice.

Control: Blast injury during early growth of the rice plants may be reduced by submergence of the land as soon as the leaf spots become evident. The breaking of the stalk and of the RACHIS may best be controlled by developing varieties that will not tend to break-over even though infected. The excessive use of fertilizers of high nitrogen content should be avoided in fields in which blast has been found and on land that has not recently been cropped to rice. Resistant varieties, such as *Zenith, Fortuna, Bluebonnet,* or *Nira*, should be planted on new land as a first crop in areas in which blast is known to occur. Second-year crops are rarely severely affected.

Reference: T.2.

BLIGHT is a general term used to describe such symptoms of plant diseases as spotting, sudden wilting, or death of leaves, flowers, stems, or entire plants.—*See also* SEEDLING BLIGHT; LEAF BLIGHT; VICTORIA BLIGHT; HALO BLIGHT; SEPTORIA BLIGHT OF WHEAT; POD AND STEM BLIGHT; BACTERIAL BLIGHT OF SOYBEAN; ASCOCHYTA BLIGHT OF PEA; BACTERIAL LEAF BLIGHT; SOUTHERN CORN LEAF BLIGHT; NORTHERN CORN LEAF BLIGHT; BACTERIAL BLIGHT OF PEA.

BLIND SEED DISEASE, caused by the FUNGUS *Phialea temulenta,* affects PERENNIAL RYEGRASS.

Heavily diseased crops have many seeds that fail to germinate; such dead seeds are referred to as *blind seeds*. It is hard to tell infected seeds from healthy ones unless the LEMMA and PALEA are removed. Then one can see the shriveled, soft, pasty appearance of diseased seeds.

The blind seeds remain dormant during the winter. In spring, when perennial ryegrass flowers, the cup-shaped spore-producing organs arise from the over-wintered blind seeds and forcibly discharge the primary spores, which are showered on the ryegrass flowers and infect the developing seeds. These secondary spores can infect other developing seeds when rain and insects spread them from head to head. The entire life cycle is confined to the seed. Infected seeds are not toxic to livestock.

Control: Disease-free seed should be selected for planting. The fungus dies after twenty-four months in dry storage; aged seed therefore is safe to plant. The planting of any seed more than $\frac{1}{4}$ in. deep, with complete soil covering, prevents emergence of the spore-producing organs.

It is helpful also to destroy infested perennial ryegrass screenings, prevent heading in pastures until July, plow clean to bury infected seeds deeply, have good soil drainage, and plow all ryegrass on a farm at the same time to prevent spread of disease from old fields to new plantings nearby.

Reference: H.40.

BLIND GUT = CECUM.

"BLIND STAGGERS."

See SELENIUM POISONING.

BLISSUS. *B. leucopterus* = CHINCH BUG.

BLOAT, or *tympany*, is chiefly a spring problem for livestock producers. It is common on succulent (juicy) PASTURES, especially when feed is eaten too rapidly by ruminants. Feedlot bloat is also common on BARLEY and ALFALFA rations. If pastures are properly planned, there is less danger from bloat. Improved pastures with high LEGUME content have been causing increasingly severe livestock losses from bloat. Exceptions are LESPEDEZAS and, to a lesser degree, the BUR-CLOVERS and SWEETCLOVERS.

Cattle and sheep should not be turned on the lush pasture when they are hungry. Animals vary greatly in their susceptibility to bloat. Animals that bloat readily should be taken off pastures with a high percentage of CLOVER or alfalfa.

Cool weather may greatly reduce, though by no means eliminate, bloat; some observers report that the most frequent and most severe cases of bloat occur on bright, warm, dry afternoons; others believe that early mornings are as dangerous.

Mowing a pasture that is causing bloat is helpful; very close grazing after mowing will greatly reduce bloat, presumably because the animals must feed a great deal longer to get a "fill."

In a GRASS-clover pasture, at least 50 percent by weight of grass is necessary to give reasonable protection against bloat. TALL FESCUE, due to its fibrous nature and its growth characteristics, is an excellent grass for bloat control. Even with fescue, care must be taken not to overgraze early in the spring, or the percentage of grass in the pasture will be reduced. The judicious use of nitrogenous fertilizer will assist in maintaining a satisfactory balance of grass and clover.

Free feeding of good-quality HAY or STRAW reduces bloat but also reduces the amount of forage consumed. In spite of hay feeding, animals which are susceptible to bloat must not be left continuously on clover pastures.

Cattle must be removed from a dangerous pasture in many cases. If an animal is held in the dry lot over night, it will usually not bloat until the afternoon; if the animal is held out until noon, it will usually not bloat after the first fill of herbage in the afternoon.

Mixtures containing minerals, yeast, or organic acids are without value in preventing bloat. SILICONES and ANTIBIOTICS may be helpful in reducing bloat.

References: H.39; G.8; S.10.

See also GRAZING; SMOOTH BROME; MOUNTAIN BROME; CRIMSON CLOVER; STRAWBERRY CLOVER; PERSIAN CLOVER; ALSIKE CLOVER; RED CLOVER; VETCHES; HAIRY VETCH; COWPEA; ALFALFAS; SORGHUMS; RAPE; BACTERIUM; HYDROCARBONS.

BLOATING is the state of being bloated (swelled).—*See also* BLOAT.

BLOOD, the red, circulating body fluid, is the main transport system whose composition varies, depending on location, time, and activity. It consists of the white and red *blood cells* (the latter containing the HEMOGLOBIN which colors the blood (red), and *plasma* (the colorless liquid portion of the blood). The blood plasma contains fat droplets, minerals and gases in solution, and proteins; it also contains vitamins, hormones, and waste materials. The blood stream is important for the ABSORPTION of the NUTRIENTS.—*See also* ACID-BASE BALANCE.

BLOOD FLOUR. *See* ANIMAL PRODUCTS.

BLOOD MEAL. *See* ANIMAL PRODUCTS.

BLOOMLESS BUCKWHEAT
        = TARTARY BUCKWHEAT.

BLOTCH.
See SOOTY BLOTCH; YELLOW LEAF BLOTCH.

BLUE BARLEY.
      See SIX-ROWED BARLEY (variety).

BLUE BELL. *See* FIELD PEA (variety).

BLUEBONNET. *See* RICE (variety).

BLUEBUNCH WHEATGRASS (*Apropyron spicatum*), or *bearded bluebunch wheatgrass*, is a native, perennial, drought-resistant bunchgrass. It is found chiefly in the dry, open areas of the Pacific Northwest and Intermountain States where it forms as much as 60 percent of the vegetative cover in many localities.

The vigorous plant growth starts rather early in the spring if enough moisture is available. The volume of forage produced is usually high and dependable. The leafage remains green throughout the growing season and is nutritious and palatable even after growth ceases, although the stems become wiry late in the season. Plants may reach a height of 4 ft., the leaves are flat, 6 to 10 in. long, and about ¼ to ½ in. wide.

Bluebunch wheatgrass. (U.6.)

This species of wheatgrass is propagated only from seed. Deferring grazing until seed maturity, thus utilizing livestock for scattering and trampling the seed into the ground, has been practiced to good advantage in some places. Many successful plantings have been made in revegetation work on range and farm lands. On millions of acres of range land, however, where unrestricted grazing has prevailed, bluebunch wheatgrass has succumbed to overstocking and too early grazing.

References: H.1; B.1.

See also BEARDLESS BLUEBUNCH WHEATGRASS; GRASSES; WHEATGRASSES; RANGE PLANTS.

BLUE GRAMA (*Bouteloua gracilis*) is a low-growing, long-lived, native perennial that is found throughout the Great Plains. The flowering stems are 12 to 18 in. tall; each stem usually has two purplish spikes that extend flaglike at a sharp angle from the stalk. The leaves are 3 to 6 in. long and less than $\frac{1}{8}$ in. wide.

Blue grama grows on all soil types, including alkaline soils, but is most abundant on the heavier rolling upland soils.

Growth begins fairly late in the season and depends on how much moisture is available. The forage is relished by all kinds of livestock. Growth ceases during long droughts but begins again upon the return of favorable moisture and temperature. Because of its wide distribution, high quality, drought resistance, hardiness, and growth habits, blue grama is one of the most important range species.

Under heavy grazing, blue grama often persists in nearly pure stands after associated grasses disappear. To insure range maintenance, the stubble height of the grass should be not less than 2 in. at the end of the grazing season and, in addition, 25 to 30 percent of the flower stalks should be left ungrazed.

Excellent stands of blue grama have been obtained by broadcast or solid-drill seeding. The seedbed should be well prepared, but the seedlings are relatively persistent and compete with weeds and other grasses if they are not grazed until they become well established. The seed usually matures in August.

Although its erosion-control properties are effective when blue grama is seeded alone, it is generally planted in mixtures with other grasses. Most revegetation

seedings have been made on range land and abandoned cropland.

Plantings of North Dakota blue grama seed have not been productive when planted in the Southern Great Plains and vice versa. This lack of plant adaption is important with many of the native grasses and has led to the general caution that locally grown seed is used whenever possible.

Reference: H.1.

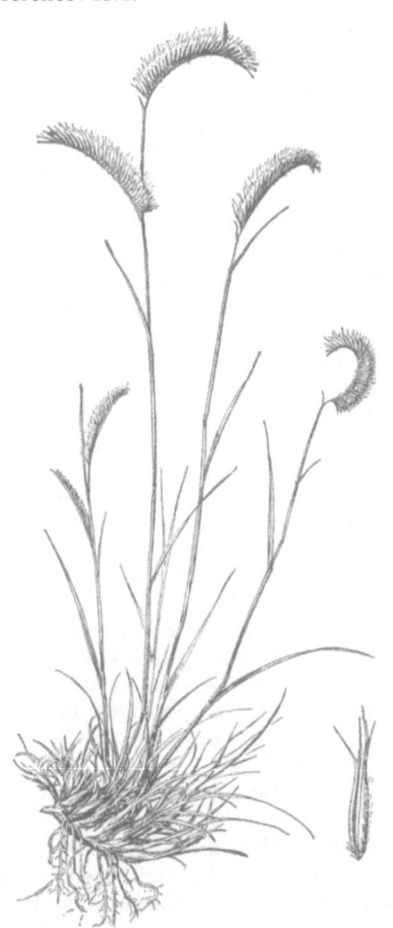

Blue grama. Plant and glume. (H.26.)

See also BLACK GRAMA; GRAMA GRASSES; GRASSES; RANGE PLANTS; PASTURE MANAGEMENT.

"BLUEGRASS."

See KENTUCKY BLUEGRASS; BLUEGRASSES.

BLUEGRASSES (*Poa* spp.) are distinguished by small, awnless SPIKELETS; LEMMAS with a heavy midnerve (like the keel of a boat); GLUMES, one- to three-nerved; and leaf blades with boat-shaped tips.

The bluegrasses are valued primarily for pasturage, hay, and lawn. They rank as the most palatable of range and pasture grasses and are suited for many special agricultural uses.

Generally, the bluegrasses should be planted in autumn when temperature and moisture are conducive to germination and promote good growth. For best growth and seed production, phosphorus, potassium, and nitrogen must be plentiful. Unless the soil is quite deficient in calcium, lime is not necessary.

Some of the bluegrasses are considered WEEDS.

Dangers: Bluegrasses may be attacked by such insects as ARMYWORMS and BILLBUGS.

Species: The most important bluegrasses—because of their forage value—are KENTUCKY BLUEGRASS and CANADA BLUE-GRASS; other species that are well known are ANNUAL BLUEGRASS, BIG BLUEGRASS, BULBOUS BLUEGRASS, FOWL BLUEGRASS, MUTTON BLUEGRASS, NEVADA BLUEGRASS, ROUGHSTALK BLUEGRASS, SANDBERG BLUE-GRASS, and TEXAS BLUEGRASS.

Reference: H.1.

See also SHEEP FESCUE; SAGEBRUSHES; ALFALFAS; GRASSES; PASTURE PLANTS; GRAZING; STRAW MEAL; VITAMIN A.

Illustrations: See KENTUCKY BLUEGRASS; MUTTON BLUEGRASS.

BLUE JACKET.
                        See COMMON WHEAT (variety).

BLUEJOINT (*Calamagrotis canadensis*) is a GRASS which grows in marshes, wet places, open woods, and meadows, from Alaska to California, New Mexico, Kansas, Missouri, and West Virginia. It is useful as forage, furnishing much of the wild hay of Wisconsin and Minnesota. If cut before maturity, bluejoint hay is almost as valuable as timothy hay, especially if fed to horses and mules.

BLUE LUPINE (*Lupinus angustifolius*) is the most important of the annual LUPINES.

The upright plants attain a height of about 3 ft. They require neutral or slightly acid soils of at least moderate fertility and do best on Coastal Plain soils east of the Mississippi. The plants branch quite freely in thin stands, but branching is somewhat restricted in thick stands. As the Latin name implies, the leaves of blue lupines are fine, rather small and digitate (fingerlike). The plants are taprooted with many large nodules. Flowers are in terminal SPIKES and are blue or pink. Seeds are borne in pods about ½ in. in diameter and 3 to 4 in. long. Blue lupine seed is nearly round and usually mottled gray.

Dangers: Among the diseases of blue lupine are BOTRYTIS STEM CANKER, BROWN SPOT OF LUPINE, and SKLEROTINIA STEM ROT.

Strains: The species is represented by *sweet* and *bitter* strains. In general, pink flowers in the blue lupine are correlated with bitter (high alkaloid), and blue flowers with sweet (nonalkaloid) plants. However, blue-flowering plants high in alkaloid do occur; plants with pink flowers and low alkaloid content have not been found. In the high-alkaloid strains the pods remain straw color, while in the nonalkaloid strains they are much darker and become blackish gray.

References: S.7; M.12.

See also WHITE LUPINE; YELLOW LUPINE.

BLUE PANICGRASS (*Panicum antidotale*), also called *giant panicum*, is adapted to warm areas, both dry and moist, but not promising in the southeastern states (possibly owing to low competitive ability in the seedling stage). It is cultivated in Arizona and Texas.

Blue panicgrass is a coarse, vigorous perennial, 5 to 9 ft. tall, with a heavy basal growth of thick, short, bulbous rootstocks (RHIZOMES). The root system is extensive and reproduction is by tillering. Seed production is abundant. The plant requires fertile soil and good drainage.

Abundant moisture is needed for germination; seedlings establish slowly and do not withstand competition. When established, blue panicgrass can recover from extensive heat and wind injury. It needs frequent cultivation to prevent sod-binding. The grass is good for grazing, has a high protein content, and is very palatable if not too mature; it provides earlier grazing than the available annuals. Production is best under rotational grazing. MADRID SWEETCLOVER or a dry-land ALFALFA are suitable companion legumes.

This PANICGRASS species is recommended for planting in washes and arroyos, on flood plains and desilting areas, and for stream-bank protection. It is also being effectively used as a windbreak in the drier parts of the Rio Grande Valley.

Reference: W.16.

See also GRASSES.

BLUE PRUSSIAN PEA = *Blue Bell.*

See FIELD PEA (variety).

BLUE ROSE. *See* RICE (variety.

BLUESTEMS (*Andropogon* spp.) comprise a large group which is well represented throughout the world's warmer regions.

The stems of these GRASSES are solid or contain pith, differing in this respect from most other grasses, the stems of which are hollow.

Species: Two widely distributed species are regarded as good forage grasses, namely BIG BLUESTEM and LITTLE BLUESTEM; BROOMSEDGE also has a wide distribution, but it ranks low as livestock forage. *King Ranch bluegrass*, a variety of TURKESTAN BLUESTEM, is of increasing importance.

References: H.1; H.2; O.1.

See also CAUCASIAN BLUESTEM; ASIATIC BLUESTEMS; BEARDGRASS; ANDROPOGON; RANGE PLANTS; PASTURE PLANTS.

Illustration: See LITTLE BLUESTEM.

BLUESTEM WHEAT is a name applied to *Fultz* and to *Purplestraw*, two varieties of COMMON WHEAT.

BLUESTEM WHEATGRASS

= WESTERN WHEATGRASS.

BLUESTONE = COPPER SULFATE.

BLUE WILD-RYE (*Elymus glaucus*) is a native perennial bunchgrass that occurs throughout the western states, particularly on old burns and cut-over areas in the Northwest. It commonly grows in small tufts and rarely forms dense, pure stands. It is the most widely distributed species of WILD-RYE GRASSES.

Seedstalks may grow to 5 ft. high, with broad, flat, smooth leaves nearly 12 in. long. Roots are vigorous and penetrate deeply. The plant derives its name from the bluish bloom on leaves and stems.

It is most abundant on moist soils, but it stands considerable drought. The foliage, although rather coarse, is grazed by cattle and horses, especially during the early part of the season.

Blue wild-rye produces good growth during the cool season in parts of California; it also persists very well there under limited rainfall and shows promise for use as dry-land pasture, hay, or range.

In experimental plantings in wood lots in Washington, blue wild-rye was shade-tolerant, provided excellent ground protection, and gave a high forage yield. It may be attacked by such diseases as GRASS SMUTS.

Reference: H.1.

See also GRASSES.

BODY WASTES, consisting of the undigested residues of feed, are eliminated from the animal's body as *feces*. They contain chiefly cellulose (crude fiber), mineral matter, remainders of the digestive fluids, mucus, bacteria, etc.

BOEHMERIA. *B. nivea* = RAMIE.

BOER'S LOVEGRASS (*Eragrostis chloromelas*), or *Boer lovegrass*, is a perennial LOVEGRASS with long basal leaves and erect stems. It is used for grazing, hay, and dune control in southern Arizona, New Mexico, and Texas.—*See also* GRASSES.

BOKHARA CLOVER

= WHITE SWEETCLOVER.

BOLL is the capsule which develops from the enlarged ovary as fruit of COTTON and FLAX. The boll splits open at maturity.—*See also* LINSEED.

BOLLWORM = CORN EARWORM.

BOLTING means: *sifting*—e.g., for sizing ground feedstuffs, for separating bran from flour, or for removing hulls.

BONDA. *See* COMMON OAT (variety).

BONE ASH. *See* MINERAL FEEDS.

BONE BLACK = BONE CHARCOAL.

BONE BUILDING VITAMIN

= VITAMIN D.

BONE CHARCOAL, or *bone black*, is one of the MINERAL FEEDS. It consists chiefly of calcium phosphate and the remains of burnt bones and blood.

BONE MEAL is a term often rather vaguely used; finely ground bone meal is sometimes called *bone flour.*

All bone meals are valued primarily for their calcium and phosphorus contents; e.g., RAW BONE MEAL and four of the officially recognized MINERAL FEEDS: *steamed bone meal, special steamed bone meal, cooked bone meal,* and *dicalcium phosphate from bone.—See also* GRAZING.

BONE PHOSPHATE OF LIME, or *BPL,* is a term which was once used to describe BONE MEAL. Not now accepted for use on labels of feeds.—*See* MINERAL FEEDS.

BONITA and *Combine Bonita* are grain SORGHUMS.

BOONE SOYBEAN.

. *See* SOYBEAN (variety).

BOONE COUNTY WHITE.

See CORN (variety).

BOOT is the upper leaf SHEATH of a grass.

BORAX, or *sodium borate,* forms colorless crystals or a white powder. It is only slightly soluble in cold water. Borax is often used to prevent houseflies from breeding in manure piles and as fertilizer to increase the BORON content of the soil. Bur-clovers, kudzu, and other LEGUMES often require borax fertilizers.

BORDEAUX MIXTURE is a fungicide consisting of COPPER SULFATE, HYDRATED LIME, and water. LEAF SPOT OF PEANUT can be effectively controlled by spraying with bordeaux mixture. This treatment controls at the same time tobacco thrips.

Bordeaux 4-2-100 means that the mixture consists of 4 lb. copper sulfate and 2 lb. hydrated lime in 100 gal. water.

BORDERED SHEATH SPOT is a FUNGUS disease of RICE caused by *Rhizoctonia* spp., especially *R. oryzae, R. solani,* and *R. zeae.* It occurs rather commonly in Louisiana and Texas and to some extent in Arkansas and California. The disease appears chiefly on the sheaths above the water line and may occur also on the leaves. The spots are irregular in outline. Their centers, particularly between the veins, are cream-colored, and their outer borders are rather broad and dark reddish brown. The inner edges of the borders are usually jagged. The disease is of minor importance on rice.

Reference: T.2.

BORER. *See* COMMON STALK BORER; SUGARCANE BORER; SOUTHWESTERN CORN BORER.

BORON is a chemical element which occurs in some minerals, plants, and animals. It is not one of the essential elements, but certain LEGUMES and other plants require boron-containing fertilizers, such as BORAX.

BOSS COWPEA = *Wonderful.*

See COWPEA (variety).

BOT. (bot.) is the abbreviation for *botanical* VARIETY.

BOTANICAL VARIETY.

See VARIETY; SUBSPECIES.

BOTANY is the science of the vegetable kingdom—from its lowest, microscopic forms to the flowery and seed-bearing plants.

Systematic botany = TAXONOMY.

BOTRYTIS. The FUNGUS *Botrytis cinerea* causes BOTRYTIS STEM CANKER and GRAY MOLD.

BOTRYTIS STEM CANKER is caused by the soil-borne FUNGUS *Botrytis cinerea* which attacks LUPINES. It is found in varying abundance in BLUE LUPINE fields. The fungus attacks plants that have been injured by freezing; it can, however, infect healthy plants under suitable conditions. The disease is most evident on the stems and large branches, where it produces large brown cankers that, under high-moisture conditions, may be covered with a fluffy, moldy growth containing the fruiting bodies of the fungus. The pods are sometimes affected by the disease and may be covered by the same moldy growth. For the most part botrytis stem canker is not serious. The disease is seen most often in the spring.

Control: No effective control measure is known.

Reference: W.8.

See also SCLEROTINIA STEM ROT.

BOUTELOUA. The *Bouteloua* spp., or GRAMA GRASSES, include *B. curtipendula* = SIDE-OATS GRAMA, *B. grácilis* = BLUE GRAMA, *B. eriopoda* = BLACK GRAMA, and *B. barbata* = SIXWEEKS GRAMA.

BOVINE is an animal belonging to the cattle group.

BPL = BONE PHOSPHATE OF LIME.

BRABHAM. *See* COWPEA (variety).

BRACKEN (*Pteridium aquilinum*) is a perennial fern containing toxic substances which have a cumulative effect; one toxic factor is *thiaminase*, an enzyme that destroys thiamine (vitamin B_1). When horses or cattle eat about 5 lb. fronds (leaves) daily for a month or so, they begin to show symptoms of poisoning: the horses by becoming unable to control the legs and weakening; the cattle by developing hemorrhages in various parts of the body.—*See also* POISONOUS PLANTS.

GRASSHOPPERS; RED HARVESTER ANT; OFFAL; WHEATS; RICE.

BRAN BUGS. *See* GRAIN BEETLES.

BRAND NAMES of feeds are widely used; they must comply with certain *labelling* requirements. When a brand name carries a percentage value, it is understood to signify PROTEIN content only, even though it may not explicitly modify the percentage with the word "protein." Other percentage values are permitted if they are followed by the proper description. When a figure is used as part of a brand name (except in MINERAL, VITAMIN, or other products where the protein guaranty is nil or unimportant) it shall be

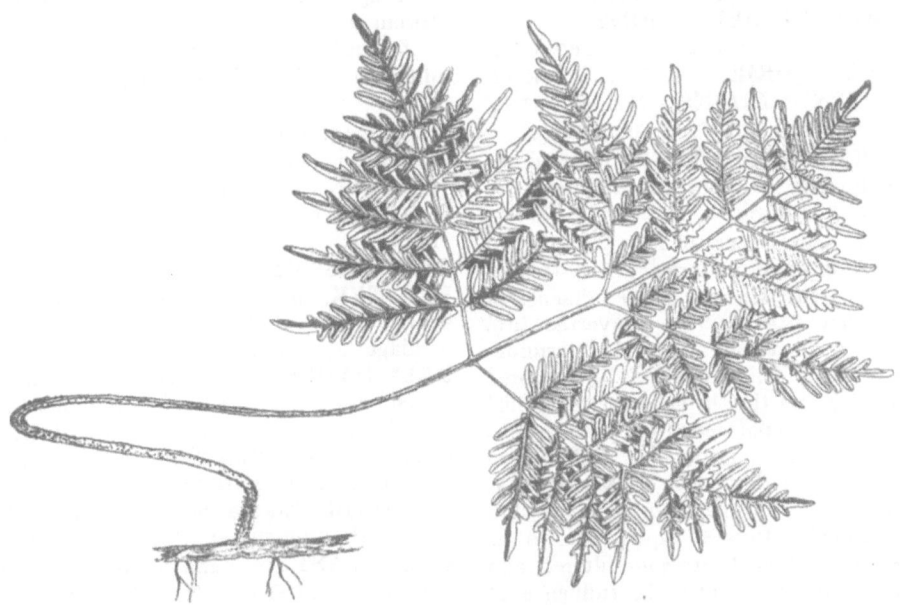

Bracken. Leaf attached to rootstock. (H.48.)

BRACT is a modified leaf or a scalelike organ subtending (i.e., embracing in its AXIL) a flower or flower branch.

Bracted means: provided with, or composed of, bracts.—*See also* LEMMA; GLUME; PALEA; SCALE.

BRAN, or *seed coat*, is the coat of a grain (of a cereal); it is removed in milling and widely used as feed. WHEAT BRAN, for instance, is rich in protein, fat, and phosphorus, and does not contain too much fiber.—*See also* POISONED BRAN MASH;

preceded with the word "number" or some other suitable designation.

Reference: F.6.

See also DEHYDRATED ALFALFAS; LABELS.

BRASSICA. *B. napus* = RAPE; *B. caulorapa* = KOHLRABI; *B. napobrassia* = RUTABAGA; *B. oleracea* var. *acephala* = KALE, var. *asparagoides* = BROCCOLI, and var. *capitata* = CABBAGE.

Many of the brassicas contain goitrogens (antithyroid substances) which reduce the metabolic rate of animals. Such plants are therefore more valuable for fattening animals than for growing or lactating animals.

BREAD, when *stale*, is useable as replacement of part of the grain in the normal stock ration. Its composition is similar to that of grain, but its water content is much higher (usually over 30 percent), so it has only about 75 percent of the value of CORN for pig feed.

BREAKFAST FOOD BY-PRODUCTS are obtained in the manufacture of corn or wheat flakes by sifting or cracking. They are used in feeds for calves and pet animals.—*See also* BY-PRODUCT FEEDSTUFFS.

BREEDING RATIONS.

See POULTRY RATIONS.

BREWERS' DRIED GRAINS.

See BREWERS' PRODUCTS.

BREWERS' DRIED YEAST. *See* YEAST.

BREWERS' GRAINS. *See* BREWERS' PRODUCTS; WET BREWERS' GRAINS.

BREWERS' PRODUCTS are by-products of the brewing industry; most of them are important as feeds.

In the manufacture of beer the grain of selected BARLEY varieties is so treated that part of its starch—under influence of its enzyme DIASTASE—is converted into *malt sugar* during an accelerated germination process. This process is interrupted by heating and the shriveled roots—called *sprouts*—are separated from the *malted grain* which then is crushed and *mashed* with water, so that most of the starch is transformed into malt sugar. This and other solubles form a liquid, known as *wort*, which is boiled with *hops*, filtered, and finally fermented under the influence of added YEAST. The beer is separated from the residues which are used for feeding purposes.

Officially recognized are the following brewers' products:

Brewers' dried grains—the dried extracted residue of barley malt alone or in mixture with other cereal grain or grain products resulting from the manufacture of wort. Brewers' grains are predominantly a cattle feed to be used interchangeably with other feeds of similar type as to bulk and crude-fiber and protein contents.

Malt sprouts—obtained by the removal of the sprouts from malted barley together with the malt hulls, other parts of malt, and foreign material unavoidably present. It shall contain not less than 24 percent protein. (The term "malt sprouts" when applied to a corresponding portion of other malted cereals must be used in qualified form, as for example: rye malt sprouts, wheat malt sprouts, etc.). Dried malt sprouts are used chiefly in mixed feeds for dairy cattle, frequently along with brewers' grains.

Malt cleanings—obtained from the cleaning of malted barley or from the recleaning of malt which does not meet the minimum protein standard of malt sprouts. It must be designated and sold according to its protein content.

Malt hulls—consisting almost entirely of hulls obtained in the cleaning of malted barley.

Dried spent hops—obtained by drying the material filtered from hopped wort.

References: F.6; E.12.

See also FERMENTATION INDUSTRY BY-PRODUCTS.

BREWERS' RICE. *See* RICE.

BRIDGER. *See* COMMON OAT (variety).

BRISTLE is a short, stiff, hairlike appendage; e.g., an AWN.

BRIX HYDROMETER is an instrument which indicates the *sugar* content of a solution in *Brix degrees*. Each Brix degree equals 1 gm. SUCROSE per 100 cc. (at 15.6° C.).—*See also* MOLASSES.

BROAD BEANS are the *large-seeded* varieties of the HORSEBEAN.

BROADCAST (by hand or by machine) means: to scatter or sow seed on the surface of the land.

BROAD-LEAVED HERB. *See* WEED.

BROAD-LEAVED TREFOIL is a type of the BIRDSFOOT TREFOIL.

BROCCOLI, like the closely related *cauliflower*, is a variety of *Brassica oleracea* var. *asparagoides*. The leaves and stems of this vegetable are PLANT BY-PRODUCTS which are used as feedstuffs.

Dehydrated broccoli leaf meal may be used instead of ALFALFA meal in diets for poultry. It contains vitamins A and B_2 and protein.

Reference: E.12.

BROILER RATIONS.
                    See POULTRY RATIONS.
BROMAR.
        See MOUNTAIN BROME (variety).
BROME GRASSES (*Bromus* spp.), or
bromes, are found mostly in the North
Temperate Zones. Most bromes are highly
palatable during their period of most
active growth.

Leaf blades of the bromes are characteristically flat, and the edges of the
sheath grow together to form a tube. The
seed heads are usually open and spreading,
forming PANICLES. The tip of the rather
rigid LEMMA is notched into two teeth,
between which the awn arises.

Species: About 43 species of brome
grasses are native to the United States.
Some of the most troublesome weeds belong to this genus—and also some of the
most important forage grasses, e.g., CALI-
FORNIA BROME; MOUNTAIN BROME; RESCUE-
GRASS; and SMOOTH BROME.

Of minor importance as forage plants,
and sometimes even considered weeds, are
FOXTAIL BROME, CHESS, and CHEATGRASS.

Reference: H.1.

See also MUTTON BLUEGRASS; GRASSES;
ALSIKE CLOVER; ALFALFAS; PASTURES;
PASTURE PLANTS; RANGE PLANTS; SILAGE
CROPS; HAY; ERGOT.

Illustration: See SMOOTH BROME.

BROME HAY. *See* SMOOTH BROME.

BROMES = BROME GRASSES.

BROMUS. The genus *Bromus* includes
the following BROME GRASSES: *B. carinatus*
= CALIFORNIA BROME; *B. catharticus* =
RESCUEGRASS; *B. inermis* = SMOOTH
BROME; *B. marginatus* = MOUNTAIN
BROME; *B. rubens* = FOXTAIL BROME; *B.
secalinus* = CHESS; *B. tectorum* = CHEAT-
GRASS.—*See also* RANGE PLANTS.

BRONCOGRASS = CHEATGRASS.

BROOD SOW RATIONS.
                        See SWINE RATIONS.

BROOM. *See* WITCHES'-BROOM.

BROOMCORN belongs to the SORGHUMS.
See also SILAGE CROPS; DUST STORMS.

BROOMCORN MILLET = PROSO.

BROOMCORN SILAGE.
                        See SILAGE CROPS.

BROOM GROUNDSEL. *See* GROUND-
SELS; SENECIO; POISONOUS PLANTS.

BROOMSEDGE (*Andropogon virginicus*)
is widely distributed throughout the
eastern half of the United States. Its range
has probably moved westward considerably from its original natural distribution.

This undesirable, weedy perennial is a
bunchgrass with a fibrous, shallow root

Broomsedge. Plant and (upper left) spikelet
with rachis joint. (H.26.)

system. The stems may reach 2 to 4 ft. in height; they become very hard with maturity. Broomsedge can be easily distinguished from the other BLUESTEMS in fall and winter by its conspicuous bright yellow color and the shiny appearance of the stems. Eradication of this plant is one of the major problems in meadows and pastures.

Broomsedge is considered a poor-land plant. It makes its appearance quickly on land that has been abandoned. It usually appears first on wet, poorly-drained soils or seepy slopes where the nitrogen supply is extremely low. However, this grass is now invading soils of all fertility levels.

As palatability is very low after hard stems form, broomsedge ranks low as a livestock forage. Cattle will eat the young plant in the early part of the grazing period.

The roots of broomsedge can be destroyed by one thorough plowing, preferably in late fall or early winter. For best control, destroy the old plants by cultivation, then fertilize properly and establish good pasture grasses and LEGUMES that will withstand heavy grazing.

References: E.1; H.1.

See also GRASSES; SERICEA.

BROOM SNAKEWEED (*Gutierrezia sarothrae*), or *broomweed*, is a half shrub with woody roots, crowns, and stem bases. It occurs abundantly in the western states, primarily in the plains, foothills, and mountain slopes. Throughout most of its range this plant is considered worthless as forage. Horses eat the plant during the winter; it ranks as fair forage for sheep directly after growth begins in the spring.

Broom snakeweed plays an important role on ranges depleted of better vegetation in that it protects the soil against wind and water erosion.

Dangers: Heavy utilization of this plant by livestock would probably result in sickness or death.

Reference: U.6.

See also RANGE PLANTS.

BROWN-BORDERED LEAF SPOT is a disease caused by the FUNGUS *Phyllosticta glumarum*. It occurs on RICE in Arkansas, Louisiana, Texas. The spots are light brown at first and their centers become lighter as they enlarge and extend from the midrib to the margin of the leaf. The light-brown fungus bodies in which the spores are produced are visible only under a magnifying lens. Most commercial varieties of rice are rather resistant to this leaf spot; it has not caused any great damage.

Reference: T.2.

See also NARROW BROWN LEAF SPOT; BROWN SPOT OF RICE.

BROWN COTTON BUG (*Euschistus impictiventris*), or *Arizona brown stink bug*, occurs from Texas to Utah, Nevada, and southern California. The adults of this STINK BUG species average ½ in. in length. Their yellowish-brown color is remarkably uniform.

Reference: R.6.

BROWN DURRA, or *Brown "Gyp,"* is one of the grain SORGHUMS.

BROWN HAY is HAY that turned brown because it was stacked when half cured and allowed to ferment in the stalk. In the fermentation process, up to 40 percent of the dry matter may be lost in form of gases and water vapors. The hay becomes less digestible and has less protein available. The feeding value decreases as the color of the hay gets darker.

See also BROWNING REACTION.

BROWNING REACTION, or *Maillard reaction*, occurs when PROTEINS are heated in the presence of sugar, and certain AMINO ACIDS (e.g., lysine) are made unavailable. If feedstuffs are to have high value, they should be processed to hold this reaction to a minimum.

BROWN KAOLING. *Manchu - Brown kaoling* is one of the grain SORGHUMS.

BROWN PATCH, caused by the FUNGUS *Pellicularia filamentosa*, is a disease which affects ST. AUGUSTINEGRASS and all varieties of COLONIAL BENT. Where these bentgrasses are grown alone (i.e., not in mixtures with other grasses), they are likely to be injured by brown patch. It may attack susceptible plants wherever high temperatures and humidity prevail. It makes roundish areas that are 1 in. to 3 ft. across or may extend over an entire field.

Control: Brown patch can be controlled with TERSAN.

BROWN RICE. *See* RICE.

BROWN SPOT OF LUPINE, caused by the FUNGUS *Ceratophorum setosum*, does damage to the LUPINE crop in the northern part of the Lupine Belt.

On BLUE LUPINE the disease is most abundant on the leaflets. The lesions may be small, just visible to the unaided eye, or as big as the width of the leaflets. Usually the spots are circular; some lesions consist of fine dark streaks that form an irregular netlike pattern. The spots and streaks are blackish brown to almost black. Centers of the larger lesions are brown. The brown spot fungus also attacks the PETIOLES, stems, blossoms, pods, and seeds of blue lupine. It causes a great deal of defoliation, often leaving the stems bare, except for a few young leaves at the top of the plant. It may kill large plants as well as seedlings only a few inches tall.

The brown spot fungus may also attack the YELLOW LUPINE and WHITE LUPINE and destroy much, if not all, of the stand. The fungus also attacks VETCH, ROUGHPEA, CROTALARIA, and probably other plants. As some of these plants volunteer, they may keep the brown spot fungus alive in the absence of its lupine host and thereby offset some of the value of rotation as a method of control. It is not known how long the brown spot fungus lives in the seed, but certainly it survives for more than two years.

Control: At present the most effective method is to plant disease-free seed on land that has not grown lupines for three to four years.

Reference: W.8.

BROWN SPOT OF RICE, caused by the FUNGUS *Helminthosporium oryzae*, is one of the most serious RICE diseases found in Louisiana and Texas and to a lesser extent in Arkansas. It attacks the seedlings, leaves, "necks," hulls, and kernels of rice. It may attack the seedlings and cause SEEDLING BLIGHT until they attain a height of about 4 in. Brownish discolorations first appear on the sheaths between the germinated seed and the surface of the

soil or on the roots. Badly affected seedlings die because their upper parts are cut off from the roots by the girdling action of the fungus.

On the leaves the spots are circular to elongate and usually are most abundant in August and September. The small spots are dark reddish brown, and the larger ones have dark reddish-brown margins and grayish centers. On severely affected plants the leaves may dry up before the crop is fully mature, and the yield and quality of the rice may be affected seriously.

The brown spot fungus also causes the condition known as "*rotten neck*," which is similar to that caused by the fungus of BLAST OF RICE. The infected tissues turn brown, shrivel, and die; the weight of the heads (or parts of the heads) then breaks the stalks at the weakened points, making the heads hang down. If the "necks" break before the rice is mature, the kernels are light in weight and of inferior quality.

Spots on the hulls when they are still green look much the same as those on the leaves, and the brown edges of the spots are still visible after maturity. Spots may occur on the kernel either under the spots on the hulls or along the sides of the kernel. Discolored kernels are called *pecky rice* (which may be due also to injury of the kernels by other fungi or by insects).

Control: Seedling blight resulting from the brown spot disease may be controlled in part by treatment with ARASAN, CERESAN M, and other fungicidal dusts. The fungus, however, is carried also on plant refuse in the soil and occurs on wild grasses in or near the fields; SEED TREATMENT, therefore, gives only temporary protection to the young plant. The use of resistant varieties of rice probably is the best means of controlling the disease.

Reference: T.2.

See also BROWN-BORDERED LEAF SPOT; NARROW BROWN LEAF SPOT.

BROWN SPOT OF SOYBEAN is caused by the FUNGUS *Septoria glycines*. It is one of the earliest diseases to appear on SOYBEANS in the spring. The symptoms are angular, reddish-brown spots on the first pair of leaves to come out. The reddish-brown color is more pronounced on the under than on the upper leaf surface. As the plant grows, the disease moves upward to the younger leaves. Infected leaves gradually turn yellow and are shed prematurely. In badly infected fields the lower half of the stem may loose all its leaves. Brown spot also causes brown discolorations on stem, branches, and pods. The spots vary in size from pin point to about ⅛ in. When these small dots join one another, the spots become much larger.

The fruiting bodies of the fungus are borne in the tissues of the leaf and stem. They can be seen only with the aid of a hand lens. The spores, discharged to the surface through a comparatively large pore, spread infection to new leaves or other plants. Warm, moist weather and poor drainage favor the spread of the disease. The fruiting bodies overwinter on diseased leaves and stems and serve as a source of infection the next year. The disease is seed-borne.

Control: Brown spot must be combated through crop rotation.

Reference: C.9.

BROWN STEM ROT is caused by the FUNGUS *Cephalosporium gregatum.* It is one of the most serious SOYBEAN diseases. The fungus enters the plant through the root and lower stem and travels slowly upwards. Infected plants at first show no outward symptoms; if the stem is split lengthwise, a brown discoloration can be detected inside the root and at the base of the stem. The progress of the disease up the stem is favored by cool weather. The leaf symptoms appear late, the tissues between the veins turning brown; near the veins the natural green stays for a while, but soon the whole leaf withers. The over-all appearance of the field takes on a brownish cast. This late phase of the disease starts so abruptly that the plants may look as if they had been damaged by frost. Brown stem rot may also cause lodging.

Control: The disease is soil-borne; the only way to combat it is to use a rotation in which soybeans are grown on the same ground every four to five years.

Reference: C.9.

BROWN STRIPE is caused by the FUNGUS *Scolecotrichum graminis* which forms water-soaked lesions on the leaves; these are first olive gray (when wet) or dull gray (when dry), and turn later brownish-purple to ocher with gray centers. As the leaves slowly die, the spots tend to form streaks. The spore-bearing bodies of the fungus can be seen as black dots arranged in parallel rows.

ORCHARDGRASS and many other GRASSES are affected with brown stripe.

Control: Careful burning of dead grass reduces the spores which otherwise infect new leaves.

Reference: H.44.

BROWSE is a term applied to the twigs with their shoots and leaves, which are cropped by the livestock from shrubs, small trees and woody vines. In emergencies, large trees may be felled and the stock allowed to eat the leaves. Soft wooded varieties such as maple, poplar, and linden are preferred.

Browse is recognized as a class of RANGE FORAGE.—*See also* PASTURE PLANTS; PASTURES; GRAZING.

BRUCHOPHAGUS. *B. funebris* = CLOVER-SEED CHALCID.

BRUCHUS. *B. pisorum* = *pea weevil.*

BRUNKER. See RED OAT (variety).

BRUSH is the hairy formation at the end opposite the germ of WHEAT (and other seeds).—*See also* COMMON WHEAT; BRUSHES.

BRUSHES—the different kinds of low growing woody plants—should not be disturbed in some areas. Removal of this natural cover on steep slopes or on stony or sandy land would result in severe soil erosion. On some RANGE land, the cost of killing brush may be too great to offset returns from increased grass production.

In some of the brush land, native GRASSES are present and ready to cover the soil when the brush is killed. Where good grasses are not present to take over, the brush must be removed to allow planting and fertilizing machinery to operate.

Where brush control is required, a method best adapted to the individual brush problem must be selected. Some species are easy to control with chemicals, others must have several treatments, and some are resistant to all chemical treatments now available for this purpose.

2,4-D, and 2,4,5-T are effective on certain kinds of brush. Both chemicals can be used as foliage sprays or for dormant treatments. Other useful chemicals are AMMONIUM SULFAMATE ("Ammate") and arsenic compounds (ARSENICALS).

Reference: E.15.

See also PASTURE MANAGEMENT; RANGE MANAGEMENT.

BRUSH PASTURES. *See* PASTURES.

BRYOBIA. *B. praetiosa* is one of the CLOVER MITES.

BUCHLOE. *B. dactyloides* = BUFFALO-GRASS.

BUCKWHEAT BRAN. *See* BUCKWHEATS.

BUCKWHEAT FEED. *See* BUCKWHEATS.

BUCKWHEAT GRAIN. *See* BUCK-WHEATS.

BUCKWHEAT GROATS. *See* BUCK-WHEATS.

BUCKWHEAT HONEY. *See* BUCK-WHEATS.

BUCKWHEAT HULLS. *See* BUCK-WHEATS.

BUCKWHEAT MIDDLINGS. *See* BUCK-WHEATS.

BUCKWHEATS (*Fagopyrum* spp.) do not belong to the true cereals or to the grains (because they do not belong to the grass family), but are classed commercially among grain crops. They are unable to compete with other grain crops and are usually found growing back in the hills. Although the crop has little national importance in the United States, it is of considerable value in certain limited areas of the country. Because of its short growing season it may be sown later than any regular grain crop and for this reason is often used as a catch crop, seeded where other crops are failing or cannot be planted in season because of wet weather or other reasons.

The plant, an annual, grows 2 to over 5 ft. in height and has a single stem with branches. The stem is grooved and green or reddish; on aging it turns brown. The leaf blades are 2 to 4 in. long and heart shaped. The taproot has many short laterals and may extend 3 to 4 ft. in the soil. The flowers have no petals and are white or tinged with pink. The fruit (an ACHENE) is brown, gray-brown, or black.

The buckwheat species are grown principally in the northeastern part of the United States, the leading buckwheat states being Pennsylvania, New York, Minnesota, Ohio, West Virginia, and Michigan. The buckwheats can be grown with some success almost anywhere north of the Cotton Belt. In the more southern states they are best adapted to the uplands and mountainous sections.

Buckwheat.

Buckwheat plants do best in a moist, cool climate. They are very sensitive to cold, being killed quickly when the temperature falls below freezing. They can be grown rather far north and at high altitudes because their growing season is only ten to twelve weeks and because they require only a small amount of heat for development. Rather high temperatures during the day are not destructive if the nights are cool, the wind not excessive, and the soil not too dry. Unfavorable weather in the principal flowering period tends to reduce materially the set of seed and the yield of grain.

If the climatic conditions are favorable, the buckwheats will produce a better crop on infertile, poorly tilled lands than any other grain. They are suited to light, well-drained soils such as sandy loams and silt loams. The plant needs but little lime, growing well where RED CLOVER and ALFALFA would not succeed.

Best results can be obtained by plowing the land early in the spring and keeping it in condition by occasional harrowing until the crop is sown. Good yields are often obtained on land plowed and harrowed just before sowing. If plowing is done late, the land should be well worked to firm the seedbed. The preparation of the seedbed should be in general the same as for CORN. Old meadow and pasture lands usually are very suitable for the buckwheats.

The buckwheat species are often grown year after year on the same land with little or no apparent ill effect, but the plants will exhaust the soil sooner or later if not properly fertilized.

A good rotation for much of the area suited to this crop is: First year, ALSIKE clover or red clover; second year, buckwheat; third year, POTATOES; fourth year, RYE, OATS, or WHEAT seeded to clover. Another rotation is as follows: First year, corn with CRIMSON CLOVER sown at the last cultivation; second year, buckwheat followed by rye; third year, SOYBEANS.

Phosphorus is the only commercial fertilizer usually necessary for the buckwheat crop on soils of fair to good fertility; 150 to 200 lb. superphosphate per acre is sufficient. Potassium is used by the buckwheats in comparatively large quantities. If the land has been poorly farmed, it may be necessary to add a small amount of potash to secure the best yields. A little nitrogen in the soil is indispensible, and on poor land its application will increase the harvest. The growing of other crops, such as clovers or other LEGUMES, in the rotation may make nitrogen available in sufficient quantity for the buckwheat plants. Excess nitrogen results in the development of straw rather than grain. On very poor soil that has not been well farmed, 100 to 300 lb. per acre of a complete fertilizer furnishing principally phosphorus, probably is best. Lime is beneficial occasionally, but it is not necessary to neutralize the soil acidity in order to secure good buckwheat crops. It seems that 500 lb. lime per acre is the maximum even on soils most deficient of lime.

Buckwheat seed more than one year old should not be used for seeding. The seed germinates best when the soil temperature is about 80° F., but it will germinate at any temperature between 45° and 105°. The plants usually come up in about a week. Blooming begins in about five or six weeks. The first grain begins to ripen about three to four weeks after flowering starts, and ripening continues until frost.

Buckwheat seed may be sown with a grain drill or broadcast and harrowed in. The crop may be harvested with a cradle, scythe, self-rake reaper, grain binder, or combine. The plants are ordinarily shocked in the field and left to dry until threshing time. This requires a week to ten days of good weather. The crop may be stacked or stored in a mow if thoroughly dry. It is easily threshed with either a hand flail or a thresher.

The heavy growth made by buckwheat plants makes them useful in fighting QUACKGRASS and other weeds.

The buckwheats have also great value as soil renovators. They can utilize relatively insoluble mineral soil constituents to good advantage; when plowed under as green manure, they render this plant food available, besides furnishing humus to the soil. A heavy growth plowed under decays more quickly and completely than most other green manure crops, making the residues soon available for a succeeding crop.

Buckwheats are important fall honey plants. Numerous flowers are produced on each plant, and a plant may continue to bloom for a month or more. The flowers are usually well supplied with nectar, each one having eight nectaries.

Buckwheat honey is dark and has a distinct flavor. It usually is highly regarded in sections where it is produced, but commands a lower price than some of the lighter colored clover honeys.

Buckwheat grain may be mixed with oats and BARLEY as a livestock feed. In some cases buckwheat is seeded as a mixture with barley or oats, and the resulting crop is used as a grain feed. The whole grain may be used as a poultry feed. For this purpose Tartary buckwheat is better than the other types, because of the smaller size of the seed; it is used in many mixed poultry feeds, being popular with feed dealers, because the kernels show up well in the mixture.

Buckwheat straw sometimes is used for feed and is eaten readily by livestock if well preserved; it is rich in minerals and carbohydrates. It makes good bedding for cattle but does not last well. The straw makes a good manure.

NOTE: Buckwheat grain, in each 100 lb., contains on an average, 10.8 lb. crude protein, 72.5 lb. carbohydrates, and 2.5 lb. fat. The same quantity of straw contains 5.2 lb. crude protein, 78.1 lb. carbohydrates, and 1.3 lb. fat. Thus, buckwheat grain contains about the same amount crude protein as corn, more fat than wheat and rye, and about the same amount digestible carbohydrates as oats; buckwheat straw contains about the same amount crude protein as corn stover, or wheat or rye straw, is superior in digestible fat to corn stover or wheat, rye, or oat straw, but contains less digestible carbohydrates than corn stover or the straw of these other grains.

Buckwheat products find many uses. *Buckwheat flour* is used in pancake mixtures. The *buckwheat groats*—i.e., the kernels with the hulls removed—are used as breakfast food, porridge, and thickening for soups, gravies, and dressing, much in the same manner as corn meal. Buckwheat flour and groats must be used fresh, as they soon become rancid, owing to their high fat content.

When buckwheat products are eaten steadily, or in too large quantities, then sometimes cause a rash on the skin of mey and white-colored animals. The substance that produces these effects apparently is located in the buckwheat hulls, particles of which are contained in the flour.

Buckwheat hulls are the outer layers of the kernels; they are removed in milling and are used to some extent as an ingredient in mixed livestock feeds. The hulls contain considerable carbohydrate and other nutrient material and are of some small value as feed.

Buckwheat bran is a mixture of the middlings and the hulls. It varies in composition but is usually much poorer as a feed than the middlings alone.

Only two of the buckwheat products are officially recognized:

1. *Buckwheat middlings*—the portion of the buckwheat grain immediately under the hull after separation of the flour. It shall contain no more hulls than is obtained in the usual process of milling and not more than 10 percent crude fiber.

Buckwheat middlings is commercially listed among the CEREAL GRAIN BY-PRODUCTS. It contains considerable protein, carbohydrates, fat, and mineral matter and is considered good feed for cattle and hogs; it is used principally by dairymen. The middlings apparently have no bad effects on the animal or the dairy products if not fed in excess or as the only concentrate. The manure from livestock fed with middlings has a high fertilizing value, because of the nitrogen, phosphorus, and potassium contained in it.

2. *Buckwheat feed*—a mixture of buckwheat middlings and buckwheat hulls—must contain not more than 30 percent crude fiber.

Dangers: The buckwheat species are particularly free from destructive plant diseases or insect enemies, and serious losses occur rarely from these causes. The blasting of the flowers often results from unfavorable climatic conditions but is not due to disease.

WIREWORMS have been reported as damaging buckwheat seed and roots and under certain conditions APHIDS will attack the plants. (The buckwheats are not injured by CHINCH BUGS or WHEAT JOINT-WORMS.) Birds and poultry, when numerous, may consume a considerable quantity of grain before it can be threshed.

Species: The buckwheat species grown in the United States are COMMON BUCKWHEAT, TARTARY BUCKWHEAT, and WINGED BUCKWHEAT.

References: Q.1; E.12; F.6.

See also STRAW; WILD BUCKWHEAT.

BUCKWHEAT STRAW.

See BUCKWHEATS.

BUD BLIGHT is caused by the tobacco ring spot virus. It is the most serious VIRUS DISEASE of SOYBEANS and may attack plants throughout the growing season. The first symptom appears on young plants. It is a browning and curling of the terminal (topmost) bud to form a crook. The bud becomes dry and brittle, while the leaf immediately below it shows a flecking with rusty-brown specks. Sometimes the inside of the stem at the upper nodes (joints) turns brown. The infected plant is stunted and produces no seed. Soybeans infected later in the season may produce no pods or small, undeveloped ones. Plants affected in either of these ways are known as *duds*. In the fall they remain green after normal plants have matured.

Late symptoms of the blight are pods that are poorly filled and drop prematurely, or. that are covered with purple blotches and remain on the plant. The disease always appears first at an edge of a field and progresses inward. It is not seed-borne.

No control method is known at the present time.

Reference: C.9.

BUD ROT is a disease which affects SAFFLOWER; its cause is not known. The bud rot is serious only in areas of high rainfall and high humidity.

Reference: T.3.

BUD SAGEBRUSH (*Artemisia spinescens*) is a low spiny shrub 3 to 12 in. high with small and finely divided leaves. The new growth of the leaves forms a dense green cover over the spines as soon as the weather becomes warm. Later in the spring

tiny yellow flowers, in small budlike clusters, are produced. During the summer the leaves turn brown and fall to the ground. In winter the spiny stems do not have any foliage but are covered between the buds with short white hair.

Bud sagebrush ordinarily grows on slightly alkaline soils, especially in Nevada. It is one of the most palatable sheep forages on the *winter range;* it is highly relished late in winter when the sheep strip the bark and tender buds from the twigs. It is one of the first plants to be injured and destroyed by heavy grazing.

Reference: H.41.

See also RANGE PLANTS.

BUDWORM

= SOUTHERN CORN ROOTWORM.

Budworm injury, also called *ragworm injury*, is caused by the CORN EARWORM.

BUFFALO ALFALFA belongs to the COMMON ALFALFAS.

BUFFALOGRASS (*Buchloe dactyloides*) is a fine-leaved, native sod-forming perennial. Generally it grows 4 to 6 in. high and produces leaves less than ⅛ in. wide and 3 to 6 in. long. It spreads rapidly by surface runners and forms a dense, matted turf. During the growing season the foliage is grayish green which turns to a light straw color when the plants cease growth.

Growth of this grass begins in late spring and continues through the summer. Livestock like its forage. Its palatability, prevalence, and adaption to a wide range of soil and climatic conditions make it the dominant forage species on large areas of upland on the central part of the Great Plains where it withstands long, heavy grazing better than any other grass native to that region. On ranges consistently subjected to severe use it often survives as a nearly pure stand.

It is, however, advisable to grow a LEGUME in combination with Buffalograss to keep the grass vigorous and the soil productive.

Because of its excellent ground cover, aggressive spread under use, and relative ease of establishment, buffalograss is ideally suited for erosion control on range and pasture lands where the soil does not contain too much sand.

The seed of the grass is enclosed in hard burs, one or more grains in a bur. The plants are UNISEXUAL; about half are female and produce the seed burs; the others are male and produce pollen grains only.

Buffalograss. Plant and spikelet. (H.26.)

Revegetation by use of sod pieces is effective. Sod pieces about 4 in. in diameter are placed at 3- to 4-ft. intervals on a well-conditioned seedbed. Usually this results in a complete sod cover by the end of the second growing season.

References: H.1; D.6.

See also GRASSES; RANGE PLANTS; PASTURE PLANTS; PASTURE MANAGEMENT; PRAIRIE HAY.

BUFF CATJANG. *See* COWPEA (variety).

BUFFELGRASS (*Pennisetum ciliare*) is grown in southern states. It is readily established; the roots grow as much as 1 in.

a day. This grass, characterized by a high protein content, provides a large amount of palatable forage on relatively poor soils but must be fertilized and managed properly to maintain high production. Buffelgrass is not adapted where winter temperatures are as low as minus 3°–5° F. Seeding commences sixty days after emergence and continues profusely until frost. Seeds are fuzzy and difficult to harvest.

Reference: W.16.

See also GRASSES; PENNISETUM.

BUG is a term applied popularly to any BEETLE-like insect.—*See also* GREEN BUG; CHINCH BUG; JUNE BUG; BROWN COTTON BUG; SAY STINK BUG; RED-SHOULDERED PLANT BUG; LYGUS BUG; STINK BUGS; BILLBUGS.

BULB is a (usually subterranean) leaf bud composed of fleshy scales—not a root (as it often is thought to be).—*See also* BULBLET; BULBOUS; CORM.

BULBIL = BULBLET.

BULBLET, or *bulbil*, is a small, bulblike bud. It is capable of developing into a new plant, when planted.—*See also* BULB.

BULBOUS means: swollen at the base like a BULB; or containing one or more bulbs; or growing from bulbs.

BULBOUS BLUEGRASS (*Poa bulbosa*) has not reached great importance as pasture grass; it is grown most extensively in southern Oregon and northern California. It forms true bulbs at the base and bulblets in the PANICLE. The plants rarely produce perfect flowers and true seeds on the culms or the seedstalks. Growth usually begins early in October and ceases in May when the bulblets are mature. The seedstalks reach a height of 10 to 18 in., depending on the fertility of the soil.

Bulbous bluegrass is best adapted to regions that have a dry summer, a mild winter, and winter rainfall.

Reference: H.1.

See also BLUEGRASSES; GRASSES.

BULGARIAN WHEAT = TURKEY.
                   See COMMON WHEAT.

BULL is the male BOVINE.

BULL CLOVER = *American single-cut clover. See* RED CLOVER.

BULL RATIONS. Herd bulls should be kept in thrifty condition, but over-fattening should be avoided. Basic rations should contain good quality roughage with a light feed of grains other than corn. In summer, the bull is best consigned to pasture where he can obtain exercise and sunshine. It will sometimes be necessary to add limited amounts of grain to the ration of a mature bull. Younger bulls should receive from 3 to 5 lb. of CONCENTRATES daily depending on number of services and other conditions. —*See also* HAY; SOILING CROPS.

BULRUSH (*Scirpus mucronatus*), or *rough-seed bulrush*, is a WEED pest in RICE fields. It produces numerous tuber-like rootstocks and viable seeds and can therefore spread rapidly.

Reference: J.5.

BUMBLEBEE is a large relative of the honeybees. It is especially important in the cross fertilization of RED CLOVER.—*See also* BEES.

BUNCHGRASSES are GRASSES which differ from the sod-forming grasses in that they send up tillers from the plant's crown, thus forming dense clumps. Among the many bunchgrasses are INDIAN GRASS, most of the WHEATGRASSES and BLUE-STEMS, and certain BLUEGRASSES and perennial FESCUES, GIANT WILD-RYE, NEEDLE-AND-THREAD GRASS, SQUIRREL-TAILS, SACATON, etc.

Feather bunchgrass = GREEN NEEDLE-GRASS.—*See also* PACIFIC BUNCHGRASS RANGE.

BUNCH VELVETBEAN = *bush velvetbean. See* DEERING VELVETBEAN (variety).

BUNDLE CORN = *shock corn. See* CORN.

BUNDLE FEED, or *fodder*, is a term used when the whole plants—stalks and heads with the seeds—of grain crops, especially grain SORGHUMS, are harvested together. When the heads have been removed, only the so-called *butts* are left. A bundle of grain sorghum fully matured and field-cured, will weigh 10 to 12 lb. and will contain from 20 to 40 percent heads and from 15 to 30 percent grain. It will therefore contain about 8 lb. dry matter, about 2 lb. of which will be grain. Hegari and kafir are used extensively as bundle feeds, but other grain sorghums—because of the low palatability of their stalks—are seldom fed in bundles.

Good bundle feed, with a little protein supplement, will maintain livestock in good condition throughout the winter with little or no grain supplement. It is nearly equal to corn fodder in food value and, if sufficiently fine-stemmed and leafy, is more palatable.

References: M.1; S.3.

BUNT is one of the SMUTS of WHEAT. It occurs in two distinct forms: COMMON BUNT and DWARF BUNT.—*See also* DURUM WHEAT; CLUB WHEAT; COMMON WHEAT.

BUR is a spiny or prickly fruit or fruit envelope; e.g., the burs of BUR-CLOVER and BUFFALOGRASS.

BUR-CLOVERS (*Medicago species*) are annual LEGUMES, much like the true clovers; they differ from them in that their small yellow flowers are in clusters of 5 to 10, and the coiled pods are commonly beset with spines, forming the so-called bur. The roots are fibrous and do not extend very deep. Most of the plants have 10 to 20 or more spreading branches 6 to 20 in. long. When in fruit, they are thickly beset with burs. Well-developed plants may contain more than 1,000 pods.

In the Pacific Coast area, from Oregon south to Arizona, and in the Great Plains area of eastern Texas and southern Oklahoma, bur-clover is a valuable winter annual pasture plant, well thought of for its high feeding value.

Bur-clover is commonly used as a green-manure crop in the orchards of California.

In North Carolina bur-clover may be pastured by the middle of February, and near the Gulf Coast it furnishes practically continuous winter pasture. Few legumes will make more growth in that area during cool weather.

The plant will succeed in practically all types of soil, but loams are most suitable. It is very tolerant of alkali. In the South it grows best in soils rich in lime but will also succeed in somewhat acid soils. As a rule it prefers moist, well-drained soils.

Bur-clover does poorly on soils of low fertility. In establishing stands, the use of liberal quantities of stable manure or commercial fertilizer is recommended. Usually superphosphate is the most essential fertilizer ingredient, and 400 lb. per acre of it is all that is needed. In some cases potash and boron are beneficial, and on very poor soils the addition of nitrogen will give increased growth.

If hulled seed is used, sowing may be done from two to four weeks later than if seed in the bur is used.

On the Pacific Coast INOCULATION occurs naturally. In other areas, however, lack of inoculation has often been the cause of failure in establishing bur-clover. Farmers can now purchase legume inoculants prepared with superior and selected strains of bacteria for bur-clover.

Bur-clover is utilized mostly as *pasture* for hogs, cattle, sheep, and poultry. Farm animals do not eat it readily at first, but they soon acquire a taste for the plant and then eat it freely. When bur-clover is growing in cultivated lands it is best not to pasture continuously, but to put the stock on the land for only a few hours each day, as this reduces very much the injury by trampling. Not only do animals eat the herbage, but sheep, especially, are very fond of the ripe pods and will lick them from the ground. The burs are eaten more readily when they have been softened by rain.

An objection to bur-clover is that the burs become entangled in the wool of sheep and reduce its value. The value of the forage, however, far outweighs any small damage done by the burs. On cultivated land the difficulty is easily avoided by removing the sheep from the pasture before the burs are ripe, or by pasturing the bur-clover so heavily that few burs are formed. (It is mainly on account of this objection that spineless varieties of bur-clover are considered desirable on cultivated land.)

For permanent pasture in the South, a combination of bur-clover and BERMUDA-GRASS is very satisfactory. The Bermuda-grass furnishes pasture during the warm weather until further growth is stopped by frost, whereas the bur-clover begins to grow with cool weather and provides pasture during the winter and spring. In the South, undoubtedly the greatest value of bur-clover is that it is the cheapest and most easily handled legume that can be used as a combination cover and green-manure crop. Even a growth only a few inches in height will prevent to a large degree the washing of the land in winter and, when plowed under, add sufficient humus and nitrogen to improve materially the following cotton crop.

California bur-clover with pods. (M.6.)

Under favorable conditions bur-clover will make a dense stand 18 to 24 in. high. From such stands, yields of about 12,000 lb. and more green weight, or of 2 to 3 tons of *hay* per acre have been recorded. Unless the stand is very dense, however, bur-clover plants lie close to the ground. The hay is not regarded very highly and is seldom used.

Dangers: Few cases have been recorded of bur-clover causing BLOAT; but when the growth is lush, care should be exercised.

The only insect that does any serious damage to bur-clover is the CLOVER-SEED

CHALCID; it destroys considerable quantities of seed but does no harm to the herbage. Of minor importance are HARVESTER ANTS, STINK BUGS, and such plant diseases as SOUTHERN ANTHRACNOSE.

Species: Two species of bur-clover are commonly cultivated in the United States, the SPOTTED BUR-CLOVER (or *southern bur-clover*) and the CALIFORNIA BUR-CLOVER (or *toothed bur-clover*). TIFTON BUR-CLOVER and LITTLE BUR-CLOVER are more recently introduced species which are gradually spreading. There are about 35 other species, among them several with large spineless burs, e.g., BUTTON-CLOVER.

References: M.6; M.3.

See also CLOVERS; SPINELESS BURCLOVERS; ALFALFAS; KUDZU; INOCULATION.

BURNT LIME = LIME.

BURT. *See* RED OAT (variety).

BUSH-AND-BOG is a heavy cut-away disk tillage implement, useful on brushy or rough pasture land.

BUSH CLOVERS = LESPEDEZAS.

BUSH VELVETBEAN.

See DEERING VELVETBEAN (variety).

BUTTERMILK. *See* MILK PRODUCTS.

BUTTONCLOVER (*Medicago orbicularis*) is one of the species of BUR-CLOVERS with large spineless burs.

Reference: M.6.

See also CLOVERS; SPINELESS BURCLOVERS.

BUTTS is BUNDLE FEED from which the heads have been removed.

BUTYLATED HYDROXYANISOLE, or *BHA*, is one of the officially recognized ANTIOXIDANTS; it is used for stabilizing animal fats.

BUTYLATED HYDROXYTOLUENE, chemically closely related to BUTYLATED HYDROXYANISOLE, is also an ANTIOXIDANT.

BUTYRIC ACID is a colorless, oily liquid with an unpleasant, rancid odor. The acid is produced in the fermentation process of SILAGE and by rumen bacteria. Large amounts of butyric acid make the silage unpalatable and may increase the danger of *ketosis*, or *acetonemia*, a disease often affecting cattle and sheep.—*See also* BACTERIA.

BYNG. *See* SIX-ROWED BARLEY (variety).

BY-PRODUCT FEEDSTUFFS are important in feeding poultry and swine, as well as beef and dairy cattle.

Research in animal nutrition includes a continual search for new by-products to use in feeds and for new ways of combining by-products with other feedstuffs into rations that give a better balance of nutrients and improve the efficiency of use of feeds by livestock.

Cost is important in determining the value of a by-product in animal feeding. Each year new by-products, resulting from research and from technological development, become available to feed-manufacturers and feeders. Increased use of by-products reduces not only the cost of animal production but also the competition of livestock with humans for food supply.

Almost every food industry furnishes some by-products for animal feed, but the most important sources are the milling of grain, the processing of oilseeds, the manufacture of dairy products, the slaughter of meat animals, and brewing and distilling.

The by-product feeds can be grouped according to their origin as follows: (1) CEREAL GRAIN BY-PRODUCTS; (2) OILSEED MEALS; (3) FERMENTATION INDUSTRY BYPRODUCTS; (4) DAIRY BY-PRODUCTS; (5) PACKING-HOUSE BY-PRODUCTS (including by-products from poultry dressing plants, whale fisheries, and the products obtained from rendering plants); (6) FISHERY BYPRODUCTS; (7) PLANT BY-PRODUCTS (from vegetables or fruits).

Reference: E.12.

See also WOOD MOLASSES; COCOA MEAL; PUFFED CEREALS.

C

C = CARBON.

C 11 is one of the inbred lines of CORN hybrids.

Ca = CALCIUM.—*See also* MINERAL FEEDS.

CABBAGE (*Brassica oleracea* var. *capitata*) is a biennial valued as succulent forage for sheep and dairy cattle; it should be fed after milking.—*See also* VITAMIN A.

CACHE. *See* COMMON WHEAT (variety).

CaCO₃ = CALCIUM CARBONATE. *See also* MINERAL FEEDS.

CACTI (sing.: *cactus*) belong to the Cactaceae. They vary in size from small to treelike growths. The cacti are succulent perennials which are able to withstand drought. They occur in the Southwest, especially in arid areas, and are sometimes useful as emergency forage. All cacti are low in protein and cause scours when fed without sufficient dry feed.—*See also* RANGE PLANTS, PRICKLY PEAR CACTI.

CADET. *See* COMMON WHEAT (variety).

CAESALPINIODEAE is a subfamily of the LEGUMES.

CAFETERIA SYSTEM.
                                        See SELF-FEEDING.

CAL. = CALORIE.

CALAMAGROTIS. *C. canadensis*
                                        = BLUEJOINT.

CALCIFEROL = VITAMIN D_2..
                                        See also VITAMIN D.

CALCIFYING VITAMIN = VITAMIN D.

CALCINED means: heated to a high temperature; e.g., calcined CALCIUM PHOSPHATE.

CALCITE is a rock-forming mineral consisting of CALCIUM CARBONATE.—*See also* MINERAL FEEDS.

CALCIUM—the CHEMICAL ELEMENT *Ca*—is widely distributed in nature in the form of numerous salts. It is essential to normal nutrition of plants and animals.—*See also* MINERAL FEEDS; GRAZING; VITAMIN D; MOLASSES; FERTILIZERS; BLACK POD; LIME; OYSTER SHELL FLOUR; MINERALS.

CALCIUM ARSENATE is a white powder which is only slightly soluble in water. It is a violent poison and must be handled with great care.

Applied at the rate of 10 lb. dust or, as spray, 4 lb. in 100 gal. water per acre, it protects ALFALFA against GARDEN WEBWORMS. However, fields so treated should not be used for hay or pastured with livestock. To kill the larvae of the ALFALFA WEEVIL, dust or spray the crop with 2 lb. calcium arsenate per acre.

Calcium arsenate is also used in POISONED BRAN BAIT against ARMYWORMS and CUTWORMS. It is contained in *poison bait pellets* used for the control of the common garden SLUG, one of the pests of SUBCLOVER.

Calcium arsenate, as well as other AR-SENICALS, may be used as dust or spray on SOYBEAN plants to prevent serious damage from *rabbits*.

CALCIUM BIPHOSPHATE
                                = MONOCALCIUM PHOSPHATE.

CALCIUM CARBONATE, the chemical compound $CaCO_3$—better known as *limestone, chalk,* or *whiting*—is one of the MINERAL FEEDS available in various natural and prepared forms. It is a white powder, which is practically insoluble in water, but which dissolves readily in the acids of the stomach.

The leaves of sorghums, especially milos, grown on soils rich in calcium carbonate may become yellow (CHLOROTIC).—*See also* NONPARASITIC SORGHUM-LEAF DISEASES; LIME.

CALCIUM CYANAMIDE="CYANAMIDE."

CALCIUM DEFICIENCY is the cause of many plant and animal diseases. While legumes are an excellent CALCIUM source for cattle, grasses (in form of grazing or hay crops) and grains may not supply sufficient calcium for heavy milking cows and thus cause a withdrawal of calcium from the cows' bones (*osteomalacia*). In young animals the bones fail to harden (*rickets*), and sometimes the blood calcium becomes depressed (*tetany*).

CALCIUM HYDROXYDE
                                = HYDRATED LIME; LIME.

CALCIUM OXALATE is a white, water-insoluble powder.—*See also* SILAGE CROPS.

CALCIUM OXIDE = LIME.

CALCIUM PANTOTHENATE is a commercially available form of PANTOTHENIC ACID, one of the B-COMPLEX VITAMINS. It is a white, water-soluble powder; the aqueous solution is not stable.—*See also* VITAMINS.

CALCIUM PHOSPHATES are officially recognized as mineral ingredients of feedstuffs and include MONOCALCIUM PHOSPHATE, DICALCIUM PHOSPHATE, TRICALCIUM PHOSPHATE, ROCK PHOSPHATE, and SOFT PHOSPHATE WITH COLLOIDAL CLAY.

Calcined, fused, or precipitated calcium phosphate of low fluorine content is called DEFLUORINATED PHOSPHATE.—*See also* MINERAL FEEDS; BONE CHARCOAL.

CALCIUM-PHOSPHORUS RATIO, ranging from 1:1 to 2:1—i.e., one or two parts

Ca for each part P (provided that both minerals are available in sufficient quantities)—is considered ideal for animal rations. A great excess or deficiency of one element, even though the other one is available in adequate amounts, may cause various disease conditions.

CALCIUM PROPIONATE is a white, water-soluble powder. It is widely used as a MOLD INHIBITOR.

CALCIUM SULFATE. *See* GYPSUM.

CALCUTTA BUCKWHEAT
                    = TARTARY BUCKWHEAT.

CALENDRA. The *Calendra* spp. are known as BILLBUGS.

CALEY-PEA = ROUGHPEA.

CALF RATIONS. *See* DAIRY RATIONS.

CALIFORNIA BROME (*Bromus carinatus*) is a vigorous, short-lived native of the Rocky Mountain and Pacific Coast regions.

Plants grow 3 to 4 ft. tall. Leaf blades are flat, 6 to 8 in. long, and about $\frac{1}{2}$ in. wide. The seed is rather strongly awned.

The species is characterized by its capacity to produce large quantities of leafy forage that is relished by all classes of livestock. The mature foliage is harsh and less palatable, a condition that is offset somewhat by the fact that the seed heads are palatable and nutritious.

This grass would have extensive use for revegetation if adequate seed supplies were available.

Reference: H.1.

See also BROME GRASSES; GRASSES.

CALIFORNIA BUR-CLOVER (*Medicago hispida*), or *toothed bur-clover*, is one of the BUR-CLOVERS and therefore not a true clover. It resembles the SPOTTED BUR-CLOVER, but is readily distinguished from it by the lack of the large reddish-brown spot in the center of the leaf. California bur-clover occurs on the Pacific Coast and in the less humid parts of eastern Texas and southern Oklahoma.

Reference: M.6.

See also CLOVER; SPINELESS BUR-CLOVER; PASTURE PLANTS.

CALIFORNIA COAST BARLEY.
                    See SIX-ROWED BARLEY (variety).

CALIFORNIA COMMON 49 is one of the COMMON ALFALFAS.

CALIFORNIA FEED = *California Coast barley*. *See* SIX-ROWED BARLEY (variety).

CALIFORNIA NEEDLEGRASS (*Stipa pulchra*) is a conspicuously awned GRASS sometimes called *southwestern porcupine grass*. It has numerous slender stems 24 to 40 in. high. The awns are 2 to $3\frac{1}{4}$ in. long. This needlegrass species ranks high in forage value, being palatable to all classes of livestock, and particularly to cattle and horses. The seeds mature in early summer; the awns, thereafter, may be somewhat troublesome to sheep. When not grazed down during the summer, the species provides a very good fall and winter forage.

This valuable bunchgrass depends chiefly upon seed for reproduction. On many ranges it has been largely killed off because it was grazed so closely that seed could not mature.

Reference: U.6.

See also RANGE PLANTS.

CALIFORNIA No. 23 SUDANGRASS.
                    See SUDANGRASS (variety).

CALIFORNIA OATGRASS (*Danthonia californica*) is a leafy, fairly tall perennial which typically grows in small tufts. It ranges from British Columbia to Montana, Colorado, and California. While immature, this GRASS is considered good forage for cattle and horses, but somewhat less palatable to sheep. Like other species of oatgrass, it usually occurs in scattered locations; but in California, for instance, it forms stands which are dense enough to be cut for hay.

Reference: U.6.

See also RANGE PLANTS.

CALIFORNIA RED.
                    See RED OAT (variety).

CALIFORNIA WATERHEMLOCK.
See WATERHEMLOCKS; POISONOUS PLANTS.

CALIFORNIA WHEAT, better known as *shallu*, is one of the grain SORGHUMS.

CALIVERDE belongs to the COMMON ALFALFAS.

CALLISTEPHUS. *Callistephus* spp. = *China asters*. *See* ASTERS.

CALLOSOBRUCHUS. *C. maculatus*
                    = COWPEA WEEVIL.

CALORIE is the unit of quantity of heat (or ENERGY). The heat necessary to raise the temperature of 1 gm. water 1° C. is 1 (small) calorie; 1 (large) Calorie raises the temperature of 1 kg. water 1° C. Thus, *1 Cal.* = 1000 *cal.*; and 1000 *Cal.* = 1 *therm.*

Calorific value, or *energy value,* of a substance is the amount of heat each 1 gm. gives out on complete combustion. This value is of importance for feedstuffs and foodstuffs. It is (per gm.), 4.2 cal. for cellulose or starch, 3.8 cal. for sugar, and 9.5 cal. for FATS; for PROTEINS it is about 5.7 cal.

> NOTE: The calorific value of *sugar* differs, depending on whether simple sugars (e.g., GLUCOSE—3.76) or disaccharides (such as SUCROSE—3.96) are involved.

—*See also* ARMSBY FEEDING STANDARDS; CALORIMETER.

CALORIMETER is an apparatus that measures the heat produced by burning a substance.

CALORO. *See* RICE (variety).

CALOSOMA. *C. calidum* is an INSECT enemy of the ARMYWORM; *C. scutator* = CATERPILLAR HUNTER.—*See also* GROUND BEETLES.

CALOSOTA. *C. metallica,* a WASP, is a parasite of the WHEAT STRAWWORM.

CALROSE. *See* RICE (variety).

CALYX (pronounced: *kay*-liks) is the outer group of floral leaves of a plant; it is often small and green in color, as contrasted to the inner, more showy part, the COROLLA.—*See also* PERIANTH.

CAMELLIA OAT. *See* RED OAT (variety).

CAMNULLA. *C. pellucida* = *clear-winged grasshopper.* See GRASSHOPPERS.

CANADA BLUEGRASS (*Poa compressa*), which is extensively naturalized in this country, resembles KENTUCKY BLUEGRASS, but is different from it because of its blue-green foliage, distinctly flat culms, and short and much contracted PANICLES. It spreads by underground rootstocks (RHIZOMES).

Canada bluegrass is adapted to open, rather poor, dry soils, and in such situations competes with Kentucky bluegrass as a pasture grass. For lawns and golf links it can be used to advantage under conditions too dry for, or otherwise not entirely favorable to, Kentucky bluegrass.

Reference: H.1.

See also BLUEGRASSES; GRASSES; PASTURE PLANTS.

CANADA FIELD PEAS is a name applied to *Canadian Beauty* and other white- and yellow-seeded FIELD PEA varieties grown in the North Central States.—*See also* SILAGE CROPS; ALFALFAS; OATS; HAY.

CANADA-THISTLE (*Cirsium arvense*) is a perennial 1 to 4 ft. tall. The hollow stems are branched at the top. The leaves are deeply cut into lobes and tipped with frequently sharp spines. The flowers are rose-purple or white and the ⅛ in. long seeds are tan and have a white downy tuft.

This WEED is widely distributed in the northern half of North America and difficult to control.—*See also* ALFALFAS; WHEATS; RYE.

CANADA WILD-RYE (*Elymus canadensis*) is a vigorous, widely distributed perennial bunchgrass. It is most abundant in the Great Plains, the Pacific Northwest, and the Rocky Mountain states.

The seed heads grow from 3 to 5 ft. and many may be green or green-blue. The leaf blades are broad, flat, and rough. They are 6 to 12 in. long and usually ½ in. or more wide. The mature SPIKES, dark purple in color, average nearly 6 in. in length and have sharp awns that emerge from the SPIKELET parts.

Canada wild-rye begins its growth in the spring. If moisture conditions are favorable, it usually continues to grow through the summer. After a summer drought it may resume growth in the fall, if enough moisture is available then.

The palatability of the forage is fair but becomes less so as the plants become harsh and woody at maturity. The young seedlings are exceptionally vigorous and quickly form a good protective cover; Canada wild-rye therefore is useful in mixtures, especially with grasses that do not produce ground cover rapidly.

Hay of good quality may be had if the wild-rye is harvested just as the seed heads are emerging from the boot.

Dangers: Among the plant diseases which attack Canada wild-rye are the GRASS SMUTS.

Variety: *Mandan wild-rye,* an improved variety developed through selection and breeding, produces more forage and seed than Canada wild-rye.

Reference: H.1.

See also WILD-RYE GRASSES; GRASSES.

Canada wild-rye. Plant, spikelet, and floret. (H.26.)

CANADIAN BEAUTY.

See FIELD PEA (variety).

CANADIAN VARIEGATED = *Ontario Variegated. See* VARIEGATED ALFALFAS, (variety).

CANARYGRASSES (*Phalaris* spp.) found in the South, are winter annuals; those that occur in the North are summer annuals. HARDINGGRASS and REED CANARYGRASS are two perennials which grow most rapidly during the cool season and since they are nutritious and palatable to livestock, are cultivated for forage.

Reference: H.1.

See also GRASSES.

Illustration: See REED CANARYGRASS.

CANAVALIA. *C. ensiformis* = JACKBEAN.

CANDIDA is a genus of a FUNGUS; some of the *Candida* spp. are used for the production of *dried candida yeast.—See also* YEAST.

CANE is a term properly belonging to SUGARCANE, but sometimes also used for forage (or *sweet*) SORGHUMS, especially those used for making sorghum molasses.—*See also* DUST STORMS; MOLASSES.

CANE MOLASSES, used for feeding purposes, is more correctly called *feeding cane* (or *blackstrap*) *molasses.—See also* MOLASSES.

CANE SUGAR. *See* SUCROSE.

CANKER is a localized diseased area which is characterized by destruction of tissue and open wounds. This is usually found on woody or semi-woody stems; e.g., ASCOCHYTA STEM CANKER and BOTRYTIS STEM CANKER.—*See also* SOUTHERN BLIGHT; HEAT CANKER.

CANNABIS. *C. sativa* = HEMP.

CAPILLARY means: hairlike, or a very small tube.

CAPITAL SOYBEAN.

See SOYBEAN (variety).

CAPROCK is one of the intermediate-type grain SORGHUMS.

CAPSELLA. *C. bursa-pastoris*
= SHEPHERD'S-PURSE.

CAPSICUM, or *red pepper*, is the dried fruit of various *Capsicum* spp. It is officially recognized as a FEED INGREDIENT and often used as a CARMINATIVE.

CAPSULE = POD.

CARAWAY SEED is the dried fruit of *Carum carvi*. This officially recognized FEED INGREDIENT is used as a stomachic STIMULANT, CARMINATIVE, and CONDIMENT.

CARBAMIDE = UREA.

CARBOHYDRATES contain, in addition to carbon, the molecules of hydrogen and oxygen in the ratio 2:1. They include CELLULOSE, STARCHES, and SUGARS, and are the chief energy-producing constituents of all plants, vegetable food, and feed materials. In the small intestine of animals they split into simple sugars for ABSORPTION.

The term carbohydrates is not to be used on TAGS; the percentage of carbohydrates in feedstuffs can easily be calculated by the addition of percentages of crude FIBER and NITROGEN-FREE EXTRACT.—*See also* MOLASSES; CALORIE; PENTOSANS.

CARBOLIC ACID = PHENOL.

Straw-colored carbolic acid = CRESYLIC ACID.

CARBON, or *C*, is a CHEMICAL ELEMENT which occurs in nature in varying degrees of purity as charcoal, lampblack, etc. It is a constituent of all ORGANIC compounds and of the INORGANIC carbonates and cyanides.—*See also* ORGANIC COMPOUNDS.

CARBONACEOUS means: containing CARBON.

Carbonaceous feedstuffs are those whose dry matter contains a high percentage of CARBOHYDRATES; e.g., the *carbonaceous concentrates* (beet pulp, cereal grains, molasses, etc.).

CARBONATE is any salt of *carbonic acid*, i.e., the solution of CARBON DIOXIDE in water.

CARBON BISULFIDE
= CARBON DISULFIDE.

CARBON DIOXIDE is an odorless and colorless, incombustible gas.—*See also* YEAST; SILOS; OXYGEN.

CARBON DISULFIDE, or *carbon bisulfide*, is a colorless, poisonous, and inflammable liquid characterized by its disagreeable odor. It may be used for preventing damage to the seed of the FIELD PEA by the PEA WEEVIL; it is best to fumigate the seed as soon as it is threshed. The seed must be placed in a tight container and exposed from thirty to forty-eight hours to the fumes of the carbon disulfide.

This liquid should be contained in a shallow dish placed on top of the seed, since the vapor is heavier than air; 1 lb. (or a little more) carbon disulfide is sufficient to fumigate 100 bu. seed.

Carbon disulfide is also used as FUMIGANT for the control of both the RED HARVESTER ANTS and the WESTERN HARVESTER ANTS, when these pests are in the nest. For a colony that has cleared an area of not more than 4 ft. in diameter, use 4 fl. oz. carbon disulfide. Pour the liquid into the entrance hole and then stamp dirt into it. Larger colonies must first be prepared by removing a 6-in. layer of soil from an area 3 to 6 ft. in diameter surrounding the entrance. Thus, the vertical tunnels which lead to the chambers will be exposed. On the following day pour 8 fl. oz. carbon disulfide into the exposed tunnels. Then close the openings with firm soil and replace the layer of dirt that was removed the day before. Re-treat from time to time, using 4 fl. oz. carbon disulfide per colony.

Apply the carbon disulfide early in the morning or late in the evening, never at midday.

Caution: Carbon disulfide is a poison, inflammable and explosive. It must be stored in a cool place and handled carefully. Never open a container where there is poor air circulation; do not expose to fire.

CARBON TETRACHLORIDE is a colorless, noninflammable liquid. It possesses INSECTICIDAL properties.

CARBON TETRACHLORIDE MIXTURES are used as FUMIGANTS in the control of COWPEA WEEVIL in stored seed. A mixture of 3 parts ETHYLENE DICHLORIDE with 1 part CARBON TETRACHLORIDE or a mixture of 1 part CARBON DISULFIDE with 4 parts carbon tetrachloride is recommended. These fumigants are available as ready-mixed liquids under various trade names. (Their preparation at home should not be attempted.)

The seeds are placed in tight bins or receptacles and fumigated for at least forty-eight hours with one of the mixtures mentioned above, using 2 lb. of it to 100 bu. seed. The fumigant is poured into shallow pans set on top of the seed or sprinkled directly upon it. As the fumigant evaporates, it forms a gas heavier than air which sinks down among the seeds. After fumigation the seed must be thoroughly aired, or the germination and flavor may be affected. Fumigation should be conducted when the temperature is above 60°F., and preferably over 70°F.

Reference: M.23.

CAREX. *Carex* spp. = SEDGES.—*See also* CYPERACEAE.

CARLETON is the name of a COMMON OAT variety and a white DURUM WHEAT variety.

CARMEL. *Mount Carmel* is a SOYBEAN variety.

CARMINATIVES are drugs which stimulate the movements of stomach and intestines; they are used for relieving flatulence and colic. Among the carminatives are ANISE SEED, CARAWAY SEED, FENNEL, GINGER, and CAPSICUM.—*See also* FEED INGREDIENTS.

CAROB BEAN, also called *St. John's bread* or *locust bean*, is the fruit of the LEGUME tree *Robina pseudacacia*. The beans are imbedded in fleshy pods; because beans and pods are very rich in sugar, they are ground and used as nutrients in calf rations and other mixed feeds.—*See also* FEED INGREDIENTS.

CAROLINE CLOVER (*Trifolium carolinianum*) is a TRUE CLOVER; it is only of local importance.

CAROLINA RUNNER is a runner-type PEANUT.

CAROTENE, or *provitamin A*, has VITAMIN A value. It is a yellow pigment which accompanies CHLOROPHYLL in green leaves and other plant parts. A rough estimate of the carotene content can be made on the basis of the green color: fresh forage is green and rich in carotene, while old hay is bleached and contains no carotene.

The ability of animals to convert carotene into physiologically active vitamin A varies; poultry require only 0.6 mcg. *beta carotene* for conversion into 1 I.U. vitamin A; farm animals require 1 to 2 mcg. beta carotene for 1 I.U. vitamin A, while calves (according to recent work) may convert ½ or less and some carnivora are unable to

convert carotene.—*See also* ALFALFAS; ALFALFA PELLETS; ALFALFA-HAY STANDARDS; VITAMINS.

CARPETGRASS (*Axonopus affinis*) has spread over the Coastal Plain from Virginia to Texas and inland to Arkansas.

A perennial grass, it makes a dense sod and is distinguished by its compressed, two-edged, creeping stems (which root at each joint) and its blunt leaf tips. The slender flower stems grow 1 ft. high—rarely 2 ft. (in fertile soil). Because it has no underground stems, this grass never has become a pest in cultivated fields.

Carpetgrass is especially adapted to sandy or sandy loam soils, particularly where the moisture is near the surface most of the year, but it does poorly in swamps or where seepage is continuous. Carpetgrass is probably most valuable for permanent PASTURE. It also has value for firebreaks in forests, for lawn and turf, for use along roads, and for open areas in the pine forests. The cheapness and abundance of seed and ease of establishment make it popular in the South.

Because its sod is dense and its habit of growth is aggressive, LEGUMES are maintained with difficulty when carpetgrass is used in a cultivated pasture. On fertile soils it makes a good growth, but generally it is not high enough in feed nutrients to furnish a balanced diet.

Seed can be sown on a well-prepared seedbed or broadcast on burned-over open areas in timberland. It is spread quite easily by grazing animals and by natural reseeding. Seeding is best done in spring; but even summer or midsummer will not be too late.

Reference: H.1.

See also GRASSES; LAPPA CLOVER; LESPEDEZAS.

CARROT (*Daucus carota*) is a root which is too valuable to be used for stock feeding. Where available cheaply, carrots—which are rich in CAROTENE (vitamin A)—make an excellent crop for poultry and horses.

Carrot top is a VEGETABLE WASTE PRODUCT used as stock feed.

Dehydrated carrot leaf meal contains protein and vitamins A and B_2; it can be used to advantage in diets of poultry in place of ALFALFA meal.—*See also* PLANT BY-PRODUCTS; GRAZING.

CARTER GRASS = NAPIERGRASS.

CARTHAMUS. *C. tinctorius*
= SAFFLOWER.

CARTIER. *See* COMMON OAT (variety).

CARUM. The dried fruit of *C. carvi* is known as CARAWAY SEED.

CASEIN, a yellow-white, odorless powder, is the major protein of milk. In cow's milk it is present as a neutral calcium salt (*calcium caseinate*).—*See also* MILK PRODUCTS; RENNIN; HYDROCARBONS; PARACASEIN.

CASEIN WHEY is the liquid milk residue which is obtained in cheese manufacture, after most of the nitrogenous matter is removed from the milk. *See also* MILK PRODUCTS.

CASH CROP is a crop which, being sold off the farm, brings in ready cash.

CASSAVA (*Manihot utilitissima*) is a root crop grown in Florida and other southeastern states as raw material for the starch industry or as feed for cattle and other stock. The same purposes may be accomplished by using *cassava meal*, or *manihot meal*, the by-product of starch factories; it is low in protein, fat, and fiber.

CASTLE VALLEY CLOVER
= GARDNER SALTBUSH.

CASTOR BEAN (*Ricinus communis*) is a plant which grows very high (in the North, over 12 ft.; in the tropics, up to 30 ft.) and may therefore be used to provide shade on poultry ranges.—*See also* GRAZING.

CASTOR SEED MEAL, the CASTOR BEAN residue obtained in castor oil factories, is poisonous. It may be used as fertilizer, but must not be used in feeds.

CATALUNA. *See* RYE (spring variety).

CATCH CROP is a crop planted on land where other crops have failed or another crop has already been harvested.

CATERPILLAR is a term popularly used for the larval stage of moths and butterflies. The caterpillar has a distinctly set-off head, a pair of appendages (jointed legs) on each of the first three body segments, and prolegs (short, fleshy, unjointed legs) on some other segments.—*See also* LARVAE; GRUB; MAGGOT; ALFALFA CATERPILLAR; VELVETBEAN CATERPILLAR.

CATERPILLAR HUNTER (*Calosoma scrutator*) is one of the GROUND BEETLES. This beneficial insect is about 1¼ in. long and has green wing covers with rosy margins.

Reference: J.7.

CATJANG, or *Hindu cowpea*, is the name of a group of COWPEA varieties.

CATKIN is a SPIKE with scalelike BRACTS.

CATTAIL (*Typha latifolia*), or *tule*, is a WEED which grows on poorly drained land and in RICE fields. It is a perennial which spreads by seeds and creeping rootstocks. —*See also* BLACK SHEATH ROT.

CATTAIL MILLET = PEARLMILLET.

CATTLE RATIONS. *See* BEEF CATTLE RATIONS; DAIRY RATIONS.

CAUCASIAN BLUESTEM (*Andropogon caucasian*) is one of the newer BLUESTEMS now being cultivated in the southern areas of the United States. It is a vigorous bunchgrass which is easily established from seed. The high-yielding plants have fine texture, are drought-resistant, offer good weed competition, and give a good ground cover. This species volunteers freely and does well when seeded alone on SORGHUM stubble. It gives satisfactory results in the southern Great Plains area when used for dry-land grazing and for hay.

Reference: W.16.

CAULIFLOWER is a vegetable closely related to BROCCOLI. It may be used like the latter, but ordinarily it is too valuable to be fed to animals.

CAUSTIC means: corrosive.

Caustics is a term generally applied to the water-soluble hydroxide compounds of the ALKALIES; e.g., sodium hydroxide.

CAYUGA. *See* SOYBEAN (variety).

CC 1 is one of the inbred lines of CORN hybrids.

CECUM, or *blind gut*, is a part of the DIGESTIVE TRACT.

CEIBA. *C. pentandra* = KAPOK TREE.

CELATORIA (*C. diabroticae*) is a two-winged fly which attacks the SOUTHERN CORN ROOTWORM in the beetle stage. It places a maggot (larva) in the abdomen of the beetle and the maggot feeds on the vital organs of the host and finally kills it.

After death of the beetle the parasite larva enters the ground and forms a PUPARIUM in which it passes into the pupal stage. In due time, the adult, or fly, emerges from the puparium which starts the cycle over again.

Reference: L.9.

CELL is the structural (usually microscopic) unit of all living organisms. It is surrounded by a cell wall and consists largely of PROTOPLASM; its most essential part is the NUCLEUS which is fundamental in growth, METABOLISM, and reproduction.

CELLULASE is an ENZYME.

CELLULOSE, a CARBOHYDRATE, is the major component of the cell wall of plants. Almost all vegetable tissues contain cellulose. It is a white fibrous material, insoluble in water, and makes up one part of crude FIBER.—*See also* BACTERIA; RESTING SPORE; HEMICELLULOSES; LIGNIN.

CENTAUREA. *C. solstitialis*

= YELLOW STAR-THISTLE.

CENTIPEDEGRASS (*Eremochloa ophiuroides*) grows in the southern states and as far west as the Pacific Coast. It is a perennial which spreads by STOLONS and makes a dense mat of creeping stems and leaves.

Centipedegrass will grow on clay soil and the poorest Norfolk sand, if enough moisture and plant food are available for it to get started. It has withstood temperatures of 12° F. or slightly lower, but is not adapted to conditions in the North.

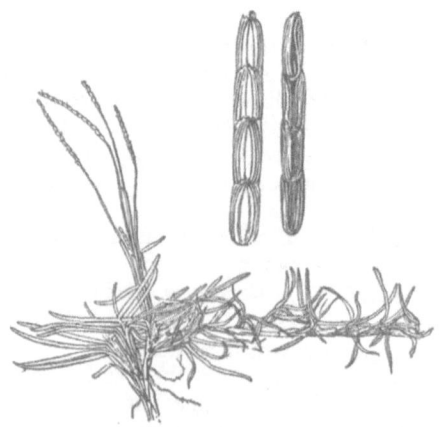

Centipedegrass. (H.1.)

Because of its low nutritive value, its best use is for lawns and in erosion control. In pastures it generally has given poor livestock gains. Because of its dense and aggressive growth it crowds out desirable LEGUMES. Chemical analysis of grass samples, even in young vegetative growth, resembles the analysis of cereal straws.

Reference: H.1.

See also GRASSES.

CEPHALOSPORIUM. *C. gregatum* is a FUNGUS which causes BROWN STEM ROT.

CEPHUS. *C. cinctus*

= WHEAT STEM SAWFLY.

CERATOPHORUM. The FUNGUS *C. setosum* causes BROWN SPOT OF LUPINE.

CEROCARPUS. *C. intricatus*

= LITTLELEAF MOUNTAIN-MAHOGANY.

CERCOSPORA is the imperfect stage of the FUNGUS *Mycosphaerella*. The species *C. zebrina* is the cause of SUMMER BLACK-STEM; *C. sojina*, also known as *C. daizu*, causes FROG-EYE; *C. oryzae*, NARROW BROWN LEAF SPOT; *C. sorghi*, GRAY LEAF SPOT; and *C. cruenta*, RED LEAF SPOT.

CERCOSPORA LEAF SPOT

= SUMMER BLACKSTEM.

CERCOSPORINA. The FUNGUS *C. kikuchii* is the cause of PURPLE SEED STAIN.

CEREAL is a GRASS cultivated for its edible seeds, or grains. The term cereal is applied to both, the grain and the plant itself.

Cereal crops are also called *grain crops*. Among the *true cereal crops* are the SMALL GRAINS (oats, rye, barley, wheat), as well as CORN, PROSO, grain SORGHUMS, and RICE.

NOTE: BUCKWHEATS, which do not belong to the true cereals, are commercially classified among the grain crops.

See also SILAGE CROPS; STRAW; PUFFED CEREALS.

CEREAL FORAGE.

See HAY; PASTURES; SILAGE.

CEREAL GRAIN BY-PRODUCTS are obtained from such cereal grains as WHEAT, CORN, RYE, OATS, BARLEY, RICE, and BUCKWHEAT; commercially, SORGHUM products are also listed among them. When these cereal grains are processed for human food, part of the grain is removed and becomes a by-product which is used mainly in feeds for farm animals. Most of the grain by-products have higher levels of protein, fat, and fiber than do the original grains; some contain more vitamins. The higher fiber content limits their use in feeds for swine and poultry.

Reference: E.12.

See also BY-PRODUCT FEEDSTUFFS; SCREENINGS.

CEREAL GRAIN SILAGES. *See* SILAGE; CEREAL SILAGE.

CEREAL GRAINS. *See* CEREAL; LEGUMES.

CEREAL GRASSES.
                    See DRIED CEREAL GRASSES.

CEREAL HAY from SMALL GRAINS is widely used in the Pacific Coast states, Montana, and North Dakota. In compostiion and feeding value, good cereal HAY is similar to good timothy hay; both, therefore, may be used to replace each other in livestock rations.

CEREAL PASTURES are of great value in the northern part of the United States where there exists a need for forage in late fall (RYE or WHEAT) and in early summer (OATS or BARLEY). In the South, they are important as winter PASTURES.

CEREAL SILAGE from CORN or grain SORGHUMS is easily made because these crops are rich in sugar; therefore, neither molasses nor any other preservatives need to be added. Green cereals, such as OATS, WHEAT, and BARLEY are also widely used in SILAGE making. A better product is obtained if PRESERVATIVES are used. 40 to 60 lb. of MOLASSES per ton of fresh material is satisfactory.—*See also* SILAGE CROPS.

CERES. *See* COMMON WHEAT (variety).

CERESAN is a fungicide which contains as active ingredient 2 percent ETHYL MERCURIC CHLORIDE. It is the most effective material for the SEED TREATMENT of PEANUTS, when used as dust at the rate of 3 oz. per 100 lb. shelled seed; it is also an effective protectant against insect and rodent damage.—*See also* CERESAN M.

CERESAN M is a powder which contains 7.7 percent of the somewhat volatile FUNGICIDE *ethyl mercury para-toluene sulfonanilide*. For SEED TREATMENT of SORGHUM it should be applied at the rate of not more than ½ oz. per bushel.

Seed-treatment of OATS tends to reduce the seed-borne fungus diseases VICTORIA BLIGHT, COVERED SMUT OF OATS, BLACK LOOSE SMUT, OAT LEAF SPOT, CULM ROT, and HALO BLIGHT. The seed oats are treated with ½ oz. Ceresan M in slurry form per bushel. The same quantity of Ceresan M and the same treatment is recommended for the control of BACTERIAL STRIPE OF OATS; WHEAT SCAB, FLAG SMUT OF WHEAT, COMMON BUNT, and DWARF BUNT of wheat; COVERED SMUT OF BARLEY; and BROWN SPOT OF RICE.

Application of 1 oz. Ceresan M dust to 1 bu. seed is needed for the prevention of GRASS SMUTS.

Mix the fungicide and the seed in a special treater or the mixing equipment described for COPPER CARBONATE. After treatment, put the seed in a cloth sack and let it stand in a dry place for not less than one day and not more than three days before planting.

Caution: Ceresan M is poisonous. Care should be taken to avoid inhaling the dust or the fumes or having it come in contact with the skin, especially when moist, as it will cause blisters. Treatment of the seed should be done in the open air or in a well-ventilated place. If dust is present in the air, a dust mask should be worn. Long sleeves should be worn, and gloves used to cover hands and wrists. Treated seed should not be used for feed or food.

Reference: L.1.

CERTIFICATION OF SEED, grown and sold under specified conditions and inspection, is important to insure the maintenance of varietal purity and freedom from seed-borne pests.

CERTIFIED DYES are prepared in accordance with federal government specifications. It is illegal to use other than certified (F.D.C.) dyes for coloring feedstuffs.—*See also* COLORS.

CHACODERMUS. *C. aeneus*
                    = COWPEA CURCULIO.

CHAETOCNEMA. *C. pulicaria*
                    = CORN FLEA BEETLE.

CHAFF is (1) a thin SCALE, GLUME, or BRACT, or (2) a fragment material of light plant-tissues broken in threshing.—*See also* SCREENINGS.

CHAFF AND/OR DUST.
See SCREENINGS.

CHALCIDS are insects which parasitize other insects; an exception is the CLOVER-SEED CHALCID.

CHALK = CALCIUM CARBONATE.

CHALK ROCK is a source of CALCIUM CARBONATE.—*See also* MINERAL FEEDS.

CHAMISE (*Adenostoma fasciculatum*), or *chamiso*, is a Californian evergreen shrub which forms dense brushfields. The sprouts produced by chamise the first season are palatable to livestock, but older sprays are largely inedible.

Reference: U.6.

See also RANGE PLANTS.

CHAMIZA = FOURWING SALTBUSH.

CHANCELLOR PEA.
See FIELD PEA (variety).

CHANG PEA. *See* FIELD PEA (variety).

CHAPARRALS are thickets formed by often tangled shrubs, such as CHAMISE; they are found on ranges in the western states.

CHARCOAL consists chiefly of the element *carbon* and small amounts of ash or minerals. It is obtained by incomplete combustion of organic substances, such as wood, blood, or bones. The black powder, granules, or lumps absorb odors and coloring matter. Charcoal is officially recognized as one of the MINERAL FEEDS. When mixed in feed, it passes through the animal's body undigested, but adsorbs gases and carries them off.

Charcoal is often added to mash feeds in amounts of 1 to 2 percent; it may also be mixed in pigeon feed or scratch feed of poultry, or fed to poultry cafeteria style (in hoppers).

CHARCOAL ROT, caused by the FUNGUS *Macrophomina phaseoli*, has resulted in serious losses in fields of SORGHUMS (but not the kafir-type Quadroon) throughout Texas, Oklahoma, Kansas, Nebraska, and New Mexico; it occurs, furthermore, in California, Illinois, Indiana, Maryland, and other states. This STALK ROT also affects SUDANGRASS. It is sporadic in its appearance, and is associated with crop sequence at the critical stage of development and certain soil and weather conditions that subject a crop to periods of stress such as extreme heat or drought. The rot is likely to occur in dry spots in a field, such as terrace crowns, knolls, or areas underlain by coarse sand or gravel.

Injury due to charcoal rot usually does not become evident until the sorghum plants approach maturity. Close examination at that time reveals many poorly filled heads with lightweight kernels, a premature ripening and drying of entire stalks, and the presence of lodged stalks. Further examination shows that many stalks are soft and discolored at the base, with the pith disintegrated at this point and the separated VASCULAR fibers in the stalks presenting a shredded appearance.

An abundant moldlike growth of a pink or white *Fusarium* fungus is found frequently at this stage and probably, along with the BACTERIA present, assists in the destruction of the stalk. Soon thereafter the affected stalks break over at the base. If a period of dry, warm weather follows this stage of the disease, the vascular fibers in the interior of the stalks become covered with small black bodies visible to the naked eye; these compact masses, known as SCLEROTIA, are formed by the fungus causing the disease.

When the roots and lower parts of the stalks decay in the field, the sclerotia become incorporated in the soil. Here, under proper temperature and moisture conditions, they may germinate and infect the roots of any one of more than 30 different kinds of crops. Among the latter are CORN, RED CLOVER, ALFALFA, LESPEDEZA, COTTON, SUGAR BEETS, POTATOES, SWEETPOTATOES, SUNFLOWERS, COWPEAS, SOYBEANS, and several species and varieties of peas and beans.

Control: Varietal resistance to charcoal rot exists among sorghums. The milos and milo derivatives, including Martin, Midland, and Westland, suffer the greatest damage; feterita, hegari, and Sweet Sudangrass are less seriously affected. Most varieties of sorgo and kafirs are rather resistant.

Reference: L.1.

See also FUSARIUM STALK ROT; RHIZOCTONIA STALK ROT.

CHARD. *See* SWISS CHARD.

CHARLEE. *See* SOYBEAN (variety).

CHARLOTTETOWN 80. *See* TWO-ROWED BARLEY (variety).

CHEAT = CHESS.

CHEATGRASS (*Bromus tectorum*), also called *cheatgrass brome, downy chess,* and *broncograss,* is not to be confused with the related "cheat," or CHESS. It is widely distributed in the United States. Cheatgrass is found as a weedy roadside annual in eastern states and covers extensive areas in the West. It is a rapidly growing, short-lived grass that is very aggressive under moist conditions and may deter

Cheatgrass. Plant at maturity and awned seed. (F.10.)

establishment of perennial species. Herbage production is high and provides good forage when young, but fluctuates widely in yield from year to year. Awns, when mature, cause mechanical injury to stock. Cheatgrass, when dry, is a great fire hazard.

Reference: W.16.

See also GRASSES; BROME GRASSES; ALFALFAS; SAGEBRUSH; RANGE PLANTS; WEED.

CHECKROW is a row of plants (or seeds) which is spaced equally in both directions.

CHEESE MEAL, a by-product of cheese manufacture, consists chiefly of the cheese trimmings. It is an excellent source of protein and fat and at its best is as valuable as best-grade tankage.

CHEESE RIND. *See* MILK PRODUCTS.

CHEMICAL BARRIERS for the control of migrating CHINCH BUGS include CREOSOTE BARRIERS (*dirt-ridge creosote barrier* and *creosote-treated paper-fence barrier*), TAR BARRIER, DINITRO-DUST BARRIER, DDT-DUST BARRIER, and OIL BARRIERS.

CHEMICAL ELEMENT is any one of about 100 simple substances composed only of one kind of atom; e.g., ARSENIC, BORON, CALCIUM, FLUORINE, IODINE, MAGNESIUM, MOLYBDENUM, NITROGEN, OXYGEN, PHOSPHORUS, SELENIUM, SILICON, SULFUR, etc.—*See also* TRACE ELEMENTS.

CHENOPODIUM. *C. album* = LAMBSQUARTERS.

CHEROKEE. *See* COMMON OAT (variety).

CHERRY.
See WILD CHERRY; POISONOUS PLANTS.

CHESS (*Bromus secalinus*), also called *cheat* or *chess brome,* is a weed which often causes trouble in MEADOW FESCUE and WHEAT stands; its seeds closely resemble those of the fescue.—*See also* GRASSES.

Downy chess = CHEATGRASS.

CHESS BROME = CHESS.

CHEWINGS FESCUE. *See* FESCUES; SUBCLOVER; RED FESCUE; GRAZING.

CHEYENNE KAFIR is not a true kafir, but one of the intermediate-type grain SORGHUMS.

CHEYENNE WHEAT. *See* COMMON WHEAT (variety).

CHICKASAW PEA = MUNGBEAN.

CHICK PEA (*Cicer arietinum*)—also known as *gram* or *garbanza*—is a bushy, hairy LEGUME widely grown in California. The seeds are sometimes used as feed for horses and other livestock; the herbage should not be fed to animals since it may cause poisoning.

CHICKWEED (*Stellaria media*) is a WEED which grows .to 2 ft. height. It is very troublesome in rich soils and spreads very fast during the cool growing season.—*See also* ALFALFAS; WHEATS.

CHIEF SOYBEAN.
        See SOYBEAN (variety).

CHIEFKAN.
        See COMMON WHEAT (variety).

CHILEAN ALFALFA is a variety of the nonhardy ALFALFAS.

CHILEAN DODDER (*Cuscuta racemosa chileana*), one of the DODDER species, is a parasitic plant which attacks ALFALFA and CLOVER. It occurs occasionally in the Mississippi Valley and California.
Reference: H.19.

CHILE SALTPETER = SODIUM NITRATE.

CHILTEX is one of the intermediate-type grain SORGHUMS.

CHINA GRASS = RAMIE.

CHINCH BUG (*Blissus leucopterus*) is one of the most destructive insects, attacking the grain and grass crops of the United States. It is most abundant in the regions drained by the Mississippi, Ohio, and Missouri Rivers. The chinch bugs prefer to feed on BARLEY, WHEAT, RYE, and OATS. After the spring flight, probably 90 percent are in the SMALL GRAINS; where the acreage of small grains is low, the bugs may be found on TIMOTHY, KENTUCKY BLUEGRASS, ST. AUGUSTINEGRASS, and several wild GRASSES.

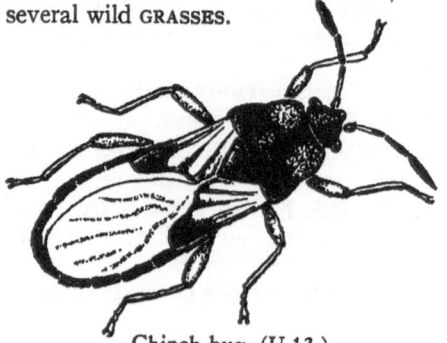

Chinch bug. (U.13.)

Throughout the CORN-growing area the second generation of chinch bugs feeds on grains and grasses that are succulent late in the summer, such as timothy, FOXTAIL, and BENTGRASS, but mainly on corn and various SORGHUMS. They also attack RICE, PROSO, and SOYBEANS.

Most of the varieties of forage sorghums are not injured extensively by the chinch bug unless infestation occurs when the plants are small. Among the sorgos, Honey, Leoti, Black Amber, and Red Amber are the most susceptible. SUDANGRASS, also, is attacked by this insect pest. Hegari, Early Kalo, milos and their hybrids (especially Sooner and Colby) are more susceptible than other sorghums. But the grain sorghums darso, Gurno, Midland (when early planted), and many of their derivatives, are able to resist injury from the chinch bug.

Fortunately, the chinch bug does not develop on any of the legumes. Other common crops which prevent its development are sunflower, flax, rape, buckwheat, beets, and all the truck or garden crops, except sweet corn.

The full-grown chinch bug is about $\frac{1}{6}$ in. long, and black with white markings. The *long-winged* form prevails throughout the Central States and is capable of flying long distances when the wind is favorable. The *short-winged* form is unable to fly.

In fields of small grain the bugs will feed a few days before the females begin laying eggs. The female chinch bug lays several hundred eggs at the rate of 15 to 20 a day. The eggs, which are about $\frac{1}{32}$ in. long, hatch in seven to forty-five days, depending mainly upon the temperature. A very young bug, or NYMPH, is about half the size of a pinhead, and bright red with a transverse band of white. This bug sheds its skin five times, and with each change it becomes darker until, in the last stage before acquiring wings, it is grayish-black with a conspicuous white spot on the back. In all immature stages the insect is wingless. In the sixth, or adult, stage the insect has wings.

There are at least two generations of chinch bugs a year, throughout their entire range in this country. A third generation usually develops in the extreme South, and occasionally a partial third generation as far north as Iowa.

Chinch bugs overwinter in the adult stage in tufts of perennial grasses or in clump-forming grasses such as timothy, orchardgrass, and sedges. Many bugs pass the winter under leaves and litter in the borders of woodlands and under hedges. Chinch bugs may sometimes be found under large leaves of weeds, in standing cornstalks, and in sorghum stubble; also under the bark of dead trees, boards, and shingles of houses.

The spring flights of overwintering bugs take place when the temperature remains at 70° F. or more for several hours.

Weather is the chief factor governing the number of chinch bugs, which are most susceptible to weather conditions while they are hatching. The adults are less able to survive an open, wet winter than a cold one with heavy snow cover.

Probably the most destructive natural enemy of the chinch bug is the disease caused by the white fungus, *Beauveria globulifera*, which is generally present in the fields throughout the country.

Next in importance is the tiny parasitic insect *Eumicrosoma benefica*. This wasp lays an egg in the chinch bug egg; the maggot hatching from the wasp egg consumes the content of the chinch bug egg. A number of birds are also known to feed on the chinch bug.

Control: Since first-generation bugs feed mainly on small grains and those of the second generation on corn and sorghum, a good way to hold both in check is to make their food supply as scarce as possible. This can be done by reducing the acreage of small grains where corn and sorghum are the leading crops, or of corn and sorghum where small grains predominate, and by planting legumes or other immune crops in their place. In recent years much progress has been made in the development of hybrid corns and sorghums distinctly resistant to chinch bugs.

In the more western areas, where the native bunchgrasses form the principal shelter, burning over these grasses while the bugs are in them—some time between November 1 and March 15—may help to reduce their numbers. However, permanent pastures and hayfields should not be burned over, and the burning of small-grain stubble and cornstalks is not warranted, because few bugs overwinter successfully in such cover.

Another way of reducing chinch bug damage is to avoid adjacent plantings of small grain (particularly barley and wheat) and corn. Where this is not possible, a barrier should be run between the plantings. A *dust barrier* around the field to be protected is satisfactory during dry weather. It is generally made by plowing a furrow, throwing the dirt both ways, and then dragging a log back and forth in this furrow until the sides and bottom have been worked down to a fine dust.

A *chemical barrier*, such as CREOSOTE BARRIERS (*dirt-ridge creosote barrier* and *creosote-treated paper-fence barrier*), OIL BARRIERS, DINITRO-DUST BARRIERS, DDT-DUST BARRIERS, and TAR BARRIERS are of great value. At harvest time (except in the South) only a few chinch bugs have reached the winged stage, and most of them have to migrate on foot; thus, in one instance 8 bu. (at least 60 million) bugs were caught along ½ mile of creosote barrier.

Insecticide dusts or *sprays* have not yet been found practical for reducing chinch bug infestations. Since chinch bugs do not eat plant tissue, they cannot be killed by poisons merely sprayed or dusted on the plants, but must be hit by such insecticides as SABADILLA POWDER, NICOTINE, ROTENONE, DIELDRIN, or TOXAPHENE. These insecticides are recommended for use on plantings of valuable seed corn or cornfields that have been invaded suddenly by chinch bugs.

References: P.5; M.1.

CHINESE MUNGBEAN is a variety of the green-type MUNGBEAN.

CHINESE SUGARCANE (*Saccharum sinense*) is one of the SUGARCANE species grown in America; JAPANESE SUGARCANE is one of its varieties.

CHINESE VELVETBEAN (*Mucuna nivea*), also called *White Chinese*, is in nearly all respects like the LYON VELVETBEAN, but less vigorous and about six weeks earlier, usually ripening before frost south of central Georgia, Alabama, and Mississippi.

Reference: P.10.

See also LEGUMES; INOCULATION; VELVETBEANS.

CHISEL is a tillage implement with points, used to stir the soil 10 to 18 in. deep.

CHLORAMPHENICOL
= CHLOROMYCETIN.

CHLORANIL, or *tetrachloro-para-benzoquinone,* forms yellow plates which are practically insoluble in water. It has FUNGICIDAL properties and is therefore used as seed protectant. Chloranil is the active ingredient of SPERGON.

CHLORATES, especially sodium chlorate, are used as herbicides. If a chlorate-containing weed killer falls on leaves, it causes spots.—*See also* NONPARASITIC SORGHUM-LEAF DISCOLORATION.

CHLORDANE is one of the *chlorinated hydrocarbons* which have become so useful as modern INSECTICIDES. It is a heavy, amber-colored liquid, insoluble in water, but soluble in kerosene and other organic solvents. Chlordane is available in form of emulsifiable concentrates, solutions, wettable powders, and dusts.

STINK BUGS can be controlled by the application, per acre, of 10 gal. emulsion spray containing 2 lb. chlordane.

For the complete eradication of ANT colonies, apply a 5-percent chlordane dust over the mound of their nest; where definite colonies cannot be located, ant infestation can be controlled by spraying with a mixture containing 4 lb. of 40 percent chlordane wettable powder in 100 gal. water.

For the control of the RED HARVESTER ANT and the WESTERN HARVESTER ANT apply (during warm periods when there is little wind) a 5-percent chlordane dust at the rate of 8 oz. in a 4 to 6 in. wide band, making a circle 5 to 6 ft. in diameter with the center at the nest entrance. The treatment has to be repeated if necessary.

To control GRASSHOPPERS, 1 lb. chlordane is required per acre; it has a residual action of two to four weeks.

To kill the adults of the ALFALFA WEEVIL spray 1½ to 2 lb. chlordane per acre when the spring growth of alfalfa is 1 to 2 in. tall.

Spraying or dusting infested plants with chlordane kills the SPOTTED CUCUMBER BEETLE and its larva, the SOUTHERN CORN ROOTWORM. The application must be repeated every two or three weeks.

Caution: Keep chlordane away from eyes, nose, and mouth; wear respirator while applying dust; bathe and change clothing after dusting.

Do not dust vegetation that may be eaten by man, dairy animals or animals being finished for slaughter.

Antidotes: Wash contaminated areas thoroughly with soap and water. Take 1 tablesp. salt in a glass of warm water to induce vomiting and repeat until the vomit fluid is clear. Lie down and remain quiet; call a doctor.—*See also* DRY BAIT.

CHLORINATED   HYDROCARBONS are often useful as *insecticides;* e.g., ALDRIN, BENZENE HEXACHLORIDE (LINDANE), CHLORDANE, DDT, DIELDRIN, HEPTACHLOR, TOXAPHENE, and related compounds.

CHLORINE is a suffocating, greenish-yellow gas which is valuable as disinfectant. However, this chemical element is essential for normal nutrition when consumed in form of its inorganic salts, e.g., SODIUM CHLORIDE. Organic compounds are widely used as modern insecticides; e.g., the CHLORINATED HYDROCARBONS.

CHLORIS. The *Chloris* spp. are known as FINGERGRASSES. *C. gayana* = RHODESGRASS.

CHLOROCHROA. *C. sayi*
= SAY STINK BUG.

CHLOROMYCETIN, or *chloramphenicol,* is an ANTIBIOTIC obtained from the soil BACTERIUM *Streptomyces venezuelae* or produced synthetically. It forms needles or plates which are only very slightly soluble in water.

CHLOROPHYLL is a complex nitrogenous substance which occurs (with other pigments, such as CAROTENE) in plant cells that contain MAGNESIUM and iron and are exposed to light. It is green in color and responsible for the prevalent green hue of the vegetable kingdom. Chlorophyll is essential to the formation of starch and other carbohydrates in plants. With the help of the energy of the sunlight it converts water and carbon dioxide of the air into these foodstuffs. The conversion process is called *photosynthesis.—See also* VITAMIN A; CLOVER MITES.

CHLOROPICRIN is a poisonous liquid with an intense odor which causes tears and headache. It is almost insoluble in water. Chloropicrin is a disinfectant and insecticide and can be used for sterilizing soil infested with PERICONIA ROOT ROT.

CHLOROSIS is the yellowing or blanching of leaves and other green plant parts due to partial failure of CHLOROPHYLL to develop.—*See also* WITCHES'-BROOM.

CHLOROTIC means: pertaining to CHLOROSIS; as the yellow or yellow-striped appearance of leaves.—*See also* NONPARASITIC SORGHUM-LEAF DISCOLORATIONS.

CHLORTETRACYCLINE
= AUREOMYCIN.

CHOLESTEROL, or *cholesterin*, is a STEROL which occurs in fish liver oils, and such animal products as fat, egg yolk, brain, and bile. One of the cholesterol derivatives, when irradiated, gives VITAMIN D₃.—*See also* VITAMIN D.

CHOLINE, one of the B-COMPLEX VITAMINS, is a viscid, water-soluble liquid that is widely distributed in plant and animal cells. It is important in nutrition.

Because *betaine* is an oxydation product of choline and the animal body can reduce betaine, it is possible to replace choline with BETAINE HYDROCHLORIDE in the ration.—*See also* YEAST.

CHOLINE CHLORIDE forms deliquescent crystals which are very soluble in water. This nutritional factor is often used (in place of the liquid CHOLINE) in rations and may be expressed as such on labels.—*See also* VITAMINS.

CHOLINE PANTOTHENATE, which forms cyrstals, is used in feed rations.—*See also* VITAMINS.

CHOPPED ALFALFA. See ALFALFAS.

CHOPPED COWPEA HAY. See COWPEA.

CHOPPED HAY. See HAY.

CHOPPING ROUGHAGE for the purpose of saving animals "work" in mastication and digestion is a waste of money (for equipment) and of time. *Chopping of hay* or *other roughages*—like *grinding of grains* or *other seeds*, or *soaking of grains* (such as oats or corn) is advisable whenever animals are to be induced to eat more of a not too palatable feedstuff by mixing it with other feeds; or when very young, sick, or old animals are unable to properly chew other than prepared feed.

CHOP SUEY BEAN = MUNGBEAN.

CHORIZAGROTIS. *C. auxiliaris*
= ARMY CUTWORM.

CHRYSOTHAMNUS. *C. nauseosus* = *rubber rabbitbrush; C. stenophyllus* = SMALL RABBITBRUSH.—*See also* RABBITBRUSH.

CHUFA SEDGE (*Cyperus esculentus*) is a WEED occurring in the southern parts of the United States. Its small tubers are low in protein, but when supplemented with protein-rich feed and corn (to harden the carcass), they may be used successfully for fattening swine.—*See also* CYPERACEAE.

CICER. *C. arietinum* = CHICK PEA.

CICUTA. The *Cicuta* spp., or WATERHEMLOCKS, are POISONOUS PLANTS, e.g., ‣ *C. californica* = *California waterhemlock, C. occidentalis* = *western waterhemlock, C. maculata* = *spotted waterhemlock*, and *C. vagans* = *tube waterhemlock*.

CINQUEFOIL. See SHRUBBY CINQUEFOIL.

CIRPHIS. *C. unipuncta* = ARMYWORM.

CIRSIUM. *C. arvense* = CANADA-THISTLE.

CITRIC ACID forms colorless crystals or a white powder and is readily soluble in water. This organic acid, which occurs in the juices of citrus fruits, is one of the officially recognized ANTIOXIDANTS.—*See also* FERMENTATION PRODUCTS.

CITROVORUM FACTOR, which occurs in liver, is related to FOLIC ACID (now called *pteroylglutamic acid*) and acts similarly, especially as a growth stimulant and preventive of *macrocytic anemia* in chicks.

CITRUS. *C. limonum* = LEMON.

CITRUS MEAL = DRIED CITRUS PULP.

CITRUS MOLASSES. *See* MOLASSES.

CITRUS PRODUCTS, used as feedstuffs, are the officially recognized DRIED CITRUS PULP and CITRUS SEED MEAL as well as CITRUS PULP and *citrus molasses*.—*See also* MOLASSES; PLANT BY-PRODUCTS.

CITRUS PULP is a term used collectively for peel, seeds, and pulp, left in large quantities from the canning of citrus juices and other CITRUS PRODUCTS. These citrus by-products are fed to ruminants either fresh, or ensiled, or dried; *citrus molasses* is also a citrus by-product.

Reference: E.12.

See also PLANT BY-PRODUCTS; MOLASSES; DRIED CITRUS PULP.

CITRUS SEED MEAL, one of the officially recognized CITRUS PRODUCTS, consists principally of the seeds of oranges and grapefruits, singly or mixed, from which the major portion of the oil has been removed; it is composed mostly of kernel with such portion of the hull as is necessary in the manufacture of citrus seed oil. It may contain small amounts of the seed of other varieties of citrus, and must be designated according to its protein content.

Reference: F.6.

CLAM SHELLS are very rich in CALCIUM (in form of carbonate) and are therefore used as calcium supplement in feed rations.

CLARAGE. *See* CORN (variety).

CLARKAN.

See COMMON WHEAT (winter variety).

CLARK'S BLACK HULL = *Blackhull wheat. See* COMMON WHEAT.

CLARK'S NO. 40 = *Clarkan. See* COMMON WHEAT (variety).

CLAVATE means: Club shaped. —

See also CLUB WHEAT.

CLAVISEPS. *C. purpurea* = ERGOT.

"CLAY" is a term applied to a group of COWPEA varieties.

CLAYS are hydrated aluminum silicates. They consist of very small colloidal mineral soil particles. Finely ground clay is used as diluent in insecticidal dusts.—*See also* BENTONITE; CUBE; DERRIS; CRYOLITE.

CLEAR-WINGED GRASSHOPPER.

See GRASSHOPPERS.

CLEMSON and *Clemson Nonshattering*, which is better known as *C.N.S.*, are SOYBEAN varieties.

CLICK BEETLES are slender, ½ in. long BEETLES of brown, gray, or black color. They are the adults of the WIREWORMS, various species of which feed on potatoes as well as cereal and forage crops.

CLIMATE is the total long-time characteristic weather of a region.—*See also* ARID; HUMID CLIMATE; SUBHUMID CLIMATE; SEMIARID.

CLIMAX is a variety of KOREAN LESPEDEZA.—*See also* LESPEDEZAS; ANNUAL LESPEDEZAS.

CLINTON. *See* COMMON WHEAT (variety).

CLIPPED OAT BY-PRODUCT.

See OATS.

CLIPPED OATS. *See* OATS.

CLOVER is a name applied not only to the TRUE CLOVERS, or *Trifolium* spp., but also to other LEGUMES and even some plants of other groups. Among these so-called clovers, which resemble the true clovers in some respects, are ALYCECLOVER; SWEETCLOVERS (and their many species and varieties); BUR-CLOVERS (including *Buttonclover*); LESPEDEZAS, or *bush clovers* (including *Japan clover*); WHITE SWEETCLOVER, formerly called *Bakhara clover* (including *Hubam clover*); and DESMODIUM, commonly called *tick clover*, of which FLORIDA BEGGARWEED is a species.

Dangers: Clovers are attacked by FUSARIUM ROOT ROT; BLUESTEM OF CLOVER; CLOVER-SEED CHALCID; CLOVER ROOT BORER; LYGUS BUGS; PEA APHID; YELLOW CLOVER APHID, and other APHIDS; and STEM NEMATODE.

Many of the clovers, if grazed when succulent, may cause BLOAT in ruminants.

See also HAY; STRAW; ALFALFAS; SOYBEAN; WHEATS; OATS; RICE; COWPEA; CORN; SORGHUMS; REDTOP; SMOOTH BROME; FIELD BEANS; VETCHES; TREFOILS; CLOVER DODDER; CHILEAN DODDER; PASTURES; PASTURE MANAGEMENT; RANGE PLANTS; LEGUME BACTERIA.

Illustrations: See BUR-CLOVERS; RED CLOVER; SPOTTED BUR-CLOVER; WHITE SWEETCLOVER.

CLOVER APHID (*Anuraphis bakeri*) is a small green or pink plant louse that may become very abundant in CLOVER (especially RED CLOVER and ALSIKE CLOVER) blossoms. This APHID secretes a sticky honey-dew that lowers the quality of the seed and often greatly reduces the yield.

Control: Probably the best means of avoiding loss is to produce seed from the second crop. Clip the first crop when the infestation becomes heavy, remove the hay as quickly as possible, and allow the field to remain without water until the plants are dry—about ten days in good sandy loam soil. If the sun is hot and the ground dry, good control on second crop clover is attained.

Reference: M.27.

See also YELLOW CLOVER APHID.

CLOVER DODDER (*Cuscata epithymum*), one of the DODDERS, is a plant parasite showing a decided preference for CLOVER and ALFALFA. It occurs in parts of the West and is found in the East from Maine to Pennsylvania. The stems of this species are distinctly reddish in color. Clover dodder rarely seeds in the United States.

Reference: H.19.

CLOVER FAILURE, erroneously called *clover sickness,* may be due to one or more causes, such as unfavorable soil conditions, poor seed, poor seeding methods, wrong seed treatment, attacks by diseases or insects, etc.—*See also* RED CLOVER.

CLOVER-FLOWER MIDGE (*Dasyneura leguminicola*), also called *clover-seed midge,* is one of the insects that may reduce the amount of RED CLOVER seed produced. The adult can be easily recognized by its red abdomen. Its body length is approximately $\frac{1}{12}$ in.

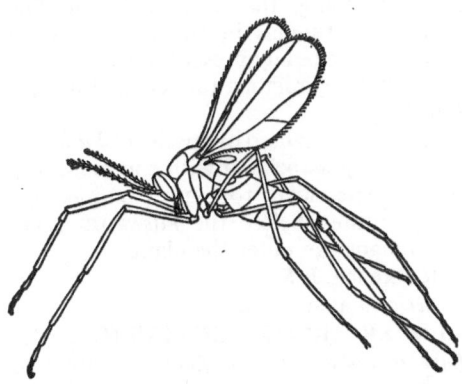

Clover-flower midge. (C.7.)

The female midge deposits eggs on the FLORETS; almost immediately after hatching, the larvae penetrate to the developing ovaries where they feed until they reach maturity. The pink or salmon-colored full-grown larvae drop to the ground from the clover head and form cocoons in which they pupate.

Control: Close pasturing, early cutting, clipping, and soiling are the best farm practices by which this insect pest can be controlled.

References: E.5; C.7.

CLOVER HAY, in general, is rich in good quality proteins; it is particularly indicated for feeding to ruminants. It is also rich in calcium and contains a fair amount of phosphorus, especially if grown on fertile soil.—*See also* HAY GRADING.

CLOVER-HEAD WEEVIL (*Phytonomus meles*) is approximately the same size as the LESSER CLOVER-LEAF WEEVIL but its color ranges from dark brown to dark gray. Adults of this species are observed in greatest numbers during the early part of the summer. The larvae are similar to those of the lesser clover-leaf weevil and both have similar feeding habits. They reach a population peak early in the summer.

Reference: E.5.

See also INSECTS.

CLOVER LEAFHOPPER (*Aceratagallia sanguinolenta*) is an insect pest which feeds preferably on the young leaves of SAF-FLOWER. The insect can be effectively controlled with DDT.

CLOVER LEAF WEEVIL (*Hypera punctata*) is a pest of CLOVER and ALFALFA. The affected plants have a ragged appearance caused by small holes in the leaves and by irregular patches eaten from the margins. If a careful search is made around the base of the plants, the greenish worm-like, or larval, stage of the weevil may be found. A crop is rarely entirely lost, but often considerable injury may result (especially in backward seasons) before the larvae become full grown or are killed by an almost universally prevalent FUNGUS disease to which they are susceptible.

The adult weevil varies somewhat in size, averaging about $\frac{1}{4}$ in. in length and $\frac{1}{8}$ in. in width. It is covered with small brown, yellow, and gray scales, which give it a mottled appearance, and has a short but distinct snout.

The eggs are oval, about 1/25 in. long and about half as wide. They are yellowish when first laid but darken with age and finally turn black. The beetles deposit their eggs in various places about the host plant—in cavities gnawed in fresh stems, in hollows of old stems, or on the leaves and stems of fresh plants. The eggs are usually laid during the fall of the year, but some of the weevils live through the winter and may deposit eggs during mild periods in the winter or spring.

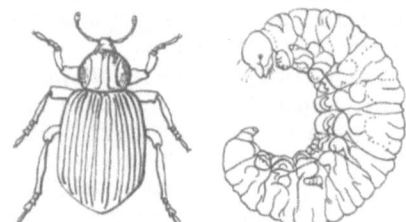

Clover leaf weevil and larva in typical curved position. (D.20.)

Most of the eggs hatch in the fall, although some remain through the winter. The small, green larvae begin feeding at once, eating the tender clover leaves and continuing their feeding on mild days throughout the winter and early spring. The full-grown larvae are from 3/8 to 1/2 in. long, usually green but sometimes yellowish, with a white or pinkish line down the center of the back.

When it is full-grown the larva spins an oval cocoon. After about eleven days the adult beetle emerges from the cocoon, feeds for a short period, and then becomes inactive for much of the summer.

Control: A disease caused by the fungus *Empusa sphaerosperma* usually keeps the clover leaf weevil from becoming a serious pest on clover and alfalfa. This fungus is well distributed over the clover- and alfalfa-growing regions of the United States. Widespread epidemics of this disease occur during periods of high humidity.

Good control has also been obtained by treating the crop with a spray containing LEAD ARSENATE.

Reference: L.3.

See also TRUE CLOVERS; WHITE CLOVER; RED CLOVER; LESSER CLOVER-LEAF WEEVIL.

CLOVER MITES (*Bryobia praetiosa*) and the other clover MITES, are exceedingly small animals, which belong to the SPIDER (and not to the INSECT) group. They occur commonly in clover fields and can be observed during the entire growing season but are generally most common during July and August. Heavy infestations may be noted on dry locations.

Clover plants injured by mites are usually stunted; the leaves are discolored and often distorted. The mites are able to pierce the leaf tissues and destroy CHLOROPHYLL. Mite colonies surrounded by webby material can be observed easily on the under-surface of infested leaves.

Large mite populations can seriously reduce the hay yield.

Reference: E.5.

CLOVER PASTURE may cause BLOAT in ruminants; there is less danger from bloat with mixed clover-grass PASTURES and from RED CLOVER pastures.—*See also* PASTURE PLANTS; PASTURE MANAGEMENT.

CLOVER ROOT-BORER (*Hylastinus obscurus*), one of the most injurious of CLOVER pests, is one of the principal factors limiting the life of a RED CLOVER stand after the first crop year.

The eggs of this INSECT are deposited in grooves on the surface of roots or in the walls of galleries made by the female insect. After hatching, the larva begins feeding and forms tunnels in the roots. Several larvae are generally present in the taproot; these make numerous tunnels and usually kill the plant.

A ROOT ROT caused by species of *Fusarium* is generally associated with the root-borer injury. The clover root must be mechanically injured before the FUSARIUM ROOT ROT fungus can enter the plant.

Reference: E.5.

See also ALSIKE CLOVER.

CLOVER ROOT CURCULIOS. *Sitona hispidula* and *S. flavescens* are two species of a common INSECT pest.

Adult curculios, which belong to the BEETLES, cause relatively little damage as they usually feed on the leaf margin. They are not common until later in the summer, when they often can be observed feeding on the plants. It is in the larval stage that this insect is most destructive. The young larvae eat the small tender roots and later attack the larger roots. Large cavities are excavated along the main roots. Taproots of RED CLOVER or ALFALFAS may show numerous scars indicating previous injuries that have healed over. The insect winters in both the adult and egg stages.

Several species of *Fusarium* have been isolated from wilted plants showing injuries on the root caused by the curculios. This would indicate that the injuries caused by the clover root curculios also facilitate entry of FUSARIUM ROOT ROT organisms.

References: E.5; G.6.

CLOVER RUST of the CLOVERS is caused by different varieties of the FUNGUS *Uromyces trifolii*. The rusts usually are not spread over entire fields but appear in patches, which may be seriously reduced in forage quantity by loss of leaves and stems. First crops are seldom heavily infested with rust. If a second crop is threatened, it should be harvested early to avoid continuous loss (which usually is not too great).

Reference: T.1.

See also TRUE CLOVERS; WHITE CLOVER; RED CLOVER; ALSIKE CLOVER; SUBCLOVER.

CLOVER-SEED CHALCID (*Bruchophagus funebris*) is one of the worst of the CLOVER (especially RED CLOVER and CRIMSON CLOVER) and ALFALFA pests. Most of the chalcids parasitize other INSECTS, but this species feeds on clover or alfalfa seed and may reduce the seed crops of these plants materially.

All stages of development of this insect are completed within the infested seed. The adult deposits a single, minute egg into the soft developing seed. The egg hatches in from three to twelve days, depending upon the temperature at the time the development is taking place. Under average temperature conditions the larvae begin their feeding a day or two after hatching. These larvae are white, quite plump and footless. They continue feeding within the tender growing seed until nothing remains but the thin outer shells. When the larvae have completed

their development, they transform to the pupal stage, and after resting for a period of from five to forty days, the insects emerge as winged adults. When the temperature has fallen, due to the approaching fall, the larvae may remain for an indefinite period within the hollowed-out seed until both—temperature and moisture—are favorable for their changing to the pupal stage.

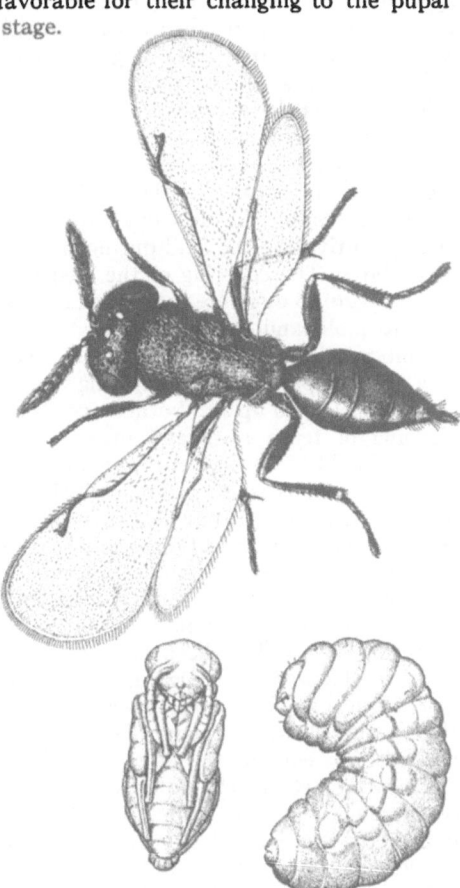

Clover-seed chalcid. Adult, larva, and pupa. (W.11.)

The adult chalcid emerges from the pupal stage as a small, black insect not quite $\frac{1}{12}$ in. in length and resembles a tiny wasp. It is then ready for feeding, mating, and egg-laying in other susceptible seeds, thus continuing the destruction indefinitely.

In southern Arizona as many as six

generations of this insect have been reported, while in the colder northern areas there are only one or two, possibly three generations, during the season.

Control: The chalcid larvae in many cases are destroyed by one of ten kinds of insects of about the same size as the chalcid itself. These little insects lay their eggs within the seeds already infested with the chalcid larvae, and the parasitic larvae entirely consume the developing seed pest.

Much loose seed and many seed pods containing chalcid larvae fall to the ground during the process of harvesting, and these overwintering forms give rise to the following year's infestation. The larvae in these seeds can readily be destroyed by thoroughly cultivating the field during the fall and winter. The covering of the seed and the exposure to moisture cause the infested seed to mold and decay, thus destroying the hibernating larvae. When this cultivation is completed, all remaining areas should be cleaned up and burned over. The cleaning of fence rows and other waste places should be continued throughout the season. A good method and one which not only destroys chalcid but also saves much time and labor (besides killing weed seeds and other insect pests) is the practice of fencing all ditches and ditch banks and then destroying uncut alfalfa or clover by pasturing with either sheep or cattle.

In alfalfa-seed growing districts one may often see alfalfa straw stacks. Before the weather is warm enough to allow the chalcids to emerge, all the screenings around these stacks should be cleaned up so thoroughly that no sign can be found of even the location of last year's threshing operations.

In many localities the chalcids develop an entire generation in the seeds of BUR-CLOVER before seed-pods have started to develop on alfalfa in the fields. Under such conditions all bur-clover should be destroyed before its seed pods complete their development. Frequently a rancher may discover that a field is severely infested and may attempt to save a portion of the forage value of the crop by pasturing the field. This should not be done, because it results in scattering the infested seed and seed pods by the animals in the field; instead, the crop should be harvested for hay and well stacked, so that the sweating process will kill a great majority of the chalcids that are still within the seed.

In sowing alfalfa or clover, it is important to use only the best recleaned seed.

References: E.5; W.11.

CLOVER-SEED MIDGE
= CLOVER-FLOWER MIDGE.

CLOVER SEED SCREENINGS have a satisfactory feeding value; when finely ground they may be used together with other concentrates in livestock rations.

CLOVER SICKNESS = CLOVER FAILURE. —See also RED CLOVER.

CLOVER SILAGE.
See SILAGE CROPS; SILAGE.

CLOVER STRAW is lower in digestible nutrients than clover hay, especially if it does not contain the protein-rich leaves. It should be fed together with good hay or silage to ruminants rather than to horses.

CLOVER TYCHIUS (*Tychius picirostris*) is a pinhead-sized gray weevil which can be collected from RED CLOVER flower heads at any time during the blooming season.

The larvae are difficult to recover from flower heads. Since they feed on the developing pod and ovary, the flower must be dissected to determine if the larvae are present.

Adults are found on the flower heads from the time they emerge from the sheath until the heads are approaching a brown condition after blooming.

Both the adult and the larvae damage the seed.

Reference: E.5.

See also INSECTS.

CLOVERWORM.
See GREEN CLOVERWORM.

CLUBHEAD is a forage SORGHUM.

CLUB KAFIR is one of the grain SORGHUMS.

CLUB MARIOUT.
See SIX-ROWED BARLEY (spring variety).

CLUB WHEAT (*Triticum compactum*), or *dwarf wheat*, is one of the WHEAT species grown in the United States. The plants may be of either winter or spring habit and vary widely in height. The stems are stiff and strong. The SPIKES usually are awnless but may be awned. The kernels are small and laterally compressed ("pinched"), because of crowding in the compact spikes. Most club wheat kernels have a small, short brush, and a narrow, very shallow crease. The grain may be either white or red and is used largely for cake and pastry flours.

> NOTE: The varieties of wheat grown in the eastern part of the United States and often referred to as "club," because of having clavate spikes, do not belong to this species but are COMMON WHEATS.

Varieties: *Utac* is a spring variety with awned spikes; it is grown in Utah.

Among the more important varieties with awnleted spikes are the following: *Alicel* (has winter habit and is very susceptible to BLUNT); *Big Club 43* (has spring habit and is resistant to HESSIAN FLY and some races of BUNT and STEM RUST; grown in California and Utah); *Elgin* (similar to Alicel and replacing it; has winter habit and is grown in Oregon, Washington, and Idaho); *Hybrid 128* (has winter habit and is very susceptible to BUNT); *Hymar* (very similar to Hybrid 128, but resistant to several races of BUNT; it is grown in Washington, Idaho, Montana, and Oregon); and *Jenkin* (has spring habit).

Of local interest are such awnleted varieties as *Albit* (resistant to some BUNT races); *Elmar* (practically identical with Elgin in plant characters, but more resistant to BUNT races); and *Poso 48* (has spring habit and is resistant to several races of BUNT and STEM RUST and to the HESSIAN FLY; grown in California and Utah).

Reference: B.15.

CLUPEA. *C. pilchardus* = SARDINE.

CLUSTER CLOVER (*Trifolium glomeratum*), locally known as *McNeill clover* (because the plant experiments were carried out at McNeill, Miss.) is recommended for use on sandy soils.

This TRUE CLOVER is of local importance in the longleaf pine area of southern Mississippi and on similar soils in adjoining states. It is an annual, the seed of which germinates in the fall. The plant lives over winter and grows rapidly the following spring. It is somewhat similar to HOP CLOVER. The leaves are small and broad and the pale-lavender to rose bloom is very inconspicuous, being borne on a small, round seed head. The plant has no main stem, but branches just above the ground and rebranches to form a great number of very fine stems. When the stand is thin, single plants spread out flat on the ground to form a circle which may be more than 2 ft. in diameter. In thick broadcast stands, the erect plant reaches a height of 12 to 15 in. and the branches interlace to form a dense mat.

Cluster clover is relished by cattle, sheep, and hogs, both as pasturage and as hay, but its chief use appears to be for grazing, since it gives very early pasturage, lasting until June when LESPEDEZA is ready to graze. When grazed, the stems lie close to the ground and produce seed, which, because it is very small, will sift through the closest grass turf and maintain a stand indefinitely. If manure chips in the pasture are chopped up with a hoe and a small pinch of seed dropped in each one, a good start of cluster clover may be obtained. In the South the plant will provide grazing from late February until June.

On the best crop land, cluster clover will give a satisfactory yield of early hay. Small plots have yielded 2.7 tons of field-cured hay per acre. For hay the clover should be cut just after it has come into bloom, though it will not reseed the land under such conditions.

It appears that a stand of cluster clover can be maintained indefinitely when followed by late crops of CORN, SORGHUMS, SOYBEANS, or COWPEAS. From small, unfertilized plots on good land that has been manured in previous years, a green weight yield of 15.8 tons per acre has been obtained in May when the plants were in bloom.

This clover is either immune or highly resistant to the ROOT KNOT NEMATODE and the POWDERY MILDEW which are serious pests on many other LEGUMES.

Reference: G.2.

See also INOCULATION.

C.N.S., short for *Clemson Nonshattering*, is a SOYBEAN variety.

COAGULATION VITAMIN
= VITAMIN K.

COAL OIL BRUSH = LITTLELEAF HORSEBRUSH.—*See also* HORSEBRUSHES.

COAL TAR is a thick liquid obtained (besides gas and coke) in the distillation of coal. It is used for making TAR BARRIERS which serve as protection against migrating CHINCH BUGS.

COAL-TAR ACIDS are produced by distillation of COAL TAR as fractions yielding at high temperatures; one of these fractions is CRESYLIC ACID.

COAL-TAR CREOSOTE, also called *creosote oil* or *creosote*, is distilled from COAL TAR. This fraction forms a yellow to brown liquid of varying viscosity, depending on contents of phenol and other COAL-TAR ACIDS. It is used for the preparation of the CREOSOTE BARRIERS, i.e., the *dirt-ridge creosote barrier* and the *creosote-treated paper-fence barrier* which are recommended for the destruction of CHINCH BUGS. For making barriers, the light creosotes—particularly the grade *T T-W-556A*—are preferred to heavy oils.

Caution: Coal-tar creosote and its fumes have a caustic effect on the skin. Coating the hands and face with petrolatum or cup grease helps to prevent *creosote burns*.

COASTAL BERMUDA.
See BERMUDA-GRASS (variety).

COAST BARLEY.
See SIX-ROWED BARLEY (variety).

COB is the woody AXIS of CORN to which the seeds are attached.

COBALT is an important TRACE ELEMENT; to prevent or treat cobalt deficiency, various cobalt salts are used.—*See also* VITAMIN B_{12}.

COBALT ACETATE, or *cobaltous acetate*, forms red crystals which are water-soluble. It contains 25 percent elemental COBALT and is used in MINERAL FEEDS.

COBALT CARBONATE, or *cobaltous carbonate*, a rose-colored powder containing 49.5 percent elemental COBALT, is insoluble in water.—*See also* MINERAL FEEDS.

COBALT CHLORIDE, or *cobaltous chloride*, available in form of dark-red crystals

of 24.7 percent COBALT content, is soluble in water. It is one of the officially recognized cobalt salts used in MINERAL FEEDS.

COBALT SULFATE, or *cobaltous sulfate*, forming red crystals, is water-soluble; it contains not quite 21 percent elemental COBALT and is widely used in the treatment of cobalt deficiency of ruminants.—*See also* MINERAL FEEDS.

COB MEAL. *Corn-and-cob meal = ear corn chop.*—*See also* CORN; SWEETCLOVER.

COCCUS (plural: *cocci*) is a BACTERIUM having round or ovoid form.

COCK is a small conical pile of hay.

To cock means: to put into cocks.

COCKLEBUR (*Xanthium echinatum*) is a POISONOUS PLANT containing a GLUCOSIDE which is dangerous to pigs and cattle. Toxic effects may show up within a few minutes if the animals eat the green plants (particularly the first leaves of seedlings) at the rate of 12 oz. per 100 lb. body weight. Symptoms of poisoning are prostration and inflamed stomach.

COCKSFOOT = ORCHARDGRASS.

COCOA MEAL is obtained as a byproduct when extracted *cocoa beans* are manufactured into cocoa in the chocolate industry. It is used as a feed ingredient, but its digestibility and value for feeding purposes are low.—*See also* COCOA TREE.

COCOA SHELLS have an agreeable chocolate flavor and are sometimes used as feed ingredient; they form the thin covering of *cocoa beans* and must be ground before they can be mixed with feed.—*See also* COCOA TREE.

COCOA TREE (*Theobroma cacao*), or *cacao tree*, is an evergreen which grows 25 ft. high. It is cultivated in Central and South America. The large seeds, called cocoa beans, are obtained in seed pods up to 1 ft. long; they are used in the manufacture of cocoa butter, cocoa, and chocolate, and give such by-products as COCOA MEAL and COCOA SHELLS.

COCONUT OIL MEAL, or *copra oil meal*, is the ground residue obtained after the extraction of part of the oil from the dried meat of the *coconut*. It must be designated and sold according to its protein content.

This officially recognized OILSEED MEAL —largely an imported by-product obtained in the manufacture of oil from the fruit of the COCONUT PALM—is extensively used in feeding dairy cattle and, to a lesser extent, of beef cattle and lambs.

References: F.6; E.12.

See also COCONUT PALM.

Cocklebur. Leaf, fruiting branch, and two seedlings. (W.36.)

COCONUT PALM (*Cocos mucifera*) is a tree whose fruit is known as *coconut*. The dried meat of the nut is called *copra*. From the latter, *coconut oil* is produced and, as a by-product, COCONUT OIL MEAL.

COCOS. *C. mucifera* = COCONUT PALM.

COD LIVER OIL.

See VITAMINS; HAY GRADING.

COD LIVER OIL MEAL, the residue of the cod livers obtained in the manufacture of cod liver oil, is rich in protein, fat, and VITAMIN D. It is occasionally used in poultry mashes.

COD LIVER OIL WITH ADDED VITAMIN A AND D CONCENTRATES.

See VITAMINS.

CODY OAT. *See* COMMON OAT (variety).

CODY SORGHUM is one of the grain SORGHUMS.

COEFFICIENT OF DIGESTIBILITY.

See DIGESTIBILITY.

COELOMIC means: relating to *coeloma*, or *coelom*, i.e., the body cavity of fish.— *See also* MARINE PRODUCTS.

COENZYMES are produced by living cells and are necessary for certain ENZYME actions.—*See also* VITAMINS.

COES and *Highland Improved Coes* are intermediate-type grain SORGHUMS.

COKER'S VICTORGRAIN.

See RED OAT (variety).

COLBY, a *milo* variety, is one of the grain SORGHUMS.

COLIAS. *C. philodice eurytheme*

= ALFALFA CATERPILLAR.

COLLAGEN is a protein substance closely related to GELATIN; it is present in connective tissues, bones, and cartilages.

COLLETOTRICHUM. Some species of the FUNGUS *Colletotrichum* attack many different hosts, whereas others are selective. *C. villosum* causes VETCH ANTHRACNOSE; *C. pisi*, PEA ANTHRACNOSE; *C. graminicola*, ANTHRACNOSE OF GRASS and COLLETOTRICHUM STALK ROT; *C. falcatum* is the cause of *red rot*, a term which should be restricted to SUGARCANE STALK ROT. *C. trifolii* causes SOUTHERN ANTHRACNOSE; *C. crotalariae* is the cause of CROTALARIA ANTHRACNOSE, but there is a strong possibility that more than one species of this fungus can attack crotalaria.—*See also* GLOMERELLA.

COLLETOTRICHUM STALK ROT is caused by *Colletotrichum graminicola*, the same FUNGUS that causes ANTHRACNOSE OF GRASS, occurring on leaves of the sorghums. This condition is also called *red rot* but since true red rot is caused by another fungus, *C. falcatum*, which infects sugarcane and not the sorghums, this term should be restricted to the SUGARCANE STALK ROT.

Colletotrichum stalk rot affects broomcorn in Illinois and in other states, and other sorghums in many areas of the South, especially where sorgos are grown intensively for sirup production. The STALK ROT phase of the disease is usually preceded by the anthracnose stage.

The fungus apparently enters the stalk directly through the rind and spreads rapidly in the interior of the stalk. This results in poor development of heads and seeds of sorghum plants. In diseased broomcorn plants the brush is dead and of very poor quality.

The lesions that form on the outside of diseased stalks usually have reddish to purplish margins and whitish centers, although the colors vary somewhat with different varieties. The fruiting bodies of the fungus are in the whitish centers of the lesions. Under moist conditions they produce spores in abundance.

When infected stalks are split, the pith is found to be red or purplish red in most varieties. In Leoti and some of its derivatives, however, the pith is yellowish or orange.

Control: Clean culture, rotation, and the use of resistant varieties are recommended as control measures. Atlas and Planter are practically the only commercial varieties of sorgos known to be resistant. Most of the varieties planted for sirup almost completely escape serious losses when harvested early, or when the heads are in the soft-dough stage; however, in the more susceptible varieties, such as the Honey sorgo, the stalk rot may be destructive even when they are harvested early.

Among the grain sorghums, hegari, Western Blackhull kafir, Club kafir, and Early Kalo are resistant to colletotrichum stalk rot.

Reference: L.1.

COLLIER SORGO, and its selection, *Kansas Collier 704-D*, are forage SORGHUMS.

COLLOID is a state of subdivision of matter which consists of single large MOLECULES (such as proteins) or of aggregations of smaller molecules; e.g., gelatin, glue, starch, or colloidal clay.

COLLOIDAL means: pertaining to COLLOID.

The colloidal systems include *suspensions* (solid, finely distributed in liquid) and *emulsions* (liquids distributed in liquids, e.g., milk).

"COLLOIDAL PHOSPHATE" = SOFT PHOSPHATE WITH COLLOIDAL CLAY.

COLMAN SORGO—erroneously called *Red Orange*—is one of the forage SORGHUMS.

COLON. *See* DIGESTIVE TRACT.

COLONIAL BENT (*Agrostis tenuis*), or *colonial bentgrass*, also called *Rhode Island bent*, does not creep as extensively as the other fine bentgrasses do; like all of them, it has been found well adapted for putting greens and lawn mixtures in much of the northern half of the United States.

Fields of colonial bent are grazed after seed harvest in seed-producing areas of the Northwest.

Dangers: Colonial bent is often affected by BROWN PATCH, a fungus disease.

Reference: H.1.

See also BENTGRASSES; GRASSES.

COLOR is a term used loosely to indicate material substances (such as inorganic *pigments* or organic *dyes*), rather than their appearance.

The use of artificial colors in feedstuffs is disapproved; however, FDC CERTIFIED DYES may be used to indicate the distribution of valuable ingredients, or to increase or aid in proper intake of a feedstuff. But dyes shall not be used to enhance the natural color of a feed or feed ingredient, and thereby conceal inferiority.

Reference: F.6.

See also FDC.

COLORADO 37.
See COMMON OAT (variety).

COLORADO 152 is a CORN hybrid.

COLORADO BARBERRY.
See BARBERRY BUSHES.

COLORADO BLUESTEM
= WESTERN WHEATGRASS.

COLORADO RUBBERWEED. *See* RUBBERWEEDS; ACTINEA; POISONOUS PLANTS.

COLOSTRUM is the first, thicker milk secreted soon after parturition. It is rich in VITAMINS and is important for the newborn since it gives the animal some resistance to bacterial and other diseases.—*See also* CATTLE RATIONS.

COLSESS.
   See SIX-ROWED BARLEY (variety).

COLT RATIONS.
   See HORSE AND MULE RATIONS.

COLUMBIA OAT.
   See RED OAT (variety).

COLUMELLA is a long, dark, pointed structure left in the place of a GALL after its spores have been blown away.—*See also* LOOSE KERNEL SMUT OF SORGHUM.

COLUSA. *See* RICE (variety).

COLZA OIL. *See* RAPE.

COLZA OIL MEAL = RAPESEED MEAL.

COMANCHE.
   See COMMON WHEAT (variety).

COMBINE is a machine that combines harvesting and threshing into a single operation.

COMBINE-7078 is one of the intermediate-type grain SORGHUMS.

COMBINE BONITA is a hegari variety which belongs to the grain SORGHUMS.

COMBINE HEGARI = *Early hegari.* —
   See also SORGHUMS.

COMBINE KAFIR-60 is one of the grain SORGHUMS.

COMBINE-TYPE GRAIN SORGHUMS, or *combine sorghums*, are *dwarf* and *double dwarf* varieties that can be harvested with the ordinary grain combine by elevating the cutting bar to the highest position. These SORGHUM varieties do not exceed a height of 48 to 56 in.; the best range is from 40 to 46 in., and true double dwarfs usually range from 30 to 36 in. Combine sorghums are, e.g., *Bonita, Caprock, Combine-7078, Combine kafir-60, Combine Waxy kafir, Double Dwarf Yellow Sooner, Dwarf Feterita, Dwarf White Durra, Early hegari, Martin, Midland, Plainsman milo, Quadroon,* and *Redbine-60.*
   Reference: W.2.

COMBINE WAXY KAFIR is one of the grain SORGHUMS.

COMBINE WHITE KAFIR OKLA. 44-14 = *Dwarf kafir 44-14.* See SORGHUMS.

COMFREY. *See* PRICKLEY-COMFREY.

COMMERCIAL FERTILIZERS.
   See FERTILIZERS.

COMMERCIAL MIXED FEEDS, or *formula feeds*, are mixtures of *processed grains* (e.g., ground, rolled, flaked, cracked) and of various *by-products* from milling, distilling, edible oil production, and fat rendering operations; they are properly fortified with VITAMINS and MINERALS and designed to perform a specified job. They may contain added DRUGS, MOLD INHIBITORS, or conditioning agents.

The formula feeds may be classed on the basis of the use for which they are intended, the species to which they are to be fed, or their physical state.

Protein supplements usually include feeds which contain at least 20 percent crude PROTEIN and were formulated primarily for supplementing farm grains. Many of the protein supplements are also excellent sources of minerals and vitamins and sometimes contain added amounts of these substances. There is a tendency to call these feeds "*supplements*," but the word protein must be retained because of registration procedures.

Mineral supplements, or *mineral mixtures*, usually contain several of the more important minerals and may contain a complete mixture of TRACE ELEMENTS.

Vitamin supplements include feeds which have been prepared to contain high amounts of vitamins; alfalfa meals are sometimes sold on the basis of their CAROTENE content.

Complete feeds are those which have been formulated to meet all of the NUTRIENT requirements of the animal. Examples of complete feeds are a 15 percent laying mash or a dog food. When farmers buy the complete feed, they should be cautioned against diluting it with farm-grown grains and thereby reducing the protein, mineral and vitamin levels enough to lower the performance of the animal.

More than half of all the commercial feeds sold in the United States are fed to *poultry*. Many of the poultry feeds will contain 25 or 30 separate ingredients, assembled from many areas, which are blended to produce the very high performance levels demanded by the poultry industry, especially in the sections producing market birds.

Dairy cattle in this country consume the second largest amount of commercial feed. Most of this feed is 14, 16, or 18 percent dairy feed, intended for feeding with farm ROUGHAGES. There are a number of excellent dairy feeds available but in this highly competitive area of the feed trade some feeds are bagged rather with the idea of underselling the competitor than with the idea of providing what the dairy cow needs.

About 10 percent of the total feed is used by *swine*. Most of this is in the form of protein supplements which are used to balance farm grains. There is an increasing volume of specialty feeds such as pre-starter and *creep feeds* and feeds for pregnant sows.

Something over 5 percent of the total feed sold is *beef cattle* and *sheep* feed, with sheep accounting for only a very small percentage of the total. Most of the beef cattle feed is in the form of protein supplements. The introduction of STEROID hormone-like materials in beef supplements will probably continue to increase the supply of commercial feed for beef cattle.

Miscellaneous feeds account for about 5 percent of the total product. These feeds include *horse* and *mule* feeds, pet feeds, and fur and game animal feeds. In some areas, however, they make up a considerable volume.

Most mixed feeds were originally sold as MASHES. More recently feeds in other physical forms have become popular. One widely used process is *pelleting*. The mash mixture is put through a high pressure pelleter which in the presence of suitable binding agents (which may have nutrient value) and heat causes the formation of a PELLET. Some manufacturers break the pellets down to form *crumbles*. There is also demand for feeds which are prepared from coarse grains or flaked products which result in a very coarse-textured feed. These are chiefly sold to dairy men.

The sale of commercial feed is regulated by the states. Interested parties should consult their State Department of Agri-

culture for details regarding registration, labeling, and sale of feed.—*See also* LABEL; TAG; F.D.A.

(COMMERCIAL) TURKESTAN ALFALFA. *See* ALFALFAS (variety).

COMMON ALFALFAS, also called *purple alfalfas* and (in Europe) *lucernes*, are often distinguished by the name of the state where they were grown for many years: e.g., Kansas Common, Montana Common, or Dakota Common, which are *Northern Commons*, and Arizona Common, Oklahoma Common, or other *Southern Commons*. Among the newer varieties are Buffalo, California Common 49, Caliverde, and Williamsburg.—*See also* ALFALFAS.

COMMON BAHIAGRASS
　　　　　　= BAHIAGRASS.

COMMON BEAN (*Phaseolus vulgaris*) is one of the FIELD BEANS. Its more than 50 varieties are either of the bushy or trailing type; they include Navy bean (also known as *White Pea*), *Pinto*, *Great Northern*, and *Red Kidney*.

COMMON BICOLOR
　　　　　　= BICOLOR LESPEDEZA.

COMMON BUCKWHEAT (*Fagopyrum esculentum*) is the most important BUCKWHEAT species found in the United States. It is highly self-sterile and must be cross-pollinated to effect fertilization. Bees and other insects distribute the pollen. Hybrids thus occur in the mixture and the following crops are of an intermediate type with characteristics of both parents.

Varieties: *Japanese buckwheat* is probably the most extensively grown and widely adapted type. The seed is brown to dark-brown in color and usually large in size. In cross section it is nearly triangular. The plants are tall, have large leaves and coarse stems, and are inclined to lodge.

Silverhull usually is distinguished from Japanese buckwheat by its smaller, glossy, silvery-gray seeds. The seed is more nearly round than triangular in cross section. The plants are smaller than those of the Japanese, the stems not so coarse, and the leaves smaller. The stems of Silverhull show a tendency to be more reddened at maturity than those of Japanese.

Common Gray is very similar to Silverhull but has smaller seeds.

NOTE: *"Common buckwheat"* is a term often applied to mechanical and hybrid mixtures of Japanese and Silverhull.

Reference: Q.1.

COMMON BUNT, or *stinking smut*, is caused by two *Tilletia* spp., namely *T. caries* and *T. foetida.* The FUNGUS species *T. carles* also causes the related DWARF BUNT. Losses from this wheat disease are of two kinds—field losses, due to reduction in yield; and market losses, due to the lower price paid for smutty wheat because of its discoloration and foul odor. Occasional smut explosions, which may occur during threshing or while handling smutty seed in elevators, also cause losses.

Common bunt stunts plants a few inches to as much as half the height of healthy plants, depending upon which of the two fungi is involved. Symptoms of the disease usually are not apparent until heading time. At the time of emergence from the boot, smutted wheat heads are bluish-green (as compared to the yellowish-green color of healthy heads) and contain smutted kernels, or "smut balls," in place of normal kernels. Each smut ball contains a mass of sooty, black powder, the individual particles of which are the spores that perpetuate the smut fungus. Soil-borne spores of common bunt remain infective only in areas like the Pacific Northwest where the soil remains dry from threshing until after seeding. It is advisable to sow winter wheat late in order to avoid common bunt infection.

The smut has a distinctive fishy odor.

Infection by common bunt is favored by soils rich in humus, and it is inhibited somewhat in heavy clay, peat, and highly acid soils. Potassium and phosphate fertilizers seem to stimulate BUNT infection, while lime, nitrogen fertilizers, superphosphate, and calcium cyanamide repress it. These factors may account for the difference in infection seen at times in different fields of wheat that is grown from the same lot of seed.

Control: The most effective methods of controlling common bunt are (1) the use of smut-free seed or of well-cleaned and CERESAN M-treated seed from which the smut balls have been separated and (2) the use of varieties immune from or highly resistant to the disease.

Reference: L.10.

See also SEED TREATMENT.

COMMON CHOKECHERRY.
    See WILDCHERRY; POISONOUS PLANTS.

COMMON DODDER (*Cuscata gronovii*), one of the DODDERS, resembles FIELD DODDER in its indifference to the plants which it may attack.

Reference: H.19.

COMMON EMMER = VERNAL. —
                        See also EMMER.

COMMON FOXTAIL MILLET.
        See FOXTAIL MILLET (variety).

COMMON GRAY BUCKWHEAT.
        See COMMON BUCKWHEAT (variety).

COMMON LEAF SPOT, one of the best known of the foliage diseases of ALFALFA, is caused by the FUNGUS *Pseudopeziza medicaginis.*

Spots produced on the hosts are round and 3 to 4 mm. ($\frac{1}{8}$ to $\frac{1}{6}$ in.) in diameter when fully developed. The edge of the spot is toothed, a feature that distinguishes it from all other leaf spots. At the center of the spot is a tiny structure, which is flesh-colored when it is open under moist conditions and from which great numbers of spores are discharged. The airborne fungus is soon found on alfalfa no matter how far away from old fields.

The fungus is rarely abundant in early spring in northern regions and depends on moisture for development and spread. The injury caused by spots is local, but the spots, when they are abundant, cause yellowing of foliage and dropping of leaflets. Only a few individuals in common alfalfa varieties are highly resistant. The only variety in which resistance has been reported to be incorporated is CALIVERDI.

Spots rarely become abundant enough in old leaves to cause much loss of foliage.

Reference: J.3.

COMMON LESPEDEZA (*Lespedeza striata*), often called *Japan clover* or *"Jap clover,"* has spread widely throughout the South of the United States. This ANNUAL LESPEDEZA species is prostrate. For this reason it is able to persist with GRASSES in closely grazed permanent pastures better than the more upright-growing lespedeza species.

Common lespedeza. (M.3.)

Varieties: Two improved varieties of the unimproved common lespedeza are of importance:

1. *Kobe:* This is one of the improved LEGUMES for southern pastures; its large growth assures increased yields of hay,

and its late maturity provides nutritious feed at a time of scarcity. Kobe grows more upright than unimproved common lespedeza, but not as upright as Tennessee 76. Kobe is considered superior to Tennessee 76 and to unimproved common lespedeza for general use under cultivation.

In temporary pastures or in pastures used in rotation Kobe is decidedly superior to unimproved common lespedeza. It fits well into a rotation with small grain, and such a rotation provides a continuous ground cover that prevents leaching of plant food and reduces loss of soil by erosion. Such a rotation also fits in well with livestock production.

A SMALL GRAIN crop and lespedeza may be used in rotation with COTTON, CORN, PEANUTS, SOYBEANS, and other summer crops.

Kobe, being a large-growing variety, gives maximum yields on hay.

2. *Tennessee 76:* This variety has a more erect habit of growth than other annual lespedezas. It matures a little later than Kobe; its seed is distinctly smaller than that of Kobe or unimproved common lespedeza.

References: M.7; E.7; M.8.

See also LESPEDEZAS; PASTURE PLANTS.

COMMON OAT (*Avena sativa*), also called *spreading oat* or simply *oat*, is commonly cultivated in the United States. This oat species differs from COMMON WILD OAT in having mostly two-flowered SPIKELETS, the FLORETS not readily separating from the GLUMES.

> NOTE: A group of late, large-strawed, large-grained, thick-hulled varieties with wide panicles, called *side oats*, includes *Canada Cluster*, *Mammoth Cluster*, *Storm King*, and *Tartar King*. They are sometimes grown in the northeastern states, but they are much less productive than common oat and give a high percentage of hull and a low feeding value.

Varieties: An alphabetical listing of common oat varieties follows. The particular parts of the country in which respective varieties are grown are indicated by references to numbered regions, such as: Region 1—*North Central* oat region; 2—

Central red oat region; 3—*Northeastern* oat region; 4—*Southern* red oat region; 5—*Great Plains* oat region; 6—*Rocky Mountain* oat region; 7—*Pacific Coast* oat region.

Abundance: region 7; midseason oat.

Ajax: regions 1 and 3; spring-sown; disease-resistant.

Andrew: regions 1 and 3; spring-sown; disease-resistant.

Arlington: region 5; fall-sown; disease-resistant.

Banner: region 7; midseason oat.

Bannock: region 6; midseason oat; SMUT-resistant.

Benton: region 1; spring-sown; early maturing; similar to *Clinton.*

Bonda: region 1; spring-sown; early maturing; disease-resistant.

Bridger: region 6; midseason oat; SMUT-resistant.

Carleton: region 6; midseason oat; white kernels; SMUT-resistant.

Cartier: regions 1 and 3; spring-sown; early maturing.

Cherokee: region 1; spring-sown; disease-resistant.

Clinton: regions 1 and 3; spring-sown; early-maturing; yellow kernels; resistant to CROWN RUST, STEM RUST, SMUTS, and VICTORIA BLIGHT.

Cody: region 7; spring-sown.

Colorado 37: region 6; midseason oat; white kernels.

De Soto: region 5; fall-sown; disease-resistant.

Golden Rain: regions 1 and 3; spring-sown.

Gopher: regions 1 and 3; midseason oat.

Green Russian: regions 1 and 3; spring-sown.

Lee: region 5; fall-sown.

Maine 340: region 3; midseason oat.

Marida: region 7; spring-sown.

Marion: region 1; spring-sown; early maturing; white grain; resistant to VICTORIA BLIGHT and SMUTS.

Markton: regions 6 and 7; midseason oat; grown under irrigation; yellow-grained; highly productive; SMUT-resistant.

Mission: region 6; midseason oat; SMUT-resistant.

Mohawk: region 3; spring-sown; early maturing; similar to *Clinton.*

Nemaha: region 1; spring-sown; resistant to VICTORIA BLIGHT and SMUTS.

Osage: region 1; spring-sown; early maturing; yellow kernels.

Overland: regions 1, 6, and 7; spring-sown; low hull percentage.

Rainbow: region 1; midseason oat.

Silvermine: regions 1 and 3; spring-sown.

Southland: region 5; midseason oat.

Swedish Select: regions 1 and 3; spring-sown.

Uton: region 6; midseason oat; SMUT-resistant.

Victory: regions 1, 3, 6 (grown under irrigation), and 7; midseason oat; white kernels.

Wayne: region 3; midseason oat.

White Tartar: regions 1 and 3; spring-sown.

Winter Turf, also called "*Winter Grazing,*" *Gray Winter,* or *Oregon Gray:* region 7; fall-sown; grows tall with narrow, dark brown leaves and gray kernels; it is often better adapted to pasture and hay than to grain production.

Wintok: region 4; fall-sown; hardiest common oat variety.

References: R.7; F.7; S.19; S.21.

See also RED OATS; OATS.

COMMON OLEANDER (*Nerium oleander*) is a very POISONOUS PLANT, affecting all species of animals. The intake of only small amounts of the leaves soon causes such symptoms as stupor, trembling, convulsions, paralysis, vomiting, and diarrhea.

Illustration: See page 108.

COMMON RAGWEED. *See* RAGWEED.

COMMON RED CLOVER
= *Double-cut clover. See* RED CLOVER.

COMMON RYE, which includes *winter rye* and *spring rye* types, is the most important RYE variety.—*See also* GRAZING.

COMMON RYEGRASS, as grown in the United States, is often referred to as *domestic ryegrass, Oregon ryegrass, western ryegrass, native ryegrass,* and *Pacific ryegrass.* These names designate mechanical and hybrid mixtures of PERENNIAL RYEGRASS with ITALIAN RYEGRASS. It is also often called the *South American type of*

Common oleander. Flowering shoots. (W.35.)

Italian ryegrass since its conglomerate make-up contains numerous types, including that prevailing in the South American seed.

The importance of common ryegrass in the United States as a forage, lawn, and seed crop is rapidly increasing. Plant and seed characteristics of the common ryegrass resemble those of the Italian ryegrass very closely, though the common ryegrass usually does not grow as tall and its stems are somewhat stiffer and heavier; its seeds, too, are plumper and the awns are shorter and weaker.

Common ryegrass grows from 2 to 3 ft. tall, is leafy and tender, and when used as pasturage is very palatable to all classes of livestock. Since it furnishes early grazing and acts as a nurse-crop to the more permanent grasses that are generally slow in becoming established, it makes an excellent addition to a permanent pasture mixture.

It also gives very good fall, winter, and early spring grazing when seeded alone. The grass is a heavy yielder and, when properly handled, gives a high grade of very palatable hay.

For lawns and putting greens in the southern states it is very satisfactory.

Reference: S.1.

See also RYEGRASSES; LOLIUM; GRASSES.

COMMON VETCH (*Vicia sativa*), sometimes called *spring vetch* or *tares*, belongs to the VETCHES. The plants are semiviny, with slightly larger leaves and stems than HAIRY VETCH. They cannot be grown as a winter crop except in regions that have a mild climate. In western Oregon and western Washington common vetch is hardy in most winters, but it often winterkills in the northern part of the Cotton Belt. In the Pacific Coast states it is grown as a hay and seed crop, as well as for green manure, silage, and pasture. In the southeastern states it has been used largely for green manure.

Varieties: *Willamette vetch* is quite vigorous, winter-hardy, and well adapted to the more fertile soils of the southeastern states.

Pearl vetch is a variety with light-pink seed, and is grown occasionally in western Oregon as a spring-sown crop.

Reference: M.18.

See also LEGUMES; INOCULATION; HUNGARIAN VETCH; NARROWLEAF VETCH.

COMMON SALT = SODIUM CHLORIDE.

COMMON ST. JOHNSWORT.

See ST. JOHNSWORT.

COMMON SUDANGRASS.

See SUDANGRASS.

COMMON WHEAT (*Triticun vulgare*), the most widely cultivated of all WHEAT species, is distinguished from CLUB WHEAT, which it most closely resembles, by a

SPIKE that is long in proportion to its thickness. The spike is usually dorsally compressed. The SPIKELETS are two- to five-flowered, far apart, and nearly erect. The GLUMES are shorter than the LEMMAS, firm, and either glabrous or pubescent. The lemmas are awnless or have awns less than 4 in. long. The blades of the leaves are usually narrower than those of the DURUM WHEAT and POULARD WHEAT. The kernel may be either soft or hard and white or red.

Common wheat is adapted to widely varying climatic conditions:

Varieties: The 204 varieties of common wheat cultivated in the United States can be classified according to certain plant characteristics. The more important varieties are listed in the following subdivisions:

1. GLABROUS, WHITE GLUMES; WHITE KERNELS

Idead: soft to semihard spring wheat; awnless; grown in the Northwest.

Lemhi: soft spring wheat; awnless; grown in the Northwest.

White Federation 38: hard spring wheat; awnless; grown in California.

Yorkwin: soft winter wheat; awnleted; the leading variety in New York and Michigan.

2. GLABROUS, WHITE GLUMES; RED KERNELS

Cache: hard winter wheat; awnleted; resistant to many races of BUNT and moderately resistant to DWARF BUNT; grown in Utah, southern Idaho, Montana, and Washington.

Cadet: hard spring wheat; awnleted; resistant to STEM RUST and to some races of LEAF RUST, LOOSE SMUT, and MILDEW; grown in North Dakota, South Dakota, and Minnesota.

Chiefkan, also called *Kanhull* or *Newchief:* hard winter wheat; awnleted and swaybacked, moderately resistant to LEAF RUST and STEM RUST but very susceptible to LOOSE SMUT OF WHEAT and BUNT; grown chiefly in Kansas, Oklahoma, Texas, and Colorado.

Clarkan, or *Clark's No. 40:* soft winter wheat; awnleted; moderately resistant to FLAG SMUT OF WHEAT but susceptible to

MOSAIC; grown largely in Missouri, Kansas, and Kentucky.

Fairfield: soft winter wheat; awnleted; resistant to MOSAIC and LOOSE SMUT OF WHEAT, and somewhat resistant to LEAF RUST; grown in Indiana, Illinois, and Ohio.

Forward: soft winter wheat; awnleted; resistant to several races of LOOSE SMUT OF WHEAT; grown in the East.

Fulhio, or *Ohio No. 127:* very similar to *Fultz.*

Fultz, also called *Bluestem, Hickman, Posey, Slickhead,* and *Snow:* soft winter wheat; awnleted; grown mainly in Illinois, Missouri, and Indiana.

Hardired: semihard, intermediate wheat; awnleted; moderately resistant to LEAF RUST and POWDERY MILDEW; grown in the southern states, especially in the Carolinas.

Leap, also known as *Hastings Prolific, Leap's Prolific,* or *Woods Prolific:* soft winter wheat; awnleted; resistant to LOOSE SMUT OF WHEAT grown in Virginia and other eastern states.

Marquis: hard spring wheat; awnleted; grown mainly in Montana and Canada.

Newhatch: hard spring wheat; awnleted; resistant to STEM RUST, LOOSE SMUT OF WHEAT and BUNT, but susceptible to LEAF-RUST and WHEAT SCAB; grown largely in the Dakotas, Montana, and Minnesota.

Purplestraw, also known as *Bluestem, Georgia Red,* and *Ripley:* soft, intermediate spring wheat; awnleted; grown in the southeastern states.

Redhart: semihard, intermediate spring wheat; awnleted; grown chiefly on the East Coast.

Redman: hard spring wheat; awnleted; resistant to STEM RUST, BUNT, and to some races of LEAF RUST and LOOSE SMUT OF WHEAT; grown in Minnesota, the Dakotas, Montana, Wisconsin, and Canada.

Regent: hard spring wheat; awnleted; resistant to STEM RUST, BUNT, and some races of LEAF RUST; grown in the Dakotas and Minnesota.

Rescue: hard spring wheat; awnleted; resistant to the WHEAT STEM SAWFLY, somewhat resistant to STEM RUST, but susceptible to LEAF RUST and BUNT; has solid stems and low protein content;

grown in Montana, North Dakota, nad Canada.

Sanford or *Sanett:* very similar to *Purple-straw.*

Thatcher: hard spring wheat; awnleted; resistant to STEM RUST, but susceptible to LEAF RUST; grown extensively in Montana, the Dakotas, and Canada.

Trumbull: similar to *Fultz;* grown in Ohio and Indiana.

Vigo: soft winter wheat; awnleted; resistant to LEAF RUST, MOSAIC, and some races of LOOSE SMUT OF WHEAT; grown chiefly in Ohio and Indiana.

3. GLABROUS, BROWN GLUMES;
WHITE KERNELS

Cornell 595: soft winter wheat; awnleted; resistant to MOSAIC, some races of LOOSE SMUT OF WHEAT, POWDERY MILDEW, and SEPTORIA BLIGHT OF WHEAT; grown largely in New York, Michigan, and Ohio.

Federation: soft spring wheat; awnleted; grown in the western states.

Goldcoin, also called *Fortyfold, Junior No. 6, Klondike,* or *White Clawson:* soft winter wheat; clavate spike, awnleted; purple straw; collared brush; grown in the Northwest.

Golden: differs from *Goldcoin* in being slightly later and having shorter and stronger stems.

Ramona 44: hard spring wheat; awnless; resistant to several races of BUNT, STEM RUST, and LEAF RUST; grown in California, Arizona, and Nevada.

Rex: soft winter wheat; awnless; resistant to BUNT and to lodging and shattering; grown in the Northwest.

4. GLABROUS, BROWN GLUMES;
RED KERNELS

Red Chief, or *Superred:* hard winter wheat; awnleted; grown mainly in Kansas, Oklahoma, Colorado, and Texas.

Red May, also called *Michigan Amber, Beachwood, Early May, Jones Longberry, Michigan Wonder, Orange, Purdue No. 4,* and *Red Republic:* soft winter wheat; awnleted; grown chiefly in Indiana, Michigan, and Missouri.

Thorne, or *T. N. 1006:* soft winter wheat; awnleted; resistant to LOOSE SMUT OF WHEAT and MOSAIC; grown in Ohio, Pennsylvania, New Jersey, Indiana, and Illinois.

5. PUBESCENT, WHITE GLUMES;
RED KERNELS

Reward: hard spring wheat; awnleted; grown mainly in Montana, Colorado, and South Dakota.

6. PUBESCENT, BROWN GLUMES;
WHITE KERNELS

Galgalos, also known as *Russian Red* or *Velvet Chaff:* soft spring wheat; awnleted; grown in the Northwest.

7. GLABROUS, WHITE GLUMES;
WHITE KERNELS

Baart: semihard spring wheat; awned; it can be distinguished from all other wheat varieties by its large yellowish pear-shaped kernels; grown in the Northwest.

Baart 38: very similar to *Baart;* resistant to some races of STEM RUST and BUNT; grown in California and Arizona.

Orfed: soft to semihard spring wheat; awned; resistant to BUNT and FLAG SMUT OF WHEAT; grown in the Northwest.

8. GLABROUS, WHITE GLUMES;
RED KERNELS

Blackhull, also known as *Black Chaff, Superhard,* and *Clark's Black Hull:* semihard to hard winter wheat; awned; has black stripes on the glumes; grown chiefly in Kansas, Oklahoma, Texas, and Colorado.

Blue Jacket: hard winter wheat; awned; grown in Texas, Kansas, Oklahoma, New Mexico, and Iowa.

Ceres: hard spring wheat; awned; moderately resistant to STEM RUST and to drought; grown mainly in Montana, the Dakotas, and Colorado.

Cheyenne, or *Nebraska No. 50:* hard winter wheat; awned; grown in Nebraska, Wyoming, Colorado, Kansas, Oklahoma, and Texas.

Comanche: hard winter wheat; awned; early maturing; resistant to BUNT, and somewhat resistant to LEAF RUST and STEM RUST; grown extensively in western Kansas, Oklahoma, Texas, and Colorado.

Early Blackhull, also called *Early Hardy, Early Kansas, Early Russian,* and *Haeberle:* differs from *Blackhull* principally in being eight days earlier and somewhat shorter.

Fulcaster, known under many names, e.g., *Dietz, Miracle, Stoner, Bearded Bluestem, Cumberland Valley, Duffy, King, Lancaster, Marvelous, Peck,* or *Red Wonder:*

soft winter wheat; awned; has purple straw and orange-colored stripes on the glumes; grown in the eastern half of the United States.

Henry: semihard to hard spring wheat; awned; resistant to STEM RUST, BUNT, and LEAF RUST; grown chiefly in Wisconsin, Minnesota, and North Dakota.

Kanred, or *P-762:* hard winter wheat; awned; resistant to some races of LEAF RUST and STEM RUST; grown chiefly in Colorado, Kansas, Texas, and New Mexico.

Karmont: similar to *Turkey;* grown largely in Montana.

Kawvale: semihard to hard winter wheat; awned; resistant to LOOSE SMUT OF WHEAT, somewhat resistant to LEAF RUST, STEM RUST, and HESSIAN FLY; grown in Kansas, Missouri, and Illinois.

Mida: hard spring wheat; awned; resistant to STEM RUST, some races of LEAF RUST, and to BUNT; grown in the Dakotas and neighboring states.

Nebred: differs from *Turkey* in being slightly earlier, shorter, and stronger and in being resistant to BUNT; grown largely in Nebraska and South Dakota.

Nigger, or *Winter King:* soft winter wheat; awned; grown mainly in Ohio.

Nittany, or *Penn. No. 44:* soft winter wheat; awned; grown in Maryland and Pennsylvania.

Pawnee: hard winter wheat; awned; highly resistant to LOOSE SMUT OF WHEAT, moderately resistant to LEAF RUST, BUNT, and HESSIAN FLY; grown widely in the Midwest.

Pilot: hard spring wheat; awned; resistant to STEM RUST and some races of LEAF RUST, BUNT, and POWDERY MILDEW; grown in the Dakotas, Montana, and Wyoming.

Premier: very similar to *Mida;* grown in North Dakota.

Rio: differs from *Turkey* only in having slightly shorter stems and in being resistant to BUNT; grown in the Pacific Northwest.

Rival: hard spring wheat; awned; resistant to STEM RUST and some races of LEAF RUST and BUNT; grown in the Dakotas and Minnesota.

Rudy, or *Black Mediterranean:* soft winter wheat; awned; distinct in having long kernels and black stripes on the glumes; grown in Indiana, Illinois, and Ohio.

Tenmarq: hard winter wheat; awned; resistant to some races of LEAF RUST and STEM RUST; grown chiefly in Colorado, Kansas, Texas, and New Mexico.

Triumph, also called *Dane's Early Triumph, Early Dain,* and *Premium:* hard winter wheat; awned; very early maturing; grown largely in Kansas, Oklahoma, Texas, and Colorado.

Turkey, known under many other names, e.g., *Alberta Red, Argentine, Bulgarian, Crimean, Hundred-and-One, Malakof, Kharkof, Hungarian, Minnesota Reliable, Red Russian, Romanella, Taruanian, Theiss, Ulta,* and *Zuni:* hard winter wheat; awned; drought-resistant; widely grown in the United States, especially in the western half of the country.

Wasatch: hard winter wheat; awned; resistant to DWARF BUNT and (ordinary) BUNT; grown in Idaho, Utah, Montana, and Washington.

Westar: hard winter wheat; awned; resistant to some races of LEAF RUST; grown in Texas, Oklahoma, and New Mexico.

Wichita: hard winter wheat; awned; very early maturing; grown in Kansas, Colorado, Oklahoma, and Texas.

Yogo: semihard to hard winter wheat; awned; the most winter-hardy variety known in the United States; resistant to some races of BUNT; grown in Montana.

9. GLABROUS, BROWN GLUMES;
WHITE KERNELS

Requa: soft winter wheat; awned; grown largely in Ohio.

10. GLABROUS, BROWN GLUMES,
RED KERNELS

Austin: soft intermediate wheat; awned; resistant to STEM RUST, LOOSE SMUT OF WHEAT, and some races of LEAF RUST; grown in Texas and Oklahoma.

Goens, also known under many other names, e.g., *Baldwin, Cummings, Dunlap, Early Red, Early Ripe, Going, Hall, Owen, Red Chaff, Red Rudy,* or *Shelby Red Chaff:* soft winter wheat; awned; grown in Ohio, Indiana, Illinois, Missouri, and Kentucky.

Mediterranean, sometimes called *Acme, Farmers Trust, Key's Prolific, Lancaster Red, Lehigh, Miller, Missouri Bluestem,*

Mortgage Lifter, *Red Chaff*, *Red Sea*, *Red Top*, *Standby*, or *Swamp:* soft winter wheat; awned; grown largely in Texas, but widely distributed in the United States.

References: B.15; R.9.

See also TALL WHEATGRASS.

COMMON WHITE CLOVER.

See WHITE CLOVER (variety).

COMMON WILD OAT (*Avena fatua*), also called *wild oat*, differs from COMMON OAT in having usually three-flowered SPIKELETS. It occurs from Maine to Pennsylvania, in Missouri, and westward, and is a common WEED on the Pacific Coast.

Reference: H.26.

See also OATS; AVENA; RYE; PASTURE PLANTS; SCREENINGS.

COMMON WINTERFAT.

See WINTERFAT.

COMMON YARROW. *See* YARROWS.

COMPANA.

See TWO-ROWED BARLEY (variety).

COMPLETE FEEDS.

See COMMERCIAL MIXED FEEDS.

COMPLETE FERTILIZERS.

See FERTILIZERS.

COMPLEX SUGARS. *See* SUGARS.

COMPOSITE is a member of the Composite family (Compositae) which is characterized by its composite flowers, each one consisting of an assembly of many small flowers surrounded by, or even surmounting, BRACTS. To this family belong the SUNFLOWERS, ASTERS, and many weeds.

COMPOUND LEAF is a leaf composed of separate LEAFLETS.

CONCEALED DAMAGE IN SEED, or *internal seed decay*, is an early stage of seed decay that occurs in PEANUTS principally after harvest. Seeds with concealed damage may contain a variety of FUNGI, primarily *Diplodia theobromae;* others are species of *Fusarium, Aspergillus, Penicillium, Rhizopus,* and *Rhizoctonia.* Occasionally, *Sclerotium rolfsii* and *S. bataticola* are found.

Peanuts harvested during humid weather and stacked without sufficient wilting or while still damp from rain or dew, and those that have been windrowed, quickly become covered with a moldy fungus growth. The fungi pass through the pegs into the interior of the seed where they continue development and discolor the inner surface of the cotyledons before any damage is noticeable on the outer surface of the seed.

When peanuts are picked with a moisture content above 9 percent, mold growth continues in the bins. Concealed damage and complete decay of the seed may develop rapidly.

There is considerable difference among peanut varieties in regard to the occurrence of concealed damage. The latter is most serious in runner types, grown in the lower South. The more susceptible varieties, though, when grown in the northern part of the Peanut Belt and harvested during cooler weather of the late fall, do not show any appreciable amount of concealed damage.

Control: Late planting of the runner type peanuts in the southeastern states, so that they may be harvested and cured during cooler weather and after the tropical storm period, is important. Greater care in harvesting, curing, and stacking will also reduce losses from this condition.

Reference: B.10.

CONCENTRATED BUTTERMILK.

See MILK PRODUCTS.

CONCENTRATED CULTURED SKIMMED MILK. *See* MILK PRODUCTS.

CONCENTRATES, or *concentrate feeds*, contain large amounts of ENERGY in small bulk. Thus, they allow animals to consume large amounts of raw materials needed for production. Swine and poultry have limited intestinal capacity and therefore are more dependent on concentrates than ruminants. Common concentrates are all grain crops, cottonseed meal, linseed oil, meal, gluten feeds, tankage, etc. They can be divided into several groups:

1. **Farm grains.**
2. **By-product feeds** purchased in bulk.
3. **Special mixtures,** such as

a) *Feeding concentrates*, or *supplement feeds*—these are added in small quantities to home-grown grains to increase the amount of proteins, minerals, and vitamins of the ration. These concentrates are available in a great variety of formulations as *hog supplements, dairy concentrates, poultry concentrates*, etc.

b) *Mixing concentrates*—primarily for formulated poultry mash, also used occasionally as base for hog feeds. They are often used in large quantities to balance rations prepared from locally grown grains. —*See also* ALFALFA PELLETS; GRAZING; RANGE MANAGEMENT.

CONDENSED BEET SOLUBLE PRODUCT (tentatively accepted as feedstuff) is made (after partial removal of glutamic acid) from the condensed filtrate that is obtained when recovering sugar from beet molasses.

Reference: F.6.

See also MISCELLANEOUS PRODUCTS; PLANT BY-PRODUCTS.

CONDENSED BUTTERMILK.
See MILK PRODUCTS.

CONDENSED CULTURED SKIMMED MILK. *See* MILK PRODUCTS.

CONDENSED CULTURED WHEY and *condensed cultured whey solids* are MILK PRODUCTS.

CONDENSED FERMENTATION SOLUBLES. *See* VITAMINS.

CONDENSED FISH SOLUBLES.
See MARINE PRODUCTS.

CONDENSED HYDROLYZED WHEY.
See MILK PRODUCTS.

CONDENSED SKIMMED MILK.
See MILK PRODUCTS.

CONDENSED WHEY FERMENTATION SOLUBLES. *See* VITAMINS.

CONDENSED WHEY-PRODUCT and *condensed whey-product solids* are MILK PRODUCTS.

CONDENSED WHEY SOLUBLES.
See MILK PRODUCTS.

CONDENSED WHOLE WHEY.
See MILK PRODUCTS.

CONDIMENT is a spicy vegetable product used to flavor feed; e.g., CARAWAY SEED; FENNEL; GINGER.—*See also* FEED INGREDIENTS.

CONGO COWPEA.
See COWPEA (variety).

CONIUM. *C. maculatum*
= POISONHEMLOCK.

CONTARINIA. *C. sorghicola*
= SORGHUM MIDGE.

CONTOUR FURROWS are plowed at the level and at right angles to the slope, so as to retain runoff water.

CONVOLVULUS. *C. arvensis*
= FIELD BINDWEED.

CONWAY. *See* RICE (variety).

COOKED BEANS. *See* FIELD BEANS.

COOKED BONE MEAL.
See MINERAL FEEDS.

COOKING FEED, or *steaming feed*, often decreases the digestibility of proteins and destroys vitamins. Feeds that may be cooked advantageously are FIELD BEANS, SOYBEANS, and POTATOES to be fed to swine or poultry. Each of these feeds contains an undesirable compound that is destroyed by heat.

COPPER must be present in the animal body for the conversion of *iron* into the *hemoglobin* of blood. Deficiency of copper in pasture plants and other feedstuffs may cause various diseases in farm animals.

Copper-deficient *cattle* often scour excessively. The hair coats of black and red animals fade, especially around the eyes. Copper-deficient *sheep* also show a fading of color. An early symptom of deficiency is "steely wool," or wool with no crimp.

Excess *molybdenum* in the feed will make copper unavailable and produce signs of copper deficiency.

Copper is one of the TRACE ELEMENTS and may be supplied to feedstuffs in small amounts in form of COPPER CARBONATE, COPPER HYDROXIDE, or COPPER SULFATE.— *See also* MINERAL FEEDS.

COPPER ACETATE, or *cupric acetate*, is a poisonous, greenish-blue, water-soluble powder with 31.8 percent COPPER content. —*See also* PARIS GREEN.

COPPERAS = IRON SULFATE.

COPPER CARBONATE is more correctly called *basic copper carbonate* or *basic cupric carbonate*. It forms a dark green or blue powder and is used to supply the TRACE ELEMENT copper to animals. Copper carbonate dust for SEED TREATMENT comes in two grades, one containing about 50 percent and the other 18 to 25 percent COPPER. The higher grade, which is to be preferred, should be applied at the rate of 2 to 3 oz. per bushel of SORGHUM; the lower grade at twice the rate. The dust should be thoroughly mixed with the seed in a special treater, or the mixing may be done in an old milk can, churn, or other dust-tight container. Treated seed can be stored indefinitely. Meanwhile, the dust protects the seed from insects and rodents.

Copper carbonate dust controls KERNEL SMUTS and also improves emergence and stand of sorghums by combating harmful FUNGI in the soil.

Caution: Like most copper salts, copper carbonate is poisonous and may cause extreme nausea and vomiting if inhaled or swallowed. Therefore, an effective dust mask should be worn over the nose and mouth when applying the dust or when handling treated seed, even when the work is done in the open air. Treated seed should not be used for feed or mixed with untreated seed that is to be so used. Sacks that have contained treated seed should be thoroughly cleaned before being used for grain that is to be fed.

Reference: L.1.

See also MINERAL FEEDS.

COPPER HYDROXIDE, also called *copper hydrate* or *cupric hydroxide*, is a light-blue powder and insoluble in water. It contains 65.1 percent COPPER and is used in rations as a source for this TRACE ELEMENT.—*See also* MINERAL FEEDS.

COPPER META-ARSENITE is a greenish powder, insoluble in water and very poisonous.—*See also* PARIS GREEN.

COPPER SUBSULFATE, better known as *basic copper sulfate* or *basic cupric sulfate*, is a blue-green powder containing about 50 percent COPPER; it is a FUNGICIDE which is sold under various trade-names. It should be applied in the same manner as COPPER CARBONATE, and all other statements made about the use of that preparation apply to it.

Caution: Copper subsulfate is poisonous and must be handled with care, for instance in the SEED TREATMENT of SORGHUMS.

Reference: L.1.

COPPER SULFATE, or *curpic sulfate*, is commonly called *bluestone*. It is commercially available in form of water-soluble blue crystals, granules, or powder.

TADPOLE SHRIMPS in *rice* fields can be controlled by applying granular copper sulfate from an airplane at the rate of 10 lb. per acre. A spray of a 5-percent copper sulfate solution is recommended for the eradication of DODDERS.

Copper sulfate is also used to supply rations with COPPER, one of the essential TRACE ELEMENTS. In proper dosage, it possesses anthelmintic (worm-expelling) action against a number of widely distributed intestinal parasites.—*See also* MINERAL FEEDS.

COPPERWEED (*Oxytenia acerosa*) is a plant containing an as yet unknown poison; its leaves, when eaten by cattle and sheep in the fall, often cause poisoning; its characteristic effects are vomiting, frothing, and weakness.—*See also* POISONOUS PLANTS.

COPRA OIL MEAL
                                = COCONUT OIL MEAL.

CORDOVA. *See* SIX-ROWED BARLEY (variety).

CORM is an enlarged or "swollen," rounded, and fleshy (mostly subterranean) stem base—like a BULB in shape and appearance, except that it is solid.

CORN (*Zea mays*), more correctly called *Indian corn* or *maize*, is one of the coarse annual GRASSES and belongs to the true cereal crops. It is by far the most important cultivated crop grown in the United States and is consumed largely on the farms where it is grown, furnishing the basic concentrate feeds for much of the American livestock.

The great variety of vegetative types of corn makes it possible for this crop to be grown under such widely diverse conditions as can be found in an area extending from central Canada to the tropical regions of the southern hemisphere, as well as from sea level to a 10,000 ft. altitude.

The corn stalk ranges from 2 to 25 ft. in length and from $\frac{1}{2}$ to 2 in. in diameter. At the base of the alternately grooved internodes of the plant develops a bud, which later produces an ear shoot; buds below the soil surface produce the tillers which must not be injured because they contribute greatly to grain formation and good yield. The roots spread in all directions approximately $3\frac{1}{2}$ ft. from the stalk, but may penetrate 5 to 9 ft. deep. The leaf is large and consists of the broad blade, the overlapping sheath which clasps the stalk, and the collarlike LIGULE; hairs are scattered over the upper surface of the leaf.

Corn is a MONOECIOUS plant, i.e., its flowers are borne on the same stalk—the female flowers (which, when mature, become the *ear*) in the axils of the leaves and the male flowers in the *tassel* at the top of the stalk. Cross-pollination occurs by simple pollen shedding. The tassel contains numerous SPIKELETS which produce millions of pollen grains to fertilize the hundreds of long styles (*silks*) of each ear. The ear is a SPIKE with a thickened and semi-woody axis (known as *cob*). The *kernels* are the pistillate (female) SPIKELETS arranged in 8 to 16 rows or even as many as 30. Each row contains 20 to 70 kernels—a good ear yields about 600 kernels, but a large ear may contain as many as 1,000.

Corn grows best on fertile, well-drained, medium loams. In addition to the prairie soils, well-drained bottom or second-bottom lands (if not too sandy), as well as the loess soils are admirably suited to corn production.

Crop rotation was once considered essential for corn growing. But now many farmers are trying to grow corn on the same field for a continuous number of years. They expect to return corn stalks to maintain organic matter, replace removed nutrients with commercial fertilizers, and control insects and diseases with recently developed pesticides. It is too early at present to evaluate this new approach but there are indications that new hazards are developing. Corn requires lots of water and when rainfall is limited, the yields of fields in continuous corn may be reduced more than those of the fields on which a standard rotation has been maintained.

The chief rotations in the Corn Belt and surrounding regions are a 2-year rotation of corn and SMALL GRAIN, a 3-year rotation of corn, small grain, and CLOVER; and a 4-year rotation of corn, OATS, WHEAT, and clover. The small grain may be oats, wheat, or BARLEY; RED CLOVER or SWEET-CLOVER also may be used, depending upon the locality. The 2-year rotation of corn and small grain can be improved materially by growing sweetclover with the oats, to

be turned under the following spring as a green-manure crop.

Probably no crop is more generally beneficial in preparation for corn than ALFALFA. Where this can be grown profitably, it should by all means be used in the rotation.

In much of the South corn ranks second only to COTTON. On some of the rich alluvial soils two-thirds or more of the land is used for cotton, with corn as the only other major crop. The planting of SOYBEANS in the corn and a follow-up with RYE, WINTER PEAS, or some other crop to be turned under early in the spring, will help maintain soil fertility.

The growing and plowing under of green manure will increase the productivity of most soils in any area. In the North, an increase in the percentage of LEGUMES for hay or pasture and a better conservation of crop by-products and manure will frequently be sufficient to maintain organic matter in moderately productive soils. In the southern states, winter-grown green-manuring crops serve also as a cover crop and check winter erosion and the loss of plant food.

The liberal application of well-rotted manure adds both needed nutrients, especially nitrogen and potash, and organic matter, thereby increasing the fertility and improving the physical condition of the soil. Well-rotted manure may be applied to the land for corn harvesting in the fall or spring, unless the soil is quite sandy; on sandy soils manure should be applied in the spring to avoid undue loss from leaching. There is little difference in the effect of well-rotted manure, either plowed under or applied as a top dressing after plowing.

Nitrogen fertilizers promote quick and vigorous growth of stalks and leaves. The nitrogen supply of the soil generally can be maintained more economically by turning under legumes such as COWPEAS, clover, alfalfa, and the like. Nitrogen fertilizers, such as sodium nitrate, ammonium nitrate, or anhydrous ammonia, may be used advantageously to supplement this supply.

Phosphatic fertilizers tend particularly to increase the yield of grain. Superphosphate is the most practical form in which to apply phosphorous for immediate utilization. This may be applied after the ground is plowed and ready for planting.

Corn. Ear (pistillate inflorescence); two branches of tassel (staminate inflorescence) and staminate spikelet; pistillate spikelets, one soon after flowering, and a pair attached to cob (rachis). (H.26.)

Potash fertilizers contribute generally to the health of the plants and the quality of the grain. Both potassium chloride and potassium sulfate are good sources.

Except where corn is to be listed, the first step in seedbed preparation is plowing. Land that has been in a cultivated crop—corn, cotton, small grain, soybeans, POTATOES, etc.—can be plowed about equally well in the fall or spring, so far as the yield of the following corn crop is concerned. Fall plowing in the southern states is undesirable because it promotes soil washing and because vegetation turned under in the fall may decay so rapidly as to be of little value to the following corn crop. Fall plowing also should be avoided on soils subject to blowing.

In the Corn Belt, it usually is necessary that at least part of the cultivated acreage be fall-plowed if the crop is to go in on time in the spring. In general, the fields to be fall-plowed should be those that have been in alfalfa, clover, or pasture.

Plowing on all soils should be deep enough to turn all surface growth under to a point where it will remain moist and start to decay. Plowing less than 5 in. deep usually is insufficient to turn under surface growth and trash properly. Plowing more than 7 to 8 in. deep, or subsoiling, usually is wasted energy. Between these limits, plowing should be somewhat deeper on heavier soils, when heavy vegetation or sod is turned under, and when done in the fall. Conversely, it may well be at the lower limits in the spring when plowing lighter soils that have been in cultivated crops.

The final steps in preparing land for corn consist in obtaining a surface layer of some 2 in. of reasonable fine soil in which to plant, and, of even more importance, putting the entire 5 to 8 in. of soil that have been turned by plowing in the best condition to promote full development of the plants throughout their growth. This preparation should precede planting as shortly as possible to avoid both excessive drying and the chance of rain ruining the seedbed.

Land that has been plowed in the fall can frequently be put in excellent condition by simply double disking and harrowing. Spring-plowed land sometimes can be fitted satisfactorily by similar treatment. Finishing the preparation with a plank drag leaves the soil surface smooth and even for planting.

Most corn nowadays is drilled because more seed is planted on an acre. Weeds are controlled with herbicides.

NOTE: With tractor operations it takes too long to get off at the end of the field and move the check wire. Checking is now obsolete.

On many of the more poorly drained soils of the South it is customary to plow the land into beds the width of a corn row and to plant in these beds, or "lists." Toward the western part of the Corn Belt and in the Great Plains listing without previous plowing is a common practice.

Corn planting begins in southern Texas, usually before February 1, and progresses northward through the eastern two thirds of the United States at an average rate of about 13 miles a day. Planting is relatively early in the central part of this area, and is delayed toward the west by higher elevations and toward the east by heavier, colder, and wetter soils. The spread between start and finish of the planting is smallest in the extreme north and increases toward the south.

Corn should be planted deep enough to put the seed in moist soil where it will not be in danger of drying out. In soils that tend to be cold, 1 or 1½ in. is deep enough. On sandy soils, it may be necessary to plant 3 in. deep, or even more.

Weeds can be killed most easily when the field is being prepared for planting and before the corn is up or while it is still small enough to stand harrowing without injury. At such times the land can be gone over quickly with a harrow, weeder, or rotary how. Harrowing from just before the corn seedlings emerge to just after should be avoided unless it is essential. Later cultivation can be made satisfactorily with any of the usual cultivators. Corn should be cultivated often enough to control weeds. Shallow cultivation is less likely to injure the corn roots and does not bring many weed seeds to the surface.

One of the objects of listing corn is to obtain deeper rooting of the plants. This can be accomplished only if the furrows are kept open until the crown roots are established. A lister-cultivator may be used in the first cultivation of listed corn; its disks are set to cultivate the sides of the furrows without pulling the soil down into them. Subsequent cultivations may be made with the same implement, or with any of the ordinary cultivators.

Most of the corn acreage is harvested for grain, a small part is hogged down, grazed, or cut for fodder or silage. Corn usually is husked from the standing stalk by hand or with a mechanical corn picker; in the South corn is often snapped from the standing stalk with the husk left on; in other parts of the country, corn is sometimes cut and *shocked* before being husked mechanically or by hand.

The grain of freshly gathered corn (except the moisture-rich sweet corn) contains 16 to 30 percent moisture; before shelling, the ears are stored in cribs until the moisture content is reduced to less than 14 percent; then the *shelled* grain can be stored safely in bins. Care must be taken to avoid spoilage; sufficient air circulation in the cribs or artificial drying may be necessary to avoid losses.

Soft corn is a term which refers to moisture-rich, immature grain killed by freezing; it has a high water content. After drying, it looks chaffy or shrunken, but has almost the same feeding value as sound corn. The frozen immature corn can be made into satisfactory "ear corn silage."

Shock corn, or *bundle corn*, is a term referring to corn fed without husking; such corn is usually called *"corn fodder."*

Corn stover, also called *corn stalks*, is cured shock corn which remains after removal of the ears from shock corn. It consists mainly of leaves and stalks. Corn stover may be put through a shredder which reduces the stems and allows more feed to be stored in a given area.

Corn silage is very nutritious—it consists of corn plants cut before their leaves fall off. In areas where corn is well adapted, farmers can produce more total livestock feed from an acre of corn made into silage than they can by any other use of the land.

Corn fields are sometimes hogged down or they are grazed by sheep, or by cattle; hogs should follow cattle in the pasture and thus pick up grain the cattle may waste.

Seed corn must be carefully dried and stored to retain its viability.

While most of the corn crop is used—in this order—for hog feed, cattle feed, poultry feed, horse and mule feed, and sheep feed, a part of the harvest is used industrially for the production of alcohol, and for food—*corn meal, hominy grits, breakfast cereals, popcorn, corn sirup, corn sugar, starch, corn oil,* etc. Sweet corn is widely used for *canning* and *freezing* purposes.

There is considerable variation in the composition of corn grown in the United States. The average kernel contains approximately 61.0 percent starch, 2.0 percent fiber, 7.4 percent other carbohydrates (sugars, etc.), 9.0 percent protein, 4.0 percent oil, 1.2 percent ash, 15.0 percent water, and 0.4 percent other substances; it is rich in VITAMIN A, but relatively poor in B-complex vitamins.

Corn must be cleaned before it can be milled. In the milling process, the bran (hull), germ, and flour are sifted from the degermed grain which is used for hominy or ground into grits or corn meal. *Corn by-products* rank next to wheat by-products as the most important CEREAL GRAIN BY-PRODUCTS of this country.

The following are the officially recognized feed products obtained from corn:

Corn meal—the finely ground, unbolted corn.

Corn bran—the outer coating of the kernel, with little or none of the starchy part or germ.

Corn feed meal—the fine siftings obtained in the manufacture of screened corn chop, screened ground corn, or screened cracked corn.

Corn chop, ground corn, or *cracked corn*—the entire product made by chopping, grinding, or cutting the grain; it may be fine, medium, or coarse and must contain not more than 4 percent foreign material.

Screened corn chop, screened ground corn, or *screened cracked corn*—the coarse portion of corn chop, ground corn, or cracked corn, from which most of the fine particles have been removed; it must contain not more than 4 percent foreign material.

Corn grits, also called *hominy grits*—consisting of the hard flinty portions of sound corn and containing little or none of the bran or germ.

Ear corn chops, or *corn-and-cob meal,* is corn and cob chopped, or run through a hammer mill, without the husk and with no greater proportion of cob than occurs in the ear corn in its natural state.

Ear corn chops with husks is corn, cob, and husk chopped, with no greater proportion of cob or husks than occurs in the ear corn in its natural state.

Flaked corn is cracked corn that has been steamed and run over smooth flaking rolls and subsequently dried and cooled.

Toasted corn flakes is the product obtained by running cracked corn—which has been properly tempered—over smooth flaking rolls; subsequently it is dried, cooled, and toasted.

Corn gluten feed—the part of shelled corn that remains after the extraction of most of the starch and germ in the wet milling manufacture of corn starch or sirup.

Gluten feed is used in livestock feeding, especially for dairy cattle; it also serves as a protein supplement in the fattening of beef cattle and sheep, but is usually not equal to the oil meals. It is not used extensively in rations of swine because, like corn, it is deficient in lysine; however, small amounts may be included with other supplements which contain proteins of high biological value, e.g., fish meal.

Corn gluten meal—the part of shelled corn that remains after the extraction of most of the starch and germ, and the separation of bran in the wet milling manufacture of corn starch or sirup.

Gluten meal is fed under much the same conditions as gluten feed and is considered somewhat more valuable in keeping with its higher protein content. *Yellow corn gluten meal* is valued as a constituent of the diet of growing chickens because it provides vitamin A and gives the desirable yellow color to the shanks and skin of the chickens.

Maltose processed corn—the dried residue of cleaned whole ground corn, obtained after the removal of starch in the manufacture of malt sirup.

Maltose process corn gluten feed—the dried residue from degermed corn, obtained after removal of starch in the manufacture of malt sirup.

Hominy feed—a mixture of corn bran, corn germ, and part of the starchy portion of a kernel, as produced in the manufacture of pearl hominy, hominy grits, or table meal; it must contain not less than 5 percent crude fat. If prefixed with the words "white" or "yellow," the product must correspond thereto.

Hominy feed is generally used as a replacement for part or all of the corn in the rations of livestock; it can also replace grain in poultry rations.

Corn oil cake and *corn oil flakes* consist of the corn germ from which most of the oil has been removed. The designating name for each product (obtained in the wet milling process of manufacturing corn starch, corn sirup, and other corn products) must include a term descriptive of the manufacturing process, such as: *Hydraulic, expeller,* or *solvent extracted.*

Corn oil meal is ground oil cake or corn oil flakes; its name must include a descriptive term (hydraulic, expeller, or solvent extracted) to indicate the method of manufacture. It is a satisfactory protein supplement.

Corn germ cake consists of corn germ with other parts of the corn kernel from which part of the oil has been pressed. It is obtained in the dry milling process of manufacturing corn meal, corn grits, hominy feed, and other corn products.

Corn germ meal is ground corn germ cake.

Corn screenings consists of small light grains of corn, parts of grain of corn and other cereals, and other materials with feeding value; this product is obtained by screening shelled corn, excluding sand, dirt, and similar inert materials.

NOTE: The following definitions of corn products are as yet not official: *Dried corn sirup*—the product resulting from spray-drying or roll-drying regular conversion corn sirup. It shall have a dextrose equivalent of not less than 42 percent.

Hydrolized corn protein—the product resulting from complete hydrolysis of isolated corn gluten, after part of the glutamic acid has been removed.

Dangers: Corn is subject to many diseases; some of these are largely local in their distribution or importance, whereas others are most widespread. CORN SMUT probably is the most common disease of corn in the United States, followed by HEAD SMUT, CHARCOAL ROT and the many other ROT diseases of the roots, stalks, and ears. Corn is also attacked by WHEAT SCAB.

Among the leaf diseases are LEAF BLIGHT, BACTERIAL LEAF BLIGHT, SOUTHERN CORN LEAF BLIGHT, STEWART'S WILT, ZONATE LEAF SPOT, GRAY LEAF SPOT, BACTERIAL SPOT, HELMINTHOSPORIUM LEAF SPOT, and NORTHERN CORN LEAF BLIGHT.

More than 350 different species of insect pests are known to attack corn; the most destructive group includes the CORN EARWORM, CHINCH BUG, WHEATSTEM SAWFLY, GREENBUG, GARDEN WEBWORM, CORN ROOTWORM, CORN ROOT APHID, CORN LEAF APHID, EUROPEAN CORN BORER, SOUTHWESTERN CORN BORER, CORN FLEA BEETLE, SOUTHERN CORN ROOT WORM, SUGARCANE BORER, GRASSHOPPERS, ARMY CUTWORM and other CUTWORMS, BILLBUGS, FALL ARMYWORM and other ARMYWORMS, GRAIN WEEVILS, WHITE GRUBS, WIREWORMS, STINK BUGS, and many other kinds of beetles, bugs, and borers.

Varieties: More than a thousand varieties of corn have been grown in the United States, but their names are of little interest to farmers, particularly since most corn is hybrid and local conditions greatly influence the performance of the very same varieties, even in neighboring areas.

Hybrids are just as definitely adapted to various localities as are varieties, and hybrid seed corn should be used only when it is known that the particular hybrid is adapted and productive in the particular locality.

Possibly the simplest way to give a general idea of what hybrid corn is, is to compare it to the mule. A mule is the first-generation hybrid between a mare and an ass, and partakes of the better qualities of both parents; it does not reproduce, but must be produced anew each generation, not for reproduction but for its own value. A corn hybrid is the first-generation hybrid between two strains of corn, and it, too, must be produced anew each generation, not for reproduction but for its own value. During that generation good hybrids produce larger yields of high-quality corn per acre than do the best commercial varieties.

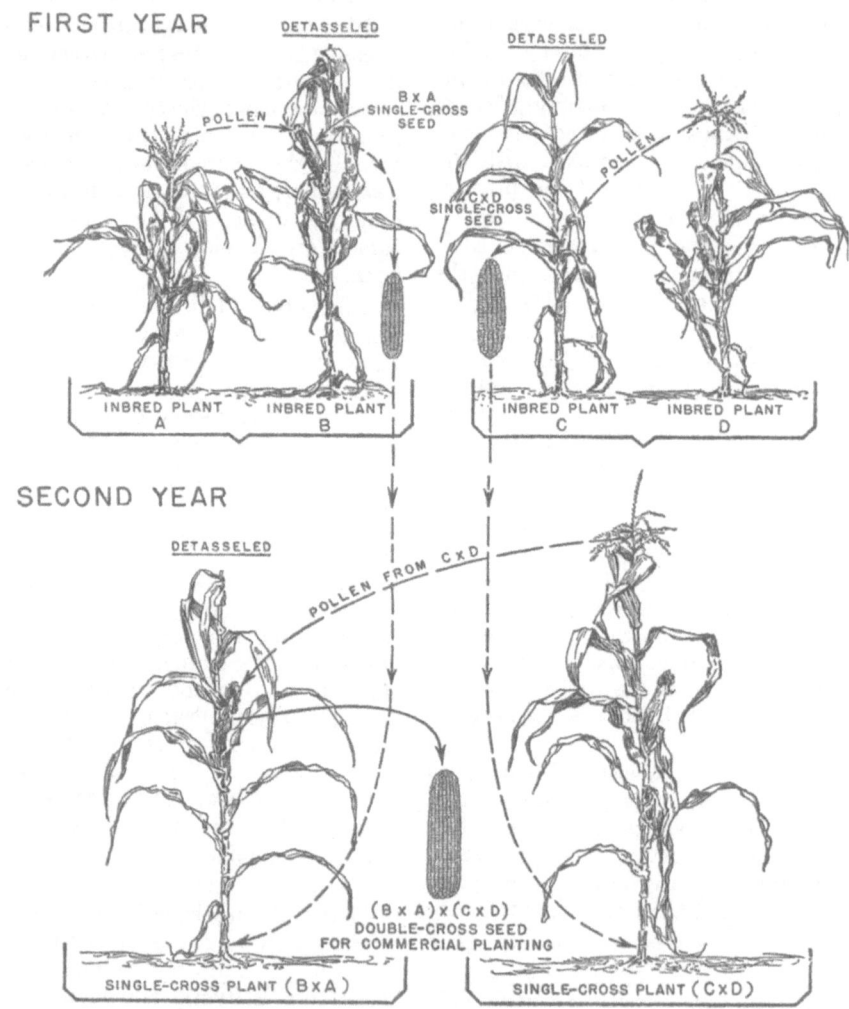

Corn. Diagram of method of crossing inbred plants and the resulting crosses to produce double-cross hybrid seed. (R.8.)

NOTE: These are steps for hybrid seed corn production:

1. Select a number of prospective parents.
2. Inbreed by placing a bag on the shoot to prevent other plants from pollinating; place another bag over the tassel to collect pollen which is then transferred to the shoot. This is repeated for several generations.
3. Select the inbred corns that have the desired characteristics and re-produce.
4. Cross two inbred lines to get a single-cross hybrid. Several rows of one line are planted next to the second line. The female plants are detasseled to prevent them from shedding pollen.
5. Cross two single-cross hybrids to get seed for planting.

At present several hundred different corn hybrids are being grown in this country. They are produced from a large number of outstanding *inbred lines*, e.g., those designated as *38-11, A, C 11, CC 1, Cornell 11, Hy, Kys, L 317, Os 420, Pr, R 4, Tr,* and *WF 9.*

Probably the best known corn hybrid is *U.S. 13,* a double cross produced from four inbred lines, namely (WF 9 × 38-11) × (Hy × L 317). Other leading hybrids are *Colorado 152, Indiana 844, Iowa 939, Indiana 909A, Kentucky 103, Missouri 8, Ohio C 38, Oklahoma 311, Tennessee 101, Texas 8,* and *Wisconsin 416.*

The species Indian corn, *Zea mays,* consists of seven types, each of which includes numerous varieties:

1. DENT CORN

This is the most important corn-type grown in the United States, especially in the Corn Belt. The grain shows a wide range of colors, with yellow or white predominating; the kernel forms a dent or wrinkle when it dries.

Among the best-known varieties of dent corn are *Boone County White, Clarage* (yellow), *Leaming* (yellow), *Northwestern Dent* (red-kernelled), *Reid Yellow Dent, Silver King* (white), various southern *Prolific* varieties (white), and *St. Charles White.*

2. FLINT CORN

The kernels are rounded at the tip and mature early. If grown in northern areas, the ear has 8 rows of kernels; in the South it may have 12 or 14 rows. This type includes the varieties *King Philipp Flint* (yellow), and *Longfellow Flint* (yellow).

3. FLOUR CORN

The kernels may have any color, but white, blue, and variegated predominate. The blue and variegated types are called *squaw corn,* or *native corn;* they are much richer in niacin content than ordinary white flour corn.

4. POD CORN

Each kernel is enclosed in a pod (husk). This type is of no commercial importance.

5. POPCORN

The small, very hard kernels of this type differ in color and size; when the kernels are pointed, the varieties are called *"rice";* when rounded, they are known as *pearl.*

Among its leading varieties are *South American* and *Supergold* (or *Yellow Pearl*). From inbreds developed from them, many hybrids have been produced; e.g., *K 4* and *Purdue 31.*

6. SWEET CORN

The kernels are nearly transparent and wrinkled when mature. Because they contain some sugar before ripening, these kernels taste sweeter than those of other types. Popular are such varieties as *Stowell's Evergreen* (white), *Golden Bantam* (yellow), *Country Gentleman* (white), and *Golden Sunshine* (yellow). Leading hybrids are *Golden Cross Bantam* (yellow), *Spancross* (yellow), and *Marcross* (yellow).

7. WAXY CORN

This plant is grown for its special starch which consists of AMYLOPECTIN only, while the starch of other plants also contains AMYLOSE.

NOTE: Three grain SORGHUMS are named "corns":

Jerusalem rice corn, also called *Mexican desert wheat corn;* this is a shallu.
Rice corn, or *Jerusalem corn,* is *durra.*
White Egyptian corn, or *White durra,* is a durra selection.

References: R.8; F.6; M.32; L.8; B.12; B.13; B.14; U.3; U.7; U.8; E.11; W.22; P.14; D.10; E.12.

See also VELVETBEANS; FIELD BEANS; BUCKWHEATS; CROTALARIAS; LESPEDEZAS; COMMON LESPEDEZA; SPELT; KUDZU; PROSO; FLORIDA BEGGARWEED; FIELD PEA; ROUGH-PEA; PEANUT; CRIMSON CLOVER; ALSIKE CLOVER; CLUSTER CLOVER; TEOSINTE; DODDERS; BY-PRODUCT FEEDSTUFFS; YEAST DRIED GRAIN OR VINEGAR DRIED GRAIN; DISTILLERS' PRODUCTS; SILAGE; SILAGE CROPS; WHITE-FRINGED BEETLES; ROOT KNOT; STEM NEMATODE; VITAMIN E; PETERSEN'S RELATIVE FEED VALUES; SOAKING FEED; SHREDDING; SCREENINGS; GRAZING; PASTURE MANAGEMENT; SEED TREATMENT; ARASAN.

CORN-AND-COB MEAL
= *ear corn chops. See* CORN.

CORN BORER. *See* SOUTHWESTERN CORN BORER; EUROPEAN CORN BORER.

CORN BRAN. *See* CORN.

CORN CHOP. *See* CORN.

CORN COBS, when ground finely, can be used as roughage for ruminants. Some states permit the sale of corn cobs in mixed feeds, if the package is labelled "For ruminants only."

Corn cobs are useful for bedding and for absorbing MOLASSES before incorporating it in dry feed.

CORNCOCKLE (*Agrostemma githago*) is a tall branching plant with purplish-red flowers. It is a WEED very often found in WHEAT and other grain fields.

CORN DISTILLERS' DRIED GRAINS
and *corn distillers' dried grain with solubles* are DISTILLERS' PRODUCTS.

CORN DISTILLERS' DRIED SOLUBLES. *See* DISTILLERS' PRODUCTS.

CORN EARWORM (*Heliothis armigera*) is the most destructive insect enemy of ear CORN in the United States. When it feeds on COTTON it is called *bollworm*. Under the name "*tomato fruitworm*" it is a very destructive enemy of early tomatoes.

Corn earworm occurs throughout this country wherever corn is grown, but it is most destructive in the southern states where the breeding season is longer. Sweet corn is especially attractive to the earworm, which in a single year has caused an estimated loss of $75 million to the corn crop of the United States.

In addition to cotton and tomatoes, it attacks tobacco, SOYBEANS, grain SORGHUMS, VETCHES, PEANUT foliage, ALFALFAS, COWPEA, SUNFLOWER, many vegetables, and other plants.

The damage to corn is caused entirely by the worms, or larvae. In early plantings they feed on the tender, unfolding leaves; when these injured leaves unfold, they present a ragged appearance, which is often called "*ragworm injury*," or "*budworm injury*." When tassels appear, the worms immediately attack them, too. Corn silk is an attractive food for the larvae only while it is fresh; after it dries out, they feed upon the developing kernels. The principal injury, however, occurs when the worms reach the ears, where many of the kernels are destroyed while still soft.

The indirect injury caused by the earworm is sometimes as important as the direct loss, since other insects, such as the GRAIN BEETLES and WEEVILS, and MOLDS gain admission to the damaged ears. Ears severely injured by molds are unsafe for feed, especially for horses.

The egg of the corn earworm is about half the size of the head of a pin. It is shaped like a flattened ball and is ribbed. When first laid it is a light yellow but it soon darkens and when ready to hatch, is a dusky brown. Hatching occurs from two to eight days after the egg is deposited, depending on the temperature.

The newly hatched larva is whitish with a black head, and very small. Growth is rapid, and the worm attains full size in thirteen to twenty-eight days after hatching. The increase in size is accomplished by molting: every two to five days the old, hard skin is cast off, and the worm expands greatly before the new skin has become hardened. Usually, five such molts occur during the process of growth.

When full-grown the earworm is about 1½ in. long and very robust. The coloration at this time varies greatly. Many of the larvae are marked with conspicuous stripes of varying shades of cream, yellow, brown, slate, and black. A few are without stripes and may be pink, green, cream, or yellow. The larvae stage is the only destructive stage of the insect.

The full-grown larva then leaves the ear and drops to the ground. It enters the soil and bores down 1 to 9 in., depending on soil and weather conditions. The larva then forms a cell, where it transforms into a light-brown pupa about ¾ in. long. While in the pupal stage it changes into a moth which emerges and makes its way to the surface.

In midsummer the period from the time the larva leaves the ear until the moth emerges may be as short as fourteen days, but it may be several months if the insect passes the winter in the pupal stage.

The moth is about ¾ in. long and has a wingspread of 1½ in. The coloration is dull, from a light olive-green to a rather dark reddish-brown. The moths become active early in the evening, feed on the nectar of various flowers, and then fly in search of suitable plants on which to lay their eggs. The females live about twelve days, and each may deposit 400 to 3,000 eggs, the average being about 1,000.

In hot weather the full life cycle can be completed in a month. There may be as many as seven generations a year in some of the southern areas. In the most northern part of the country, however, there is usually only one generation each year. Throughout the greater part of the Corn Belt there are three or four generations annually. In the East pupae do not survive much farther north than an imaginary line from central Virginia through St. Louis, Mo., to Topeka, Kans.; along the Pacific Coast, pupae survive at least as far north as southern Washington.

Time of planting affects the severity of earworm damage. In general, in the northeastern areas and North Central States corn planted the earliest is injured the least and that planted the latest is damaged the most. In the southern states early planted corn is more likely to be infested by the earworm.

Control: The most important factor in helping reduce earworm damage to corn is the earworm's habit of cannibalism:

Whenever two worms come into contact with each other, they fight until one or both are injured beyond recovery.

A tiny wasplike insect known as *Trichogramma minutum* is the most important egg parasite. Earworms feeding upon vetch and alfalfa are easy prey to several parasites, particularly the two-winged fly *Winthemia quadripustulata*. The most important predatory insect enemy of the earworm egg and small larva is a small blackish bug, *Orius insidiosus*.

Corn earworm. (U.13.)

At least 21 species of birds feed on the corn earworm. Most important are the redwinged blackbirds, the boat-tailed grackle, the English sparrow, and the downy woodpecker.

MOLES destroy many corn earworm pupae. During prolonged periods of wet weather large numbers of larvae and pupae die of certain diseases.

Usually it is not practical to use insecticides for the control of the corn earworm in field corn. However, protecting the ears by applying DDT sprays may be desirable in the case of valuable seed corn.

Reference: B.9.

See also ALFALFA CATERPILLAR.

CORNELL 11 is one of the inbred lines of CORN hybrids.

CORNELL 595.

                See COMMON WHEAT (variety).

CORN FEED MEAL. *See* CORN.

CORN FIELD. *See* CORN.

CORNFIELD ANT (*Lasius niger alienus americanus*) is a small brown ANT, which, in the fall, carries the CORN ROOT APHID eggs to its nest and cares for them; in spring, when the eggs hatch, the ant places the helpless aphids on the host plant. The aphids are cared for in the same way during the summer months. In return, the ant obtains for its work a sweetish fluid, the predigested sap of the corn or other plants, which is given off by the aphid.

Reference: D.10.

CORN FLEA BEETLE (*Chaetocnema pulicaria*) feeds on CORN leaves and so causes wounds into which it introduces the bacteria that are the cause of STEWART'S WILT. This shiny black beetle is not much larger than a flea and jumps like one when disturbed. The beetle usually picks up bacteria when it feeds on leaves of diseased plants. The bacteria may live over winter within the bodies of the adult beetles.

Control: The corn flea bettle can be readily controlled with DDT.

Reference: P.16.

CORN FODDER. *See* CORN.

CORN GERM CAKE. *See* CORN.

CORN GERM MEAL. *See* CORN.

CORN GLUTEN FEED. *See* CORN.

CORN GLUTEN MEAL. *See* CORN.

CORN GRITS. *See* CORN.

CORN HYBRIDS. *See* CORN.

CORN LEAF APHID (*Aphis maidis*) is an insect pest which attacks CORN—especially those lines which are susceptible to the EUROPEAN CORN BORER—and SORGHUMS. These APHIDS winter in the South and migrate north during spring and summer; they are difficult to control.

CORN LEAF BLIGHT. *See* NORTHERN CORN LEAF BLIGHT; SOUTHERN CORN LEAF BLIGHT.

CORN MEAL.

See CORN; BUCKWHEATS; COWPEA.

"CORN MEAL BY-PRODUCTS."

See LABELS.

CORN MOLASSES = *feeding corn sugar molasses. See* MOLASSES.

CORN OIL CAKE. *See* CORN.

CORN OIL FLAKE. *See* CORN.

CORN OIL MEAL.

See CORN; MOLASSES.

"CORN PRODUCT." *See* LABELS.

CORN ROOT APHID (*Anuraphis maidiradicis*) causes serious injury to growing CORN each year. It is distributed generally throughout the United States east of the Rocky Mountains, but it is an especially destructive pest in the Corn Belt and in southern Wisconsin. It also is injurious to COTTON.

The corn root aphid is a small, soft-bodied insect not larger than a pinhead, of a bluish-green color, but dusted with a fine whitish powder which makes it appear grayish-green. The aphids cluster on the corn roots and suck the plant juices. Infested plants are dwarfed, the leaves become brown or otherwise discolored, but even heavily infested plants are seldom killed outright. Infestations later in the season are seldom recognized, but ant hills and the common brown CORNFIELD ANTS found near corn plants often indicate the presence of the root aphids.

There are four forms of the corn root aphid—the males, the egg-laying females, and the winged and wingless females that bear living young. In the spring and during the summer only the winged and wingless females are to be found.

The true sexes—the males and the egg-laying females—occur only in the fall. These females lay pale yellowish-green eggs, which later turn to jet black. The eggs, which are kept by the cornfield ant and other species of ants in their nests over winter, begin to hatch usually the latter part of March; the young aphids are transferred by the attendant ants to the roots of convenient weeds. The young mature in about fifteen days. Members of this and the succeeding 15 or 16 generations that are born between spring and fall give birth to living young, which they produce without fertilization by the male.

Since each female gives birth to 40 or 50 young, and they in turn mature and produce young in six to eight days during the summer months, the corn root aphids increase to enormous numbers.

The first two or three generations live on the roots of weeds, but as soon as the corn sprouts, the ants transfer the aphids to the more succulent corn roots. After the second or third generation, a considerable number of the aphids may be winged, and many of these come out of the ground through the ant tunnels and fly away to a new field. If they chance to alight near an ant hill, they are seized immediately by the watchful ants and placed on a convenient root, where they will give rise to another infestation.

The males and the egg-laying females begin to appear about the first of October, and the eggs laid by these females are immediately stored by the attendant ants.

As cold weather approaches, the ants carry the eggs with their own young deeper into the soil, and usually, by the middle of November, (in the latitude of Illinois and Wisconsin) all will be found 8 in. or more below the surface, which is below the ordinary plow furrow. Similarly in summer, during periods of drought, the ant colonies may be found 8 to 12 in. below the surface.

Control: A rotation that avoids having two successive crops of corn on the same land not only prevents injury by the corn root aphid, but also controls other serious pests, especially the CORN ROOTWORM. In the Cotton Belt it is also important not to follow cotton with corn, or vice versa.

One of the most effective means of controlling the corn root aphid is by thorough stirring of the soil before planting; this procedure disturbs the ant colonies and scatters and kills the aphids so that the corn and cotton plants may make a substantial growth before the ant and aphid colonies become re-established.

Infested fields that are to be replanted to corn should—in the spring—be plowed to a depth of 6½ to 7 in.; they should be plowed after March 15 in the latitude of central Indiana and Illinois. Subsequently, they should have three or four diskings to a depth of 4 to 5 in. with a 16- or 20-in. disk.

Plowing in the fall before the ant colonies go below the plow line is sometimes useful; however, additional spring diskings are essential.

One of the greatest obstacles in the control of insects injurious to field crops is lack of community co-operation. It is important that every farmer in a community rotate his crops and cultivate his old cornfields to destroy the root aphid and the attendant ant colonies, whether he plans to replant them to corn or not.

Reference: D.10.

CORN ROOTWORM (*Diabrotica longicornis*), sometimes also referred to as the *northern corn rootworm*, is a close relative of the SOUTHERN CORN ROOTWORM; it attacks CORN to some extent in the Middle West and also is found from Maine to the Gulf of Mexico and westward to Minnesota, South Dakota, and New Mexico. —*See also* CORN ROOT APHID.

CORN SCREENINGS. *See* CORN.

CORN SILAGE.

See SILAGE CROPS; SILOS; CORN.

CORN SMUT is caused by the FUNGUS *Ustilago maydis*. It is one of the most widely distributed CORN diseases and occurs wherever corn is grown. At times it causes relatively heavy losses, especially in the warmer and somewhat drier corn-growing areas.

The smut galls may develop on any of the above-ground parts of the plant where young tissues are exposed. Infection takes

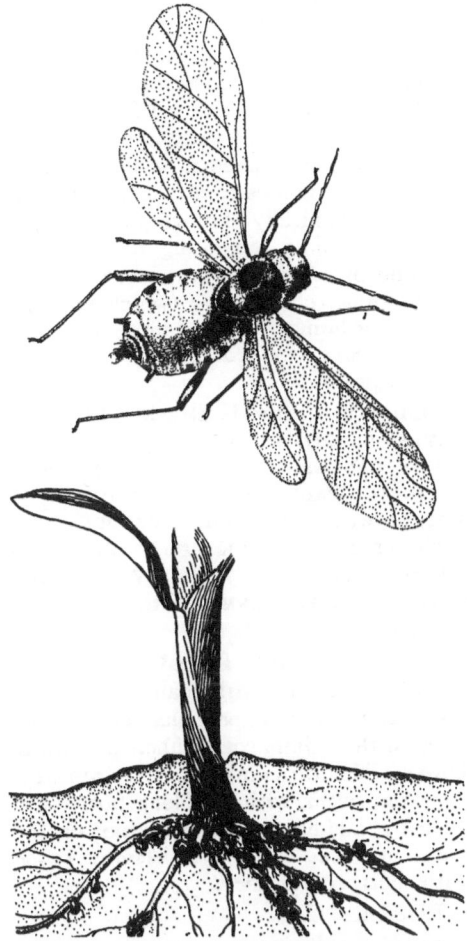

Corn root aphid. A winged aphid; wingless aphids attacking corn roots. (D.10.)

place only in young succulent tissue, unless entrance is gained through wounds. Young leaf tissue is usually the first to be attacked, next in order are tassels and ears. Stalks may be attacked through openings made by insects, implements, or other agencies. The galls formed are irregular in shape and may vary in size from ½ to 6 in. in diameter. At first the smut gall is white inside; later the closely packed fungus threads break up into black spores. The gall is covered with a white membrane until the spores are mature; then the membrane breaks and the spores are spread by the wind to the soil and to decaying plant refuse, where they overwinter. The following summer these spores produce smaller spores (sporidia), which are carried by air currents to the cornfields and infect the next crop.

Unlike other grain smuts, corn smut is not seed-borne and is not systemic; that is, the fungus does not invade the young seedling and does not develop unobserved inside the plant. Each smut gall is a separate local infection caused by spores carried to that part of the plant by air currents.

Control measures are limited to sanitation, crop rotation, and the development and use of resistant corn hybrids. The removal and burning of the smut may be practicable where infection is not too severe. A heavy stand of grain or some forage crop following corn tends to prevent the spread of the sporidia from the soil to nearby cornfields.

Reference: L.10.

CORN STALKS = *Corn stover. See* CORN.

CORN STOVER. *See* CORN.

CORN SUGAR. *See* DEXTROSE.

COROLLA is the inner part (often showy) of the floral envelope; it is composed of PETALS. The outer part is called CALYX.

CORYNEBACTERIUM. The BACTERIUM *C. insidiosum* causes BACTERIAL WILT, an important ALFALFA disease.

COSSACK ALFALFA.

See VARIEGATED ALFALFAS (variety).

COTTON is a term applied to various *Gossypium* spp. and to the downy substance growing in the capsules, or pods, of the cotton plant which is cultivated in large quantities in the southern part of the United States. Cotton is the most important fiber crop of this country.

While cotton is a perennial plant in the tropics, it is an annual in the United States. The plant grows 2 to 5 ft. high, has long taproots and a main stem with many branches and lobed leaves. Its flowers are yellow, purple, red, or white, and are followed by fruits which consist of seeds covered with fibers: they are called BOLLS.

The cotton is picked from the bolls and hauled to the *gin* where the fibers are separated from the seeds. The separated fibers are known as *lint*; the lint is packed in bales which are shipped to the spinner to be made into yarn.

On the other hand, a small percentage of the COTTONSEED produced is taken back to the farms where it is used for feed; the rest is shipped to the cottonseed oil mills. At the mills cottonseed is accumulated in large storage houses from which it is conveyed mechanically through the cleaning, delinting, hulling, separating, rolling, cooking, and pressing apparatus—according to need. Thus, *cottonseed oil* (which must be refined for human use) and its by-products (such as cottonseed cake, meal, hulls, and linters) are obtained.

Dangers: Cotton is attacked by ASCOCHYTA BLIGHT OF COTTON, ASCOCHYTA STEM CANKER, CHARCOAL ROT, RHIZOCTONIA STALK ROT, and COTTON ROOT ROT, as well as meadow NEMATODES, and such insects as STINK BUGS, CORN EARWORM, CORN ROOT APHID, VELVETBEAN CATERPILLAR, FALL ARMYWORM, and BROWN COTTEN BUG.

Species: In the United States four species of cotton—with many varieties—are of commercial importance; i.e., *upland cotton* with medium coarse fibers and white flowers which turn pink and red; *American-Egyptian cotton*, with fine, long fibers, and yellow petals having a purple spot on the base; and two species of *Asiatic cotton* with short, coarse fibers.

Reference: C.12.

See also OATS; CORN; LIGHTNING INJURY; BUR-CLOVERS;   LESPEDEZAS;   KUDZU;

VETCHES; ROUGHPEA; FIELD PEA; COWPEA; PEANUT; COMMON LESPEDEZA, WHITE-FRINGED BEETLES.

COTTON BUG.

See BROWN COTTON BUG.

COTTON ROOT ROT is caused by the

FUNGUS *Phymatotrichum omnivorum* which flourishes on more than 2,000 plant species, including COTTON, TALL FESCUE, and SERICEA lespedeza, as well as on vegetables, fruits, shrubs, and trees.

The first symptom on a plant is a yellowing or bronzing of the leaves, then a wilting of the leaves occurs which is rapidly followed by dying and browning of the foliage. By midseason and later, patches of brown, dead plants present a characteristic symptom of the disease.

Roots of infected plants are overrun with whitish or tan threads of the fungus mycelium, causing lesions which are destructive to the outer cells of the root. Then the fungus penetrates into the central parts of the plant.

Heavy rain and gullying of the soil may spread the fungus to new areas of infections.

Control: Rotation with nonsusceptible crops, in which cotton is grown one year in four, gives best control of the fungus, especially if supported by a rigid weed-control program, use of nitrogen fertilizer, early-fall plowing, and application of organic manure to the soil.

Reference: B.27.

COTTONSEED, obtained from COTTON, the great crop of the South, is sent through the cleaning, delinting, hulling, separating, cooking, and pressing apparatus of the oil mills.

Many cottonseed products, concentrates as well as roughages, are used as livestock feeds. All concentrates have the same general characteristics and qualities; their chemical composition depends mainly on the form of manufacture and the thoroughness of separating out the hulls.

Untreated cottonseed contains several compounds that are toxic, at least to nonruminants; one of these compounds is GOSSYPOL. This substance is made inactive by cooking the seed after the addition of water. Similarly, the cooking and ex-

pelling process to which cottonseed meal is subjected in order to remove the oil, is in large measure destructive to the toxic principle in the raw seed.

Cottonseed is a good source of protein, potash, and phosphorus, but is deficient in calcium and carotene (vitamin A). An adequate quantity of calcium is especially important to milking or nursing animals and to young stock, and may be supplied satisfactorily by LEGUME hays or by a mineral supplement. Some nutritional failures attributed to cottonseed meal have been the result of using it with poor-quality roughages, such as hulls or straw, and without carotene or vitamin-A supplement.

Cottonseed (uncrushed) was formerly used extensively as a feed for livestock. Feeding tests have indicated that 1 lb. of good-quality cottonseed meal is equal to 2 lb. cottonseed as a feed for fattening steers. Large rations of cottonseed tend to produce scours, but quantities up to 5 or 6 lb. give no trouble of this sort or only a very mild laxative action.

Cottonseed contains about 20 percent each of fat and crude protein; 1 ton cottonseed will yield approximately the following quantities of products: Linters or short fiber, 110 lb.; hulls, 514 lb.; cake or meal, 954 lb.; crude oil, 303 lb.; with dirt and loss in manufacture amounting to 119 lb.

Cottonseed hulls are the roughage product of cottonseed-oil manufacture. The hulls are removed from the cottonseed before the oil is extracted. They have a very low protein content and should be fed only in connection with protein-rich feeds. As a roughage the hulls have about the same energy value as has oat straw or corn stover but they contain less protein and calcium. They are valuable where no other roughage is available but can be hauled economically only as far as 50 miles. Cottonseed hulls are used extensively in the South, especially for steer feeding. They can also be used for fattening lambs or cattle, if carotene, protein, and minerals are supplied adequately by pasture forage.

Cottonseed-hull bran consists of good cottonseed hulls from which the lint has been removed. The feeding value of the bran is not appreciably greater than that of ordinary cottonseed hulls.

Officially recognized are the following cottonseed products:

Cottonseed cake—a product of the cottonseed, composed principally of the kernel with such portions of the hull as are necessary in the manufacture of oil; it must contain at least 36 percent crude protein. Cottonseed cake is classed as follows:

1. *Cottonseed cake, prime quality*—must be firm, but not flinty in texture, free of mold, not have a sour or musty odor, and when ground into meal must produce cottonseed meal, prime quality. Cottonseed cake with 43 percent crude protein must be termed "43 percent *protein cottonseed cake, prime quality*," and lower grades similarly designated.

2. *Cottonseed cake, off quality*—does not fulfill the above requirements as to odor or texture. Cotton-seed cake with 43 percent crude protein must be labeled "43 percent protein cottonseed cake, off quality," and lower grades similarly designated.

NOTE: The following cottonseed cake sizes must be designated and sold according to their protein content:

Nut-size cottonseed cake, prime or off quality, will pass through 1½ in. round perforation and over ⅞ in. round perforation. It shall be free from meal, pea-size and pebble-size cakes, and shall not contain in excess of 10 percent sheep-size cake.

Sheep-size cottonseed cake, prime or off quality, will pass through ⅞ in. round perforation and over ⅝ in. round perforation. It shall be free from meal and pebble-size cakes and shall not contain in excess of 10 percent nut-size and pea-size cakes.

Pea-size cottonseed cake, prime or off quality, will pass through ⅝ in. round perforation and over ⅜ in. round perforation. It shall be free from meal and nut-size cake and shall not contain in excess of 10 percent of sheep-size and pebble-size cakes.

Pebble-size cottonseed cake, prime or off quality, consists of fine particles and small pieces of cottonseed cake capable of passing through a ⅜ in. round perforation.

Cottonseed meal is a product of the cottonseed; it is composed principally of the kernel with such portion of the hull as is necessary in the manufacture of oil, provided it contains at least 36 percent protein. Cottonseed meal is classed as follows:

1. *Cottonseed meal, prime quality*—must be finely ground, not necessarily bolted, free from extensive lint, not have a sour, musty, or burnt odor. Cottonseed meal with 36 percent protein must be termed "36 percent *protein cottonseed meal, prime quality*," and higher grades similarly designated by changing "36" to appropriate protein content.

2. *Cottonseed meal, off quality*—does not fulfill the above requirements as to odor or texture. It must be graded "36 percent protein cottonseed meal, off quality," and higher grades similarly designated.

The following guarantees of crude fat and crude fiber are recommended for other than solvent extracted cottonseed meal:

Protein	Crude fat (minimum)	Crude fiber (maximum)
36%	3%	17%
41%	3%	14%
43%	3%	13%

For color standard refer to NOTE under cottonseed cake.

Cottonseed meal is heated during processing. Proper heating reduces toxicity to the extent that such cottonseed meal can serve as the only protein supplement in diets for swine and growing poultry. However, even the highest quality cottonseed meals at present are not recommended for laying hens because the minute, nontoxic levels of gossypol in the ration cause egg yolks to develop a green color in storage.

The proteins of cottonseed meal contain comewhat less lysine than those of properly heated soybean meal and animal-protein supplements but they are very satisfactory for ruminants.

Solvent extracted cottonseed meal—the product resulting from grinding solvent extracted cottonseed flakes. It must be sold according to its protein content.

For solvent extracted cottonseed meal, the following guarantees of crude fat and crude fiber are recommended:

Protein	Crude fat (minimum)	Crude fiber (maximum)
36%	0.5%	17%
41%	0.5%	14%
43%	0.5%	13%

Solvent extracted cottonseed flakes—the product obtained after solvent extraction of the oil from cottonseed kernels and such portions of the hull as are necessary in the manufacturing process.

Degossypolized cottonseed meal is cottonseed meal from which the gossypol has been deactivated so as to contain not more than 0.04 percent free gossypol.

Cottonseed cubes, prime quality, or *cottonseed pellets, prime quality*—processed by a cubing or pelleting machine, must be firm, but not flinty, free of mold, not have a sour or musty odor, and when ground into meal, must produce cottonseed meal, prime quality. This product shall contain not less than 36 percent crude protein. It must be sold according to the protein content.

Cottonseed feed—consisting of cottonseed meal and cottonseed hulls, must contain not less than 22 percent protein.

Whole pressed cottonseed, prime quality, is the product that results from subjecting the whole, sound, mature, clean, not decorticated cottonseed to pressure in order to extract the oil; it includes the entire cottonseed, except for the extracted oil and the removed lint. It must be sold according to its protein content.

Ground whole pressed cottonseed, prime quality, is whole, pressed cottonseed, ground. It must be sold according to its protein content.

References: F.6; E.12; E.13; G.11.

See also OILSEED MEALS; VITAMIN E.

COTTONSEED BY-PRODUCTS.
See COTTONSEED; RANGE MANAGEMENT.

COTTONSEED CAKE. *See* COTTONSEED; PRUSSIC ACID POISONING.

COTTONSEED CUBES.
See COTTONSEED.

COTTONSEED FEED. *See* COTTONSEED.

COTTONSEED-HULL BRAN.
See COTTONSEED.

COTTONSEED HULLS.
See COTTONSEED; POISON BRAN MASH.

COTTONSEED MEAL. *See* COTTONSEED; GRAZING; YEAST DRIED GRAINS OR VINEGAR DRIED GRAINS; VELVETBEANS; PETERSEN'S RELATIVE FEED VALUE.

COTTONSEED PELLETS.
See COTTONSEED.

COTYLEDON (pronounced: kot-e-*lee*dun is the first leaf produced when a seed germinates. The number of cotyledons is of importance in the classification of flowering (seed) plants. Thus, a GRASS or SEDGE is always a *monocotyledon;* LEGUMES are *dicotyledons.*

COUCH GRASS = BERMUDA-GRASS.

COUMARIN forms colorless crystals and is only slightly soluble in water. It is contained in SWEETCLOVER; when mold action transforms it into DICOUMAROL, it becomes the cause of SWEETCLOVER HAY POISONING. —*See also* VITAMIN K.

COUNTY (AGRICULTURAL EXTENSION) AGENT is an employee of a State Land Grant College who acts at the county level. In addition to his own knowledge he can draw on and request the aid of subject-matter specialists in giving professional advice to farmers. He may be able to supply lists of dependable sources of seeds, feedstuffs, insecticides, etc.

COUNTRY GENTLEMAN CORN.
See CORN (variety).

COVER CROP is a crop grown between the trees of orchards or on fields between cropping seasons. Left on the land, it prevents erosion by water and wind. In spring it is usually turned under, thus becoming a fertilizer.—*See also* DUST STORMS.

COVERED KERNEL SMUT, one of the so-called HEAD SMUTS, is caused by *Sphacelotheca sorghi*, a FUNGUS. It attacks all groups of SORGHUM, including JOHNSON-GRASS, and is probably the most destructive sorghum disease in the United States. Although it occurs wherever sorghum is grown, it is most prevalent in the Kansas-Oklahoma-Texas area. Usually, all of the kernels on smutted plants are destroyed. In smutted heads, enlarged cylindrical or cone-shaped smut galls are formed instead of the kernels. These smut galls are initially covered with a light-gray or brown membrane which may later break and release dark-brown spores. Some of these spores are scattered in the field to nearby healthy heads, but most of them remain in the galls until the crop is threshed. Threshing breaks up the galls and spreads the spores to the healthy seeds.

When this smutted seed is planted, the spores germinate along with the seed. The growing fungus then invades the developing seedling and continues to grow undetected inside the plant until after heading, when the smut galls become evident.

At least *five races* of covered kernel smut are known; they differ in their ability to attack different varieties of sorghums. All commercial varieties of sorgos, kafirs, durras, broomcorns, and SUDANGRASSES are susceptible to all five races, as are also such varieties as darso, Schrock, and Dwarf Freed.

Hegari and the true milos are resistant to *race 1*, the most common of covered kernel smut races, but are readily attacked by *race 2*. Race 2 frequently does not affect the entire head in milo and hegari; nevertheless, it often causes severe losses.

Gurno and common feterita are generally resistant to all races of covered kernel smut except *race 3*, to which they are moderately susceptible. Certain varieties or crosses that are not susceptible to race 1 and differ in their susceptibility to races 2 and 3, are attacked by *races 4* and *5*. Spur feterita and certain hybrids derived from it, are highly resistant to, or immune from, all races of covered kernel smut.

Control: Covered kernel smut can be controlled effectively by proper SEED TREATMENT and by planting only smut-free seed, or growing resistant varieties.

Reference: L.1.

See also SULFUR; PHYGON; SPERGON; ARASAN; FORMALDEHYDE SOLUTION; COPPER CARBONATE; GRASS SMUTS; SPELT.

COVERED SMUT OF BARLEY is

caused by the FUNGUS *Ustilago hordei.* Heads of BARLEY affected by this smut appear a little later than the healthy heads. Smutted heads may emerge completely or remain half-enclosed in the boot. The mass of smut spores is first covered with a thin grayish membrane, but this soon splits and permits an early spread of spores and inoculation of the developing seed into the healthy heads. This continues while the grain is in the field and is completed with

threshing. Spores often lodge beneath the glumes and under favorable conditions germinate there and form a fungus growth on the surface of the seedcoat, beneath the GLUME.

When infected barley seed is sown, the smut spores germinate at the same time that the seed germinates and the smut germ tube invades the young seedling before the latter emerges from the soil.

Infection of the seedling by covered smut is favored by certain environmental seeding conditions, such as a somewhat acid soil, deep sowing, medium moisture content of the soil, and soil temperatures between 50° and 70° F.

Control: Covered smut of barley is readily controlled by SEED TREATMENT with CERESAN M. Losses can be prevented mainly by growing covered-smut resistant varieties of barley; but varieties that are resistant in one region may be susceptible to the smut in other regions since there are 13 or more races of the covered smut fungus.

Reference: L.10.

COVERED SMUT OF OATS is caused

by the FUNGUS *Ustilago kolleri.* The smut mass, which replaces the kernel, is covered by a thin grayish membrane that usually does not rupture until harvest or threshing time. Then the spores are spread over the grain and some lodge beneath the GLUMES, where they germinate and spread a mat of fungus mycelium ("threads") over the seedcoat. The glumes, however, are left undamaged.

When the seed is sown in moist soil at the right temperature (60°–75° F.), these spores germinate with the seed and send infection "threads" into the young seedling, where they keep pace with the growing point of the plant. The mycelium under the glumes also invades the seedling. When the head forms, a mass of spores replaces the kernels within the glumes.

As the related BLACK LOOSE SMUT, covered smut now causes only small losses, because of the ease of control by SEED TREATMENT.

Control: CERESAN M is recommended for effective seed treatment. This disease can also be prevented by growing resistant varieties.

References: L.10; M.30.

COWPEA (*Vigna sinensis*) is one of the most extensively grown leguminous crops in the southern states, but it can be grown profitably much farther north, for instance, in the southern parts of Ohio, Illinois, Indiana, and New Jersey, and in parts of Michigan.

In general, the cowpea is adapted to about the same climatic conditions as CORN, but it requires somewhat more heat. It succeeds under a greater diversity of climatic, soil, and cultural conditions than most other LEGUMES. It will withstand a considerable degree of drought, but under very dry conditions will produce only a moderate quantity of hay and a very small number of seeds, if any. The leaves are injured by the least touch of frost, and a heavy frost is always fatal. The cowpea withstands a moderate shade sufficiently well to be valuable in orchards. In heavy shade the plants are usually attacked by mildew.

The cowpea succeeds practically on all types of soil that are well drained, properly inoculated, and moderately rich. It does quite as well as CLOVER or ALFALFA on sandy soils and heavy clays, but will do better than those crops on thin soils or soils poor in lime. The cowpea will do best on good cornland.

Soils that are naturally unproductive, or badly run down by continuous cropping, should be properly fertilized in order to obtain the best results with cowpeas. On very poor soils, 40 to 50 lb. sodium nitrate to the acre or its equivalent in any other nitrogenous fertilizer, about 300 lb. super-phosphate, and 50 lb. potassium chloride to the acre should be applied when preparing the seedbed. As with most other legumes, lime almost invariably increases the yield of cowpeas.

If the cowpea has been grown in a locality for many years, INOCULATION is not necessary. However, when cowpeas are grown for the first time, the soil should be inoculated by using a pure culture of the proper bacteria or by dusting the cowpea seed with soil obtained from an old cowpea field.

The cowpea will give fair results upon a poorly prepared seedbed, but best results are obtained when the soil receives as careful a preparation as it does for corn.

Cowpeas should not be sown until the soil has become thoroughly warmed and all danger of cold weather is past. After that time they can be sown whenever moisture conditions are favorable. The latest date for profitable sowing, however, is at least ninety days before the first killing frost.

Cowpea. (M.3.)

The time of sowing will also depend largely upon the purpose for which the crop is grown. If grown for seed or hay, the seed should be sown shortly after the corn crop.

When cowpeas are grown for forage or green manuring, the seeds are sown broadcast or in drill rows 6 to 8 in. apart. If grown for silage with corn, the corn and cowpeas can be sown in one operation, using the ordinary corn planter. The crop may be cultivated in two to three weeks after sowing. About three cultivations at intervals up to the time the blossoms appear will usually be sufficient. If grown for hay, cowpeas and corn are sometimes sown thickly together—a method that has brought excellent results.

The need for systematic rotations is quite apparent on most types of soil in the cowpea region. The place of the cowpea in the rotation will depend largely on whether the crop is to be plowed under as a green manure or to be harvested for grain or hay. In one case it is sown as a catch crop after small grains, and in the other in the spring as a regular crop.

A system of cropping practiced quite generally throughout the Cotton Belt allows the largest possible area to COTTON, three years being given to this crop, the fourth year to corn and cowpeas. In the SUGARCANE district of Louisiana excellent results are obtained by taking three crops of cane off the land and sowing to cowpeas, or corn and cowpeas the fourth year. On the Black Lands of Texas one of the most successful rotations is: first year, WHEAT or OATS; second year, cowpeas; third year, corn, MILO, or SORGHUM; fourth year, cotton. In Oklahoma, the following rotations are recommended: first year, cotton, corn, or KAFIR; second year, cowpeas; third year, wheat or oats (or, if not used in the first year, kafir or corn). On Missouri, Arkansas, and Tennessee cotton farms, where livestock is kept, the following rotation may be used: first year, cotton; second year, corn with cowpeas at the last cultivation; third year, WINTER OATS or wheat, with a catch crop of cowpeas for hay or seed after the grain has been removed. In Virginia the following 7-year rotation is used: tobacco, wheat, GRASS, grass again, corn with CRIMSON CLOVER as a cover crop, cowpeas, and RED CLOVER. A good rotation in the Corn Belt consists of corn, SOYBEANS, wheat, and clover, with cowpeas for hay when clover fails. Cowpeas are also to be recommended as a crop to precede alfalfa.

Although the cowpea can be satisfactorily grown alone, it is more advantageously grown in combination with other crops.

The cowpea is an excellent crop to grow with corn for silage, and it is being used extensively for this purpose on many dairy farms, especially in the northern part of the cowpea area. If grown with corn for other than silage purposes, cowpeas are allowed to ripen a fair percentage of pods, which are gathered for seed; the remainder is pastured.

Cowpeas, grown in combination with SORGOS or kafirs, make an excellent hay or silage crop. As a hay crop this mixture is more easily cured than cowpeas alone, constitutes a well-balanced ration, and is relished by all kinds of farm stock. The AMBER (sorgo) is more generally favored; however, when grown in rows, the SUMAC and ORANGE varieties of sorgo are fully as good as Amber.

Whenever JOHNSONGRASS and cowpeas are used, excellent results are obtained both in the yield and the quality of hay produced.

SUDANGRASS also is an excellent crop for growing in combination with cowpeas if hay is wanted, since Sudangrass is easily harvested and cured. The mixture is cut about the time the cowpeas are ready for hay.

FOXTAIL MILLET is sometimes grown in mixture with early cowpea varieties. The millet aids to a considerable extent in curing the hay and improves the quality by adding variety.

Cowpeas and tall, strong-growing varities of soybeans afford a very satisfactory combination, either for hay or for pasture, and the yield is nearly always greater than that of either crop alone. The hay of this mixture is of high feeding value, as both plants are rich in protein. In harvesting for hay, the best results are obtained if the mixture is cut when the soybean seed is about full grown and the first pods of the cowpeas are mature.

For feed the cowpea is especially valuable, because it will grow on all types of soil, requiring little attention and producing most excellent forage. Greater use of this crop for HAY and pasturage increases the production of livestock, an essential factor in securing the maximum returns in any system of agriculture. It does not, however, yield as well as some clovers. In many areas it has been replaced by Korean lespedeza.

Cowpea seed has a high feeding value, but is rarely cheap enough to use as feed. It is fed to some extent to poultry. Good seed can be stored for a considerable length of time without much danger of losing its vitality.

Cowpea straw, obtained from threshing the cowpea for seed, can be used as feed for all kinds of stock.

Cowpea hay is an excellent roughage for all kinds of stock. As a rule cowpeas should not be cut for hay before the pods begin to turn yellow. The best quality is produced and the hay cured most readily if the vines are cut when most of the pods are full grown and a considerable number of them are mature. If cut before this stage, the vines are watery and difficult to cure; if cut much later, there will be an unnecessary loss of leaves in handling and the stems will be tough and woody. Cowpea hay has proved satisfactory for work stock and for beef or milk production and has given good results when fed to poultry, hogs, and sheep. It is better suited as a feed for cows than for horses.

Under favorable conditions cowpeas, after being cut for hay, will sprout again from the base. Considerable pasturage or even a second crop of hay or seed is sometimes produced, especially in the Gulf Coast region.

The hay is somewhat difficult to cure, but with attention to the stage of growth and to weather conditions, little more trouble will be experienced in obtaining well-cured cowpea hay than red-clover or alfalfa hay.

The cutting of the plants should begin as soon as the dew is off. The vines should be left in the swath until well wilted on top; then they should be raked into small windrows and after one or two days the hay should be thrown into small cocks until the vines are well cured. A good rule to follow is to consider cowpea-vine hay ready for stacking or storing in the barn when it is impossible to wring moisture out of the stems by twisting them forcefully. Cowpea hay should be cured with as little exposure to the sun as possible. Too

long exposure will cause the loss of the leaves, the most nutritious part of the plant.

Well-cured cowpea hay is nutritious and fully as valuable as feed, pound for pound, as red-clover hay, and its value nearly equals that of alfalfa hay or WHEAT BRAN.

Under average conditions the cowpea will yield from 1 to 2 tons of hay to the acre, and frequently, under very favorable conditions, much larger yields are secured.

Chopped cowpea hay, if moistened, is a winter feed for turkeys, ducks, geese, and chickens; it may be substituted for clover and alfalfa hay and for green feed with good results. Chopped cowpea hay, mixed with CORN MEAL and moistened, has proved quite satisfactory for brood sows; it may also be used as a substitute for roots or green feed in winter.

The cowpea alone has not given good results as a *silage* crop; the best silage is obtained when cowpea is mixed with corn or sorghum. *Mixed silage* has a greater feeding value than either corn or sorghum silage alone.

In many sections alternate rows of cowpeas and corn are put in the silo, but the most common method is one load of cowpeas to two or three loads of corn. This mixture is easily handled, packs quite satisfactorily, keeps well, and makes a superior quality of palatable silage. Although the vines may be put in the silo without cutting, they will pack much closer if run through a silage cutter.

Pasturing cowpeas is not considered the best farm practice, but under certain conditions it is advisable and quite profitable. When cowpeas are grown in corn the grazing should be deferred until the corn has been gathered. If stock are turned on when the cowpea plants have not attained full size, there is more waste from trampling.

Good results are obtained by pasturing cowpeas with any kind of livestock. The most common practice is to hog them down. After pasturing hogs on a field for some time, cattle or sheep may be used profitably to pasture off the leaves and vines which the hogs leave. Sheep may be used on cowpeas in the same way as hogs.

When sown in corn, the stover blades and cowpea seed make a fine ration for fattening fall lambs and wethers. Dairy cows show the effect of such pasturage in much increased flow of milk.

There is danger of BLOAT when sheep and cattle are first turned on cowpeas, especially in wet weather. The danger, however, is far less than with alfalfa, and decreases as the cowpeas become mature.

As a *soiling* crop the cowpea can be advantageously used to supplement crops with less protein, such as corn, sorghum, and FOXTAIL MILLET. The wide variation in the maturity of the varieties makes it possible to have an abundance of green forage throughout the greater part of the summer and fall.

The cowpea has been used more as *green manure* for soil improvement than any other legume because it is so easily grown, has such a marked effect upon succeeding crops, and succeeds under such a great diversity of conditions.

Dangers: ROOT KNOT and COWPEA WILT are the most serious diseases of the cowpea in the United States. Several other troubles of minor importance are WHITE SPOT, CHARCOAL ROT, RED LEAF SPOT, and POWDERY MILDEW.

Among the insect enemies of the cowpea are the CORN EARWORM, STINK BUGS, the COWPEA CURCULIO, the VELVETBEAN CATERPILLAR, the COWPEA WEEVIL, and other species of WEEVILS.

Varieties: At present about 15 varieties of cowpeas are in common cultivation, and there are about three times as many that are grown less frequently.

The most valuable American varieties of cowpeas for forage are the following:

Whippoorwill, also known as *Shinney* and *Speckled,* is most commonly grown. It is a good general-purpose variety and suitable for seed or hay production. The plants are tall, half bushy, and prolific; the pods mature in about eighty-five days.

New Era is one of the important commercial varieties of cowpeas. It is especially valuable on account of its earliness, its erectness, and the smallness of the seed.

The plants are tall, half bushy, and very prolific; the pods mature in about seventy-five days.

Groit (*cowpea*) is a natural cross between the New Era and Whippoorwill varieties. It has often been confused with the New Era, but in general is from 20 to 25 percent better than the latter and has largely replaced it. The plants are suberect, half bushy, and very prolific; the pods mature in about eighty days.

Iron (*cowpea*) is especially valuable on account of its immunity to ROOT KNOT and COWPEA WILT and is largely grown where these diseases prevail. The Iron volunteers more readily and retains its vitality better than other varieties. The plants are tall, half bushy, and moderately prolific; the pods mature in ninety to 100 days.

Brabham (*cowpea*) originated as a natural cross between the Iron and Whippoorwill varieties. It has the tall habit and prolificacy of the Whippoorwill with the resistance to COWPEA WILT and ROOT KNOT of the Iron variety. The Brabham is especially adapted to the sandy soils of the South. The pods mature in about ninety days.

Victor is an artificial cross between the Brabham and Groit varieties. It has the tall habit of the Brabham and has given better results in yields of hay and seed than either parent. Victor is valuable for both forage and seed, especially in the Piedmont and Coastal Plain areas of the cowpea region. Its resistance to COWPEA WILT and NEMATODE attacks (causing ROOT KNOT) is high. The plants are tall, half bushy, and very prolific; the pods mature in about ninety days.

"Clay" is a term applied commercially to a group of medium-late varieties with buff-colored seed rather than to any one sort. Those sorts which mature their first pods in about 110 days or less make up most of the seed sold as Clay, while the sorts requiring 111 days or more to mature the first pods are called *Wonderful, Unknown, Boss,* or *Quadron.* Because of its large, vigorous growth, the Clay group is grown to a considerable extent for forage and soil improvement.

"Red Ripper" is a group name (like

Clay) which is applied to all varieties with maroon seeds rather than to any one sort. In a few sections of the South they are grown for soil improvement and forage.

"*Black*" (*cowpeas*) is the name of a group which includes many similar black-seeded sorts. The varieties of this group are viny and bear the pods near the ground. If grown in corn they are excellent, and in some sections they are used for pasturing with hogs, as the seeds do not decay readily. A variety quite often grown as *Congo*, or *Early Black*, differs from the ordinary Black sorts in being about ten days earlier. Other varieties of local importance for forage use are *Taylor*, also called *Gray Goose*, *Gray Crowder*, *Whittle*, and *Speckled Java; Early Buff* (maturing in about sixty days); *Early Red* (maturing in about seventy-five days); *White cowpeas* (a group valued for table use rather than for forage use); *Catjang*, or *Hindu cowpea* (a group to which the *Buff catjang* belongs which has given favorable results in some sections of the southern states); and *asparagus bean*, or *yard-long bean* (a group used for food, forage, and seed production).

References: M.23; M.25.

See also DODDERS; CLUSTER CLOVER; FIELD PEA; PEANUT; MUNGBEAN; VELVET-BEANS; LEGUME BACTERIA; GRAZING; SILAGE CROPS.

COWPEA CURCULIO (*Chacodermus aeneus*) is distributed throughout the States east of the Mississippi River and south of Tennessee and Virginia. In the southern part of this region there are two generations annually, and the insect hibernates in the adult stage. It infests the COWPEA seeds during their development in the field, but does not breed in dry seeds.

Control: SODIUM FLUOSILICATE has proved to be one of the most satisfactory materials for the control of the curculio on cowpeas. Insecticidal control should be supplemented by farm practices that include the location of all new plantings in areas isolated from other cowpeas, the frequent picking of ripe pods, and the rotation of crops to prevent infestation by

insects coming from hay and other fields of cowpeas.

Reference: M.23.

COWPEA HAY.
                        See COWPEA; HAY GRADING.

COWPEA PASTURE. *See* COWPEA.

COWPEA SEED. *See* COWPEA.

COWPEA STRAW. *See* COWPEA.

COWPEA WEEVIL (*Callosobruchus maculatus*) causes much injury to COWPEAS. This WEEVIL is generally distributed throughout the southern states and California. The adult lays its eggs on the cowpea pods in the field and continues to breed for successive generations in the stored seed. The food value of the crop is often entirely destroyed or its value for planting purposes seriously impaired or lost.

Control: All lots of dried beans, cowpeas, and bean or cowpea straw on farms and in warehouses serve as breeding places for cowpea weevils and as sources of infestation of the new plantings in the field. These sources should be eliminated prior to planting time by fumigating stocks of dry peas, beans, or cowpeas with CARBON TETRACHLORIDE MIXTURES or by soaking the seed for one minute in boiling water; by feeding the seed stalk and bean or cowpea straw to livestock; and by plowing the straw under deeply or burning it.

Reference: M.23.

COWPEA WILT, caused by the FUNGUS *Fusarium oxysporum* f. *tracheiphilum*, is probably the most serious disease of COWPEAS. The wilt is characterized by a premature yellowing and falling of the leaves and the final death of the plants. The stems become yellow, the plants are stunted, and usually no fruit is set. Diseased stems are brown to black inside. The dead stalks of plants killed by wilt usually turn pink, due to the presence of masses of spores of the fungus which causes the disease. Cowpea wilt usually begins to make its appearance about midseason, and by August and September the plants in infested areas may all be dead. The disease starts in spots in the field; these spots enlarge from season to season until eventually the entire area may be infected. The wilt may be spread from field to field by drainage water, cultivation, animals, or any other agency which will carry some of the diseased soil.

Control: Wilt may be controlled by using varieties of cowpeas which are immune. Fortunately the varieties known to be immune to ROOT KNOT are also highly resistant to cowpea wilt, especially *Brabham*, *Iron*, and *Victor*, as well as *Groit*.

References: M.23; W.15.

COWPER is one of the forage SORGHUMS.

COW RATIONS.

See BEEF CATTLE RATIONS; DAIRY RATIONS.

CRABGRASS (*Digitaria sanguinalis*) is a branching and spreading GRASS with ascending flowering shorts. It is a troublesome WEED throughout the United States.

Control: Crabgrass is hard to control. Best results are obtained with KEROSENE or a solution containing 1 lb. SODIUM CHLORATE in 24 gal. water.

See also ALFALFAS.

CRAB MEAL. *See* MARINE PRODUCTS.

CRACKED CORN. *See* CORN.

CRACKED OAT GROATS. *See* OATS.

CRAZY CHICK DISEASE is due to VITAMIN-E deficiency.

CRAZYWEED = *White locoweed.*

See LOCOWEEDS.

CREATINE is a nitrogen containing organic compound which occurs in muscle tissues.

CREATINURIA is the excretion of CREATINE in the urine.—*See also* VITAMINS.

CREEP is a pen or sheltered area which allows calves, lambs, or pigs to enter (creep), but is so constructed as to keep out older animals. The creep should be comfortable and easily accessible and contain palatable feed and water.

Creep feeding results in heavier calves at weaning time; their mothers remain in good flesh, thus requiring less feed during the winter months. If good nursing cows are grazing improved pastures, creep feeding causes only slight increase in gain and is not justified. However, on native or poor pasture, creep feeding will result in calves weighing 100 lb. more at weaning time.

Creep feeding is advisable for early lambs and for suckling pigs after they are two weeks old.

CREEP FEEDING. *See* CREEP; GRAZING; SWINE RATIONS.

CREEPING FOXTAIL
= REED FOXTAIL.

CREEPING RED FESCUE = RED FESCUE.—*See also* FESCUES; GRAZING.

CREOLE SOYBEAN.
See SOYBEAN (variety).

CREOSOTE. *See* COAL-TAR CREOSOTE.

CREOSOTE BARRIERS. Two types of this CHEMICAL BARRIER are used for the destruction of CHINCH BUGS:

1. Dirt-ridge creosote barrier. COAL-TAR CREOSOTE is the best barrier material because of its repellent and lasting qualities. Light creosote oils are easily applied and soak into soil and paper readily. The grade TT-W-556A is the best of any thus far tested. Heavy creosote oils soon become ineffective through loss of odor and drying.

The foundation for a creosote barrier is made by throwing up a ridge of earth with a plow. Only a small quantity of creosote is necessary for one application, as a line ½ in. wide is just as effective as one 2 to 3 in. wide. Fresh creosote should be applied to the original line at least once a day for the first few days; thereafter it need be renewed only once every two or three days, unless the weather is extremely hot and dry. When properly applied, 50 gal. creosote will maintain ¼ mile of barrier for about three weeks, which ordinarily is longer than the barrier is required.

The bugs trapped in the holes should be killed every afternoon at sundown. This may be done easily by sprinkling 1 to 2 tablesp. kerosene into each hole. The kerosene should not be ignited, but the bugs should be allowed to work it around themselves.

2. Creosote-treated paper-fence barrier. It is made by setting upright in the ground a strip of creosote-soaked paper about 4 in. wide, with half its width above the surface.

Single-faced corrugated paper, tarred (not asphalt-treated) felt paper of the 14- or 15 lb.-grade, red-rosin building paper of the 30-lb. grade or heavier, and heavy chip board or chip strawboard from 20 to 40 points thick have been used successfully. After being cut, the 4-in. rolls are soaked for at least twelve hours in a container with enough creosote to keep them covered, and then allowed to drain for an hour or more before the fence is built. A handy tool for use in unrolling and installing the paper strips may be made by fitting a broom handle into a hole in the side of a piece of two-by-four about 1 ft. long, so as to form a T-shaped carrier; the roll of paper may be slipped down over the handle until it rests on the crosspiece at the bottom.

If the paper fence has been treated properly, it should repel the bugs for two to three days. Then it will have to be freshened by applying more creosote close to the top edge. From 2 to 3 gal. creosote is needed to treat ¼ mile of paper fence. After it is well soaked, the paper barrier will not need retreatment as frequently as a creosote line on the ground. About 30 gal. creosote is sufficient to maintain ¼ mile of paper fence barrier for the season.

Reference: P.5.

CREOSOTE OIL = COAL-TAR CREOSOTE.

CREOSOTE-TREATED
PAPER-FENCE BARRIER is used for the control of migrating CHINCH BUGS.— *See also* CREOSOTE BARRIERS.

CRESOL. The technical grade of cresol is known as CRESYLIC ACID.

CRESTED WHEATGRASS has for many years—erroneously—been considered to be an *Agropyron* spp. Two types of crested wheatgrass are now grown commercially:

1. STANDARD CRESTED WHEATGRASS, or *desert wheatgrass*, which is of great importance for seeding the Northern Great Plains, the Intermountain region, and the Rocky Mountain states.

2. FAIRWAY CRESTED WHEATGRASS, or *true crested wheatgrass*, which is more extensively planted in western Canada than in the United States, is identified as *A. cristatum*.

Crested wheatgrass is a perennial bunchgrass that produces an abundance of both basal and stem leaves. The leaves are flat, 6 to 10 in. long, and about ¼ in. wide. The stems are fine and develop dense tufts about 2 to 3 ft. high. The plant has a wide-spreading, deeply penetrating root system.

This is one reason why it can survive cold and drought, withstand grazing, and compete with weeds and associated GRASSES.

Crested wheatgrass usually begins growth in early spring, when succulent feed is so important. It ceases to grow during long, hot, dry periods of summer, but it again makes growth when moist, cool weather returns. High palatability, good quality, and good volume of forage, combined with hardiness, drought resistance, and adaptation to widely different soil types, make crested wheatgrass one of the most valuable forage grasses.

If the grass is to furnish a good quality of *hay*, it is advisable to cut it shortly

Crested wheatgrass. (H.47.)

after it comes into full head and before it flowers. Young crested wheatgrass large enough for *grazing* has a high protein content (20 to 30 percent) both in spring and in fall. As it matures it becomes harsh and the protein content decreases rapidly, and by the time it is mature the content may be as low as 2 or 3 percent. The HAY cures readily, and with favorable weather can be stacked and stored soon after being mowed. It can be used successfully in mixture with other grasses and with ALFALFA and SWEETCLOVER; however, the alfalfa and sweetclover should not make up more than 20 percent of the seed by weight.

The grass can also be used effectively for the control of wind and water erosion and its tough, fibrous root system adapts it for use in rotations.

In reseeded areas, livestock must be controlled. A new stand of crested wheatgrass should not be grazed until the plants develop strong root systems; this requires at least two growing seasons.

Crested wheatgrass is long-lived; sometimes a stand is utilized as long as twenty-five years. Where the grass is adapted, yields have ranged from ¾ to 1 ton of cured hay per acre over a period of years, with considerably higher yields in the favorable seasons.

Dangers: Among the diseases attacking this grass are the GRASS SMUTS.

References: S.1; W.1; H.1; R.1; W.16.

See also WHEATGRASSES; RANGE MAN-AGEMENT.

CRESYLIC ACID, or *straw-colored carbolic acid*, is the name of the technical grade of *cresol*. It is one of the COAL-TAR ACIDS obtained from COAL TAR.

CRICKET. *See* MORMON CRICKET.

CRIMEAN WHEAT = *Turkey wheat.* —
                    See also COMMON WHEAT.

CRIMPED BARLEY
                    = *rolled barley. See* BARLEYS.

CRIMPED OATS. *See* OATS.

CRIMSON CLOVER (*Trifolium incarnatum*), one of the TRUE CLOVERS, is also called *German clover* and *scarlet clover*, and is the most important winter annual LEGUME of the central section of the eastern states. This crop may also be grown

successfully as a summer annual in northern Maine, Michigan, and Minnesota. Winter culture can be extended into Kentucky, southern Missouri, southern Indiana, and Ohio, provided varieties are grown that are adapted to these sections and the seed is sown in fertile soils early in August. Besides being an excellent pasture plant and furnishing plenty of hay, crimson clover protects the soil during fall, winter, and spring.

In general, the leaves and stems of crimson clover resemble those of RED CLOVER, but are distinguished by the rounded tips of the leaves and by more hair on both leaves and stems. The common name of this clover is derived from the bright crimson color of the blossoms.

Crimson clover does well in cool, humid weather and is tolerant of winter conditions where the temperature does not become severe or too changeable. It may be planted from midsummer to late fall. Enough soil moisture to sprout the seed and establish the seedlings is the greatest factor in obtaining a stand. Poor soils can be improved by adding phosphate and potash fertilizers and manure or by turning under such crops as cowpeas, soybeans, or lespedeza.

In many areas where crimson clover has been grown successfully for several years, it is not necessary to inoculate the seed with bacterial cultures for the production of nodules. But when crimson clover is grown for the first time, an additional INOCULATION treatment is recommended if weather conditions are dry and hot after seeding. This supplemental inoculation consists of mixing commercial cultures with sand, soil, or cottonseed meal and broadcasting the mixture over the soil surface during cloudy, rainy weather as the young seedlings are emerging. A bushel-size culture mixed with 60 lb. of the above-mentioned material is sufficient for an acre.

RYE, VETCH, RYEGRASS, and fall-sown grain crops are often seeded with crimson clover. Seeding is done at the same time, but since a greater depth is required for the seeds of most of the companion crops, two seeding operations are necessary.

Farmers often use a mixture of 5 lb. red clover and 10 lb. crimson clover per acre with excellent results.

Crimson clover grows rapidly in fall and spring and furnishes an abundance of GRAZING. With early planting and a good fall growth, the clover may also be grazed during the fall and winter months. Crimson clover combined with small grains of ryegrass has been most widely used for winter grazing.

The clover makes excellent *hay* when cut at the early-bloom stage, although the yield may be slightly reduced. For best yields it should be harvested in full bloom. The hay is easily cured either in the swath or in the windrow. Yields as high as 2½ tons per acre are not uncommon on fertile soil.

Crimson clover may be made into *silage* by the same methods as are used for other legumes and grasses. Crimson clover is an ideal green manure crop. For best results it should be plowed under two to three weeks before the succeeding crop is planted.

Dangers: Animals grazing on crimson clover seldom BLOAT; however, it is advisable not to turn them into clover fields for the first time when they are hungry. As crimson clover reaches maturity, the hairs of the heads and stems become hard and tough. When it is grazed continuously or when it is fed as hay at this stage, large masses of the hairs are liable to form into HAIR BALLS in stomachs of horses and mules, and occasionally with fatal results. If small quantities of other feeds, particularly roughages, are fed along with the clover, the formation of these balls will be reduced. Cattle, sheep, and swine do not seem to be affected.

The most serious diseases that affect crimson clover are the CROWN ROTS; in the cooler sections of North America, it is also affected by NORTHERN ANTHRACNOSE; while SOUTHERN ANTHRACNOSE may affect it in all parts of the country. Sandy soils in the southern part of the crimson clover belt are often infested with NEMATODES. While the CLOVER-SEED CHALCID, the PEA APHID, and other insects sometimes become numerous in crimson clover, they do not ordinarily cause appreciable damage.

Variety: *Dixie crimson clover* grows well in the Gulf Coast section; it is more widely adapted and appears more winter-hardy than common crimson. Dixie has successfully volunteered to good stands when grown in pastures with BERMUDA GRASS, JOHNSONGRASS, SMALL GRAINS, and in rotation with such cultivated crops as SORGHUM or late planted CORN. The seed and plant of Dixie cannot be distinguished from common crimson clover, and the variety may be readily contaminated. The farmer should buy only certified seed.

Reference: H.11.

See also OATS; BUCKWHEATS; ALFALFAS; BERMUDA-GRASS; KUDZU; ALSIKE CLOVER; PEANUT; COWPEA; SILAGE CROPS; PASTURE PLANTS; PASTURES.

Crimson clover. (L.4.)

CROPS in the United States. *See* table next page.

CROP VARIETY

= *agronomic variety. See* VARIETY.

CROSS. *See* HYBRID.

CROSSBREED. *See* HYBRID.

CROSS-DRILL means: to DRILL two or more kinds of seeds crosswise, usually at right angles to each other.

CROPS HARVESTED IN THE UNITED STATES, 1954

	Million acres	Acre yield
Corn.................	79.9	37.1 bu.
Hay, total harvested..	72.8	1.43 ton
Wheat...............	53.7	18.1 bu.
Oats................	42.2	35.6 bu.
Cotton..............	19.2	339 lb.
Soybeans............	17.0	20.1 bu.
Barley..............	13.0	28.5 bu.
Grain sorghums......	10.8	19.0 bu.
Sorghum forage......	7.0	1.1 ton
Flax................	5.7	7.3 bu.
Rice................	2.4	2447 lb.
Rye.................	1.7	13.8 bu.
Tobacco.............	1.6	1337 lb.
Field beans.........	1.6	1200 lb.
Peanuts.............	1.4	763 lb.
Potatoes............	1.4	252.8 bu.
Sugar beets.........	0.9	16 tons
Velvet beans........	0.9	820 lb.
Sweet potatoes......	0.35	86.5 bu.
Sugar cane..........	0.34	20 tons
Cowpeas.............	0.28	6.0 bu.
Field peas..........	0.27	1300 lb.
Broomcorn..........	0.24	
Deciduous fruits.....	2.1	
Citrus fruits........	0.8	10 tons
Commercial vegetables	3.9	6.5 tons

CROSS-POLLINATION.

See POLLINATION.

CROTALARIA. Among the *Crotalaria* spp., or CROTALARIAS, are numerous *poisonous* species; e.g., *C. spectabilis* = SHOWY CROTALARIA; *C. juncea* = SUNN CROTALARIA; *C. sagittalis* = RATTLEWORT; and species without common names, such as *C. burkeana* and *C. dura*.

Among the *nonpoisonous* species are *C. mucronata* (or *C. striata*) = STRIPED CROTALARIA; *C. intermedia* = SLENDER-LEAF CROTALARIA; *C. lanceolata* = LANCE CROTALARIA; *C. incana* = SHAK CROTA-LARIA; and *C. grantiana* (without a common name).

Reference: M.14.

CROTALARIA ANTHRACNOSE is an ANTHRACNOSE caused by the FUNGUS *Colletotrichum crotalariae* which attacks the stems of both showy and striped CROTALARIAS and kills their barks, which then slough away.

CROTALARIAS (*Crotalaria* species), also called *rattle boxes*, belong to the LEGUMES. Only a few of the more than 600 species of crotalaria are of economic importance.

While the principal use of crotalaria in this country has been for soil improvement, the value of the nonpoisonous species for forage has been recognized. These are upright summer annuals or short-lived perennials. The stems are coarse; the central stem is upright and branches quite freely, except in very thick stands. The leaflets are borne singly or in threes in the axis of the leaf and vary in shape from linear to broad ovate. The plants in general are leafy, bloom freely, and set seed in abundance. The yellow flowers are showy and the seed pods as a rule are quite conspicuous. The seed color varies from straw-yellow through brown to black.

Most species of crotalaria are treated as summer annuals in the United States. Warm temperatures with moderate humidity seem to favor their growth.

Crotalaria is better suited to poor sandy soils than most crops. It is not adapted to heavy clay; it will make good growth on most soils of the Coastal Plains region of the southeastern states.

Seeds of crotalaria remain viable for a number of years. Since the seed is comparatively small, thorough seedbed preparation is necessary for the first seeding to insure a prompt and good stand. Subsequent stands, which will be largely volunteer, can be assured by ordinary cultivation or disking to check other plant competition.

Crotalaria will grow at a comparatively low soil-fertility level; however, when a heavy crop is desired, fertilizing with superphosphate and potash is recommended. On acid soils a moderate quantity of lime benefits crotalaria.

Seeding should be done after all danger of frost is past. Crotalaria makes good growth and develops many nodules, so that artificial INOCULATION is not necessary.

The fact that NEMATODES cannot live in the roots of crotalaria makes the crop especially good for lands where nematode-susceptible crops are being grown. On lands not growing a cash crop, a winter cover crop of RYE or a WINTER LEGUME should follow crotalaria in the fall. Crotalaria will not choke out BERMUDA-GRASS, as is sometimes suggested.

On poor lands that cannot be cropped every year one can well afford to grow crotalaria in the years the land does not produce a cash crop. When crotalaria is seeded in OATS or other SMALL GRAIN and grown after the harvesting of the grain, allow it to mature some seed each year to induce volunteering. Crotalaria grown with CORN should be seeded several weeks before the corn is laid by.

In experimental plantings increased yields of corn and other crops have followed the use of crotalaria in rotations and inter-plantings, with maximum increases equaling 100 percent.

No species of crotalaria is used extensively as forage; however, it is known that animals will eat both the *poisonous* and the *nonpoisonous* species. As crotalaria is a coarse, fibrous plant, it is not well adapted for *hay*. When cut quite early (before bloom), a fair forage can be produced. Curing is difficult; it is more satisfactory to take freshly cut crotalaria, make it into bundles and then shock it. Its most satisfactory use for forage has been as *silage*. When cut green and put into the silo, it can be handled with an ordinary silage cutter.

On fertile soils crotalaria makes very strong growth and gives heavy yields. More than 30 tons of green weight per acre has been reported for some species. A good average yield is 10 tons per acre, which is equivalent to about 2 tons of dry forage.

Most of the chemical analyses reported for crotalaria indicate that the various species are relatively high in PROTEIN, averaging about 17 percent when cut early, but decreasing to about 10 percent as they advance toward maturity. While crotalaria is recognized as only a second-rate forage, its feeding value is increased with increase in protein and it should be pastured or cut during the early stages of development, when protein content is highest. The fiber content is high and increases to as much as 47 percent at maturity.

In tests for palatability of hay even the least palatable poisonous species was eaten in sufficient quantity to cause POISONING.

Cattle will eat some species in waste places and woods pastures when other feed is not abundant, but the plants are not as palatable as the native grasses. Cattle and mules have browsed all species grown. It is also known that livestock eat, to a limited extent, many of the crotalaria species made into silage.

Showy crotalaria. Flowering shoot and pod. (W.35.)

Dangers: No widespread epidemics of disease have ever been reported in farmers' plantings, and crotalarias can be depended upon to produce as regularly as any other crop. CROTALARIA ANTHRACNOSE attacks the stem of both showy and striped crotalarias. STEM CANKER OF CROTALARIA occurs in central Florida. SOUTHERN BLIGHT and BROWN SPOT OF LUPINE also attack the stems of the plants. GRAY MOLD has been a conspicuous disease of both showy and striped crotalarias. DAMPING-OFF of crotalaria seedlings may result in a considerable degree of loss and does not appear to be limited to any specific species of host plants. A *virus disease* characterized by general stunting of the plants, by mottling, blistering, and malformation of the leaves, and in most cases by abnormally stimulated lateral branches, or WITCHES'-BROOM, is prevalent on crotalaria grown in Virginia.

Species: Crotalaria species most extensively grown in the United States are SHOWY CROTALARIA (poisonous), STRIPED CROTALARIA, SLENDERLEAF CROTALARIA, LANCE CROTALARIA, SHAK CROTALARIA, SUNN CROTALARIA, also called *Sunn hemp* (poisonous), and RATTLEWORT (poisonous).

Other species—without common names —are *Crotalaria burkeana* (poisonous), *C. dura* (poisonous), and *C. grantiana* (which, with its fine stems and branching growth, has the appearance of a good forage plant, but is more exacting in growth requirements than other species).

> NOTE: The *poisonous* species contain an ALKALOID and should not be used in new plantings. When other feed is available, animals usually avoid the poisonous crotalarias, but in winter, when feed is scarce, they may consume a sufficient quantity to cause trouble.

References: M.14; M.10.

See also ROOT-KNOT NEMATODES.

CROWFOOT GRAMA = BLACK GRAMA.

CROWN of any perennial herbaceous plant is its base, often slightly enlarged— the portion near the surface of the ground where the stem emerges from the roots.— *See also* STEM.

CROWN ROTS, or *stem rots*, are caused by various FUNGUS species; many of these fungi can live in the soil for a long time. Crown rots—which belong to the most destructive diseases of RED CLOVER, WHITE CLOVER, ALFALFAS, and other forage LEGUMES—as well as the related ROOT ROTS, may occur any time during the growing season.

There is a very close relationship between winterkilling of CLOVERS (e.g., CRIMSON CLOVER) and crown and root rots. Plants having low vigor because of the low temperatures, or plants injured by freezing and thawing, are readily attacked by these fungi.

Root and crown rots also make plants more susceptible to winterkilling. For example, plants that are infested with these diseases in the summer are likely to go into the winter dormancy in a weakened condition.

The development of crown rots is stopped by warm spring days; however, periods of cool, wet weather later in the spring may result in a further loss of plants.

Control of crown rots is probably best accomplished by the use of adapted varieties and good management. Varieties that are adapted to a given area will survive unfavorable periods better than unadapted varieties. Second or third crops of clover should not be cut too late so that the plants can recover and store an adequate food reserve before they go into the winter.

References: T.1; E.5; G.6.

See also TRUE CLOVERS; SOUTHERN BLIGHT; SOUTHERN ANTHRACNOSE.

CROWN RUST, caused by *Puccinia coronata avenae*, is one of the most important FUNGUS diseases of OATS. This rust forms small orange-yellow pustules, principally on the upper side of leaves of oat plants. The pustules may become so numerous that they come together on the leaf to form larger irregular patterns. If an infected leaf is pulled between the thumb and forefinger, a small amount of orange-colored dust—spores—can be collected.

Crown rust spores are spread from field to field by wind. They may travel for many miles before coming to earth in another oat field to cause new infections. If weather conditions are favorable—that is, warm and damp—they germinate in a film of water on living oat leaves and form more pustules within a week to ten days.

Crown rust may become so severe that oat plants will actually be killed in the growing stage. More often, plants heavily infected with crown rust are retarded in growth, lodge badly, and ripen prematurely. This greatly reduces forage and grain production and causes any grain produced to be light and chaffy.

Control: Crown rust can be adequately controlled only by the use of resistant oat varieties.

Reference: M.30.

See also VICTORIA BLIGHT; VICTORIA.

CROWN WART, caused by the FUNGUS *Urophlyctis alfalfae*, is a stem and leaf disease of ALFALFAS. The fungus enters scales and leaf tissue of developing buds about the crowns of well-established plants and causes them to expand into rough galls often of remarkable size. The young galls are white, but they turn brown and disintegrate as the fungus spores mature within them. Infection and gall development require a long wet and cool spell in early spring and a supply of germinating spores from the previous year. Thus, there are few localities—in western and southern states—where the disease appears sporadically. The damage to infected plants appears largely from loss of shoots in the spring and from crown necrosis following decay of the galls.

Reference: J.3.

CRUDE CARBOLIC ACID. *See* PHENOL.

CRUDE FAT = ETHER EXTRACT.

— *See also* FAT.

CRUDE FIBER = FIBER. *See also* HAY.

CRUDE OIL, or *crude mineral oil*, is unrefined PETROLEUM. It has insecticidal and fly-repellent properties.—*See also* DODDERS.

CRUDE PROTEIN, as listed in ordinary feed analyses, is obtained by multiplying the nitrogen by 6.25. The term includes not only TRUE PROTEIN, but also its building blocks, the AMINO ACIDS and simpler nonprotein nitrogen compounds, such as UREA, and under certain conditions considerable inorganic nitrogen in the form of AMMONIA, nitrates, or nitrites.—*See also* HAY; MILK PRODUCTS.

CRUMBLES are broken PELLETS.—*See also* COMMERCIAL MIXED FEEDS.

CRUSHED OATS. *See* OATS.

CRYOLITE, a sodium-aluminum fluoride, is a pale-gray mineral.

Applied in the late afternoon at the rate of 15 lb. per acre, it controls young VELVET CATERPILLARS; a second application may be necessary about ten days later to destroy the newly hatched caterpillars.

For the control of the MEXICAN BEAN BEETLE, a spray may be used consisting of 3 lb. cryolite to 50 gal. water; or a dust prepared from 60 lb. cryolite and 40 lb. diluent (such as finely ground TALC, CLAY, SULFUR, TOBACCO, or GYPSUM) may be applied to the infested plants as soon as the beetle is found in the field. Treatment is repeated every week or ten days, but should not be used on soybeans after their pods begin to form.

CRYPTOGAM is any *nonflowering* plant (which does not produce seed); e.g., a FUNGUS, fern, or moss.—*See also* OÖSPORE.

CUBE (pronounced: *kew*-bay) is the root of the Peruvian cube plant which contains an important insecticide: ROTENONE. Cube powder is used to prepare insecticidal dusts or sprays.

For the control of the MEXICAN BEAN BEETLE, cube root containing 4 percent rotenone is mixed at the rate of $1\frac{1}{2}$ lb. to 50 gal. water or $12\frac{1}{2}$ lb. to $87\frac{1}{2}$ lb. diluent (such as finely ground TALC, CLAY, SULFUR, GYPSUM, TOBACCO, or other powder, except lime). The treatment is to be repeated every week to ten days.

Cube (pronounced: *kewb*) is a PELLET of feed especially prepared for feeding under range conditions.

CUCUMBER BEETLE. *See* SPOTTED CUCUMBER BEETLE; WESTERN SPOTTED CUCUMBER BEETLE.

CUCURBITA. *Cucurbita* spp.

= SQUASHES.

CUD. *See* RUMEN BACTERIA; RUMINATION.

CULL BEANS. *See* FIELD BEANS.

CULL POTATOES. *See* POTATO.

CULL RAISINS can replace some of the grain, especially in lamb and pig rations. —*See also* RAISIN PULP.

CULM is the jointed STALK or STEM of a plant belonging to the GRASS family; it is usually hollow (except at the nodes). The term is also frequently applied to the solid stalk of grasslike plants; e.g., SEDGES.

CULM ROT, sometimes called *foot rot*, is caused by *Helminthosporium sativum*. It attacks young seedlings, causing a rotting of the roots and basal parts of the plant. OATS retarded in growth by overgrazing, cold weather, or lack of fertility are subject to serious damage and often die of culm rot infections. Plants attacked by this FUNGUS may show reddening of the leaves. Culm rot symptoms are somewhat similar to VICTORIA BLIGHT.

True culm rotting occurs at the time the heads are starting to fill with grain, causing the plants to lodge. The stems break over near the soil line, become water-soaked, and later turn black. This black color is due to the growth and sporulation of the fungus on the stems. Culm breaking causes considerable loss of grain because of the difficulty of harvesting lodged oats.

Control: Damage from culm rot can be materially decreased by sowing seed—after it has been cleaned and treated with CERESAN M—on land which has not grown oats for at least two years. Keeping the land in good fertility and using carefully controlled grazing also helps to keep down the effects of culm rot. There exists a wide difference in resistance among oat varieties.
Reference: M.30.
See also SEED TREATMENT.

CULTIVATORS of many types are used to loosen soil around growing plants. These implements may be equipped with DISKS, sweeps, shovels, teeth, or knives.

CULTURED WHEY SOLIDS.
See MILK PRODUCTS.

CULTURE MEDIUM. *See* MEDIUM.

CULTURING is the process of growing (cultures of) FUNGI or BACTERIA on a prepared food material, or culture MEDIUM.

CUMBERLAND RED CLOVER.
See RED CLOVER (variety).

CUMBERLAND VALLEY = *Fulcaster.*
See COMMON WHEAT (variety).

CUMMINGS = *Goens.*
See COMMON WHEAT (variety).

CUPRIC ACETATE = COPPER ACETATE.

CUPRIC HYDROXIDE
= COPPER HYDROXIDE.

CUPRIC SULFATE = COPPER SULFATE.

CURCUBITO. *C. pepo* = PUMPKIN.

CURCULIOS. The CLOVER ROOT CURCULIOS and the COWPEA CURCULIO are insect pests.

CURED ALFALFA. *See* ALFALFAS.

CURLED TOE PARALYSIS is a chick disease caused by lack of VITAMIN B_2 in the feed.

CURLER is a disk implement used to cultivate crops grown in lister furrows.

CURLY MESQUITE (*Hilaria lelangeri*) is similar to BUFFALOGRASS. It grows in the Southwest and Mexico. This short grass is esteemed highly for forage and is very palatable to all classes of livestock. It is very resistant to extended drought and withstands close grazing.
Reference: U.6.
See also RANGE PLANTS; PASTURE PLANTS.

CURVULARIA. The FUNGUS *C. lunata* causes BLACK KERNEL of rice.

CUSCATA. The *Cuscata* spp., or DODDERS, include the following: *C. epithymum* = CLOVER DODDER; *C. pentagona* (also known as *C. arvensis*) = FIELD DODDER; *C. planiflora* = SMALL-SEEDED ALFALFA DODDER; *C. indecora* = LARGE-SEEDED ALFALFA DODDER; *C. gronovii* = COMMON DODDER; and *C. racemosa chileana* = CHILEAN DODDER.

CUSTER.
See SIX-ROWED BARLEY (variety).

CUT ALFALFA
= *chopped alfalfa. See* ALFALFAS.

CUTICLE is the outer waxy or corky covering of plants.

CUT OAT GROATS. *See* OATS.

CUT-OVER PASTURES. *See* PASTURES.

CUTWORMS are the larvae of dull-colored moths (millers) that fly by night. In appearance and feeding habits cutworms vary according to species; the most important ones are the VARIEGATED CUTWORM and the ARMY CUTWORM. They are stout, soft-bodied, smooth, and cylindrical. In color they may be brown, gray, or nearly black. Some are spotted, others striped. When full-grown, they are 1 to 2 in. long.

Most cutworms hide in the soil during the day and feed at night. On dark, cloudy days they may be seen above the ground. In the soil they are generally in a coiled position.

Cutworms feed on most kinds of plants, but cause great damage to cultivated crops. Some species cut off the plants just above the soil surface, others at the surface, and still others just below the surface. Some climb the stems to feed on buds, leaves, or fruit, and some remain in the soil to feed on roots and underground portions of such plants as SOYBEANS, CORN, SAFFLOWER, MEADOW FOXTAIL, VETCHES, or SWEETCLOVERS.

Cutworms are destructive only in the immature stage. The adults (moth) feed on nectar and do not injure plants.

The eggs of most species of moths that produce cutworms are laid on the stems of GRASSES and WEEDS, or behind the leaf sheath of these plants; the eggs of some species are laid on bare ground. Each female may lay a few hundred to as many as 1,500 eggs. The egg stage lasts from two days to two weeks. From the eggs emerge the larvae (cutworms).

Most cutworms pass the winter in the larval stage, hidden in the soil, or under trash, or in clumps of grass. They resume feeding in the spring and grow until early summer, when the full-grown larvae enter the next stage of development: in hollowed-out cells, or chambers, they change to pupae. Next, the pupae change to moths. Both these changes take place beneath the surface of the soil. The moths emerge from the soil, and the females soon lay eggs; another generation of cutworms is thus started.

In most of the common species there is but one generation a year. In a few species there are two to four generations, and sometimes there is so much overlapping of generations that moths may be found at almost any time from late spring to the middle of the fall. The cutworms in the northern part of the United States usually have just one generation a year.

Control: TOXAPHENE gives best control. DDT also is effective against many cutworms. Cutworms that feed above ground can also be controlled with poisoned baits, e.g., with a POISONED BRAN MASH containing CALCIUM ARSENATE, PARIS GREEN, SODIUM FLUOSILICATE, or toxaphene.

References: U.4; G.7; W.22.

See also PALE WESTERN CUTWORM.

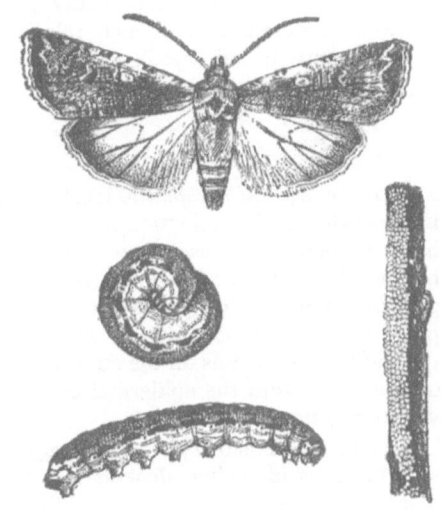

Variegated cutworm. Moth; caterpillar, side view and in curved position; egg mass on twig. (W.22.)

"CYANAMIDE," or *calcium cyanamide*, is a NITROGENOUS FERTILIZER. It forms black powder or lumps, and, in presence of water, slowly liberates ammonia.

CYANIDE POISONING

= PRUSSIC ACID POISONING.

CYANIDES are salts of the hydrocyanic acid (PRUSSIC ACID).

CYANOCOBALAMIN = VITAMIN B 12

CYANOGENETIC GLUCOSIDES are found in so-called *cyanogenetic plants;* some of them cause large losses of livestock because their GLUCOSIDES develop toxic PRUSSIC ACID. Among these plants are the chokecherry (WILD CHERRY) species, ARROW PODGRASS, and SORGHUMS.—*See also* PRUSSIC ACID POISONING.

CYMADOTHEA. The FUNGUS *C. trifolii* causes SOOTY BLOTCH.

CYNODON. *C. dactylon*

= BERMUDA-GRASS.

CYPERACEAE = *sedge family.* It includes the SEDGES (*Carex* spp.), CHUFA (one of the *Cyperus* spp.), and other genera.

CYPERUS is a genus of the Cyperaceae; *C. esculentus* = CHUFA.

CYSTINE is a dispensable sulfur-containing AMINO ACID. It can replace part of the METHIONINE in rations. Cystine forms crystals and is practically insoluble in water. It is a constituent of many plant and animal proteins and occurs in relatively large amounts in hair, wool, and feathers.

CYTOLOGY. *See* MORPHOLOGY.

D

2,4-D, or *2,4-dichlorophenoxyacetic acid,* is one of the new organic HERBICIDES which have become very useful as weed and brush killers. Its action is based on its growth-stimulating properties; it acts as a plant hormone. It is generally applied to the leaves and remains on the outer surface or penetrates into the epidermal cells and vascular system of the plant.

Such compounds of 2,4-D as (inorganic) *salts* or *esters* of higher alcohols are most

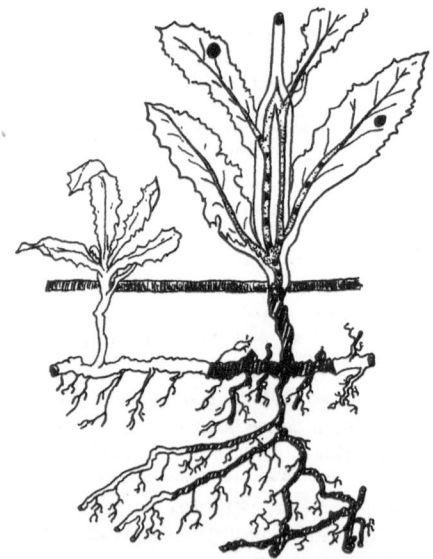

2,4-D. Lower portion of a perennial weed showing how 2,4-D works when used properly. The chemical is applied to the leaf where it slowly penetrates the plant. It gets into the translocation system (stem enlarged to show this system) and moves slowly to the roots where it kills those directly below stem. (D.18.)

effective. It is recommended to apply any of the 2,4-D compounds when growing weather is favorable.

2,4-D acid is a stable, white, water-insoluble solid not itself used as an herbicide but as raw material for the following herbicides:

2,4-D salts—among them the sodium salt, which can easily be prepared from 2,4-D (acid) by reacting with sodium hydroxide. It is somewhat water-soluble and often used in dry formulations.

2,4-D amines are more water-soluble than the inorganic salts and less volatile than the esters, which are advantages; the amines are recommended for use near susceptible crops.

2,4-D esters are insoluble in water and therefore must be applied to weeds in form of emulsions or dissolved in organic solvents. They are recommended for use against "tough" weeds.

> NOTE: When used in insufficient quantities, 2,4-D sometimes produces a leaf distortion very similar to that produced by SOYBEAN MOSAIC.

Applied from an airplane at the rate of 1 lb. of 98-percent 2,4-D acid per acre (preferably in form of an emulsion and in volume sufficient to insure complete coverage) it will control SAGEBRUSH, SKUNK-BRUSH, and many WEEDS, such as WATER-PLANTAIN and REDSTEM, RAGWEED, and SHEPHERD'S PURSE.

Application of 2,4-D at the rate of 1 lb. acid equivalent per acre (in any kind of formulation) will greatly reduce stands of FIELD BINDWEED, but even repeated applications will not, in most cases, eradicate established stands. Best results are obtained from applying 2,4-D following a short period of intensive cultivation.

To kill BARBERRY BUSHES, 2,4-D at full strength is applied with a paint brush to each of the freshly cut stubs of the brush canes. The native species may also be killed by using 2,4-D as foliage spray.

D-5 = *Pentad. See* DURUM WHEAT (variety).

D-ACTIVATED ANIMAL STEROL
= VITAMIN D_3.

D-ACTIVATED PLANT STEROL
= VITAMIN D_2.

DACTYLIS. *D. glomerata* = ORCHARD-GRASS.

DAGHESTAN SWEETCLOVER (*Melilotus suaveolens*) is a yellow-flowering SWEETCLOVER species of limited use.

Variety: *Redfield* is a biennial which is late in maturity; it is particularly adapted to sections of the northern Great Plains.

Reference: P.7.

DAIRY BY-PRODUCTS are used as feedstuffs; they are officially classified as MILK PRODUCTS.—*See also* BY-PRODUCT FEEDSTUFFS.

DAIRY RATIONS for milk production should be based on good-quality ROUGHAGES. During the *winter* season 2 lb. of average quality dry roughage should be fed, or 1 lb. of dry roughage and 3 lb. SILAGE per 100 lb. of body weight. If high quality leafy HAY and excellent silages are available, cows will readily increase their hay intake to 2½ to 3 lb. per 100 lb. of body weight. The roughages are balanced with CONCENTRATE mixtures containing 10 to 18 percent CRUDE PROTEIN, depending on the type of roughage that is used. Cattle of the small breeds receive 1 lb. concentrate mixture for each 3 lb. milk produced, and cattle of the larger breeds, producing low fat milk, receive 1 lb. feed for every 4 or 5 lb. milk produced. If high production is desired, it should be profitable to increase the amount of concentrate until the last added pound produces just enough extra milk to pay for it.

In the *summer* good quality PASTURE is the basic dairy feed in most sections of the country. Most improved pastures contain enough PROTEIN, CALCIUM, and VITAMINS to support good production, and enough energy for 20 to 30 lb. milk per cow. The limiting factors for milk production in most pastures are ENERGY and PHOSPHORUS. Therefore, cows on good pastures need extra grain to balance their RATION; they may also need a simple mineral supplement. Feed about 1 lb. concentrate for every 6 lb. milk produced, or omit the feeding of concentrate for the first 20 lb. milk and then add it at the same rate as you would in stall-feeding for milk above 20 lb.

The dairy cow rations table shows some grain rations that contain 16 percent crude protein and lists methods of adjusting these to other protein levels.

DAIRY COW RATIONS
Amounts to mix approx. 1 ton of feed

Ration:	1	2	3	4	5
Ground corn or milo, lb.	650	1000	800	400	750
Ground oats, lb.	650	600	...	400	400
Ground wheat or rye, lb.	...	...	...	200	...
Ground barley, lb.	...	...	200	...	...
Beet pulp, lb.	...	...	200	...	100
Corn gluten feed, lb.	...	...	...	200	...
Corn distillers' grains, lb.	...	...	...	200	...
Wheat bran, lb.	350	...	450	400	200
Soybean, cottonseed, or peanut meal, lb.	300	350	160	50	300
Linseed oil meal, lb.	...	...	150	50	...
Molasses, lb.	...	...	...	50	200
Salt, lb.	20	30	30	30	30
Steamed bonemeal, lb.	20	20	20	20	20
% Crude protein, approx.	16	16	16	16	14

Ration 1—University of Missouri herd ration.
Ration 2—University of Illinois herd ration.
Ration 4—Commercial ration.

The protein content of these rations may be adjusted by 2% by exchanging 110 lb. corn and 110 lb. soybean meal. To change ration 1 to an 18% dairy feed, remove 110 lb. corn and add 110 lb. soybean meal. To change ration 2 to a 12% dairy feed, remove 220 lb. soybean meal and add 220 lb. corn.

To Guernseys and Jerseys, feed 1 lb. grain mix for each 3-4 lb. milk produced/day.

To Holsteins, Ayrshires, and Brown Swiss, feed 1 lb. grain mix for each 4-5 lb. milk produced/day.

To cows on good pasture, feed 1 lb. grain mix for each 6-7 lb. milk produced/day.

CALF RATIONS

A number of satisfactory systems have been worked out for the feeding of calves:

1. *The nurse-cow method* is favored by many farmers because it produces a superior calf with less trouble. The main limitation is that cows who have suckled calves do not produce as much MILK as do similar cows who have been machine milked during their entire lactation period. However, it is possible to use a nurse cow with several calves each year, and if carefully planned, this would be a reasonably efficient operation.

The addition of suitable ANTIBIOTICS providing about 40 mg. per calf per day has resulted in improved performance if calves are not nursing cows. Research workers are not agreed as to the length of time that antibiotics should be left in the ration, but the best response is obtained until the calves are eight weeks of age, with some investigators reporting responses up to 16 weeks.

2. *Limited whole milk plan:* The calf is allowed COLOSTRUM for the first two to three days and is then placed on the regular milk feeding schedule. During the second week fine-stemmed LEGUME HAY and a grain mixture are provided. From the time the calf is weaned until it is six months old it receives hay free choice and is allowed all the grain it will consume up to a maximum of 4 lb. per day.

During this period the calf can be allowed good PASTURE if available, but the grain and hay should still be in front of it. Good quality SILAGE can be introduced into the ration after two to three months, if care is taken to remove the uneaten parts to prevent spoilage.

Suitable grain mixtures are listed in the calf rations table.

3. *The milk replacer plan:* In many areas whole milk is too valuable to feed to calves after they are two or three weeks old. Well balanced milk replacers can be used under these conditions.

Satisfactory milk replacers usually contain at least 40 percent dairy by-products. Plant by-products, apparently, are not well digested by the young calf. Ample vitamin and mineral fortification should be provided. Several milk replacers are shown in the calf rations table.

HEIFER RATIONS

The feeding of replacement heifers is often severely neglected when they are six to eighteen months old. Some people use any enclosed dry lot and call it "heifer pasture." The animal is developing its size and scale and the capacity to make maxi-

CALF RATIONS
Starter rations

Ration:	1	2	3	4	5 †
Ground yellow corn, lb.	50	49	25	30	15 (dist. sol)
Ground oats, lb.	20	20	20	30	5 (flour)
Wheat bran, lb.	..	..	15	10	..
Soybean oil meal, lb.	25	28	15	5	..
Linseed oil meal, lb.	..	..	5	10	10 (dried whey)
Alfalfa meal, lb.	2	..	7	5	..
Molasses, lb.	..	..	5	..	7 (dextrose)
Dried skimmed milk, lb.	..	..	5	5	50
Blood flour, lb.	..	..	..	2.8	10
Salt, lb.*	1	1	1	1	..
Steamed bonemeal, lb.*	1-4/5	1-4/5	1-4/5	1	..
A and D oil, lb.	1/5	1/5	1/5	1/5	½
Aureomycin or terramycin, gm.	1	1	1	1	1

† Milk replacement.
* If desired, the salt and steamed bonemeal may be replaced by 3 lb. complete mineral mixture (ground limestone 300 lb.; steamed bone meal 300 lb.; salt 350 lb.; iron sulfate 15 lb.; copper sulfate 4 lb.; manganese sulfate ½ lb.; cobalt sulfate ½ lb.; potassium iodine ¼ lb.; magnesium carbonate 20 lb.). These rations were formulated to supply 18% crude protein and small amounts of fiber. They furnish 10,000 I.U. vitamin A and 500 I.U. vitamin D per day. Calves are usually placed on these at two weeks of age and allowed all they will eat to a maximum intake of 4 lb. per day.

mum use of roughage. Good quality hay can be the bulk of the ration. Heifers will gain satisfactorily to 1½ lb. a day on good quality roughage. If poor roughage is used, additional amounts of CONCENTRATE should be added, usually at the rate of 1 lb. concentrate for every 100 to 200 lb. of live weight. If the heifers are grazing on unimproved pasture, it may sometimes be necessary to add between 2 to 4 lb. good hay and 1 to 4 lb. grain to get satisfactory gains.

SUPPLEMENT TO ADD TO FARM GRAINS WHEN CALVES ARE BEING CREEP FED ON LOW-YIELDING OR DEAD PASTURES

Ration:	1	2
Soybean oil meal, lb.*....	40	30
Linseed oil meal, lb.*.....	42½	32½
Cottonseed oil meal, lb.*..	..	20
Alfalfa leaf meal.........	10	10
Steamed bonemeal, lb.†...	5	5
Salt, lb.†...............	2½	2½

  * May be replaced by other oilseed meals if price or availability dictates.
  † May be replaced by 7.5 lb. complete mineral mixture.

DAKOLD. *See* RYE (variety).

DAKOTA AMBER SORGO and *Dakota Amber 39-30-S*, more often called *39-20-S*, are (Black)Amber-type forage SORGHUMS.

DAKOTA COMMON ALFALFA belongs to the common ALFALFAS.

DALLISGRASS (*Paspalum dilatatum*), which is an upright-growing bunching grass, is the most winter-hardy and most widely adapted of the PASPALUM GRASSES. It can be grown as far north as the Carolinas and thrives in the irrigated sections of the milder parts of the southwestern states.

Dallisgrass requires a moist but not wet soil; growth is best where organic matter is abundant. Because it seldom forms a dense sod, it is an excellent grass to mix with LEGUMES and with other GRASSES. It is used for GRAZING, HAY making, and erosion control.

Seeded alone, it often fails to make a perfect stand. Because the seed is produced throughout the summer, it is possible to harvest two or more crops in a season.

Dangers: The germination of Dallisgrass seed is often poor because of the fungus disease ERGOT which attacks or destroys the seed.

Reference: H.1.

See also DALLISGRASS POISONING; LAPPA CLOVER; RICE; ANNUAL LESPEDEZAS; PASTURES; PASTURE PLANTS.

Dallisgrass. Plant, two views of spikelets, and floret.(H.26.)

DALLISGRASS POISONING occurs in cattle grazing on PASTURES where DALLIS-GRASS grows. The term is a misnomer because the toxic substance is in an ERGOT that grows on the Dallisgrass seed rather than the grass itself; symptoms develop only when animals are eating plants carrying seeds that are infested with the ergot. Since such infestation is nearly universal in the South and Southeast, practically every pasture in which Dallisgrass is producing seed is potentially dangerous. Poisoning is usually most prevalent in the late summer and fall and occurs most frequently on good pastures which are not overgrazed. A few cases have been seen in cattle and horses during the winter when the animals were eating Dallisgrass hay.

Symptoms have been observed in cattle four or five days after they were placed on pastures containing large amounts of Dallisgrass seed. Excitability and muscular inco-ordination are the two principal symptoms. Once an affected animal begins moving by flexing the limbs with each stride, there is a tendency for the movements to become more emphasized until the animal finally falls. Sometimes a poisoned animal goes down on its knees and remains in this position for several seconds.

Unless they are severely poisoned, most affected animals will struggle for a few seconds after falling, then become quiet, and get up and walk after a minute or two. If they are not excited any more, they may move quietly and soon begin grazing, but any attempt to handle or drive them usually brings on another period of excitement followed by falling. When the poisoning is well developed, the animals, when standing, usually show muscular tremors and, occasionally, a palsy-like movement of the head. Even when moving slowly such animals may sway from side to side.

Absolute quiet and freedom from excitement seem to be the best treatment. Steps should be taken immediately to stop the eating of ergotized Dallisgrass. If it is necessary to drive the herd to a new pasture, the animals should be moved very slowly and very quietly. If animals are down, it will probably be advisable either

to move them to shade, using a drag slide or wagon, or to build a temporary shelter over them. Under this management nearly all animals will recover in from five to twenty days. Owners should not persist in trying to drive an affected animal to the barn after it has had several muscular spasms and has fallen.

If seed production is heavy and ergot is plentiful, mowing the pasture may be the best control method. If the clippings are heavy enough they may be fed as hay during the winter with little danger, especially if this hay is not the whole source of feed.

Reference: S.28.

See also GRAZING.

DAMAGED BEANS. *See* FIELD BEANS.

DAMAGED PEANUTS. *See* PEANUT.

DAMPING-OFF is a disease of seeds or young seedlings caused by FUNGI; it is most evident in young seedlings that topple over and die just after they emerge from the soil.

The fungus *Rhizoctonia solani* is the cause of damping-off of CROTALARIAS; it may result in a considerable degree of loss during periods of cool weather in spring.

Reference: M.14.

DANE'S EARLY TRIUMPH = *Triumph. See* COMMON WHEAT (variety).

DANTHONIA. *D. californica* = CALIFORNIA OATGRASS.

DARSET, the most important DARSO variety, belongs to the grain SORGHUMS.

DARSO, *Oklahoma No. 1 darso,* and *little darso* (better known as *darset*) are grain SORGHUMS.

DASIPHORA. *D. fruticosa* = SHRUBBY CINQUEFOIL.

DASYNEURA. *D. leguminicola* = CLOVER-FLOWER MIDGE.

DAUBENTONIA. *D. drummondii* = POISONVETCH.

DAWN KAFIR, sometimes erroneously called *Dwarf kafir* (a name that belongs to another variety), is a grain SORGHUM.

DAY MILO and *Sixty Day milo,* which is better known as *Sooner,* are grain SORGHUMS.

DD is the abbreviation for *Double Dwarf;* e.g., *DD milo No. 38.*

DDD = TDE.

DDT, or *1,1,1-trichloro-2,2-bis(para-chlo-rophenyl)ethane*—one of the *dichlorodi-phenyltrichloroethanes* and, thus, one of the CHLORINATED HYDROCARBONS—forms a white powder practically insoluble in water but soluble in many organic solvents.

DDT is widely used as insecticide in form of solutions or suspensions applied as sprays, or as dusts or paints.

For the control of the CORN EARWORM, solutions of 1½ lb. DDT in 25 gal. WHITE MINERAL OIL (having a viscosity 65 to 95 Saybolt seconds at 100° F.) may be applied as a fine spray to the individual ears of corn; only the silks should be wetted, but not before they are five to six days old. For protecting large acreages, 3 gal. of a 25 percent DDT emulsifiable concentrate mixed with 7 gal. white mineral oil (viscosity as above) and 90 gal. water, is to be used at the rate of 25 gal. per acre —the first time two days after the silks appear, the second time three days later, or three times at two-day intervals.

Of a 25 percent emulsifiable concentrate, 3 qt. (or 1½ lb. actual DDT)—after dilution with sufficient water—is to be applied per acre for the control of ARMYWORMS, CUTWORMS, ALFALFA WEEVILS, and LYGUS BUGS; 2 qt. of the same emulsion is sufficient for PEA APHIDS.

Good control of the CORN FLEA BEETLE can be obtained by applying a DDT solution as a very fine mist with a power blower to corn plants at the rate of 2 lb. DDT per acre.

A 3-percent dust applied in the late afternoon at the rate of 15 lb. per acre destroys newly hatched VELVETBEAN CATERPILLARS and GREEN CLOVERWORMS.

To control TOBACCO THRIPS on peanuts, a 3-percent DDT dust should be applied at the rate of 10 lb. per acre as soon as the injury appears on the young plants; the treatment is to be repeated in ten days. A DDT-emulsion spray prepared by mixing 3 qt. of a 25-percent emulsifiable concentrate in 5 gal. water can be used in place of DDT dust.

Infestations by the grubs of WHITE-FRINGED BEETLES can be controlled by mixing 10 lb. DDT per acre into the upper 3 or 4 in. of soil when preparing the land for the peanut crop.

Dust containing 5 percent DDT applied at the rate of 20 lb. per acre controls the PEA WEEVIL and the WESTERN SPOTTED CUCUMBER BEETLE. Repeat in three or four days if necessary; 40 lb. per acre are needed to control the fall ARMYWORM.

As 10-percent dust, DDT provides good control of the ARMYWORM and CUTWORM if applied at the rate of 20 lb. per acre.

Spraying or dusting infested plants with DDT every two or three weeks is recommended for the control of the SPOTTED CUCUMBER BEETLE and its larva, the SOUTHERN CORN ROOTWORM. DDT controls also the CLOVER LEAFHOPPER and the JAPANESE BEETLE.

DDT mixed with other insecticides effectively controls STINK BUGS; e.g., as dust, 1¼ lb. DDT and 8 oz. LINDANE per acre, or as emulsion sprays, 1¼ lb. DDT with 2 lb. CHLORDANE or with 2 lb. TOXAPHENE per acre give satisfactory results.

DDT-lindane mixture, a dust containing 5 percent DDT and 8 percent lindane, controls the THREE-CORNERED ALFALFA HOPPER if the dust is applied at the rate of 15 to 20 lb. per acre.

Caution: Do not feed hay or silage from crops that have been treated with DDT to dairy animals, poultry, or meat animals being finished for slaughter. Do not pasture animals on such crops. Fields treated with DDT should be closed to livestock for about two weeks after treatment.

DDT-DUST BARRIER, containing 10 percent DDT, is an effective barrier for migrating CHINCH BUGS, but the bugs may live for several days after having come in contact with the dust. About 50 lb. dust per acre is necessary for satisfactory control.

> NOTE: This CHEMICAL BARRIER is prepared in the same manner as a DINITRO-DUST BARRIER.

Caution: Do not inhale dust; protect the hands while working with it to prevent irritation of the skin. After handling the material bathe thoroughly and wash contaminated clothing with soap and water.

DEATHCAMASSES (*Zygadenus* spp.), sometimes erroneously called *lobelias*, are POISONOUS PLANTS. These grasslike, ALKALOID-containing herbs are not conspicuous until they bloom. When sheep or cattle eat leaves and stems of any deathcamas species at the daily rate of 8 oz. per 100 lb. body weight, they develop such symptoms as vomiting, frothing, and weakness.

Deathcamass. (N.6.)

DEAN. *See* RYE (variety).

DECIDUOUS means: falling off, not persistent; said of leaves which drop off in autumn or of petals which fall before the fruit is formed. Deciduous plants lose their leaves each year.

DECORTICATED means: husked; stripped of the cortex (bark) or any outer covering (of a tree or seed).—*See also* COTTONSEED.

DECUMBENT conveys the idea of weakness; applied to stems, the term means reclining on the ground but with the end ascending.

DEERING VELVETBEAN (*Mucuna deeringiana*) is the most important VELVETBEAN species. It includes a number of useful varieties which are grown widely in the Southeast of the United States.

Deering Velvetbean. (M.3.)

Varieties: *Florida velvetbean*, the oldest of the velvetbeans grown in America, makes a very rank growth of vine and requires a frost-free period of eight to nine months to mature. The purple flowers are borne in clusters 3 to 8 in. long. The pods are 2 to 3 in. long, nearly straight, blunt at each end, and covered with a black velvety pubescence. The seeds are nearly spherical, about ⅜ in. in diameter, and usually grayish, marbled with brown.

Georgia velvetbean, which has also been called "*Ninety-Day Speckled*", "*Hundred-Day Speckled*", and "*Early Speckled*"—names later transferred to the Alabama variety—matures early; it reaches maturity in 110 days and is adapted to all parts of the Cotton Belt. The Georgia variety makes a much less vigorous growth and yields somewhat less seed than the Florida variety.

Arlington velvetbean, developed from an early selection of the Georgia variety, is very similar to it in all respects, except that the Arlington matures earlier, makes less vine growth, and is one of the earliest velvetbeans. It matures seed as far north as Washington, D.C.

Alabama velvetbean, also called *"Hundred-Day Speckled"*, *"Early Speckled"*, and *"Ninety-Day Speckled"*—names originally used for the Georgia velvetbean—is similar to the Georgia variety but makes a more vigorous growth and matures about six weeks later. It is best adapted to the region south of central Georgia, Alabama, and Mississippi; in that region it is the principal variety.

Bush velvetbean, or *Bunch velvetbean,* also called *Florida Bush,* is a nonvining bushy type, branching near the base and ordinarily attaining a height of 3 ft., with an occasional branch 5 to 6 ft. in length. Clusters of pods form a dense mass near the base. The plant matures in about the same time as the Alabama variety but is less productive. The main objections to this variety are that the pods cannot be gathered so quickly as those of the vining varieties, and that, lying so close to the ground, they become watersoaked in wet weather, which causes many of them to decay.

Osceola velvetbean is an intermediate type (not a true hybrid), developed from plants belonging to two velvetbean species: the Florida and the LYON VELVETBEANS. It is adapted to the region south of a line running through the center of Georgia, Alabama, and Mississippi.

Reference: P.10.

See also LEGUMES; INOCULATION.

DEFERRED-AND-ROTATION MANAGEMENT. *See* RANGE MANAGEMENT.

DEFERRED GRAZING.

See RANGE MANAGEMENT.

DEFICIENCY.

See MINERAL DEFICIENCY; VITAMINS.

DEFLUORINATED PHOSPHATE.

See MINERAL FEEDS.

DEGERMED is a term which must precede the name of any feed product from which the germ has been wholly or partially removed.

Reference: F.6.

DEGOSSYPOLIZED COTTONSEED MEAL. *See* COTTONSEED.

DEGREASED DEHYDRATED GARBAGE. *See* DEHYDRATED GARBAGE.

DEHULLED SOLVENT EXTRACTED SOYBEAN FLAKES. *See* SOYBEAN.

DEHULLED SOLVENT EXTRACTED SOYBEAN OIL MEAL. *See* SOYBEAN.

DEHYDRATED is a term which may precede the name of any feed product, provided it has been artifically dried. When water is removed from finished mixed feed for domestic animals, the term "dehydrated" may be used in (or parenthetically added to) the brand name. However, when the ingredients of such feeds are dried separately and then mixed, or when the previously dried ingredients are mixed, moistened with water, and formed and baked into biscuits, the term "dehydrated" is not to be used in the brand name.

Reference: F.6.

DEHYDRATED ALFALFA.

See ALFALFAS.

DEHYDRATED BROCCOLI LEAF MEAL. *See* BROCCOLI.

DEHYDRATED CARROT LEAF MEAL. *See* CARROT.

DEHYDRATED GARBAGE, an officially recognized feedstuff (formerly called *processed garbage*), is composed of artificially dried animal and vegetable waste from GARBAGE that is collected sufficiently often to avoid harmful decomposition and that has been freed of crockery, glass, metal, string, and similar materials. The odor must not be suggestive of the presence of decomposition. The feedstuff shall not contain more than 0.2 percent glass, none of which shall be knifelike or needle-like particles. It shall be processed at a temperature sufficient to destroy all organisms capable of producing animal diseases. If part of the grease and fats is removed, it shall be designated as *degreased dehydrated garbage.*

Reference: F.6.

See also MISCELLANEOUS PRODUCTS.

DEHYDRATED KALE LEAF MEAL.

See KALE.

DEHYDRATED POTATOES.

See POTATO.

DEHYDRATED SORGHUM MEAL and *dehydrated sweet sorghum meal* are prepared from forage SORGHUMS.

DEHYDRATED TURNIP LEAF MEAL.

See TURNIP.

DEHYDRATER. *See* HAY DRYING.

7-DEHYDROCHOLESTEROL, or *pro-vitamin D₃*, occurs in higher animals. Irradiation with ultraviolet light changes it to VITAMIN D₃.

DELIQUESCENT means: gradually liquefying by absorbing water from the humid air.

DELPHINIUM. The *Delphinium* spp. are known as LARKSPURS.

DELSOY. *See* SOYBEAN (variety).

DELSTA. *See* SOYBEAN (variety).

DELTA-TOCOPHEROL. *See* VITAMIN E.

DENSITY is (1) the weight of any substance contained in a given unit of volume or (2) the relative degree to which vegetation covers the ground surface; the latter is often expressed in tenths, 1.0 indicating a complete ground cover.

Average density is a term used to indicate the proportion of ground surface actually covered by herbaceous or shrubby vegetation within reach of livestock.

The *density of grasses* is based on the spread when the plants are grazed to the proper extent rather than (a) on the normal plant spread when ungrazed or (b) on the reduced spread when total use has been made of the plant.

The *density of browse* is estimated from the ground surface covered by that part of the browse that is accessible to livestock.

The *density of erect weeds* is based on the amount of ground that appears covered when the vegetation is viewed directly from above.

Reference: D.9.

DENT CORN. *See* CORN.

DENTON SORGO is one of the forage SORGHUMS.

DERIVATIVE is (1) a chemical compound derived from another one by a simple chemical process or (2) a plant derived from one of the two or more parents involved in its ancestry; thus, Midland, Martin, and Plainsman may be referred to as milo derivatives, although kafir also was involved in their parentage.—*See also* STRAIN.

DERMATITIS is an inflammation of the skin. It can be due to a great variety of causes, such as contact with certain substances, infection, infestation with para-

sites, deficiencies in the ration, etc.—*See also* VITAMIN B₂, PANTOTHENIC ACID.

DERRIS (pronounced: *dare*-is) is a genus of tropical plants containing ROTENONE. *Derris root*, particularly that obtained from the species *D. elliptica*, after being finely ground, is widely used in INSECTICIDES.

For the control of the MEXICAN BEAN BEETLE, derris root containing 4 percent rotenone is mixed at the rate of 1½ lb. to 50 gal. water or 12½ lb. to 87½ lb. diluent, such as finely ground TALC, CLAY, SULFUR, GYPSUM, or TOBACCO (but not with lime). The treatment is to be repeated every week or ten days.

DESERT WHEAT CORN. *Mexican desert wheat corn*, better known as *shallu*, is one of the grain SORGHUMS.

DESERT WHEATGRASS
    = STANDARD CRESTED WHEATGRASS.

DESMODIUM, commonly called *tick clover*, is a genus of which FLORIDA BEGGARWEED, or *D. purpureum*, is a species.

DE SOTO. *See* COMMON OAT (variety).

DETERGENTS are synthetic cleansing agents which are widely used in place of soaps.—*See also* SURFACTANTS.

DEVIL GRASS = BERMUDA-GRASS.

DEVIL'S-GUTS, *devil's hair*, and *devil's ringlet*, are common names of DODDER.

DEXTRINS are products obtained by the partial chemical breakdown of STARCH by the action of heat, acids, enzymes, or a combination of these agents. The dextrins are not definite chemical compounds but usually contain a mixture of CARBOHYDRATES ranging between the simple SUGARS (such as dextrose), MALT SUGAR, and the original starch; they differ from starch in that they dissolve in water. Dextrins are constituents of SIRUPS and MOLASSES.

DEXTROSE is a crystallized SUGAR obtained chiefly by hydrolysis of starch or a starch-containing substance; when produced from corn starch, it is called *corn sugar*.—*See also* GLUCOSE; FEED INGREDIENTS.

DI- is a prefix, signifying two or double.

DIABROTICA. *D. longicornis* = CORN ROOTWORM; *D. undecimpunctata* = WESTERN SPOTTED CUCUMBER BEETLE; *D. undecimpunctata howardi* = SOUTHERN CORN ROOTWORM.

DIAPORTHE. Two varieties of the FUNGUS *D. phaseolorum*—var. *batatatis* and var. *sojae*—cause POD AND STEM BLIGHT.

DIASTASE is an ENZYME obtained from malt; it hydrolyzes starch into MALT SUGAR via DEXTRIN. It forms a light-colored powder which is water-soluble.—*See also* BREWERS' PRODUCTS; FERMENTATION.

DIATRAEA. *D. grandiosella* SOUTH-=WESTERN CORN BORER; *D. saccharalis* = SUGARCANE BORER.

DIBASIC CALCIUM PHOSPHATE = DICALCIUM PHOSPHATE.

DICALCIUM PHOSPHATE, or *dibasic calcium phosphate*, is available in form of a white crystalline powder. It contains over 23.2 percent CALCIUM and 18.0 percent PHOSPHORUS.—*See also* MINERAL FEEDS; CALCIUM PHOSPHATES.

DICALCIUM PHOSPHATE FROM BONE = *precipitated bone phosphates*. *See* MINERAL FEEDS.

DICENTRA. *D. cucullaria* = DUTCHMAN'S BREECHES.

1,1-DICHLOR-2,2-BIS (PARA-CHLOROPHENYL) ETHANE = TDE.

DICHLORODIPHENYLDICHLOROETHANE = TDE.

DICHLORONAPHTHOQUINONE, also known as *dichlone*, forms yellow needles and is practically insoluble in water. This FUNGICIDE, the active ingredient of PHYGON, is used for SEED TREATMENT.

DICOUMAROL, or *dicumarol*, is contained in spoiled SWEETCLOVER and LESPEDEZA hay or silages. It is an anticoagulant which removes VITAMIN K from the body and causes hemorrhages. If digested by cattle, it may cause the animals to bleed to death internally or following a small external wound.—*See also* SWEETCLOVER HAY POISONING.

DIELDRIN, an oxidation product of ALDRIN, is a CHLORINATED HYDROCARBON. The commercial product, containing 85 percent of the active ingredient, forms a buff, crystalline powder which is insoluble in water, but soluble in organic solvents. It is a valuable INSECTICIDE.

To control GRASSHOPPERS, ½ to ¾ oz. dieldrin per acre is needed; this amount suffices for a period of four to seven weeks.

For the control of both, the RED HAR-VESTER ANT and the WESTERN HARVESTER ANT, a dust containing 2 percent dieldrin may be spread thinly in a continuous band 4 to 6 in. wide, making a circle 5 to 6 ft. in diameter centering at the nest entrance. For small colonies, the band of dust may be placed around the edge of the cleared nest area.

To control WIREWORMS, 2 lb. dieldrin per acre is required; the material must be properly mixed into the soil; it can be applied as dust, emulsion, in granular form, or in combination with a suitable fertilizer.

Dieldrin is also recommended as crop spray to prevent migration of CHINCH BUGS from small grains to corn. The spray should be applied to the base of the plants (where the bugs congregate) and the bugs wet with the spray.

For the control of the RICE LEAF MINER, 8 oz. (actual chemical) per acre in 10 gal. water is applied by airplane.

To kill the adults of the ALFALFA WEEVIL, apply as a spray 4 oz. dieldrin per acre when the spring growth of alfalfa is 1 to 2 in. tall.

Caution: Dieldrin is poisonous. Wear a respirator while applying a dust. Bathe and change to clean clothing after dusting. Handle dieldrin-dusted clothing carefully. Do not dust vegetables that may be eaten by man, dairy animals, or animals being finished for slaughter.

Antidotes: Wash thoroughly with soap and water. Take 1 tablesp. salt in a glass of warm water to induce vomiting and repeat until vomit fluid is clear. Lie down; remain quiet; call a doctor.

DIENESTROL DIACETATE is a female sex hormone-like compound similar to STILBESTROL. It forms fine needles which are soluble in vegetable oils. Such solutions (or suspensions) are DRUGS rather than nutrients and when used in feed must be labeled in accordance with the FDA regulations.

The drug is sometimes mixed into regular poultry rations at the rate of 0.007 percent and fed for three weeks before slaughter. The treated birds produce capon-like carcasses.—*See also* HORMONES.

DIETHYLSTILBESTROL
= STILBESTROL.

DIETZ = *Fulcaster. See* COMMON WHEAT (variety).

DIFFERENTIAL GRASSHOPPER.
See GRASSHOPPERS.

DIGESTER TANKAGE.
See ANIMAL PRODUCTS.

DIGESTER TANKAGE WITH BONE.
See ANIMAL PRODUCTS.

DIGESTIBILITY of a nutrient refers to the amount of the nutrient that is absorbed from the intestine. Usually digestibilities are expressed as percentages and are called *digestion coefficients*, or *coefficient of digestibility*. Two methods for determining digestibility are used: (1) The *total collection method* which consists of feeding a known amount of a feedstuff of known chemical composition to experimental animals and analyzing the weighed feces (representing the undigested portions of the various nutrients of the feed). The difference between amount of nutrient fed and amount of nutrient excreted is the amount digested; from it one can easily calculate the digestion coefficient. (2) The *reference method* which uses an inert marker and indirect calculations.

Both methods give estimates of apparent digestibility. For careful studies, *true digestibilities*, obtained after correcting for metabolic nutrients, are more reliable.

The digestibility of feedstuffs differs with different animal species, and even within the same species (because of age and other factors). Ruminants digest feed rich in fiber better than swine, horses, or poultry. Carbohydrates, in general, are digested more easily than fat or protein. —*See also* DIGESTION; DIGESTIVE TRACT; ABSORPTION.

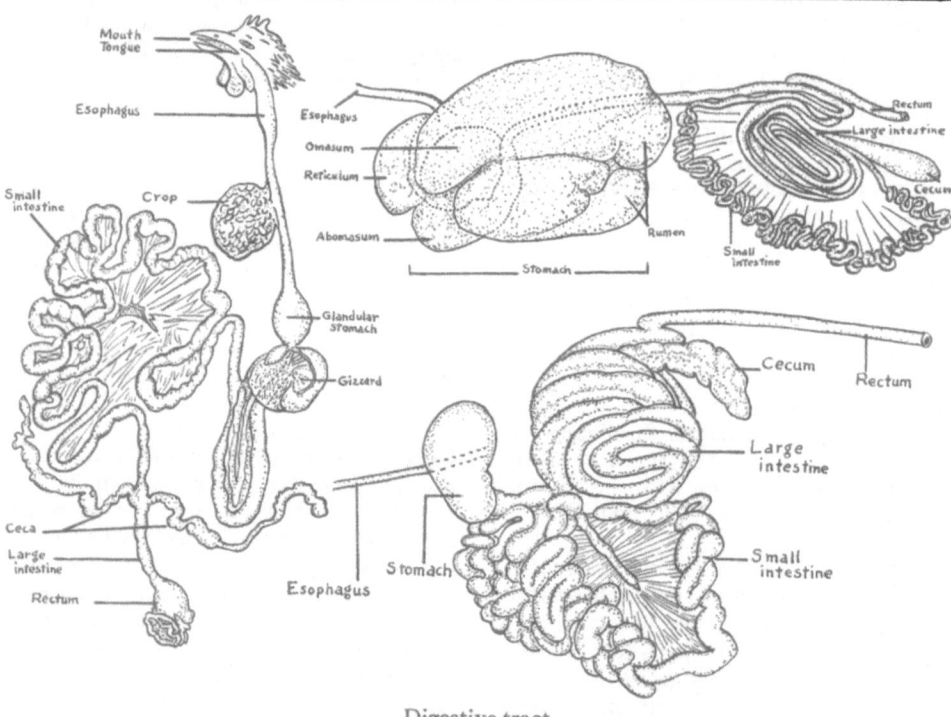

Digestive tract.

Left, fowl: Av. capacity of gizzard, 1½ oz.; av. length of small intestine, 62 in., of large intestine, 4½ in. *Top right*, cattle: capacity of stomach, 65 gal.; av. length of small intestine, 130 ft., of large intestine, 30 ft. *Bottom right*, swine: capacity of stomach, over 2 gal.; av. length of small intestine 60 ft., of large intestine, 16½ ft. (S.33.)

DIGESTIBLE ENERGY *See* ENERGY.

DIGESTIBLE NUTRIENT is the part of each NUTRIENT which can be digested and absorbed into the blood and lymph systems.—*See also* DIGESTIVE TRACT.

DIGESTION is the breakdown (by hydrolysis) of complex foods into simple compounds; e.g., the change of proteins, starches, and fats, under the influence of ENZYMES, into AMINO ACIDS, GLUCOSE, FATTY ACIDS, and GLYCERIN which are absorbed into the blood and lymph systems. —*See also* ABSORPTION; BACTERIA.

DIGESTION COEFFICIENT.
                          See DIGESTIBILITY.

DIGESTIVE JUICES.
                     See DIGESTIVE TRACT.

DIGESTIVE TRACT, or *alimentary tract*, is the place where DIGESTION occurs as the result of chemical processes following the chewing and swallowing of the feed. In the STOMACH, the *gastric juices* are added and their ENZYMES, especially the *proteolytic enzyme* pepsin, soon act upon the feed's proteins. In the duodenum, the *pancreatic juices* are mixed to the partly digested feed, and their enzymes act upon fats as well as carbohydrates and proteins: *bile* from the liver also enters the duodenum, aiding in the breaking down of the fats. In the small intestine, still more *digestive juices* with their specific enzymes are secreted into the feed mass, for the purpose of digesting the last remains of the nutrients.

The *digestible nutrients* of the feed are absorbed into the body proper, while the undigested parts are excreted from the body in the feces. The digestive tract also includes the *mouth* where saliva moistens the food and *salivary ferments* start the break-down of starch; the *esophagus* (gullet), i.e., the portion between larynx and stomach; the *large intestine* where water is absorbed; the *cecum* (blind gut) which forms the first part of the large intestine; the *colon* which extends from the cecum to the rectum; and the *rectum*, the terminal portion of the digestive tract, with the *anus* as its lower opening.—*See also* ABSORPTION; FERMENTS.

DIGITARIA. *D. decumbens* = PANGOLA-GRASS; *D. sanguinalis* = CRABGRASS.

DIGITATE, or *palmate*, means: finger-like, as the digitate leaflets of lupines—a compound leaf with five similar parts radiating from a common point.

Digitate leaf of lupine. (D.9.)

DIHYDROSTREPTOMYCIN.
                         See STREPTOMYCIN.

DILAURYL THIODIPROPIONATE is used, similarly to DISTEARYL THIODIPROPIONATE, as ANTIOXIDANT.

DILUENT is an inert ingredient added to an active ingredient so as to increase the bulk of insecticidal or fungicidal powders, etc.—*See also* CUBE; DERRIS; CRYOLITE.

4,4-DINITROCARBANILIDE.
                           See NICARBAZIN.

DINITRO-DUST BARRIER. A 4-percent dinitro dust—consisting of DINITRO-ORTHOCRESOL, DINITRO-ORTHOCYCLOHEXYL-PHENOL, or DINITRO-ORTHOISOPROPYL-PHENOL—has proved to be an effective chemical barrier for migrating CHINCH BUGS. It is advisable to use 2 lb. of the dust per rod of barrier. The dust is applied in a continuous band 1 to 2 in. wide in a shallow depression or track made by dragging a log or driving an automobile or small truck along the field. About 5 percent mineral oil (100 seconds Saybolt viscosity at 100° F.) is sometimes added to give the dust stability and prevent it from being blown away. The bugs, upon coming in contact with the dust, are poisoned and die within a few hours.

The dust may be applied from a tin can or with a special dust applicator. After the dust line is made, it should be patrolled every day during the migration period to remove leaves and other debris that may have been blown onto it. Breaks in the line must be repaired. Lines that have been destroyed by heavy rain should be completely renewed.

Caution: The dinitro dusts are injurious to plant life and therefore should not be applied directly to crops. Do not inhale the dusts; protect the hands while working with them to prevent irritation of the skin. After handling the material bathe thoroughly and wash contaminated clothing with soap and water.

Reference: P.5.

See also CHEMICAL BARRIERS.

DINITRO-ORTHOCRESOL forms yellow crystals and is only sparingly soluble in water. It is used in DINITRO-DUST BARRIERS for migrating CHINCH BUGS.

Caution: Dinitro-orthocresol is poisonous and must be handled with care. Avoid breathing the dust.

DINITRO-ORTHOCYCLOHEXYL-PHENOL is closely related to, but more effective than, DINITRO-ORTHOCRESOL and used like it to control migrating CHINCH BUGS by means of DINITRO-DUST BARRIERS.

DINITRO-ORTHOISOPROPYL-PHENOL, one of the compounds used in DINITRO-DUST BARRIERS, is recommended for the destruction of migrating CHINCH BUGS.

DINITRO - (ORTHO-SECONDARY-)-BUTYLPHENOL is reddish-brown liquid, slightly soluble in water, but forming water-soluble salts which readily liberate the parent compound. It is used as insecticide and as herbicide.—*See also* PEANUT.

DIOECIOUS (or *diecious*) means: the male and female flowers are borne on different individual plants. For example, the flowers of TEXAS BLUEGRASS are dioecious. —*See also* MONOECIUS.

DIPLODIA. The FUNGUS *D. theobromae* is the main cause of CONCEALED DAMAGE OF SEED.

DIRT. *See* SCREENINGS.

DIRT-RIDGE CREOSOTE BARRIER is used for the control of migrating CHINCH BUGS.—*See also* CREOSOTE BARRIERS.

DISACCHARIDES. *See* REDUCING SUGARS; NONREDUCING SUGARS.

DISCOLORATION. *See* NONPARASITIC SORGHUM-LEAF DISCOLORATIONS.

DISEASES of plants are the result of injury or irritation following attacks by insects, micro-organisms, or environmental conditions, such as moisture or drought, lack or surplus of nutrients, abnormal soil acidity or alkalinity, etc. Some diseases are as yet of unknown cause; e.g., BUD ROT. —*See also* BACTERIAL DISEASES; FUNGUS DISEASES; VIRUS DISEASES.

DISINFECTANT is any material that kills micro-organisms. Labeling of disinfectants must comply with regulations of the U.S. Department of Agriculture.—*See also* SEED ROT; LEGUME BACTERIA.

DISINFESTANT is an agent that removes or inactivates external pests found on plants or animals. Agents used for killing these parasites must be labeled so as to comply with regulations of the U.S. Department of Agriculture.

DISK PLOWS are important for plowing loose soils, dry and hard soils, and sticky soil. The DISKS replace the moldboard of the common moldboard plow. They measure 20 to 30 in. and make 4- to 10-in. deep furrows.—*See also* ONE-WAY DISK PLOW.

DISKS are often found on intertillage implements, such as CULTIVATORS, PLOWS, and HARROWS. On the latter, they may be set to cut weeds on the sides of the furrows. —*See also* DISK PLOW; ONE-WAY DISK PLOW.

DISODIUM ETHYLENE BIS-DITHIO-CARBAMATE. *See* DITHANE.

DISTEARYL THIO-DIPROPIONATE, an ester of the *thio-dipropionic acid*, is one of the ANTIOXIDANTS used in concentrations not exceeding 0.09 percent.

DISTICHLIS. *Distichlis* spp. = SALT-GRASSES.

DISTILLERS' PRODUCTS are obtained as by-products in distilleries. The distillery by-products are the residues after the production of alcohol from grain. Several processes are involved: (1) Grain starch is converted into sugars by enzymatic action; (2) the sugars are fermented to alcohol and carbon dioxide by the actions of YEAST and other micro-organisms; and (3) the alcohol is removed by distillation. The remaining material—the *whole stillage*—contains the unfermentable part of the grain consisting mostly of protein, fats, and minerals, along with unfermentable carbohydrates. It also includes most of the vitamins present in the original grain, together with those formed by the yeast during the fermentation process. Most distillers screen the whole stillage to remove the grains, which are then pressed and dried to produce *distillers' dried grains*. The liquid remaining is known as *distillers' thin slop*. When this slop is evaporated under vacuum and then dried on rotary driers, the resulting product is called *distillers' dried solubles*. If the solubles are dried on the grain residue, the resulting product is *distillers' dried grains with solubles*. Sometimes the by-products may be fed as WET BREWERS' GRAINS or as *slop*.

Distillers' dried grains contain relatively high percentages of protein and fiber; their contents of fat and of vitamins add to their value. *Distillers' dried solubles* contain less crude fiber and fat but include substantial amounts of the water-soluble vitamins VITAMIN B$_2$, PANTOTHENIC ACID, and NIACIN, in addition to the high contents of proteins, minerals, and lactic acid. Since distillers' dried grains with solubles are a combination of distillers' dried grains and solubles, their composition is intermediate between the two, depending on the amount of solubles recombined. When the latter is 50 to 55 percent, the distillers' grains with solubles are representative of whole stillage.

Feeding tests have shown that the distillery by-products have wide application in the feeding of all classes of livestock. In general, the dried grains and the dried grains with solubles have found their widest use through practical experience in rations for dairy cattle, beef cattle, and sheep.

Although not so high in protein as the OILSEED MEALS, distillers' grains rank well in feeding value. Fed on an equal protein basis, there usually is little choice; e.g., in the feeding of soybean meal and distillers' grains to fattening *cattle*. Distillers' dried solubles are also a satisfactory substitute for soybean oil meal as the protein concentrate for the drylot fattening of cattle.

Milo or corn distillers' dried solubles make a very satisfactory protein supplement for yearling steers being wintered on carbonaceous roughages and to be grazed the following season without grain.

The distillery by-products have had wide use as an ingredient in mixed concentrate feeds for milking cows. Both distillers' solubles and distillers' grains with solubles have been found to be excellent in feed mixtures for calves. In general the dried grains from corn and wheat have given somewhat better results than the products from rye.

As a *swine* feed, distillers' solubles are preferred to distillers' grains. Distillers' dried grains with solubles stand between them. This is largely explained on the basis of fiber and vitamin contents. The proteins, being mainly those from the original grains used in the fermentation mashes, are not of so high a quality as those in the animal-protein supplements or in most of the oilseed meals. Distillers' solubles included in the diet at levels of 5 to 10 percent make a significant contribution of vitamin factors to many swine rations.

Distillers' solubles are an excellent supplement for market-pig rations, and under certain market conditions this product is a very economical protein supplement. The product is useful in rations for growing and fattening pigs and for pregnant and lactating cows. Its greatest value lies in its vitamin content.

If adequate supplements of other high-quality protein feeds are included, distillers' solubles and distillers' grains with solubles help make up the protein deficit of corn in the ration.

Distillers' grains with solubles contribute protein and vitamins to *poultry* diets, but their high content of fiber limits their use.

Dried distillers' solubles are important sources of vitamin B$_2$ and other water-soluble vitamins for poultry; sometimes, however, these solubles produce a laxative effect if fed as 5 percent or more of the diet, especially if the diet contains other ingredients that tend to be laxative. Ordinarily the levels required to supply riboflavin are well below the laxative level.

Officially recognized are the following distillers' products:

1. *Corn* (respectively, *rye, wheat, grain sorghum*) *distillers' dried grains*—obtained in the manufacture of distilled liquors and alcohol from CORN. (respectively, RYE, WHEAT, GRAIN SORGHUM) or from a grain mixture in which corn (respectively, rye, wheat, grain sorghum) predominates, by drying that portion of the whole stillage retained by screens.

2. *Corn* (respectively, *rye, wheat, grain sorghum*) *distillers' dried solubles*—obtained in the manufacture of distilled liquors and alcohol from corn (respectively, rye, wheat, grain sorghum) or from a grain mixture in which corn (respectively, rye, wheat, grain sorghum) predominates, by condensing and drying the screened stillage.

3. *Corn* (respectively, *rye, wheat, grain sorghum*) *distillers' dried grains with solubles*—obtained in the manufacture of distilled liquors and alcohol from corn (respectively rye, wheat, grain sorghum) or from a grain mixture in which corn (respectively, rye, wheat, grain sorghum) predominates, by drying that portion of the whole stillage retained by screens to which condensed screened stillage has been added so that the final product contains not less than $3/4$ of the solids content of the whole stillage.

4. *Molasses distillers' dried solubles*—obtained by drying the residue from the yeast fermentation of MOLASSES after the removal of the alcohol by distillation.

5. *Molasses distillers' condensed solubles*—obtained by condensing to a sirupy consistency the residue from the yeast fermentation of molasses after the removal of the alcohol by distillation.

6. *Potato distillers' dried residue* is the dried product obtained after the manufacture of alcohol and distilled liquors from POTATOES or from a mixture in which potatoes predominate.

References: E.12; F.6; L.11; G.12; H.30.

See also FERMENTATION INDUSTRY BY-PRODUCTS.

DISTILLERS' BY-PRODUCTS.
                    See DISTILLERS' PRODUCTS.
DISTILLERS' DRIED GRAINS.
                    See DISTILLERS' PRODUCTS.

DISTILLERS' DRIED GRAINS WITH SOLUBLES. *See* DISTILLERS' PRODUCTS.
DISTILLERS' DRIED SOLUBLES.
                    See DISTILLERS' PRODUCTS.
DISTILLERS' THIN SLOP.
                    See DISTILLERS' PRODUCTS.
DITCH FESCUE = TALL FESCUE.
DITHANE is a water-soluble FUNGICIDE based on *disodium ethylene bis-dithiocarbamate*; it is effective for the control of NORTHERN CORN LEAF BLIGHT when dusted or sprayed on corn soon after silking.
DITROPINOTUS. *D. aureoviridis* is one of the WASPS that are parasites of the WHEAT STRAWWORMS.
DITYLENCHUS. *D. dipsaci* = STEM NEMATODE.
DIURETIC is any drug which increases urine excretion; e.g., AMMONIUM PHOSPHATES, GLAUBER'S SALT, or UREA.
DIXIE CRIMSON CLOVER.
                    See CRIMSON CLOVER (variety).
dl - 2 - HYDROXY - 4 - METHYL-THIOBUTYRIC ACID CALCIUM = *methionine hydroxy analogue calcium.* *See* AMINO ACIDS.
dl-METHIONINE. *See* METHIONINE.
DOCKAGE. *See* GRAIN GRADING.
DODDERS (*Cuscuta* spp.) are troublesome WEEDS which are known by various names; e.g., *love vine, strangleweed, devil's-guts, goldthread, pull-down, devil's ringlet, hellbind, hairweed, devil's-hair,* and *hailweed.*

When large enough to become noticeable, dodder is a slender twining parasite. The tough, curling, threadlike, leafless stems of dodder are usually yellowish or orange, sometimes tinged with red or purple, and occasionally almost white. Close examination of the stem will reveal minute rudiments of leaves.

Dodder is of economic importance principally because of the damage it causes in CLOVER and ALFALFA. A number of other plants are subject to dodder infestation; the most important of these are the LESPEDEZAS, SUGAR BEETS, FLAX, BIRDSFOOT TREFOIL, RAGWEED, wild shrubs, and various wild grasses.

The dodder obtains its nutrition by sinking minute suckers into the food-conducting tissues of the host, thus extracting food that is utilized by the dodder in growth. The dodder climbs, gaining growth and strength at the expense of the plant upon which it grows. As it reaches the top of the host, the twisting tips of the dodder reach out and attack adjacent plants,until a gradually increasing circle of infestation is formed. Meantime, the plants first attacked are usually killed, and the dodder upon them also dies, but the dodder on the newly attacked host at the edge of the infested area keeps alive.

The dodder winters over either by means of seeds which lie dormant in the soil or by means of its stems, which are not always killed by the winter's frost and may renew growth when the favorable conditions of spring arrive.

Dodder may be introduced and disseminated on the farm by impure seed, dodder-infested hay or manure, or dodder seeds, stems, and flowers carried by farm workers, wagons, farm animals, or irrigation water.

Control: Dodder may be prevented from entering the farm by observing the following rules: Do not sow dodder-infested seed. Do not use dodder-infested hay. Do not allow animals from dodder-infested fields to have access to dodder-free fields. Do not allow dodder to grow where irrigation water may carry the pest from place to place. Do not use farm manure suspected to contain dodder seed until such manure has been composted for a period of at least six weeks.

When dodder appears in small scattered patches, mow these areas before the seed has matured (1) allowing the mowed plants to dry or sprinkling them with crude oil or kerosene and burning them, or (2) feeding the cut plants for hay.

When dodder occurs in large areas, it is of extreme importance to attack it before viable seed has formed. Any of the following methods may be used: (1) Early mowing. (2) Grazing; for this purpose sheep are particularly good, although hogs or any other grazing animals may be used. (3) Plowing.

When an entire field of clover or alfalfa is infested with dodder and the weed has been allowed to go to seed, one of the following methods should be selected:

1. *Burning.*

2. *Utilizing the crop for hay.* In case the alfalfa or clover is worth saving, the crop may be cut for hay and fed directly from stacks on the infested field. It should not be forgotten that the resulting manure will contain viable dodder seeds. The following spring, the dodder-infested clover or alfalfa should be cut for hay just before the dodder has opened its flowers. During the remainder of the season, the land must be grazed closely, so that no dodder will be allowed to mature seed. Similar early-cutting and close-grazing treatment should be given any clover or alfalfa grown for the following four years, in order to prevent dodder seed from again fouling the land.

Dodder. (B.28.)

3. *Rotation.* On land known to contain the seeds of dodder it is recommended to use a 5-year cropping system consisting of plants which are immune to the attacks of the pest. The following plants are not susceptible to injury by dodder: CORN, SOYBEANS, VELVETBEANS, COWPEAS, and

SMALL GRAINS, such as OATS, WHEAT, and RYE.

4. *Chemical means of eradication.* Various chemical sprays, such as solutions of IRON SULFATE, COPPER SULFATE, SULFURIC ACID, common salt (SODIUM CHLORIDE), crude carbolic acid (PHENOL), SODIUM NITRATE, and POTASSIUM SULFATE, have been recommended for use against dodder. These chemicals are expensive to apply; for this reason, the use of chemical plant poisons as a means of controlling dodder is not ordinarily to be recommended.

Species: At least 54 species of dodder are found in North and Central America, and the West Indies. The species are very similar in appearance. Some of the dodders show marked preference for certain crops; e.g., *alfalfa dodder,* which is practically never very harmful on any crop except alfalfa, though it may grow on other plants. Conspicuous exceptions to this rule are the *common dodder* and the *field dodder,* which seem indifferent as to their hosts.

Clover dodder rarely seeds in the United States; hence it is to be feared only during the first season after sowing; under favorable conditions it may live over by means of stems.

In addition to COMMON DODDER and FIELD DODDER, the species that are most damaging in the United States are SMALL-SEEDED ALFALFA DODDER, CLOVER DODDER, LARGE-SEEDED ALFALFA DODDER, and CHILEAN DODDER.

Reference: H.19.

See also RAGWEED.

DOG'S TOOTH GRASS
                              = BERMUDA-GRASS.
DOLOMITIC LIMESTONES.
                    See GROUND LIMESTONE.
DOMESTIC HEMP. *See* HEMP.
DOMESTIC RYEGRASS
                         = COMMON RYEGRASS.
DOMINANT is the character of one of the parents of a HYBRID that is manifested in the hybrid to the apparent exclusion of the contrasted (RECESSIVE) character from the other parent. This hereditary character is controlled by one or more GENES with greater activity than the recessive. In some instances the dominant masks the recessive; in others the dominance is incomplete.

DORMAN. *See* SOYBEAN (variety).
DORMANCY is due to external factors.
                    —*See also* DORMANT.
DORMANT is the inactive state, or rest period, of a seed or plant; this state is caused by unfavorable weather conditions.
DORSAL means: upon, or relating to, the back or outer surface of an organ.
DORTCHSOY 2. *See* SOYBEAN (variety).
DOUBLE-CUT CLOVERS include the RED CLOVER varieties *Kenland, Midland, Cumberland,* and *Pennscott.*
DOUBLE DWARF MILO No. 38 is one of the grain SORGHUMS.
DOUBLE DWARF WHITE SOONER is a milo-type SORGHUM.
DOUBLE DWARF YELLOW SOONER.
                         See SORGHUMS.
DOWNY CHESS = CHEATGRASS.
DOWFUME W-40 is the trade name of a water-clear solution containing 41 percent ETHYLENE DIBROMIDE. It is a FUMIGANT particularly useful for preplanting application to the soil.

Dowfume W-40 is effective for the control of the STING NEMATODE, ROOT-KNOT NEMATODE, WIREWORM, and other soil pests. Depending on the crop, 15 to 30 gal. Dowfume W-40 is recommended per acre. The use of fumigation equipment designed for field application is necessary for best results.

Fumigated soil should not be disturbed for a week; if the soil temperature is above 50° F., planting can be done in two weeks, but if it is between 40° and 50°, planting should be delayed for three weeks.

Caution: Avoid breathing the toxic vapors. Avoid contact with clothing, eyes, and skin. Use only with adequate ventilation. Do not use containers or handling equipment made of aluminum, magnesium, or their alloys.

DOWNY MILDEW, a name applied to various FUNGUS diseases affecting leaves, is caused by *Peronospora* spp. which occur widely in temperate regions. The more important types of this disease are the following: DOWNY MILDEW OF ALFALFA; DOWNY MILDEW OF SOYBEAN; DOWNY MILDEW OF VETCH. The fungi attack also SWEETCLOVER, OATS, and other crops.

NOTE: The downy mildews attacking LEGUMES comprise a complex group of many physiological races. Some authorities are inclined to lump (former) species together, on the basis of characteristics, such as host range; hence, the former *P. pisi* is now considered as possibly a specialized race of *P. viciae*.

DOWNY MILDEW OF ALFALFA is caused by the FUNGUS *Peronospora trifoliorum* to which about a fourth of the plants of common ALFALFA varieties are susceptible. In the most susceptible plants the foliage becomes distorted and yellowed. Sometimes the growing end is killed, although vigorous shoots may outgrow the fungus. Spores are produced in succession from the underside of the leaves. Usually only single leaves are infected. Grown buds, infected in late fall, may carry the fungus over winter by invading the shoots developing from them. Probably these shoots are the first source of infection in the spring. In strains that have a high percentage of susceptible plants, the disease attacks seedlings grown under a cover crop in wet weather and damages or destroys them. Because susceptible plants sometimes appear to be the most vigorous and succulent in population, it is easy to introduce them into breeding stocks.

References: C.9; M.18; J.3.

See also DOWNY MILDEW.

DOWNY MILDEW OF SOYBEAN is caused by the FUNGUS *Peronospora manshurica* which produces small pale-green spots, visible on the upper surface of the SOYBEAN leaf. Later the spots become dark gray or brown. On the lower surface of the leaves small grayish tufts of moldy growth develop on the spots. These tufts may drop off, leaving a brown spot. The spores produced on the underside of the leaves spread the infection to other plants throughout the growing season. Resting spores (oöspores) are produced in the tissues of the leaf and carry the fungus through the winter. The resting spores may also encrust the seed and thus distribute the disease wherever the beans are planted. Among soybean varieties, differ-

ences in resistance to DOWNY MILDEW are striking.

Reference: C.9.

DOWNY MILDEW OF VETCH is caused by the FUNGUS *Peronospora viciae* which attacks not only the VETCH species but also the FIELD PEA. The disease is widespread but seldom very destructive except in certain humid areas. The fungus causes localized light-green spots, often blotched with brown, on the upper surfaces of the leaves. In the later stages of the disease the affected areas die and turn brown. On the lower surface of the leaf the affected spots are covered with a whitish, moldy appearing growth. Downy mildew is carried in the seed and lives in the soil for an indefinite period of time.

Control: Rotation should be practiced and seed should be obtained preferably from semiarid regions.

Reference: M.24.

DRIED APPLE PECTIN PULP is the sound, dried residue obtained by the removal of pectin from apple products.

Reference: F.6.

See also PLANT BY-PRODUCTS; MISCELLANEOUS PRODUCTS.

DRIED APPLE POMACE is the sound, dried residue obtained by the removal of cider from apples.

This officially recognized feedstuff, like fresh APPLE POMACE, is fed preferably to cattle.

Reference: F.6.

See also PLANT BY-PRODUCTS; MISCELLANEOUS PRODUCTS.

DRIED BEET PULP is the dried residue from sugar beets that have been cleaned and freed from crowns, leaves, and sand, and that have been extracted in the process of manufacturing sugar.

This officially recognized feedstuff is fed like wet BEET PULP to cattle and other animals.—*See also* MISCELLANEOUS PRODUCTS; PLANT BY-PRODUCTS.

DRIED BREAD. *See* BREAD.

DRIED BREWERS' YEAST. *See* YEAST.

DRIED BUTTERMILK.

See MILK PRODUCTS.

DRIED CANDIDA YEAST. *See* YEAST.

DRIED CEREAL GRASSES are obtained by dehydration of young cereal grasses, especially of WHEAT, OATS, and RYE. They are valuable feed ingredients because of their high contents of vitamins, minerals, and proteins.

DRIED CITRUS PULP, or *citrus meal*— one of the officially recognized CITRUS PRODUCTS—consists of the dried ground peel, residue of the inside portions, and occasional cull fruits of the citrus family. If calcium oxide or calcium hydroxide is added as an aid in processing, the maximum percentage present, expressed as CALCIUM (Ca), must be stated.

Dried citrus pulp is often fed to cattle, alone or as an ingredient of mixed feeds. It is also used with *citrus molasses*.

Reference: F.6.

See also MOLASSES; CITRUS PULP.

DRIED CORN SIRUP. *See* CORN.

DRIED CULTURED SKIMMED MILK.
See MILK PRODUCTS.

DRIED EXTRACTED FERMENTA-TION SOLUBLES include *dried extracted citric acid fermentation solubles, dried extracted penicillin fermentation solubles, dried extracted streptomyces fermentation solubles,* etc.; they belong to the FERMENTATION PRODUCTS.

DRIED EXTRACTED MEAL AND FERMENTATION SOLUBLES, e.g., *dried extracted citric acid meal and fermentation solubles, dried extracted penicillin meal and fermentation solubles,* and *dried extracted streptomyces meal and fermentation solubles,* are FERMENTATION PRODUCTS.

DRIED FERMENTATION SOLUBLES.
See VITAMINS.

DRIED FISH SOLUBLES PRODUCTS.
See MARINE PRODUCTS.

DRIED GRAPEFRUIT REFUSE can be used as cattle feed like DRIED BEET PULP. It has a mild laxative action.—*See also* PLANT BY-PRODUCTS.

DRIED HYDROLYZED WHEY.
See MILK PRODUCTS.

DRIED LEMON PULP.
See LEMON PULP.

DRIED MILK ALBUMIN.
See MILK PRODUCTS.

DRIED MOLASSES.
See MOLASSES; SILAGE.

DRIED MOLASSES-BEET PULP.
See MOLASSES.

DRIED OLIVE PULP. *See* OLIVE PULP.

DRIED PINEAPPLE PULP.
See PINEAPPLE PULP.

DRIED SACCHAROMYCES YEAST.
See YEAST.

DRIED SKIMMED MILK.
See MILK PRODUCTS.

DRIED SPENT HOPS.
See BREWERS' PRODUCTS.

DRIED SWEETPOTATO PULP is the residue from the manufacture of starch from SWEETPOTATOES; it is rich in carbohydrates and sometimes used as feedstuff. —*See also* SWEETPOTATO PULP.

DRIED TOMATO POMACE—an officially recognized feedstuff—is a dried mixture of tomato skins, pulp, and crushed seeds resulting from the process of extracting juice from tomatoes.

Reference: F.6.

See also MISCELLANEOUS PRODUCTS; PLANT BY-PRODUCTS; TOMATO POMACE.

DRIED WHEY. *See* MILK PRODUCTS.

DRIED WHEY FERMENTATION SOL-UBLES. *See* VITAMINS.

DRIED WHEY-PRODUCT.
See MILK PRODUCTS.

DRIED WHEY-PRODUCT LACTOSE.
See MILK PRODUCTS.

DRIED WHEY SOLUBLES.
See MILK PRODUCTS.

DRIED WHOLE MILK.
See MILK PRODUCTS.

DRIED WHOLE WHEY.
See MILK PRODUCTS.

DRIED YEAST. *See* YEAST.

DRIER. *See* HAY DRYING.

DRILL is a machine for sowing seeds in furrows.

To drill means: to sow in furrows.

DRILLROW is a row of seeds or plants sown with a drill.

DRILLWORM
= SOUTHERN CORN ROOTWORM.

DRIP. *See* SUGAR DRIP; HONEY DRIP.

DROPSEEDS (*Sporobolus* spp.) are widely distributed perennials. The 36 species native to the United States are most abundant in the southern Great Plains and the Southwest. Most of them are palatable to animals.

The common name, dropseed, refers to the prompt casting of seed as it nears maturity. The SPIKELETS are single-flowered. In most species the stems are solid or pithy, rather than hollow.

Practically all the dropseeds produce an abundance of viable, long-lived seed which —because of hard seed coat— may lie dormant for many years before germinating under natural conditions; this characteristic accounts for the appearance of seedlings after long periods of drought.

Species: The most important dropseed species for forage production are two perennials: ALKALI SACATON and SAND DROPSEED.

Reference: H.1.

See also GRASSES.

DROUGHT. Lack of rain causes crop losses for some farmers and stockmen somewhere in the United States every year. The Corn Belt, Cotton Belt, Wheat Belt, western range, and general farming sections—all are damaged by drought at times. Drought may damage crop even on irrigated land and on fields that normally require drainage.

Some areas that have droughts during one season may get more water than is needed during other seasons. Then excess runoff, soil erosion, and floods damage the crops and land.

A well-planned soil and water *conservation* program is helpful in preventing damage from drought. The conservation problems will vary according to the kind of land and the climate.

The following types of soil and water conservation are recommended:

1. *On sloping land in wet climates*, these measures will help insure crops against drought damage: (1) Use adapted plants; (2) build up the soil's organic matter so that it will absorb and hold more water—you must stop erosion to do this; (3) check runoff when the heavy rains come and store water in the soil for future use; and (4) control the runoff the soil will not absorb—keep it out of gullies; spread it on pastures, meadows, or woodlands; or store it in farm ponds.

2. *Wet land* can be made more drought-resistant by drainage. Good drainage will enable crops to make a faster early growth; the roots will grow deeper into the soil, and the crops can then use some subsoil moisture if dry weather exhausts the moisture in the topsoil.

On many wet soils waterlogging or puddling has broken down the granular structure. Most of these wet soils can be improved by plowing under green manure, crop residues, or other organic matter. This helps make the soil porous so that air can circulate through it, and it will drain better. The organic matter will hold some of the water from the wet seasons to help tide the crops through the drought. Furthermore, the land can be cultivated more easily. Most of the other soil-improving practices that are used on sloping land in wet climates will also benefit poorly drained or wet land.

3. *For dry-farming areas*, the basic principles for conservation are as follows: (1) Do not try to grow crops that are not adapted to a dry climate; (2) do not try to cultivate land that is suited only for PASTURE or RANGE; (3) hold all the rain water on the land where it falls until it soaks into the ground; (4) improve the soil so that it will absorb and hold more water; (5) use for crops the water that you store in the soil—do not let weeds use it up, and do not lose it through evaporation; (6) use a flexible cropping system that you can change according to the season; (7) keep extra feed and cash on hand to tide you through drought years.

4. *To semiarid range land of the West*, drought comes very frequently and it may last for months or even years. To eliminate practically all the damage to range and much of the financial loss of livestock that usually comes with the droughts, a good range-conservation program is needed. The main conservation measures are proper grazing, maintaining feed reserves, water conservation, and reseeding depleted ranges.

5. *On irrigated land*, where the farmer does not have to depend on the amount of soil moisture at planting time or on the amount of rain and snow that falls during the growing season, a general drought over the watershed may seriously deplete the

amount of water available through irrigation canals for the next crop season. The farmer can usually find out about how much irrigation water he can get for a coming season. If the advance reports show that the water supply for an area is going to be short, crops that do not require too much water should be planted; a part of the land may be left fallow for the season.

By land leveling, some irrigation farmers have saved more than half the water formerly used. The substitution of borders and corrugations for flooding and improving irrigation canals and water-diversion systems may also bring large savings in water. The types of soil-building practices that help to make nonirrigated lands more drought-resistant will do the same thing for most irrigated lands.

Reference: D.12.

See also DUST STORMS; RANGE MANAGEMENT.

DRUGS are substances used in medicine; to make sure that the same amount of active ingredients is contained in each dose, drugs must comply to certain official standards of purity and safety.—*See also* MINERAL FEEDS; MEDICAMENTS.; F.D.A.

DRY BAIT is well adapted to spreading by aircraft, but not to spreading by ground equipment. It consists of 100 lb. standard bran (no shorts or middlings), 1 lb. TOXAPHENE (or ½ lb. CHLORDANE) and 2 qt. KEROSENE or FUEL OIL. Dissolve the insecticide in the latter and spray onto bran while mixing. This bait is applied at the rate of 10 lb. per acre for the control of MORMON CRICKETS.

DRY COW RATIONS. During her dry period the mature cow stores nutrients in her body to replace those lost during her last lactation. The calf makes most of its growth during the last three months of pregnancy. It is, therefore, especially important that adequate amounts of MINERALS, PROTEINS, and good quality energy feeds be provided. The gains obtained during this period will show up later in terms of increased milk production. Cows that are fed well during the dry period may produce 1000 lb. more milk than similar cows not well fed.

When pastures are not available, feed legume hay or silage; add sufficient grains to the feed, especially if no silage is fed to the animals. It is often necessary to add phosphorus supplement, such as BONE MEAL, to the ration.

A week to ten days before calving time, bulky concentrates, having laxative action, should be fed; e.g., a ration consisting of equal parts of wheat bran and ground oats and maybe also linseed oil meal.

The danger from milk fever can be reduced by feeding sun-cured LEGUME HAYS or by giving VITAMIN D for a week before the cow calves

DRYMARIA. *D. holosteoides* = THICKLEAF DRYMARY.

DRY MATTER IN FEEDS represents the amount of material left after the water has been removed. It is usually expressed as a percentage of the original weight.

Fresh green grass may contain only 20 to 30 percent dry matter; after curing, the hay produced may contain 85 to 90 percent dry matter. Beets and other root crops have a very high water content, often more than 90 percent; on the other hand, cereal grains contain often 90 percent and more of dry matter.

DRY MIXING. *See* FEED MIXING.

DRY-RENDERING.

See ANIMAL PRODUCTS.

DRY ROT = SOIL ROT.

DRY WEIGHT OF FORAGE.

See SILAGE.

DUAL-PURPOSE SORGHUMS are grain SORGHUMS that produce satisfactorily grain and forage.

DUCK RATIONS. *See* POULTRY RATIONS.

"DUCK WHEAT"

= TARTARY BUCKWHEAT.

DUCTLESS GLANDS.

See HORMONES; GLANDS.

"DUD" is a stunted plant; e.g., a soybean that does not produce seeds.—*See also* BUD BLIGHT.

DUFFY = *Fulcaster. See* COMMON WHEAT.

DUMP HOPPERS. *See* STORING FEED.

DUNFIELD. *See* SOYBEAN (variety).

DUNLAP = *Goens. See* COMMON WHEAT.

DURRA. Among the durras, formerly known as *Jerusalem corn* and *rice corn*, are the varieties *White durra* (also called

White Egyptian corn or "*Gyp*"), *Dwarf White durra*, and *Brown durra* (or *Brown Gyp*); they belong to the grain SORGHUMS.

DURUM WHEAT (*Triticum durum*) is one of the WHEAT species. It is similar to certain varieties of POULARD WHEAT. The plants are of spring habit. The SPIKES are compact and laterally compressed. The GLUMES are sharply keeled, and the LEMMAS are awned (except in a few awnless forms which are not in commercial production). The awns are long and coarse and are white, yellow, brown, or black. The kernels are white or red and usually rather long and pointed; they are very hard and translucent, making the white-kerneled forms appear amber-colored. The kernels always have a short brush and are the hardest of all known wheats.

The center of durum production is in northeastern North Dakota; relatively small durum areas are found in Montana, South Dakota, Minnesota, and other northern states. The durums furnish the great bulk of the wheat for the manufacture of semolina, which is made into macaroni, spaghetti, and similar products.

Dangers: Durum is attacked by ERGOT and other wheat diseases.

Varieties: Only one *red durum* (spring) variety is commercially important—*Pentad*, or *D-5* (red kernels and white awns; used largely for feed).

Of the *white durum* varieties the following are preferred (all have spring habit, white kernels, and—with the exception of Peliss—white awns): *Carleton* (late maturing, resistant to STEM RUST and many LEAF RUST races); *Kubanka*, also known as *Beloturka, Gharnovka, Pererodka*, and *Taganrog* (late maturing, somewhat resistant to STEM RUST); *Mindum* (late maturing); and *Stewart* (late maturing, resistant to LEAF RUST and somewhat resistant to BUNT; the most important durum variety).

Of minor importance are these white durums: *Arnautka*, also called *Goose, Johnson, Nicaragua*, and *Pierson* (late maturing); *Nugget* (very early maturing); *Peliss*, or *Blackbearded durum* (black awns); and *Vernum* (early maturing, RUST-resistant).

Reference: B.15.

See also COMMON WHEAT; TALL WHEATGRASS.

DUST. *See* SCREENINGS.

DUST BARRIER is a simple protection against migrating CHINCH BUGS; it is satisfactory only during dry weather. During recent years the dust barrier has been replaced by more dependable CHEMICAL BARRIERS, including the DDT-DUST BARRIER and the DINITRO-DUST BARRIER.

DUST STORMS often occur during DROUGHTS, when vegetation no longer protects the soil from the erosive action of high winds. The worst dust storms occurred in the thirties—they are well remembered as the *Dust Bowl* of the Great Plains. However, there are precautions that can be taken to prevent another Dust Bowl.

Drought and high winds do not always cause dust storms; it takes loose dirt to make dust. Soil that is protected with ground cover will not blow away. Soil that retains its original structure and is bound together with organic matter and crop residues is not likely to blow.

Enough rain falls, even during most drought years, to grow some sort of crop if all the water is saved and used for the crop. Soil blowing can be stopped, even on a bare field, if one gets to it in time.

Deep listing is probably the best and quickest way to stop an unprotected field from blowing. Lister furrows as much as 2 rods apart may check blowing. In fact, any implement that leaves a ridged or rough and cloddy surface should help stop soil blowing. All emergency tillage should be on the contour if the land is sloping.

If a drought lasts for many months and if the land has been blowing it should be covered with a quick growing crop as soon as possible.

Conserve every drop of water that falls —store it in the soil until you have enough to plant and grow a *cover crop*. GRAIN SORGHUMS, CANE, SUDANGRASS, and BROOMCORN are the best cover crops if land has been blowing. If blowing has not been severe, WHEAT, BARLEY, RYE, and other SMALL GRAINS make good cover for the land when enough moisture is available for them to get a good start. Even a cover of weeds is better than bare land.

Once a cover crop is established on the land, keep it. Do not graze the stalk and stubble heavily. Start stubble-mulch farming. Stubble mulch will protect the land and help conserve moisture. It is also a good time to start strip cropping; strips of SORGHUM will help protect land that you wish to plant to wheat. If there is idle or abandoned land next to a farm, border strips of drilled sorghums around your farm may help protect it from nearby blowing fields.

Reference: D.15.

DUST TREATMENT OF SEEDS.
See SEED TREATMENT.

DUTCH BOY, better known as *Dwarf Ashburn,* is one of the forage SORGHUMS.

DUTCH CLOVER. *White Dutch clover,* more correctly called *common white clover,* is a WHITE CLOVER variety.

DUTCHMAN'S BREECHES (*Dicentra cucullaria*) is toxic to beef cattle because of its ALKALOID content. Trembling, frothing, and convulsions are symptoms of poisoning due to feeding on leaves and

Dutchman's breeches. Plant with tuberous roots at base of stem. (H.48.)

stems, particularly in spring and early summer.—*See also* POISONOUS PLANTS.

DWARF ASHBURN belongs to the forage SORGHUMS.

DWARF BUNT is a SMUT disease caused by the FUNGUS *Tilletia caries*. It is not as prevalent as COMMON BUNT.

Dwarf bunt stunts infected WHEAT plants severely. Its principal region of distribution is the Pacific Northwest, but it occurs also in Wyoming, Colorado, and New York.

Plants with dwarf bunt are one-half to one-fourth the height of healthy plants. The smutted wheat heads are bluish green when they emerge from the boot. They tend to be long and lax and to ripen sooner than healthy ones. The smut balls protrude beyond the GLUMES as they enlarge. Dwarf-bunt infected heads usually are more compact than those infected with common bunt, and the glumes are spread apart so that the smutted heads have a feathery look.

Dwarf bunt stimulates excessive tillering of infected plants. Bunt-infected flowers have longer PISTILS and longer and broader ovaries than do healthy flowers. Diseased ovaries are green; STAMENS in diseased flowers are reduced in length and breadth, and the ANTHERS have a pale-yellow color.

The spore ball shape tends to conform to the shape of the wheat kernel. The spore balls are broken in threshing and the grain becomes contaminated with spores. If it is used for seed, the spores germinate and penetrate the young seedling while it grows from the seed to the surface of the soil. The growth of the parasite keeps pace with plant development; at maturity, bunt balls are formed in place of wheat kernels.

Wheat seedlings also may become infected by soil-borne spores. Spores of dwarf bunt may remain viable for seven years. Even so, dwarf bunt does not attack spring wheat. The greatest infection occurs at soil temperatures of 40°-60° F., with moisture contents ranging from 15 to 60 percent of field carrying capacity.

Control of dwarf bunt depends on the use of clean seed of a smut-resistant wheat variety properly treated with CERESAN M or another appropriate fungicide. Also effective, when it is practical, is the seeding of wheat when soil temperatures are unfavorable to bunt development.

References: H.29; L.10.

See also SEED TREATMENT; COMMON WHEAT.

DWARF DISEASE, or *alfalfa dwarf*, is a VIRUS DISEASE, which is known to occur only in California. There it often causes a rapid thinning of second- and third-year stands in a large part of the ALFALFA acreage.

LEAFHOPPERS and SPITTLEBUGS carry the virus of dwarf disease from plant to plant. When the leafhoppers feed upon dwarf-diseased plants, they take the virus into their bodies, and retain it for several months, during which time they may spread the virus from field to field.

Alfalfa plants infected with the dwarf virus gradually lose vigor for several months. Stems are short and spindly. The leaves get smaller and often darker in color than leaves of healthy plants. Internally, gum forms in the water-conducting elements and the woody portion of the roots and crown becomes yellow or brown. Susceptible plants usually die six to eight months after infection.

Dwarf disease may be confused with BACTERIAL WILT of alfalfa which it resembles both in the dwarfing of foliage and in the discoloration in the root; but it is distinguished by the normal green color of the dwarfed foliage in contrast to the pale-green or yellow color characteristic for the wilt.

The variety CALIFORNIA COMMON 49 is highly tolerant to the virus of dwarf disease.

References: J.3; J.2.

DWARF FETERITA is one of the grain SORGHUMS.

DWARF FREED is a grain SORGHUM variety.—*See also* COVERED KERNEL SMUT.

DWARF HEGARI = *hegari*. See GRAIN SORGHUMS.

DWARF KAFIR 44-14, is not a true kafir variety, but one of the intermediate-type grain SORGHUMS.

NOTE: The name Dwarf kafir is sometimes erroneously applied to DAWN KAFIR.

DWARF SHUNTUNG KAOLIANG belongs to the grain SORGHUMS.

DWARF WHEAT = CLUB WHEAT.

DWARF WHITE DURRA is a grain SORGHUM.

DWARF WHITE MILO is one of the grain SORGHUMS.

DWARF YELLOW MILO belongs to the grain SORGHUMS.

DYES. *See* COLORS; CERTIFIED DYES.

E

EAR CORN CHOPS, or *corn-and-cob meal*, and *ear corn chops with husks* are feedstuffs; if not finely ground, animals may refuse them.—*See also* CORN; HAY GRADING; SILAGE CROPS.

EARLY AMBER, now commonly called *Black Amber*, is one of the forage SORGHUMS.

EARLYANA. *See* SOYBEAN (variety).

EARLY BLACKHULL.
        See COMMON WHEAT (winter variety).

EARLY BUFF. *See* COWPEA (variety).

EARLY DAIN = *Triumph.*
        See COMMON WHEAT (variety).

EARLY FORTUNE. *See* PROSO (variety).

EARLY HARDY = *Early Blackhull.*
        See COMMON WHEAT.

EARLY HEGARI is one of the grain SORGHUMS.

EARLY KALO is an intermediate-type grain SORGHUM.

EARLY KANSAS = *Early Blackhull.*
        See COMMON WHEAT.

EARLY KOREAN.
        See KOREAN LESPEDEZA (variety).

EARLY MAY = *Red May.*
        See COMMON WHEAT.

EARLY PROLIFIC. *See* RICE (variety).

EARLY RED COWPEA.
        See COWPEA (variety).

EARLY RED WHEAT = *Goens.*
        See COMMON WHEAT.

EARLY RIPE = *Goens.*
        See COMMON WHEAT.

EARLY RUSSIAN = *Early Blackhull.*
        See COMMON WHEAT.

EARLY SPECKLED is better known as *Alabama*, but the term was originally used to identify the *Georgia* variety of DEERING VELVETBEAN.

EARLY SUMAC SORGO and *Extra Early Sumac* are forage SORGHUMS.

EARLY WHITE MILO is one of the grain SORGHUMS.

EARLY WOODS YELLOW = *Arksoy*. See SOYBEAN (variety).

EARTHNUT = PEANUT.

EARWORM. See CORN EARWORM.

EAST INDIA BLUESTEM = TURKESTAN BLUESTEM.

EBONY. See SOYBEAN (variety).

ECHINOCHLOA. *E. crusgalli* = BARN-YARDGRASS (and its variety, *Japanese millet*).

ECOTYPE is a VARIETY or STRAIN adapted to a particular environment.

ECTOPARASITES are PARASITES which live on the surface of plants or animals; e.g., the STING NEMATODE (but not the other nematodes).

EDEMA is a swelling due to accumulation of fluid in the tissue spaces.

EDIBLE BEANS. See FIELD BEANS.

EDSOY, which has been renamed *Delsoy*, is a SOYBEAN variety.

EELWORMS = NEMATODES.

EGG is the female reproductive cell.— See also VITAMIN D.

EGYPTIAN CLOVER = BERSEEM.

EGYPTIAN CORN. See WHITE EGYPTIAN CORN.

EGYPTIAN COTTON. *American - Egyptian cotton* is a species of COTTON.—See also GOSSYPIUM.

EGYPTIAN WHEAT, better known as *shallu*, is one of the grain SORGHUMS.

ELAEIS. Palm trees of the genus *Elaeis*, particularly *E. guineensis* and *E. melanococco*, develop a fruit from whose rind (outward covering) as well as kernel, or nut, an oil can be produced: the former gives *palm oil*, the latter, *palm kernel oil* and PALM KERNEL OIL MEAL.

ELEMENT. See CHEMICAL ELEMENT; TRACE ELEMENT.

ELEOCHARIS. *E. palustris* = SPIKE RUSH.

ELEPHANTGRASS = NAPIERGRASS.

ELGIN. See CLUB WHEAT (variety).

ELKAN BLUESTEM. See TURKESTAN BLUESTEM (variety).

ELLIPTICAL means: oblong and rounded at the ends.

ELLIS is one of the forage SORGHUMS.

ELMAER. See CLUB WHEAT (variety).

ELRENO SIDE-OATS GRAMA, also called *El Reno*, is a variety of SIDE-OATS GRAMA.

ELSBERRY. See SMOOTH BROME (variety).

ELYMUS. The genus *Elymus* includes these species of WILD-RYE GRASSES: *E. canadensis* = CANADA WILD-RYE; *E. condensatus* = GIANT WILD-RYE; *E. glaucus* = BLUE WILD-RYE; *E. junceus* = RUSSIAN WILD-RYE; and *E. salinus* = SALINA WILD-RYE.

EMBRYO, or *germ*, is the undeveloped plant in a seed, resulting from the union of a male and female cell. It becomes a SEEDLING in the process of GERMINATION. In animals, the embryo develops either in the uterus, or outside of the body, e.g., in eggs of birds or insects.

Embryonic means: undeveloped; resembling an embryo or pertaining to the embryo state.—*See also* ENDOSPERM.

EMERALD. See RYE (variety).

EMERGENCE is a growth outward from beneath the EPIDERMIS; e.g., the LIGULE of a grass.

EMMER (*Triticum dicoccum*), often incorrectly called "*speltz*," is one of the WHEAT species grown to a limited extent in the United States. It may be of either winter or spring habit and is usually awned, with its leaves pubescent. The kernels, which remain enclosed in the GLUMES after threshing, are red or white, and long as well as slender, with both ends acute.

Emmer is distinguished from SPELTZ by the shorter, denser SPIKES, which are laterally compressed. The pedicel of emmer is shorter and narrower and is usually attached to the face of the next lower SPIKELET. The inner side of the spikelet is flat instead of arched, and the kernel usually is of a darker red than that of spelt.

Emmer is used as feed for livestock.

The states leading in the production of emmer are South Dakota, North Dakota, Nebraska, Minnesota, and Colorado.

Varieties: The *Vernal*, or *common emmer*, is the most productive spring variety; *Khapli* is preferred in some sections. Both of these varieties are very resistant to STEM RUST.

Black Winter (emmer) is not very winter-hardy, nor resistant to RUST or drought.

References: B.15; M.33.

Emmer. Spikelets, showing short and pointed pedicel attached to base. (M.33.)

EMPOASCO. *E. fabae*
                    = POTATO LEAFHOPPER.

EMPUSA. *E. sphaerosperma* is a FUNGUS well distributed over the clover- and alfalfa-growing regions of the United States; it keeps the CLOVER LEAF WEEVIL from becoming a serious pest. The fungus also attacks the larvae of the LESSER CLOVER WEEVIL.

EMULSIONS are liquid-in-liquid COL-LOIDAL systems.

Emulsifiable is a concentrated solution —e.g., of an insecticide—which, after mixing with water, forms an emulsion.

Emulsifiers are substances, mostly inert, which are mixed in emulsifiable formulations to aid in the formation of stable emulsions when two not mixable liquids, such as oil and water, are mixed.

ENCELIA (*Encelia farinosa*) is one of the forbs (weeds) which is useful as forage on ranges.—*See also* RANGE PLANTS.

ENDOCRINE GLANDS = *ductless glands.* See HORMONES; GLANDS.

ENDOSPERM is the starchy main portion of the grain kernel. With the exception of

the small space occupied by the germ (EMBRYO), the endosperm fills the interior of the kernel. It is a nutritive material on which the embryo feeds when germinating. —*See also* ALEURONE.

ENERGY—the ability to produce action, effect, or to do work—is expressed in calories.

Gross energy of a feedstuff is the total energy released; it is determined by burning in a bomb CALORIMETER.

Digestible energy is gross energy less energy excreted in feces.

> NOTE: Most air-dry feedstuffs—except those rich in fat—have a gross energy of about 4.5 Cal./gm. Silages, grasses, and roots are low in energy. The digestible energy of a specific feedstuff will vary to some extent with species, age, and individuality of animal fed, the feeding plan, and the other feeds which are present in the ration.

Metabolizable energy, or *available energy,* is that portion of the feed which is of use to the animal; i.e., the amount left after deducting from the gross energy the energy lost in the feces (undigested material), urine (formed from proteins), and gases (formed by fermentation of carbohydrates).

Net energy is the amount of feedstuff left for utilization by the animal after deducting its heating effect from metabolizable energy. Net energy is used to meet the requirements for normal body functions (breathing, movements, etc.), and any surplus over this *maintenance energy* is available for the production of fat, wool, milk, work, etc. Thus, net energy can be defined as gross energy minus metabolizable energy minus energy needed for the work of digestion (also called *heat increment*).

> NOTE (example): Timothy hay, containing 100 Cal. gross energy, is fed to a steer; 47.5 Cal. were lost in the feces, leaving 52.5 Cal. digestible energy; the loss in methane gas was 7.5 Cal. and in urine 4 Cal., leaving 41 Cal. metabolizable energy; the heat increment was 17 Cal., leaving a remaining net energy value of 24 Cal.,

or 24 percent of the original gross energy. The net energies of other feedstuffs determined under the same circumstances are: wheat straw, 5 percent; peanut oil (fat), 54 percent; starch (carbohydrates), 53 percent; and wheat gluten (protein), 37 percent.

See also ARMSBY FEEDING STANDARDS; MORRISON FEEDING STANDARDS; KELLNER FEEDING STANDARDS; FRAPS FEEDING STANDARDS; SCANDINAVIAN FEED-UNIT SYSTEM; FERMENTATION; DIGESTION; ABSORPTION NUTRIENTS; FEED EFFICIENCY.

ENGLISH CLOVER = *American single-cut clover.* See RED CLOVER.

ENGLISH RYEGRASS
= PERENNIAL RYEGRASS.

ENGLISH WILD WHITE.
See WHITE CLOVER (variety).

ENHEPTIN = AMINO-NITROTHIAZOLE.

ENICOSPILUS. *E. purgatus* is an insect enemy of the ARMYWORM.

ENLARGED HOCKS in poultry are caused by the lack of VITAMIN E in the feed.

ENSILAGE = SILAGE.

ENSILE means: to make into SILAGE.

ENSILING PROCESS. *See* SILAGE.

ENTERITIS is an inflammation of the intestines. It may be caused by infections, internal parasites, vitamin deficiency, poisoning, etc.—*See also* ARSONIC ACIDS.

ENTYLOMA. The FUNGUS *E. oryzae* causes LEAF SMUT OF RICE.

ENZYMATIC means: relating to ENZYMES.

ENZYMES (pronounced *enn*-zymes) are chemical products of plant or animal cells that bring about changes in organic materials, like ripening, fermentation, or digestion. Enzymes are protein substances, but will retain their activity after being separated from living organisms; they induce or speed up chemical reactions in substances of plant or animal origin.

The enzymes acting during digestion are the following:

Amylase—hydrolyzes STARCH to dextrin and maltose.

Lactase—hydrolyzes LACTOSE to glucose and galactose.

Lipase—hydrolyzes FATS to fatty acids and glycerin.

Pepsin—hydrolyzes PROTEINS (therefore called a *proteolytic enzyme*) to proteoses and peptones.

Peptidases—hydrolyzes PEPTIDES to amino acids.

Sucrase—hydrolyzes SUCROSE to fructose and glucose.

Trypsin (another proteolytic enzyme) —hydrolyzes PROTEOSES to peptides.

Urease—hydrolyzes UREA to ammonia (compounds).

Mammals have no enzymes to digest CELLULOSE; they must depend therefore, on bacteria which have the enzyme *cellulase.*

Labeling, which implies that the presence of added enzyme-bearing materials improves the utilization of a feed product, is officially condemned.

NOTE: Recent investigations indicate that the newborn of some animal species do not have enough enzymes to digest certain types of feed. For this reason, *pepsin* is now an approved ingredient in the rations of baby pigs.

See also SILAGE; ALCOHOL; DIGESTIVE TRACT; DIASTASE; THIAMINASE; INVERTASE; FERMENTS; COENZYMES.

EPARGYREUS. *E. tityrus*
= LOCUST SKIPPER BUTTERFLY.

EPHEDRA. *E. nevadensis*
= NEVADA EPHEDRA.

EPIDERMIS is the thin outer-cell layer in the higher plants or animals; e.g., the epidermis present in leaves or herbaceous stems.

EPILACHNA. *E. varivestis*
= MEXICAN BEAN BEETLE.

EPSOM SALT = MAGNESIUM SULFATE.

EQUINES are a related group of mammals represented by the domestic horse, mule, and ass.

EQUISETUM. *Equisetum* spp. = HORSE-TAILS; *E. hyemale* = SCOURING-RUSH.

EQUIVALENT PROTEIN.
See NONPROTEIN NITROGEN.

ERAGROSTIS. The genus *Eragrostis* comprises 250-odd species of which about 40 occur in the United States, among them *E. intermedia* = PLAINS LOVEGRASS; *E. curvula* = WEEPING LOVEGRASS; *E. trichodes* = SAND LOVEGRASS; *E. lehman-*

niana = LEHMANN'S LOVEGRASS; *E. chloromelas* = BOER'S LOVEGRASS; and *E. obtusiflora* = ALKALI LOVEGRASS.

EREMOCHLOA. *E. ophiuroides*
= CENTIPEDEGRASS.

ERGOSTEROL, also called *provitamin D_2* or *ergosterin*, is a STEROL commonly obtained from yeast or ergot. It forms colorless crystals which are not water-soluble.

Irradiation or electronic bombing changes ergosterol to VITAMIN D_2, or calciferol.—*See also* VITAMIN D.

ERGOT of grains and grasses is caused by the FUNGUS *Claviseps purpurea*. BAHIA-GRASS is attacked by *C. phasphali*. Loss in yield from sterility is always associated with this FUNGUS DISEASE. Resting bodies (SCLEROTIA) of the fungus replace some of the kernels, and the adjacent kernels do not develop. The disease is common and destructive on GRASSES (e.g., MEADOW FOXTAIL), RYE, BARLEY, and WHEAT, particularly DURUM WHEAT and some varieties of hard spring wheat. The damage occurs in the humid sections and extends into the subhumid regions of the United States, notably in the spring grain areas of Nebraska, the Dakotas, and Montana.

The disease is recognized first by the sticky fluid (*honeydew stage*) on parts of the SPIKES soon after heading and later by the purplish-black sclerotia in the ripening heads. The sclerotia are shaped somewhat like a rye kernel. They are usually longer than a grain and thus protrude from the chaff. The sclerotia thresh out with the grain.

> NOTE: Any grain containing more than 0.3 percent ergot sclerotia by weight is graded as *ergoty* and is discounted at the markets. The ergot content of milled products is limited by law.

The life cycle of the fungus is well synchronized with that of the grain or grass host. The sclerotia falling on the soil or planted with the seed germinate when the grains and grasses are flowering.

The sexual spores are wind-borne to the flowers of the grain or grass, where they invade the young kernels and replace them with fungus growth. The first fungus growth forms a folded mat, which bears millions of spores in a sticky, sweet, honey-dew-like mass. The minute spores are carried by insects or are spattered by rain to infect numerous other kernels. The sclerotia develop following this spreading stage.

Ergot. Heads of rye and smooth brome grass bearing ergot kernels. (D.19.)

Dangers: The ergot sclerotia contain several chemical compounds, some of which cause acute injury to animals and man. Ergot sclerotia invaded by *Fusarium* spp. contain compounds affecting the respiratory and other automatic muscular action. Milk flow in livestock is reduced greatly by continuous feeding of small quantities of ergot sclerotia. Relatively small quantities of ergot may cause ERGOT POISONING and thus serious losses among animals. (Bright, hard ergot sclerotia free from molds command a high price for medicinal use.)

For animals on pasture, the danger of *ergotism* can be avoided by keeping the seed heads of grasses mowed off in late summer. It is suggested that the mower be set high to avoid cutting off the foliage of the grass.

Control: Neither crop rotation nor ergot-free seed controls the disease. Destroying grasses in the grainfields, particularly QUACKGRASS, BROME GRASS, and WHEATGRASSES, and mowing grasses near grainfields before they head helps in control.

No varieties of barley, rye, or wheat resistant to ergot have been found.

References: D.11; K.1.

See also REE WHEATGRASS; DALLISGRASS; DALLISGRASS POISONING; TURKESTAN BLUEGRASS.

ERGOT POISONING is due to the ERGOT fungus found on forage plants and hay or straw fed to livestock, especially cattle, over a prolonged period of time. It may cause serious losses among farm animals, particularly in late fall and winter.

Affected animals—because of dry gangrene—may lose part of their ears, hoofs, or tails; or they may develop severe sores on mouths and teats; pregnant animals may abort.

ERIDONTOMERUS. *E. isosomatis* is a useful WASP; it destroys WHEAT STRAWWORMS.

ERIGERON. *E. strigosus* = WHITETOP FLEABANE.

ERODIUM. *E. circutarium* = ALFILERIA.

EROSION is the wearing away of the land surface by running water, wind, etc.—*See also* BRUSHES; RANGE MANAGEMENT.

ERYSIPHE. The FUNGUS *E. polygoni* causes POWDERY MILDEW.

ESOPHAGUS, or *gullet*, is a part of the DIGESTIVE TRACT.

ESSENTIAL OILS, or *volatile oils*, occur in plants in very small amounts and give them their characteristic odors.—*See also* AROMATICS.

ESTAFIATA = FRINGED SAGEBRUSH.

ESTERS are ORGANIC compounds formed from ALCOHOLS and ACIDS by the elimination of water; e.g., WAXES.

ESTROGENS are a group of sex HORMONES which produce estrus (heat) in female mammals. STILBESTROL is a synthetic (hormone-like) product with estrogenic properties.—*See also* SUBCLOVERS.

ESTRUS, or *heat*, is the period of intense sexual urge in female animals.—*See also* HORMONES; VITAMINS.

ETHER EXTRACT is the total of *crude* FATS, oils, waxes, and similar products extracted with ether from plant or other feed (food) material.

ETHOXY-TRYMETHYL DIHYDRO-QUINOLINE is an ANTIOXIDANT, the use of which is limited to poultry feeds.

ETHYL ALCOHOL. *See* ALCOHOL.

ETHYLENE DIBROMIDE is a very heavy liquid (1 gal. weighs 18 lb. 3 oz.) which is slightly soluble in water. It is the active ingredient of the soil fumigant DOWFUME W-40.

ETHYL MERCURIC CHLORIDE forms silvery leaflets which are practically insoluble in water. It possesses fungicidal properties and is the active ingredient of CERESAN.

ETHYL MERCURY PARA-TOLUENE SULFONANILIDE has fungicidal properties; it is the active ingredient of CERESAN M.

EUCHLAENA. *E. mexicana* = TEOSINTE.

EUMICROSOMA. The tiny WASP *E. benefica* is a parasite which, by laying an egg in the CHINCH BUG egg, aids in destroying this insect pest.

EUPATORIUM. *E. rugosum* = WHITE SNAKEROOT.

EUPELMUS. *E. allynii*, a WASP, is one of the WHEAT STRAWWORM parasites.

EUROPEAN BARBERRY. *See* BARBERRY BUSHES.

EUROPEAN CORN BORER (*Pyrausta nubilalis*) attacks many cultivated crops and weeds, but CORN is its favorite host plant. The eggs are laid on the underside of the corn leaves, and hatch in four to nine days. The tiny borers crawl immediately to protected places on the plants, where they feed on the tissues of the immature leaves and tassels, and eventually bore inside the stalks and into the ears of corn. They become full grown in about a month and change to pupae inside the burrows. In ten to fourteen days the adult moths emerge from the pupal cells and lay about 400 eggs each. The moths live from ten to twenty-four days. They are active fliers during the evening or night and may migrate several miles. The insects pass the winter in the borer stage inside infested stems of corn and other plants, and change to moths late in the spring or early in the summer. There are one or more generations a year, depending on the length of the growing season in different latitudes.

Control: Community application of the following methods is necessary for most effective control:

In order to destroy overwintering borers dispose of infested cornstalks by (1) feeding them to livestock direct, or as silage, or in finely cut or shredded form; (2) plowing them under in the fall or early spring before the moths emerge; (3) burning infested plants completely.

Plant, as late as practicable, resistant or tolerant kinds of *hybrid corn* (no immune strains are available).

Avoid sowing fall WHEAT or other SMALL GRAIN in standing corn or corn stubble. Dispose of all early *sweet cornstalks* in fields immediately after harvesting the ears, by feeding, ensiling, or plowing the cornstalks under.

Reference: U.3.

See also SUGARCANE BORER; CORN LEAF APHID.

European corn borer. (U.13.)

EUROTIA. *E. lanata* = WINTERFAT.
EUSCHISTUS. *E. impictiventris*
                    = BROWN COTTON BUG.
EVAPORATED BUTTERMILK.
                    See MILK PRODUCTS.
EVAPORATED CULTURED SKIMMED MILK. *See* MILK PRODUCTS.
EWE RATIONS. *See* SHEEP RATIONS.
EXCELSIOR. *See* RYE (variety).
EXPELLER SOYBEAN OIL CHIPS.
                    See SOYBEAN.
EXPELLER SOYBEAN OIL MEAL.
                    See SOYBEAN.
EXSERTED means: protruding from or projecting beyond the surrounding organs; said, for instance, of PISTILS and STAMENS.
EXTRACTED ALFALFA MEAL.
                    See ALFALFAS.
EXTRACTED ANIMAL LIVER MEAL.
                    See ANIMAL PRODUCTS.

EXTRACTED MEALS, such as *extracted citric acid meal, extracted penicillium meal, extracted streptomyces meal,* etc., are FERMENTATION PRODUCTS.
EXTRACTED PRESSCAKE—e.g., *extracted citric acid presscake, extracted penicillium presscake,* or *extracted streptomyces presscake*—belongs to the FERMENTATION PRODUCTS.
EXTRA EARLY SUMAC is one of the forage SORGHUMS.
EYE. *See* TUBER.

F

F = FLUORINE.—*See also* MINERAL FEEDS.
FABACEAE = LEGUMES.
FABOIDEAE, or *Papilionoideae,* are the most important subfamily of the LEGUMES (Leguminoseae).
FAGOPYRUM. The genus *Fagopyrum* includes the following BUCKWHEAT species: *F. esculentum* = COMMON BUCKWHEAT; *F. emarginatum* = WINGED BUCKWHEAT; and *F. tataricum* = TARTARY BUCKWHEAT.
FAIRFIELD.
                    See COMMON WHEAT (variety).
FAIRWAY CRESTED WHEATGRASS (*Agropyron cristatum*), or *true crested wheatgrass,* is a type of the CRESTED WHEATGRASS. It is characterized by bright green color and great uniformity. Compared to STANDARD CRESTED WHEATGRASS (which contains some Fairway crested wheatgrass) it is shorter (about 22 in. high), more leafy, smaller stemmed, smaller seeded, has horizontally spreading SPIKELETS and more pronounced awns.

Reference: W.16.

FALL ARMYWORM (*Laphygma frugiperda*), known mainly as an enemy of growing CORN, feeds also on many other cultivated crops (ALFALFA, SOYBEAN, COTTON, WHEATS, VETCHES, PEANUTS, GRASSES) and wild plants. The eggs are laid on plants at night and hatch in about five days. The young larvae, also called caterpillars, or "worms," feed at first near the ground, become full-grown in about twenty days,

Fall armyworm. (R.13.)

and then enter the soil for a few inches and change into pupae. The inactive pupal stage lasts about ten days. After the moths emerge from the pupal cases, they often fly many miles before the females lay eggs.

The fall armyworm—not to be confused with the (true) ARMYWORM—may have as many as six generations a year in the Gulf States, but does not survive the winter farther north.

Control: The fall armyworm can be controlled with TOXAPHENE, DDT, TDE, or with POISONED BRAN MASH containing PARIS GREEN.

Reference: U.3.

FALLOW is land that has not been planted to a crop, so as to build up its water resources and productive capacity or to destroy weeds, diseases, or pests. It may be tilled to kill weeds.

FALL-SOWN OATS.
> See OATS; COMMON OAT; RED OAT.

FALSE ANTHRACNOSE OF VETCH, caused by the FUNGUS *Kabatiella nigricans*, is different from (true) VETCH ANTHRACNOSE, but resembles it closely. It is prevalent on hairy vetch in the South. The disease causes brown discoloration and girdling of stems; spots on the leaves are small and circular but tend to form elongated streaks. When the pods are heavily spotted, the fungus penetrates the seed, causing it to become infected.

FALSE SAFFRON = SAFFLOWER.

FAMILY, in TAXONOMY, is a group of closely related TRIBES and GENERA; e.g., the grass family. The names of plant families regularly end in—*aceae*; as Poaceae (rather than the older term Gramineae) for GRASSES.—*See also* ORDER.

FANWEED = PENNYCRESS.

FARGO BROME is a northern-type SMOOTH BROME.

FARGO MILO, or *Straightneck milo*, is not a true milo, but one of the kafir-milo derivatives which belong to the grain SORGHUMS.

FARM ANIMAL AND MEAT PRODUCTION. The graphs on pages 177 and 178 show the trends of farm production of livestock and meat.

FARMERS TRUST = *Mediterranean wheat. See* COMMON WHEAT.

FAT MIXING. *See* FEED MIXING.

FATS are important NUTRIENTS. The true fats are composed of FATTY ACIDS combined with GLYCERIN and may be emulsified by bile and changed to mono- and di-GLYCERIDES in the small intestine before ADSORPTION into the blood stream. When not needed immediately for ENERGY production, they are deposited in the *fatty tissues*.

Most feedstuffs of animal and plant origin contain at least small amounts of fat; feeds rich in fat may become rancid and may cause digestive disturbances. Small amounts of *animal fats* or *vegetable fats* are needed in the ration because they carry the important fat-soluble VITAMINS in the system. Swine and poultry need at least 1 percent fat, and dairy cattle need at least 3 percent in the ration.

> NOTE: *Oils* are fats in liquid form. The term "fat," when used in connection with feedstuffs or feeding, refers to either the liquid or solid form.

True fats as well as natural fatlike substances—such as ESSENTIAL OILS, WAXES, STEROLS (e.g., CHOLESTEROL and ERGOSTEROL), CAROTENE, certain fatty acids, etc.—are recovered in the *ether extract* of feedstuff analyses (because their content in feedstuffs is determined by extraction with ether); sometimes the terms crude fat, lipids, or lipoids are used instead of ether extract.

The fat content of such feeds as pet feeds, ANIMAL PRODUCTS, and MARINE PRODUCTS, shall not exceed the guaranteed amount by more than 50 percent, according to the tentatively accepted resolution of the Association of American Feed Control Officials.

As synthetic detergents have replaced soaps, considerable fat that had been used for soap has become available at reasonable prices. Experiments indicate that rations with added fat increase the rate of gain and feed efficiency of swine and poultry. The use of fat allows the feed industry to prepare rations with rather high energy and has caused a re-evaluation of other nutrient requirements. General requirements are now stated in terms of *energy levels*.

The use of fat has other advantages for

Cattle. *Heifers and calves not for milk, and all steers and bulls. °2 years and older not for milk. †Cows and heifers 2 years and older for milk.

Hogs.

Stock sheep and lambs. *Eleven western states and South Dakota.

Horses and mules.

Livestock on farms, 1866-1954. (U.S.D.A.)

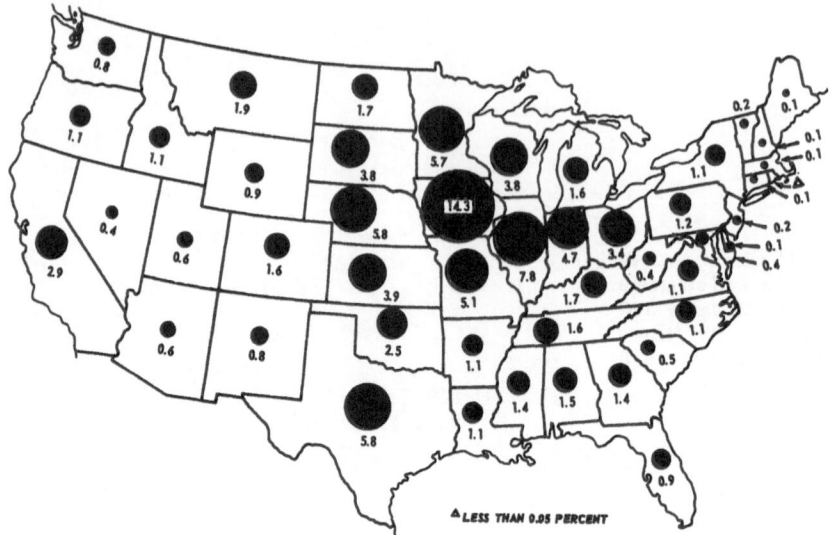

Farm production of meat animals. By states, as percent of U.S. total, 1955: 49,987 million lb. live weight. (U.S.D.A.)

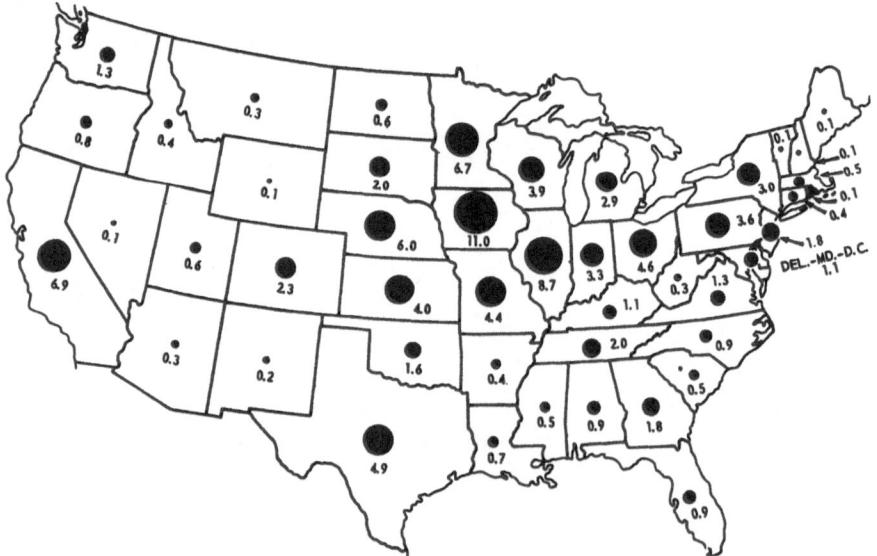

Meat production (from all slaughter, including farm). By states, as percent of U.S. total, 1955: 25,333 million lb. (U.S.D.A.)

the manufacturer. It reduces the dust in the mill and the wear on the machinery, and it improves PELLET formation. Calculations indicate that one can pay from two to three times as much for fat as for farm grains if the fat is used for ENERGY. Many manufacturers are willing to pay more for

a small amount of fat to improve the physical character of their feeds.

See also NITROGEN-FREE EXTRACT; LABELS; SOAPS; CALORIE; FATTENING; ANIMAL PRODUCTS; MARINE PRODUCTS.

FATTENING of animals has the result that their lean meat is improved in tenderness, flavor, and juiciness, because the FAT is stored not only in fat layers (which are of little value) but also between the fiber bundles of the muscles. Long fattening periods increase the amount and percentage of fat stored while the percentage of protein, minerals, and water is decreased although their amounts may increase slightly. These gains are very expensive in terms of feed energy. More feed is needed for maintenance of a fat animal than for a thinner animal. The fattening process is, therefore, relatively more expensive than the growing process.

The successful husbandryman will feed his animals to obtain enough fat deposition for desired flavor and tenderness but will avoid the wasteful production of excess fat that the consumer discards.

FATTENING LAMBS.
See SHEEP RATIONS; FATTENING.

FATTENING RATIONS. *See* BEEF RATIONS; SWINE RATIONS; FATTENING.

FATTY ACIDS are organic acids formed when FATS split in the small intestine. Fatty acids are sometimes neutralized to SOAPS before ABSORPTION; only a small percentage of free acids is absorbed into the blood stream.

F.D.A. is an abbreviation used for *Food and Drug Administration*, the federal agency which enforces the FDC ACT. The activities of the F.D.A. have greatly improved conditions in the *interstate* trade of feedstuffs and drugs; federal authorities, however, have no right to prevent selling worthless or misbranded products within the border of the state in which they are prepared.

The *intrastate* trade of feeds in most of the United States is regulated by state laws.—*See also* TAG; VITAMIN F.

F.D.C. ACT means: *Food, Drug, and Cosmetic Act.—See also* COLORS; CERTIFIED DYES; F.D.A.

Fe = IRON.

FEATHER BUNCHGRASS
= GREEN NEEDLEGRASS.

FEATHERING. *See* FOLIC ACID.

FECES. *See* BODY WASTES.

FEDERATION WHEAT.
See COMMON WHEAT (variety).

FEEBAR.
See SIX-ROWED BARLEY (variety).

FEED BARLEY = *Stravropol*.
See SIX-ROWED BARLEY.

FEED CONTROL OFFICIALS. *See* ASSOCIATION OF AMERICAN FEED CONTROL OFFICIALS.

FEED EFFICIENCY is a measure of the effectiveness of farm animals in converting feed to animal tissues. It is expressed as the ratio of the weight of feed required per unit of gain and does not consider cost of feed. In the United States, the common expression is pounds of feed per pound of gain or 100 lb. feed per 100 lb. gain. If a pig weighing 40 lb. required 485 lb. corn, 65 lb. supplement, and 5 lb. minerals to reach a weight of 225 lb., his feed efficiency is

$$\frac{485+65+5}{225-40} = \frac{555}{185} = 3.00$$

or 3 lb. feed per lb. gain. The more efficient animals and the better feeds result in lowest *feed efficiency ratios* and best feed efficiency. *Egg* or *milk efficiency ratios* are sometimes figured in as much as they are reflections of the ability of the animal to convert feed into eggs or milk.

In general, younger animals are more efficient than older ones. Animals fed moderate to high amounts of feed are more efficient than those fed minimum or maintenance RATIONS or those fed so much feed that they become wasteful.

Rations that are well-balanced usually promote better efficiency than those that are poorly balanced. In fact, it may be said that the greatest improvement that could be made in standard rations used on farms today lies not in promoting faster growth rates of animals but rather in improving the efficiency with which the animal can make gains. Rations high in energy are more efficient, although not necessarily less expensive, than rations formulated from bulky feeds.

Well-bred animals, properly managed, have superior feed efficiencies to those which are poorly managed, or of poor ancestry, or are subjected to undue stress from diseases and parasites.

Good efficiency ratios at the present time are considered to be as follows:

Poultry broilers..2.3 lb. feed per lb. gain
Swine..........3 lb. feed per lb. gain
Beef cattle
    and sheep.....7 lb. feed per lb. gain
Lower values have been obtained experimentally and may eventually become common in the feed lot.—*See also* NUTRITION; GROWTH; ENERGY.

FEEDING. *See* SLOP FEEDING; RATIONS; SELF-FEEDING; SWINE RATIONS; CREEP.

FEEDING BEET MOLASSES.
                            See MOLASSES.

FEEDING CANE MOLASSES.
                            See MOLASSES.

FEEDING CONCENTRATES.
            See CONCENTRATES.

FEEDING CORN SUGAR MOLASSES.
            See MOLASSES.

FEEDING OAT MEAL. *See* OATS.

FEED INGREDIENTS that are officially recognized among the MISCELLANEOUS PRODUCTS, are listed in the table.
    Reference: F.6.

FEED INSPECTION. *See* F.D.A.; TAG; ASSOCIATION OF AMERICAN FEED CONTROL OFFICIALS.

FEEDING STANDARDS have been established to express systematically the *feed requirements* of farm animals and the *feeding values* of the different feedstuffs. Among them are the following: the ARMSBY FEEDING STANDARDS; the KELLNER FEEDING STANDARDS, the WOLFF-LEHMANN FEEDING STANDARDS, the FRAPS FEEDING STANDARDS, and the MORRISON FEEDING STANDARDS.— *See also* SCANDINAVIAN FEED-UNIT SYSTEM; PETERSEN'S RELATIVE FEED VALUES; NUTRIENT REQUIREMENTS; ENERGY.

FEEDING TANKAGE = *digester tankage.*
                        See ANIMAL PRODUCTS.

FEEDING TANKAGE WITH BONE
= *digester tankage with bone. See* ANIMAL PRODUCTS.

FEED LOTS are enclosures provided with feed, water, and shelter where animals are detained prior to shipment, during fattening, or before slaughter. *Farm feed lots* are ordinarily used for winter feeding and fattening livestock and may vary in size from a few feet to many acres. *Commercial feed lots* often operate the year around with large groups of animals coming through at all seasons of the year. Railways and other commercial carriers often provide feed lots along their rights-of-way for feeding and watering livestock that are in transit.

Successful operation of feed lots usually depends upon a convenient supply of economical feed, especially a basic roughage such as ALFALFA, BEET PULP, CORN cobs, or some by-product that is difficult to move, such as wet distillers' grain, CITRUS

Recognized English name and English synonym, if any	Article of substance indicated by the name and synonym
Anise seed	The dried ripe fruit of *Pimpinella anisum.* Stimulant, carminative, and flavoring agent.
Capsicum (or red pepper)	The dried ripe fruit of any species of *Capsicum.*
Caraway seed	The dried fruit of *Carum carvi.* Stimulant, carminative, and condiment.
Dextrose *	A refined, crystallized sugar obtained chiefly by the hydrolysis of starch or of a starch-containing substance.
Fennel	The dried, ripe fruit of *Foeniculum vulgare.* Stimulant, carminative, galactagogue, and condiment.
Fenugreek seed	The dried fruit of *Trigonella foenumgraecum.* Aromatic.
Gentian	The dried rhizome and roots of *Gentiana luten.* Simple bitter.
Ginger	The dried rhizome of *Zingiber officianalle.* Condiment, carminative, and stimulant.
Glucose *	A sirup obtained by the incomplete hydrolysis of starch or of a starch-containing substance.
Kelp	Any of the various large brown seaweeds of the families Fucaceae and Laminariaceae. Source of iodine.
Kamala	The hair of the capsules of *Mallotus philippinensis taenifuge.*
Lactose (milk sugar)	A sugar obtained from cow's milk.
Locust bean (carob bean)	The fruit of *Robina pseudacacia.* A nutrient.

* The terms glucose and dextrose, as defined above by the A.A.F.C.O., apply to feed ingredients only; otherwise glucose is used synonymously with dextrose and refers to the crystallized sugar defined above as dextrose.

PULP, and refuse from various canning factories.

Commercial feed lots are designed to make maximum use of labor saving equipment and must be set up to control diseases and parasites without undue losses from them. If feed lots have concrete floors, they are easy to clean and lead to a considerable increase in the amount of MANURE which can be returned to land.

In areas where there is not a great deal of cold moisture falling in the winter, the only thing that is needed in the way of shelter is a wind break; in wet areas it is desirable to have sheds which are open, away from the prevailing winter wind so that the animals can remain dry.—*See also* DISTILLERS' PRODUCTS; PLANT BY-PRODUCTS.

FEED MIXING can be done in different ways:

Dry mixing—uses fast vertical, slower horizontal, or (occasionally) trunnion-type *batch mixers* of 1,000 lb. to 5 ton capacity; or it uses *continuous mixers* which are so arranged that the raw material enters the equipment at one end and the mixed feed comes out at the other.

Molasses mixing—usually a continuous process performed in large mixers of a capacity from 5 to 40 tons per hour. Using warm MOLASSES greatly reduces the relatively large amount of power needed for the mixing of dry feed with the molasses.

Fat mixing—the process of evenly distributing animal, fish, or vegetable fats and oils in various types of feeds for the purpose of increasing their fat content and, thus, their energy value and possibly also their palatability. Good results have been obtained by adding fat at 160°F. ANTIOXIDANTS are mostly added to the fats and oils; copper and alloys must not be used in fat mixing equipment.

Manual mixing on the farm is possible with a good scale, a clean, smooth barn floor, and a scoop shovel. Weigh or measure (or use bags containing known amounts of) the ingredients that go into 500 or 1000 lb. or 1 ton of the formula. When no mixer is used, the main ingredient should be spread on the floor first, forming a 2-to 5-in. layer, followed by layers of the other ingredients. The ingredient used in the smallest amount may be premixed before it is put on top of the pile. The feed is then turned three or four times with the scoop, each time forming new piles of the same size as the original one. With only a few ingredients this method is very satisfactory. For larger and more efficient operations, wagons or trucks with mixing equipment are available. The various layers of feed are run into the wagon, and the feed is mixed as the farmer drives to the pens.

FEED MOLASSES. *See* MOLASSES.

FEEDSTUFF COMPOSITION. The approximate composition of the common feedstuffs is given in the tables of the N.R.C. One of them is an alphabetical list of the more important feedstuffs with complete analyses, the other a compilation of analytical data for dry roughages, silages, and roots tubers, and for concentrates. *See* pages 182-188 for tables.

FEEDSTUFFS. *See* RATIONS; FLEXIBLE RATIONS; NUTRIENTS.

FEED SUPPLEMENTS. *See* VITAMINS.

FEED-UNIT SYSTEM.
        See SCANDINAVIAN FEED-UNIT SYSTEM.

FEED VALUES.
        See PETERSEN'S RELATIVE FEED VALUES.

FEELER = ANTENNA.

FENCE SILOS. *See* SILOS.

FENDLER BLUEGRASS
                        = MUTTON BLUEGRASS.

FENDLER CLOVER (*Trifolium fendleri*), one of the TRUE CLOVERS, is a productive native species of the West. It is of only minor importance as pasture plant.
Reference: M.3.

FENDLER THREE-AWN (*Aristida fendleriana*) is a small, tufted perennial GRASS which inhabits dry, sandy soils of deserts, plains, foothills, and mountain parks in the West. It has some forage value while green, but in midsummer or later becomes unpalatable.
Reference: U.6.
See also RANGE PLANTS.

FENNEL, the dried fruit of *Foeniculum vulgare*, is a stomach STIMULANT, CARMINATIVE, CONDIMENT, and GALACTAGOGUE.— *See also* FEED INGREDIENTS.

AVERAGE COMPOSITION OF FEEDSTUFFS [1]

Feeding Stuff	Moisture %	Protein %	Fat %	Fiber %	N-free extract %	Ash %	Calcium %	Phosphorus %	Iron %	Manganese mg/lb	Copper mg/lb	Cobalt mg/lb	Carotene mg/lb	Thiamine mg/lb	Niacin mg/lb	Riboflavin mg/lb	Pantothenic acid mg/lb
Alfalfa meal, dehydrated, 17% protein	6.7	17.8	2.8	24.2	39.7	8.8	1.7	0.2	0.033	15.0	3.1	0.05	36.0[2]	1.5	8.7	7.3	12.3
Alfalfa meal, sun-cured, 17% protein	7.8	17.6	2.0	23.8	38.9	9.9	1.4	0.2					24.0[2]	1.1	16.1	5.0	12.7
Alfalfa meal, dehydrated, 20% protein	7.1	20.9	2.9	19.8	38.2	11.1	1.7	0.3	0.039	28.6	7.1	0.17	60.0[2]	3.1	17.3	7.4	18.5
Alfalfa meal, sun-cured, 20% protein	8.1	20.3	3.2	17.8	40.3	10.3	1.4	0.2					48.0[2]	2.0	22.7	7.2	13.6
Barley, excluding Pacific Coast	10.6	12.7	1.9	5.4	66.6	2.8	0.09	0.5	0.005	8.3	5.1	0.01		1.7	24.1	0.8	3.7
Barley, Pacific Coast	6.4	9.7	2.2	6.2	68.7	2.2	0.06	0.4	0.007	7.8	5.0	0.05	1.8	1.8	20.0	0.6	3.3
Bone meal, raw	11.0	26.0	5.0	1.0	2.5	59.1	21.7	10.0	0.044	1.9	8.5			0.1	1.9	0.5	1.0
Bone meal, steamed, special	6.4	13.4	9.7	1.1	1.2	71.3	29.3	15.1	0.080	5.1	9.0	0.03		0.9	2.0	0.4	0.8
Buttermilk, dried	3.3	32.4	6.4	0.3	43.3	10.0	1.4	0.9		1.5				1.7	2.8	15.8	13.5
Corn, dent, yellow	15.0	8.9	3.9	2.0	68.9	1.3	0.02	0.3	0.002	2.3	0.9	0.01	1.33	1.7	9.8	0.5	2.6
Corn gluten meal, 41% protein	8.6	42.9	2.0	3.9	40.1	2.5	0.2	0.4	0.047	4.4	13.1	0.04	10.0	0.1	24.8	0.7	3.8
Cottonseed oil meal, 41% protein	7.8	41.2	6.2	11.2	27.7	5.9	0.2	1.2	0.008	12.9	7.7	0.06		1.8	13.0	2.5	4.4
Crab meal	7.9	31.5	2.2	10.5	6.0	41.9	14.5	1.5								2.3	3.0
Distillers dried corn grains, with solubles	6.9	28.8	8.9	9.0	41.7	4.7	0.2	0.7		18.2				2.1	36.3	3.4	5.2
Distillers dried solubles	6.9	28.0	6.7	3.3	47.6	7.5	0.4	1.4	0.050	45.4	36.3	0.09		2.7	54.3	5.2	8.9
Fish meal, menhaden	6.4	62.2	8.5	0.7	4.2	18.0	5.0	3.4	0.057	10.0	3.9	0.08		0.2	25.9	2.4	

[1] From a compilation of the Committee on Feed Composition of the National Research Council.
[2] Rough approximations since carotene content is too variable for dependable averages.

AVERAGE COMPOSITION OF FEEDSTUFFS—(Continued)

Feeding Stuff	Moisture %	Protein %	Fat %	Fiber %	N-free extract %	Ash %	Calcium %	Phosphorus %	Iron %	Manganese mg/lb	Copper mg/lb	Cobalt mg/lb	Carotene mg/lb	Thiamine mg/lb	Niacin mg/lb	Riboflavin mg/lb	Pantothenic acid mg/lb
Fish meal, sardine.	6.9	67.2	5.0	0.6	5.4	14.9	4.2	2.5	0.03	10.3	9.2	0.08		0.2	26.0	2.5	1.3
Fish meal, white-fish.	9.6	63.0	6.7	0.1	0.1	20.5	6.8	3.7		5.0				0.4	36.0	4.0	1.4
Hominy feed, white, 5% fat.	10.1	10.8	5.7	4.7	66.0	2.7	0.05	1.0	0.004	8.2	9.9			5.9	25.1	1.0	3.1
Hominy feed, yellow.	9.5	11.1	7.1	5.8	63.9	2.6	0.05	0.7	0.024	6.9	5.7	0.02	0.5	3.8	22.6	1.0	3.5
Limestone.							38.3										
Linseed oil meal, o. p., 33% protein.	8.8	35.0	5.6	8.1	36.9	5.6	0.4	0.9	0.024	19.1	11.8	0.17		3.9	18.9	1.9	7.5
Linseed oil meal, o. p., 37% protein.	9.1	38.0	5.9	7.7	33.7	5.6	0.5	0.9	0.028	18.6	11.9	0.18		2.7	14.7	1.7	8.9
Meat scrap, 52% protein.	6.9	52.9	7.3	2.2	4.3	26.4	8.7	4.4	0.050	4.0	5.5	0.09		0.1	27.1	2.4	2.1
Meat scrap, 60% protein.	6.2	60.9	8.8	2.4	1.1	20.6	6.3	3.5	0.041	5.4	3.7	0.04		0.1	23.7	2.5	2.3
Meat and bone scrap, 50% protein.	6.4	50.6	10.0	2.0	2.0	29.0	9.7	4.2		5.3					21.4	2.1	1.5
Milk, skimmed, dried.	5.8	34.7	1.2	0.2	50.3	7.8	1.3	1.1	0.005	1.2	5.2	0.03		1.5	5.7	10.0	16.0
Molasses, beet.	19.5	8.4	0.0	0.0	62.0	10.1	0.08	0.02							22.0	1.1	17.9
Molasses, cane.	26.0	2.9	0.0	0.0	62.1	9.0	0.7	0.08						0.4	20.9		
Oats, excluding Pacific Coast.	9.8	12.0	4.6	11.0	58.6	4.0	0.09	0.4	0.008	19.2	2.4			2.9	8.2	0.4	6.8
Oats, Pacific Coast.	9.8	9.0	5.4	11.0	61.1	3.7	0.07	0.4							4.5		
Oatmeal, feeding.	9.3	15.0	7.4	2.0	64.4	1.9			0.006	13.6	1.6	0.02		3.5	4.5	0.6	6.6
Oyster shell, ground.							37.9										
Peanut oil meal, o. p., 43% protein.	7.2	43.1	7.6	13.9	23.0	5.2	0.1	0.6						3.3	77.5	2.4	24.1
Phosphate, defluorinated rock.							28.3	12.3		40.9	29.5						
Phosphate, dicalcium.							26.5	20.5		70.9							
Rice bran.	9.0	12.8	13.1	12.7	41.7	10.7	0.08	1.4						10.3	129.1	1.4	10.3

AVERAGE COMPOSITION OF FEEDSTUFFS—(Continued)

Feeding Stuff	Moisture %	Protein %	Fat %	Fiber %	N-free extract %	Ash %	Calcium %	Phosphorus %	Iron %	Manganese mg/lb	Copper mg/lb	Cobalt mg/lb	Carotene mg/lb	Thiamine mg/lb	Niacin mg/lb	Riboflavin mg/lb	Pantothenic acid mg/lb
Rice polishings	9.7	12.7	11.4	3.5	56.6	6.2	0.04	1.1						8.8	325.0	0.9	5.5
Rye, grain	10.5	12.6	1.7	2.4	70.9	1.9	0.01	0.3	0.008	37.0	3.4			2.0	7.1	0.7	4.2
Sorghum, kafir	10.5	11.2	2.9	2.3	71.1	2.0	0.04	0.3	0.008	7.5	3.0	0.03		1.6	18.3	0.5	5.7
Sorghum, milo	10.6	11.3	2.9	2.2	71.3	1.7	0.03	0.3	0.005	5.9	7.8	0.02		1.8	13.1	0.4	5.0
Sorghum, milo head chop	10.0	10.1	2.5	6.8	67.1	3.5	0.1	0.3									
Soybean oil meal, 41% protein	10.2	41.1	5.3	5.5	30.4	7.5	0.3	0.6	0.017	12.3	11.1					0.9	
Soybean oil meal, 43% protein	9.4	43.1	5.6	5.7	30.4	5.8	0.3	0.6		12.7							
Soybean oil meal, 44% protein	9.2	44.2	5.3	5.6	29.9	5.8	0.3	0.7	0.015	14.0	7.7	0.05		0.8	16.7	2.5	6.1
Soybean oil meal, solvent, ext.	9.4	46.1	1.0	5.9	31.8	5.8	0.05	0.6	0.01	13.8	6.8			1.4	17.1	1.5	6.2
Wheat, hard red winter	10.4	15.2	1.8	2.6	68.3	1.7	0.05	0.4	0.005	18.0	2.0			2.3	24.1	2.0	6.3
Wheat, northern spring	9.9	15.8	2.2	2.5	67.8	1.8	0.05	0.4	0.005	33.0	4.9	0.03		2.3	28.8	1.4	6.4
Wheat, soft, Pacific Coast	10.8	9.9	2.0	2.7	72.7	1.9		0.3	0.007	27.7	4.4			2.2	26.8	0.5	5.2
Wheat bran	10.3	16.4	4.3	9.9	53.0	6.1	0.1	1.3	0.016	56.0	4.9	0.05		3.9	63.5	0.5	13.6
Wheat flour middlings	10.3	18.1	4.6	4.9	58.5	3.6	0.07	0.6	0.002	39.0	2.1			6.0	44.2	0.5	4.5
Wheat standard middlings	9.9	17.6	5.0	6.7	56.6	4.2	0.1	0.8	0.009	53.6	9.8	0.04		5.8	44.3	0.8	9.3
Whey, dried	6.5	12.2	0.8	0.2	70.4	9.9	0.9	0.8	0.02	1.1	24.2	0.05		1.8	5.1	13.0	22.4
Yeast, brewers, dried	6.3	46.8	1.2	2.8	35.7	7.2	0.1	1.5	0.014	2.4	15.1	0.08		43.0	213.6	14.0	49.1

1 pound = 454 grams.   1 ounce = 28.4 grams.
1 microgram per gram is the same as parts per million.
1 I.U. (U.S.P.) unit of vitamin A equals the activity of 0.6 micrograms of B-carotene.
1 microgram of B-carotene equals 1.666 I.U. (U.S.P.) unit of vitamin A.

Conversion table for calculating feed formulas
1 gram = 1000 milligrams.
1 milligram = 1000 micrograms.

AVERAGE COMPOSITION AND DIGESTIBLE NUTRIENTS[1]

Feedstuff	Total Dry Matter	Dig. Protein	Total Dig. Nutrients	Calcium		Phosphorus		Carotene
	%	%	%	%	gm./lb.	%	gm./lb.	mg./lb.
Dry Roughages								
Alfalfa hay, all analyses.....	90.5	10.5	50.3	1.47	6.67	0.24	1.09	11.4
Alfalfa hay, before bloom....	90.5	13.7	53.4	2.22	10.08	0.33	1.50	....
Alfalfa hay, ⅒ to ¼ bloom....	90.5	11.2	51.7	1.26	5.72	0.22	1.00	19.4
Alfalfa hay, ¼ to full bloom..	90.5	10.3	50.1	....	....	....	....	9.0
Alfalfa hay, past bloom......	90.5	9.2	47.6	....	....	....	....	3.2
Alfalfa meal, good...........	92.7	11.8	53.6	1.69	7.67	0.25	1.14	16.6
Alfalfa meal, dehydrated*....	93.6	13.8	54.7	1.50	6.81	0.35	1.59	43.4
Alfalfa straw*..............	92.6	4.5	42.6	....	....	0.13	0.59	....
Barley hay..................	90.8	4.0	51.9	0.26	1.18	0.23	1.04	....
Barley straw................	90.0	0.7	42.2	0.32	1.45	0.11	0.50	....
Bromegrass hay, all analyses.	88.1	5.0	48.9	0.20	0.91	0.28	1.27	....
Clover hay, red, all analyses.	88.1	7.1	52.2	1.35	6.13	0.19	0.86	8.6
Clover and mixed grass hay, high in clover.............	89.7	5.5	52.2	0.90	4.09	0.19	0.86	....
Clover and timothy hay, 30 to 50% clover.............	88.1	4.8	51.2	0.68	3.09	0.20	0.91	
Corn cobs, ground...........	90.4	0.0	45.7	....	....	0.02	0.09	....
Corn fodder, very dry.......	91.1	3.8	58.8	0.24	1.09	0.16	0.73	1.8
Corn stover, very dry.......	90.6	2.1	51.9	0.29	1.32	0.05	0.23	....
Cowpea hay, all analyses....	90.4	12.3	51.4	1.37	6.22	0.29	1.32	....
Grass hay, mixed eastern states, good quality.......	89.0	3.5	51.7	0.48	2.18	0.21	0.95	....
Kafir fodder, very dry.......	90.0	4.5	53.6	0.35	1.59	0.18	0.82	2.0
Kafir stover, very dry.......	90.0	1.9	51.3	0.54	2.45	0.09	0.41	1.1
Lespedeza hay, annual, all analyses.................	89.2	6.4	47.5	0.98	4.45	0.18	0.82	....
Lespedeza hay, annual, before bloom................	89.1	7.2	49.2	1.04	4.72	0.19	0.86	22.4
Lespedeza hay, annual, in bloom....................	89.1	6.5	47.5	1.02	4.63	0.18	0.82	....
Lespedeza hay, annual, after bloom....................	89.1	3.6	39.6	0.90	4.09	0.15	0.68	....
Mixed hay, good, less than 30% legumes*.............	88.0	4.4	49.8	0.61	2.77	0.18	0.82	....
Oat hay....................	88.1	4.9	47.3	0.21	0.95	0.19	0.86	....
Oat straw..................	89.7	0.7	44.7	0.19	0.86	0.10	0.45	....
Pea hay, field..............	89.3	10.6	55.1	1.22	5.54	0.25	1.14	....
Pea-and-oat hay............	89.1	8.6	52.9	0.72	3.27	0.22	1.00	....
Peavine hay, sun cured......	86.3	8.6	54.2	1.48	6.72	0.16	0.73	....
Prarie hay, western, good quality...................	90.7	2.1	49.6	0.36	1.63	0.18	0.82	9.3
Prairie hay, western, mature..	91.7	0.6	46.7	0.28	1.27	0.09	0.41	....
Sorghum fodder, sweet, dry..	88.8	3.3	52.4	0.34	1.54	0.12	0.54	1.1
Rye hay*...................	91.3	2.8	44.7	....	....	0.18	0.82	....
Rye straw..................	92.8	0	42.2	0.26	1.18	0.09	0.41	....
Soybean hay, good, all analyses....................	88.0	9.6	49.0	0.94	4.27	0.24	1.09	....
Soybean hay, in bloom or before...................	88.0	12.0	52.4	1.53	6.95	0.27	1.23	....
Soybean hay, seed developing*...................	88.0	9.8	48.5	1.35	6.13	0.25	1.14	13.6
Soybean hay, seed well developed...................	88.0	10.8	52.5	1.14	5.18	0.27	1.23	....
Soybean straw..............	88.8	1.2	38.5	....	....	0.13	0.59	....
Sudan grass hay, all analyses.	89.3	4.3	48.5	0.36	1.63	0.26	1.18	....

AVERAGE COMPOSITION AND DIGESTIBLE NUTRIENTS[1]—Continued

Feedstuff	Total Dry Matter	Dig. Protein	Total Dig. Nutrients	Calcium		Phosphorus		Carotene

Dry Roughages (Continued)

Feedstuff	%	%	%	%	gm./lb.	%	gm./lb.	mg./lb.
Timothy hay, all analyses...	89.0	2.9	48.9	0.23	1.04	0.20	0.91	5.3
Timothy hay, before bloom..	89.0	5.4	56.8	....	....	....	....	9.2
Timothy hay, early bloom...	89.0	4.1	50.8	0.41	1.86	0.21	0.95	....
Timothy hay, full bloom.....	89.0	3.3	48.1	0.23	1.04	0.20	0.91	4.2
Timothy hay, late seed.....	89.0	2.1	42.2	0.14	0.64	0.15	0.68	2.5
Timothy and clover hay, ¼ clover*................	88.8	3.7	49.4	0.51	2.32	0.20	0.91	....
Vetch and oat hay, over half vetch..................	87.6	8.4	52.7	0.76	3.45	0.27	1.23	....
Wheat hay..............	90.4	3.3	46.7	0.14	0.64	0.18	0.82	....
Wheat straw...............	92.5	0.3	40.6	0.21	0.95	0.07	0.32	....

Silages, Roots and Tubers

Feedstuff	%	%	%	%	gm./lb.	%	gm./lb.	mg./lb.
Alfalfa silage, wilted before being ensiled.............	36.0	4.1	21.3	0.51	2.32	0.12	0.54	21.1
Alfalfa—molasses, not wilted.	26.8	2.7	15.4	0.41	1.86	0.08	0.36	14.5
Alfalfa—molasses, wilted....	36.6	4.0	21.4	0.56	2.54	0.11	0.50	....
Beet top silage, sugar.......	31.6	2.6	14.9	0.31	1.41	0.07	0.32	5.1
Corn, dent, well matured, all analyses.................	27.4	1.2	18.1	0.10	0.45	0.06	0.27	6.4
Corn, dent, well matured, well eared...............	28.4	1.3	20.0	0.08	0.36	0.06	0.27	....
Corn, dent, well matured, fair in ears...............	26.0	1.1	17.1	0.09	0.41	0.06	0.27	....
Corn, dent, immature, before dough stage..............	20.4	0.9	13.0	....	....	....	....	....
Corn, sweet, stover (ears removed...................	24.0	0.9	13.0	....	....	....	....	....
Corn—canning factory waste (husks, cobs, and waste ears)...................	22.4	1.1	16.1	....	....	....	....	....
Grass silage, some legumes, wilted, molasses added.....	35.0	2.7	19.8	0.32	1.45	0.12	0.54	....
Grass silage, small proportion legumes, wilted slightly, molasses added...	29.0	1.6	18.0	....	....	0.07	0.32	....
Grass silage, small proportion legumes, not wilted*..	26.5	1.6	16.2	....	....	....	....	....
Mangel-wurzel.............	9.2	0.9	7.0	.0.01	0.05	0.03	0.14	....
Pea-vine, from canneries.....	24.5	1.9	14.0	.0.32	1.45	0.06	0.27	21.0
Potato, tubers.............	21.2	1.3	17.9	0.01	0.05	0.05	0.23	....
Sorghum silage, sweet.......	25.3	0.8	15.2	0.08	0.36	0.04	0.18	2.7
Soybean, not wilted........	24.8	2.9	14.6	0.34	1.54	0.09	0.41	14.6
Soybean, wilted..............	33.2	3.9	19.1	0.45	2.04	0.12	0.54	....

Grains, Seeds, and By-product Concentrates

Feedstuff	%	%	%	%	gm./lb.	%	gm./lb.	mg./lb.
Barley, excluding Pacific coast......................	89.4	10.0	77.7	0.09	0.41	0.47	2.13	....
Barley, Pacific coast.........	89.0	7.7	78.3	0.06	0.27	0.41	1.86	....
Beans, field or navy........	90.0	20.2	78.7	0.15	.68	0.57	2.59	....
Beans, lima*.................	89.7	18.7	77.8	0.09	0.41	0.37	1.68	....
Beet pulp, dried............	90.1	4.3	67.8	0.71	3.22	0.12	0.54	....
Beet pulp, molasses, dried...	91.9	7.1	72.1	0.62	2.81	0.09	0.41	....
Beet pulp, wet..............	11.6	0.8	8.8	....	....	....	....	....
Bone meal, steamed.........	96.3	....	....	25.22	14.50	11.9	54.03	....

AVERAGE COMPOSITION AND DIGESTIBLE NUTRIENTS[1]—Continued

Feedstuff	Total Dry Matter	Dig. Protein	Total Dig. Nutrients	Calcium		Phosphorus		Carotene
	%	%	%	%	gm./lb.	%	gm./lb.	mg./lb.
Grains, Seeds, and By-product Concentrates (Continued)								
Brewers' grains, dried, below 25% protein	92.3	16.8	61.8	....	....	....	....	....
Brewers' grains, dried, 25% protein	92.9	22.1	67.1	0.25	1.14	0.49	2.22	....
Citrus pulp, dried*	90.1	3.0	73.2	2.28	10.35	0.17	0.77	....
Coconut oil meal, expeller process	92.6	17.4	80.8	0.12	0.54	0.62	2.81	....
Corn, dent, yellow, No. 2....	85.0	6.8	80.0	0.02	0.10	0.27	1.23	1.3
Corn and cob meal*	86.1	5.3	73.2	....	....	0.22	1.00	....
Corn, gluten feed, 23% protein	91.4	21.3	76.3	0.68	3.09	0.82	3.72	6.1
Corn gluten feed, 25% protein	91.1	22.9	76.5	0.40	1.82	0.82	3.72	....
Corn gluten meal, 41% protein	91.4	36.5	80.2	0.20	0.91	0.41	1.86	10.0
Cottonseed, whole pressed, 28% protein	93.5	20.3	59.8	0.15	0.68	0.77	3.50	....
Cottonseed meal, 36% protein	92.2	30.4	71.9	0.19	0.86	1.13	5.13	....
Cottonseed meal, 39% protein	91.5	32.9	72.6	0.11	0.50	1.30	5.90	....
Cottonseed meal, 41% protein	92.2	34.2	73.6	0.18	0.82	1.14	5.18	....
Cottonseed meal, 43% protein	92.5	35.4	74.6	0.23	1.04	1.14	5.18	....
Cottonseed meal, 45% protein	92.4	37.1	76.4	0.27	1.23	1.09	4.95	....
Distillers' dried corn grains..	92.9	20.7	82.4	0.11	1.04	0.47	2.71	....
Flaxseed screenings	91.1	9.2	57.9	0.37	1.68	0.43	1.95	....
Flaxseed screenings oil feed*.	91.9	14.0	55.1	....	....	....	....	....
Hominy feed, white, 5% fat..	89.9	7.7	82.9	0.05	0.23	1.03	4.68	....
Hominy feed, yellow*	90.5	7.9	85.1	0.05	0.23	0.66	3.00	4.1
Linseed oil meal, O.P., 31% protein	91.0	28.2	77.0	0.36	1.63	0.90	4.09	....
Linseed oil meal, O.P., 33% protein	91.2	30.4	77.1	0.44	2.00	0.94	4.27	....
Linseed oil meal, O.P., 37% protein	90.9	33.1	77.5	0.49	2.22	0.89	4.04	....
Molasses, beet	80.5	4.4	60.8	0.08	0.36	0.02	0.09	....
Molasses, cane	74.0	0	54.0	0.74	3.36	0.08	0.36	....
Oats, excluding Pacific coast.	90.2	9.4	70.1	0.09	0.41	0.43	1.95	....
Oats, Pacific coast	90.2	7.0	71.4	....	....	....	....	....
Orange pulp, dried	87.9	6.1	78.4	....	....	....	....	....
Peanut oil meal, O.P., 41% protein	93.8	38.0	83.2	....	....	....	....	....
Peanut oil meal, O.P., 43% protein	92.7	39.2	82.0	0.16	0.73	0.54	2.45	....
Peanut oil meal, O.P., 45% protein	93.4	41.5	83.7	....	....	....	....	....
Potato meal, or dried potatoes	92.8	3.0	70.6	0.08	0.36	0.22	1.00	....
Rice bran	91.0	8.7	68.4	0.08	0.36	1.36	6.17	....
Rice polishings	90.3	9.6	82.6	0.04	0.18	1.10	4.99	....
Rye	89.5	10.0	76.1	0.01	0.05	0.33	1.50	....
Safflower seed oil meal from hulled seed	91.0	32.7	55.5	....	....	....	....	....

AVERAGE COMPOSITION AND DIGESTIBLE NUTRIENTS[1]—Continued

Feedstuff	Total Dry Matter	Dig. Protein	Total Dig. Nutrients	Calcium	Phosphorus		Carotene

Grains, Seeds, and By-product Concentrates (Continued)

Feedstuff	%	%	%	%	gm./lb.	%	gm./lb.	mg./lb.
Safflower seed oil meal from unhulled seed*	91.0	15.7	39.1	....	....	....	....	....
Safflower seed*	93.1	14.0	86.8	....	....	....	....	....
Sorghum, Kafir	89.5	9.1	80.7	0.04	0.18	0.33	1.50	....
Sorghum, milo	89.4	8.8	80.1	0.03	0.14	0.27	1.23	....
Sorghum, milo, head chops	90.0	7.7	77.2	0.14	0.64	0.26	1.18	....
Soybeans	90.0	33.7	87.6	0.27	1.23	0.62	2.81	....
Soybean oil meal, 41% protein	89.8	34.5	76.4	0.26	1.18	0.59	2.68	....
Soybean oil meal, 43% protein	90.6	36.2	78.7	0.28	1.27	0.61	2.77	....
Soybean oil meal, 44% protein	90.8	37.1	78.6	0.30	1.36	0.66	3.00	....
Soybean oil meal, solvent extracted	91.6	42.4	78.5	0.29	1.32	0.63	2.86	....
Wheat, hard, red, winter	89.6	12.8	80.0	0.05	0.23	0.41	1.86	....
Wheat, northern, spring	90.1	13.3	80.7	0.05	0.23	0.41	1.86	....
Wheat, soft, Pacific coast	89.2	8.3	80.0	....	....	0.29	1.32	....
Wheat bran	89.7	13.3	66.4	0.14	0.64	1.30	5.90	....
Wheat flour middlings	89.7	15.9	79.0	0.07	0.32	0.65	2.95	....
Wheat standard middlings	90.1	14.6	78.0	0.14	0.64	0.78	3.54	....

* Where no digestion coefficients were available or where the available data were too few to be reliable, the digestion coefficients for comparable feeds were used.

[1] From a compilation of the Committee on Feed Composition and the Committee on Animal Nutrition of the National Research Council.

FENUGREEK SEED is the dried fruit of *Trigonella foenumgraecum*. It is used as an AROMATIC to flavor feeds (rather than as a drug).—*See also* FEED INGREDIENTS.

FERBAM = FERMATE.

FERGUSON 922. *See* RED OAT (variety).

FERGUSON BARLEY is a winter variety of the SIX-ROWED BARLEY.

FERMATE, or *ferbam*, is *ferric dimethyl dithiocarbamate;* it is used for control of BLACKPATCH and other FUNGUS DISEASES.

FERMENT. *See* FERMENTS.

FERMENTATION (sometimes referred to as *sweating*) is any transformation of complex organic substances (such as CARBOHYDRATES, e.g., various sugars) by the action of the ENZYMES of micro-organisms, such as YEAST or bacteria, into more simple compounds. Thus, DIASTASE hydrolyzes starch to sugar; INVERTASE converts sucrose (cane sugar) into dextrose and levulose; in *alcoholic fermentation*, alcohol and carbon dioxide are formed from starch (of grain)

via sugars by the action of ferments contained in yeast.—*See also* HAY; INVERT SUGAR; FERMENTS.

FERMENTATION INDUSTRY BY-PRODUCTS include DISTILLERS' PRODUCTS, BREWERS' PRODUCTS, so-called FERMENTATION PRODUCTS, YEAST, DRIED GRAINS OR VINEGAR DRIED GRAINS, and such VITAMIN products as *dried fermentation solubles* and *condensed fermentation solubles.*

FERMENTATION PRODUCTS is a broad term which includes all FERMENTATION INDUSTRY BY-PRODUCTS, but in the terminology of the feed industry is limited to products obtained in the manufacture of *antibiotics* and organic acids, especially *citric acid.*

Officially recognized are the following:

Extracted . . . presscake—the filtered and dried MYCELIUM obtained from . . . fermentation. (For label identification the source must be indicated in the blank spaces as

PENICILLIUM, STREPTOMYCES, CITRIC ACID, etc.).

Extracted . . . meal is ground . . . press-cake. (The source must be indicated as penicillium, streptomyces, citric acid, etc.).

Dried extracted . . . fermentation solubles —the dried extracted broth obtained from penicillium, streptomyces, citric acid, etc. fermentation, as the case may be.

Dried extracted . . . meal and fermentation solubles—the filtered and dried mycelium and dried extracted broth obtained from . . . fermentation. (The source shall be indicated as penicillium, streptomyces, citric acid, etc.).

References: F.6; E.12.

FERMENTS—a term often used synonymously with ENZYMES—consist of enzymes in association with the living organisms or with substances produced by the latter. Ferments are capable of causing *fermentation* of other substances, apparently without undergoing chemical changes themselves.—*See also* VITAMINS; INVERTASE; DIGESTIVE TRACT.

FERRIC DIMETHYL DITHIOCARBA-MATE is a white powder, slightly soluble in water. It is the active ingredient in the fungicide FERMATE.

FERRIC OXIDE. *See* RED IRON OXIDE.

FERRIC PHOSPHATE
= IRON PHOSPHATE.

FERROUS CARBONATE
= IRON CARBONATE.

FERROUS SULFATE
= IRON SULFATE.

FERTILE means: fruit-producing or capable of proper functioning in reproduction.

FERTILITY is the ability (1) of plants and animals to reproduce sexually or (2) of soils to provide the chemical compounds needed for the growth of specified plants.

Self-fertility and *pseudo-self-fertility* are terms applied to any plant possessing complete units of reproduction; e.g., RED CLOVER.

FERTILITY VITAMIN. *See* VITAMIN E.

FERTILIZATION means (1) the union of the male nucleus (pollen) of a plant with the female cell (egg); (2) the union of sperm and ovum (egg) following insemination of animals; and (3) the addition to the soil of those elements or compounds that are needed for the nutrition of plants.— *See also* PASTURE MANAGEMENT; FERTILIZER.

FERTILIZERS are chemical compounds, added to the soil for the purpose of improving its nutrient content; they enable plants to grow faster and bigger, thus increasing the crop yield.

LIME is often not considered a fertilizer because it contributes rather to the improvement of the soil than to the nutrition of the plants. Similarly, MANURE is generally not included in the term fertilizer, even though it possesses plant-food and soil-improving values; it is rather considered an *"indirect fertilizer."*

So-called *commercial fertilizers* are either *organic* (by-products from packing houses, fisheries, oil mills, breweries, etc.) or *inorganic*. The latter may contain many constituents, but their value depends chiefly on three essential plant-food elements: NITROGEN, PHOSPHORUS, and POTASSIUM.

The *mixed fertilizers* consist of more than one of these three essential elements. Their content is expressed in three numbers, with the first referring to the percentage of nitrogen, or N; the second to that of phosphorus, expressed as P_2O_5 (phosphorus pentoxide); and the third to potassium, expressed as K_2O (potassium oxide—often misnamed *"potash"*).

Mixed fertilizers, widely used on farms, are the following: (a) *without nitrogen*— 0-9-27, 0-10-20, 0-12-20, 0-14-10, 0-14-14, etc.; (b) *complete fertilizers*—2-12-6, 2-12-12, 3-12-12, 4-8-4, 4-12-4, 6-6-6, etc. Fertilizers containing only one plant food, too, are expressed in the same manner— e.g., 0-45-0, indicating that the particular product contains 45 percent P_2O_5, but neither nitrogen nor K_2O.

The chemical analysis, as expressed in the three numbers, is not sufficient: the *qualitative* composition of the mixed fertilizers, too, must be known: the same percentage of P_2O_5 may refer to practically insoluble (natural) rock phosphate, to more readily soluble superphosphate, or to very soluble ammonium phosphate. Thus, when buying mixed fertilizers, it is necessary to consider the actual cost of their total plant-food ingredients rather than the cost per bag or per ton.—*See also* CALCIUM; AMMONIUM; BASIC SLAG; GYPSUM.

FESCUE FOOT = FESCUE LAMENESS.

FESCUE LAMENESS, or *fescue foot*, is a disease condition affecting cattle. It has been recognized in Colorado, Missouri, Kentucky, New Zealand, and Australia. The disease is chiefly a problem connected with blizzard conditions and injudicious management. Fescue lameness affects hungry, probably vitamin-A starved animals, that have access to TALL FESCUE pasture at its peak stage of growth. Eating large amounts of any strain of this grass may in a short time cause the condition, which is characterized by a lameness and apparently circulatory disturbance of, especially, the hind feet, ears, and tails. If the animals are taken off the fescue as soon as the first signs appear, complete recovery occurs.

> NOTE: Recent cases were seen in Missouri where cattle were running to silage and hay and had access to fescue pasture that was chiefly stand-over from last season's growth. The animals lost ears, tails, as well as hooves.

Reference: C.5.

FESCUES (*Festuca* spp.) compose a very large genus of which there are about 100 species in temperate or cool zones. They vary in texture and growth and comprise a versatile group of varying uses.

The annual species are weedy, but the perennials are excellent for forage and turf. The fescues cultivated as pasture GRASSES can be classified as the *broad-leaf* and the *fine-leaf* species.

Of the broad-leaf species that currently are most widely used, MEADOW FESCUE and TALL FESCUE (also known as *reed fescue* and under numerous common names, such as *King fescue, giant fescue, ditch fescue,* or *tall bank fescue*)—and the latter's two strains *Alta fescue* and *Kentucky 31*—are outstanding. Of the fine-leaf species, SHEEP-FESCUE is perhaps the most useful; *red fescue* (or *creeping red fescue*) and its not-creeping variety, *chewings fescue,* are used mainly for lawns and turfs.

References: H.1.

See also MUTTON BLUEGRASS; ANNUAL LESPEDEZAS; SAGEBRUSHES; BLOAT; PAS-TURE PLANTS; RANGE PLANTS; BUNCHGRASS. *Illustrations: See* MEADOW FESCUE; RED FESCUE; SHEEP FESCUE.

Fescue. Plant and spikelet. (H.47.)

FESTUCA. The *Festuca* spp., commonly known as FESCUES, include *F. elatior* = MEADOW FESCUE; *F. arundinacea* = TALL FESCUE; *F. ovina* = SHEEP FESCUE; *F. rubra* = RED FESCUE (and its not creeping variety *F. rubra* var. *commutata,* or *chewings fescue*).

FETERITA. *Standard feterita, Spur feterita,* and *Dwarf feterita* are grain SORGHUMS.

FIBER is threadlike, tough tissue derived from plants, animals, or minerals.

In feed analysis, the term *crude fiber* (or, for short, fiber) indicates the content of CELLULOSE and related carbohydrates which form the cell walls of the plants with protective LIGNIN and which animals—and particularly swine—can digest only partly, and this under waste of much energy. No mammal can digest an appreciable amount of lignin.

The higher the fiber content of a feed, the lower its feeding value. STRAWS are very high in fiber, as are ROUGHAGES in general; CONCENTRATES are lower in fiber content; CORN contains only about 2 percent fiber, wheat slightly more, but BARLEY and OATS have tough hulls that cause a fiber content two to four times that of corn.—*See also* NITROGEN-FREE EXTRACT; LABELS.

FIBROUS means: composed of FIBERS.

FIELD BEANS are LEGUMES and resemble in many respects the SOYBEAN. They belong to a number of genera, the most important of which includes the *edible beans* of the *Phaseolus* spp., e.g., the COMMON BEAN with its more than 50 varieties (among them the Navy bean and Pinto), the TEPARY BEAN, MUNGBEAN, and LIMA BEAN.

These and other edible beans are successfully grown on both irrigated and non-irrigated lands under a variety of soil and climatic conditions in 21 states of the United States, but a major part of the crop is in fairly distinct areas of 7 states.

Beans are a concentrated direct food and are high in protein, phosphorus, iron, and vitamin B_1. They fit well into rotations on many farms and are an important cash crop; both the straw and the cull beans can be used as feed for livestock.

Cleaning and grading of beans are almost always done at the "bean house." The cull beans are either returned to the grower for livestock feeding or sold to feeders or feed dealers.

Cull beans—the *damaged beans* which are sorted out from the first-quality beans —can be used satisfactorily for livestock feeding, but their feeding value is less than would be expected from the chemical analysis. They are not very palatable to stock, and their digestibility is not high when they are fed raw, especially to hogs. Cull beans often include such waste as coarse bits of stem, small stones, and dirt.

Cull beans can well be fed to sheep for they will sort out the beans from trash. *Whole cull beans* are satisfactory for lamb fattening, if they are not more than 20 to 25 percent of the grain mixture; a larger quantity may be unpalatable and cause scouring. It is especially difficult to

keep the animals eating large amounts of beans in the summer, possibly because cracked beans soon become rancid.

Pods and leaflets of the mungbean which belongs to the field beans.

If the cull beans are not too "trashy," they can be fed to other livestock. They should be cooked or steamed and mixed with grain if fed to hogs or steers. *Ground cull beans* can be fed to dairy cows as a substitute for other protein supplements, but they should not be more than a fifth of the concentrate mixture. If they are mixed with more palatable feeds—such as ground CORN, ground OATS, or LINSEED OIL meal—the cows are more likely to clean up the feed. More *cooked beans* can be fed to dairy cows, but cooking is rather expensive and may not be justified.

Bean straw can be used for feeding

purposes. The bean land is often rented to sheepmen after the harvest, and the sheep are *pastured* a few days on the straw.

Bean straw, in many bean-growing districts, is used for feeding sheep, cattle, or horses. Its feeding value varies widely, but straw of a good grade, especially if fed with some legume hay, may be considered worth about half as much per ton, in terms of net energy, as ALFALFA or CLOVER hay. In the West it is used mainly for wintering livestock and to a limited extent in fattening rations.

Bean straw, when fed as the sole roughage in a lamb-fattening ration, does not produce satisfactory gains, since the lambs are inclined to eat so much bean straw that they scour severely.

Cracked and unthreshed beans in the straw add to the feed value and unused or surplus straw can be thrown into the feed lot to help build up the manure.

Dangers: Ewes eating cull beans have given birth to lambs that became stiff.

Reference: M.45.

See also STRAW; VITAMIN E.

FIELD BINDWEED (*Convolvulus arvensis*), commonly called *bindweed*, is one of the bindweeds which in many respects is similar to the closely related morning glory.

Field bindweed is a perennial with bell-like flowers and arrow-shaped 1 to 2 in. long leaves. The plants spread rapidly—the creeping roots may grow 1 in. a day—and often penetrate the soil to a depth of more than 20 ft. The smooth stems form vines which grow up to 6 ft. in length, spreading and twining over the ground or climbing up WHEAT, OAT, BARLEY, SOR-

Field bindweed. (B.28.)

GHUM, or other plants in the field, thus becoming noxious weeds, especially in the Midwest and the western states. The encapsulated seeds of bindweeds may remain in the ground for several years before they germinate.

Control: Intensive fallow usually eliminates the bindweed in two seasons or less, provided cultivations to a depth of 4 or 5 in. are carried out at the proper time and with the proper implements which will cut completely all bindweed shoots and thus prevent additional food storage in the root system.

Cultivation operations should be every two weeks until the bindweed has been weakened and emerges more slowly. The interval then may be safely lengthened to three weeks.

SODIUM CHLORATE is a most satisfactory soil sterilant. Bindweed stands may be greatly reduced (but rarely eradicated) with 2,4-D.

References: H.42; P.23.

FIELD DODDER (*Cuscata pentagona*, also known as *C. arvensis*) is one of the most destructive of all the DODDERS. It shows little preference as to the plants which it attacks. The stems of this species are pale yellow in color, a characteristic which is useful when determining the parasitic plant in the field. It grows in most parts of the United States, but causes the greatest damage east of the Mississippi River. Its seed is a common impurity in RED CLOVER and ALFALFA seed.

Reference: H.19.

FIELD MICE. If a growth of ALFALFA is left standing in the fall, a large population of mice of different species may be attracted to the field. They attack also PROSO, SUBCLOVER, shocked grain (corn, grain sorghums), and many other crops.

Control: Pieces of hollow tile or tin cans laid on their sides may be placed along fence rows and baited with STRYCHNINE-POISONED GRAIN to protect trees or other crops from mice that may be living in fields.

References: G.7; R.3.

See also RODENTS.

FIELD PEA (*Pisum arvense*) is a LEGUME. The plants are annual and have slender succulent stems 2 to 4 ft. long, which ascend with support.

Field pea. (M.3.)

A cool growing season is necessary for the field pea, as high temperatures are much more injurious than frosts and are most disastrous when they occur when the pods are setting. These climatic requirements limit the successful production of the field pea as a summer crop to the northern states and Canada. The legume may, however, be grown as a winter crop in the southern states. Its moisture requirements are less exacting than its temperature requirements, but it does best where rainfall is fairly abundant. In the northern Great Plains a rainfall of 15 in. is sufficient to produce a good crop, while 20 in. rain in Kansas, Nebraska, or Colorado is inadequate.

Field peas are grown for forage and seed in New York, Michigan, Wisconsin, Minnesota, eastern North Dakota, South Dakota, Montana, Idaho, Oregon, and Washington, and for a cover crop and green manure in the southeastern part of the Cotton Belt and in the Pacific Northwest.

Well-drained clay loams of limestone origin, or soils that are neutral or of low acidity, are best suited to the field pea, although it does well on fertilized sandy loam soils. Heavy, black soils, rich in humus, tend to produce a heavy growth of vines with comparatively few pods and result in a large tonnage of hay.

In the northern United States it is advantageous to fall-plow the land for field peas because of the necessity for early seeding. Fall-plowed land should be disked as early as possible in spring and smoothed down with a drag harrow in case the seed is to be planted with a drill.

In the southern states the largest acreage of winter legumes follows COTTON, in which case very little or no preparation of the soil is necessary. Other crops, such as tobacco, PEANUTS, COWPEAS, SOYBEANS, and melons, leave the soil in good condition, and only little preparation of a seedbed is necessary following these crops.

In the North, field peas must be planted early enough in the spring to set pods before the warm weather of summer arrives. The young plants are not harmed by light frosts. In the northern part of the Cotton Belt, seeding should be done, if possible, during the last half of September, and in the southern part, early in October.

Field peas are best sown with a grain drill. Where that is not available, they may be sown broadcast and covered with a disk, spike-tooth harrow, or cultivator. The peas should be planted from 2 in. deep (in clay loam) to 4 in. (in sandy soil). When field peas are seeded with GRAIN,

the common practice is to mix and sow them in one operation.

Seed more than two years old is liable to have low germination. In purchasing seed the grower should endeavor to obtain seed of the current season's crop that is accompanied by analysis tags, showing the purity and germination of the seed.

The use of fertilizers with field peas is not recommended for most regions of the northern United States. In the Cotton Belt of the Southeast it is often essential. If the soil is poor the use of 200 to 400 lb. superphosphate or 300 to 600 lb. basic slag and 50 lb. sodium nitrate or ammonium sulfate is advisable. In the Pacific Northwest gypsum applied at the rate of 50 to 100 lb. an acre, increases yields on most soils.

For best results, field peas must be inoculated. This is especially true in the more recently developed farming districts. INOCULATION can be accomplished by the use of commercially available pure cultures or by the use of soil from a field that has recently grown field peas.

> NOTE: Commercial fertilizer, unless it is basic slag, should not come in contact with inoculated seed, as it may injure the inoculating organism. However, barnyard manure is effective in inducing inoculation, and its use is recommended.

When used as a *green manure* to precede annual crops, field peas should be turned under about two weeks before the planting of the annual crop. In the South, where field peas are used most largely for this purpose, earlier plowing is advised if the field peas have made sufficient growth.

When field peas are planted for *pasturage*, either alone or with SMALL GRAIN, they give best results when allowed to mature before the animals are turned in, so the entire plant may be utilized. This is the common practice in parts of the Rocky Mountain states, where such pastures are usually grazed by sheep. Animals pasturing on field peas should be confined—by movable fences—to one portion of the field at a time. A good crop will usually fatten from 10 to 15 lambs per acre, each animal gaining about 8 lb. a month; hogs, if not obliged to gather their food over too large an acreage, will make an average daily gain of 1 lb. More rapid gains may be made by hogs when a limited quantity of CORN or BARLEY and a supplementary protein are fed in addition to the peas. Access to a mineral mixture should also be provided.

To assure continued substantial gains, the fattening animals must be moved to a new field as soon as peas become scarce, and stock animals must be used for cleaning up the field.

The proper time to cut field peas for *hay* is when most of the pods are well formed. When they have been seeded with grain, the varieties of field peas and grain should be so chosen that the crop can be harvested at the most favorable period of maturity for both.

When grown for hay, field peas are usually mixed with OATS, RYE, or barley. These mixtures stand up much better and make harvesting easier. The presence of grain in the crop also causes it to cure more quickly. Hay yields of field peas alone, or in mixture with grain, range from 1 to 3 tons per acre.

When grown for hay, the field pea works into a rotation very nicely, because it is removed from the field early in the year, thus allowing ample time for a thorough preparation of the soil during the fall. The feeding value of field pea hay is about the same as that of ALFALFA.

Field peas make good *silage* when grown in mixture with a small grain and cut when the grain is nearly mature. Such silage has a high feeding value and has given excellent results when used for fattening cattle and sheep.

The field pea should be cut for *seed* when the pods are mature and the seed is firm. It is not well, however, to wait until the vine and pods both are dry, since then the loss of shattering and WEEVIL damage is sure to be large.

The threshing of field peas is usually done with a grain separator or combine. Where the field peas are not sold for seed purposes, but are intended wholly for livestock feeding, no precautions are neces-

sary since cracked seed then is not objectionable.

Dangers: The *fungus* diseases ASCOCHYTA BLIGHT COMPLEX—namely ASCOCHYTA FOOT ROT and ASCOCHYTA BLIGHT OF PEA—and the closely related LEAF AND POD SPOT, LEAF BLOTCH, POWDERY MILDEW, PEA ANTHRACNOSE, FUSARIUM WILT OF FIELD PEA, various ROOT ROTS (especially FUSARIUM ROOT ROT), and the same fungus which causes DOWNY MILDEW OF VETCH affect field peas as do BACTERIAL BLIGHT OF PEA and the *virus* disease MOSAIC. Among the *insect* enemies are PEA WEEVIL, PEA APHID, and the PEA MOTH. A nematode causes ROOT KNOT.

Varieties: The most important field peas grown for feed follow:

Agnes and *Canadian Beauty* are recommended for pasture and hay in Colorado. Canadian Beauty is also suited for the Great Lake states.

Alaska and *Blue Bell*, or *Blue Prussian*, are grown in the Pacific Northwest.

Austrian Winter is the most winter-hardy variety; it is grown in the Pacific Northwest and (almost exclusively) in the southern states where fall planting is practiced.

Chancellor, Chang, French Gray (which matures earlier than the Austrian Winter), *Marrowfat, Multipliers,* and *Scotch* are recommended in the Great Lake states. Reference: M.24.

See also STRAW; CANADIAN FIELD PEAS.

FILAMENT is the STALK of a STAMEN on which is borne the pollen-bearing ANTHER.

FILAREE = ALFILERIA.

FINE BENTGRASSES include COLONIAL BENT.—*See also* BENTGRASSES.

FINETOP SALTGRASS
                        = ALKALI SACATON.

FINGERGRASSES (*Chloris* spp.) form part of the forage for grazing animals on ranges in the Southwest and in the Hawaiian Islands. RHODESGRASS is an important representative of this genus.—*See also* RANGE PLANTS.

FINISHING RATIONS.
        See BEEF CATTLE RATIONS; RATIONS.

FINNEY MILO is one of the grain SORGHUMS.

FIORIN. *See* REDTOP.

FIRE ANTS are useful because they feed on VELVETBEAN CATERPILLARS and on the eggs of this insect pest.

FISHER BROME is a southern-type SMOOTH BROME.

FISHERY BY-PRODUCTS include (1) the various MARINE PRODUCTS obtained in canneries and other fish-packing factories and (2) *whale products* which—because the whale is a mammal—are officially listed as ANIMAL PRODUCTS.—*See also* BY-PRODUCT FEEDSTUFFS; SHARK MEAL.

FISH FACTOR. *See* U.G.F.

FISH-LIVER AND GLANDULAR MEAL. *See* MARINE PRODUCTS.

FISH-LIVER OILS are good sources of VITAMINS A and D. VITAMIN A is obtained from the cod, halibut, salmon, and shark; VITAMIN D_3, from the cod, sardine, and other fish. Oils from different fish vary in vitamin content. The amount of fish-liver oil to be added to rations (especially in winter months), depends on the potency of the type of fish oil available.

The vitamin A in fish oils is not stable— it is easily destroyed on oxidation. Buy only fresh oils and do not store unused rations for long periods of time.

Dangers: Rancid fish-liver oil not only contains little or no vitamin A but also destroys VITAMIN E. If rancid oil is fed to young ruminants, *white muscle disease* of calves or *stiff lamb disease* may result.

FISH MEAL. *See* MARINE PRODUCTS.

FISH RESIDUE MEAL.
                    See MARINE PRODUCTS.

FISSION means: splitting into parts.
                            See BACTERIA.

FLAG SMUT OF WHEAT—one of the most virulent STRIPE SMUTS—is caused by the FUNGUS *Urocystis tritici.* In the United States the range of the disease is gradually increasing. Although it has not become severe in any locality in this country, its persistence in certain major WHEAT-producing areas is a potential danger.

Flag smut appears as long black stripes running lengthwise on the leaf blades and on the upper parts of the stems of the plants. Infected plants have dwarfed stalks that seldom produce heads. Usually the entire plant is affected, although partly infected plants are not uncommon in certain varieties. When infected leaves dry, they split along the black streaks and free the black powdery spores of the smut fungus. The spores are blown to nearby plants or fall to the ground, where they may be spread by the feet of animals and men and by machinery. When infected wheat is being harvested or threshed, flag smut spores are carried to the sound wheat kernels from infected plants. This smut lives over in the soil, or may be carried on wheat seed. Infection of the young wheat seedling takes place when the seed germinates. After entering the seedling, the fungus MYCELIUM grows inside the stem and leaves of the plants. Infected plants can be detected early by the black streaks on the leaves before the jointing of the plant begins.

Control: Flag smut can be controlled by crop rotation, seed treatment with CERESAN M, and growing of resistant varieties.

The flag smut spores in the soil usually do not survive when wheat is not grown on the land for one year. Flag smut persists only in winter-wheat areas where a mild winter climate prevails and continuous wheat culture is practiced.

Reference: L.10.

See also COMMON WHEAT.

FLAKED CORN. *See* CORN.

FLAMBEAU. *See* SOYBEAN (variety).

FLAVOR. *See* MEAT QUALITY.

FLAVORING AGENTS are sometimes added to feedstuffs because it is thought that they increase the palatability of the feed; ANISE SEED is widely used for this purpose. Many experts doubt, however, that flavoring agents have any value in making animals perform better.—*See also* FEED INGREDIENTS.

"FLAX" (*Linum usitatissimum*) is the name under which the cultivated *Old World Flax* is generally known. It is the most important of all the FLAXES; it is grown for fiber (linen yarn) or for seed.

Flax is a winter annual in warm climates. It reaches 1 to 3 ft. height, has a short taproot, a main stem, and branches with leaves and flowers; the latter develop into bolls which include the seeds.

Flaxseed, more often called LINSEED, contains about 40 percent linseed oil. After the oil has been extracted from flaxseed, the residue is pressed into cakes or ground into linseed oil meal for livestock.

Reference: U.6.

See also ALFALFAS; DODDERS; CHINCH BUGS; WHEAT STEM SAWFLY; FLAX STRAW; FLAX PLANT BY-PRODUCTS.

FLAXES (*Linum* spp.) are common sights in the West and occur on well-drained to dry soils. They are, as a class, low in forage value and their palatability rates from worthless to fair for sheep, somewhat lower for cattle; and horses seldom touch them.

Species: There are about 100 flax species, some of which are known to be poisonous to livestock, e.g., the *New Mexican flax* and the *stiffstem flax*. Most important among all flaxes is the *Old World flax*, commonly referred to as (cultivated) "FLAX."

Reference: U.6.

See also STRAW; FLAX STRAW.

FLAX PLANT BY-PRODUCT,—i.e., the leftover portion of the FLAX plant after removal of seeds, bast fiber, and part of the shives (defibered stems)—has only little feed value.

FLAX PLANT FEED. *See* LINSEED.

FLAXSEED = LINSEED.

"FLAXSEED" is the PUPARIUM of the HESSIAN FLY; this term is used because the puparia of this insect pest resemble flax-seeds in size and color.

FLAXSEED SCREENINGS OIL FEED. *See* LINSEED.

FLAX STRAW of good quality can be used in place of oat straw as roughage for wintering cattle.

Caution: Do not feed flax straw if it contains immature seeds because they yield PRUSSIC ACID.

FLEABANE. *See* WHITETOP FLEABANE.

FLEA BEETLE = GARDEN FLEA HOPPER.

—*See also* CORN FLEA BEETLE.

FLEXIBLE PROTEIN SUPPLEMENT. *See* SWINE RATIONS.

FLEXIBLE RATIONS. Animals require NUTRIENTS, not *feedstuffs*, and should receive RATIONS designed to meet their needs for all known nutrients and recognized "factors." In addition, the rations should be palatable, properly balanced physically and chemically, suitable for the animal that eats it, free of dust, free of objectionable odor, nontoxic, easy to feed and handle, and economical.

Naturally most sample formulas will be based on feedstuffs that are common to the area where the rations will be formulated.

After a feedstuff has appeared in recommended rations for several years, many people begin to regard it as essential. Thus, many people believe that dairy cows must have some bran; horses some oats; fat steers some oil meal, and poultry some milk products. Each of these feeds is good for the animal indicated, but with our present knowledge of nutrition we can replace them by other feedstuffs, or combinations that will give equally satisfactory results.

Purified VITAMINS, MINERALS, and AMINO ACIDS as well as large quantities of FATS and SUGARS also can be effectively used to make the ration formulator less dependent on natural feeds.

Thus, the discoveries in the area of nutritional requirements and those in the area of production can be utilized to give great flexibility to rations. The feed manufacturer and the livestockman do not have to be bound by obsolete formulas which require certain feedstuffs that may be scarce or expensive during certain times of the year.—*See also* RATION FORMULATION.

FLIES are two-winged INSECTS; they are dangerous, because they carry many animal diseases, but only a few flies are dangerous to plants; e.g., the HESSIAN FLY. Some flies are beneficial to plants; e.g., the parasitic WINTHEMIA, CELATORIA, and PHOROCERA species.—*See also* RICE LEAF MINER; SORGHUM MIDGE; GARDEN WEB-WORM; ROBBER FLIES.

FLINT CORN. *See* CORN.

FLOAT is (1) a plank clod masher or (2) a land leveler.

FLORAL means: of, or pertaining to, a flower or other plant.

FLORET is a diminutive flower, especially one of the readily detachable flowers of a grass SPIKELET.

FLORIDA BEGGARWEED (*Desmodium purpureum*)—one of the so-called *tick clovers*—is a plant attaining a height of 4 to 7 ft. The main stem is branched rather sparsely and is noticeably pubescent (hairy). In thick stands the branches are reduced and the lower ones absent. The leaves are trifoliolate with large ovate leaflets. Racemes of rather inconspicuous flowers terminate the main stem and lateral branches. The small seed is borne in a joined pod (loment).

Florida beggarweed is a short-lived perennial in frostless regions; but it behaves as an annual in most parts of the United States. It is particularly adapted to the Coastal Plains area of the southern states. In Florida, southern Georgia, and Alabama it has been used as a regular and volunteer crop; under favorable conditions, it will make a fair vegetative growth as far north as the Great Lake states.

In composition, Florida beggarweed is comparable to most of the common LEGUME hay crops. It is of considerable value for grazing in late summer and when fed in this way, is very fattening. For green manure it serves best on sandy loam soils in rotations where a volunteer crop can be handled. The seed of beggarweed is readily eaten by quail.

On poor sandy soil or soil low in fertility an application of manure or commercial fertilizer greatly increases the growth of Florida beggarweed. On lands not previously fertilized 200 lb. superphosphate, 75 lb. potassium chloride, and 50 lb. sodium nitrate or ammonium sulfate can be used to advantage.

A moderate rainfall is essential to good growth even though the plants will stand as much drought as the average farm crop. Florida beggarweed is of particular value because of its ability to grow on moderately acid soils. Its lime requirement is low; few other legumes are as well adapted to the soils of the southeastern United States as this crop.

INOCULATION is unnecessary. The organism that inoculates Florida beggarweed seems to be present in most, if not all, southern soils.

Best time for seeding is late spring. Scarified seed should be used for quick stands.

Harvesting for forage must be undertaken when the plants are in early bloom or when the first seed pods form; i.e., about July in Florida. Hay is hard to cure and spoils easily. Forage yields amount to 1 to 2 tons per acre.

Florida beggarweed can be used in rotation with CORN, giving a full season to the beggarweed, or it can be volunteered in the corn after last cultivation.

The plants are immune to root knot and no serious damage from diseases or insects has been reported.

Reference: M.13.

FLORIDA BUSH = *Bush velvetbean.*
>   See DEERING VELVETBEAN.

FLORIDA NAPIERGRASS is a strain of NAPIERGRASS.

FLORIDA VELVETBEAN.
>   See DEERING VELVETBEAN (variety).

FLOUR CORN. *See* CORN.

FLOWER GLAND. *See* NECTARY.

FLOWERING GLUME = LEMMA.
>   —*See also* GRASSES.

FLOWERS OF SULFUR, *sulfur flowers,* or *sublimed sulfur,* is used as a FUNGICIDE for SEED TREATMENT.—*See also* SULFUR.

FLUORINE—the very reactive chemical element *F*—is a gas but it is so reactive that it does not occur *free* in nature; its solid compounds are contained in minerals which, when fed to animals, may cause FLUORINE POISONING.--*See also* MINERAL FEEDS.

FLUORINE POISONING is caused by FLUORINE, a cumulative poison. It is dangerous to feed to livestock rations containing ROCK PHOSPHATE because this mineral usually contains 2 to 4 percent fluorine. Other minerals as well as water, soil, and plants in some areas of this country—e.g., in Arkansas, California, South Carolina, and Texas—contain high enough fluorine content to cause trouble with livestock by mottling teeth and interfering with normal calcification.

Mineral mixtures fed directly to livestock must not exceed 0.30 percent F for cattle, 0.35 percent for sheep, and 0.45 percent for swine. The fluorine level in dry feed for all species of animals must not exceed 0.003 percent.

Industrial contamination is becoming

more common; fumes containing fluorine or fluorides often settle on plants, etc.

FLUSHING EWES. See SHEEP RATIONS.

FLY. See FLIES.

FLYNN and *Flynn 37* are SIX-ROWED BARLEY varieties.

FODDER is any dry feed for livestock; the term is applied more specifically to such feedstuffs as CORN, SORGHUMS, and coarse GRASSES harvested whole and cured in an erect position.

Pulled fodder consists of leaves of corn or sorghum stripped by hand from the stalk and then cured.

Topped fodder refers to the tops of corn stalks cut off above the ears and then cured.—*See also* SHREDDING.

"FODDER" = BUNDLE FEED.

FODDER CORN = *corn fodder.* See CORN.

FOENICULUM. The dried ripe fruit of *F. vulgare* is known as FENNEL.

FOLGER'S SORGO, or *Folger's Early,* is one of the forage SORGHUMS.

FOLIAGE DISEASES = LEAF DISEASES.

FOLIATE means: leaved.

FOLIC ACID is a term which has been changed to *pteroylglutamic acid.* It is one of the B-COMPLEX VITAMINS and occurs in leafy plant material and grasses, in yeast, liver, and kidney. It is important to animal life.

A shortage in chick diets leads to *macrocytic anemia,* poor growth, and poor feathering. A related compound, known as CITROVORUM FACTOR, has similar activity.

Pure folic acid forms yellowish-orange crystals that are only slightly soluble in water.—*See also* VITAMINS; ANTIBIOTICS.

FOOD AND DRUG ADMINISTRA- TION = F.D.A.

FOOT ROT of cattle and sheep may occur at any time of the year, but is more prevalent when livestock is on PASTURE during wet seasons. The symptoms are usually soreness of the feet, lameness, swelling, redness, heat, and a definite indication of pain. After the disease has progressed for a few days, necrosis (sloughing of the tissues) and a distinctive foul odor will be noticed.

References: S.28; S.10.

See also GRAZING.

FOOT ROTS, *foot-and-root rots,* and ROOT ROTS are terms applied to a variety of FUNGUS DISEASES which are usually found to be most severe on dwarfed, undernourished plants or on plants receiving an excess of NITROGEN fertilizer. Among the more important foot rots are ASCOCHYTA FOOT ROT, CULM ROT, and LEAF AND POD SPOT.

Reference: M.37.

See also FUNGUS; ROTS.

FORAGE, in a broad sense, is the vegetable matter that is fed to animals, but the term is usually applied to the leaves, flowers, stems, and twigs. Forage and ROUGHAGE are high in fiber and contain relatively small amounts of net ENERGY per unit volume.

Green forage is fresh forage.

Fermented forage = SILAGE.

FORAGE BLOWER. See SILOS.

FORAGE CROPS and PASTURES supply the most important feeding stuffs for all species of farm animals, except swine and poultry. HAY is among the most valuable FORAGE crops.—*See also* FORAGE; SILAGE CROPS; STRAW MEAL; RANGE PLANTS; PASTURE PLANTS.

FORAGE ELEVATORS. See SILOS.

FORAGE POISONING is a loosely used term to indicate poisoning of animals due to their eating spoiled (moldy) cornstalks, pastures, silage, and other FORAGE. The affected animals may show a variety of symptoms—from staggering gait to delirium and death.—*See also* POISONOUS PLANTS; PRUSSIC ACID POISONING.

FORAGE RANGE.

See RANGE PLANTS; RANGE MANAGEMENT.

FORAGE SORGHUMS. See SORGHUMS.

FORAGE SPECIES.

See RANGE PLANTS; PASTURE PLANTS.

FORBS are nongrass-like herbs; in the range stockman's sense, they are WEEDS which are more harmful than beneficial. Forbs are LUPINES, dandeloins, BASSIA, and many others.—*See also* RANGE PLANTS; SAGEBRUSHES.

FORKEDEER. See RED OAT (variety).

FORMALDEHYDE GAS is a pungent, very irritating, colorless gas which is occasionally used as a FUMIGANT. The gas is soluble in water; this solution is known as FORMALDEHYDE SOLUTION.

FORMALDEHYDE SOLUTION, or *formalin*, is a colorless liquid, the vapors of which are very irritating. The solution is used for SEED TREATMENT because it is very effective in preventing seed-borne diseases, but it has no effect on those caused by soil-borne FUNGI. It frequently injures the seed. It is therefore not favored, unless used on SMUT-infested seed. Formaldehyde solution may be used as follows for SORGHUM seed:

Mix 1 part of commercial formaldehyde solution (containing 37 percent FORMALDEHYDE GAS by weight) with 240 parts of water in a tub. The seed, first thoroughly cleaned, is placed in a loosely woven burlap bag, half filled and tied at the top. Immerse the sack of seed in the formaldehyde solution for half an hour, lift it out, and let it drain a few minutes; then spread the seed in a thin layer on a clean floor or canvas in a well-aired place to dry. Stir it occasionally to hasten drying and plant on the same day—as soon as it is dry enough to feed through the planter. Treated seed, however, should not be planted in dry soil.

Caution: Formaldehyde solution is poisonous. Keep it out of eyes and do not breathe the fumes.

Reference: L.1.

See also PERICONIA ROOT ROT; GRASS SMUTS.

FORMALIN = FORMALDEHYDE SOLUTION.

FORTUNA RICE. *See* RICE (variety).

FORTYFOLD = *Goldcoin*.
See COMMON WHEAT.

FORWARD WHEAT.
See COMMON WHEAT (variety).

FOUR-ROWED BARLEY.
See SIX-ROWED BARLEY.

FOUR-WINGED POISONVETCH.
See POISONVETCHES.

FOURWING SALTBUSH (*Atriplex canescens*), or *chamiza*, is a freely branching shrub attaining a height up to 10 ft., though it is usually lower. It prefers deep, sandy soil.

Fourwing saltbush is one of the most valuable forage shrubs in the Southwest and Intermountain regions. Its importance is due to its abundance, accessibility, size, evergreen habit, high palatability, and nutritive value. Moreover, it exhibits hardiness to cold and a remarkable ability to withstand drought because of its tremendous root development (sometimes penetrating to a depth of 20 ft.). The leaves, stems, flowers, and fruits are cropped by all classes of range livestock, except horses, which graze the species only in winter when other forage is sparse.

Reference: U.6.

See also RANGE PLANTS.

FOWL BLUEGRASS (*Poa palustris*), also called *fowl meadow grass*, occurs in meadows and moist open grounds in Alaska, Canada, and to the south in Virginia, Missouri, New Mexico, and California.

This GRASS yields up to 2½ tons of hay per acre; the hay is liked by all farm animals.

FOXTAIL is a term applied to different genera of plants; some foxtail GRASSES of the genus *Alopecurus* are WEEDS. Of importance as forage plants are MEADOW FOXTAIL, REED FOXTAIL, and MILLET FOXTAIL. Foxtail species which are succulent late in summer are often attacked by CHINCH BUGS.—*See also* ALFALFA; COWPEA; FOXTAIL BARLEY.

FOXTAIL BARLEY (*Hordeum jubatum*), also called *squirreltail grass*, is a perennial WEED from 8 to 10 in. high. It grows in tufts or bunches and is widely distributed, occurring from Alaska to New Jersey, Texas, and California. This weed is very common, especially along roadsides and other waste places as well as in grain and hay fields.

While young, foxtail barley is palatable to livestock; after the bearded heads form, the plants become troublesome—their SPIKELETS, with the stiff awns, easily become inbedded in the mouth tissues, nostrils, and eyes of livestock and game animals.

Control: Once established, foxtail barley is difficult to eradicate. It is a prolific seeder. Seeding plowable meadows or pastures after thorough cultivation to a grass which will quickly form a dense stand, is effective in reducing the amount of foxtail barley. On the range, where cultivation is seldom practical, conservative grazing which will facilitate the establishment of the native, palatable, perennial grasses, is the most feasible method of reducing this weed.

Reference: U.6.

See also BARLEYS; WILD BARLEYS.

Foxtail barley. Plant and flower heads (spikes). (U.6.)

FOXTAIL BROME (*Bromus rubens*), or *foxtail chess*, is one of the BROME GRASSES; it occurs on dry hills and in waste or cultivated ground in the western part of the United States and in Massachusetts. Foxtail brome is considered a poor forage grass.—*See also* RANGE PLANTS.

FOXTAIL MILLET (*Setaria italica*)—not to be confused with MEADOW FOXTAIL—has erect or ascending stems. It grows $2\frac{1}{2}$ to 5 ft. high and bears broad, flat leaves. The seeds are borne in a rather dense, cylindrical spike.

Foxtail millet is grown throughout the Great Plains, as far south as northern Texas, east through Missouri, southern Iowa, and northern Arkansas, and across Tennessee, Kentucky, southern Illinois, and Indiana. It can be grown in almost any area that has warm weather during the growing season and enough rain for any other crop. In fact, it has a lower rain requirement than most other crops but is seriously damaged by severe drought.

Foxtail millet is used as hay, pasture, and green fodder; the seed is used for bird feed. It is useful as a catch crop to supply supplemental feed when pastures fail or the hay crop is short.

Foxtail millet. (H.1.)

Nitrogen and phosphate particularly have given increased yields, but as a general rule, fertilizer should be applied to other crops in the rotation rather than to the foxtail millet.

The seed deteriorates rapidly; seed more than two years old usually has low germination. Therefore, only fresh seed should be used.

A good seedbed should be prepared by plowing, harrowing, and cultipacking or otherwise firming the seedbed. The seed is sown from shortly after corn-planting time to the middle of summer. A mixture of foxtail millet and SOYBEANS or COWPEAS may be broadcast by hand or drilled separately.

Foxtail-millet *hay* is fed to horses and cattle. The feeding value is greatest from the time of the bloom until the seed reaches the milk stage. Hay yields are from 1 to 3 tons. The hay is less palatable and does harm to horses if it is fed as the sole roughage, but makes fairly good hay for cattle and sheep and good roughage for growing stock.

Foxtail millet is often used as a cash crop in a regular rotation and as a catch crop following SMALL GRAIN, or other late-spring- or early-summer-maturing crops.

Varieties: A number of varieties are recognized; the ones most generally grown are *common* and *German foxtail millet;* other varieties are the *Hungarian, Siberian,* and *Kursk foxtail millets.*

Reference: H.1.

See also BACTERIAL SPOT; GRASSES.

FRANKLINIELLA. *F. Fusca*
= TOBACCO THRIPS.

FRAPS FEEDING STANDARDS are expressed in terms of dry matter, digestible crude protein, and productive value (in THERMS of net energy). Because Fraps considers not only the true proteins but all nitrogenous substances in feeds, his standards are preferred by some to the ARMSBY FEEDING STANDARDS, particularly by poultry men who use them to calculate the energy of their rations.

> NOTE: Many of Fraps' values were determined on feedstuffs that are not representative of products with the same name used today; e.g., most

soybean meal contains much less fat (and therefore energy) and more protein than Fraps' values indicate. His values were also determined on slow-growing birds that may have had to use more feed for maintenance and less for production than present-day broilers.

FREED and *Dwarf Freed* are grain SORGHUMS.

FREMONT SORGO is one of the forage SORGHUMS.

FRENCH GRAY PEA.
See FIELD PEA (variety).

FRENCHWEED = PENNYCRESS.

FRIABLE means: easily crumbled or broken to powder.

FRINGED SAGEBRUSH (*Artemisia frigida*), also known as *fringed wormwood, estafiata,* and *pasture sagebrush,* is a half-shrub with a low, perennial, woody base. Its leaves are finely dissected and the flowers have a strong, camphor-like odor. This sagebrush species extends from Mexico, through the western United States and Canada, into Alaska; it prefers dry, coarse, or shallow loam soils. The plant basks in the glowing sunshine of southwestern summers and also withstands the frigid vigors of the Arctic.

Fringed sagebrush varies considerably in forage value in different places; it is highest in the Southwest, but is considered practically worthless, except during late fall and winter, on the cattle ranges of the northern plains and prairies.

Reference: U.6.

See also RANGE PLANTS.

FRINGED WORMWOOD
= FRINGED SAGEBRUSH.

FROG-EYE is caused by the FUNGUS *Cercospora sojina,* also known as *C. daizu.* This leaf spot disease appears on SOYBEANS rather late in the season. Although it affects other parts of the plant, the conspicuous phase of the disease is the "eye-spot" composed of a gray-to-tan central area, usually bordered by a narrow, darker margin. The leaves, when badly infected, fall prematurely. The disease also affects the stems and pods and is carried on the seed.

Diseased leaves and stems serve to carry

the fungus over the winter and lead to new infections in the spring.

Control: When the disease appears in a field, soybeans should not be planted on the land the following year.

Reference: C.9.

FROZEN SILAGE. *See* SILAGE.

FRUCTOSE = LEVULOSE.

FRUIT is the ripened PISTIL of a seed plant; a pea pod, a grain of wheat, a huckleberry, and a rose haw are all, botonically speaking, FRUITS.

FRUITING BODY is a complex FUNGUS structure that contains or bears spores and disseminates them.—*See also* NONPARASITIC SORGHUM-LEAF DISCOLORATIONS.

FRUITS, such as apples, peaches, plums, pears, and other surplus fruits, may be used to replace a part of the regular rations fed to livestock. The fruits contain SUGARS as main nutrients, but they are very low in protein.—*See also* FRUIT.

FRUIT SUGAR = LEVULOSE.—*See also* FRUITS.

FT. COLLINS SUDANGRASS.
                    See SUDANGRASS (variety).

FUCACEAE are a seaweed family.
                    —*See also* KELP.

FULCASTER, also known under many other names, is a soft winter wheat variety of COMMON WHEAT.

FULGHUM. *See* RED OAT (variety).

FULGRAIN. *See* RED OAT (variety).

FULHIO. *See* COMMON WHEAT (variety).

FULTEX. *See* RED OAT (variety).

FULTZ. *See* COMMON WHEAT (variety).

FULWIN. *See* RED OAT (variety).

FUMIGANTS are liquid or solid substances that form vapors which destroy insects, micro-organisms, etc. They are often used in soils or in closed structures. Widely used fumigants are CARBON DISULFIDE, METHYL BROMIDE, FORMALDEHYDE GAS, CARBON TETRACHLORIDE, and DOW-FUME W-40.—*See also* SOIL FUMIGATION; RED HARVESTER ANT; WESTERN HARVESTER ANT.

FUMIGATION is a method of disinfection or disinfestation using FUMIGANTS as DISINFECTANTS or DISINFESTANTS.—*See also* SOIL FUMIGATION.

FUNGI. *See* FUNGUS.

FUNGICIDAL means: of, or pertaining to, FUNGICIDES; e.g., fungicidal dusts.

FUNGICIDES are chemical substances which kill FUNGI. They may be applied to plants or seeds as spray or dust; e.g., for SEED TREATMENT.

ARASAN, BORDEAUX MIXTURE, CERESAN, CERESAN M, COPPER CARBONATE, COPPER SUBSULFATE, DICHLORONAPHTHOQUINONE, DITANE, FORMALDEHYDE SOLUTION, PANOGEN, PARIS GREEN, PARZATE, PHYGON, SPERGON, SULFUR, and TERSAN, are widely used fungicides.—*See also* FUNGUS.

FUNGI IMPERFECTI, or *imperfect fungi*, are a major FUNGUS group for which no sexual production of SPORES is known.

FUNGUS (plural: *fungi*), a very primitive plant, is a micro-organism. The fungi, lacking CHLOROPHYLL, live on living or decaying plant or animal matter.

The body of a fungus consists of delicate threads, or *hyphae*, many of which form branched systems called *mycelia*. Many fungi multiply by forming SPORES at the ends of, within, or on specialized hyphae.

Classification of fungi is based on the manner in which the spores are produced and the appearance of the FRUITING BODIES from which the spores are disseminated. The sexual activity of some fungi consists in the fusion of male and female cells; others produce sexual spores; and one group, consisting of the FUNGI IMPERFECTI, is characterized by the nonsexual type of spore formation. Some fungi produce both sexual and nonsexual spores. A fungus in its *perfect*, or *sexual* stage, sometimes has a name different from the one in its *imperfect*, or *unsexual* stage; e.g., GLOMERELLA (perfect stage) and COLLETOTRICHUM (imperfect stage), or MYCOSPHAERELLA (perfect) and CERCOSPORA (imperfect).

NOTE: The Botanical Congress of 1905 set up certain rules in regard to botanical nomenclature; one of them states that a plant could have only one proper Latin binominal, and that in the case of fungi, possessing two or more morphological forms, the name applying to the *perfect* stage should have preference. However, this rule is not always observed because of long association of a disease with the name of the imperfect stage; e.g., *Cercospora* leaf spot instead of *Mycosphaerella* leaf spot of peanut.

Higher fungi are the *Saccharomyces* spp., commonly known as YEASTS. MOLDS are fungi, found on damp or decaying matter.

Among the more important fungus species, causing FUNGUS DISEASES, are the following: *Alternaria* spp.; *Aristastoma oeconomicum; Ascochyta* spp.; *Aspergillus* spp.; *Beauveria globuliferia; Botrytis cinera; Candida* spp.; *Cephalosporium gregatum; Ceratophorum setosum; Cercospora* spp.; *Cercosporina kikuchii; Claviseps* spp.; *Colletotrichum* spp.; *Curvularia lunata; Cymadothea trifolii; Diaporthe phaseolorum; Diplodia theobromae; Empusa sphaerosperma; Entyloma oryzae; Erysiphe polygoni; Fusarium* spp.; *Gleocercospora sorghi; Glomerella cingulata; Helicoceras oryzae; Helmintosporium* spp.; *Kabatiella* spp.; *Leptosphaeria* spp.; *Macrophomina phaseoli;* MOLDS; *Mycosphaerella* spp.; *Neovosia horrida; Ophiobolus* spp.; *Pellicularia filamentosa; Penicillium* spp.; *Periconia circinata; Peronospora* spp.; *Phialea temulenta; Phoma trifolii; Phyllosticta glumarum; Phytophthora drechsleri; Piricularia oryzae; Pleospora herbarum; Pseudopeziza* spp.; *Pseudoplea trifolii; Puccinia* spp.; *Pythium* spp.; *Ramulispora sorghi; Rhizoctonia* spp.; *Rhizopus* spp.; *Rhynchosporium* spp.; RUSTS; *Saccharomyces* spp.; *Sclerotinia* spp.; *Sclerotium* spp.; *Scolecotrichum graminis; Septoria* spp.; SMUTS; *Sphacelotheca* spp.; *Stemphylium sarinaeforme; Tilletia* spp.; *Torulopsis* spp.; *Urocystis* spp.; *Uromyces* spp.; *Urophlyctis alfalfae;* and *Ustilago* spp.—*See also* CLOVER LEAF WEEVIL; CHINCH BUG; BLACK POD.

FUNGUS DISEASES, caused by a great variety of FUNGI, include these plant diseases: ALTERNARIA LEAF SPOT; ANGULAR LEAF SPOT; ANTHRACNOSES; ASCOCHYTA BLIGHT COMPLEX; ASCOCHYTA BLIGHT OF COTTON, ASCOCHYTA BLIGHT OF PEA, ASCOCHYTA FOOT ROT, and ASCOCHYTA STEM CANKER; BARLEY STRIPE; BLACK KERNEL; BLACK LOOSE SMUT; BLACKPATCH; BLACK SHEATH ROT; BLACK STEM OF CLOVER and BLACK STEM OF OATS; BLAST OF RICE; BORDERED SHEATH SPOT; BOTRYTIS STEM CANKER; BROWN-BORDERED LEAF SPOT; BROWN PATCH; BROWN SPOT OF LUPINE, BROWN SPOT OF RICE, and BROWN SPOT OF SOYBEAN; BROWN STEM ROT; CHARCOAL ROT; CLOVER RUST; COLLETOTRICHUM STALK ROT; COMMON BUNT; COMMON LEAF SPOT; CONCEALED DAMAGE IN SEED; CORN SMUT; COVERED SMUT OF BARLEY and COVERED SMUT OF OATS; COWPEA WILT; CROTALARIA ANTHRACNOSE; CROWN ROTS, CROWN RUST, and CROWN WART; CULM ROT; DAMPING-OFF; DOWNY MILDEWS, e.g., DOWNY MILDEW OF ALFALFA and DOWNY MILDEW OF SOYBEAN; DWARF BUNT; ERGOT; FALSE ANTHRACNOSE OF VETCH; FLAG SMUT OF WHEAT; FOOT ROTS; FROG-EYE; FUNGUS LEAF DISEASES; FUSARIUM ROOT ROT, FUSARIUM STALK ROT, FUSARIUM WILT OF ALFALFA, and FUSARIUM WILT OF FIELD PEA; GRASS SMUTS; GRAY LEAF SPOT; GRAY MOLD; HEAD DISEASES; HELMINTHOSPORIUM LEAF SPOT; KERNEL SMUT OF RICE and other KERNEL SMUTS; LARGE BROWN PATCH; LEAF AND POD SPOT; LEAF BLIGHT; LEAF BLOTCH; LEAF DISEASES; LEAF RUSTS; LEAF SMUT OF RICE; LEAF SPOT OF PEANUT and other LEAF SPOTS; LESPEDEZA ANTHRACNOSE; LOOSE SMUTS OF BARLEY and LOOSE SMUT OF WHEAT; LUPINE ANTHRACNOSE; MILDEWS; NORTHERN ANTHRACNOSE; NORTHERN CORN LEAF BLIGHT; OAT LEAF SPOT; PEA ANTHRACNOSE; PEPPER SPOT; PERICONIA ROOT ROT; PLEOSPORA LEAF SPOT; POD AND STEM BLIGHT; POWDERY MILDEW; PSEUDOPEZIZA LEAF SPOT; PURPLE SEED STAIN; PYTHIUM ROOT ROT; REDDISH-BROWN SHEATH ROT; RED LEAF SPOT; RHIZOCTONIA STALK ROT; ROOT AND STEM ROT, ROOT ROT OF SAFFLOWER, and other ROOT ROTS and ROTS; "ROTTEN NECK"; ROUGH SPOT; RUSTS, e.g., SAFFLOWER RUST; SCALD; SCLEROTINIA STEM ROT; SEEDLING BLIGHT; SEED ROT; SEPTORIA BLIGHT OF WHEAT; SHAKES; SMUTS; SOIL ROT; SOOTY BLOTCH; SOOTY STRIPE; SORGHUM RUST; SOUTHERN ANTHRACNOSE, SOUTHERN BLIGHT, and SOUTHERN CORN LEAF BLIGHT; SPOTS; STALK DISEASES, such as STALK ROTS; STEM CANKER OF CROTALARIA and other STEM CANKERS; STEMPHYLIUM LEAF SPOT; STEM ROT OF RICE; STEM RUST OF OATS and other STEM RUSTS; STEM SMUTS; STRIPE RUST and STRIPE SMUTS; SUGARCANE STALK

ROT; SUMMER BLACKSTEM; TAKE-ALL; TAR-
GET SPOT; VETCH ANTHRACNOSE; VICTORIA
BLIGHT; VIOLET ROOT ROT; WEAK NECK;
WHEAT SCAB and WHEAT STEM RUST;
YELLOW LEAF SPOT; YELLOWS; and ZONATE
LEAF SPOT.—*See also* POTATO LEAFHOPPER;
VELVETBEAN CATERPILLAR.

FUNGUS LEAF DISEASES are distinguished from BACTERIAL LEAF DISEASES by
roughened leaf spots which are due to the
presence of fungal FRUITING BODIES.—*See
also* FUNGUS DISEASES.

FUSARIUM. The FUNGI *F. avenaceum, F.
culmorum, F. graminearum,* and other
Fusarium spp. cause WHEAT SCAB. Various
species,—e.g., *F. oxysporum, F. solani, F.
moniliforme,* and *F. roseum*—are associated
with FUSARIUM ROOT ROT. *F. oxysporum, F.
medicaginis,* and other *Fusarium* spp.
cause FUSARIUM WILT OF ALFALFA; *F.
oxysporum* f. *cheiphilum* causes COWPEA
WILT; *F. orthoceras* var. *pisi* causes
FUSARIUM WILT OF FIELD PEA.

Fusarium spp. are among the causes of
CONCEALED DAMAGE IN SEED and of SEED
ROT; *F. culmorum* and *F. moniliforme*
cause SEEDLING BLIGHT; and the last-
named species also causes FUSARIUM STALK
ROT.—*See also* CHARCOAL ROT; ERGOT;
CLOVER ROOT CURCULIOS; CLOVER ROOT-
BORER; BAHIAGRAS.

FUSARIUM BLIGHT
                = FUSARIUM WILT OF ALFALFA.

FUSARIUM HEAD BLIGHT
                            = WHEAT SCAB.

FUSARIUM ROOT ROT plays a major
role in CLOVER failures. Several *Fusarium*
spp. are associated with this bad disease;
these include *F. oxysporum, F. solani, F.
moniliforme* and *F. roseum*. These FUNGI
enter the host plants through wounds in
the crown and the tap root and are responsible for a severe thinning of plants
during the second year. These fungi also
kill some seedling plants during the first
year.

Infected plants have several symptoms:
The top growth may be stunted or wilted
and the leaflets of large plants often show
yellow or red discoloration of the margins;
and at least a part of the vascular system
of the tap root is dark in color. This
disease is generally associated with in-

juries to the root system caused by the
CLOVER ROOT CURCULIOS or by the CLOVER-
ROOT BORER.

Fusarium root rot also attacks other
plants, e.g., LUPINES and FIELD PEAS.

Control of this disease depends on the
control of insects that feed on the root
system of the plant.

References: E.5; W.8.

FUSARIUM STALK ROT, caused by
the FUNGUS *Fusarium moniliforme,* occurs
in the northern part of Texas. This STALK
ROT produces symptoms almost the same
as those of CHARCOAL ROT. Instead, however, of small black SCLEROTIA which are
visible to the naked eye, this fungus produces within the dried rotted stalks of
SORGHUMS a powdery mass of white spores
that can be seen individually only with
the aid of a microscope. The fungus invades the sorghum plant through openings
made by insects or mechanical injuries. It
grows rapidly after it once gets into the
stalk and soon causes it to break over at
the base.

Control: Some, varieties seem less susceptible than others to this fungus, but no
definite statements on varietal reaction
can be made at present.

Reference: L.1.

FUSARIUM WILT OF ALFALFA, or
fusarium blight, is a FUNGUS DISEASE caused
by *Fusarium* spp., especially *F. oxysporum*
f. *medicaginis*. It may be easily confused
with BACTERIAL WILT of alfalfa. In fusarium
wilt, which occurs from northeastern
Mississippi as far north as Virginia and in
the southern part of California, the root is
more deeply discolored than in bacterial
wilt, but wilting and death are not preceded by dwarfism or discoloration of the
foliage.

Reference: J.2.

FUSARIUM WILT OF FIELD PEA,
caused by the FUNGUS *Fusarium orthoceras*
var. *pisi,* is characterized by a rapid
wilting of the vines without a conspicuous
rotting of the roots.

Control: Rotation of field peas with
other crops is recommended.

Reference: M.24.

FUSED is a mineral or other inorganic material sintered and then cooled to form a compact mass; e.g., fused CALCIUM PHOSPHATE.

G

GADIDAE is a family of fish to which the *Gadus* spp. belong.—*See also* GADUS; VITAMINS.

GADUS. Cod liver oil is obtained from livers of species of the family Gadidae, especially of *Gadus morrhuae.*—*See also* VITAMINS.

GALACTAGOGUE is any substance which promotes the flow of milk; e.g., FENNEL.—*See also* FEED INGREDIENTS.

GALACTOSE is a water-soluble sugar derived from milk sugar (LACTOSE).—*See also* REDUCING SUGARS.

GALGALOS is a soft spring variety of COMMON WHEAT.

GALL = KNOT.—*See also* CROWN WART; ROOT KNOT NEMATODES; WHEAT JOINT-WORM.

GALLETA (*Hilaria jamesii*), often called *curly grass* (or—erroneously—*curly mesquite*), grows in patches forming an open turf. It reproduces and increases largely by means of short rootstocks beneath the ground. Ordinarily the leaves are 2 to 5 in. long and they curl up when dry. The flower stalks are often 8 to 15 in. in height. In some years when precipitation is favorable, two crops of seed stalks and seed are produced.

Galleta, one of the most common GRASSES on the winter range, grows on the valley slopes, the best growth occurring with summer rains.

Galleta is one of the desirable forage species on the *winter range*. It is readily utilized by sheep and on properly grazed range about half of the herbage is eaten. It withstands heavy grazing because of the rootstocks.

Reference: H.41.

See also PASTURE PLANTS; RANGE PLANTS.

GAMBLE OAK (*Quercus gameblli*) is one of the poisonous OAK species.—*See also* POISONOUS PLANTS.

GAMMA-TOCOPHEROL.

See VITAMIN E.

Galleta plant and single spike. (H.26.)

GARBAGE is animal and vegetable refuse; it is usually collected by contractors from kitchens or from cooking and feeding operations. Depending on its origin, garbage varies widely in its composition and feeding value; best quality garbage comes from restaurants. It must be fresh and free from injurious material. Most states now have laws that require boiling of garbage for thirty minutes for the prevention of *V.E.* (vesicular exanthema) *disease;*

it is also required that swine fed garbage be immunized against *hog cholera*.

Garbage is essentially a source of EN-ERGY. It usually contains considerable FAT and FIBER. Best results are obtained if pigs weigh at least 100 lb. when they are started on garbage. A good PROTEIN supplement and MINERALS should be available. Garbage-fed hogs often produce soft pork and are discounted by discriminating markets.—*See also* DEHYDRATED GARBAGE.

GARBANZO = CHICK WEED.

GARDEN FLEA HOPPER (*Halticus citri*), also called *flea beetle*, is a minute black bug which sucks up the sap of WHITE CLOVER and other useful plants.

GARDEN SLUG. *See* SLUGS.

GARDEN WEBWORM (*Loxostege similalis*) is a serious pest of ALFALFA in California, Nebraska, Iowa, Missouri, New Mexico, Kansas, Oklahoma, and Texas. The webworm also occurs in South America and Mexico. The second and third annual cuttings of alfalfa may be entirely destroyed by the pest.

The larva, or caterpillar, is the stage that causes plant injury. Soon after hatching the caterpillars begin feeding on the undersurfaces of alfalfa leaves. As the larvae grow they web together the tops of the plants and eat the portions within the web masses, until nothing is left except the skeletons of leaves and stems. Many fields show damage only in irregular spots, usually on sandy upland, but a sufficient number of larvae may ruin an entire cutting of the crop. The larvae also feed on CORN, SORGHUM, SOYBEAN, WHEAT, PIG-WEED, LAMBSQUARTERS, RUSSIAN-THISTLE, and many truck-crop and garden plants.

When fully grown the larvae are greenish-brown and about ¾ in. long. Upon each body segment on each side of the back are three tiny black spots arranged in triangles, and on each spot is a short, stiff hair. The larvae descend to the ground and enclose themselves in silken cocoons in trash and leaves or in the soil just beneath the surface. In these cocoons they transform into light-brown pupae. After about ten days the pupal cases split open and the adult moths emerge.

The moths are reddish-brown to dark gray, with markings on the wings. After emerging they remain quiet for several hours until their wings become fully expanded. They then have a wing spread of about ¾ in. When disturbed the moths make short flights from 10 to 25 yd. and usually alight on some hidden part of the foliage. The moths may migrate several miles. They are most active at night and are strongly attracted to light.

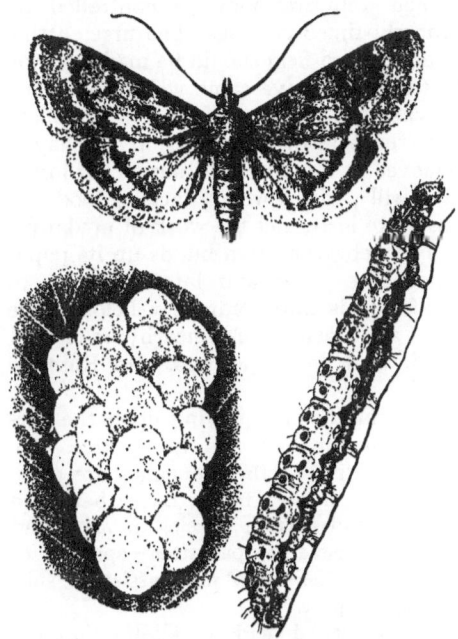

Garden webworm. Eggs, caterpillar, and moth. (P.8.)

Within a few days after the females emerge they mate and begin to lay eggs. They deposit their eggs in masses of 40 to 50 on the undersides of the leaves, usually those near the top of the plant. One moth may deposit as many as 300 or 400 eggs. The moths live only a few days. They feed on the nectar in the alfalfa blossoms.

The eggs are oblong, oval and about ¹⁄₄₀ in. in diameter. At first the egg masses are almost transparent and are difficult to detect against the green background of the leaves, but about a day before hatching they become darker. After four or five days the eggs hatch into tiny larvae.

In the latitudes of Kansas and Oklahoma there are at least four generations annually. The first moths appear in March or April. The insects overwinter as larvae or pupae within cocoons in the soil.

In the latitude of southern Texas the insects are active in the fields during the entire year. They complete their development in thirty to thirty-five days in warm weather, but during the cooler months they require a longer time.

Control: In fields of hay alfalfa the garden webworm may be controlled by properly timed cuttings. The first cutting in an infested field should be made as soon as the plants have come well into bloom, or when the young shoots begin to appear near the base of the plants. If the hay is removed at this time, the larvae left in the field will perish through lack of food, exposure to heat, and the work of predators.

The webworm often builds up its populations on weeds and later migrates to alfalfa. It is important that fields, fence rows, and waste ground nearby be cleared of pigweed, lambsquarters, and Russian-thistle.

Natural enemies preying upon the garden webworm assist in its control. The more effective of these enemies are common toads, horned toads, birds, barnyard poultry, and certain species of ANTS and BEETLES. Several parasites—species of wasps and flies—also may assist in subdueing the pest.

Newly seeded fields of alfalfa that became infested with garden webworms have been protected by dusting or spraying with CALCIUM ARSENATE; more recently, TOXAPHENE has been used to control this insect.

Reference: P.8.

GARDNER SALTBUSH (*Atriplex gardneri*), also called *saltsage* or *Castle Valley clover*, is a low grayish or whitish, narrow-leaved shrub which grows 1 to 2 ft. high. The basal portions are woody and the branches often spread along the ground. New growth each year is produced from the woody basal stems. The flowers are inconspicuous; the seeds produced in large quantities are nutlike, about 1/8 in. across the broader end.

Gardner saltbush grows on clay soils in the valley bottoms that are strongly alka-line. It covers extensive areas where free water sometimes accumulates temporarily, especially in Colorado, Utah, and Wyoming.

Gardner saltbush is readily eaten by sheep and is fairly nutritious, especially when the seeds and seed stalks are available.

Reference: H.41.

See also RANGE PLANTS.

GAS-TIGHT STEEL TOWER SILO.
See SILOS.

GASTRIC JUICE, or *stomach juice*, plays an important role in the DIGESTION of feed in animals with single stomachs, such as horses or swine. The juice is a clear liquid secreted by the stomach glands and consists of water, HYDROCHLORIC ACID, salts, and PEPSIN, one · of the proteolytic ENZYMES; sometimes another enzyme, RENNIN, is also present.

The first three stomachs of ruminants contain water, BACTERIA, and PROTOZOA that digest CELLULOSE and PENTOSANES as well as more soluble compounds; the fourth, or true, stomach contains gastric juice which acts upon the partially broken down plant material and upon the bacteria and protozoa.—*See also* ABSORPTION.

GATAN. *See* SOYBEAN (variety).

GELATIN is a protein-rich, jelly-like substance obtained from animal tissues. It has low nutritive values, being markedly deficient in the AMINO ACIDS methionine and cystine and devoid of tryptophan.

GEM is a SIX-ROWED BARLEY variety.

GENE is the factor in any germ-cell that determines a certain character in an individual produced by the fusion of two cells.

GENERA is the plural form of GENUS.

GENERIC means: of, or pertaining to, a GENUS.

GENOTYPE = Sort.

GENTIAN, the dried RHIZOME and roots of *Gentiana luten*, is a simple BITTER used as stomachic STIMULANT.—*See also* FEED INGREDIENTS.

GENTIANA. *See* GENTIAN.

GENUS (plural: *genera*) is a group of related SPECIES showing similar characteristics and appearing to have common ancestry.—*See also* SECTION; SUBGENUS.

GEORGIA NAPIERGRASS is a strain of NAPIERGRASS.

GEORGIAN SOYBEAN. *See* SOYBEAN (variety).

GEORGIA RED = *Purplestraw.*
    See COMMON WHEAT.

GEORGIA RUNNER is a runner-type PEANUT.

GEORGIA VELVETBEAN. *See* DEERING VELVETBEAN (variety).

GERM is (1) an EMBRYO or (2) a vegetable micro-organism, especially one of the pathogenic BACTERIA.—*See also* GERM CELL; PEANUT.

GERMAN CLOVER = CRIMSON CLOVER.

GERMAN FOXTAIL MILLET.
    See FOXTAIL MILLET (variety).

GERM CELL is a CELL capable of, or sharing in, reproduction.

GERMINATION is the sprouting of seed; i.e., the beginning of seed growth.

GHARNOVKA = *Kubanka.* See DURUM WHEAT.

GIANT BROWN SOYBEAN = *Mammoth Brown.* See SOYBEAN (variety).

GIANT FESCUE = TALL FESCUE.

GIANT PANICUM = BLUE PANICGRASS.

GIANT SPOTTED BUR-CLOVER.
    See SPOTTED BUR-CLOVER (variety).

GIANT STRIATA.
    See STRIPED CROTOLARIA (variety).

GIANT WILD-RYE (*Elymus condensatus*), the largest of all the native WILD-RYE GRASSES, is a coarse, robust perennial with thick, short rootstocks. It occurs in all western states. Individual plants often grow 10 ft. high and form clumps several feet thick. The erect flower SPIKES may grow 1 ft. long. Leaves are flat and coarse, nearly ¾ in. wide and 2 ft. long. The extensive root system has short, thick, perennial rootstocks.

This Pacific bunchgrass is abundant on moist or wet saline soils; it occurs also on moderately dry, fertile soils, but moderate grazing, especially during early spring, is essential to good stands.

Giant wild-rye is grazed readily while young. Later the foliage becomes coarse and harsh, and livestock leave it if they can get more palatable forage. If it is left standing, the grass provides a considerable amount of winter feed for cattle and horses. Fairly good hay can be had from the young growth.

Good stands of giant wild-rye have been obtained by drilling the seed about 1 in. deep. It is valuable for range reseeding on flood plains.

Reference: H.1.
See also GRASSES; RANGE PLANTS.

GIBSON is a SOYBEAN variety.

GILT RATIONS. *See* SWINE RATIONS.

GINGER, the dried RHIZOME of *Zingiber officianalle,* is used as CONDIMENT, CARMINATIVE, and stomachic STIMULANT.—*See also* FEED INGREDIENTS.

GLABROUS means: hairless, devoid of pubescence, or smooth.

GLACIER BARLEY. *See* SIX-ROWED BARLEY (variety).

GLANDS are secreting organs of the animal body which, when stimulated to activity by HORMONES or by nerves, pour out substances which are used in the body (e.g., digestive ENZYMES) or eliminated from it (e.g., sweat). The glands are divided into two types: (1) *endocrine glands,* or *ductless glands,* which empty their secretions into the blood stream, and (2) *duct glands* which pass their secretions through a series of small tubes, e.g. mammary glands.

Among the glands are pancreas, thyroid, spleen, gall bladder, and liver.

Glandular means: relating to a gland.— *See also* NECTARY.

GLANDULAR MEAL. *Animal liver and glandular meal* is one of the ANIMAL PRODUCTS.

GLAUBER'S SALT is hydrous SODIUM SULFATE with almost 56 percent water. It forms colorless crystals or white granules which are very soluble in water.

Glauber's salt is used as a purgative and diuretic.—*See also* MINERAL FEEDS.

GLEOCERCOSPORA. The FUNGUS *G. sorghi* is the cause of ZONATE LEAF SPOT.

GLOBEMALLOWS (*Sphaeralcea* spp.) are herbs and shrubs with mallow-like (5-parted) flowers; they occur in warm regions and are considered good forbs (weeds) on ranges.—*See also* RANGE PLANTS.

GLOBIN. *See* HEMOGLOBIN.

GLOMERELLA. A strain of the FUNGUS *G. cingulata*—which is the name of a

Colletotrichum spp. in its perfect, or sexual, stage—causes LESPEDEZA ANTHRACNOSE; another strain causes LUPINE ANTHRACNOSE.—*See also* COLLETOTRICHUM.

GLUCOSE is a CARBOHYDRATE that is a constituent of many animal and vegetable fluids. It occurs naturally in plants, especially in fruits, and is obtained by the incomplete hydrolysis of STARCH; it is an important NUTRIENT.

This simple SUGAR comes in form of *glucose sirup* (which is the "glucose" of feed people) and as white cyrstalline powder—which is often called DEXTROSE. It is water-soluble.

Glucose injections are used to supplement the treatment of PRUSSIC ACID POISONING with such antidotes as SODIUM NITRITE and SODIUM THIOSULFATE.—*See also* FEED INGREDIENTS; ABSORPTION; GLUCOSIDE; REDUCING SUGARS.

GLUCOSIDES are compounds which occur in plants and consist of GLUCOSE and some other substance. Some form soapy solutions with water, and therefore have been called SAPONINS; others are called CYANOGENETIC GLUCOSIDES because they develop compounds that under certain circumstances release hydrocyanic acid, or PRUSSIC ACID; still others are outside these two categories that form the toxic principles of many POISONOUS PLANTS.

GLUMES are the two lowest BRACTS of a grass SPIKELET. They are empty, i.e., they do not bear STAMENS or PISTILS in their AXILS. The lower one is known as the *first glume* and the upper one as the *second glume*.

Flowering glumes are called LEMMAS.—*See also* GRASSES.

GLUTAMIC ACID is a dispensible AMINO ACID; it has recently been reported, however, to be an essential for chicks and laying hens. It forms colorless, water-soluble crystals and occurs in many seeds and in beets. Glutamic acid is produced by hydrolysis of such protein substances as GLUTEN and CASEIN.—*See also* CONDENSED BEET SOLUBLE PRODUCTS.

GLUTEN is a protein-rich product obtained from cereal grains. It is the sticky, yellowish mass that remains after the STARCH is washed out of wheat flower.

Gluten feed and *gluten meal feed* are terms often applied to *corn gluten feed* and *corn gluten meal*, two important CORN feedstuffs.—*See also* VITAMIN A.

GLUTINOUS means: gluelike, sticky; i.e., of the nature of cooked starch pastes of certain WAXY grain SORGHUMS, RICES, and other cereal grains.

GLYCERIDES are "salts" (esters) of glycerin; e.g., FATS, the FATTY ACID esters of glycerin.

GLYCERIN, or *glycerol*, is a highly viscous (sirupy), water-soluble, sweet-tasting liquid; it exists in combined form in FATS and oils. Fats split into glycerin and FATTY ACIDS which then are absorbed from the small intestine into the blood stream.—*See also* ABSORPTION.

GLYCINE means different things; among them: (1) *Glycine max* = SOYBEAN. (2) One of the dispensible AMINO ACIDS which forms white, sweet, water-soluble crystals.

GLYCOGEN, or *animal starch*, is a CARBOHYDRATE which has the same general composition as vegetable STARCH. It is stored in the LIVER and in muscles of animals.

GOAT RATIONS. *Dairy goats* may be fed rations of a quality similar to those that are satisfactory to cows. A ration of 5 lb. of good LEGUME HAY and 2½ to 3½ lb. CONCENTRATES should support a production level of from 6 to 8 lb. milk. The concentrate mixture may be relatively simple, but should be free of dust and finely ground feeds.

A satisfactory grain mixture can be prepared from 73 lb. rolled barley or a mixture of cracked corn and oats, 10 lb. black strap molasses, 15 lb. cottonseed or linseed meal, 1 lb. bone meal, and 1 lb. salt.

Angora goats usually obtain their feed from browse or by grazing poor PASTURES. In emergencies they may be fed grass HAY. —*See also* RATIONS.

GOATWEED = ST. JOHNWORT.—*See also* POISONOUS PLANTS.

GOENS is a soft winter variety of COMMON WHEAT.

GOING = *Goens. See* COMMON WHEAT.

"GOING CRAZY" is a condition sometimes observed in animals fed *ammoniated molasses.—See also* MOLASSES.

GOITER. *See* IODINE.

GOITROGENS are goiter-causing anti-thyroid substances; some of them occur in BRASSICAS. Goitrogens are valuable for fattening animals; e.g., THIOURACIL.

GOLDEN BANTAM and *Golden Cross Bantam* are CORN varieties.

GOLDEN GRAM = *golden mungbean.* See MUNGBEAN (variety).

GOLDEN MARIOUT = *Club Mariout.* See SIX-ROWED BARLEY.

GOLDEN MUNGBEAN is a MUNGBEAN variety.

GOLDEN RAIN. *See* COMMON OAT (variety).

GOLDENROD. RAYLESS GOLDENROD is one of the POISONOUS PLANTS.

GOLDEN SUNSHINE is a CORN variety.

GOLDEN WHEAT. *See* COMMON WHEAT (variety).

GOLDSOY. *See* SOYBEAN (variety).

GOLDTHREAD = DODDER.

GOOBER = PEANUT.

GOOSE DURUM = *Arnautka. See* DURUM WHEAT.

GOOSENECK SORGO is one of the forage SORGHUMS.

"GOOSE STEPPING."
                    See PANTOTHENIC ACID.

GOPHER OAT.
                    See COMMON OAT (variety).

GOPHERS are crop-damaging, burrowing animals which belong to several RODENT species. They do much damage to agriculture and in gardens west of the Mississippi River.

Control: Gophers are controlled by trapping, poisoning, or gassing.—*See also* FIELD MICE; POCKET GOPHERS.

GORDURA GRASS = MOLASSES GRASS.

GOSSYPIUM. Among the *Gossypium* spp. are the following COTTON species (and their varieties): *G. hirsutum,* or *Upland cotton; G. barbadense,* which includes *Sea Island* and *American-Egyptian cottons; G.arboreum* and *G. herbaceum,* two *Asiatic cottons.*

GOSSYPOL is a toxic agent found in *Gossypium* spp. (COTTON species). It is a fat-soluble, dark-colored substance: the poisonous principle present in untreated COTTONSEED.

GRAIN is (1) the fruit of CEREALS and other GRASSES or (2) any cereal crop.—*See also* CEREAL HAY; CRIMSON CLOVER; ROSE CLOVER; RED CLOVER; STRAWBERRY CLOVER; VETCHES; ROUGHPEA; FIELD PEA; SCREENINGS; HAY GRADING; CHINCH BUG; RED HARVESTER ANT; WESTERN HARVESTER ANT; GRAINS.

GRAIN BEETLES, which belong to various genera, as well as other insects are often referred to as *bran bugs;* they are found in stored grain, milled products, and broken grains.

GRAIN BIN. *See* STORING FEED.

GRAIN BINDER. *See* BINDER.

GRAIN BUG = SAY STINK BUG.

GRAIN CROPS = CEREAL CROPS. *See* CEREAL.

GRAIN CRUSHING.
                    See GRAIN GRINDING.

GRAIN DISTILLERS' DRIED YEAST.
                    See YEAST.

GRAIN ELEVATOR. *See* STORING FEED.

GRAIN GRADING in accordance with the *U.S. Grain Standards Act* is undertaken by official inspectors who sample, test, and grade bulk grain. Grain grading is of great practical importance for regulating marketing, market value, and financing of the grains.

A variety of factors determine the grades of the different types of grains. Exact grain inspection and laboratory procedures have been established for obtaining the proper grain samples and for determining such factors as moisture content, density, plumpness (fullness), and shape of the kernels, all of which influence the "test weight per bushel."

Unless the grain is adulterated, high *test weight* should indicate cleanliness of grain and, thus, high feeding value because of low content of crude fiber (hull and/or bran).

NOTE: *Dockage* is the foreign material removed from the grain by means of testers or of screening devices; when it amounts to more than 1 percent of the total weight of the grain, it is deducted from the weight. Thus, clean grain has practically no dockage.

Official standards exist for the grains listed in the table. In addition, there exist

GRAIN GRADING—OFFICIAL STANDARDS *

Grain	Minimum test weight per bushel	Maximum limits of damaged kernels	foreign material
	lb.	%	%
Wheats.................	60–50	2–15	1–7
Corn...................	54–44	2–7	3–15
Barleys................	47–35	0.1–3	0.5–6
Oats..................	32–24	0.1–6	2–10
Rye...................	56–49	2–15	3–10
Grain sorghums..........	55–49	2–15	1–10
Flaxseed (linseed).........	49–47	20–30	.....
Soybean...............	56–49	2–8	2–6

* This is a condensed tabulation showing only the value limits for all classes of each grain listed. In general, it can be said that the higher the minimum weight of a grain, the lower the maximum limit allowed for damaged kernels and foreign material.

standards for *mixed grains*, i.e., mixtures of those grains for which standards have been established.

Reference: U.11.

GRAIN GRINDING and *grain crushing* are operations that break up the outer hull and reduce the particle size. The ground grain is more easily attacked by digestive juices.

All animals, except the very young, the sick, and those with poor or missing teeth, can do a reasonably good job of chewing (or grinding) their own feeds. *Sheep* do the most thorough job and only very hard or very small seeds should be ground for them. *Swine* do a reasonably good job, but grinding will improve the digestibility of CORN some 3 to 5 percent. Usually the value of the grain saved will be less than the cost of grinding. Grains other than corn can be economically ground for hogs.

Grains should be coarsely ground for *dairy cows*.

Beef calves, six to ten months old, chew corn very well. Other *beef animals* should be fed ground corn unless they are followed by hogs. Hogs will recover enough undigested feed from the cattle to make 1 to 2 lb. pork gains for each bushel of corn fed to cattle. Other grains should be ground for beef cattle.

Grains other than OATS and corn should be rolled, crimped or crushed for *horses*.

Poultry can do an efficient job of grinding their own feeds and may be fed whole grains; however, many feeds are ground so

that they may be properly mixed in a complete ration.—*See also* GRINDING OF FEED MATERIAL; SOAKING FEED.

GRAINS belong to the GRASS family. However, commercially, BUCKWHEATS are classed among the GRAIN crops.—*See also* CEREAL; SILAGE; GRAZING; RANGE MANAGEMENT; GRASS TETANY.

GRAIN SCREENINGS. *See* SCREENINGS.

GRAIN SORGHUM CHOP.
See SORGHUMS.

GRAIN SORGHUM DISTILLERS' DRIED GRAINS. *See* DISTILLERS' PRODUCTS.

GRAIN SORGHUM DISTILLERS' DRIED GRAINS WITH SOLUBLES. *See* DISTILLERS' PRODUCTS.

GRAIN SORGHUM DISTILLERS' DRIED SOLUBLES. *See* DISTILLERS' PRODUCTS.

GRAIN SORGHUM FEED MEAL.
See SORGHUMS.

GRAIN SORGHUM GLUTEN FEED.
See SORGHUMS.

GRAIN SORGHUM GLUTEN MEAL.
See SORGHUMS.

GRAIN SORGHUM HEAD CHOP.
See SORGHUMS.

GRAIN SORGHUM HEAD STEMS.
See SORGHUMS.

GRAIN SORGHUM MILL FEED.
See SORGHUMS.

GRAIN SORGHUM OIL CAKE.
See SORGHUMS.

GRAIN SORGHUM OIL MEAL.
See SORGHUMS.

GRAIN SORGHUMS. *See* SORGHUMS.

GRAIN SORGHUM SILAGE.
>*See* SILAGE CROPS.

GRAIN WEEVILS are various insect pests found in stored CORN and other grains and seeds. The true grain weevils cannot breed in grain that has a moisture content below 9 percent.

GRAM, abbreviated *gm.,* is a weight unit in the metric system: 1 gm. = 15.432 grains; 1,000 gm. = 1 kilogram (or approx. 2.2 lb.).—*See also* GRAMS.

GRAMA GRASSES (*Bouteloua* spp.) or *gramas,* of which eighteen species occur in the United States, are well represented throughout the Great Plains and Western States. They are reliable producers of good forage on range and pasture land.

The gramas are summer growers, and the amount of forage they produce depends upon the moisture available during the growing season. In years of extreme drought they make little or no new top growth. Most species cure naturally, however, and standing growth from previous seasons makes very satisfactory and palatable forage for most classes of livestock.

Individual SPIKELETS are small and single-flowered, a characteristic that greatly simplifies the identification of gramas.

Dangers: Among the diseases which attack grama grasses are the *sausage smut,* one of the STRIPE SMUTS.

Species: The most important species of grama grasses are SIDE-OATS GRAMA, the BLUE GRAMA, BLACK GRAMA, and SIXWEEK GRAMA.

Reference: H.1.

See also GRASSES; PRAIRIE HAY; RANGE PLANTS; PASTURE PLANTS.

Illustrations: See BLACK GRAMA; BLUE GRAMA; SIDE-OATS GRAMA.

GRAMICIDIN is an ANTIBIOTIC substance —one of the components of tyrothricin.

GRAMINEAE = GRASSES.

GRAM PEA = CHICK PEA.

GRAMS. *Golden gram* (or *golden mungbean*) and the *green grams* (or *green mungbeans*) are MUNGBEAN varieties.

GRANGER. *See* SOYBEAN (variety).

GRANULATED FEEDS can be prepared by breaking or cutting PELLETS with gran-

ulating machines and by sifting the fine material out of the finished product. The *granules* are often mixed with ground mash for baby chicks.

GRAPEFRUIT REFUSE.
>*See* DRIED GRAPEFRUIT REFUSE.

GRAPE POMACE, or *grape-seed meal*— consisting of seeds and skins of pressed grapes—has a low feeding value; it is sometimes used in feed rations to counteract the laxative action of other feed ingredients.

GRAPHOGNATUS. *G. leucoloma* is one of the WHITE-FRINGED BEETLE species.

GRASS. *See* GRASSES.

"GRASS EGGS." *See* SILAGE.

GRASSES are the plants belonging to the botanical family of Gramineae, or (better)

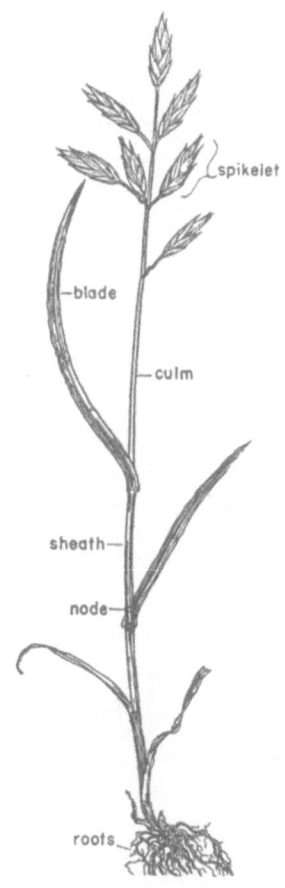

Grasses. Their general habit. (H.47.)

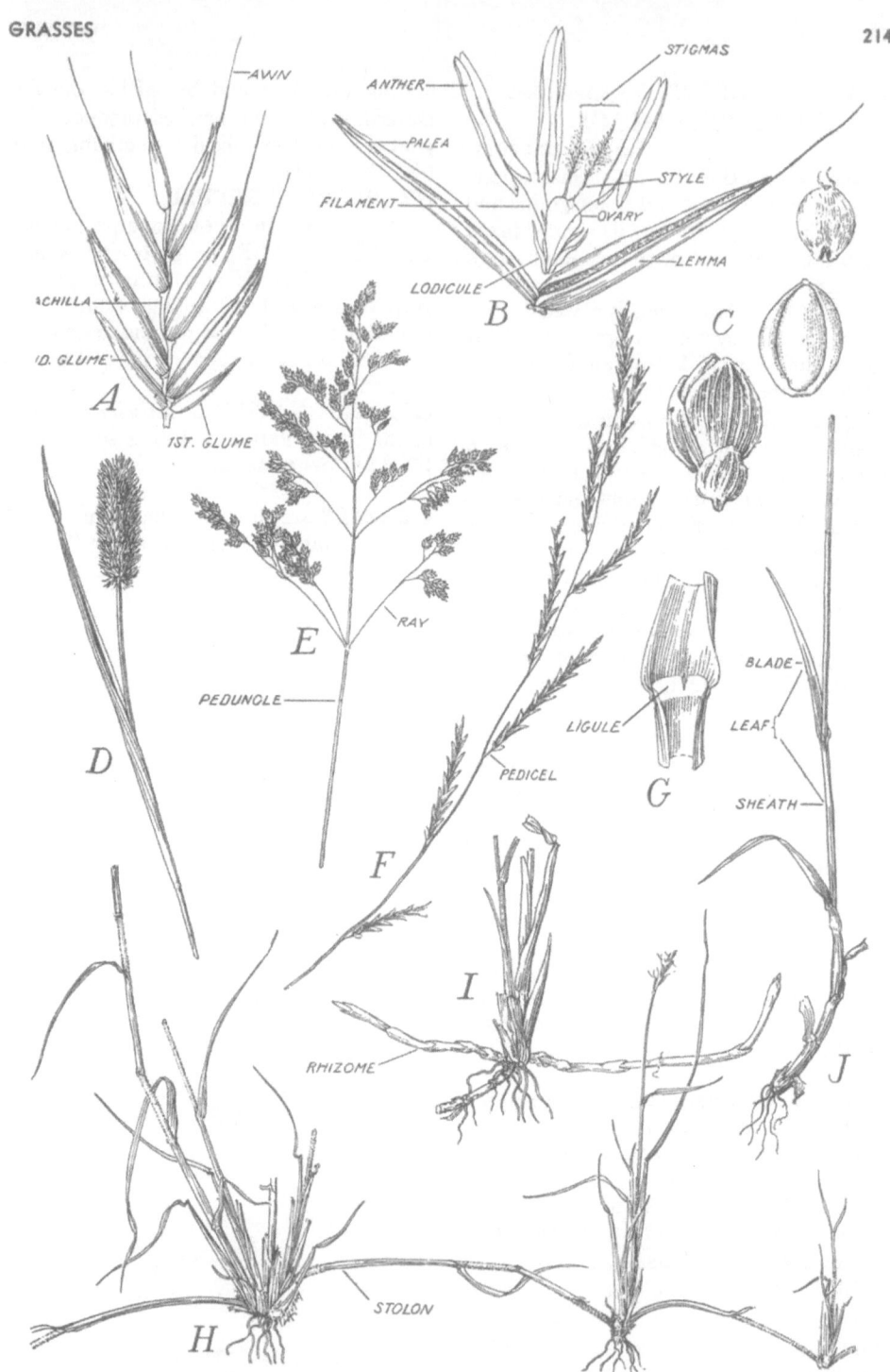

Grasses. A and B, parts of the flowers; C, fruit; D, E, and F, spikelets aggregated into terminals spikes, panicles, and racemes; G and J, stem; H and I, root formations. (D.1.)

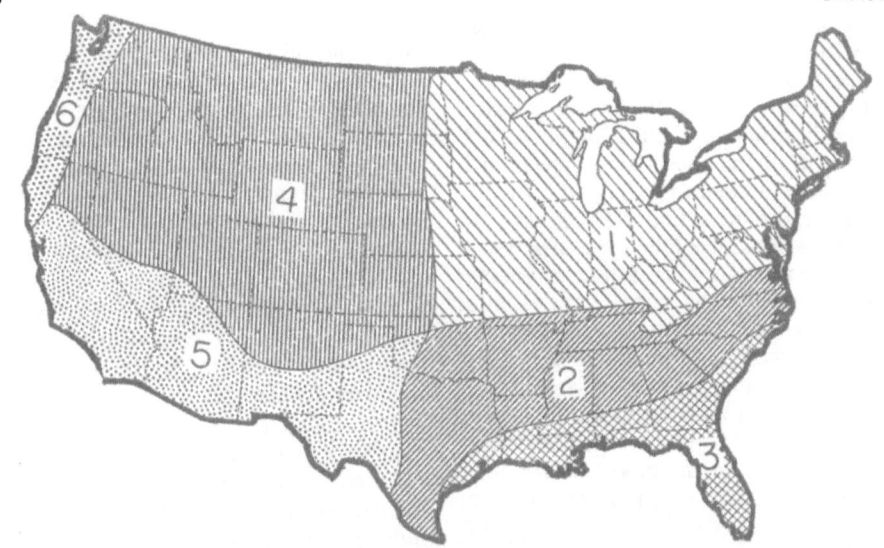

Regions of general adaptation of grasses. *Region 1* ÷ Colonial bentgrass, Canada bluegrass, Kentucky bluegrass, smooth brome grass, chewings fescue, creeping red fescue, quackgrass, and redtop. *Region 2* ÷ Bermuda-grass and redtop. *Region 3* ÷ Bermuda-grass, carpetgrass, and centipedegrass. *Region 4* ÷ Buffalograss, smooth brome grass (Eastern), blue grama grass, crested wheatgrass, and western wheatgrass. *Region 5* ÷ Bermuda-grass (under irrigation), Indian ricegrass, sand dropseed, and weeping lovegrass. *Region 6* ÷ Colonial bentgrass, Kentucky bluegrass, chewings fescue, creeping red fescue, redtop, and perennial ryegrass. (U.S.D.A.)

Poaceae. Most of them have hollow stems, or CULMS, with closed and solid joints, or NODES; however, the number of solid-stemmed grasses is considerable.

The leaves, alternate in two ranks, are parallel-veined; they consist of (1) a basal portion, called SHEATH, which envelopes the stem; (2) the LIGULE, the projecting, often tonguelike, membranous end of the lining of the leaf-sheath, seen at the base of the leaf-blade, between it and the stalk (sometimes it is reduced to a mere fringe of hairs or to a hardened ring); (3) the BLADE.

The flowers of grasses are mostly perfect —i.e., with both male and female floral organs—although occasionally they are one-sexed, arranged primarily in small clusters, or SPIKELETS, on a central axis (RACHILLA). The spikelets are aggregated into characteristic terminal arrangements, distinguished as SPIKES, PANICLES, and RACEMES.

On the central axis are also two or more two-ranked BRACTS: the two lowest bracts, termed GLUMES, are empty (occasionally one or both of these may be absent); above

the glumes are from one to many flowering glumes, or LEMMAS; opposite each lemma is often a second, two-nerved bract which is named the PALEA.

The male floral organs, or STAMENS, vary from one to six in number and have two-celled ANTHERS. The female floral organ, or PISTIL, has a one-celled OVARY and usually two STYLES, ending in feathery (plumose) receptive organs (for the pollen grains), called STIGMAS. The LODICULES, usually two in number, are small, thin scales, which help to force open the lemma and palea at the time of fertilization.

The seed and fruit of grasses are united.

The grasses are widely distributed and range in height from less than 1 in. to more than 100 ft (in the Tropics).

Dangers: Among the more important grass enemies are the ANTHRACNOSE OF GRASS; STRIPE RUST; ERGOT; WHEAT SCAB, BACTERIAL LEAF BLIGHT; GRASS SMUTS, STRIPE SMUT and other SMUTS; the STEM NEMATODE; the FALL ARMYWORM and other ARMYWORMS; GRASSHOPPERS; the CHINCH BUG; WHEAT STEM SAWFLY; ARMY CUT-

WORM and other CUTWORMS; the POTATO LEAFHOPPER; and the MORMON CRICKET.

Genera and Species: Grasses form a large natural family comprising about 600 genera. Of these, about 150 genera with 1,500 species occur in the United States.

Some of the more important grass genera and species are as follows: BAHIAGRASS; BARLEY; BARNYARDGRASS; BERMUDA-GRASS; BLUEGRASSES; BLUEJOINT; BLUESTEMS; BROME GRASSES; BUFFALOGRASS; BUFFEL-GRASS; CALIFORNIA NEEDLEGRASS; CALIFORNIA OATGRASS; CANARYGRASSES; CARPETGRASS; CENTIPEDEGRASS; CHESS; COLONIAL BENT; CORN; DALLISGRASS; DROPSEEDS; FENDLER THREE-AWN; FESCUES; FOWL BLUEGRASS; FOXTAIL MILLET; GALLETA; GRAMA GRASSES; GUINEAGRASS; INDIAN GRASS; INDIAN RICEGRASS; JOHNSONGRASS; LOVEGRASSES; MEADOW FOXTAIL; MOLASSES GRASS; NAPIERGRASS; NATALGRASS; NEEDLEGRASSES; NEVADA BLUEGRASS; OATS; ORCHARDGRASS; PANGOLAGRASS; PANICGRASS; PARAGRASS; PEARLMILLET; PEPPERGRASS; PORCUPINE GRASS; REDTOP; RHODESGRASS; RICE; RYE; RYEGRASSES; SACATON; SALINA WILD-RYE; SALTGRASSES; SIDE-OATS GRAMA; SORGHUMS; SQUIRRELTAILS; ST. AUGUSTINEGRASS, STIPA GRASSES; SUGARCANES; the SWEET VERNALGRASS; TALL OATGRASS; TEOSINTE; TIMOTHY; TOBOSA; VELVETGRASS; WHEATGRASSES; WHEATS; WILD GRASSES; WILDRICES; WILD RYE-GRASSES; and ZOYSIA.

Reference: D.1.

See also BLOAT; SELENIUM POISONING; LESPEDEZAS; ANNUAL LESPEDEZAS; COMMON LESPEDEZA; KUDZU; SOYBEAN; BIRDFOOT TREFOIL; SUBCLOVER; RED CLOVER; BLACK MEDIC; MUNGBEAN; VELVETBEANS; COWPEA; BRUSHES; SAGEBRUSHES; GRASSLIKE PLANTS; CEREALS; TRUE CEREALS; SMALL GRAINS; STRAW; PRAIRIE HAY; HAY; ALFALFA-HAY STANDARDS; GRASS MEAL; SILAGE CROPS; GRAZING; PASTURE PLANTS; PASTURES; PASTURE MANAGEMENT; RANGE PLANTS; RANGE MANAGEMENT; POISONOUS PLANTS; WEEDS; U.G.F.

GRASS FACTOR. *See* U.G.F.

GRASS HAY. *See* HAY GRADING.

GRASSHOPPERS, which destroy crops and cause losses totaling millions of dollars,

are found all over the United States, but serious outbreaks seldom develop in the East. Outbreaks occur mostly in the western two-thirds of the United States, in the northern Great Plains, including eastern Montana, North Dakota, South Dakota, Nebraska, and Kansas.

Of the 142 grasshopper species collected on range plants in the western states, only a few are known well enough to have common names. At least 90 percent of all grasshopper damage is caused by the following five species:

1. *Lesser migratory grasshopper*, or *Melanoplus mexicanus*, the most widely distributed and most destructive species.

2. *Clear-winged grasshopper*, or *Camnulla pellucida*, which is most common in mountain meadows, grassy openings in timbered land, and well-sodded, closely grazed pastures on the open plains.

3. *Differential grasshoppper*, or *Melanoplus differentialis*, like the next two species, prefers heavy soil and lush vegetation.

4. *Two-striped grasshopper*, or *M. bivittatus*.

5. *Red-legged grasshopper*, or *M. femurrubrum.*

Under outbreak conditions all five species spread far from favored habitats and feed on a variety of crops and vegetation.

In parts of the United States where grasshoppers are abundant they do considerable damage to WHEAT, BARLEY, CORN, OATS, RYE, SAFFLOWER, KUDZU, SWEETCLOVER, ALFALFA, SOYBEANS, BEANS, the VETCHES, LUPINES, as well as GRASSES, e.g., RICE, OATS, SUDANGRASS, MEADOW FOXTAIL, and KING RANGE BLUESTEM, a variety of TURKESTAN BLUESTEM. The grasshoppers have food-plant preferences; they damage crops mainly by gnawing and eating the leafy parts of the plants.

The female grasshopper feeds for several days to several weeks, then mates and begins to lay eggs. The eggs are laid well below the surface of the ground. While the eggs are being laid, a sticky, frothy liquid is added, which hardens as it dries. This aids in holding the eggs in a compact mass. The number of eggs laid in each mass varies from 20 to 100 or more, depending on the species. Most female grasshoppers lay from one to three egg masses.

After the eggs hatch, the young grasshoppers push to the surface of the ground and begin feeding on the nearest vegetation. The immature (NYMPH) stages of the grasshoppers are similar to the adult ones except that in the first they do not have wings and the head is proportionately larger compared to the rest of the body. All grasshoppers increase in size by molting. Usually six such molts occur before the insect reaches the adult stage.

Grasshoppers are most active on warm, sunny days. They seldom move any great distance from their breeding areas, except when a lack of food material and other unfavorable factors force them to do so.

Control: The cultivation of fields and fence rows late in the fall or early winter destroys large quantities of grasshopper eggs and helps to keep this insect pest in check.

The practice of "burning-over" grassy areas to control grasshoppers is satisfactory only when it is delayed until after the eggs are hatched. To destroy the nymphs before they leave the breeding areas is the cheapest method of controlling grasshoppers.

Among the most effective methods are the use of POISONED BRAN MASH or of any one of a number of newer insecticides, namely TOXAPHENE, CHLORDANE, DIELDRIN, PARATHION, or HEPTACHLOR.

References: V.1; A.2; K.5; P.6.

See also LYGUS BUGS; MORMON CRICKET; RANGE MANAGEMENT.

GRASSLAND FARMING.
                                        See PASTURES.

GRASS-LEGUME SILAGE.
                                        See SILAGE CROPS.

GRASSLIKE PLANTS—the so-called "*water grasses*" of the stockman—are also called "*marsh grasses*" or "*wet meadow grasses*" because the majority of these plants grow on wet sites. They resemble true GRASSES superficially, but do not belong to that family and the use of the term "grasses" for them is, therefore, inaccurate and misleading. As a rule, these plants are inferior to true grasses as forage plants. However, they form one of the four classes of RANGE FORAGE.

The most characteristic grasslike plants are the SEDGES and RUSHES.

GRASS MEAL—one of the officially recognized feedstuffs—is the product obtained from drying and grinding GRASS which has been cut before formation of the seed. If the name of a grass species is used, the product must correspond thereto.

Reference: F.6.

See also MISCELLANEOUS PRODUCTS.

GRASS SILAGE.
                        See SILAGE CROPS; SILAGE; SILOS.

GRASS SMUTS occur on nearly all GRASSES, both wild and cultivated. However, SMUT is a serious problem on only a comparatively few. These include MOUNTAIN BROME, SLENDER WHEATGRASS, CRESTED WHEATGRASS, TALL OATGRASS, CANADA WILD-RYE, BLUE WILD-RYE, and BIG BLUEGRASS.

Control: SEED TREATMENT is recommended whenever any of the above grasses are planted. The following seed disinfectants give good results: CERESAN M, ARASAN, TERSAN, PANOGEN, or FORMALDEHYDE SOLUTION. Smut-resistant varieties of grasses should be used when they are available.

Among the more important grass smuts are STRIPE SMUTS, HEAD SMUT; COVERED KERNEL SMUT; and LOOSE KERNEL SMUT.

References: M.26; F.4.

GRASS SORGHUMS include SUDANGRASS and JOHNSONGRASS.—*See also* SORGHUMS.

GRASS TETANY is a disease often observed soon after recently freshened cows are turned on a lush PASTURE—RYEGRASS, ORCHARDGRASS, OATS, etc. This pasture disease often causes death losses. The incidence of the disease can be reduced by feeding MAGNESIUM OXIDE in concentrates or salt, since grass tetany may be caused by the precipitation of MAGNESIUM in the intestine through AMMONIA formed from nitrates in forage. When cattle have been given hay or grain the losses have stopped, proving that cattle should not be grazed exclusively on cereal grains in the winter.

References: S.28; S.10.

See also GRAZING; MAGNESIUM.

GRAY CROWDER = *Taylor. See* COWPEA.

GRAY GOOSE COWPEA = *Taylor. See* COWPEA.

GRAYIA. *G. spinosa* = SPINY HOPSAGE.

GRAY LEAF SPOT, caused by *Gercospora sorghi*, a FUNGUS, occurs on grain and forage SORGHUMS, JOHNSONGRASS, and CORN. This disease is of minor importance, but occasionally it causes considerable spotting of sorghum leaves in limited areas of the more humid and warmer sections, particularly the Gulf Coast areas.

The spots usually are reddish-purple, and in some varieties tan. When small they are indistinguishable from other LEAF SPOTS, but as they enlarge they become long and narrow, being limited somewhat by the leaf veins. These long, narrow spots may grow together and thus destroy large areas of the leaves. As the spots enlarge, they usually become covered with grayish-white fuzz, made up of the fruiting structures of the fungus.

Control: No control measures and no sorghum varieties resistant to gray leaf spot are known at the present.

Reference: L.1.

See also FUNGUS LEAF DISEASES.

GRAY MOLD, or *gray-mold leaf spot*, caused by the FUNGUS *Botrytis cinerea*, attacks both showy and striped CROTOLARIAS; on the former it may become serious enough to cause much defoliation by late summer. Gray mold also occurs on VETCHES in the southern United States.

Reference: M.14.

GRAY SUMMER-CYPRESS (*Kochia vestita*) is a fast-growing, fair shrub with dense foliage which sometimes serves as forage on ranges.—*See also* RANGE PLANTS.

GRAY WINTER OAT = *Winter Turf. See* COMMON OAT.

GRAZING of PASTURES and RANGES by livestock and poultry is of great economical importance to farmers and rangers.

The plant species on *pastures* differ in palatability and some are more readily grazed than others, which tends to jeopardize the more palatable and to protect the less palatable species. Likewise, some species are naturally more persistent than others, so that even though equally palatable, the more persistent tend to survive while the less persistent tend to disappear under grazing. Again, species differ in their ability to recover after grazing, some requiring more time than others. The time and intensity of grazing with respect to the stage of growth also affects persistency, especially that of species dependent upon natural reseeding for reproduction; if these are grazed in a manner to prevent seed formation they do not survive.

The specific height to which the vegetation should be eaten down depends on the crops which constitute the pasture and the location of the pasture. BLUEGRASSES and volunteer WHITE CLOVER should not be grazed shorter than 1 or 2 in.; 2 in. has been recommended as the minimum to which pastures containing LADINO CLOVER should be grazed in Connecticut and pastures containing BERMUDA-GRASS in Georgia. Hay-type GRASSES and LEGUMES will not withstand close grazing. It is recommended in Wisconsin that Ladino clover be pastured not shorter than 4 to 5 in. In order to maintain a well-balanced mixture of ALFALFA and BROME GRASS through three or four seasons, it is considered advisable in Michigan to maintain a growth 8 to 10 in. tall during May and June, and not less than 4 in. high during the summer.

Intense grazing may result in the total elimination of certain forages from the stand during the course of a single growing period. Grassy swards occasionally are grazed closely during certain specified periods and with success—they reduce the vigor of the grass component sufficiently so that short-lived legumes (such as LESPEDEZA, DOUBLE-CUT CLOVERS, and Ladino clover) are more readily re-established from natural reseeding. But if that is not the objective, close and continuous grazing is generally harmful.

Unfortunately, no exact rules of *grazing management*, applicable to all forages under all conditions, can be made, but it can be stated as a broad generalization that the ideal plan of grazing management involves maximum use of pastures at immature stages of growth while at the same time safeguarding the forage stand and providing a reserve for emergencies.

Grazing *rotationally* may increase yields approx. 10 percent above those obtained

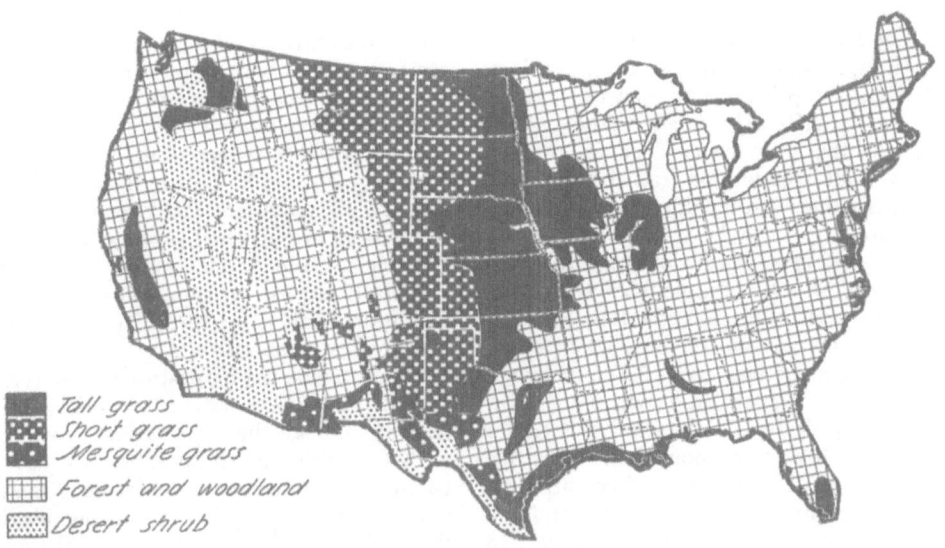

Tall grass
Short grass
Mesquite grass
Forest and woodland
Desert shrub

Natural vegetation. Short-grass lands characterize a large part of the subhumid lands of the Temperate Zone. The rainfall of the short-grass lands keeps the upper soil layers moist during the warm season in most years, enough to support the shallow-rooted short grasses. Near the humid border of the short-grass land, several years of above-normal rainfall may encourage tall grasses and convert short-grass land to tall-grass land.Rainfall varies greatly—years occur when rainfall is no greater than might be expected in the desert, and so do years when it equals that in the humid forested regions. Tall-grass land or prairie is found where soil moisture is deeper than in short-grass lands. (U.S.D.A.)

by continuous grazing. Such increases are often not large enough to justify the extra cost of fencing and of providing water.

Various types and parts of *range* are best fitted for different kinds and classes of livestock. Level and rolling grasslands can grow high-quality forage, especially well adapted for production of high-grade calves and for growing out yearling steers. Such lands, too, are used in the production of feeder lambs. The higher mountain ranges with their cool summers and heavy growth of palatable grasses and range "weeds" are especially valuable for the production of high-quality, grass-fat lambs. Browse ranges are often adapted to Angora goats. Ranges that produce an abundance of mast are sometimes used for grazing hogs. Many arid and semiarid mesa and valley lands have proved to be well adapted to winter grazing by sheep. Also, if adequate water is available, they may be satisfactorily grazed all year or seasonally by cattle.

Supplementing low-value range forage, especially in the winter, with protein concentrates or GRAIN, keeps young animals growing and breeding herds in good productive condition.

Chemical analyses of range forage show that the native forage is often deficient in phosphorus and other minerals, particularly while dormant during winter months. Feeding of BONE MEAL or other mineral supplements is usually necessary. Range forage is often deficient in CAROTENE for six months of the year.

Adequate distribution of grazing can be facilitated by proper salting, development of additional water places, construction of drift fences, and construction of trails.

PASTURES AND RANGES
FOR LIVESTOCK

Dairy cattle: Productive pastures as a rule are a cheaper source of the nutrients needed for milk production than any other crop, largely because the cost of establishing the pasture is spread over a period of years and the grazing animals reduce the labor and cost of harvesting.

The largest volume of milk is produced in the spring when the pasture grass is at its best; the price of milk, however, is low in spring. After the flush of the pasture season, the production declines rapidly. More milk production can be expected when legumes are included in the pasture mixture.

On lush grass and clover pasture the average cow will consume from 100 to 125 lb. forage daily. A large cow may consume 150 lb. a day. On short over-grazed pasture a cow might be limited to 50 lb. or less of forage daily. Under these conditions a cow consuming daily 100 lb. grass and clover pasture will receive about 3 lb. digestible protein and 15 lb. total digestible nutrients. This amount of feed will maintain a 1000-lb. cow and enable her to produce 15 to 18 lb. milk daily.

When the grass becomes tougher, less palatable, and more fibrous, the cows do not eat as much as they did earlier in the season. This fact is the principal explanation of the rapid decline of production during the summer.

Aside from common salt (SODIUM CHLORIDE), which is needed with all rations, young pasturage grown on a fertile soil provides a suitable ration for cows producing a medium or small quantity of milk. Some dairy cows, however, have been developed to such a state of productivity that they must have supplementary ENERGY feeds. Concentrates, being high in energy, are preferred, although CORN or SILAGE is useful. Good quality HAY may simply decrease pasture consumption.

Deficiencies of vitamins A and D in a cow's ration are the only ones that have been shown to be serious enough at times to affect health. In most areas cows get sufficient sunshine to provide vitamin D, but in northern states some vitamin D source may be needed.

Beef cattle: It is generally agreed that beef cattle make gains on high-quality pasture at less expense than by lot-feeding, yet much more time is usually required to get cattle fat enough for market by the exclusive use of pasture. A combination of mother's milk and good pasturage is usually sufficient for beef calves. One or more concentrates are often recommended for fattening calves on poor pasture prior to weaning. This practice involves what is known as *creep feeding*.

It is possible to speed up the rate of finishing 3-year-old steers by feeding a mixture of CORN and COTTONSEED MEAL while they are on good pasture; it is also possible to put a marketable finish on 2-year-old steers by supplementing their pasturage with a mixture of protein and carbohydrate concentrates.

Hogs: Swine fed on pasture rather than in a dry lot are usually more thrifty and healthy; unless the best rations fortified with VITAMINS and ANTIBIOTICS are used, pasture-fed animals may be fattened more rapidly. Pasture gains are economical. With pigs weighing 75 lb., a satisfactory rate of gain can be obtained from good pasture, farm grains, and a mineral mixture containing calcium, salt, and phosphorus. (Feeding hogs on pasture also helps to maintain soil fertility.)

Sheep: For fattening lambs, average quality pasture will give better results when supplemented. Good-quality temporary pastures supplement permanent pastures in midsummer as well or better than concentrates.

To provide for reasonable stability in winter grazing, and to assure an adequate forage supply on winter ranges in most years, a basic stocking rate that will utilize 75 percent of average forage production is recommended. On ranges in fair to good conditions 1.0 to 3.5 acres are required per sheep per month, whereas on those in poor condition 2.3 to 5.7 acres are required.

Certain management practices are of benefit to both ranges and sheep: they include subdividing large grazing allotments so that grazing use may be rotated from year to year; using open herding and 1-night bedgrounds; leaving the range early to avoid grazing during the late winter period; and providing water for sheep each day.

Horses and mules: Permanent pastures, and most kinds of temporary pasture when not overgrazed, are important factors in keeping horses and mules healthy and maintenance costs at a minimum. Such pastures do not, as a rule, need to be supplemented with hay or grain in order to nourish idle mares and young stock adequately. In the Corn Belt, work horses are pastured about six months of each year, but much of this time they are turned out only at night on Sundays or during other short periods of idleness; under these practices, pasture may be considered a supplement to the work-horse ration.

PASTURE AND RANGES
FOR POULTRY

Chickens, turkeys, and *other* kinds of poultry are natural consumers of a certain amount of roughage, and the cheapest and easiest way to supply this is to let them range on a green crop.

Many poultrymen now rear birds in confinement and replace green forage with legume meals and vitamin supplements. If one plans to rear poultry on ranges, houses should be near well drained, fertile soil which is suitable for growing grazing crops. The yards should be large enough to furnish grazing over a reasonably long period. This will require about 1 acre of land for each 400 to 500 pullets during the growing period, or for 200 to 300 laying hens. Turkeys are better grazers than chickens and will require considerably more space; at least 1 acre, and preferably 2 acres, should be provided for each 100 breeder hens or 100 growing turkeys.

Feed costs of chickens or turkeys can be reduced from 10 to 20 percent if an adequate supply of green feed is available. The saving in feed costs will depend upon the quantity and quality of the green feed and on how the birds are rotated.

There are some objections to using irrigated crops or pastures for poultry inasmuch as wet soil provides an ideal place for many disease organisms and parasites to multiply, especially during warm weather. If irrigated crops or pastures are used, sufficient acreage should be available so that the birds can be rotated to avoid contact with the soil during and immediately following irrigation.

Forage crops for poultry should be low in fiber and high in digestible nutrients. CLOVERS, especially Ladino clover, or other fine-stemmed, leafy legumes are preferred.

Cereal grains can be used to advantage in spring and fall months. Poultry raisers sometimes grow BEANS, or ARTICHOKES to provide shade for the birds.

Alfalfa makes an ideal crop for the summer; it makes the best grazing when it is cut several times. When once established, alfalfa will tolerate considerable grazing unless the weather is unseasonably dry. If succulent green feed is not available, a good grade of baled alfalfa or alfalfa leaf meal can be used for poultry.

Ladino clover is probably the best crop for an irrigated green feed plot where it is cut and carried to the birds. It is also a good crop for chickens and turkeys to graze if sufficient moisture is available. Lawn clippings are frequently used as green feed for birds in confinement. They should be short and fresh; if lawn clippings are too long and tough they may cause impaction of the crop. SWISS CHARD is an easy crop to grow on well-drained soil; the best way to feed it is to cut and carry it to the birds.

When the cereals WHEAT, OATS, and BARLEY are used for feeding, the birds are often allowed to harvest them as they ripen. Turkeys are more effective in harvesting a grain field than chickens.

Permanent grass pastures are frequently used for poultry range. Some of the tall growing grasses, such as ORCHARDGRASS, ALTA FESCUE, and TALL OATGRASS, and some of the short grasses (MEADOW FOXTAIL, CREEPING RED FESCUE, CHEWINGS FESCUE) and some BENT GRASSES as well as PERENNIAL RYEGRASS (an intermediate), are useful. One or more of the grasses should be mixed with a legume, e.g., SUBCLOVER, LOTUS, HOP CLOVER, RED CLOVER, or ALSIKE CLOVER.

Some of the best grasses to make a heavy sod pasture are creeping red fescue, chewings fescue, alta fescue, perennial ryegrass, and COMMON RYEGRASS.

COWPEAS and SOYBEANS may be used for summer grazing; turkeys or chickens will eat either readily. For turkeys it is best to plant cowpeas and soybeans in rows so that they can walk down the rows and pick off the leaves.

KALE is also used as a green feed crop,

especially during the fall and winter months. It is usually cut and carried to birds in confinement.

CARROTS are a good, succulent feed for chickens and turkeys. They are high in vitamin A and make a desirable green feed substitute, especially for breeding stock during the fall and winter months. They should be allowed to dry and be placed in a cool, dark, well ventilated storage room.

In the Cotton Belt, *geese* are often turned into cotton fields to eat JOHNSON-GRASS. This is good practice because geese are not bothered by wet weather, and their gains represent profit.

Other crops which may be used to supply grazing are bluegrass, REDTOP, Bermuda grass, DALLISGRASS, ITALIAN RYE GRASS, RYE, CRIMSON CLOVER, WHITE CLOVER, and RAPE.

Silage may also be fed, at the rate of 3 to 4 lb. per 100 laying birds per day; turkeys can be fed from 10 to 12 lb. per 100 birds per day.

> NOTE: Poultry, and particularly turkeys, do not care for VETCH, soybeans, or cowpeas when other green feed is available.

An important advantage of a "poultry range" i.e., any wide expanse of countryside, is that young birds can be separated from older birds on the farm which is very important from the standpoint of disease and parasite control. The out-door range provides also other well-recognized advantages, such as fresh air, direct exposure to sunshine, and pasturage which includes succulent green feed, bugs, insects, worms, and the like. The developing pullets on the range should receive, in addition, a good growing mash or a grain, oyster shell, and plenty of good water.

Poultry *manure* is very high in nitrogen. If it is properly distributed, it provides a good fertilizer for pastures and green feed crops. Chickens and turkeys will produce from 1 to 2 lb. fresh droppings for 1 lb. feed consumed. Market turkeys are on range from four to five months, and range-reared pullets about three months. During these periods, 100 growing turkeys or about 400 pullets will produce a minimum of 3½ tons poultry manure.

Dangers: Some of the dangers of grazing animals on pastures or ranges are BLOAT; PRUSSIC ACID POISONING; POISONOUS PLANTS; WILD WINTER PEA POISONING; SLAG POISONING; and DALLISGRASS POISONING; such conditions as FOOT ROT, SHIPPING FEVER, GRASS TETANY, and losses from internal parasites; and a great variety of diseases and insect pests attacking the various species of pasture and range plants. Predators are more likely to kill pigs, lambs, and birds on the range than in confinement.

References: C.15; H.35; W.29; E.14; H.36; U.9; B.21; C.16; A.7; S.28; H.37; S.27; W.28; P.21; M. 43; G.16; G.17; M.44; L.12; D.14; K.10; B.22.; S.11.

See also DROUGHT; DUST STORMS; RANGE MANAGEMENT.

GRAZING MANAGEMENT.
 See GRAZING; PASTURE MANAGEMENT.

GRAZING REGIONS.
    See PASTURE PLANT.

GRAZING SEASON.
    See PASTURE MANAGEMENT.

GREASEWOOD (*Sarcobatus vermiculatus*), also called *black greasewood*, is a perennial spiny shrub which grows in the arid areas of the West. When eaten in small amounts, it has some value as forage. However, when sheep eat 1½ lb. leaves in a few minutes, they soon become depressed and develop kidney lesions.—*See also* POISONOUS PLANTS; RANGE PLANTS.

GREAT BEARDLESS BARLEY
 = *Horsford. See* SIX-ROWED BARLEY.

GREAT NORTHERN BEAN—one of the COMMON BEAN varieties—is grown on irrigated land. The trailing plant has white flowers and white seeds.—*See also* FIELD BEANS.

GREELEY is one of the intermediate-types grain SORGHUMS.

GREENBUG (*Toxoptera graminum*) is a little plant-louse that sometimes causes great losses to WHEAT, CORN, OATS, grain SORGHUMS, and other small grains.

GREEN CLOVERWORM (*Plathypena scabra*) frequently causes damage to SOYBEANS by eating the foliage.

Control: Apply DDT dust to the infested plants.

Greasewood. Twig and flowers. (D.19.)

GREEN CROPS. *See* GRAZING.

GREEN-FLOWERED PEPPERGRASS
= PEPPERGRASS.

GREEN GRAMS = *green mungbeans. See* MUNGBEANS.

GREEN MANURE is any crop (prefer-ably of leguminous plants) grown and plowed under to improve the soil by supplying it with primarily organic matter.

Widely used green manure crops are CANADA FIELD PEAS, VETCHES, BUR-CLO-VERS, LESPEDEZAS, CRIMSON CLOVER, SOY-BEAN, and RYE.

GREEN MUNGBEANS are varieties of the MUNGBEAN.

GREEN NEEDLEGRASS (*Stipa viri-dula*), known also as *feather bunchgrass*, is a rather coarse, leafy, native perennial adapted to the western states. It is most abundant on the upland prairie and ranges of the northern Great Plains. Green needle grass is well adapted to most soil types but

makes its best growth on the sandier soils.

Green needlegrass grows to a height of 3 ft. Its leaves, mostly basal, vary in width from ¼ to ½ in., and in length from about 8 to 12 in. The seed heads are com-pact PANICLES about 4 to 8 in. long. Seed SPIKELETS have bent awns about 1 in. long that are conspicuous, but not nearly so troublesome to grazing animals as the awns of other species of NEEDLEGRASSES, especially NEEDLE-AND-THREAD GRASS. Green needlegrass has rather fibrous, deep penetrating roots, which accounts for its wide adaption.

Green needlegrass. (H.1.)

Growth starts early in the spring and continues into the fall if enough moisture is available. This grass makes excellent recovery after grazing or clipping and pro-vides good pasture forage for all classes of livestock. Hay of excellent quality may be produced. If the plants are permitted to stand, fairly good winter grazing is fur-nished.

Green needlegrass is a good native species for use in revegetation. The young seedlings are vigorous and fairly resistant to drought and insects. Customarily, the grass is seeded in mixture with one of the adapted species.

Variety: Plant selection and breeding with this species has resulted in the development of a new variety, named *green stipa grass*, which has excellent seedling vigor, a high degree of disease resistance, and good yields of forage and seed.

Reference: H.1.

See also GRASSES.

GREEN RUSSIAN OAT.
                              See COMMON OAT (variety).

GREEN SILAGE. *See* SILAGE.

GREENSTEM PAPERFLOWER (*Psilostrophe sparsiflora*) is poisonous to sheep; it contains an as yet unknown toxic principle. If sheep eat the leaves or flowers of this plant early in spring or late in fall, when other feed is scarce, such symptoms of poisoning develop as depression, weakness, and emaciation.—*See also* POISONOUS PLANTS.

GREEN STIPA GRASS.
                See GREEN NEEDLEGRASS (variety).

GREEN WEIGHT FOR FORAGE.
                                        See SILAGE.

GRIMM ALFALFA is one of the VARIEGATED ALFALFAS.

GRINDING OF FEED MATERIAL is performed with various types of grinding machines, i.e., *grinders* or such *mills* as hammer, roller, and attrition mills, as well as old French burr mills.—*See also* GRAIN GRINDING.

GRIT is obtained from insoluble, inert material, especially from granite, fine gravel, or pebbles. Since birds do not have teeth, they must grind the swallowed feed in their gizzards. Grit greatly aids in the grinding process, thus increasing the digestibility of other than all-mash feed.—*See also* MINERAL FEEDS.

GROATS are the hulled kernels of grain. *Oat groats, cut oat groats, cracked oat groats, ground oat groats,* and *undried oat groats* (better known as *hulled oats*) are used as feedstuffs.—*See also* OATS.

GROHOMA is one of the grain SORGHUMS.

GROIT. *See* COWPEA (variety).

GROSS ENERGY. *See* ENERGY.

GROSS MORPHOLOGY.
                              See MORPHOLOGY.

GROUND ALMOND HULLS, or *almond hull meal,* is a tentatively accepted feedstuff; it is obtained by drying and grinding that portion of the almond fruit which surrounds the nut. It shall be reasonably free of nutshell and other foreign material.

Reference: F.6.

See also PLANT BY-PRODUCTS; MISCELLANEOUS PRODUCTS.

GROUND BARLEY. *See* BARLEYS.

GROUND BEETLES form an INSECT family with many beneficial species; one of them is the CATERPILLAR HUNTER. These beetles have long legs; most of them are shining black, but some species are blue, green, brown, or spotted. Both the adults and the larvae hide during the day under stones and debris and come out during the night to feed on other insects, e.g., on VELVETBEAN CATERPILLARS and on their pupae and eggs.

GROUND CORN. *See* CORN.

GROUND CULL BEANS.
                              See FIELD BEANS.

GROUND GREEN MUNGBEAN SEED, the ground beans of MUNGBEAN, is used in concentrate mixtures fed to dairy cows.

GROUND LIMESTONE is mostly CALCIUM CARBONATE. *Dolomitic limestones* contain also MAGNESIUM, and most limestones contain varying amounts of TRACE ELEMENTS.—*See also* MINERAL FEEDS.

GROUNDNUT = PEANUT.

GROUND OAT GROATS. *See* OATS.

GROUND OATS. *See* OATS.

GROUND ROUGH RICE. *See* RICE.

GROUNDSELS, like RAGWORT, belong to the genus *Senecio.* Many of them contain poisonous ALKALOIDS; e.g., *broom groundsel, Riddell groundsel, threadleaf groundsel,* and *lambstongue groundsel.*—*See also* POISONOUS PLANTS; SENECIO.

GROUND SOYBEANS. *See* SOYBEANS.

GROUND VELVETBEAN AND POD.
                              See VELVETBEANS.

GROUND WHOLE PRESSED COTTONSEED. *See* COTTONSEED.

GROWER RATIONS.
                              See POULTRY RATIONS.

Groundsel. (D.19.)

GROWTH, i.e., the increase in muscle and skeleton size and length of farm animals, is the basis of meat production. The limit of growth is fixed by inherited characteristics, but proper feeding enables the animal to reach its full size. ANTIBIOTICS, HORMONES, and some drugs (e.g., ARSONIC ACIDS), may also increase growth rate by helping to reduce disease or parasites, or by acting on the cells. Of the NUTRIENTS needed for growth, the most important are the PROTEINS or their building blocks, the AMINO ACIDS. CALCIUM, PHOSPHORUS and other MINERALS are needed for other structures, especially for the development of the skeleton. VITAMINS, too, must be supplied in the ration to assure normal growth.—*See also* PABA; RUMEN BACTERIA; BACITRACIN; THYROPROTEIN; VITAMIN C; VITAMIN F; TRYPTOPHAN; TYROSINE; THIOURACIL; ZINC; FOLIC ACID; CITROVORUM FACTOR; FEED EFFICIENCY; STILBESTROL.

GROWTH FACTORS, UNIDENTIFIED. *See* U.G.F.

GROWTH-PROMOTING VITAMIN
= VITAMIN A.

GRUB is the LARVA of a BEETLE. It resembles a CATERPILLAR in that it usually has a pair of appendages (jointed legs) on each of the first three body segments and also in its distinctly set-off head, but it can be quickly distinguished by its lack of prolegs (short, fleshy, unjointed legs).

"*Legless grubs*" and "*footless grubs*" are terms popularly used for MAGGOTS.—*See also* WHITE GRUBS; WILD GRUBS.

GRYLLOTALPA. *Grillotalpa* spp. = MOLE CRICKETS.

GUAIAC. *See* RESIN GUAIAC.

GUAIACUM (pronounced: *gwy*-ah-*kum*). The *Guaiacum* spp. are hardwood trees which grow in tropical America. They produce RESIN GUAIAC.

GUINEAGRASS (*Panicum maximum*) is one of the PANICGRASSES and grows chiefly in Florida, but also in Texas and California. It is a tall, vigorous, moderately coarse bunchgrass which spreads slowly from creeping rootstocks. This GRASS does best on heavy, rich soil; it spreads rapidly and reseeds itself where well adapted.

Guineagrass is used for fodder and for pasturage when young, but it is too coarse for forage when mature. It may be cut once a month during the growing season in Florida.

Reference: W.16.

GULLET = ESOPHAGUS.

GURNO is one of the intermediate-type grain SORGHUMS.

GUTIERREZIA. *G. sarothrae* = BROOM SNAKEWEED.

GYP. *Brown Gyp,* or *Brown durra,* and *White Gyp,* or *White durra,* are selections of *durra* varieties which belong to the grain SORGHUMS.

GYPSUM, or *land plaster,* is native *calcium sulfate.* It occurs in lumps or as a powder. Gypsum is very slightly soluble in water. It is occasionally used as FERTILIZER to supply SULFUR (of which it contains about 18 percent); to improve the physical condition of ALKALINE soil; or, when finely ground, as a DILUENT in insecticidal dusts. Applied to peanut plantings (when needed), it may prevent BLACK POD.—*See also* CUBE; DERRIS; CRYOLITE.

H

HABARO. *See* SOYBEAN (variety).

HABERLANDT. *See* SOYBEAN (variety).

HABIT is the manner of growth of a plant.

HAEBERLE = *Early Blackhull.* *See* COMMON WHEAT.

HAILWEED = DODDER.

HAIR BALLS may form in the stomachs of equines from the hairs of the stems and heads of CRIMSON CLOVER; they are found also in ruminants that have licked loose hair. Hair balls are sometimes fatal.

Reference: H.11.

HAIRGRASS DROPSEED

= ALKALI SACATON.

HAIRWEED = DODDER.

HAIRY INDIGO (*Indigofera hirsuta*) is a relatively new crop adapted to the Coastal Plain area from Florida to Texas. In this region, where other summer-growing LEGUMES are limited, hairy indigo is of special value because it produces an abundant growth of high-quality forage that can be cut for hay, pastured by livestock, used as a cover crop, or plowed under as green manure.

Hairy indigo is resistant to ROOT-KNOT NEMATODES and grows best on sandy loam soil. It is a summer annual crop, but can be volunteered from year to year with proper attention and management. This saves the cost of annual seeding. Seed crops mature in the fall. Combine harvesting is practical. Seed inoculation is unnecessary.

The hairy indigo plants are coarse when grown singly, but in thick stands they make good forage. The leaves resemble vetch leaves, except that they are larger and covered with short hairs. The stems are moderately coarse and become woody with age. If used for hay, hairy indigo should be cut when still young.

Strains: A *late-maturing* strain of hairy indigo has been grown in southern Florida, but the need for a long growing season prohibits its use in other areas. Plantings of an *early-maturing* strain of hairy indigo are expanding rapidly in the Coastal Plain areas of the United States.

Reference: M.17.

See also INOCULATION.

HAIRY-LEAVED TREFOIL is a type of BIG TREFOIL.

HAIRY PERUVIAN ALFALFA is a strain of Peruvian ALFALFA.

HAIRY VETCH (*Vicia villosa*), also known as *sand vetch*, is probably the most important among the VETCHES. It is very winter-hardy and will stand the winter temperatures of the northern part of the United States. Hairy vetch is a winter annual LEGUME which may be planted either in the spring or in the fall. When sown in the spring it acts like a biennial, that is, it does not set seed until the second season of growth. It can be seeded in the spring either alone or with OATS for pasture, hay, or silage.

It is a viny, weak-stemmed plant. The stems attain a length of 2 to 5 ft. The stems and leaves are covered with a heavy pubescence which gives the plant a silvery appearance when it reaches the bloom stage. The flowers are blue-violet, borne in dense one-sided clusters. The pods are light green, and smooth; they contain from two to eight small spherical seeds. When ripe the seeds are nearly black.

Hairy vetch succeeds especially well on sandy loam soils, but can be grown on most well-drained soils. It often succeeds on soils where SWEETCLOVER and ALFALFA fail, because it is more tolerant to acid (lime-deficient) soils than most leguminous crops.

Hairy vetch is relished by all farm livestock. It makes good pasture, hay, and silage and is an excellent cover crop for sandy soils. As a pasture crop, hairy vetch is very valuable for sheep and cattle.

INOCULATION is essential and should be supplied particularly where hairy vetch is grown for the first time.

Dangers: Hairy vetch should not be pastured when wet because of danger of BLOAT.

Among the diseases of hairy vetch is FALSE ANTHRACNOSE OF VETCH.

Variety: *Smooth vetch* differs from hairy vetch in having fewer hairs on stems and leaves. The seed is very much like that of hairy vetch, and the two kinds are often being sold in mixture under the name "hairy vetch".

In the southern states, smooth vetch is perfectly hardy during the winter months. In the Cotton Belt it is well adapted as a winter green-manure and forage plant.

References: G.8; M.18.

See also WOOLYPOD VETCH; COMMON VETCH; HUNGARIAN VETCH; PURPLE VETCH; MONANTHA VETCH; CARD VETCH; ROUGH-PEA; WEEPING LOVEGRASS.

weather disease which disappears quickly as warm weather returns.

Halo blight forms water-soaked spots on leaves; as the spots become larger, they form irregular, clear areas. The disease is spread by insects, especially APHIDS and LEAFHOPPERS.

Control: SEED TREATMENT and crop rotation reduce the primary phase of this disease.

Reference: M.30.

HALOGENS are the nonmetallic elements CHLORINE, FLUORINE, bromine, and IODINE. —*See also* HALIDES.

HALOGETON (*Halogeton glomeratus*), also called *barilla*, is an annual which grows 2 to 24 in. high. It prefers the finer

Hairy vetch. (M.3.)

HALF-BREED. *See* HYBRID.

HALIDES are compounds of metals or organic groups (radicals) with HALOGENS.

HALL = *Goens*. *See* COMMON WHEAT.

HALO BLIGHT is a bacterial disease caused by *Pseudomonas* spp. The BACTERIUM *P. coronafaciens* affects OATS, and the *P. phaseolica*, KUDZU. Like the related BACTERIAL STRIPE OF OATS, it is a cool-

Halogeton. (D.17.)

heavily alkaline soils. Halogeton has red stems, blue-green leaves and foliage, and fleshy finger-like leaves tipped with a single hair. In the fall the plants take on a reddish coloration.

Halogeton, rich in oxalate, is a stock poisoning plant and not very palatable to sheep except in unusual circumstances. It occurs on the *winter ranges* in Utah, Nevada, Idaho, Wyoming, Montana, California, and western Colorado.

Halogeton is one of the forbs which rapidly invade disturbed areas along roadways and depleted ranges.

Reference: H.41.

See also RANGE PLANTS; POISONOUS PLANTS; OXALIC ACID.

HALTICUS. *H. citri* = GARDEN FLEA HOPPER.

HANNCHEN.

    See TWO-ROWED BARLEY (variety).

HANSEN WHITE SIBERIAN.

    See PROSO (variety).

HARBIN is a variety of KOREAN LESPE-DEZA.—*See also* ANNUAL LESPEDEZAS; LES-PEDEZAS.

HARDIGAN. *See* VARIEGATED ALFALFAS (variety).

HARDING GRASS (*Phalaris tuberosa* var. *stenoptera*), an outstanding forage grass in Australia, is a perennial which gives good yields of forage in places where winter weather is relatively mild and soils are fertile. This canarygrass resembles REED CANARYGRASS in general appearance—it is a tall bunchgrass with a loose, branching base—but differs from it principally in that the seed heads are more compact and the rootstocks spread less. Harding grass, a winter grower, is drought-resistant, but grows best on moist, heavy soils.

Reference: H.1.

See also CANARYGRASSES; GRASSES.

HARDIRED.

    See COMMON WHEAT (variety).

HARDISTAN is one of the Turkestan ALFALFAS.

HARDPAN is a hardened or "cemented" subsoil, often composed of clay and sand; it prevents the penetration of plant roots and absorption of water.

HARD RED SPRING WHEAT.

    See WHEATS; COMMON WHEAT.

HARD RED WINTER WHEAT.

    See WHEATS; COMMON WHEAT.

HARMAN. *See* SOYBEAN (variety).

HARMEL PEGANUM (*Peganum harmala*) is a plant whose fruits, leaves, and stems are toxic to sheep and cattle. Characteristic effects of poisoning with this ALKALOID-containing POISONOUS PLANT—which is consumed only when there is a scarcity of desirable feed—are nervousness, inco-ordination, and paralysis.

HARMOLITA. *H. grandis* = WHEAT STRAWWORM; *H. tritici* = WHEAT JOINT-WORM.

HARROW is an implement used for seed-bed preparation. It breaks clods, stirs the soil, smooths the field, and kills weeds.

To harrow means: to till the soil with a harrow.

HARVESTED CROPS. *See* CROPS.

HARVESTER ANT. *See* RED HARVESTER ANT; WESTERN HARVESTER ANT.

HASTINGS PROLIFIC = *Leap.* *See* COMMON WHEAT.

HAWKEYE. *See* SOYBEAN (variety).

HAY is among the most valuable forage crops. It is herbage made from GRASSES, small grains, LEGUMES, and a few other fine-stemmed plants by cutting and curing.

Efficient farm and ranch management, and farm or ranch practice, include the cutting and storing of hay at the proper time for use during winters and in times of drought or feed shortage.

Young grasses and the hays made from them provide more NUTRIENTS per acre than mature grasses. Seasonal trends and major differences in composition occur as the season advances; crude protein and phosphorus contents decline, while crude fiber and manganese contents increase. Other constituents are variable.

The period in which hay is cut affects the digestibility of the nutrients of the hay. This is particularly true of protein and to a lesser degree of carbohydrates. In general, early cut hay is more palatable and digestible than late cut hay; but quantity is often as important as quality. Hay may be richer than needed (by an animal) to obtain a satisfactory total production. In order to secure the largest amount of all nutrients, without seriously sacrificing the quality of the hay crops, grasses and legumes had best be cut at the following periods:

TIMOTHY—when the bloom first shows and not later than full bloom.

Mixtures of timothy and other grasses—not later than full bloom.

Cereals—milk to early dough stage.

LESPEDEZAS—not later than full bloom.

ALFALFAS—first crop, starting not before the plants are one-tenth in bloom and finishing not later than when they are in the one-fourth-bloom stage; bloom may be somewhat more advanced (to three-fourth) for the second and third cuttings.

SUDANGRASS and JOHNSONGRASS—as soon as the first seed heads begin to form, or even a little earlier.

COWPEAS—when the first seed pods begin to ripen.

CLOVER (such as RED CLOVER, ALSIKE CLOVER, LADINO CLOVER, and other TRUE CLOVERS)—closer to the half- than the full-bloom stage.

SWEETCLOVER—before it blooms. Because of its growth habits, biennial sweetclover should not be cut closer than 5 or 6 in. to the ground, except for the last cutting.

PRAIRIE GRASSES (for prairie hay)—at pre-bloom stage.

Other important hay crops are SOYBEANS, PEANUT VINES, SMALL GRAINS (cut green, especially OATS), and sorgo (forage SORGHUMS); in addition, such tame hays as BARNYARDGRASS, BROME GRASSES, CRESTED WHEATGRASS, RYEGRASSES, VETCHES, and CANADA FIELD PEAS are widely used.

Farmers have never given as much attention to the production of hay as they give other farm crops; losses in quality, therefore, have been high. High-quality hay cannot be produced unless the meadows are kept clean of weeds (such as VELVETGRASS) and other trash, unless hay plants are cut at the proper stage of maturity, and the crop is cured and stored as quickly as possible.

Foreign material is usually rejected by livestock and remains uneaten in the manger. It is therefore just and fair that the hay grade and thus, indirectly, its price should be lower when the hay contains much of this kind of material.

Fermentation, often referred to as *sweating*, occurs in hay that is baled, stacked, or mowed before it is completely cured. In fact, all hay will sweat a little after it is stored, unless it is overdried in the field. The sweating of newly harvested hay is, therefore, a natural process and may increase its palatability by softening the stems and improving the aroma. Hay that has been properly field-cured before being stored will go through a moderate sweat in the bale, stack, or mow with little or no loss of green color.

If hay is stored when it contains too much moisture (either sap, rain, or dew), excessive heating will take place and the hay will lose much of its green color and may become musty and moldy. Hay that is severely discolored from stack or mow sweating is known as *stack-burnt hay* or *mow-burnt hay*. Baled hay which sweats excessively may have discolored cores in the center of the bale, whereas the outside, where the air has prevented heating, will not be discolored.

Fermentation of wet or undercured hay may cause high temperatures to develop. Hay that is warm to the touch and gives off a strong, sour odor, is called *heating hay*, or *hot hay*. The temperature of hot hay occasionally becomes high enough to cause spontaneous ignition.

Hay-making methods necessarily vary with the kind of hay, the local climatic conditions, the farm conditions and buildings, and the labor and machinery available. A few hints on how to avoid some of the most common errors made in curing and storing hay, which are reflected in the low quality of the hay produced, are discussed below:

Right *moisture content* is important to preserve hay quality. Hay containing 22 to 25 percent moisture can be stored without danger of heating or spoiling. Two rule-of-thumb methods are often used by farmers to determine whether the hay is cured sufficiently for storage: (*a*) Twist a wisp of hay in your hands. If the twisted hay is tough and there is evidence of moisture where the stems are broken, the hay is considered too sappy for safe storage; if the stems are slightly brittle when broken and there is no evidence of moisture when the stems are twisted, the hay can be stored without danger of spoilage. (b) Scrape the outside of the stems of clover hay with your fingernail. If the outside bark can be peeled from the stem, the hay is considered undercured; if it does not peel off, the hay is dry enough for storage.

Rain damage, while the hay is in the swath or windrow, is the hay producer's greatest "bugbear." Summer showers or sudden, heavy rains sometimes spoil the plans of the best farmer. One practice that should always be followed is to cut only that quantity of hay which can be handled with the available crew before it becomes overdried.

In many areas it is common practice to cure hay completely in the *swath* and to rake it into windrows shortly before it is baled or hauled. On bright, hot days this method is conducive to swath bleaching (a loss of green color and carotene) and overdrying of the leaves of clover that are shattered and lost when the dry hay is raked. These losses can be prevented to a considerable extent by use of the side-delivery rake, especially when heavy mixtures of clover are being handled.

Hay must be somewhat drier to be baled from the windrow than to be stored loose. The period during the day when the hay is just right for *windrow baling* without serious shattering is rather short. Hay that is baled before it is dry enough is likely to heat and sweat in the bale; hay that is too dry when it is baled will be broken and the leaves will be badly shattered. Hay that is baled from the windrow should be piled on edge and spaces left between the bales so that air can circulate around them. Baled hay that is heating may be reconditioned through aeration, but in most cases where it is distinctly hot, it will become musty and moldy.

Hay is often stacked by hand, in tall, narrow *stacks* about a central pole. These stacks usually contain 2 to 4 tons of hay. It is poor practice to store hay in such small stacks because so much of it is ex-

posed to weather. Stacks holding 10 to 15 tons can be built easily with any of several types of hay stackers. The percentage of weather-damaged hay to total stack tonnage is much less in large than in small stacks. The risk of damage to the center is small if the stack is built with side-wall bulges 3 to 4 ft. above the ground level, and is drawn to a peak at the top. In areas of heavy rainfall, straw or grass hay may be spread over the top to assist in shedding the rain. A layer at least 1 ft. thick should be used for this purpose.

In some localities, hay is stored in *barns* that are used to house the livestock or that are used entirely for hay. Barn-stored hay will not be damaged further by weather if the barn has a good roof, but great care must be taken to control the moisture content and thus avoid loss from fire caused by spontaneous ignition.

When artificial HAY DRYING is used—e.g., barn curing by placing partly cured hay in barns equipped with air ducts on the floor of the mow through which air can be forced—the hay should be partly cured in the field unless it must be taken up to prevent rain damage.

The practice of *chopping* hay at the time of storage to reduce storage space and labor involved requires power machinery. Two to two and one-half times as much chopped hay can be stored in the same space as that required for long hay; chopped hay, also, can be removed from the mow for feeding much easier, but barns that have been built for long hay must be strengthened if they are to be used for chopped hay.

In good hay management the following five points are important: (1) Availability of a fertile soil sown to an adapted species; (2) cutting of hay at proper stage of maturity; (3) curing so as to avoid leaf loss through shattering and leaching from rain; (4) storing at proper moisture content to prevent heating; (5) feeding hay to an animal that can make good use of it—a full feed of U.S. No. 1 leafy alfalfa is wasted on a mature beef cow, and a dairy cow cannot produce milk on No. 4 Timothy.

References: H.34; N.4; M.41; M.42.

See also HAY MEASURING; HAY GRADING;

BLOAT; OAT HAY POISONING; SWEETCLOVER HAY POISONING; GRAZING; PASTURES; PASTURE MANAGEMENT; MARSH HAY; HAY YIELDS.

HAY CROPS.
See HAY; SILAGE CROPS.

HAY CROP SILAGE.
See SILAGE CROPS.

HAY DRYING (*artificial*)—with the help
of driers—is one means of greatly reducing the serious losses from rain damage of HAY feed value suffered throughout the country each year. Leaching, bleaching, and leaf shattering can prevent the farmer from putting more than one-half the original feed value of the hay crop in the mow. Heavy rains on hay, waiting to be cured in the field, often result in nearly the total loss of the crop. Although artificial drying in the manner that is practical on the farm cannot eliminate all weather hazards, it can, by proper management, reduce field losses to a minimum.

> NOTE: Tests run by experts of the U.S. Department of Agriculture show that with good weather field-cured hay is as good as barn-cured hay; the big difference was found with hays made on cloudy or wet days.

Many types of driers and drying methods are used commercially.

A *drier* is usually any mechanical arrangement for removing moisture from the crop by the use of induced or forced ventilation of unheated air. A *dehydrator* employs the use of supplemental heat, applied either directly to the product itself or to the air forced through or over it. Since most farm driers are used to finish the drying process after most of the water

has been removed in the field, they are aften referred to as *"hay finishers."*

Where *buildings* are available, it is possible to use them for mow drying with few modifications. They should have a good, tight roof for rain protection, and large windows or doors for exhausting the moisture-laden air. It is essential that the floor be reasonably airtight, in order to keep the air in the *ducts* from escaping downward instead of being forced up through the hay. Where a barn is less than 20 ft. wide, a single open-type center duct would normally require the least labor and materials. In case the mow is 20 ft. to 30 ft. wide, a side-main system is suggested.

Hay can be dried artificially either long, chopped, or baled. Many drying systems will handle it in any of the three forms without alteration. All driers must be loaded evenly over the air distribution system. Heavy fork loads or sling loads of long hay must be torn apart if they are to dry properly. Chopped hay should be leveled off without tramping.

Various types of hay drying *fans* are used to provide sufficient push to overcome the resistance of air movement through the duct system and through the hay.

Whether *heating systems* for supplemental drying *heat* (in addition to natural heat generated by the hay) should be provided depends upon such factors as how much field wilting or curing is done, how rapidly the hay is to be dried, and how much drying is to be done in humid or rainy weather. There are numerous types of heating and blower units on the market. No heating unit should be used unless it is equipped with a high-temperature cut-

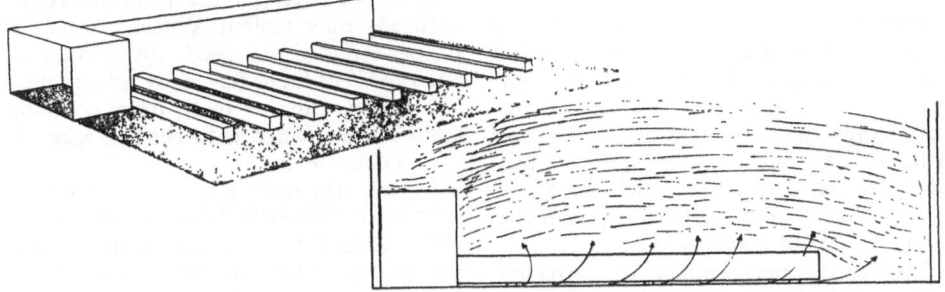

Hay drying. Left, lateral-duct system with side main-air distribution duct. Right, section view showing entrance of air into the hay stack. (K.9.)

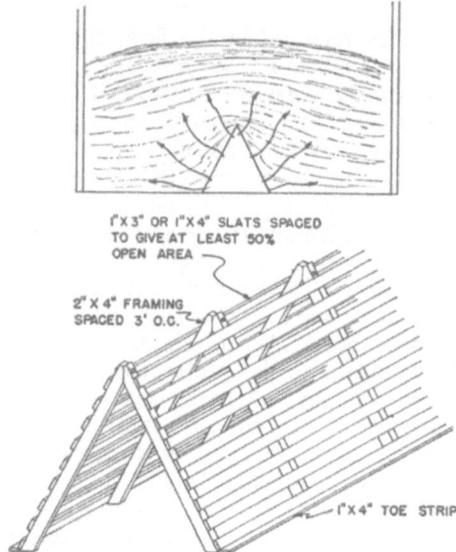

I"X 3" OR I"X 4" SLATS SPACED
TO GIVE AT LEAST 50%
OPEN AREA

2"X 4" FRAMING
SPACED 3' O.C.

I"X 4" TOE STRIP

Hay drying. Top, cross section of A-frame duct showing its relative position in the mow and the region of air discharge. Bottom, a section of the A-frame duct showing method of construction; first 6 ft. of A-frame should be tightly sheathed. (K.9.)

off which will shut off the fuel supply in case of fan failure.

The most important drying systems are as follows:

1. Mow drying, a process in which the major portion of the water originally in the green forage is evaporated in the field during a short wilting period, with only the remaining 10 to 20 percent of excess moisture being removed by the drier. Hay should be field-wilted to 35 to 40 percent moisture (from the 75 to 80 percent moisture content at the time of cutting) before it is placed in a mow drier.

Under average conditions long hay can be safely stacked about 8 ft. high in the drier and chopped hay about 6 ft. high. After the first stack is dried, a second batch may be placed over it to a depth of about 6 ft. for long hay and 4 ft. for chopped hay. The recommended depth for baled hay is 7 to 9 ft. and the total depth of the stack should not exceed 12 ft.

Mow-type driers should be provided with from 15 to 20 cu. ft. air per minute for each square foot of drying area. Air

velocities should be held below 1,000 ft. per minute. High-velocity air tends to bypass the hay near the main duct. Where the hay is all cured from one cutting, a simple "A" duct, with air openings over its entire surface, is adequate.

2. Circular hay keeper is a useful wooden silo-type drier, generally 24 ft. in diameter and either 20 or 30 ft. high. Capacities will vary from 20 to 50 tons. Hay is blown or elevated through a dormer provided in the roof. The central flue or shaft serves as a chute when the hay is pitched from the top of the stack into a feed cart at the base of the shaft.

The keeper is able to do a fair job of drying by natural draft. As heating takes place in the hay, the air in the center duct becomes warmer and tends to rise up and out of the duct, which acts as a chimney. The air rising up the chimney creates a pull which slowly draws the warm air out of the hay. Fresh air in turn is pulled into the hay through the cracks left between the outside vertical wall boards.

The natural-draft keeper can be converted into a forced-draft drier, using either cold or heated air. The blower and motor are located in the doorway, with the blower or fan so arranged as to blow air into the short alley and up the center chute. A simple plug made of plywood is placed in the chute about 4 ft. below the level of the hay to prevent the air from escaping up the chute and force it to pass horizontally through the hay and out the cracks in the outer wall.

If the air is to be heated, a heating unit can be placed in a duct or tunnel ahead of the fan.

3. Stack drying in the field can be done with any duct system. Where the hay is to be threshed for seed, the temporary low-type "A" frame has many advantages. The frame can be built from common lumber and covered with chicken wire.

Reference: K.9.

See also HAY GRADING; HAY MEASURING.

"HAY FINISHERS." *See* HAY DRYING.

HAY GRADING, on the basis of the Official Standards for Hay, is done by licensed inspectors who have the right to issue official hay grade certificates.

The chief factor influencing the established grades is the *foreign material* in HAY. Other factors influencing the official grades are: *color, maturity, fineness,* and (for legume hays) *leafiness.* A maximum of 10 percent of foreign material is permitted in grade *U.S. No. 1;* 15 percent in grade *U.S. No. 2;* and 20 percent in grade *U.S. No. 3;* hay that contains more than 20 percent but not more than 35 percent foreign material is placed in the *Sample grade.* Sample grade also includes good hay which has an objectionable odor, is undercured, hot, wet, musty, moldy, caked, badly weathered, badly broken, badly overripe, or very dusty.

Special grade designations—such as Extra Leafy, Leafy, Extra Green, Green Fine, and Coarse—are added to the numerical grades for some of the officially recognized types.

The official standards include 11 groups of hay; i.e., (1) *alfalfa* and alfalfa mixed hay; (2) *timothy* and *clover* hay; (3) *prairie hay;* (4) *Johnson* and Johnson mixed hay; (5) *grain, wild oat, vetch,* and grain mixed hay; (6) *lespedeza* and lespedeza mixed hay; (7) *soybean* and soybean mixed hay; (8) *cowpea* and cowpea mixed hay; (9) *peanut* and peanut mixed hay; (10) *grass* hay; and (11) *mixed hay.*

Choice hay is more valuable when converted to pounds of beef than as a cash crop, except where other conditions cause livestock feeding to be unprofitable. *Poor hay,* on the other hand, can be utilized after seasons of unfavorable weather when the crop is overmature or damaged by rain. Good results have been obtained with a system that calls for full feeding of EAR CORN CHOPS, OIL MEAL, MOLASSES, VITAMIN A or CAROTENE, MINERALS, SALT, and all the poor hay the cattle will eat.

References: N.4; K.8.

See also HAY DRYING; HAY MEASURING.

HAY MEASURING is the basis of estimating the weight of HAY: farmers, ranchers, and hay dealers buy and sell large quantities of hay in the stack. Hay is sold by measure because in most of the valleys where range stock is wintered, no scales are available; even if they were, it would sometimes be impracticable to weigh the hay. Loading it onto a wagon or sled, and hauling it to the scales, would often add an expense to the feeding operations that might be greater than any gain from buying the hay by weight instead of by measure. In most instances the hay is either fed directly from the stack in racks arranged about the stack, or scattered from a wagon or sled about the pasture or feed lot.

The volume of an *oblong* or *rectangular stack* is equal to its average length = L, multiplied by the area of the cross section; the latter is computed from two measurements: W = width of the stack at the ground and O = "over"—the distance from the ground on one side over the stack to the ground on the other side.

Hay stacks can be divided into three types based on shape; the volume formula for each type is as follows:

For *low, round-topped stacks*
$$(0.52 \times O - 0.44 \times W) \times WL;$$
for *high, round-topped stacks*
$$(0.52 \times O - 0.46 \times W) \times WL;$$
for *square, flat-topped stacks*
$$(0.56 \times O - 0.55 \times W) \times WL.$$

Example: The volume of a rectangular stack of the high, round-topped type that is 20 ft. wide, 45 ft. over, and 50 ft. long is $(0.52 \times 45 - 0.46 \times 20) \times (20 \times 50) =$ 14,200 cu.ft.

A simple formula does not exist for round stacks, but when the circumference is between 45 and 98 ft. and the over between 25 and 50 ft., it can be read from an official table of which every fifth column is reprinted here.

The volume of stacks having circumferences or overs greater or smaller than those given above can be calculated by using the following formula: Volume = $(0.04 \times O - 0.012 \times C) \times C^2$. In this formula C equals the circumference (distance around the stack at the ground) and O equals the over. It is usually advisable to take two over-measurements at right angles to each other and then average them.

Example: The volume of a round stack that is 100 ft. in circumference and has an average over of 60 ft. is $(0.04 \times 60 - 0.012 \times 100) \times (100)^2 = 12,000$ cu. ft.

VOLUME OF ROUND STACKS OF HAY OF SPECIFIED DIMENSIONS

Circumference (ft.)	Indicated volume in cubic feet when the over is:					
	25 ft.	*30 ft.*	*35 ft.*	*40 ft.*	*45 ft.*	*50 ft.*
45	825	....	....	....	....	....
50	900	1,560	2,215	....	....	....
55	975	1,650	2,345	3,060	....	....
60	1,050	1,740	2,480	3,280	4,150	....
65	1,125	1,830	2,615	3,495	4,490	5,485
70	1,200	1,925	2,745	3,715	4,825	5,955
75	....	2,010	2,875	3,935	5,165	6,430
80	....	2,105	3,010	4,150	5,510	6,905
85	....	....	3,140	4,365	5,850	7,375
90	....	....	....	4,585	6,195	7,845
95	....	....	....	4,805	6,540	8,320

WEIGHT OF HAY IN A STACK
Showing number of cu. ft. of hay that weigh a ton

Kind of hay	Length of time in stack	
	30 to 90 days	over 90 days
Alfalfa............	485 cu. ft.	470 cu. ft.
Timothy and timothy mixed...	640 cu. ft.	625 cu. ft.
Prairie............	600 cu. ft.	450 cu. ft.

Many factors affect the *density* of hay in the stack and therefore the number of cubic feet required for a ton of hay. The factor that causes the greatest variation probably is moisture in the hay at stacking time. Tough or slightly undercured hay will settle and become more compact than very dry and overcured hay. Other factors, like texture and foreign material, may also affect the density. However, the figures shown in the table can be used with fairly good results when figuring the *weight of hay in a stack*.

Reference: H.33.

HAYSEED. *See* SOYBEAN (variety).

HAY YIELDS (for 1954) by type and leading states are listed in the table.

HEAD is a headlike formation, especially a rounded, congested inflorescence or seed cluster.

HEAD BLIGHT. *Fusarium head blight* = WHEAT SCAB.

HEAD DISEASES of the SORGHUMS are caused by FUNGI. The most important head diseases are COVERED KERNEL SMUT, HEAD SMUT, LOOSE KERNEL SMUT OF JOHNSONGRASS, and LOOSE KERNEL SMUT OF SORGHUM.

HEADER is a machine which cuts the stems of grain crops a few inches below the heads and elevates the latter into the *header box* (also called *header barge*).

HEADLESS DEFECT prevents the formation of heads, and sometimes also of side branches, on the stalks of milo, darso, and their hybrids. It is a hereditary defect which often is mistaken for PERICONIA ROOT ROT.—*See also* SORGHUMS.

HEAD RICE. *See* RICE.

HEAD SMUT is a term loosely applied to a variety of smuts that occupy all or part of the flowering structures, but do not replace the seeds themselves; examples are COVERED KERNEL SMUT, LOOSE KERNEL

HAY YIELDS

Hays	Million acres	Acre yield tons	Leading states
Alfalfa..................	22.7	2.21	Wis., Neb., Minn., S. D., Iowa.
Clover and timothy.......	19.7	1.42	N. Y., Iowa, Ohio, Penn., Wis.
Wild hay...............	14.4	.85	Neb., S. D., N. D., Mont., Minn.
Lespedeza..............	5	1.05	Mo., Tenn., Ky., N. C., Va.
Cereal grains............	3.1	1.2	Calif., Mo., Mont., Ore., Okla.
Soybean hay............	1.1	1.2	N. C., Miss., Ark., Ill., Mo.
Cowpea hay............	.5	.8	
All other types of hay.....	7.6	1.1	

SMUT, and (more specifically) the true head smut caused by the FUNGUS *Sphacelotheca reiliana* which attacks forage and grain SORGHUMS, CORN, and RUSSIAN WILD RYE. It is not common in the United States, but causes occasional damage in Kansas.

Head smut, caused by *S. reiliana*, destroys the entire head, transforming it into a large mass of dark-brown powdery spores. The smut becomes evident at heading time, when the large gall bulges out of the boot. At first the gall is covered with a whitish membrane, which soon breaks and allows the spores to be scattered by wind and rain to the soil and plant refuse, where they pass the winter. The following spring and summer the spores germinate and produce smaller spores of another type, which in turn, infect the sorghum plants. The fungus grows within the plant until the latter reaches the heading stage.

The head smut also destroys the tassels and ears of corn in a manner similar to the destruction of the sorghum heads. However, it seems likely that corn and sorghum are attacked by different *races* of this fungus. Head smut seems to be severely damaging only to the sorgos, durras, and varieties developed from hybrids with these two sorghum groups. Kafirs and Sudangrass show only moderate susceptibility; milo, feterita, and broomcorn are seldom, if ever, attacked. Among the sorgos, Red Amber, certain strains of Black Amber, Ellis, and varieties apparently derived from Amber crosses (including Leoti and Colman) seem to be particularly susceptible.

Control: Sanitation is the chief means of controlling head smut. If head smut of sorghum is discovered in a field, the infected plants—usually only a few—or at least the galls, should be removed and burnt before the spores are scattered. If seed from a field containing plants infected with head smut must be used for planting in an uninfested area, it should first be treated with a good FUNGICIDE.

References: L.1; F.4.

See also SEED TREATMENT; FORMALDE-HYDE SOLUTION; GRASS SMUTS.

HEAT = ESTRUS.

HEAT CANKER appears to be the pri-mary cause of the most serious losses during the seedling stage of PEANUTS. On light sandy soils the stand is sometimes so reduced as to require replanting. Radiant heat from the sun during a clear, bright afternoon is sufficient to burn the tender tissues of the stems during or shortly after emergence.

Sometimes enough tissue may collapse to cause the plant to topple over and die; more frequently only the west or southwest side of the stem is injured. The lesions may heal and the seedlings develop into a normal, or nearly normal plant. Often, however, the injured tissue becomes infected by one or more weakly parasitic fungi that may complete the destruction of the seedling at once or by and by.

Control: Running the rows from east to west, and spacing the plants closely, affords more shade. A rough, slightly lumpy surface soil about the plants reflects less heat than a flat surface. Damage may frequently be reduced by going over the field with a weeder just before the seedlings emerge.

Reference: B.10.

See also SEEDLING DISEASES.

HEAT INCREMENT. *See* ENERGY.

HEATING HAY. *See* HAY.

HEAVING is the lifting action of the soil caused by alternate freezing and thawing; some soils may heave to such an extent that they lift plants out of the ground.

HEGARI, *Early hegari*, *Hi-hegari*, Bonita (or *Little hegari*), and Combine Bonita are forage-type grain SORGHUMS.

HEIFER RATIONS. *See* DAIRY RATIONS; BEEF CATTLE RATIONS.

HEINRICH. *See* RYE (variety).

HELENIUM. *H. hoopesii* = WESTERN SNEEZEWEED.

HELIANTHUS. The *Helianthus* spp. are known as SUNFLOWERS; most of the common sunflower varieties, and the wild sunflower, belong to the species *H. annuus*.

HELICOCERAS. The FUNGUS *H. oryzae* causes REDDISH-BROWN SHEATH ROT.

HELIOTHIS. *H. armigera* = CORN EAR-WORM.

HELLBIND = DODDER.

HELMINTHOSPORIUM. The FUNGUS *H. maydis* causes SOUTHERN CORN LEAF

BLIGHT; *H. carbonum*, HELMINTHOSPORIUM LEAF SPOT; *H. gramineum*, BARLEY STRIPE; *H. sativum*, CULM ROT; *H. avenae*, OAT LEAF SPOT; and *H. victoriae*, VICTORIA BLIGHT. *H. turcicum* is the cause of SEED ROT, SEEDLING BLIGHT, LEAF BLIGHT, as well as NORTHERN CORN LEAF BLIGHT; *H. sorghicola* causes TARGET SPOT; *H. sigmoideum* var. *irregulare* is one of the causes of STEM ROT OF RICE (the other is *Leptosphaeria salvinii*); *H. oryzae* causes BROWN SPOT OF RICE as well as "ROTTEN NECK" and "PECKY RICE."—*See also* BERMUDA-GRASS.

HELMINTHOSPORIUM LEAF SPOT

is a disease of CORN and occasionally WHEAT, caused by two races of the FUNGUS *Helminthosporium carbonum*. Race I often completely destroys certain susceptible inbred lines of corn. In spite of this, farmers are not troubled by this disease because it does not occur in commercial single- or double-cross hybrids, unless two or four susceptible inbreds are combined. The disease caused by race II has never reached severe proportions. Helminthosporium leaf spot occurs in the eastern half of the United States; it thrives under humid conditions.

The disease caused by *race I* is recognized by tan lesions on the leaves. The lesions, ranging in size up to 1 in. long by ½ in. wide, are oval or circular in outline and show concentric zones within the affected area. When the disease is severe, all the leaves on a plant are virtually covered with these lesions. Spores are produced in damp weather; they are especially abundant on the leaf sheaths where they form black, velvety masses. The disease also attacks the ears of corn, causing a black rot which imparts a charred appearance to the kernels.

Symptoms produced by *race II* are recognized by oblong, chocolate-colored spots, ranging up to 1 in. by ¼ in. in size. Ear infections caused by race II are indistinguishable from those caused by race I.

Corn appears to be the only host for the fungus. No sexual stage in the life cycle of the parasite has been observed.

Control: Helminthosporium leaf spot can be controlled easily by using resistant inbred lines.

Reference: U.7.

HEMATIN. See HEMOGLOBIN.

HEMATINIC TONICS improve the blood quality; ARSENICALS and IRON compounds belong to these TONICS.

HEMICELLULOSES are complex CARBOHYDRATES, often of gummy consistency. They contain SUGARS, especially PENTOSES, and URONIC ACIDS. Hemicelluloses are ill-named, because they are not related to cellulose, except that they are fairly insoluble and not digested by non-ruminants. They are often found as reserve foods in plants, especially in root crops; CORN cobs, many HAYS, and WHEAT bran, too, are rich sources of hemicellulose.

HEMLOCK.

See POISONHEMLOCK; WATERHEMLOCK.

HEMOGLOBIN is the pigment of the red blood corpuscles. It consists of *hematin* (an organic iron compound) and *globin* (a protein) and is important because it is the oxygen carrier of the BLOOD.—*See also* OAT HAY POISONING.

HEMORRHAGE is *bleeding*, i.e., the escape of blood from its vessels due to disease or accident.—*See also* SWEETCLOVER HAY POISONING; DICOUMAROL.

HEMORRHAGIC SEPTICEMIA, also known as *shipping fever, shipping pneumonia*, and *stockyard disease*, is one of the ailments related to cattle movement. It is a widespread infectious disease that is becoming a serious problem to buyers of grazing and feeding cattle. It appears more often during the colder months of the year. Animals infected with it stop eating and have the following symptoms: a high temperature; a discharge from the nose; rapid, pneumonia-like breathing; coughing; and sometimes diarrhea.

> NOTE: All animals purchased through sales barns or traders should be vaccinated with hemorrhagic septicemia bacterin.

Sick animals should have prompt and early treatment by a veterinarian.

References: S.28; S.10.

See also GRAZING.

HEMP (*Cannabis sativa*) is an erect,

dioecious annual; only the female plants produce seed. Hemp grows 5 to 15 ft. high. The hallow stem produces branches.

While the fiber-type hemp is useful because of its bast fibers, the *drug*-plant type —also called *domestic hemp* or *Kentucky hemp*—is rich in *marijuana*.

> NOTE: The grower of hemp needs a license from the Commissioner of Narcotics.

The seeds of hemp, after being modified by killing its germination (in accordance with federal regulation), can be fed to cage birds or used for the production of oil and its by-product, HEMPSEED MEAL.

HEMPSEED MEAL is one of the OILSEED MEALS. It is obtained as by-product in the manufacture of oil from hempseed. The crude fiber content of this protein supplement is high and the palatability limited, except if mixed with other feeds.

Reference: F.8.

See also HEMP.

HEPTACHLOR, a close relative of CHLORDANE, is an INSECTICIDE of the CHLORINATED HYDROCARBON group. It forms white crystals, but the technical material commonly used contains only about 72 percent heptachlor and about 28 percent related compounds and therefore is of a soft, waxy consistency and of a tan color. It is water-insoluble, but soluble in most organic solvents.

The recommended amount of heptachlor for GRASSHOPPER control is 2 to 4 oz. per acre; the residual action lasts two to four weeks. Applied as 5-percent dust at the rate of 40 lb. per acre or as 2-percent granulated material at the rate of 100 lb. per acre, heptachlor controls the SOUTHERN CORN ROOTWORM.

Mixed into the soil at the rate of 2 lb. per acre, heptachlor (as dust, emulsion, in granular form, or in combination with fertilizers) controls WIREWORMS.

When applied by airplane at the rate of 8 oz. (of actual chemical, suspended in 10 gal. water) per acre, heptachlor controls the RICE LEAF MINER.

HERB is a flowering plant whose stem is free from woody tissue; it perishes when flowers and fruit are matured. An herb

may have an annual, biennial, or perennial root, but the stem is ordinarily annual.

HERBACEOUS means: having the characteristics of, or pertaining to, HERBS.

HERBICIDE is a chemical substance which kills WEED or any herbaceous plant; e.g., SODIUM TCA; 2,4-D; 2,4,5-T; TCA; AMMONIUM SULFAMATE; DINITRO-(ORTHO-SECONDARY-)BUTYLPHENOL.

HERBIVOROUS ANIMALS are those feeding on grass and other plants; particularly ruminants and equines.

HERD'S GRASS = TIMOTHY. This name should not be used for REDTOP.

HERMAN. *See* SOYBEAN (variety).

HERRING, a fish about 10 in. long and widely distributed over the North Atlantic, is often used in *fish meal.—See also* MARINE PRODUCTS.

HERRING OIL. *See* VITAMINS.

HERSHEY MILLET = PROSO.

HESSIAN FLY (*Phytophaga destructor*) is without doubt the most formidable insect enemy of soft red winter WHEAT. It attacks also other wheats, BARLEY, and RYE.

The flies lay their eggs in the grooves of the upper surface of the leaves of young wheat. One female may deposit several hundred eggs and infest many plants as she flies over the grain field. The eggs hatch in three to twelve days, and the small red maggots make their way down the leaf and behind the sheath, where they begin feeding on the tender tissues of the plant. The maggots are full-grown in two to four weeks. They are then glistening white, but soon turn brown, forming puparia, or "flaxseeds." Adults emerge from the overwintering "flaxseeds" in early spring to lay their eggs.

Small tillers of infested plants die; jointed tillers often break over and fall to the ground before harvest. Adults that emerge from the "flaxseeds" in stubble and in volunteer plants of harvested fields infest early fall-seeded fields, with resultant stunting and death of tillers and plants.

Control: The most important means of controlling this pest is to plant winter wheat at a date that will delay the appearance of the young wheat above ground until after the main brood of flies has emerged and died. Sound cultural practices that contribute to the vigor of the growing crop are necessary in combating the hessian fly. Chief among these practices are plowing under the stubble of the preceding wheat crop before the flies emerge from it in the early fall; prompt destruction of volunteer wheat; the sowing of good seed; and the promotion of rapid, vigorous growth by proper fertilization and planting in a well-prepared, firm seed-bed. Improved wheat varieties resistant to the hessian fly are available.

References: B.16; U.3.

See also CLUB WHEAT.

HETERODERA. *H. marioni* is a name no longer used for ROOT-KNOT NEMATODES; they are now classified as various species of the genus *Meloidogyne.*

HETEROECIOUS (pronounced: het-er-*ee*-shus), a term pertaining to the rust fungi, means: passing through the various stages of the life cycle on more than one kind of plant.—*See also* MONOECIOUS; RUSTS.

HEVEA. *H. brasiliensis* = PARA RUBBER TREE.

HEXANE, the chief fraction of the most volatile fraction of PETROLEUM, is acceptable as solvent in the manufacture of extracted SOYBEAN products.

HEXESTROL is a HORMONE-like compound; its action is that of natural female hormones. Hexestrol forms needles or thin plates which are practically insoluble in water, but soluble in vegetable oils. This drug must be used carefully so as not to cause side-reactions.

Daily oral administration of 10 mg. hexestrol, well distributed in a fattening ration, results in weight gain of cattle.

HEXOSES are SUGARS.—*See also* PENTOSANS.

HICKMAN = *Fultz.* See COMMON WHEAT.

HIGEARY = *hegari* (grain sorghum). *See* SORGHUMS.

"HIGH GEAR" = *hegari. See* SORGHUMS.

HIGHLAND IMPROVED COES is one of the intermediate-type grain SORGHUMS.

HI-HEGARI is one of the grain SORGHUMS.

HILAND. *See* SIX-ROWED BARLEY (variety).

HILARIA. *H. belangeri* = CURLY MES-

QUITE; *H. jamesii* = GALLETA; *H. mutica* = TOBOSA.

HILL RICE. *See* RICE.

HIMALAYA. *See* SIX-ROWED BARLEY (variety).

HINDU COWPEA = *Catjang. See* COWPEA.

HIRSUTE means: hairy with rather coarse, stiffish, straight, beardlike hairs.

HISTIDINE is an essential AMINO ACID which forms colorless, water-soluble crystals.

HISTOLOGY. *See* MORPHOLOGY.

HISTOSTAT
= 4-NITRO-PHENYLARSONIC ACID.

HOE DRILL is a grain DRILL best suited for use in loose soil; it turns up clods as well as trash, but does not pulverize the soil.

HOG BARLEY = *Stravropol. See* SIX-ROWED BARLEY.

HOG CHOLERA. *See* GARBAGE.

HOGGING DOWN means: turning pigs into a field to harvest the crop; e.g., hogging down (standing) CORN, (ripe) grain, such as WHEAT, RYE, or BARLEY, and stubble fields. This fattening method saves labor. However, to obtain best results in the shortest time, it is necessary in many farm areas to simultaneously supply the animals with protein-rich feed, either by allowing them access to RAPE or LEGUMES (especially SOYBEAN) pastures or to PROTEIN supplements.

HOG MILLET = PROSO.

HOHENHEIM SYSTEM.
See PASTURE MANAGEMENT.

HOLCUS. *H. lanatus* = VELVETGRASS.

HOLLAND JUMBO is a Virginia-type PEANUT.

HOLLAND STATION RUNNER.
See PEANUT.

HOLLYBROOK. *See* SOYBEAN (variety).

HOMINY FEED. *See* CORN; VITAMIN A.

HOMINY GRITS = *corn grits. See* CORN.

HOMOGENIZED CONDENSED FISH.
See MARINE PRODUCTS.

HOMOGENIZED CONDENSED FISH WITH ADDED SALT. *See* MARINE PRODUCTS.

HONEYBEE. *See* BEES.

HONEYCOMB = RETICULUM. *See* STOMACH.

HONEYGRASS = MOLASSES GRASS.

HONEY MESQUITE and *western honey mesquite* are MESQUITE varieties.

HONEY SORGO, also known as *Honey Drip*, is one of the forage SORGHUMS.

HONGKONG. *See* SOYBEAN (variety).

HOP (*Humulus lupulus*) is a tall vine producing flowers used in breweries to give beer its characteristic flavor.—*See also* BREWERS' PRODUCTS; SPENT HOPS.

HOP CLOVERS are sometimes called *yellow clovers;* they include SMALL HOP CLOVER and LARGE HOP CLOVER.—*See also* RICE; CLUSTER CLOVER; WEEPING LOVEGRASS; PASTURES; PASTURE PLANTS; GRAZING.

HOPPER. *See* THREE-CORNERED ALFALFA HOPPER; LEAFHOPPERS.

HOPSAGE. *See* SPINY HOPSAGE.

HORDEUM. Among the *Hordeum* spp., or BARLEYS, are two major species: *H. distichum* = TWO-ROWED BARLEY and *H. vulgare* = SIX-ROWED BARLEY; *H. jubatum* = FOXTAIL BARLEY.

HORMONES are organic substances produced by *ductless* GLANDS. The blood carries hormones to other organs. Although present in the organism only in minute amounts, hormones are able to accelerate or retard internal processes, especially METABOLISM. The table shows the most important hormones and their physiological action in mammals.

Of increasing economical importance are the hormone-like materials that are produced chemically and used to stimulate the GROWTH rate of farm animals and/or increase efficiency of feed utilization. Among them are DIENESTROL DIACETATE, STILBESTROL, and HEXESTROL, all of which possess sex-hormone activity. *Iodinated casein* (a thyroxine-like material) has been used to increase growth rate of swine and milk production; it should be used only in winter months, however, and extra feed must be used. THIOURACIL and other GOITROGENS have been used to increase fat deposition of hogs.

Plant hormones are substances that promote growth of roots or other parts of plants; e.g., 2,4,-D.—*See also* THYROXINE; THYROPROTEIN.

HORN. *See* TWO-ROWED BARLEY (variety).

HORSE AND MULE RATIONS. In addition to supplying the needed nutrients, horse feeds should be of high quality. HAYS should be bright, free from dust and mold. GRAINS should be of good weight

HORMONE FUNCTIONS

Hormone	Secreting gland	Principal actions
Growth (somatotrophic)	Anterior pituitary	Stimulates growth
Gonadotrophic (LH; FSH)	Anterior pituitary	Stimulates gonads
Thyrotrophic	Anterior pituitary	Stimulates thyroid
Corticothrophic (ACTH)	Anterior pituitary	Stimulates adrenal cortex
Lactogenic	Anterior pituitary	Stimulates secretion of milk
Pitocin (oxytocin)	Posterior pituitary	Stimulates contraction of uterus
Pitressin (antidiuretic)	Posterior pituitary	Causes contraction of blood vessels; controls H_2O balance
Thyroxine	Thyroid	Stimulates metabolism
Parathormone	Parathyroid	Regulates calcium and phosphorus metabolism
Insulin	Pancreas (Islets of Langerhans)	Involved in carbohydrate metabolism
Secretin	Duodenal mucosa	Stimulates flow of pancreatic juice
Epinephrine (adrenalin)	Adrenal medulla	Regulates blood pressure and carbohydrate metabolism
Corticosterols (cortin)	Adrenal cortex	Involved in metabolism of carbohydrates, water, and minerals; enables animal to overcome effects of stress
Estrogen	Ovarian follicle and placenta	Maintains female sex accessories and secondary sex characteristics; influences sexual behavior
Progesterone	Corpora lutea and placenta	Prepares female for pregnancy and lactation
Testosterone (androgen)	Testicles	Maintains male sex accessories and secondary sex characteristics

HORSE AND MULE RATIONS

	Young stock	Light horses					Draft stock	
		Idle ration	Moderate work			Lacta-tion	Med. work	Hard work
			1	2	3			
Grass hay, lb.	4–6	20	12	6	15	8	18	16
Legume hay, lb.	1–3	..	..	5	..	8	..	..
Oats, lb.	2–4	..	8	..	3	6	8	8
Corn, lb.	2	..	3	10	5	..	3	10
Bran, lb.	1	..	..	1	1	1	..	..
Oil meals, lb.	0–1	..	..	..	..	½	..	..

Iodized salt, free choice.
Mineral mixture that is high in Ca., free choice.

and free from dust. Attempts to skimp on feeds will usually lead to digestive disturbances and lowered performance.

It is customary to feed some grain to all horses, except idle mature horses from which no work is expected. All horses are fed about 1 lb. of hay per 100 lb. of body weight, and, depending on the animal's work, an additional pound of feed (either grain or hay) is added for each 100 lb. of body weight. Horses at *very hard work* will require as much as 1½ lb. grain per 100 lb. of body weight. Hay consumption of *light horses* is limited to keep them trim and active.

Since a large amount of horse feed is grain, the ratio of CALCIUM to PHOSPHORUS is usually rather low and for this reason it is important to supply additional calcium to many rations to insure formation of sound bones. It is believed that the proper balance of TRACE ELEMENTS is extremely important in the development of sound bones. It is probably no accident that the blue grass areas are noted for producing sound bones.

It is usually customary for *brood mares* to continue to do some light work or at least be exercised daily until foaling time. They will need limited amounts of grain to keep in good condition. Unless the mares are young it will probably not be necessary to add additional PROTEIN to a ration of good quality grass hay and farm grain. MINERAL supplement should be available and unless grass is available, special provisions should be made to provide green hay as a source of CAROTENE.

Some rations are given in the table.

If given the opportunity, the *young colt* will soon begin nibbling dry feed. Make sure than an adequate supply of clean grain and fine-stemmed forage is available.

After weaning, colts are continued on a ration of mixed hay and grain or if available, good quality PASTURE. Mineral supplement should be allowed free choice and a small amount of protein may be used during the first year.—*See also* RATIONS; PASTURE.

HORSEBEAN (*Vicia faba*), is related to the VETCHES, but differs from them decidedly in habit of growth.

The horsebean requires a cool season for its best development, and it is grown as a winter annual in the South where it will not winterkill. In the North it is not winter-hardy, and even in the South it cannot be grown successfully where the temperature fluctuates rapidly.

Dangers: Among its enemies is the VELVETBEAN CATERPILLAR.

Varieties: There are many varieties of horsebeans, most of which are grown for their seed. The *small-seeded* varieties are sometimes grown for green manure, but are more generally used for stock feed; the *large-seeded* varieties, usually referred to as *broad beans*, are used as a vegetable.

Reference: M.18.

See also LEGUMES; INOCULATION.

HORSEBRUSHES (*Tetradyma* spp.), particularly LITTLELEAF HORSEBRUSH and SPINELESS HORSEBRUSH, are POISONOUS PLANTS. While being trailed, hungry sheep may eat the leaves and small stems of these plants, which contain an as yet unknown poison. Then, as a result of sensitization to light, white or light-colored animals may develop a disease called *bighead*, which is due to edema (swelling), affecting principally the head.

Horsebrush. (N.6.)

HORSE MEAT. *See* ANIMAL PRODUCTS.
HORSE MEAT BY-PRODUCTS.
                See ANIMAL PRODUCTS.
HORSETAILS (*Equisetum* spp.)—e.g., *scouring-rush*—are perennial, stiff, RUSH-type herbs with creeping rootstocks. When horses eat the tops of these POISONOUS PLANTS, to be found in hay, they develop such symptoms of poisoning as weakness, craving for the plant, diarrhea, loss of flesh, and lack of control of the legs.
HORSFORD is a spring variety of SIX-ROWED BARLEY.
HORT. (hort.) is an abbreviation for *hortensis*, meaning *horticular* VARIETY.
HORTICULAR VARIETY = *agronomic variety*. *See* VARIETY; HORT.
HOST is the organism from which a PARA-

SITE derives its sustenance. Thus, clover is a frequent host for DODDER.
HOT HAY = *heating hay*. *See* HAY.
HOT-WATER SEED TREATMENT. *See* LOOSE SMUT OF WHEAT; LOOSE SMUTS OF BARLEY.
HUBAM is an annual variety of the biennial WHITE SWEETCLOVER.
HULL is an organ which encloses a seed or fruit; e.g., a GLUME, LEMMA, PALEA, or POD.
    To hull means: to remove hulls from a seed.—*See also* SCREENINGS.
HULLED OATS, or *undried oat groats*, is a feedstuff.—*See also* OATS; HULL; HULLING MACHINERY.
HULLING MACHINERY is a piece of equipment found in many feed mills; it is particularly useful for making such feedstuffs as HULLED OATS and ground OATS.
HULL-LESS BUCKWHEAT
                = TARTARY BUCKWHEAT.
HUMID CLIMATE has sufficient precipitation—usually 30 to 40 in. annually—to support a forest vegetation.
HUMULUS. *H. hupulus* = HOP.
HUMUS is the well-decomposed, relatively stable portion of organic matter of the soil; it is the residual substance that remains after the action of BACTERIA on plant (or animal) products.

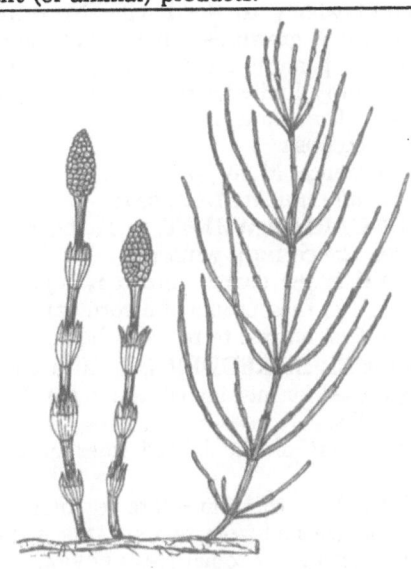

Horsetail. **Fruiting stalks and a vegetative stalk.** (D.19.)

HUNDRED-AND-ONE = *Turkey wheat.* See COMMON WHEAT.

"HUNDRED-DAY SPECKLED" is better known as *Alabama velvetbean;* originally, the term was used to identify the *Georgia* variety of the DEERING VELVETBEAN.

HUNGARIAN CLOVER (*Trifolium pannonicum*) is a TRUE CLOVER of little, if any, importance as forage crop.

HUNGARIAN FOXTAIL MILLET.

See FOXTAIL MILLET (variety).

HUNGARIAN VETCH (*Vicia pannonica*) is one of the VETCHES. The plants are less viny than HAIRY VETCH or COMMON VETCH and tend to be erect when the growth is short or when the plants have some support.

Both the stems and the leaves are covered with medium-long hairs, which give them a decidedly grayish color. A stem length of from 3 to 4 ft. is attained under favorable conditions, but under average conditions 2 to 2½ ft. is more common. The plants are rather winter-hardy.

This LEGUME is especially well adapted to heavy clay soils and will do better in wet situations than other vetches. In the southern states it has done poorly on sandy land. Hungarian vetch is confined almost wholly to the Pacific Northwest, where it is grown as a hay, silage, green-manure, pasture, and seed crop. The seed is used to a limited extent in mixed ground feeds. The plant is resistant to VETCH ANTHRACNOSE.

Reference: M.18.

See also INOCULATION; OATS.

HUNGARIAN WHEAT = *Turkey wheat.* —See also COMMON WHEAT.

HUSK is the coarse outer envelope of a fruit; e.g., the GLUMES of a corn ear.

To husk means: to remove the husks.

HUSKER-SHREDDER is a farm equipment used for husking CORN and shredding its stalks.

HY is one of the inbred lines of CORN hybrids.

HYBRID is the immediate (or first generation) *cross* between two STRAINS, VARIETIES, SPECIES, or sometimes GENERA.

Many authorities prefer to limit the term hybrid to a cross between different species; crosses between varieties are then termed *crossbreeds*, and those between strains, *half-breeds*. When the ancestry is mixed, the term *mongrel* is often applied.— See also CORN; WHEAT CLOVER.

HYBRID 128 is a CLUB WHEAT.

HYBRID CLOVER = ALSIKE CLOVER.

HYDRATED LIME, a technical grade of *calcium hydroxide*, forms a fine powder; it readily absorbs carbon dioxide from the air, thus changing to CALCIUM CARBONATE. It is only sparingly soluble in water and very alkaline. Hydrated lime is sometimes use to "sweeten" (neutralize) acid soils.

Hydrated lime dust may be applied to plants when JAPANESE BEETLES first appear; the treatment should be repeated as needed to maintain a protective coating of all parts of the plants until the beetles disappear.

HYDRAULIC means: pertaining to liquids, especially water.

Hydraulic extraction is extraction performed with a hydraulic press which uses water to transmit pressure; this method is widely used to extract oils from seeds.— See also SOYBEAN.

HYDRAULIC SOYBEAN OIL CAKE.

See SOYBEAN.

HYDRAULIC SOYBEAN OIL MEAL.

See SOYBEAN.

HYDRELLIA. *H. griseola* var. *scapularis* = RICE LEAF MINER.

HYDROCARBONS are organic compounds consisting only of two elements: carbon and HYDROGEN.

The use of such hydrocarbons as MINERAL OIL, PETROLEUM JELLY, PETROLATUM, and PARAFFIN in commercial feeds is disapproved, except temporarily as a dust control agent in mineral mixtures in an amount which will result in an intake not to exceed 0.06 percent of the total ration and not to exceed 3 percent in the mineral mixture.

The hydrocarbons are sometimes used to treat digestive disturbances, such as BLOAT and *impaction*.

Reference: F.6.

HYDROCHLORIC ACID, the technical grade of which is called *muriatic acid*, is a solution of the gaseous *hydrogen chloride* in water. It is a colorless, fuming, corrosive, and poisonous liquid.—*See also* GASTRIC JUICES.

HYDROCYANIC ACID is also called *diluted hydrocyanic acid* or PRUSSIC ACID.

HYDROGEN is a gaseous chemical element contained in many compounds; e.g., in water, CARBOHYDRATES, and HYDROCARBONS.—*See also* pH.

HYDROGEN CHLORIDE.
See HYDROCHLORIC ACID.

HYDROGEN CYANIDE is a very toxic, colorless gas; its water solution is known as PRUSSIC ACID.

HYDRO KAFIR is one of the grain SORGHUMS.

HYDROL = *feeding corn sugar molasses.* See MOLASSES.

HYDROLYSIS is a decomposition reaction caused by water, diluted acids and alkalis, or ENZYMES.—*See also* DEXTROSE; GLUCOSE; CORN; MILK PRODUCTS.

HYDROLYZED means: produced by HYDROLYSIS.

HYDROLYZED CORN PROTEIN.
See CORN.

HYDROLIZED SAWDUST and *hydrolized straw* can be prepared by HYDROLYSIS of sawdust and straw; even though the digestibility and nutritive value of these hydrolyzed feedstuffs are much higher than those of the raw materials—due to the conversion of crude FIBER into SUGARS and other soluble products—they are still far below those of grain; yet, the cost of these hydrolyzed feedstuffs is higher than that of grain, except in war emergency.

HYDROLYZED STRAW.
See HYDROLYZED SAWDUST.

HYDROLYZED WHEY. *Dried hydrolyzed whey* is one of the feedstuffs obtained from MILK PRODUCTS.

2 - HYDROXY - 4, 6 - DIMETHYLPYRIMIDINE. See NICARBAZIN.

4 - HYDROXY - 3 - NITROBENZENEARSONIC ACID = 3-NITRO-4-HYDROXYPHENYLARSONIC ACID.

HYGROSCOPIC (pronounced: hye-gro-*skopp*-ik) means: readily absorbing and retaining moisture. SUGAR and table salt are examples of hygroscopic substances.

HYLASTINUS. *H. obscurus* = CLOVER ROOT-BORER.

HYMAR. See CLUB WHEAT (variety).

HYPERA. *H. punctata* = CLOVER LEAF WEEVIL; *H. postica* = ALFALFA WEEVIL.

HYPERICUM. *H. perforatum* = ST. JOHNSWORT.—*See also* POISONOUS PLANTS.

HYPERKERATOSIS is an overgrowth of the horny layer of the epidermis (outer skin).—*See also* VITAMINS.

HYPHA (pronounced: *high*-fa) is one of the threadlike strands (filaments) that constitute the body of a FUNGUS. Some hyphae are specialized for producing SPORES or for penetrating host tissues.—*See also* MYCELIUM; SCLEROTIA.

HYSSOP. See WATER HYSSOP.

I

I = IODINE.—*See also* MINERAL FEEDS.

IDEAD. See COMMON WHEAT (variety).

ILLINI. See SOYBEAN (variety).

IMAGO is an adult INSECT.
— *See also* PUPA.

IMPACTION. See HYDROCARBONS.

IMPERFECT is a term with two meanings: (1) *unsexual*, or nonsexual; e.g., the *imperfect stage*, i.e., the life period of a FUNGUS during which SPORES are produced nonsexually, in contradistinction to the PERFECT STAGE, (2) *unisexual*; e.g., flowers lacking either STAMENS or PISTILS as distinguished from bisexual flowers which possess both the female and male organs.

IMPERFECT FUNGI
= FUNGI IMPERFECTI.—*See also* IMPERFECT.

IMPERIAL KAFIR is indistinguishable from *Ajax*, one of the intermediate-type grain SORGHUMS.

IMPERIAL RYE. See RYE (variety).

IMPERVIOUS means: not to be penetrated. An *impervious subsoil* cannot be penetrated by moisture.

IMPLANTATION of PELLETS is the insertion of a medication in pellet form under the skin, usually in the neck or ear. Hormone-like substances such as STILBESTROL are implanted in poultry, sheep, and cattle to finish them for market. This treatment, especially if used with male birds, improves the appearance of the carcass. There are usually no noticeable effects on beef carcasses.—*See also* HORMONES.

IMPROVED PASTURES
= *permanent pastures.* See PASTURES.

IMPROVED PERMANENT PASTURE.
See PASTURE MANAGEMENT.
IMPROVED SPANISH PEANUT is a Spanish-type PEANUT.
INDIANA 844 and *Indiana 909A* are CORN hybrids.
INDIAN ALFALFA.
See ALFALFAS (variety).
INDIAN CORN = CORN.
INDIAN GRASS (*Sorghastrum nutans*), or *Yellow Indian grass*—not to be confused with INDIAN RICEGRASS—is one of the best all-purpose tall GRASSES. It is palatable and nutritious in summer, has a wide range of adaptation, and the seed is readily harvested and prepared for planting. The grass has a wide range throughout the eastern United States and parts of Canada and Mexico.

Indian grass is a robust bunchgrass, 3 to 8 ft. tall, with wide leaves and a plume-like seed head. It prefers fertile bottom land and is uncommon in degenerated grasslands.

In late spring and early summer, Indian grass is one of the most palatable grasses. The seed heads are often relished by stock in the fall. Indian grass cures well enough on the stalk to make fair winter roughage, provided sufficient protein is fed as a supplement.

A firm seedbed covered with a non-competitive mulch or crop residue is preferred for pasture seedlings.

Most tall bunchgrasses, such as SWITCH-GRASS, BIG BLUESTEM, LITTLE BLUESTEM, CAUCASIAN BLUESTEM, or King Ranch bluestem (a TURKESTAN BLUESTEM variety), are satisfactory companion species.

A general rule-of thumb for tall grass grazing is "take half and leave half." In other words, do not use more than half of the season's production during the growing season. One half of the summer's growth should be left by frost. Additional winter grazing is not likely to be detrimental, but heavy use early in the spring can cause quite serious damage.

When tall grass is harvested for hay, one mid-summer cutting followed by very light use of the aftermath will probably yield maximum sustained production.

Cutting two or three times is usually very detrimental to the stand.

Dangers: Indian grass is subject to several plant diseases, but none appears to be of practical importance at present.

Reference: O.1.

Indian grass. Plant, spikelet with pedicel, and rachis joint. (H.26.)

INDIAN RICEGRASS (*Oryzopsis hymenoides*), a densely tufted perennial bunchgrass, is widely distributed over the western states. It is not to be confused with INDIAN GRASS.

The seeds are nearly round, black, tipped with a short awn, and thickly covered with white hairs.

The plants grow from 1 to 2 ft. tall. The slender leaves are nearly as long as the stems. The spreading PANICLE has long SPIKELETS and LEMMAS with silky hairs.

Indian ricegrass. Plant and spikelet. (H.47.)

Indian ricegrass occurs mainly on dry, sandy soils and frequently is important on sand dune areas. It is drought-resistant and somewhat tolerant of alkali. Once it was widely distributed over the western ranges, particularly in semidesert areas, but overgrazing on much of the land there has almost eliminated it. Now it grows in abundance mainly in places that have been ungrazed or conservatively grazed. Stockmen regard the grass highly as a winter feed for animals and prize the areas where it grows.

Initial thin stands will increase in density if natural reseeding is permitted by careful management.

Reference: H.1.

See also GRASSES; RANGE PLANTS.

"INDIA WHEAT"
= TARTARY BUCKWHEAT.

INDICATORS OF RANGE CONDITIONS. *See* RANGE PLANTS.

INDIGO. *See* HAIRY INDIGO.

INDIGOFERA. *I. hirsuta*
= HAIRY INDIGO.

INDIRECT FERTILIZERS.
See FERTILIZERS.

INERT (pronounced: in-*ert*) means: not having active properties; especially, not having any chemical action.

Inert mineral matter, such as GRIT, has no nutritive value.—*See also* MINERAL FEEDS.

INFECTION is the invasion of plant or animal body tissues by pathogenic microorganisms, such as FUNGI or BACTERIA.

INFLORESCENCE is the flowering part of a plant, and especially, the mode of its arrangement.

INOCULANT = INOCULUM.
— *See also* INOCULATION.

INOCULATE means: to bring infectious material—the INOCULUM—in contact with a host plant; as to inoculate legume seeds with nitrogen-fixing bacteria.

INOCULATION of LEGUMES means the introduction of nitrogen-fixing bacteria into the soil for the purpose of increasing the production of successful legume crops, which also improve the soil.

Well-inoculated legumes have nodules (small lumps) on the roots, produced by effective nitrogen-fixing LEGUME BACTERIA. Farmers cannot be sure before they plant the legume that sufficient bacteria of the proper kind are present in their soils.

Only certain strains of bacteria will live on a particular legume, making it necessary to inoculate with the right bacteria culture to secure satisfactory results. Certain PEA species and VETCH are inoculated by the same organism; TRUE CLOVERS are mostly, though not all, grouped together. LUPINES require a special inoculant.

Special strains of inoculants are also required for each of the TREFOIL species. It has been shown that some strains of bacteria are—as are plants—much more efficient than others in performing their work.

Nitrogen-fixing bacteria need air from which to gather nitrogen; waterlogged soils, therefore, are usually inefficient and not suitable for legume crops. Lupine and CRIMSON CLOVER inoculant is very inefficient in wet soils. Organic matter and soil acidity influence bacterial activity. Sunlight, dryness, heat, or contact with superphosphate fertilizer (which has an excess of free sulphuric acid), can be fatal to the bacteria.

Commercial inoculation preparations may be jelly-like materials, but more often appear in powder form. The date of preparation and directions for its use are given on the *labels*. It is important to use fresh cultures, and to use them within the time specified on the package. Such cultures should be kept in a cool place and away from sunlight.

A convenient way to inoculate seed is to dissolve or mix the inoculating material in a small amount of water (the smaller the seed the more water is required for wetting). For example, 30 to 35 lb. lupine seed will require ⅓ of a can (100 lb. size) of inoculating material and about 1 pt. water. Some authorities recommend adding 2 or 3 tablesp. sirup to the water as an adhesive (sticking) ingredient. Mix the inoculant immediately with the seed; the mixture should be agitated rapidly for one or two minutes or until all seeds are moist.

A widely used soil-transfer method of inoculating consists of moistening 1 bu. seed with a solution of 3 oz. glue or sugar dissolved in 1 qt. water and thoroughly mixing 2 qt. finely sifted, inoculated soil with the moistened seed.

Soon after the legumes begin to grow, the legume bacteria invade the tiny root hairs and multiply in large numbers, forming growths called nodules. A definite partnership is established—a true *symbiosis* or living-together of two organisms to the advantage of both—with the legume plant furnishing the necessary sugar or energy,

and the bacteria using this energy material to fix the free nitrogen of the atmosphere and giving it directly to the plant. This process is called *nitrogen fixation*.

A green color in inoculated legumes darker than that of uninoculated ones, is a sure sign of nitrogen fixation by the bacteria. As a matter of fact, color differences are more reliable in judging nitrogen fixation than are numbers of nodules.

ESTIMATED AMOUNT OF NITROGEN FIXED FROM THE AIR BY LEGUME BACTERIA

Legume	Pounds of nitrogen per acre
Alfalfa	194
Ladino clover	179
Lupines	151
Sweetclover	119
Alsike clover	119
Red clover	114
Kudzu	107
White clover	103
Sourclover	98
Crimson clover	94
Cowpeas	90
Lespedezas (annual)	85
Vetch	80
Bur-clover	78
Peas	72
Velvetbeans	67
Soybeans	58
Peanuts	42
Beans	40

The clustering of *nodules* around the taproot, at the point where the inoculated seed is planted, generally indicates that the nodules were formed by the bacteria added with the inoculant. If the inside of the nodule is red, this indicates high nitrogen-fixing activity. Nodules scattered over the side roots are usually formed by the legume bacteria naturally present in the soil. In order to note the effects of seed inoculation when examining legumes it is well to dig plants at different stages of growth, since nodules come and go with varying moisture levels in the soil.

As the legumes mature, the nitrogen compounds formed in the nodules furnish nitrogen for the building of proteins in the leaves, stems, and seed. In the early stages the nodules may contain 5 to 8 percent nitrogen, but at seed maturity they are no richer in nitrogen than the rest of the root. The nodules disintegrate rapidly at the time of seed formation.

Although it is impossible to give a final answer to the question of how much nitrogen is fixed by the bacteria in the nodules on legume plant roots, the table compiles the findings of different investigators.

References: E.4; S.7; M.19.

INOCULUM, or *inoculant*, is the "infectious" material introduced by INOCULATION.—*See also* SILAGE CROPS.

INORGANIC is a term pertaining to substances of other than animal or vegetable origin; particularly, chemicals that do not contain CARBON—carbonates and cyanides excepted.

Inorganic acids—e.g., SULFURIC ACID or HYDROCHLORIC ACID—should not be used in silos as SILAGE conditioners.—*See also* MINERAL FEEDS.

INSECT. *See* INSECTS.

INSECTICIDAL means: possessing insect-killing properties.

INSECTICIDES control insect pests; however, many factors must be considered —in addition to the increase in hay and seed yield—when insecticides are used on forage crops. Insecticide residues carried on the harvested hay may have a harmful effect when fed to livestock. Insecticide applied to the seed crop presents a problem of another type: Control of many of the insects that reduce the quantity of seed, and are most active and abundant in fields during the flowering period, appears to be impossible without also reducing the bee population which is required to pollinate RED CLOVER, ALFALFA, and many other forage crops.

To the newer organic insecticides belong the CHLORINATED HYDROCARBONS, such as ALDRIN, BENZENE HEXACHLORIDE or LINDANE, CHLORDANE, DDT, DIELDRIN, HEPTACHLOR, METHOXYCHLOR, TDE, and TOXAPHENE; PARATHION, a phosphorus-containing chemical, also belongs to this group. Older, but still widely used, are such natural insecticides as DERRIS, NICOTINE, NICOTINE SULFATE, PYRETHRINS, RYANIAS, ROTENONE, and SEBADILLA POWDER, PINE-TAR and PINE-TAR OIL. The arsenicals SODIUM ARSENITE and PARIS GREEN, as well as METHYL BROMIDE and NAPHTHALENE DRAIN OIL, are also used as insecticides.

Reference: E.5.

See also GRASSHOPPERS; CHINCH BUG; BAIT; POISONED BRAN MASH; DRY BAIT.

INSECTS. The injury caused by insect pests is undoubtedly one of the major factors of crop failures. No parts of plants are immune to injury; some insect species feed on the parts above ground, some on the roots, and others may even be restricted to individual organs, such as the ovaries in the flower heads.

It is difficult to obtain even a rough estimate of the injury caused to plants by certain insects. In some instances there is visual evidence of injury, such as blasted flower heads and damaged leaves, that can be assigned to the feeding of specific pests. In most cases, however, the damage to individual plants is the result of feeding by several species. MITES do not belong to the insects, but to the SPIDER group.

The rapid development of improved INSECTICIDES, especially the residual insecticides, has resulted in numerous studies on the effectiveness of these materials in controlling certain of the insect pests.

Among the most important insect pests of field crops are the following: ALFALFA LOOPER; ANTS; APANTHELES; APHIDS; ARMY CUTWORM; BEES; BEETLES; CHINCH BUG; CLOVER APHID; CLOVER-FLOWER MIDGE; CLOVER-HEAD WEEVIL; CLOVER LEAFHOPPER; CLOVER LEAF WEEVIL; CLOVER ROOT-BORER; CLOVER ROOT CURCULIOS; CLOVER-SEED CHALCID; CLOVER TYCHIUS; CORNFIELD ANT; CORN FLEA BEETLE; CORN LEAF APHID; DITROPINOTUS; ENICOPHILUS; FLIES; GRAIN BEETLES; GRAIN WEEVILS; GRAPHOGNATUS; GRASSHOPPER; GREENBUG; GROUND BEETLES; JAPANESE BEETLE; LADY BEETLES; LEAFHOPPERS; LESSER CLOVER-LEAF WEEVIL; LOCUST SKIPPER BUTTERFLY; LYGUS BUGS; MAY BEETLE; MEADOW SPITTELBUG; MELANOPLUS; MERISOPORUS; MEXICAN BEAN BEETLE; MOLE CRICKETS; MORMON CRICKET; MOTHS; ORIUS; PALE WESTERN CUTWORM; PEA APHID; PEA MOTH; PEA WEEVIL; POTATO LEAFHOPPER; RED HARVESTER ANT; RED-SHOULDERED PLANT BUG; RICE LEAF MINER; RICE WATER

WEEVIL; RHODOGYNE; ROBBER FLIES; SITONA; SOUTHWESTERN BORER; SPOTTED CUCUMBER BEETLE; SPHEX; TELENOMUS; TRICHOGRAMMA; WASPS; WINTHEMIA.

Reference: E.5.

See also MEADOW FOXTAIL; NEMATODES; CELATORIA.

INTERMEDIATE GRAIN SORGHUM

TYPES, often causing great confusion as to classification and nomenclature, are divided (in this book) into four groups (which are discussed under grain SORGHUMS): *kafir-feterita derivatives; kafir-Freed derivatives; kafir-milo derivatives;* and *other intermediate types.*

INTERMEDIATE WHEATGRASS (*Agropyron intermedium*) is a sod-forming perennial. It has been tested extensively in the northern and central parts of the Great Plains and the Pacific Northwest. Under a wide variety of soil and climatic conditions it shows great promise for use as a pasture and forage species.

The plants begin to grow in early spring and reach a height of 3 to 4 ft. before growth ceases in midsummer because of scarcity of moisture. The return of moisture and cool temperatures in the late summer brings good growth recovery.

The abundant leafy foliage is relished by all classes of livestock. Plant growth is vigorous.

Planting intermediate wheatgrass with ALFALFA is good practice. Solitary stands eighteen years old have yielded from 1,600 to 2,000 lb. field-cured hay per acre.

References: H.1; B.1.

See also WHEATGRASSES; GRASSES; RANGE MANAGEMENT.

INTERNAL PARASITES cause livestock producers tremendous losses every year. Practically all farm flocks of sheep have some parasites; calves, foals, young cattle, horses, and young pigs are also seriously affected. The affected animals are unthrifty in appearance and cough frequently; membranes of the mouth and eyes are pale, and a soft swelling under the jaw may sometimes be observed.

Livestock should be watched for these symptoms.

References: S.28; S.10.

See also GRAZING; PASTURES.

INTERNAL SECRETION.
See HORMONES.

INTERNATIONAL CHICK UNITS.
See VITAMIN D; A.O.A.C.

INTERNATIONAL UNITS = I.U.

INTERNODES are the portions of a stem or branch between the NODES (joints).

INTESTINE, the tubelike part of the DIGESTIVE TRACT, extends from the stomach to the anus.

The *intestinal wall* of the *small intestine* secretes juices which contain a number of ENZYMES; these *intestinal secretions* are important for proper DIGESTION.

The average *length* of the small intestine of various livestock species has been reported as follows: swine 60 ft.; horses 70 ft.; sheep 80 ft.; and mature cattle, 130 ft.

The large intestine of sheep holds 1½ gal.; of cattle, 10 gal.; and of horses, 15 to 30 gal. This large *volume* is needed for the digestion of large amounts of roughage consumed by these animals. BACTERIA found in the *large intestine* are important for the digestion of feeds.—*See also* STOMACH; ABSORPTION.

INTRAVENOUS means: within or into a vein; e.g., intravenous injection.

INULIN, a CARBOHYDRATE, is the reserve food stored (in place of starch) in a few plants; e.g., in ARTICHOKES. It forms a white, starchlike, water-soluble powder.

INVERTASE is an ENZYME which occurs in YEAST. In the presence of water, it is capable of *inverting* (splitting) SUCROSE (cane sugar) into INVERT SUGAR.—*See also* FERMENTATION; FERMENT.

INVERT SUGAR is a mixture of the simple sugars DEXTROSE and LEVULOSE. It is produced by the *inversion* (splitting) of SUCROSE, for instance with the help of INVERTASE.—*See also* CANE MOLASSES.

INVOLUTE means: inrolled; i.e., with both edges rolled toward the middle (of a leaf).

IODINATED CASEIN is a hormone-like substance that shows action similar to that of THYROXINE.—*See also* HORMONES.

IODINE—the chemical element *I*—belongs to the HALOGENS. It forms bluish-black scales of metallic luster which are somewhat volatile even at room temperature and practically insoluble in water.

It is found combined, as in *iodides*, and is distributed widely in nature, but sparingly in quantity.

Iodine is essential for normal nutrition of animals; it is concentrated in the thyroid gland in form of an organic compound known as THYROXINE.

Iodine deficiency results in an enlargement of the thyroid gland, a condition called GOITER. (Goiter is seen most often in young animals: calves and lambs are big-necked, pigs are hairless, and foals are weak.) Certain areas have a deficiency of iodine and are referred to as *goiter belts;* in the United States, these include the northeastern section of the country, the Great Lakes, and Rocky Mountain areas. —*See also* MINERAL FEEDS; ALGAE; TRACE ELEMENTS.

"IODIZED." *See* MINERAL FEEDS.

IODIZED SALT. *See* MINERAL FEEDS.

IOWA 939 is a CORN hybrid.

IPOMOEA. *I. batatas* = SWEETPOTATO.

IREDELL.

See SIX-ROWED BARLEY (variety).

IRKUTSK is a spring RYE variety.

IRON, the chemical element *Fe*, forms two main series of compounds: the *ferrous* and the *ferric* salts; some of them are used for the prevention and treatment of iron deficiencies that show up as *anemia* or *thumps* in baby pigs.—*See also* TRACE ELEMENTS; HEMATINIC TONIC; TONICS; MINERALS.

IRON CARBONATE, or *ferrous carbonate*, which yields approximately 50 percent IRON, forms greenish-brown, water-soluble crystals; it is also available as *"ferrous carbonate mass,"* a brown, soft mass containing about 60 percent honey and sugar. Iron carbonate is used to prevent or treat iron deficiencies in animals.—*See also* MINERAL FEEDS; MOLASSES.

IRON COWPEA is a COWPEA variety.

—*See also* FIELD BEANS.

IRON DEFICIENCY.

See MINERAL DEFICIENCIES.

IRON PHOSPHATE, or *ferric phosphate*, yielding 25 percent IRON and 13.9 percent PHOSPHORUS, is a yellowish-white, water-insoluble powder used in MINERAL FEEDS.

IRON SULFATE, also known as *ferrous sulfate* or *copperas* (which is the name of the technical product), contains 20 percent IRON and over 45 percent water. (The dried product, which is commercially available, yields 63.5 percent iron). It forms bluish-green crystals or granules which are efflorescent in dry air and soluble in water.

Iron sulfate is widely used in MINERAL FEEDS and as a 15-percent solution for the eradication of DODDERS.

IRRADIATED YEAST. *See* YEAST.

IRRADIATION, the subjection to *radiation* is, for example, the exposure of ERGOSTEROL or YEAST to the action of ultraviolet rays and their change to VITAMIN D_2. —*See also* VITAMIN D; CHOLESTEROL.

IRRIGATED PASTURES.

See PASTURES; PASTURE MANAGEMENT.

ISOLEUCINE is one of the essential AMINO ACIDS. It is obtained in natural (L-isoleucine) form from beets, other plants, and milk products, or it is produced synthetically as DL-isoleucine). The leaflets or plates are slightly water-soluble.

ISOMERES are substances of identical chemical composition (formula), but, of different properties.—*See also* BENZENE HEXACHLORIDE; LINDANE.

ITALIAN RYEGRASS (*Lolium multiflorum*) is a hardy, short-lived grass, usually an annual. When seeded in the spring, late summer, or early fall it makes rapid growth and soon covers the ground, furnishing GRAZING in a remarkably short time if adequate moisture is available. It is tender and very palatable to livestock and has excellent carrying capacity. The plants grow from 2 to 4 ft. in height and make excellent hay. Italian ryegrass is distinguishable from PERENNIAL RYEGRASS by the presence of awns on the seed; by the cylindrical culm, or stem; and by the leaves which are rolled in the bud. The plants of Italian ryegrass are yellowish green at the base while those of the perennial ryegrass are commonly reddish, pure Italian ryegrass may be just that, but more often is a mixture with COMMON RYEGRASS predominant.

Since Italian ryegrass is a heavy seeder, if not kept closely grazed or mowed, many new plants appear from volunteer seeding. If moisture and fertility are sufficient for rapid growth, the grass may be cut twice during the season. It has a crude protein content of 12 to 18 percent.

Italian ryegrass. Plant, spikelet, and floret. (H.26.)

Italian ryegrass has many uses. On the Pacific Coast it is more valuable as a hay plant than anywhere else in the United States. It makes an excellent nurse crop for spring-seeded permanent pastures and lawns, and gives a quick cover for early grazing in pastures. When sown in combination with winter grains for temporary pasture it makes a desirable bottom grass and increases the length of the grazing season. Italian ryegrass is a suitable grass for temporary poultry ranges. In the South it is used extensively for fall seeding on permanent lawns and as winter grazing crop. Yields of over 13 tons of green, or 3 tons of dry, forage per acre have been reported. The grass is relatively disease-free, although LEAF RUST may cause severe damage.

References: S.2; H.1; H.9.

See also RYEGRASSES; GRASSES.

I.U. is the abbreviation for *international units*, which are used to measure various VITAMINS in accordance with standards based on international agreements.—*See also* VITAMIN UNITS.

IVANOV. *See* RYE (variety).

IVORY NUT MEAL, also known as *vegetable ivory meal*, is one of the officially recognized BY-PRODUCT FEEDSTUFFS. It is defined as the ground waste material resulting from the manufacture of buttons and similar articles from the vegetable ivory nut. The latter is the fruit of the IVORY PALM; it is rich in carbohydrates but poor in digestible protein. Ivory nut meal is occasionally used in cattle feeds.

Reference: F.6.

IVORY PALM (*Phytelephas macrocarpa*) carries nuts which are called *vegetable ivory*, or *ivory nuts*.—*See also* IVORY NUT MEAL.

J

JACKBEAN (*Canavalia ensiformis*) is an annual LEGUME. It is sometimes used as a forage plant, but the beans are hard to digest by livestock.

JACKSON BARLEY.

See SIX-ROWED BARLEY (variety).

JAPAN CLOVER = COMMON LESPEDEZA.

JAPANESE BEETLE (*Popillia japonica*) is destructive to the leaves, blossoms, and fruits of more than 275 plants, shrubs, and trees.

These beetles are widely distributed in the states along the Atlantic seaboard, at scattered points in adjoining states, and throughout the Midwest east of the Mississippi.

The beetles spend about ten months in the soil as grubs; in late May or early June the grubs go through the short pupal stage, after which the mature insects leave the soil. They fly about in numbers by early July. During July and August, the females periodically go into the ground and lay eggs.

Control: The foliage of plants, shrubs, and trees can be protected from beetle attack with a spray containing DDT or HYDRATED LIME dust.

Reference: U.3.

See also BICOLOR LESPEDEZA.

Japanese beetle. (U.13.)

JAPANESE BUCKWHEAT.
        See COMMON BUCKWHEAT (variety).
JAPANESE LAWNGRASS. *See* ZOYSIA.
JAPANESE MILLET is a variety of BARNYARDGRASS.
JAPANESE RIBBON CANE, or *Japanese-seeded Ribbon cane*—better known as *Honey sorgo*—is one of the forage SORGHUMS. (It is not, as the name implies, one of the SUGARCANES).
JAPANESE SUGARCANE, a variety of the CHINESE SUGARCANE, is cultivated in the South, primarily for forage use.
JAPANESE WHISK DWARF belongs to the broomcorn-class of the SORGHUMS.
"JAP CLOVER" = COMMON LESPEDEZA.
JAVA. *Speckled Java*
                        = TAYLOR. *See* COWPEA.
JENKIN is a CLUB WHEAT.
JERUSALEM ARTICHOKE (*Helianthus tuberosus*) has thin-skinned tubers which

are used as food and feed. Best yields are obtained in the Pacific Northwest. Light or medium loam soil is preferred by the perennial plant which reaches a height of 6 ft. and is similar to the wild sunflower.

The tubers are planted early in the spring 4 in. deep. Since it is difficult to gather all of the small tubers in the fall, and those that are left develop in the following year, the Jerusalem artichoke easily develops into a WEED.

The artichoke is often grown for hog feed, in which case the animals are allowed to gather the tubers. At other times the tubers are harvested by pickers and then used for feed. Tubers are rich in carbohydrates which hydrolyze to LEVULOSE. The fresh tops of the plants are also used as feed, but the mature tops are un palatable and hard to digest.

JERUSALEM CORN, now called DURRA, is one of the grain SORGHUMS.
JERUSALEM PEA = MUNGBEAN.
JERUSALEM RICE CORN, better known as *shallu*, is one of the grain SORGHUMS.
JIMMYWEED = RAYLESS GOLDENROD.
JOHANNES RYE. *See* RYE (variety).
JOHNSON DURRA
                = *Arnautka*. *See* DURUM WHEAT.
JOHNSONGRASS (*Sorghum halepense*) is different from all other SORGHUMS in that it is a perennial that spreads vigorous (RHIZOME-type) rootstocks. The stems, leaves, and heads of Johnsongrass resemble those of SUDANGRASS, although it seldom grows that tall.

Johnsongrass grows wherever cotton is produced and thrives where moisture is abundant. Its seed fertility is high. Two and three crops of hay are frequently harvested a season; yields of more than 15 tons are common. It is valuable for pasture and is grown with several winter annual LEGUMES; it has a profitable place in livestock production in many sections of the South.

Dangers: Johnsongrass may contain small quantities of PRUSSIC ACID, but farm animals are rarely poisoned by it.

This grass is susceptible to BACTERIAL STREAK, BACTERIAL SPOT, ROUGH SPOT, ANTHRACNOSE OF GRASS, ZONATE LEAF

SPOT, GRAY LEAF SPOT, TARGET SPOT, SOOTY STRIPE, SORGHUM RUST, LOOSE KERNEL SMUT OF JOHNSONGRASS, LOOSE KERNEL DISEASE OF SORGHUM, and COVERED KERNEL SMUT.

Johnsongrass. Plant and terminal raceme. (H.26.)

When allowed to head, Johnsongrass provides SORGHUM MIDGES with a place to hibernate. It is also attacked by BILLBUGS, HARVESTER ANTS, STINK BUGS, and SOUTHERN CORN ROOTWORMS.

Control: Farmers condemn Johnsongrass especially if it is a WEED in cultivated fields or when it grows so abundantly that it becomes a nuisance. SODIUM TCA is most effective for controlling it on non-cropped land.

Since sodium TCA acts principally through the root system of the grass, there must be adequate rainfall or irrigation to carry the material into the soil. However, excessive rainfall or irrigation will result in poor control because the chemical will leach out of the soil. Best results will be obtained if Johnsongrass is sprayed when the growth is from 6 to 12 in. high. If plants are taller, they should be mowed if possible and the new growth sprayed.

References: H.1; C.3; D.2.

See also PRUSSIC ACID POISONING; GRASSES; CRIMSON GRASS; ALFALFA; COWPEA; HAY; PASTURES; PASTURE PLANTS.

JOHNSON HAY is obtained from JOHNSONGRASS.—*See also* HAY GRADING.

JOINT. *See* NODE.

JOINTFIR. *Nevada jointfir*
= NEVADA EPHEDRA.

JOINT GRASS is a name applied to BERMUDA-GRASS and to KNOTGRASS.

JOINTWORM. *See* WHEAT JOINTWORM.

JONES LONGBERRY
= RED MAY. *See* COMMON WHEAT.

JUNCACEAE are known as the RUSHES. This plant family includes the *Juncoides* spp. and the *Juncus* spp.

JUNCOIDES. The genus *Juncoides* belongs to the Juncaceae (RUSHES).

JUNCUS. The *Juncus* spp. form one of the genera of the Juncaceae, or RUSHES.

JUNE BEETLE = MAY BEETLE.

JUNEGRASS = KENTUCKY BLUEGRASS.

JUNE RED CLOVER
= *double-cut clover.* *See* RED CLOVER.

JUNIPERS (*Juniperus* spp.) are medium-sized evergreen trees and shrubs of the Pine family.—*See also* RANGE PLANTS.

JUNIOR NO. 6
= *Goldcoin.* *See* COMMON WHEAT.

K

K 4 is a popcorn hybrid. — *See also* CORN

KABATIELLA. *K. caulivora* is the FUNGUS which causes the disease NORTHERN ANTHRACNOSE; *K. nigricans* causes FALSE ANTHRACNOSE OF VETCH.

KABOTT. *See* SOYBEAN (variety).

KAFIR is a grain SORGHUM with many varieties, strains, and hybrids.

Among the kafir varieties are the following: *Blackhull, Club, Combine, Dawn, Hydro, Pearl, Pink, Red, Reed, Rice, Sedan, Sharon, Sunrise,* and *White.*

Kafir crosses include the following intermediate types: (a) *kafir-feterita derivatives* —e.g., *Ajax, Chiltex, Dwarf 44-14, Premo,* and *Wonder;* (b) *kafir-Freed derivatives*— e.g., *Cheyenne, Coes, Greely, Modoc,* and *Wescan;* (c) *kafir-milo derivatives*—e.g., *Beaver, Bishop, Caprock, Combine-7078, Fargo, Kalo, Martin, Midland, Plainsman, Quadron, Redbine, Redland, Westland,* and *Wheatland.—See also* COWPEA; SILAGE CROPS.

KALE (*Brassica oleracea* var. *acephala*), especially the large *thousand-headed kale* variety, is a hardy biennial plant which is grown in the northern Pacific Coast region in limited quantities for forage. Green kale is rich in vitamin A. Its leaves and stems are VEGETABLE BY-PRODUCTS which are valued as feedstuffs.

Dehydrated kale leaf meal is a good source of protein and vitamins A and B_2; it can be used advantageously in poultry feed as replacement for ALFALFA meal.

Reference: E.12.

See also PLANT BY-PRODUCTS; GRAZING.

KALMIA. The *Kalmia* spp., or *kalmias,* include *K. latifolia* = *mountain-laurel* and *K. angustifolia* = *sheep-laurel.*

KALMIAS (*Kalmia* spp.) are acid-soil evergreens with showy flowers; they are often found on ranges. Among its species are two POISONOUS PLANTS known as *mountain-laurel* (a large shrub, sometimes becoming a tree over 20 ft. high), and *sheep-laurel* or *lambkill* (up to 3 ft. high).

Cattle, sheep, and goats eating only 3 oz. green plants (especially leaves) per 100 lb. body weight a day develop such symp-

toms of poisoning as salivation, vomiting, and weakness.

Kalmias. Sheep-laurel is one of the poisonous kalmia species. (D.16.)

KALO and *Early Kalo* are intermediate-type grain SORGHUMS.

KAMALA, the hairs and glands of the the capsules of *Mallotus philippinensis taenifuge,* is one of the officially accepted FEED INGREDIENTS. It has some purgative action and in sufficient dosage removes intestinal worms, especially tapeworms.

KANHULL = *Chiefkan.*

See COMMON WHEAT.

KANOTA. *See* RED OAT (variety).

KANRED is a hard winter variety of COMMON WHEAT.

KANSAS COLLIER 704-D belongs to the forage SORGHUMS.

KANSAS COMMON ALFALFA is one of the COMMON ALFALFAS.

KANSAS ORANGE SORGO, like other varieties of *Orange sorgo,* belongs to the forage SORGHUMS.

KANSAS SOURLESS 702-H is a forage SORGHUM.

KAOLING. *Dwarf Shantung kaoling* and *Manchu Brown kaoling* are grain SORGHUMS.

KAPOK OIL MEAL, the by-product of the manufacture of oil from the seed of the KAPOK TREE, is of low palatability and, because of its high fiber content, inferior to cottonseed meal. It should therefore not be used in good livestock rations.

KAPOK TREE (*Ceiba pentandra*), grown in Malaya, has large seed-pods with cotton-like down (used in mattress manufacture) and black seeds, the size of small peas, from which an edible oil and KAPOK OIL MEAL are obtained.

KARMONT.
      See COMMON WHEAT (variety).

KAWVALE.
      See COMMON WHEAT (variety).

KEEL is a projecting ridge on a surface (like the keel of a boat). When a grass GLUME or LEMMA is compressed and boat-like, it is called a keel.

KELLNER FEEDING STANDARDS give recommendations in terms of (1) *starch value*, (2) digestible *true* (not crude) *protein*, and (3) *dry matter*. These standards, expressed in pounds, have not been widely used in this country.

Kellner's starch values are based on the assumption that 1 lb. digestible starch is the net energy unit. Thus, the starch value of dent corn is 81.5 lb. starch equivalent.

According to Kellner, 1 lb. digestible true protein has a starch value of 0.94 lb., and 1 lb. pure fat has a starch value of 2.41 lb.

> NOTE: Kellner's starch value of any feedstuff can be expressed in *Armsby's net energy value* by multiplying the amount of starch equivalent by 1.071 and expressing the result in therms. Thus, the 81.5 lb. starch equivalent of dent corn may also be expressed as 87.3 therms.

KELP, a source of IODINE, is any of the various large brown seaweeds of the families Fucaceae and Laminariaceae.—*See also* FEED INGREDIENTS.

KENLAND RED CLOVER is a *double-cut* variety of RED CLOVER.

KENTUCKY 1.
      See SIX-ROWED BARLEY (variety).

KENTUCKY 31 is a strain of TALL FESCUE.—*See also* RICE.

KENTUCKY 103 is a CORN hybrid.

KENTUCKY BLUEGRASS (*Poa pratensis*), sometimes called *junegrass* or simply "*bluegrass*," is used principally for lawns, turfs, and pastures. It is widely distributed in the northern states and the mountainous and cooler localities farther south.

Kentucky bluegrass. Plant, spikelet, and floret. (H.26.)

Kentucky bluegrass grows 18 to 24 in. tall and, under exceptionally favorable conditions, often reaches 36 in. This long-lived perennial is easily identified by its boat-shaped leaf tip. Some fields of it are known to be more than sixty years old. It spreads by underground rootstocks (RHI-ZOMES) and thus makes a dense sod. The open, pyramidal PANICLE produces much seed.

Kentucky bluegrass is dormant during exceptionally hot, dry periods of summer and turns brown unless ample water is applied. Proper fertilization is important.

Dangers: Among its enemies is the CHINCH BUG.

Reference: H.1.

See also BLUEGRASSES; GRASSES; SILAGE CROPS; PASTURE MANAGEMENT; PASTURE PLANTS.

KENTUCKY BLUEGRASS-MOLAS-SES SILAGE. *See* SILAGE CROPS.

KENTUCKY HEMP is the drug-type of HEMP.

KERATIN, the substance of which claws, nails, wool fibers, horns, and feathers are composed, is a complex PROTEIN material which contains also relatively large amounts of sulfur.

KERNEL is the mature body of an OVULE.

KERNEL SMUT OF RICE, caused by the FUNGUS *Neovossia horrida*, occurs occasionally in Arkansas, Louisiana, and Texas. It reduces the RICE yield and in some cases damages the quality materially because of the black or gray color of the milled rice. This coloration is caused by the spores of the fungus that are liberated when the rice is milled. In the most severe cases the entire starchy part of the kernel (endosperm) becomes a black mass of spores. In other cases a part of the kernel may be normal.

The smutted kernels are not noticeable until the rice is mature. Usually not more than two to six smutted kernels are found on a head. Little, if any, swelling of the kernel is caused by the smut, and the most conspicuous symptom is the dark discoloration on the hulls of mature seeds caused by the black smut spores showing through the hull on the outside. The embryo (germ) is never smutted and will germinate even when the entire endosperm is reduced to a mass of smut spores. The disease is usually of minor importance.

Control: Natural conditions in the field apparently check this disease. The commercial varieties now grown appear to differ in susceptibility to kernel smut.

Reference: T.2.

KERNEL SMUTS—LOOSE KERNEL SMUT OF JOHNSONGRASS, LOOSE KERNEL SMUT OF SORGHUM, and COVERED KERNEL SMUT—are FUNGUS DISEASES. They may be controlled by SEED TREATMENT with SULFUR, PHYGON, SPERGON, ARASAN, FORMALDEHYDE SOLUTION, or COPPER CARBONATE.

The sorghum smuts are not poisonous to livestock. Most sorghum kernel smut spores are killed by passing through the digestive tract of farm animals, so that there is little danger of spreading the smuts through the medium of barnyard manure.

Reference: M.2.

See also RICE.

KEROSENE is a light oil fraction of crude PETROLEUM. A refined grade called *deodorized kerosene* is used as INSECTICIDE, HERBICIDE, DISINFECTANT, or solvent of many organic insecticides.—*See also* CREO-SOTE BARRIERS; DODDERS; CRAB GRASS; ARMYWORMS; PETROLEUM OILS.

KEY'S PROLIFIC = *Mediterranean wheat. See* COMMON WHEAT.

KHAPLI is a wheat variety, belonging to the EMMER species.

KHARKOF = *Turkey.*

See COMMON WHEAT.

KIDDER. *See* SMOOTH BROME (variety).

KINDRED.

See SIX-ROWED BARLEY (variety).

KING FESCUE = TALL FESCUE.

KING PHILIPP FLINT.

See CORN (variety).

KING RANCH BLUESTEM.

See TURKESTAN BLUESTEM (variety).

KINGWA. *See* SOYBEAN (variety).

KING WHEAT = *Fulcaster.*

See COMMON WHEAT.

KLONDIKE = *Goldcoin.*

See COMMON WHEAT.

KNOT, is (1) a NODE or (2) a *gall,* i.e., an overgrowth of tissue, particularly one of the many irregular swellings found on the root system of diseased plants. The disease ROOT KNOT is caused by the ROOT-KNOT NEMATODES.

KNOT BINDWEED
= WILD BUCKWHEAT.

KNOTGRASS (*Paspalum distichum*), or *joint grass*, is an obnoxious perennial WEED pest found in RICE fields. It is a creeping GRASS that spreads rapidly by rooting at the joints, or nodes.

NOTE: The name joint grass should not be used since it is also applied to BERMUDA-GRASS.

KNOTWEEDS (*Polygonum* spp.) are WEEDS; e.g., *knot bindweed*, better known as WILD BUCKWHEAT. These weeds are characterized by dense SPIKES of minute flowers.—*See also* ALFALFAS.

KOBE is a variety of COMMON LESPEDEZA. —*See also* ANNUAL LESPEDEZAS; LESPEDEZAS.

KOCHIA. *K. vestita*
= GRAY SUMMER-CYPRESS.

KOHLRABI (*Brassica caulorapa*), which belongs to the CABBAGE family, is occasionally grown as forage. Its turnip-like stems and leaves are relished by sheep.

KOREAN LAWNGRASS. *See* ZOYSIA.

KOREAN LESPEDEZA (*Lespedeza stipulacea*) is a larger, coarser, and earlier maturing plant with broader leaflets than COMMON LESPEDEZA. Both are ANNUAL LESPEDEZA species. Korean makes good hay crops as far north as central Illinois and Indiana, but it is too early maturing to be useful south of northern Mississippi, Alabama, and Georgia.

Since the seed and leaves of Korean lespedeza remain on the plant well into the winter, late grazing is provided by this species as livestock readily eat the dried leaves and seed.

Dangers: BLACKPATCH may attack this lespedeza species.

Varieties: The following are important varieties of Korean lespedeza: *Harbin*—it is very early maturing and makes small growth; *Early Korean*—maturing in mid-season and of medium growth; *Late Korean* and *Climax*—later maturing and giving larger yields of hay than the other varieties.

Reference: M.7.

See also LEGUMES; LESPEDEZAS; ORCHARD-GRASS; PASTURE PLANTS.

K.R.BLUESTEM (or *K-R bluestem*), short for *King Ranch bluestem*, is a variety of the TURKESTAN BLUESTEM.

KUBANKA is a white DURUM WHEAT.

KUDZU (*Pueraria thunbergiana*) is one of the few perennial LEGUMES adapted to the southeastern states and useful both for forage and for soil improvement. The plant is of special interest to this region, since locally grown kudzu hay can be used to replace ALFALFA or other hay that must be hauled long distances.

Although not entirely winter-hardy in the northern part of the United States, it has survived in somewhat protected situations as far north as Lincoln, Nebraska, in the Great Plains area. It also survives in the Pacific Northwest where the winters are mild and the summers cool, but growth is slow and slight. In the Atlantic coastal area, in the latitude of Maryland, kudzu survives the winters but does not produce as heavily as farther south. It is best adapted in the eastern states south of Virginia and Kentucky and west to eastern Oklahoma and Texas.

Kudzu is a rapid-growing, long-lived, viny, perennial plant with a comparatively large taproot and exceedingly long runners that with age become woody at the base.

In the early stage of growth kudzu vines are soft and pliable and have a fuzzy or hairy appearance. Because kudzu vines root readily at the nodes when they come in contact with moist soil, new plants are formed which are the source of much planting material.

The leaves of kudzu are abundant, very large, and look like grape leaves. The large purple flowers are produced in multitudes late in the season and precede the clusters of densely hairy pods, which are about 2 in. long and usually contain few or no seeds. The latter is small and hard and should therefore be scarified. An average germination of about 50 percent is usual after scarification.

Kudzu is a warm-weather plant. In Georgia and adjacent territory its growth starts about the first of April and continues until checked by cold weather.

Flowering occurs late in August or in the early part of September. Farther north growth starts later, with some growth continuing until frost.

Kudzu grows best on well drained loam soil of good fertility, but can be grown on poorer soil by proper use of fertilization and manure. Although a complete fertilizer might be advisable on some soils, liberal use of manure and a light application of superphosphate about the plant when it is first set, are generally what is most needed. Borax at a rate of 20 lb. per acre has given good results in a few instances. Established stands used for grazing, or stands from which a hay crop has been removed, should receive 400 to 600 lb. per acre of superphosphate every second or third year or, if available, 10 tons of good stable manure.

When kudzu seed is available, seedlings grown in nurseries for one year are satisfactory planting stock. Kudzu crowns—i.e., the plants established by the rooting of runners at the nodes—have been the stock most commonly used in establishing commercial plantings. Best results with crowns have been attained by the use of two-year-old plants of moderate size.

In the South the best time for planting kudzu is in February and early March; farther north plantings should be done late in March and April. For best results in setting it is important to protect the plants from drying while being set and to firm the moist soil about them to prevent drying afterwards.

The land on which kudzu is to be planted should be prepared in the fall or early winter. If time and labor permit, the entire field may be broken before preparing the rows. To assure maximum production, kudzu plants should be inoculated at time of planting.

For the first year or two after planting, kudzu should be cultivated to keep down weeds. Covering the vines at the nodes with soil will insure rooting, greatly increase the number of plants, and help thicken thin stands.

Kudzu should not be cut the first year and should be cut only once the second year, or grazed lightly.

Kudzu leaves and racemes of flowers. (M.9.)

A cultivated crop, such as CORN, can be grown between the rows, so that a cash return may be obtained from the land the year the kudzu is being established.

The crop should never be closely pastured. One cutting in June or early in July and another in the fall, just before frost, can usually be safely made. If one cutting only is made, this should be before August.

The feeding value of kudzu is satisfactory. The hay is rather coarse but moderately leafy and palatable to most livestock. It can be fed with very little waste to cattle or horses.

The heavy viny growth of kudzu makes it difficult to harvest, because the long vines are caught by the divider-board of an ordinary mower. To overcome this difficulty a specially made iron rider-bar may be attached to the end of the cutter bar, which divides and frees the vines as the swath is cut. Stub guards on the mower will also help to prevent clogging.

Kudzu stands should be two years old before being harvested for *hay*. The hay should be left in the swath for several hours before windrowing. The following morning when the dew is off, the kudzu should be put into small stacks or the windrows should be turned, and in the afternoon it should be put in the barn or baled. Yields of 2 tons per acre can be expected from good stands on fertile soil.

Kudzu makes good PASTURE. No case of bloat has been reported over a period of years. It can be pastured from May until frost, or even later. It is especially valuable as a reserve feed for periods of drought.

Kudzu plants should not be grazed continuously until the third year. Once well established, fields of kudzu will withstand continuous grazing, but the plants should not be grazed closer than 12 to 18 in. high. For maximum production a field should be divided into two or more pastures, and the plants grazed in rotation to 6 to 10 in. In the fall RYE, or OATS, or a winter legume (CRIMSON CLOVER, BUR-CLOVER, or VETCH), should be seeded in the kudzu pasture to prevent loss of plant food by leaching during the winter months, and to supply late winter grazing.

Good *silage* can be made from kudzu, especially if used together with GRASS in a mixture containing about 60 percent moisture. The total moisture content of the kudzu plants at time of cutting is about 75 percent; this means that the kudzu must be handled as rapidly as possible. Cattle readily eat good kudzu silage.

Kudzu has been used but little in rotation; however, crops following a kudzu planting continue to give heavier yields for several years.

When land that is suitable for growing corn, COTTON, or other standard crops has been occupied by kudzu for a term of years, the kudzu can be plowed under to the great benefit of subsequent crops.

Finally, kudzu is being used in erosion-control work to hold banks, stop washing in gullies and diversion channels, and reduce soil losses on slopes where a permanent planting can be used to advantage.

There is a rather prevalent belief that kudzu is likely to become a serious pest if planted in or near cultivated cropland. Experience has shown that this belief is unfounded. Where kudzu is growing on land that is needed for some other purpose, it may be eradicated in one season by plowing, or in two or three seasons by close, continuous grazing.

Certain states have laws that govern the handling and shipping of kudzu plants both inside and outside their boundaries. These laws usually require inspection by an authorized inspector and the issuance of an inspection certificate to accompany the plant material in transit. Information regarding inspection can be obtained from a county agricultural agent.

Dangers: No serious diseases have so far affected kudzu. HALO BLIGHT and ANGULAR LEAF SPOT are not uncommon, but have not yet become a limiting factor in the plant's production.

Kudzu is susceptible to ROOT-KNOT NEMATODES which cause a condition commonly called ROOT KNOT; the latter reduces the vigor of plants, particularly during periods of drought. GRASSHOPPERS sometimes eat the leaves of kudzu to such an extent that the foliage is somewhat ragged, but unless the attack occurs late in fall, the plants usually produce a new crop of foliage before frost. Other insects that frequently attack the leaves of kudzu are the VELVETBEAN CATERPILLAR and the LOCUST SKIPPER BUTTERFLY whose caterpillars feed on the leaves; serious injury, though, from these causes are uncommon.

Varieties: No varieties of kudzu have been established commercially, though individual plants show wide variation in leafiness, pubescence (hairiness), length of nodes, and vigor. Since kudzu is easily propagated vegetatively, it would not be difficult to establish and maintain superior varietal strains for forage and ground cover.

References: M.9; B.5; E.8; M.10.

See also INOCULATION; SILAGE CROPS; KUDZU MEAL.

KUDZU MEAL, produced by grinding KUDZU hay, is used increasingly in poultry feeds. It is similar in composition to ALFALFA meal.

KURA CLOVER (*Trifolium ambiguum*) is an excellent nectar-producing TRUE CLOVER species.

KURSK FOXTAIL MILLET.
                    See FOXTAIL MILLET (variety).

KYS is one of the inbred lines of CORN hybrids.

L

L 317 is one of the inbred lines of CORN hybrids.

LABELS are required on all packages of feed. Acconding to the official regulations of the Association of the American Feed Control Officials, the label must be printed on one side of a *tag* attached to the package, or upon one side of the package itself. The label must be clear and distinct, printed in type of sufficient size to be easily read; the names of all ingredients must be printed in the same size. The "sliding scale" method of expressing guaranties—e.g., "protein . . . 15-18%"—is prohibited.

No advertising matter of any kind shall be printed on the label.

If to any unmixed by-product-feed there should be added SCREENINGS or SCOURINGS (either ground or unground, bolted or un-bolted), such brand shall be so registered, labeled, and sold as to clearly indicate this fact. The word "screenings", or "scour-ings"—as the case may be—shall appear as part of the name or brand.

Brand names must not tend to mislead the purchaser with respect to any quality of the feed. If the brand name indicates that the feed is made for a specific use, the character of the feed must conform there-with; a mixture labeled "dairy feed", for example, must be adapted for that purpose. A brand name of a nonmedicated feed shall not be derived from one or more ingredients of the mixture. A distinctive name shall not be one representing any component of the mixture.

Each and every *ingredient* of the feed mixture must be stated, and their common English names are to be used. Each in-gredient must be specifically named. Such general statements as "corn products," "mixed bran," "wheat mill by-products," "corn mill by-products," etc., may not be used.

Such terms as "*mill run bran*" and "*mill run shorts*" cannot be used to refer to a mixture of wheat bran and wheat shorts. A mixture of wheat bran and corn bran cannot be labeled "bran"—the proper designation is "*wheat bran and corn bran.*"

Each package of feed must show the *net weight* of the feed in the package.

If the analysis of an inspector's sample of a feed shipment shows—as compared with the label and guaranty—a deficiency in the amounts of *fat* and *protein*, or an excess of *fiber*, or, if the statement of ingredients is incomplete or inaccurate, or the shipment is found to be sold or offered for sale in violation of any requirements of law or regulations, the guarantor or person responsible for the sale of the feed and the dealer shall be fully informed regarding the objection found, and be given ten days in which to submit a written explanation unless, in the judgment of the enforcing officer, immediate action is necessary.

Reference: F.6.

See also DISINFECTANT; DISINFESTANT; DRUGS; INOCULATION; NONPROTEIN NITRO-GEN; F.D.A.

LACTASE, an ENZYME occurring in the juices of the small intestine, hydrolyzes LACTOSE to GLUCOSE and GALACTOSE.—*See also* MILK PRODUCTS.

LACTIC ACID, a colorless, sirupy, water-soluble liquid, is produced by fermentation of LACTOSE and other carbohydrates; it occurs in sour milk and SILAGE.—*See also* DISTILLERS' PRODUCTS.

LACTIC BACTERIA are used in prepar-ing cultured skimmed MILK PRODUCTS. They also cause milk to sour and produce the desirable fermentation of SILAGE.

LACTOBACILLUS is a genus of long, rod-shaped micro-organisms that produce LACTIC ACID from CARBOHYDRATES. They occur in MILK, milk products, in the in-testinal tract, SILAGE, etc.—*See also* BAC-TERIUM; BACILLUS.

LACTOSE, or *milk sugar*, is a SUGAR ob-tained from cow's milk or whey. It forms a white, crystalline powder, is mildly sweet, and slightly soluble in water. Lactose has a mild laxative action. It is used as a nutritive FEED INGREDIENT.—*See also* GALACTOSE; MILK PRODUCTS.

LADAK belongs to the VARIEGATED ALFALFAS.

LADINO CLOVER, or *Ladino white clover*, is a variety of WHITE CLOVER.—*See also* ALFALFAS; TALL FESCUE; ORCHARDGRASS; HAY; SILAGE CROPS; PASTURES; GRAZING; INOCULATION.

LADY BEETLES, also called *Ladybird beetles* or *Ladybugs*, belong to various genera. They are small INSECTS of different color but usually with conspicuous black spots. Lady beetles are most beneficial; one of the species is a natural enemy of the ALFALFA CATERPILLAR.

LAMB RATIONS. *See* SHEEP RATIONS.

LAMBSKILL = *sheep laurel. See* KALMIAS.

LAMBSQUARTERS (*Chenopodium album*) is an annual WEED which is sometimes used for greens on ranges. The plant reaches a height of from 1 to 10 ft. in rich soils; it is often attacked by the GARDEN WEBWORM.—*See also* RANGE PLANTS.

LAMBSTONGUE GROUNDSEL. *See* GROUNDSELS; SENECIO; POISONOUS PLANTS.

LAMELLA is any thin or platelike structure.

LAMINA = BLADE.

LAMINARIACEAE are one of the seaweed families.—*See also* KELP.

LANCASTER. Both, *Lancaster* = *Fulcaster* and *Lancaster Red* = *Mediterranean wheat*, are varieties of COMMON WHEAT.

LANCASTER 9 = *Pennscott red clover.* *See* RED CLOVER.

LANCE CROTALARIA (*Crotalaria lanceolata*) is one of the nonpoisonous CROTALARIAS which are grown extensively in the United States, especially for soil improvement.

LANCEOLATE means: lance-shaped; i.e., several times longer than broad and tapering from the relatively narrow base to the APEX.

LAND PLASTER = GYPSUM.

LAPHYGMA. *L. frugiperda* = FALL ARMYWORM.

LAPPA CLOVER (*Trifolium lappaceum*), also called *lappacea clover*, is a winter-annual species of TRUE CLOVER. It has become well established in a portion of the Black Belt of Alabama.

The habit of growth, and the growth period of lappa clover are similar to PERSIAN CLOVER, though somewhat later in maturity. The seeds germinate in the fall and the plants make a low rosette growth during the winter. In the spring rapid growth occurs, the plants flower profusely, set seed, and die in late spring or early summer. This clover perpetuates itself by self-seeding and spreads naturally through seed carried by animals, birds, water, wind, etc. The inconspicuous flowers, borne in burlike heads at the ends of the branches, are reddish in color.

Lappa clover is principally a spring pasture clover for growing in association with summer growing grasses, such as BERMUDA-GRASS, CARPETGRASS, and DALLISGRASS. Under favorable growth conditions and when not grazed, the plants may reach a height of 15 to 18 in.; this furnishes a good cutting of hay. When pastured the plants spread out and branch profusely. All classes of livestock seem to relish this clover.

In the northern part of the southeastern states seedings should be made during late September or October, while in the southern part seedings may be delayed until late October and November. For the first sowing, the seed should be inoculated with clover culture immediately before planting.

On most soils the addition of 200 to 400 lb. superphosphate per acre at time of seeding will aid in establishing a stand and stimulate plant growth. This clover is best adapted to the low-lying heavy, dark-colored soils.

Reference: H.13.

See also LEGUMES; INOCULATION.

LAREDO. *See* SOYBEAN (variety).

LARGE HOP CLOVER (*Trifolium procumbens*) is a winter annual, similar to SMALL HOP CLOVER, except that it is more productive. For stand establishment and good growth this TRUE CLOVER species requires a greater supply of mineral elements than small hop clover. It is valuable for early spring pasturage.

Reference: M.3.

See also LEGUMES; INOCULATION.

LARGE INTESTINE.
                    See DIGESTIVE TRACT.

LARGE LIMA BEAN. *See* LIMA BEANS.

LARGE PEPPERGRASS
                    = PEPPERGRASS.

LARGE-SEEDED ALFALFA DODDER
(*Cuscata indecora*), one of the DODDERS,
exhibits the same preference for leguminous
crops, particularly ALFALFAS, as the related
SMALL-SEEDED ALFALFA DODDER. It occurs
occasionally in the South, is common in
the West, but rarely found in the East.

Reference: H.19.

LARKSPURS (*Delphinium* spp.) belong
to the POISONOUS PLANTS. Larkspurs are
perennials which occur on western ranges.
They vary in height from a few inches to
7 ft. and contain ALKALOIDS which are
highly poisonous.

If cattle—within a few minutes—eat
leaves of larkspur plants (especially of

Larkspur. (N.6.)

young plants) at the rate of 8 oz. per 100
lb. body weight, they may soon develop
such symptoms of poisoning as weakness,
trembling, and constipation.

Control: Eradication of larkspur by
grubbing, burning, or chemical means
(particularly by spraying with SODIUM
CHLORATE) is important to prevent cattle
losses on ranges.

Reference: U.6.

LARVA (plural: *larvae*) is an INSECT which
has left the egg only recently and is now
in the first ("wormlike") stage of its complete metamorphosis. The larvae of different types of insects have different names;
i.e., CATERPILLARS (of moths and butterflies), MAGGOTS (of flies), and GRUBS (of
beetles).

LASIUS. *L. niger alienus americanus* =
CORNFIELD ANT.

LASPEYRESIA. *L. nigricana* = PEA
MOTH.

LATE KOREAN.
                    See KOREAN LESPEDEZA (variety).

LATERAL means: of, or pertaining to, a
side of an organism; e.g., lateral FLORETS.

LATE RED CLOVER = *American single-
cut clover. See* RED CLOVER.

LATE SOYBEAN = *Mammoth Yellow.
See* SOYBEAN.

LATHYRUS. *L. hirsutus* = ROUGHPEA.

LAUREL is a name applied to plants
belonging to a number of genera. *Sheep-
laurel* and *mountain-laurel* are KALMIAS.—
See also BLACKLAUREL.

LAWN CLIPPINGS may be used as
poultry green feed.—*See also* GRAZING.

LAXATIVE is a mild purgative; e.g., dried
fermentation solubles, MOLASSES, COTTON-
SEED, or LINSEED.—*See also* SORGHUMS;
VELVETBEANS; VITAMINS.

LAYING RATIONS.
                    See POULTRY RATIONS.

"L" BARLEY = *Kindred. See* SIX-ROWED
BARLEY.

LEACHING means: removing materials
by dissolving; especially, losing of plant
nutrients by the action of water seeping
through porous soil.

LEAD ARSENATE, a white, water-insoluble, heavy powder, is a poison.

It is sometimes used as a spray for the
control of the CLOVER LEAF WEEVIL. The

spray may be prepared from 2 lb. lead arsenate and 2 lb. laundry soap in 100 gal. water and used at the rate of 100 gal. per acre.

LEAF AND POD SPOT, caused by the FUNGUS *Ascochyta pisi* (which produces carrot-red spores), affects the VETCHES and the FIELD PEA. The most evident symptoms of the disease are relatively few but rather definite, sunken, tan or brown spots, which have dark-brown margins and are circular on leaves and pods. The pod lesions usually become sunken. Affected leaves eventually shrivel and dry into a blighted condition.

Stem lesions originate as separate infections or as a continuation of PETIOLE infection around the nodal area. They may girdle the entire stem, weakening it so that it is easily broken.

The disease organism can infect the seed. Such infection serves to overwinter the pathogens and is a means of transporting the disease from one region to another. When infected seed is planted, the disease first appears as a foot rot on the young seedlings at the point of seed attachment and often kills or weakens the young plant.

Control: Use of seed grown in dry regions is the best insurance against seedling infection. Treatment of seed with fungicides reduces the surface contamination but will not eliminate internal infection.

References: S.13; M.24.

See also ASCOCHYTA BLIGHT COMPLEX.

LEAF BEET = SWISS CHARD.

LEAF BLIGHT, caused by the FUNGUS *Helminthosporium turcicum*, is most prevalent in the warmer humid Atlantic and Gulf Coastal Plains of the southern and southeastern states. It causes serious losses in grain and forage SORGHUMS, and particularly in Sudangrass, which it injures in the mentioned areas and in New York, Pennsylvania, Ohio, Wisconsin, and Minnesota. Leaf blight also attacks CORN.

The fungus is carried on the seed and lives in the soil on dead and decaying plant material. It may cause SEED ROT and SEEDLING BLIGHT or sorghums, especially in cold and excessively moist soil. Small reddish-purple or yellowish-tan spots usually develop on the leaves of infested seedlings. On infected leaves of older plants the spots gradually enlarge and form elliptical areas, $\frac{1}{8}$ to $\frac{1}{2}$ in. wide and several inches long. These spots may merge sufficiently to kill large parts of the leaves, which then wither to such an extent that badly affected plants look as if they had been frosted. The center of the individual spots usually is grayish or straw-colored and is surrounded by reddish-purple or tan borders, depending upon the variety.

A greenish, moldlike growth of spores develops in the center of the leaf spots during warm, humid weather. These spores are scattered by wind or rain and infect other leaves. When weather conditions are favorable, the disease spreads rapidly and may cause serious damage.

Control: The chief hope of controlling leaf blight lies in the development of resistant varieties. The following varieties of sorghums appear to be somewhat resistant: Most kafirs and certain kafir crosses— Quadroon, Early hegari, Spur feterita, shallu, Dawn—and Atlas, McLean, Gooseneck, Norkan, Denton, and Cowper sorgos. Tift Sudangrass also shows some resistance but is not immune. Rotation is not an effective method of control, because the fungus lives in the soil for several years. SEED TREATMENT may prevent some seedling infection and spread of the LEAF DISEASE to new areas.

Reference: L.1.

See also FUNGUS LEAF DISEASES; NORTHERN CORN LEAF BLIGHT; SOUTHERN CORN LEAF BLIGHT; BACTERIAL LEAF BLIGHT.

LEAF BLOTCH, caused by the FUNGUS *Septoria pisi*, kills areas of the leaf and may run into the PETIOLES and stems of FIELD PEAS. Usually the spots appear at the margin of the leaf as yellowish areas which gradually darken and enlarge. The entire leaf and petiole may be killed and brownish, dead areas produced on the stems. Occasionally a field of peas may be injured severely, but generally the disease is not important.

Control: Rotation helps to hold leaf blotch in check.

Reference: M.24.

See also OAT LEAF SPOT; YELLOW LEAF BLOTCH.

LEAF DISCOLORATIONS. *See* NON-PARASITIC SORGHUM-LEAF DISCOLORATIONS.

LEAF DISEASES, or *foliage diseases*, may be caused by (1) FUNGI, which develop roughened leaf spots or (2) BACTERIA, which form drops or films of exudate that dry to thin, crustlike scales.

Among the more important leaf diseases are the following: YELLOW LEAF BLOTCH; BROWN STRIPE; DOWNY MILDEW; SUMMER BLACKSTEM; LEAF RUST, ALFALFA RUST; CLOVER RUST; COMMON LEAF SPOT; PLEOSPORA LEAF SPOT; ALTERNARIA LEAF SPOT; ASCOCHYTA BLIGHT OF COTTON; ASCOCHYTA BLIGHT OF PEA; PSEUDOPLEA LEAF SPOT; BACTERIAL LEAF AND STEM SPOT; ANGULAR LEAF SPOT; NORTHERN ANTHRACNOSE; and ANTHRACNOSE OF GRASS.—*See also* FUNGUS LEAF DISEASES; BACTERIAL LEAF DISEASES; LEAF SMUTS; LEAF SPOTS.

LEAFHOPPERS. RED CLOVER is attacked by several species of leafhoppers, one of the most common being the POTATO LEAFHOPPER. Leafhoppers also attack many other crops; e.g., SOYBEAN, RICE, OATS; grain SORGHUMS, and VETCHES. And they transmit VIRUS DISEASES of LEGUMES, especially of red clover.

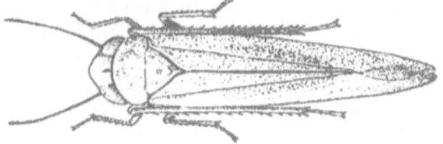

Leafhopper. (U.13.)

The INSECTS feed chiefly on the undersides of the leaves. Leafhopper injury is generally inicated by the leaf margins' changing to a red or yellow color and later browning as the margin dries. A V-shaped "tipburn" is a common symptom, but is not always present. In most cases of serious injury, the plants show a lack of vigor and growth is retarded.

Rather severe infestations of leafhoppers have been observed in fields of dry locations. Some leaf symptoms are evident on the first hay crop, but they do not reach a peak until later in the summer.

Reference: E.5.

See also HALO BLIGHT; BACTERIAL STRIPE OF OATS; CLOVER LEAFHOPPER; DWARF DISEASE; WITCHES'-BROOM.

LEAF MEAL. *See* ALFALFAS.

LEAF MINER. *See* RICE LEAF MINER.

LEAF RUSTS are FUNGUS DISEASES which affect many crops, particularly LEGUMES and GRASSES. Among these LEAF DISEASES are ALFALFA RUST and CLOVER RUST, caused by various *Uromyces* spp. A number of *Puccinia rubigo-vera* varieties possess the capacity to infect grasses, including cereals, such as ITALIAN RYEGRASS, TURKESTAN BLUESTEM, RYE, and WHEATS.—*See also* DURUM WHEAT; SPELT; COMMON WHEAT.

LEAF SCALD. *See* SCALD.

"LEAF SHATTER" is due to overdrying of HAY.

"LEAF SMUT" = (*true*) *stripe smut. See* STRIPE SMUTS.

LEAF SMUT OF RICE caused by the FUNGUS *Entyloma oryzae*, occurs in Arkansas, Louisiana, and Texas. It is found in July, August, and September on RICE leaves, leaf sheaths, and occasionally on the upper part of the stalk, where it shows as small, black spots. These spots contain the black spores of the smut fungus. The spots are at first slightly raised, but later the leaf tissues over the spots break open and allow the spores to escape. The injury to the plant in this disease consists in reducing the leaf area.

Control: Varieties differ in susceptibility to leaf smut. The use of the more resistant types, therefore, will help to control the disease.

Reference: T.2.

LEAF SMUTS include LEAF SMUT OF RICE and the STRIPE SMUTS.

LEAF SPOT OF PEANUT is caused by the fungi *Mycosphaerella arachidicola* and *M. berkeleyii*, which produce spots on leaves, stems, and pegs of the plants. The first mentioned FUNGUS attacks PEANUTS early in the season, the other late.

The causal fungi overwinter on the diseased leaves and stems in the field. During the spring and early summer months spores are developed by the fungi. These spores are spattered by raindrops or blown by wind and, falling upon a peanut leaf, produce infection which spreads rapidly. Spores of this fungi are carried long distances by air currents, and leaf spot is universally present in peanut fields late in the season.

As the disease progresses, the older leaves may show as many as 100 spots. The leaves turn yellow and drop off; spots appear on the stems, weakening them; and the plants finally die prematurely. As soon as the plants die, fungi and bacteria take over, rotting the pegs, roots, and stems. The dead stems and pegs quickly become brittle. There is no practical way to save a very large percentage of the nuts after the stems are dead. In severely infected peanuts, harvesting must be done before full maturity of the nuts.

Control: Leaf spot of peanut can be effectively controlled by dusting with SULFUR, or by spraying with BORDEAUX MIXTURE.

References: B.10; W.17.

LEAF SPOTS include SUMMER BLACK STEM; COMMON LEAF SPOT; RED LEAF SPOT; ZONATE LEAF SPOT; GRAY LEAF SPOT; PSEUDOPEZIZA LEAF SPOT; PLEOSPORA LEAF SPOT; STEMPHYLIUM LEAF SPOT; PEPPER SPOT; BROWN SPOT; BROWN STRIPE; ALTERNARIA LEAF SPOT; LEAF SPOT OF PEANUT; OAT LEAF SPOT; ANGULAR LEAF SPOT; and BROWN-BORDERED LEAF SPOT.

White leaf-spot = WHITE SPOT.

See also LEAF DISEASES; SAFFLOWER.

LEAF WEEVIL. *See* CLOVER LEAF WEEVIL.

LEAMING. *See* CORN (variety).

LEAP, or *Leap's Prolific*, is a soft winter variety of COMMON WHEAT.

LEAVES AND TWIGS of range shrubs and trees often serve as forage. Green leaves are generally preferred and may reach the feeding value of hay, but dried, dropped-off tree leaves have practically no feeding value.

LECITHIN, a complex organic compound, containing CHOLINE, 4 percent PHOSPHORUS, FATTY ACIDS, and GLYCERIN, is obtained from egg yolk or soybeans. It forms a brownish-yellow, waxy mass which swells in water.—*See also* ANTIOXIDANTS.

LEE. *See* COMMON OAT (variety).

LEGUME BACTERIA are single-celled micro-organisms that vary in size and shape with age and with the composition of the medium in which they grow. Under the microscope they may be seen in either the usual rod forms, or the irregular X, Y, or club-shaped forms.

Farmers have been accustomed to ordering cultures for the INOCULATION of leguminous plants according to group designations. Seven of these groups are now recognized, namely: ALFALFA group; CLOVER group; PEA and VETCH group; COWPEA group; BEAN group; LUPINE group; and SOYBEAN group. Very few LEGUMES require specific strains of legume BACTERIA for effective inoculation; among them are BIRDFOOT TREFOIL and BIG TREFOIL.

Once legume bacteria are in the soil, they are subject to the conditions existing there; if the chemical reaction of the soil is suitable, if sufficient moisture and plant food is available, and if temperatures are not too high, the bacteria should function normally. Acid soils tend to eliminate the bacteria, especially those of alfalfa, SWEETCLOVER, RED CLOVER. Seed treated with legume bacteria should not come in direct contact with caustic lime or mixed fertilizers.

Bacteria that dry on seed soon die. If inoculated seed must be kept as long as forty-eight hours, it is advisable to reinoculate. For this reason the purchase of pre-inoculated seed is not advisable.

Most seed disinfectants are toxic to legume bacteria. Consequently, legume seed that has been treated with disinfectant compounds should not be inoculated in the usual manner.

The more effective strains of legume bacteria can increase the yield of PROTEIN content of legumes as much as 20 percent over the average natural legume bacteria in the soil.

Reference: E.4.

See also NITROGEN; NITROGEN FIXATION.

LEGUME FEED PRODUCTS which are officially recognized may be obtained from ALFALFAS, LESPEDEZA, SOYBEANS, VELVETBEANS, or PEANUTS.—*See also* BY-PRODUCT FEEDSTUFFS.

LEGUME HAY is particularly valuable— it is very rich in VITAMIN A, VITAMIN B, and CALCIUM and is higher in PHOSPHORUS than most grasses. LEGUMES produce more hay per acre than other forage crops.

SUDANGRASS, however, is a crop that, on the same soil, might outyield most legumes.

LEGUME-POD. *See* LEGUMES.

LEGUMES (Fabaceae, more often called Leguminosae) are a large family of plants that bear PODS; e.g., beans, peas, soybeans, and cowpeas. The pod, also called *"legume,"* is a superior one-celled fruit, usually splitting into two parts, and having the seed attached along the ventral suture.

The legume family contains three subfamilies, or divisions. These are known as (1) *Mimosoideae*, the flowers of which are regular and usually in dense heads; (2) *Caesalpiniodeae*, whose flowers are irregular and fewer in a cluster or RACEME; (3) *Fabiodeae*, or *Papilionoideae*, the most important group which includes all the economic legumes of agriculture.

The flowers of the Fabiodeae are distinctive. The petals of an individual flower vary in shape. The largest and most showy petal is known as a BANNER (or standard); it is usually nearly flat and somewhat circular and is the outer one of the petals. The two inner petals which enclose the STAMENS and the PISTILS, are folded together and usually bent or curved; they are known as the KEEL. On either side of the keel are two petals, called the WINGS. When the petals are in normal position and the banner extended, the resulting flower has somewhat the appearance of a butterfly and has accordingly been called papilionaceous (butterfly-like).

The legumes, for the most part, have higher feeding value than nonlegumes. One of the main reasons why they are superior is the fact that they generally contain a higher percentage of PROTEIN than nonlegumes, and protein is an essential food constituent. The seeds of legumes are particularly high in protein, but the leaves and stems also contain a relatively higher amount than other plants, harvested at a like stage of maturity. Not only have the legumes a higher percentage of protein, but they also have a higher quality protein than nonlegumes. This is of importance and helps greatly in obtaining high nutritive value in feeds for animals (as well as food for human consumption). The quality of the protein is such as to make the legumes valuable in supplementing the CEREAL GRAINS, which do not have the proper protein for a balanced livestock feed.

Legumes are also valuable because they contain a comparatively large amount of CALCIUM and a fair amount of PHOSPHORUS, which is necessary in proper nutrition. Likewise, they are recognized as the best source of VITAMIN A and (when properly cured) of VITAMIN D for livestock feed, and are largely depended upon for suppylying these constituents.

If turned under as green manure, legumes, furthermore, are superior for soil improvement, due to the large amount of NITROGEN they are able to supply for the use of subsequent crops. However, their ability to return a large amount of nitrogen to the soil is not the only reason that makes them desirable for soil improvement; another one is the fact that much of their nitrogen content is taken from the air. (It is generally assumed that at least half the nitrogen in legumes comes from the air.) Legumes thus add nitrogen to the soil, whereas nonlegumes merely take it from the soil.

Getting nitrogen from the air is usually considered a special function of legumes, although it is known that a few other plants can do so, too. In the case of legumes, symbiotic BACTERIA known as *Rhizobia* take nitrogen direct from the air as they grow and multiply in nodules on the roots of the legumes. The nitrogen, in turn, becomes available to the legume plant and aids in its nourishment and growth.

In order to attain the advantage of SYMBIOSIS it is necessary to bring the *Rhizobia* in contact with the young, growing rootlets of the legume. This practice is referred to as INOCULATION: it is accomplished by mixing a liquid or humus culture of *Rhizobia* inoculum with the seed just before seeding.

The percentage of nitrogen in the roots of a number of legumes ranges from 1.4 to 2.3 percent, and in the tops from 2.1 to 2.8 percent.

In the prevention of erosion, legumes fill a three-fold role: they protect the soil and thus decrease run-off; they add organic matter, which increases the absorbing power of the soil for water; and they increase the store of nitrogen in the soil.

Legumes, like other crops, require proper fertilization for maximum growth. Phosphorus is needed in almost all areas, as may potash and calcium. On soils low in organic matter, a small amount of commercial nitrogen, applied at planting time, will help get the legumes off to a good start.

Application of as little as from 5 to 15 lb. BORAX per acre to certain soils greatly increases the yield of legume seeds, especially of crimson clover, red clover, white clover, and alfalfa, but it does not influence forage yields. Larger applications, particularly on the lighter, more sandy soils, may cause stand damage.

Dangers: Among the enemies of the legumes are the following: ROOT-KNOT NEMATODES, BLACKPATCH, DOWNY MILDEW, and CROWN ROT; ARMYWORM, PEA APHID, and POTATO LEAFHOPPER.

Genera and Species: Among the most important legume genera and species are the ALFALFAS; ALSIKE CLOVER; ALYCE-CLOVER; ANNUAL LESPEDEZAS; BALL CLOVER; BIG TREFOIL; BIRDFOOT TREFOIL; BLACK MEDIC; BUR-CLOVER; CAROB BEAN; CAROLINA CLOVER; CHICK PEA; CLOVER; CLUSTER CLOVER; COWPEA; CRIMSON CLOVER; CROTALARIAS; DESMODIUM; FENDLER CLOVER; FIELD BEANS; FIELD PEA; FLORIDA BEGGARWEED; HAIRY INDIGO; HORSEBEAN; HUNGARIAN CLOVER; JACKBEAN; KOREA LESPEDEZA; KUDZU; KURA CLOVER; LAPPA CLOVER; LARGE HOP CLOVER; LESPEDEZAS; MESQUITE; MUNGBEAN; PEANUT; PERSIAN CLOVER; RED CLOVER; ROSE CLOVER; ROUGHPEA; SEASIDE CLOVER; SMALL HOP CLOVER; SOYBEAN; STRAWBERRY CLOVER; STRIATA CLOVER; SUBCLOVER; SWEETCLOVERS; TORNILLO; TREFOILS; TRUE CLOVERS; VARIEGATED ALFALFAS; VELVETBEANS; VETCHES; WHITE CLOVER; WHITE SWEETCLOVER; WHITE TIP CLOVER; ZIGZAG CLOVER.

References: M.3; P.1; S.5; W.3.

See also WINTER LEGUMES; WHEATS; BUCKWHEATS; OATS; RYE; CORN; COTTONSEED; BERMUDA-GRASS; SMOOTH BROME; BUFFALOGRASS; MOUNTAIN BROME; ORCHARDGRASS; RYEGRASSES; CENTIPEDE-GRASS; RUSSIAN WILD-GRASS; WEEPING LOVEGRASS; BIG BLUEGRASS; DALLISGRASS; JOHNSONGRASS; SORGHUMS; BROOMSEDGE; STRAW; PRAIRIE HAY; HAY; LEGUME HAY; BLOAT; PASTURE PLANTS; PASTURE MANAGEMENT; PASTURES; RANGE PLANTS; GRAZING; SILAGE; SILAGE CROPS.

LEGUME SEEDS include those of the SOYBEAN (protein-richest of all feed seeds); PEANUT; FIELD PEA and other PEA species; and FIELD BEANS, VELVETBEANS, and other BEANS.

LEGUME SILAGE.

See SILAGE; SILOS; SILAGE CROPS.

LEGUMINOSAE = LEGUMES.

LEGUMINOUS means: of, or pertaining to, the legumes; also, having the characteristics of a plant of the legume family.

LEHIGH = *Mediterranean wheat. See* COMMON WHEAT.

LEHMANN'S LOVEGRASS (*Eragrostis lehmanniana*), or *Lehmann lovegrass*, is a drought-resistant perennial which is grown in the southern parts of Arizona, New Mexico, and Texas. This LOVEGRASS species gives good pastures, but is not suitable for hay.—*See also* GRASSES.

LEMHI. See COMMON WHEAT (variety).

LEMMA, or *flowering glume*, is a BRACT: it is the lower of two leaves enclosing the flower in GRASSES.—*See also* GLUMES.

LEMON (*Citrus limonum*) is a small evergreen tree and also the name of its fruit.—*See also* POISONED BRAN MASH; LEMON PULP.

LEMON PULP has a bitter taste. Therefore, *dried lemon pulp*—which consists of the extracted peel, pulp, and seeds obtained as by-product in the manufacture of citric acid from LEMONS—is impalatable for livestock.

LEOTI SORGO, or *Leoti Red sorgo*, is one of the forage SORGHUMS.

LEPIDIUM. *L. scopulorum* = PEPPERWEED; *L. apetalum* = PEPPERGRASS.

LEPTOSPHAERIA. The FUNGUS *L. salvinii* is one of the causes of STEM ROT OF RICE; *L. avenaria* causes BLACKSTEM OF OATS.—*See also* HELMINTHOSPORIUM.

LESION is a localized spot of a diseased tissue. Spots, cankers, blisters, and scabs are examples of lesions.

LESPEDEZA. Among the *Lespedeza* spp.

are the following LESPEDEZAS: *L. striata* = COMMON LESPEDEZA; *L. stipulacea* = KOREAN LESPEDEZA; *L. cuneata* (formerly called *L. sericea*) = SERICEA; *L. bicolor* = BICOLOR.

Lehmann's lovegrass. (H. 47.)

LESPEDEZA ANTHRACNOSE of AN-NUAL LESPEDEZAS and other LESPEDEZAS may be caused by a strain of the FUNGUS *Glomerella cingulata*. The disease is characterized by circular, elliptical, or angular, brownish lesions on the leaflets; a single leaf spot is sufficient to cause a leaflet to drop off. The lesions on the PETIOLES and stems are dark brown to nearly black and vary from circular to linear. They are of importance largely on young seedlings, whose stems may be completely girdled and the plants killed. The disease may be very prevalent in local areas during wet periods, especially during the late spring and early summer months, and may disappear during subsequent dry spells.

Reference: W.6.

See also ANTHRACNOSE; LUPINE AN-THRACNOSE.

LESPEDEZA HAY.
          See HAY GRADING; LESPEDEZAS.

LESPEDEZA MEAL. *See* LESPEDEZAS.

LESPEDEZAS (*Lespedeza* spp.), also formerly known as *bush clovers*, have become a crop of major importance in the eastern half of the United States; in the western half they are not so well adapted.

The *annual* lespedeza varieties are all low-growing plants with relatively small 3-parted leaves and bluish or purplish, as well as small, inconspicuous flowers without petals. Both types produce seed. In thin stands the plants are rather spreading but in thick stands they are more erect. Sericea lespedeza, the *perennial* form, is a tall, erect plant with a growth habit somewhat like ALFALFA. All lespedezas are very leafy and produce an abundance of seed.

The protein content of lespedeza hay varies from about 10 to 15 percent, depending largely upon the fertility of the soil. Young plants have less TANNIN than more mature plants, and sericea has much more than the annual species. The percentage of tannin in sericea varies from 5.1 to 8 percent for the whole plant, and in the leaves from 7.5 to 18 percent.

Lespedezas are summer-growing LE-GUMES. The annual species germinate early in spring, and although the seedlings like cool weather for their early development, they cannot survive heavy frosts and are often killed by late frosts in the spring. Sericea lespedeza is winter hardy and survives temperatures much below zero.

Seed of lespedeza of all varieties deteriorates rapidly. It is advisable to use seed that is not more than one or, at most, two years old and to have a germination test of this before purchasing or planting. The annual lespedezas have little or no hard seed and need no scarifying. Sericea has a high percentage of hard seed and must be scarified to insure prompt germination.

In the region where lespedeza is commonly grown, INOCULATION generally is not needed. On badly eroded soils, or on land that has not previously grown lespedeza, nodulation will be increased by artificial inoculation.

Lespedeza will grow on almost any type of soil. It does well on the sandy loam soils of the Coastal Plain, the clays of the Piedmont, and the limestone soils of Virginia, Tennessee, and Kentucky, and will also grow on soils too sour to grow *clover* without the use of lime. On very sour land lime has proved very beneficial to lespedeza, especially to the Korean varieties. Where soils are poor, lespedeza will respond to lime and other fertilizers. Phosphate in particular has caused increased yields and should be used generally on all the poorer soils of the Coastal Plain and upper South; from 200 to 400 lb. per acre is recommended. Although all lespedezas are fairly drought-resistant, and sericea decidedly so, good yields are dependent on adequate moisture supply.

In North Carolina, Tennessee, and farther south, annual lespedezas should be seeded in late February or the first half of March. Farther north seeding should be delayed until late March or early April. North of the Ohio River seeding in late April or early May is advised. Sericea should be seeded a little later than the annual lespedezas.

The seed of the annual lespedezas should be sown broadcast, or drilled alone, or on WINTER GRAIN. A firm seedbed is essential in obtaining a good stand. When annual lespedezas are seeded on meadows or pastures, a spring-tooth harrow or disk should be used to loosen the surface soil before the seed is sown. Sericea should be seeded alone on a firm seedbed.

Lespedeza contains less moisture than alfalfa or red clover and is consequently more quickly cured. The field-cured *hay* contains somewhat more dry matter than similarly cured alfalfa or clover hay. To make the best hay the annual lespedezas should be cut when in first bloom or just before. In the latitude of North Carolina this will usually be in August. Annual lespedezas, cut when no more than 10 in. high, should be windrowed soon after cutting; if cut early in the morning the hay can be stacked late the same day. When more than 15 in. high, the cut hay should lie in the swath longer.

If a volunteer stand of annual lespedeza is desired the following year, the hay must be cut early (about first bloom), and high enough for the second growth to have time to produce seed.

Sericea hay, cut when 12 in. high, will insure high protein and low tannin content. It should be left in the swath in bright sun for not more than an hour, then windrowed and hauled to the barn the day it is cut. On very poor soil only one cutting should be made, whereas on fertile soil as many as four cuttings are possible each season.

An average yield of hay for the annual lespedezas is about 1 ton per acre. On good soils 2 to 3 tons may be expected. Sericea yields have ranged from 1 to 4 tons per acre per season.

Annual lespedeza hay is nearly equal to alfalfa in feeding value. Sericea is both palatable and nutritious when cut at the proper stage of maturity.

The annual lespedezas afford good grazing and can be used in both temporary and permanent *pastures*. They make their best growth during the hot summer months and provide excellent grazing during the season when many grasses are producing little. The later maturing varieties usually stay green until frost. As lespedeza does not provide much grazing before June or July, early spring pasturage must be provided by the GRASSES, RYE, WINTER BARLEY, or other cereal sown early in fall.

The annual lespedezas can be maintained well with grasses that do not form a dense sod. Grasses that make a heavy, matted growth, like CARPETGRASS and

BERMUDA-GRASS, crowd out the lespedeza almost completely the second season.

Sericea lespedeza has not been so universally liked for grazing as the annual varieties. In pure stand it has not always given satisfactory results when used as a pasture for beef cattle. Some livestock refuse to eat it, but many of them eat it readily.

All lespedezas can be grazed in pure stands without danger of BLOAT.

Because lespedeza grows 'n thick stands, it affords—throughout the growing season —an excellent cover for the prevention of erosion. When a full growth of lespedeza is worked into the soil, it greatly increases soil fertility and the yield of the succeeding crops for several years. This use of lespedeza should be emphasized and is a practice that is strongly recommended.

A great many of the cropping systems in use include one of the SMALL GRAINS. In these cropping systems, lespedeza nearly always follows the small grain and precedes an intertilled crop, such as COTTON or CORN. In cropping systems that do not include a small grain, lespedeza follows an intertilled crop and occupies the land one or two years, during which time it may be cut for hay or seed.

Few, if any, changes in the crop sequence need be made when lespedeza is added to the cropping system. In many rotations lespedeza either takes the place of grass or clover, or is grown with them.

The following lespedeza products have been officially recognized:

Lespedeza meal—obtained from the grinding of the entire lespedeza hay, without the addition of its stems, straw, foreign material, or the abstraction of leaves. It shall be reasonably free from other crop plants and weeds, and must not contain more than 28 percent crude fiber.

Lespedeza stem meal—the ground product remaining after the separation of the leafy material from lespedeza hay or meal; it shall be reasonably free from other crop plants and weeds.

Lespedeza straw meal—the ground product remaining after the separation of the seed from lespedeza; it shall be reasonably free from other crop plants and weeds.

Dangers: DODDER has been the most troublesome weed in lespedeza unless the crop is used for pasturage or hay. RAGWEED is of some local importance.

Other dangers are the FUNGUS disease LESPEDEZA ANTHRACNOSE, CHARCOAL ROT, BACTERIAL WILT, and some mold formations in hay or silage.

Species: Although over 100 species of lespedeza are known, only three are of interest to the American farmer. Two are *annual species,* namely COMMON LESPEDEZA (with its varieties known as *Kobe* and *Tennessee 76*) and KOREAN LESPEDEZA (with the varieties *Harbin, Late Korean, Early Korean,* and *Climax*). One is a *perennial species;* it is known by the common name SERICEA (LESPEDEZA). BICOLOR (LESPEDEZA), another perennial, and its variety *Natob,* are popular as a source of wildlife food.

References: M.7; W.6; E.7; F.6.

See also ANNUAL LESPEDEZAS; LEGUME FEED-PRODUCTS; HAY; STRAW; PASTURES; GRAZING; SILAGE CROPS; ORCHARDGRASS; CLUSTER CLOVER; WHEATS; VELVETBEANS; DICOUMAROL; MOLDY FEED.

Illustration: See COMMON LESPEDEZA.

LESPEDEZA STEM MEAL.
                              See LESPEDEZAS.
LESPEDEZA STRAW MEAL.
                              See LESPEDEZAS.
LESSER   CLOVER-LEAF   WEEVIL (*Phytonomus nigrirostris*) is about ⅛ in. long and iridescent green or brown. During the summer months adults usually can be found on the flower heads of RED CLOVER. However, the weevil's larvae cause the most extensive damage. The larvae are generally found in the clover heads, but they may also feed in the axils of the stem. Up to 25% percent of the FLORETS may be cut off or injured by one larva.

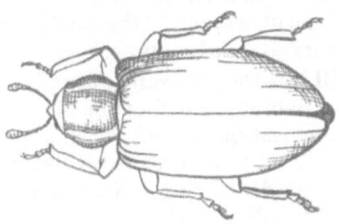

Lesser clover-leaf weevil. (D.20.)

During wet season the larvae are attacked by the FUNGUS *Empusa sphaerosperma* that also kills the CLOVER-LEAF WEEVIL.

Reference: E.5.

See also INSECTS; CLOVER-HEAD WEEVIL.

LEUCINE, one of the essential AMINO ACIDS, forms colorless leaflets which are only slightly soluble in water. CORN is an excellent source.

LEUCOTHOË. *L. davisiae* = BLACK-LAUREL.

LEVULOSE, also called *fructose* or *fruit sugar*—because it occurs in many fruits—forms colorless needles, is water-soluble and not fermentable.—*See also* INVERT SUGAR; JERUSALEM ARTICHOKE; SUGARS; REDUCING SUGARS.

LICO. See SIX-ROWED BARLEY (variety).

LIGHT is necessary to keep animals healthy. It is of particular importance in the production of VITAMIN D which aids the calcium-phosphorus metabolism.

LIGHTNING INJURY of farm crops is not uncommon. SOYBEANS are more frequently injured in the field than most other crops. They appear to rank with cotton, POTATOES, and TOMATOES in the ease with which they are hit by it.

Lightning injury can be easily mistaken for a parasitic disease. The damage, however, will not extend beyond an area that can be clearly seen a few days after lightning has struck.

Killed spots with a diameter up to 40 ft. have been observed in fields. These spots are often surrounded by a wide border of damaged plants; the lower parts of the stems of the plants in these borders may be blackened and many of the leaves may be dead.

Reference: C.9.

LIGNIN, a constituent of wood, is a complex organic substance which, together with CELLULOSE, forms the woody portions of plants.—*See also* FIBER.

LIGULE is the projecting, usually tongue-like, membranous end of the sheath lining between leaf-blade and stalk. Sometimes it is reduced to a hardened fringe of hairs, or to a hardened ring.—*See also* GRASSES.

LIMA BEAN MEAL.

See LIMA BEAN VINES.

LIMA BEANS (*Phaseolus* spp.) is a name applied to two species which belong to the FIELD BEANS; i.e., the *large lima bean* (*P. limensis*), a perennial usually grown as an annual, and the *small lima bean* (*P. lunatus*), an annual. Lima beans are chiefly grown in California.

The vines and empty pods—the PLANT-BY-PRODUCT of quick-freezing the beans—are used as SILAGE CROPS.—*See also* VELVETBEANS; LYON VELVETBEAN.

LIMA BEAN SILAGE.

See SILAGE CROPS.

LIMA BEAN VINES, in the form of SILAGE, have been used as feed for cattle. *Meal* obtained from lima bean vines can also be used to advantage in diets for poultry in place of ALFALFA meal. The meal is a good source of protein and vitamins A and B_2.

Reference: E.12.

See also PLANT BY-PRODUCTS.

LIME is a term applied to different CALCIUM compounds, especially *calcium oxide*, also called *quicklime* or *burnt lime*. It may occur in the form of white or grayish-white lumps or as a granular powder; it slowly absorbs carbon dioxide when exposed to air, thus changing to CALCIUM CARBONATE. In contact with water it changes, under development of much heat, into *calcium hydroxide* (HYDRATED LIME). It is only sparingly soluble in water.

Dusting of SOYBEAN plants with lime prevents serious damage from rabbits.

Lime is widely used in soil treatment.—*See also* FERTILIZERS.

LIMESTONE is ground (pulverized) rock composed mainly of CALCIUM CARBONATE. It is used to neutralize or better the pH value of strongly acid soils.—*See also* MAGNESIUM LIMESTONE; MINERAL FEEDS.

LIMITED WHOLE MILK PLAN.

See DAIRY RATIONS.

LIMONIUS. *Limonius* spp. = WIREWORMS.

LINCOLN BROMEGRASS is a southern-type SMOOTH BROME.

LINCOLN SOYBEAN.

See SOYBEAN (variety).

LINDANE, a white, crystalline powder, is the (most valuable) gamma isomer of BEN-

ZENE HEXACHLORIDE. It is practically odorless.

Only 2 to 3.2 oz. lindane per acre of CLOVER, or 4 oz. per acre of ALFALFA, is needed for the control of MEADOW SPITTLEBUGS.

Dust containing 2 percent lindane, if applied at the rate of 25 lb. per acre, controls STINK BUGS. Application of 8 percent lindane dust at the rate of 15 to 20 lb. per acre controls the THREE-CORNERED ALFALFA HOPPER.

LINE. *See* SELECTION.

LINEAR means: linelike, narrow and flat, with the margins parallel. Most grass leaves are linear or nearly so.

LINOLEIC ACID is an unsaturated FATTY ACID.—*See also* VITAMIN F.

LINOLENIC ACID, like the related LINOLEIC ACID, is one of the unsaturated FATTY ACIDS, occurring in linseed and other vegetable oils.—*See also* VITAMIN F.

LINSEED, or *flaxseed*, is obtained from "FLAX."

These are the officially recognized linseed (or flax) products:

Linseed oil cake, linseed oil chips, and *linseed oil flakes*—the products obtained by removal of most of the oil from flaxseed by hydraulic (old process), expeller, or solvent extraction. For each product the designated name must include the term "hydraulic," "expeller," or "solvent extracted," and the product must correspond thereto. The products shall contain not more than 0.5 percent acid-insoluble ash, and must be sold according to their protein content.

Linseed oil meal—obtained by finely grinding linseed oil cake, chips, or flakes. It must be sold according to its protein content and its name must include one of the terms "hydraulic", "expeller", or "solvent extracted" to specify the method of manufacture of the source material.

Linseed oil meal, commonly called *linseed meal*, is valued in the diets of cattle, sheep, and swine, not only as a protein supplement but also for its conditioning, appetite-stimulating, and laxative effects. The protein is somewhat inferior to that in soybean meal when used as a supplement to corn in swine rations. Linseed

meal, therefore, is used most effectively with other oil meal and animal-protein supplements. In fattening beef cattle and lambs, linseed meal is very popular and is widely used as a supplement, either alone or in combination with other protein supplements, for its protein and conditioning values. Much the same is true for fattening dairy cattle.

Linseed meal is toxic to poultry except in very low proportions. The toxicity can be largely eliminated by soaking the meal in water for twenty-four hours or by adding VITAMIN B$_6$ to the diet.

Linseed cubes or *linseed pellets* consist of linseed oil meal produced by hydraulic, expeller, or solvent extraction and processed through a cubing or pelleting machine. The product shall be firm but not flinty, of sweet odor, and free from mold. It must be sold according to its protein content and its name must include one of the terms "hydraulic," "expeller," or "solvent extracted" to specify the method of manufacture of the source material.

Linseed feed is obtained by mixing one or more linseed oil meals with flaxseed screenings oil feed; it may also contain other flaxseed by-products. It must be designated and sold according to its protein content.

Flaxseed screenings oil feed is the ground product obtained after extraction of part of the oil by crushing, cooking, and hydraulic or mechanical pressure—or by crushing, heating, and the use of solvents —from the smaller, imperfect flaxseeds, weed seeds, and other foreign materials which have feeding value and are separated in cleaning flaxseed.

NOTE: *Flax plant feed* is a term recommended for that portion of the flax plant remaining after harvesting the seed and after separation of the bast fiber and flax shives. It consists of the leaves, corticle tissues, flaxseed bolls, as well as broken and immature seeds of flax. It shall contain a minimum of 9 percent crude protein and a maximum of 35 percent crude fiber.

References: E.12; F.6.

See also OILSEED MEALS; FLAXES; FIELD BEANS.

LINSEED CUBES. *See* LINSEED.

LINSEED FEED. *See* LINSEED.

LINSEED MEAL = *linseed oil meal. See* LINSEED.

LINSEED OIL CAKE. *See* LINSEED.

LINSEED OIL CHIPS. *See* LINSEED.

LINSEED OIL FLAKES. *See* LINSEED.

LINSEED OIL MEAL. *See* LINSEED.

LINSEED PELLETS. *See* LINSEED.

LINT. *See* COTTON.

LINUM. *L. usitatissimum* = (old world) FLAX.

L. neomexicanum, or *New Mexican flax*, and *L. rigidum*, or *stiffstem flax*, are poisonous FLAXES.

LIPASE is a fat-splitting ENZYME which occurs in liver, PANCREAS, and stomach, and also in many plants.

LIPIDS, or *lipoids*, are FATS or fatlike substances characterized by their insolubility in water and their greasy feel. They readily take up oxygen from the air and become rancid, and thus may speed up the destruction of other valuable nutrients with which they are in contact, especially vitamins A and E.—*See also* ETHER EXTRACT.

LIQUID PETROLATUM
= WHITE MINERAL OIL.

LIQUID TREATMENT OF SEEDS.
See SEED TREATMENT.

LIQUOR is (1) an aqueous (watery) solution, usually of standardized strength or (2) a distilled alcoholic beverage.—*See also* DISTILLERS' PRODUCTS.

LISSORHOPTRUS. *L. simplexi* = RICE WATER WEEVIL.

LISTER is a plow with a double moldboard, used for furrowing (dry) land; it often has a planting attachment.

LITTLE BLUESTEM (*Andropogon scoparius*) is a vigorous, long-lived, native bunchgrass of wide distribution over the United States. It is most prevalent in the Great Plains, particularly in Kansas and Oklahoma, where it supplies dependable grazing and cured forage. But in many areas it is not considered of major economic value because of low palatability and poor quality of the forage obtained from mature plants.

Little bluestem is smaller than BIG BLUESTEM with which it is found in close

association; but little bluestem is more drought-resistant and therefore better adapted to sites that receive limited moisture.

Little bluestem. Plant and pair of spikelets. (H.26.)

Growth begins late in the spring and continues through the summer. Plants usually grow 1 to 3 ft. tall. The leaves, less than ¼ in. wide, from 4 to 8 in. long, and flattened at the base, are light green until the plants reach maturity, when they develop their distinctive reddish-brown color.

Because of its habit of growth and the wide range of soils on which it thrives, little bluestem is valuable in erosion control. It is suitable for use in crop rotations, and in mixtures for regrassing abandoned cultivated land. Seedings should be made on a well-prepared, firm seedbed free from weeds.

Reference: H.1.

See also BLUESTEMS; GRASSES; INDIAN GRASS.

LITTLE BUR-CLOVER (*Medicago minima*), one of the newer BUR-CLOVER species, is comparable to SPOTTED BUR-CLOVER in winter hardiness. It is readily recognized by its very small and soft spiny bur.—*See also* CLOVERS.

LITTLE DARSO = DARSET.

LITTLE HEGARI = *bonita;* it is one of the grain SORGHUMS.

LITTLELEAF HORSEBRUSH (*Tetradymia glabrata*), also called *spring rabbitbrush*, or *coal oil brush*, is a POISONOUS PLANT. It is a shrub, which, on ranges, is sometimes used as poor-quality forage.— *See also* RANGE PLANTS; HORSEBRUSHES.

LITTLELEAF MOUNTAIN - MAHAGONY (*Cercocarpus intricatus*) is an evergreen shrub occurring in dry, mountainous regions of the western states. It furnishes palatable winter feed for game, cattle, sheep, and goats.—*See also* RANGE PLANTS.

LIVER is the largest GLAND in the animal body; it secretes bile (which is needed for proper fat digestion), stores sugar as GLYCOGEN, detoxifies poisons, forms blood proteins, and exerts other important functions.—*See also* VITAMIN A; FISH-LIVER OIL.

LIVER AND GLANDULAR MEAL.

See ANIMAL PRODUCTS.

LIVER MEAL. *See* ANIMAL PRODUCTS.

LOAM is a soil composed of a mixture of two or more of the following: clay, silt, sand, and gravel.

LOBELIAS. *See* DEATHCAMASES.

LOCOWEEDS, a term which includes both *Astragalus* spp. and *Oxytropis* spp., are probably the most important stock-poisoning plants. As a matter of fact, the word "*locoed*" was for a long time used synonymously with "poisoned" in the western states, and it was difficult to educate ˙armers and rangers to distinguish

between true *locoism*—which is a definite condition caused by some locoweeds—and poisoning due to a variety of other POISONOUS PLANTS.

White locoweed, or *crazyweed*, is one of the *Oxytropis* spp.; its active principle, which has not as yet been determined and is different from other plant poisons, causes true locoism.

White locoweed. Plant and flower. (D.19.)

Some of the *Astralgus* spp. are nontoxic (e.g., *red locoweed*), while others, for instance: *A. tetrapterus*, produce poisoning, but not the true locoism; nothing is known about the chemistry of their toxic substances. Another group of this genus depends for its toxic properties upon SELENIUM which the plants take up from the soil in the form of a compound and accumulate in dangerous quantities.

Cattle, horses, sheep, and goats feeding for several days or longer on leaves and stems of locoweeds become constipated, develop a craving for the plants, and soon show such other symptoms of locoism as rough coats, inco-ordination, and various peculiar actions.

References: H.43; U.6.

LOCUST BEAN = CAROB BEAN.

LOCUST SKIPPER BUTTERFLY (*Epargyreus tityrus*) is an INSECT whose caterpillars often cause much damage. They feed on KUDZU leaves and other crops.

LODICULES are the two (occasionally three) small, thin scales found in the FLORETS of most GRASSES. They help to force open the LEMMA and PALEA, thus aiding in fertilization.

LOESS is a sticky, fine-grained soil, mostly —if not always—of wind-blown origin.

LOLIUM. The RYEGRASSES, or *Lolium* spp., comprise *L. multiflorum* = ITALIAN RYEGRASS and *L. perenne* = PERENNIAL RYEGRASS.—*See also* COMMON RYEGRASS.

LOMENT is a POD, or *legume*, constricted between the seeds.

LONGFELLOW FLINT.
                                        See CORN (variety).

LOOPER. *See* ALFALFA LOOPER.

LOOSE KERNEL SMUT. *See* LOOSE KERNEL SMUT OF SORGHUM; LOOSE KERNEL SMUT OF JOHNSONGRASS; HEAD SMUTS; and GRASS SMUTS.

LOOSE KERNEL SMUT OF JOHN-SONGRASS is caused by *Sphacelotheca holci*, a FUNGUS closely related to that causing LOOSE KERNEL SMUT OF SORGHUMS. It occurs in the southwestern states. The loose kernel smut of Johnsongrass also is able to attack certain varieties of SORGHUM. Several feteritas and feterita crosses are susceptible. The kafirs (especially Reed, Sharon, and Red) appear to be immune, although they are highly susceptible to the loose kernel smut of sorghums.

Infected plants head early and are severely stunted, frequently attaining a height of less than 1 ft., after which they sometimes die prematurely. The thin membrane covering the smut gall ruptures as soon as the gall appears, and the short-lived spores are spread at an early stage. In Johnsongrass the smut is transmitted largely through the underground RHIZOMES and by spores falling on freshly cut stubble.

Control: The control of this smut on its natural host is not important because Johnsongrass to a great extent is a noxious weed. Because of its short-lived spores, this smut is not considered a menace to sorghum. However, it can be controlled by fungicides used for the SEED TREATMENT of other groups of sorghum.

Reference: L.1.

LOOSE KERNEL SMUT OF SORGHUM, caused by the FUNGUS *Sphacelotheca cruenta*, is not very common but occurs particularly in the southern Great Plains. This fungus attacks all groups of SORGHUM, including JOHNSONGRASS, which is also attacked by the LOOSE KERNEL SMUT OF JOHNSONGRASS.

The galls formed by loose kernel smut are long and pointed and the thin membrane covering them usually breaks soon after the galls reach full size. Most of the dark-brown spores are soon blown away, leaving a long, dark, pointed structure (called a *columella*) in the central part of what was the gall. The spores of the fungus are carried on the seed and germinate soon after the seed is planted. The fungus continues to grow unobserved inside the plant until after heading when the smut galls appear in the heads in place of normal kernels. The disease stunts the infected plants and frequently induces the development of abundant side branches.

The fungus may also cause secondary infection; that is, the spores from a smutted head may infect, and cause smut to develop in the late heads of otherwise healthy plants.

Loose kernel smut comprises at least *two races* that differ in their ability to attack different groups of sorghum. Certain of the feteritas and milos, Schrock, and Dwarf Shantung kaoling, are resistant to both races; Premo, Red Amber, shallu, and Weskan are highly susceptible to both; and a large number of other varieties are susceptible to one race and resistant to the other.

Control: SEED TREATMENT and the use of smut-free seed and resistant varieties are the best control measures. Sorghum varieties that are resistant to the five races of COVERED KERNEL SMUT usually are resistant also to the two races of loose kernel smut.

Reference: L.1.

See also SULFUR; PHYGON; SPERGON; ARASAN; FORMALDEHYDE SOLUTION; COPPER CARBONATE.

LOOSE SMUT. *See* BLACK LOOSE SMUT.

LOOSE SMUT OF WHEAT, sometimes called *black head* or *black smut*, is caused by the FUNGUS *Ustilago tritici*. It is prevalent, especially, in the humid and subhumid WHEAT-producing areas of the United States.

In diseased plants the entire head, except the central axis, is replaced by a black sooty mass of spores. These spores, which appear about the time the healthy heads are in bloom, may be carried for long distances by wind, insects, or other agencies. Many of the spores lodge in the normal wheat flowers. Here they germinate and develop long slender infection threads which, under favorable conditions, grow into the developing embryo of the young wheat kernels that are forming inside the chaff. When these infected kernels are mature, they cannot be distinguished from smut-free kernels. However, if such kernels are used for seed without being properly treated, the smut fungus inside each seed starts to grow when the seed germinates and sends infection threads upward into the plant. When the head forms, a mass of black spores appears in place of the kernels. Infection by loose smut of wheat takes place only through the wheat flowers.

Wheat varieties differ widely in their resistance to loose smut of wheat. Selections that are resistant to the, at least, 11 races of loose smut in one area may be highly susceptible to races prevalent in other areas.

Control: Chemical seed disinfectants will not control loose smut of wheat. *Hot-water seed treatment*, however, is effective, and is applied as follows: The seed is first put in loosely woven burlap sacks which should be only half-filled and tied at the top. The seed is then soaked in unheated water for at least six hours.

During this period the sacks should lie on their sides and should be occasionally turned or rolled to prevent caking of the swelling seed. The presoaked seed is dipped in water at about 120° F. for a few minutes and then immersed for ten minutes in water held at 130° F. The sacks should lie on their sides also during the 10-minute treatment and should be moved about to insure an even penetration of heat. Immediately after the treatment the seed should be raked out in a thin layer to cool and dry.

Reference: L.10.

See also COMMON WHEAT.

LOOSE SMUTS OF BARLEY is a fungus disease which occurs in two forms: (1) *nuda loose smut* (caused by the FUNGUS *Ustilago nuda*, a deep-infecting species) and (2) *nigra loose smut* (caused by *U. nigra*, a shallow-infecting species). Heads smutted by these species are similar in appearance. They emerge at the time the healthy heads emerge and the loosely held spores are blown about the field. Some of the spores get into the BARLEY flowers that open in the process of blooming. At this point there is an important difference in the behavior of the two barley loose smuts. Spores of the *nuda* species develop infection threads that grow into and within the tissue of the young, developing barley kernels; spores of the *nigra* smut, however, develop infection threads that are confined to the outer layers beneath the seed hulls.

When the seed is sown, the smut resumes its growth and eventually produces a mass of spores in the heads.

There are varieties of barley resistant to either the nigra or the nuda loose smut and some are resistant to both. When such varieties, however, are grown in new localities where other races of these smuts exist, the seed may become infected and when subsequently sown, may produce smutted plants.

Control: Since the *nigra* loose smut is carried on the seed or under the seedcoat, it can be controlled by chemical SEED TREATMENT. The *nuda* loose smut, however, infects the embryo and can be controlled only by *hot-water* seed treatment, similar to that described for controlling LOOSE SMUT OF WHEAT. The exception is that after the 6-hour presoak and the tempering bath at 120° F. the seed is immersed in 126° F. water for thirteen minutes, after which it is cooled and dried.

Reference: L.10.

LOTUS. *Lotus* spp. = TREFOILS; *L. corniculatus* = BIRDSFOOT TREFOIL; *L. uli-*

ginosus = BIG TREFOIL.—*See also* GRAZING.

LOUISIANA WHITE CLOVER is a variety of WHITE CLOVER.—*See also* RICE; TALL FESCUE.

LOVEGRASSES (*Eragrostis* spp.) are represented in all temperate regions; only a few species have been recognized as of agricultural value in the United States. Several native species are considered weeds; others are used to provide vegetative cover on eroding sites.

Species: Among the approximately 40 species occurring in the United States, the following are of value as forage grasses: PLAINS LOVEGRASS, SAND LOVEGRASS, WEEPING LOVEGRASS, LEHMANN'S LOVEGRASS, BOER'S LOVEGRASS, and ALKALI LOVEGRASS.

Reference: H.1.

See also GRASSES.

Illustrations: See LEHMANN'S LOVEGRASS; WEEPING LOVEGRASS.

LOVE VINE = DODDER.

LOW FLUORINE ROCK PHOSPHATE.

See MINERAL FEEDS; ROCK PHOSPHATE.

LOXOSTEGE. *L. similalis* = GARDEN WEBWORM; *L. sticticalis* = BEET WEBWORM.

LUCERNE = COMMON ALFALFA.—*See also* ALFALFAS.

LUPINE. *See* LUPINES.

LUPINE ANTHRACNOSE, caused by a strain of the FUNGUS *Glomerella cingulata*, has become more important in recent years, particularly because BLUE LUPINE has become the major cover-crop plant in large areas of the southeastern United States. The disease is widespread in the Lupine Belt, but is usually not serious, except in wet seasons and in certain parts of southern Georgia, southern Alabama, and northern Florida. It is present, however, as far north as North Carolina and as far west as Texas.

In some lupine fields a high percentage of the plants may be diseased, and many may be killed, by the anthracnose fungus which can attack fruits of the peach, apple, and plum. Although the fungus can be carried to the fields in or on lupine seed, it may already be present in the soil, or on other plants in the field.

Usually the ANTHRACNOSE appears as small, circular, brownish pits on the seed leaves (cotyledons) of the seedling. These lesions gradually enlarge and soon spread to the stem which may become girdled. Spores of the fungus causing the disease may—under humid conditions—be formed on these early lesions and may become so abundant as to form a salmon-colored layer. These spores can be scattered by the splashing of raindrops, or by wind, insects, or any other agency that can spread contaminated drops of water. If these spores are kept moist for twenty-four hours or longer, they will germinate, grow into the plant, and start new lesions.

On stems, areas of infection appear as dark-brown to black bands, often only 1 to 2 in. wide, extending around the stem and sometimes girdling it. As the stem matures, these diseased bands become lighter brown and sometimes are several inches wide. The surface of many of the cankers may show concentric rings.

Blue lupine leaflets, usually, are not affected seriously. When present these lesions are circular to irregular in outline, slightly sunken, and dark brown to nearly black. One diseased spot may be sufficient to cause a leaflet to turn yellow and fall.

On pods the infected areas appear at first as minute, brownish pits. These gradually enlarge and often merge, involving half or more of the pod. Pod lesions are almost black, but they may have one or more salmon-colored masses of spores. As the plants mature and the leaflets fall, the large, black areas on the pods are conspicuous.

The seeds beneath the diseased parts are spotted or decayed. Many of the affected seeds are covered with white fungus mycelium (a web of interwoven threads) and may be completely decayed. Seeds less seriously affected have reddish-brown to blackish areas, sometimes with white centers.

On young *yellow lupine* plants most of the anthracnose lesions are on the leaflets and PETIOLES (leafstalks). Infected leaflets of this species do not fall off so readily as do those of the blue lupine. The lesions on the yellow lupine are similar to those on the blue species.

White lupine is also attacked by the

anthracnose fungus and sometimes seriously injured and killed.

Control: The fungus, which can survive for several months in stored seed, can be killed by exposing the seed to an air temperature of 75° C. (167° F.) for three hours. This heat treatment somewhat retards germination and may cause hardening of some seeds. In order to withstand this temperature, seed must be dried in advance to a moisture content of approximately 11 or 12 percent. Hot-water treatment is not a practical method of controlling anthracnose; water hot enough to kill the fungus within the seeds will also kill many of the seeds.

Rotation has some value in the control of anthracnose, and a 3- or 4-year rotation is recommended.

Where the disease has become widespread it seems logical to use the crop for green manure, rather than to save it for seed.

References: W.8; W.9.

See also LUPINES; WHITE LUPINE; YELLOW LUPINE; LESPEDEZA ANTHRACNOSE.

LUPINES (*Lupinus* spp.) containing an alkaloid are called *bitter;* nonalkaloid lupines are called *sweet.* In the United States plantings in the Gulf Coast area have given good results, and extensive commercial plantings are now well established in Florida, Georgia, Alabama, and Louisiana.

The plants are upright and have coarse stems and medium-sized digitate (finger-like) leaves. In thin stands they branch quite freely. Field-crop species are annuals; some ornamentals are perennials.

This LEGUME species requires cool weather for best development, and in the South commercial varieties must be grown as winter annuals. In the North they will have to be handled as summer annuals and seeded early in spring.

Lupines are used mostly for green manure. The large quantity of winter growth they produce makes them ideal for this purpose; the heavy nodulation of the roots adds a large amount of nitrogen to the soil.

Although immature plants of some species are eaten by livestock without harm, a number of species are known to be poisonous, both, in the green and dry state. The seed contains a higher percentage of poisonous alkaloids than any other part of the plant. Ordinarily livestock will not eat lupines containing poisonous alkaloids.

The nonalkaloid strains more recently developed may extend the use of both fodder and seed as stock feed. Some lupines are reported to be good honey plants.

Lupines respond to superphosphate and potash. Unless the preceding crop was heavily fertilized, apply 300 to 500 lb. superphosphate and 100 lb. potassium chloride, or 400 to 500 lb. 0-14-10 per acre. The fertilizer may injure germination if applied in direct contact with the seed.

On light, sandy lands lime is considered detrimental to lupines, but on heavier soils it may be beneficial.

INOCULATION of lupine seed is essential the first year lupines are grown in a field.

Lupines. Plant and flower. (D.19.)

It assures good nodulation and accumulation of nitrogen, even when lupine follows lupine. Commercial cultures are available.

In regions with mild winters (15° F. or above) seeding should be done from September 14 to October 15, if possible, although plantings can be made up to December 1. Later plantings sometimes give good results.

Lupines should be planted on a well-prepared, firm seedbed. The seed may be drilled or sown broadcast and covered with a cultipacker. Shallow planting, covering the seed only ½ to 2 in. deep, is essential for a good stand.

Lupines make 15 to 20 tons of green weight per acre; this amounts to 3 to 4 tons of dry matter.

Livestock will consume sweet lupines and make relatively good gains. The bitter, or high-alkaloid, lupine varieties are inferior for feed because of the danger of poisoning. They can be used, however, after having been soaked in water to remove the alkaloid, and this is sometimes done. Otherwise, consumption of only 8 oz. green plants or fruits per 100 lb. body weight in a day may develop such symptoms of poisoning as nervousness or depression in sheep, and weakness and trembling in cattle.

Sweet lupines are an excellent cover crop to boost production of BERMUDA-GRASS pastures. About every third year the Bermuda-grass should be plowed or harrowed in the fall and planted to lupines.

Lupines, like any other crop subject to disease build-up, should be rotated. Rotating them with SMALL GRAIN, WINTER PEAS, or VETCH, so that they will not be on the land more than once in two or three years, may be desirable.

Dangers: Lupines are subject to ROOT KNOT (caused by ROOT-KNOT NEMATODES). They are also susceptible to several diseases such as SCLEROTINA; STEM ROT, LUPINE ANTHRACNOSE, BROWN SPOT, PYTHIUM ROOT ROT, POWDERY MILDEW, BOTRYTIS STEM CANKER, ASCOCHYTA STEM CANKER, SOUTHERN BLIGHT, FUSARIUM ROOT ROT, RHIZOCTONIA ROOT ROT, and VIRUS DISEASES. GRASSHOPPERS often attack lupines.

Species: Of the many species recognized, the following annuals are often used commercially: WHITE LUPINE, YELLOW LUPINE, and BLUE LUPINE.

References: M.12; S.7; E.9; W.9.

See also FORBS; POISONOUS PLANTS; WHEATS; MUTTON BLUEGRASS; LEGUME BACTERIA; ROOT KNOT.

LUPINUS. Among the *Lupinus* spp., or LUPINES, are *L. angustifolius* = BLUE LUPINE. *L. albus* = WHITE LUPINE, and *L. luteus* = YELLOW LUPINE.

LYCOPERSICON. *L. esculentum* = TOMATO.

LYGODESMIA. *L. juncea* = SKELETONWEED.

LYGUS BUGS (*Lygus* spp.) thrive on a wide range of cultivated and uncultivated plants. The adults of the lygus bugs of various species, particularly the *tarnished plant bug* (*L. oblineatus*), are very abundant in CLOVER and ALFALFA fields during the summer months.

In the latitude of Utah there are three to four generations of lygus bugs a year, each taking six to seven weeks; in more southern regions, such as Arizona, a generation requires from twenty to thirty days, and the insects breed most of the year.

During cold weather the adults find protection among dormant plants or in various crop debris. With the coming of warm weather they seek early-flowering weeds. The females insert their eggs singly in the plant tissues, and the eggs hatch in ten to fifteen days. The insects become full-grown and change to adults in about three weeks. The new adults fly to more succulent plants and begin to lay eggs in about ten days.

Blasted clover buds, apparently due to the feeding of these insects, are common on the second or seed crop. On alfalfa the insects may be found from the time when the plants are in early bloom until the period of seed set. Young lygus bugs, or nymphs, are more destructive than the adults because they feed extensively on the buds of alfalfa, causing them to wilt and die. Sometimes such destruction by a large population of nymphs is severe enough to prevent flowering of the plants.

At other times the bugs feed on the flowers and cause them to drop. They will also feed on the pods, causing the injured seed to shrivel and turn brown.

Control: Successful control of lygus bugs requires properly timed applications of an insecticide. Correct timing destroys large numbers of the pest and avoids serious harm to beneficial insects.

Insecticide should be applied when population counts reveal the presence of definite numbers of lygus bugs in the field. Lygus bug counts are based on two-sweep counts taken at 10 to 20 stations over the field. At least three two-sweep counts are made at each station. The sweeps are taken with the standard sweep net, which has a 15 in. opening and a handle 26 in. long. Each sweep describes an arc of 180° with the net striking the upper 8 to 10 in. of the plants. All counts in a field are averaged, and treatment is based on average number of lygus bugs found.

Pollinators and predators—beneficial insects—must be protected. To avoid harm to them, treatments are restricted to the time in late evening when the bees leave the field, until the early morning when they return.

DDT or TOXAPHENE, at the recommended time and dosage rates, cause the least damage to the pollinators. Before spraying or dusting seed fields, it is advisable to notify beekeepers whose bees are in the vicinity.

References: E.5; A.4; U.3.

See also INSECTS.

Lygus bug. (G.20.)

LYMPH is a light straw-colored, clear fluid that contains *lymphocytes* (white cells). Lymph circulates in the lymphatic vessels of the body.

LYON VELVETBEAN (*Mucuna cochinchinensis*) makes a vigorous growth and requires a long season to mature, seldom ripening more than ten days earlier than the Florida variety of the DEERING VELVETBEAN. The white flowers are borne in pendent racemes which are often 2 to 3 ft. long. The woody pods are 5 to 6 in. long, compressed, ridged lengthwise, covered with a fine grayish pubescence, and with tendency to shatter while still in the field. The ash-colored seeds are similar in size and shape to those of the LIMA BEAN.

A hybrid between Lyon and Florida is the *Osceola velvetbean*.

Reference: P.10.

See also LEGUMES; INOCULATION; VELVETBEANS.

LYSINE is an essential AMINO ACID. It forms needles or plates and is water-soluble. Lysine is present in small amounts in cereal grains. Animal and fish by-products are good sources.

LZ is a SOYBEAN variety.

M

MAC. *See* MC.

MACOUPIN. *See* SOYBEAN (variety).

MACROCYTIC ANEMIA in chicks is a *pteroylglutamic acid* deficiency disease.— *See also* FOLIC ACID.

MACROPHOMINA. The FUNGUS *M. phaseoli* causes CHARCOAL ROT.

MACROSIPHUM. *M. pisi* = PEA APHID.

MADRID SWEETCLOVER, also called *Madrid Yellow*, is a leafy variety of YELLOW SWEETCLOVER. The foliage of this biennial is resistant to frost. It can be distinguished in the field by its dark green color. Madrid is preferred for hay production.—*See also* TALL FESCUE; BLUE PANICGRASS.

MADRID WHITE
                    = SPANISH SWEETCLOVER.

MADRID YELLOW = MADRID SWEETCLOVER.—*See also* YELLOW SWEETCLOVER.

MAGELLAN is a spring RYE variety.

MAGGOTS, also called "*legless grubs*," are the LARVAE of FLIES. The body of the maggot tapers noticeably toward the anterior end and the head is usually not distinctly set off from the rest of the body.

MAGNESIA = MAGNESIUM OXIDE.

MAGNESIUM—the chemical element Mg —is a silvery light metal which, widely distributed in nature, occurs as part of the CHLOROPHYLL molecule and in form of magnesium salts, such as *magnesite* (MAGNESIUM CARBONATE), MAGNESIUM LIMESTONE, MAGNESIUM SULFATE (*Epsom salt*), etc. Traces of magnesium are necessary for animal and plant life.

Magnesium deficiency gives rise to GRASS TETANY in cattle which, however, occurs very rarely; if it does occur, rations must contain some magnesium salts as mineral supplements.—*See also* MINERAL FEEDS.

MAGNESIUM CARBONATE — the chemical compound $MgCO_3$—occurs in nature as *magnesite* and is available commercially as a light, white, water-insoluble powder. It is used in MINERAL FEEDS.— *See also* MAGNESIUM.

MAGNESIUM LIMESTONE, or *dolomite* (or *dolomitic*) limestone, is a LIMESTONE containing more than 5 percent MAGNESIUM CARBONATE.—*See also* MINERAL FEEDS.

MAGNESIUM OXIDE, or *magnesia*, is a white powder which, when treated with water, forms *magnesium hydroxide*. It is used in the treatment and prevention of GRASS TETANY.

MAGNESIUM SILICATE. *See* TALC.

MAGNESIUM SULFATE, or *Epsom salt*, occurs in small, needlelike crystals which have a cooling, saline taste and are very soluble in water. This magnesium salt is widely used in medicine (e.g., as a purgative) and in limited amounts as an ingredient of MINERAL FEEDS.

MAGNOLIA RICE. *See* RICE (variety).

MAGNOLIA SOYBEAN.

See SOYBEAN (variety).

MAIN 340 is a COMMON OAT variety.

"MAINTENANCE ENERGY."

See ENERGY.

MAINTENANCE REQUIREMENTS must be met to supply the needs of an animal at rest. *Maintenance ration* is the amount of feed needed to enable the resting or nonproducing animal to retain its exact

weight; it is used to maintain the animal's normal body temperature, to keep up normal body functions (energy for work of the heart, etc.), and to replace waste of protein, vitamins, certain fatty acids, minerals, water, and air. Animals producing milk, meat, wool, or work need more feed than the maintenance ration. Age and differences in temperament, body size, and condition, are factors which influence the maintenance requirements of each animal. Rations that supply about 50 percent T.D.N. will usually meet the maintenance requirements of mature animals.

MAIZE = CORN.

MALAKOF = *Turkey wheat*.

See COMMON WHEAT.

MALLIARD REACTION

= BROWNING REACTION.

MALLOTUS. The hairs and glands covering the fruits (capsules) of the small evergreen tree *M. philippinensis taenifuge* are known as KAMALA.

MALT CLEARINGS.

See BREWERS' PRODUCTS.

MALTED BARLEY, contrary to some claims, is not more valuable for feeding purposes than ground BARLEY; however, it may be added to the ration for its appetizing qualities, since most farm animals like its taste.

MALTED GRAIN.

See BREWERS' PRODUCTS.

MALT HULLS. *See* BREWERS' PRODUCTS.

MALTOSE = MALT SUGAR.

MALTOSE PROCESSED CORN and *maltose processed corn gluten feed* are officially recognized feed products.—*See also* CORN.

MALT SPROUTS.

See BREWERS' PRODUCTS.

MALT SUGAR, or *maltose*, is a compound SUGAR which is formed by the action of the enzyme DIASTASE on starch: 1 molecule of malt sugar splits into 2 molecules of GLUCOSE during digestion.—*See also* BREWERS' PRODUCTS.

MAMLOXI. *See* SOYBEAN (variety).

MAMMALS, or *mammalia*, are the highest class of the vertebrates; the females possess mammae (teats). This animal class includes not only man and most species of land animals, but also some sea animals, such as whales and dolphins.—*See also* ANIMAL PRODUCTS.

MAMMOTH BROWN.
See SOYBEAN (variety).

MAMMOTH RED CLOVER = *American single-cut clover*. See RED CLOVER.

MAMMOTH WHITE is a RYE variety.
White Mammoth, better known as *White African sorgo*, is one of the forage SORGHUMS.

MAMMOTH YELLOW is a SOYBEAN variety.

MAMOTAN. *See* SOYBEAN (variety).

MAMREDO. *See* SOYBEAN (variety).

MANAGEMENT. *See* PASTURE MANAGEMENT; RANGE MANAGEMENT.

MANCHAR is a northern-type SMOOTH BROME.

MANCHU. *Manchu 3*, *Manchu 606*, and *Montreal Manchu* are important SOYBEAN varieties.

MANCHU BROWN KAOLING is one of the grain SORGHUMS.

MANCHUKOTA.
See SOYBEAN (variety).

MANCHURIA BARLEY.
See SIX-ROWED BARLEY (variety).

MANCHURIAN BEAN = SOYBEAN.

MANDAN WILD-RYE.
See CANADA WILD-RYE (variety).

MANDARIN. *Mandarin* (*Ottawa*). and *Mandarin 507*, are important SOYBEAN varieties.

MANDELL. *See* SOYBEAN (variety).

MANGANESE—the element *Mn*—is one of the TRACE ELEMENTS essential for plants and animals. Cereals, green plants, vegetables, fruits, and seeds of legumes are good sources of manganese. It occurs also in many ores. Addition of manganese salts to fertilizers is sometimes useful.

Poultry may suffer from *manganese deficiency*, and most poultry rations are fortified with 4 oz. of MANGANESE SULFATE per ton of ration. There is no convincing evidence that livestock, under farm conditions in America, suffer from such a deficiency.—*See also* HAY; MINERAL DEFICIENCIES.

MANGANESE CARBONATE forms white, rose-colored, or light-brown, water-insoluble crystals. It contains 47.8 percent MANGANESE.—*See also* MINERAL FEEDS.

MANGANESE CLOVER.
See SPOTTED BUR-CLOVER (variety).

MANGANESE DEFICIENCY.
See MINERAL DEFICIENCIES; MANGANESE.

MANGANESE OXIDE, or *manganous oxide*, is a grayish-green, water-insoluble powder which contains more than 77 percent MANGANESE.—*See also* MINERAL FEEDS.

MANGANESE PHOSPHATE, or *manganous ortho-phosphate*, is a reddish powder which is insoluble in water. It contains 11.4 percent MANGANESE.—*See also* MINERAL FEEDS.

MANGANESE SPOTTED BUR-CLOVER. *See* SPOTTED BUR-CLOVER.

MANGANESE SULFATE occurs with varying water content; the commercial grade usually contains close to 37 percent water and less than 23 percent MANGANESE. It forms pale-red, water-soluble crystals. —*See also* MINERAL DEFICIENCIES; MINERAL FEEDS.

MANGEL (*Beta vulgari* var. *macrorhiza*), or *mangel-wurzel*, a large-rooted plant, is useful as cattle and sheep feed; it contains much water, but may yield 30 tons per acre. The roots are easily harvested and can be stored without spoiling. It is recommended that only stored roots be fed, since freshly harvested, water-rich roots may cause scouring.

MANIFOLD = *omasum*. See STOMACH.

MANIHOT *M. utilitissima* = CASSAVA.

MANSOY is a SOYBEAN variety.

MANUAL MIXING. See FEED MIXING.

MANURE consists of animal or vegetable matter and is widely used to improve the soil. Fresh *stable manure* is best if first composted, watered, and repeatedly turned to avoid rapid fermentation (heating) and loss of ammonia. *Poultry manure*, for the same reason, should be mixed with soil and composted before being used in the field.

Barnyard manure, produced at the rate of 3 million tons a day by livestock in America alone, is rich in crop nutrients, but a large portion of the latter is lost through improper handling and storage. In general, 1 ton of manure is equal to 100 lb. of 3-5-10 or 4-5-10 commercial FERTILIZER.—*See also* GREEN MANURE; FIELD

BEANS; SORGHUMS; BUR-CLOVERS; KUDZU.

MANYPLIES = *omasum*. *See* STOMACH.

MARCROSS. *See* CORN (variety).

MARETT HOODED 4 is a SIX-ROWED BARLEY variety.

MARIDA. *See* COMMON OAT (variety).

MARIJUANA is a habit-forming drug, the manufacture and transportation of which are regulated by law.—*See also* HEMP.

MARINE ANIMAL OIL. *See* VITAMINS.

MARINE PRODUCTS are either manufactured from whole fish or are fishery by-products obtained in canneries and other fish-packing factories. They include the following officially recognized feedstuffs:

Fish meal is clean, dried, ground tissue of undecomposed whole fish or fish cuttings, either or both, and with or without extraction of part of the oil. If it contains more than 3 percent salt (NaCl), the amount of salt must constitute a part of the brand name, provided that in no case shall the salt content of this product exceed 7 percent.

Fish meal is widely used in poultry feeds, where it has come to be regarded as superior to meat meal, because of its greater effectiveness in supplementing a diet composed largely of grains and OILSEED MEALS. For growth and reproduction the superiority of fish meal is partially due to its higher content of LYSINE and VITAMIN B_{12}. It may also contain U.G.F.

NOTE: Among the fish widely used in the manufacture of fish meals are the menhaden, sardine (or pilchard), herring, salmon, and white fish.

Fish residue meal is the clean, dried, undecomposed residue from the manufacture of glue from nonoily fish. If it contains more than 3 percent salt (NaCl), the amount of salt must constitute part of the brand name, provided that in no case shall the salt content of this product exceed 7 percent.

Fish liver and glandular meal—obtained by drying the complete coelomic content of fish. At least 50 percent of dry weight of the product must be derived from fish liver; it must contain at least 18 mg. riboflavin (VITAMIN B_2) per pound.

Condensed fish solubles—obtained by condensing the water left from the hy-draulic extraction of oil. It is often added to poultry rations as a source of U.G.F.

Crab meal is the undecomposed, ground, dried waste of the crab, consisting of the shell, viscera, and part or all of the flesh. It shall contain not less than 25 percent protein. If it contains more than 3 percent salt (NaCl), the amount of salt must constitute a part of the brand name, provided that in no case shall the salt content of this product exceed 7 percent.

Shrimp meal is the undecomposed, ground, dried waste of shrimp consisting of the head, hull, or whole shrimp, either separately or in a mixture. If it contains more than 3 percent salt (NaCl) the amount of salt must constitute part of the brand name, provided that in no case shall the salt content of this product exceed 7 percent.

NOTE: In addition, the following (not officially recognized) marine products are used for feeding purposes:

Homogenized condensed fish—a partially dehydrated product made from fish and/or fish cuttings from which part of the oil may have been removed. It shall contain not less than 50 percent total solids, exclusive of any added salt. If salt has been added, the product shall be labeled "*Homogenized condensed fish* with . . . percent added salt."

Dried fish solubles product—obtained by drying and grinding the precipitate resulting when fish press water is adjusted to a pH of approximately 4.5. References: F.6; E.12.

See also BY-PRODUCT FEEDSTUFFS; FAT; SHARK MEAT.

MARINO = TARTARY BUCKWHEAT.

MARION. *See* COMMON OAT (variety).

MARKET HOGS. *See* SWINE RATIONS.

MARKTON. *See* COMMON OAT (variety).

MARMOR. *M. tritici*

= WHEAT MOSAIC.—*See also* MOSAICS.

MARNOBARB.

See SIX-ROWED BARLEY (variety).

MARQUIS WHEAT.

See COMMON WHEAT (variety).

MARROWFAT PEA is a FIELD PEA variety.

"MARSH GRASSES."
See GRASSLIKE PLANTS.
MARSH HAY, in general, is inferior to HAY from upland grasses; however, it may be used to feed idle horses and mules.

MARTIN BROME is a northern-type SMOOTH BROME.

MARTIN MILO, or *Martin's combine milo*, is not a true milo, but one of the kafir-milo derivatives which belong to the grain SORGHUMS.

MARVELOUS WHEAT
= *Fulcaster. See* COMMON WHEAT.
MASH is a soft, pulpy, wet feed, normally prepared from ground grain and water or milk; it is fed to young animals and chickens.—*See also* BREWERS' PRODUCTS.

MAST. *See* GRAZING; PASTURE PLANTS.

MASTICATION is the chewing process that precedes the swallowing of food. It consists of crushing and grinding the food with the teeth and moistening it with saliva; thus, a slippery mass is formed for easy passage through the gullet.

MAY BEETLE, or *June beetle*, is the adult stage of the WHITE GRUB.

McCLAVE = *Midwest soybean.*
See SOYBEAN (variety).
McLEAN SORGO is one of the forage SORGHUMS.

McNEILL CLOVER = CLUSTER CLOVER.

MEADOW is an area covered with fine-stemmed (mainly perennial) forage plants used for PASTURE or to produce HAY.—*See also* MEADOW GRASSES.

MEADOW FESCUE (*Festuca elatior*), a hardy perennial, flourishes in deep, rich soils; where it is adapted it usually grows to 15 to 30 in. The leaves are bright green and rather succulent, the leaf sheaths smooth and reddish purple at the base. The PANICLES are open. The grass flowers in June and July.

Meadow fescue produces comparatively few culms and an abundance of germinable seed.

Weeds that make stands of meadow fescue hard to maintain include CHESS, or *cheat*, and WHITETOP FLEABANE.

Meadow fescue can be grown well throughout the TIMOTHY region and also farther south and west because it can withstand more heat and drought than timothy. In regions where it is adapted, it is grown to a limited extent for pasture and hay, but it is being used less and less because of the improvement in the production of TALL FESCUE.

Reference: H.1.
See also FESCUES; GRASSES; ALSIKE CLOVER; PASTURE PLANTS.

Meadow fescue. Plant, spikelet, and floret. (H.26.)

MEADOW FOXTAIL (*Alopecurus pratensis*)—not to be confused with FOXTAIL MILLET—is a long-lived perennial grass. It resembles timothy in head so closely that it is often mistaken for it. The rootstocks are comparatively few, as are the underground branches. Old, heavy stands will produce medium dense sods. The flowering stems are erect and usually about 3 ft. high; rarely do they grow to 6 ft. The leaves are dark green and numerous.

The seeds are light in color, but occasionally plants produce brown or black seeds.

Meadow foxtail. (H.26.)

Meadow foxtail is not sensitive to heat or cold, and, therefore, succeeds in areas where summer temperatures occasionally reach 100° F. and winter temperatures drop below zero for relatively long periods. Continuous winter growth often occurs where mean minimum temperatures are 40° F. Soil moisture is the limiting factor during periods of high mean maximum temperatures. It is definitely not a southern grass but is adapted to the moist, cool climates of southern Canada and the Pacific Northwest. It also has possibilities as a forage grass in several sections of the northern half of the United States.

Meadow foxtail is naturally a wet- or moist-land grass and makes its best growth on fertile or swampy soils. It is used extensively on diked lands near the coast in the Pacific Northwest. The occasional overflow of brackish water does little or no damage to established stands. The grass is quite tolerant of saline soils, but that tolerance depends largely on soil moisture. It responds to irrigation in cool climates and is adapted to irrigated pastures when seeded alone or in mixtures of other grasses and LEGUMES.

Both fall and spring seedings are successful in the Pacific Northwest. Generally, in most sections where it is adapted, spring seeding is most satisfactory on cultivated lands.

Meadow foxtail is primarily a pasture grass. Its long life, long grazing season, winter hardiness, and succulent forage make it a valuable pasture grass. It is seldom used for hay, but the hay is leafy and palatable to all kinds of livestock. In Oregon, hay yields sometimes average about 1½ tons an acre for one cutting.

For a number of years now, meadow foxtail—in combination with various grasses and legumes—has been made into silage, especially in the coastal sections of the Pacific Northwest.

Dangers: INSECTS attack meadow foxtail only to a limited extent in humid climates. SLUGS and CUTWORMS cause some damage. In dry climates, especially late in the growing season when GRASSHOPPERS are numerous, considerable damage is caused by insects. Occasionally, APHIDS attack the green heads. No troublesome diseases have been noted, but ERGOT, SCALD, and STEM RUST have been observed.

Reference: H.1.

See also SUBCLOVER; RED FOXTAIL; GRAZING; PASTURE PLANTS; GRASSES.

MEADOW GRASSES. *"Wet meadow grasses"* is a term often applied to the GRASSLIKE PLANTS.

MEADOW NEMATODES are NEMATODES which attack PEANUTS, COTTON, and many other crops in the Southeast. Various *Pratylenchus* spp., especially the *P. leiocephalus,* cause dark-brown or black spots on the shells of peanuts. The spongy shell tissue under the discolored spots is partly or completely destroyed. The young rootlets on plants producing discolored nuts are blackened and dead, and many young pegs show discolored infested spots, usually in the tender, rapidly growing area near the tip. The root nodules, produced by nitrogen-fixing bacteria, are frequently destroyed; this destruction lowers the nitrogen-gathering power of the plants to such an extent that the plants may become yellowed and stunted.

Control: No practical methods of control have been found. The nematodes and their eggs may remain alive for several months in the cured peanut shells. Shells should be thoroughly steamed or fumigated before they are spread on fields or used as mulch.

Reference: B.10.

See also STING NEMATODE; ROOT-KNOT NEMATODE.

MEADOW SPITTLEBUG (*Philaenus leucophthalmus*). In most fields of RED CLOVER and ALFALFA, plants can be found in the spring that have frothy, whitish spittle masses somewhere on the new growth, each enclosing one or more nymphs. Eggs hatch at about the time plants break winter dormancy. Newly hatched nymphs are orange-yellow, but they change to a light green as they grow. The adult INSECTS appear generally before the first hay crop is harvested. Stunting of plants that show masses of spittle is common.

Control: Spraying with BENZENE HEXA-CHLORIDE (preferably LINDANE) effectively controls the nymphs. Early treatment is

necessary for achievement of maximum results. TOXAPHENE or METHOXYCHLOR are also recommended insecticidal sprays.

References: E.5; G.6.

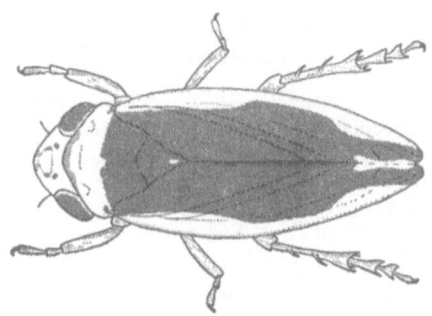

Meadow spittlebug. (Q.2.)

MEAT. *See* ANIMAL PRODUCTS.

MEAT AND BONE MEAL.
                              See ANIMAL PRODUCTS.

MEAT AND BONE MEAL DIGESTER TANKAGE. *See* ANIMAL PRODUCTS.

MEAT AND BONE MEAL TANKAGE.
                              See ANIMAL PRODUCTS.

MEAT AND BONE SCRAP.
                              See ANIMAL PRODUCTS.

MEAT BY-PRODUCTS.
                              See ANIMAL PRODUCTS.

MEAT FLAVOR. *See* MEAT QUALITY.

MEAT MEAL. *See* ANIMAL PRODUCTS.

MEAT MEAL TANKAGE.
                              See ANIMAL PRODUCTS.

MEAT QUALITY is often affected by the feed the animals consume. The term quality, as applied to meat, includes such factors as *flavor,* appearance, wholesomeness, tenderness, composition, juiciness, and nutritive value. The table on page 286 lists the effects of certain feeds.

MEAT SCRAP = *meat meal.*
                              See ANIMAL PRODUCTS.

MEDIC. *See* BLACK MEDIC.

MEDICAGO. The genus *Medicago* does not belong to the true clovers but resembles them; it includes the LEGUMES *M. arabica* = SPOTTED BUR-CLOVER; *M. hispida* = CALIFORNIA BUR-CLOVER; *M. rigidula* = TIFTON BUR-CLOVER; *M. orbicularis* = BUTTONCLOVER; *M. lupulina* = BLACK MEDIC. Also, the ALFALFAS belong to this genus; e.g., *M. satina* which includes the COMMON ALFALFAS, TURKESTAN ALFALFAS, and NONHARDY ALFALFAS; *M. media* = VARIEGATED ALFALFA; *M. falcata* = YELLOW-FLOWERED ALFALFA.—*See also* BUR-CLOVERS; SPINELESS BUR-CLOVERS; CLOVERS.

MEDICAMENTS.

See MINERAL FEEDS; DRUGS.

MELANOPLUS. Among the many GRASS-HOPPERS are a number of *Melanoplus* spp.; e.g., *M. mexicanus* = *lesser migratory grasshopper; M. differentialis* = *differential grasshopper; M. bivittatus* = *two-striped grasshopper;* and *M. femur-rubrum* = *red-legged grasshopper.*

MEAT QUALITY
Effects of feeds on wholesomeness, appearance, composition, tenderness, and flavor of meat

Feed consumed	Hog	Beef	Poultry
Rape, turnip, cabbage			Off flavor
Cod liver oil	Fishy		Fishy
Sulfur			Slows down transfer of xanthophyll to tissue
Fish oils, meals, meat scraps	Off flavor		Off flavor
Yellow corn, gluten meal, hominy feeds	Softening effect	Improve palatability. Tenderness through finish	Yellow tissue
Wheat	Firm	Firm	Lack of yellow color
Shepherd's purse, pennycress			Off-colored yolks
Silage	Same as pasture	Watery. Lack of marbling	Dark green yolks
Pasture	Meatier; less firm lecithin	Less finish	Xanthophyll
Alfalfa hay		Yellow fat carotene	Xanthophyll
Lespedeza		White fat	
Kansas bluestem		Pink-white fat	
Meat scraps, horse manure	Repulsive odor		Off flavor
Old white flour	Repulsive odor		
Molasses, beet pulp	Good up to 100 lb. live weight		
Rice polish, rice bran	Soft pork		
Brewers' rice	Very firm pork		
Potatoes	O. K.		
Barley	Firmer than when corn fed		
Rye	Firm		
Oats	As corn		
Milo, kafir, hegari	Equal to corn. White fat		
Peanuts	Soft pork		
Cow peas	Medium hard		

MEDICATED FEEDS. *See* ANTIBIOTICS; ARSONIC ACIDS; MINERAL FEEDS; DRUGS; LABELS; F.D.A.

MEDITERRANEAN WHEAT.

See COMMON WHEAT (variety).

MEDIUM (plural: *media*) is the nutrient substance used to cultivate (grow) yeast and other MICRO-ORGANISMS.

MEDIUM RED CLOVER is a RED CLOVER variety of the *double-cut clovers.*— *See also* GRAZING; TIMOTHY.

MEEKER BALTIK.

See VARIEGATED ALFALFAS (variety).

MELILOTUS. The various annual and biennial types of SWEETCLOVERS are *Melilotus* spp. and, therefore, not true clovers. They include *M. alba* = WHITE SWEETCLOVER (and its varieties *Spanish* and *Hubam*); *M. officinalis* = YELLOW SWEETCLOVER (and its variety *Madrid*); *M. suaveolens* = DAGHESTAN SWEETCLOVER (and its variety *Redfield*); and *M. indica* = SOURCLOVER.

MELINIS. *M. minutiflora* = MOLASSES GRASS.

MELOIDOGYNE. *Meloidogyne* spp. (formerly grouped under the name *Hetero-derma marioni*) = ROOT-KNOT NEMATODES. *M. hapla* attacks PEANUTS, *M. arenaria* attacks peanuts and several other plants.

MELOY 3.
     See SIX-ROWED BARLEY (variety).

MEMBRANE is a thin, soft tissue.

MEMBRANOUS means: of, or pertaining to, a MEMBRANE; or membrane-like.

MENADIONE is a synthetic substance. It forms yellow crystals, which decompose when exposed to sunlight. Menadione is insoluble in water but soluble in vegetable oil, and possesses physiological VITAMIN K properties.

MENHADEN is a salt-water fish of North America. It is a relative of the shad and valuable as a source of vitamin-rich MENHADEN OIL, fish meal, and fertilizer.— *See also* MARINE PRODUCTS.

MENHADEN OIL is obtained from the whole MENHADEN fish.—*See also* VITAMINS.

MERCURIALS are mercury-containing compounds used medicinally, in SEED TREATMENT, etc.; e.g., PANOGEN and CERESAN M.—*See also* TERSAN.

MERISOPORUS. *M. chalcidiphagus* is one of the useful WASPS that destroy the WHEAT STRAWWORMS.

MERISUS. *M. febriculosus*, a WASP, is a WHEAT STRAWWORM parasite.

MERKER GRASS is a horticultural variety of NAPIERGRASS.

MESA is an elevated table land (plateau).   —*See also* GRAZING.

MESQUITE (*Prosopis juliflora*) and the closely related TORNILLO, or *screwbean*, belong to the LEGUME family. Depending on the conditions of climate and soil, varieties of mesquite vary in character of growth—forming either trees or shrubs. Stems of the mature shrubs commonly vary in height from 2 to 5 ft. or more, whereas the trees vary from 10 to as much as 50 ft. The mesquite is deciduous, with dark-green leaves divided into numerous leaflets. The wood is hard and a reddish-brown with an outside layer of yellow sapwood. The twigs are armed with straight spines $\frac{1}{4}$ to $1\frac{1}{2}$ in. long. Flowers are small, greenish-yellow, and grow in cylindrical clusters 2 to 3 in. long near the ends of the branches. Fleshy seed pods, or

"beans," 4 to 8 in. long and containing from 10 to 20 seeds each, develop and ripen about six weeks after flowering.

Mesquite grows in the Southwest of the United States. Temperatures limit its northern extension to approximately the southern borders of Kansas, Colorado, and Utah. Within its temperature range, mesquite is a potential invader of all soil types under a wide variety of moisture conditions. The root development of mesquite varies. On deep soils that have adequate moisture, a strong taproot develops; but on shallow upland soils where moisture seldom penetrates deeply, the taproot is small and laterals may reach 50 ft. or more in all directions just beneath the surface of the soil.

Mesquite provides protective cover and a food supply for quail, doves, and other wildlife, and the flowers are an important source of nectar for honeybees. On the range it provides shelter for livestock in the winter and shade during the hot days of the summer. Its pods contain large amounts of sugar and protein and are highly relished by livestock. Its leafage also provides forage during droughts and in early spring when other forage may be scarce. However, its forage values are usually overestimated and are not great enough to offset the detrimental effect of mesquite on density and yield of perennial grasses. Once established in quantity, mesquite thickens whether the range is grazed or not.

NOTE: To preserve open grass land for profitable long-time grazing use, persistent removal of mesquite seedlings, as they invade the land, is essential. Also, on upland ranges where mesquite crowns cover 7 percent or more of the soil surface, immediate application of proven manual methods is recommended.

When the crowns of mesquite are cut off or killed, the stumps sprout. These sprouts originate from perennial dormant buds located on stem tissue in the zone immediately above the root collar. Effective chemical control methods must be used to kill these buds. Mechanical methods must uproot all tissue on which such structures occur and sever its root connection with the soil. For small plants, grubbing to a depth of 4 to 5 in. is effective and practical. Mesquite can be killed by any of several hand application methods with SODIUM ARSENITE; or by spraying or pouring light PETROLEUM OILS around the base of the plant.

Varieties: Three varieties of mesquite have been recognized; namely:

1. *Honey mesquite*, which is the common variety found in Texas. It has long, linear, glabrous (smooth), and widely spaced leaflets. The leaves and "beans" possess high feeding value. However, where range cattle are forced to subsist chiefly on mesquite "beans" and leaves, severe digestive trouble may result from compaction.

2. *Western honey mesquite*, which is common in the sand-dune areas of New Mexico, extreme western Texas, and southeastern Arizona.

3. *Velvet mesquite*, which occurs in Arizona, and has short, hairy, and closely spaced leaflets. In river bottoms in southern Arizona, velvet mesquite becomes a large tree with a well-defined trunk (up to 2 ft. or more in diameter and 40 ft. or more in height). In semiarid, sandy, windswept localities it is a many-stemmed shrub, 1 to 3 ft. in height. In upland areas of compact soil and moderate moisture it tends to form a scraggy woodland type with individual plants having one or more main stems which are intermediate in stature between the dune and large tree forms. Velvet mesquite leaves, though, are equal in feeding value to the best quality alfalfa hay and velvet mesquite "beans" are much higher in nitrogen-free extract, and lower in ash, than alfalfa hay. In digestibility mesquite pods are superior to ALFALFA.

Reference: P.11.

See also INOCULATION; VINE-MESQUITE; RANGE PLANTS; GRASSES.

METABOLISM is the total of the life processes and sums up the chemical and physical changes involved in the growth, reproduction, and aging of plants or animals. It is based on the changes taking place in NUTRIENTS after their ABSORPTION into the system for repair and formation of tissue, production of heat (body temperature), work, etc.—*See also* THYROXINE; THYROPROTEIN.

METABOLIZABLE ENERGY.

See ENERGY.

METALDEHYDE forms white prisms which are insoluble in water. It may be used in poison BAIT for the control of the common garden SLUGS, which are pests of SUBCLOVER and other crops.

METAMORPHOSIS means: transformation; i.e., any change of form, structure, or nature of a living body as result of natural development; especially the alterations of an animal from egg to adult form. —*See also* LARVA.

METHEMOGLOBIN is an abnormal blood pigment with the same composition as HEMOGLOBIN, but with the iron present in the ferric state and its oxygen more firmly bound.—*See also* OAT HAY POISONING.

METHIONINE is the only essential AMINO ACID that contains sulfur. It forms water-soluble crystals. The proteins of CORN or fishmeal are good sources of it while beans are often deficient in methionine. Part of an animal's need for methionine can be met by CYSTINE.

While *dl-methionine* is listed among the officially recognized VITAMINS, it is generally classified as one of the essential amino acids.—*See also* SESAME OIL MEAL.

METHIONINE HYDROXY ANALOGUE CALCIUM. See AMINO ACIDS.

METHOXYCHLOR forms white, water-insoluble crystals; the commercial product is about 88 percent pure, the rest consists of related materials. Generally speaking, methoxychlor has the same chemical and physical properties as the closely related DDT, but it is more stable in the presence of alkalies and has a relatively low toxicity to higher animals; it is, therefore, permitted for use on forage and food crops in form of dusts, wettable powders, or emulsified solutions.

Methoxychlor is the only INSECTICIDE now recommended for use on ALFALFA grown for hay. To protect the latter against the alfalfa CATERPILLAR, a spray containing 12 oz. methoxychlor (100 percent strength) —or 1 lb. in hot desert areas—per acre should be applied by airplane or by sprayer; if a 25 percent emulsifiable methoxychlor concentrate (properly diluted) is used, 1½ qt.—or 2 qt. in hot areas—per acre should be applied. Methoxychlor is also used in the control of the MEADOW SPITTLE-BUG; the insecticide is sprayed at the rate of 1 lb. per acre.

For the control of the POTATO LEAF-HOPPER, spray at the rate of 1 qt. 25 percent emulsifiable concentrate (properly diluted) per acre.

To kill the larvae of FALL ARMYWORMS, dust or spray the crop with 1 to 2 lb. methoxychlor per acre. For the control of VELVETBEAN CATERPILLARS or PEA WEE-VILS, use 20 lb. dust containing 5 percent methoxychlor per acre and repeat the treatment in three to four days, if necessary.

Caution: Methoxychlor is a poison. Handle it with care, according to directions on the container. Store it where children and animals cannot reach it.

METHYL BROMIDE, at ordinary temperatures, is a gas, but can be easily liquefied and is commercially available (as liquid) in cylinders. It is noninflammable, colorless, has a sweetish odor, and is toxic to animals and especially to insects; it is widely used as an insecticidal FUMIGANT.

Methyl bromide is useful for the control of both, the RED HARVESTER ANT and the WESTERN HARVESTER ANT. It shows satisfactory results in moist soils only. Use a mechanical dispenser that will release the gas into the colony entrance, 6 to 8 in. below ground level. Apply 1 to 2 fl. oz. methyl bromide from a 1 lb. can per treatment; then pack the soil tightly over the entrance hole.

WHITE TIP of rice can be controlled by fumigation with methyl bromide at the rate of 1¼ lb. per 1,000 cu. ft. seed (of 13 percent moisture content) for twelve hours.

Caution: Methyl bromide is a poisonous gas. Store it in a cool place and handle carefully. It is particularly dangerous at high temperatures. Do not store it in buildings where people live or work.

METHYLENE BLUE forms dark-green crystals with bronze luster. It is soluble in water at the rate of 1:25. Such aqueous solutions are used as antidotes; they are injected into the veins of animals affected with PRUSSIC ACID POISONING or OAT HAY POISONING.

METHYL MERCURY DICYAN DI-AMIDE is a volatile MERCURIAL; it is the active ingredient of PANOGEN, a FUNGICIDE used for SEED TREATMENT.

METHYL PARA-HYDROXYL BEN-ZOATE forms white needles which are soluble at the rate of 1:400 in water and up to 1:40 in oils. It is widely used as a preservative in concentrations of 0.05 to 0.2 percent.—*See also* MOLD INHIBITORS.

MEXICAN BEAN BEETLE (*Epilachna varivestis*) overwinters in the adult (BEETLE) stage, usually in the woodlands near bean fields. The convexly rounded adult is about ⅓ in. long and copper-colored, with 16 small black spots on its back. In the spring, the female beetles lay their eggs on the underside of SOY-BEAN and other bean leaves. These eggs hatch in five to fourteen days into larvae that feed principally on the underside of the bean leaves. The yellow larvae grow rapidly, passing through four stages. Each stage is larger than the preceding one. They reach full growth in twenty to thirty-five days. The full-grown larva attaches itself to the underside of the leaf on which it has been feeding or to some nearby

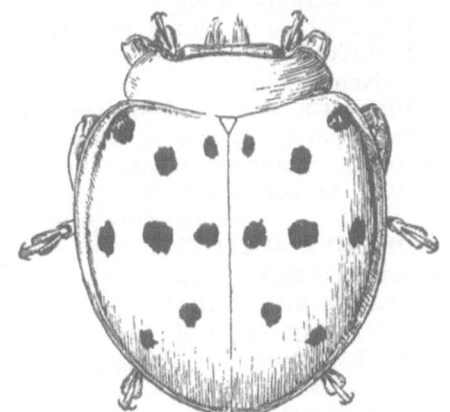

Mexican bean beetle. (U.13.)

object, and changes to a pupa. After ten days or so, the adult beetle emerges from the pupa. Within two weeks the female beetle is ready to deposit eggs for another brood.

Control: Spray or dust with DERRIS, CUBE, or CRYOLITE. The first application of the insecticide should be made when Mexican Bean beetles are found in the field. Repeat every seven to ten days if the insects are numerous.

References: U.3; M.19.

MEXICAN DESERT WHEAT CORN, better known as *shallu*, is one of the grain SORGHUMS.

Mg = MAGNESIUM.

MICE.

See FIELD MICE; RANGE MANAGEMENT.

MICHIGAN AMBER = *Red May.*

See COMMON WHEAT.

MICHIGAN WINTER BARLEY is a variety of SIX-ROWED BARLEY.

MICHIGAN WONDER = *Red May.*

See COMMON WHEAT.

MICRO- is a prefix signifying small, or minute; it is applied preferably to microscopic objects.

Micro-elements is a term that is sometimes used to indicate nutrients needed in small amounts.

MICROBE is a BACTERIUM, but often this term is applied to other MICRO-ORGANISMS as well.

MICRO-ORGANISMS are minute animals or plants; e.g., BACTERIA, FUNGI, YEASTS, MOLDS, ALGAE.—*See also* MICROBE; LEGUME BACTERIA; WEAK NECK.

MIDA. *See* COMMON WHEAT (variety).

MIDDLES is the space between the rows of cultivated crops.

MIDDLINGS is a feed derived from milling small grain, such as wheat.—*See also* OFFAL; PALMO MIDDLINGS.

MIDGE. *See* SORGHUM MIDGE; CLOVER-FLOWER MIDGE.

MIDLAND RED CLOVER is a *double-cut* variety of RED CLOVER.

MIDLAND SORGHUM is one of the intermediate-type grain SORGHUMS.

MIDVEIN, or *midrib*, is the large central ridge on a leaf.

MIDWEST SOYBEAN.

See SOYBEAN (variety).

MILDEWS are FUNGUS growths occuring on damp substances and plants.—*See also* POWDERY MILDEW; DOWNY MILDEW; DOWNY MILDEW OF ALFALFA; DOWNY MILDEW OF SOYBEAN; DOWNY MILDEW OF VETCH; COMMON WHEAT.

MILK BY-PRODUCTS.

See MILK PRODUCTS.

MILK FEVER. *See* DRY COWS.

MILK OF SULFUR

= *precipitated sulfur. See* SULFUR.

MILK PRODUCTS include *milk* and the three *milk* (or *dairy*) *by-products* which are fed in the liquid, condensed, and dried forms: (1) *Skimmed milk,* or *skim milk,* is the low-fat fraction separated from cream by centrifugal separators. (2) *Buttermilk* is the by-product remaining after butter has been produced from cream by churning. It contains slightly less fat than does skimmed milk. A little more than one-third of the solids of both consists of high quality protein. (3) *Whey* is the by-product remaining after the making of cheese from milk. It is lower in protein than skimmed milk and buttermilk.

The *liquid* milk products are fed both in *sweet* and *sour* forms, principally on the farms where they are produced; the *condensed* and *dried* forms are important commercial feedstuffs. All of them possess, like milk, high nutritive values due to their vitamin, mineral, and LACTOSE (or *milk sugar*) contents and to the excellent quality of their protein. The fresh products should be pasteurized before they are fed.

Officially recognized are the following milk products:

Condensed skimmed milk—the product resulting from the removal of a considerable portion of water from clean, sound skimmed milk. It shall contain not less than 27 percent total solids.

It is valued for its protein, VITAMIN B_2, and other water-soluble vitamins.

Dried skimmed milk—the product resulting from the removal of water from clean, sound skimmed milk. It shall contain not more than 8 percent moisture.

On many farms that sell *whole milk*, little or no skimmed milk is available for calf feeding, and the use of whole milk alone for this purpose is too expensive. In that case, feeding of dried skimmed milk is recommended.

In preparing dried skimmed milk, mix 1 part powder to 9 parts warm water by weight. Feed it in the same way, and same amount, as ordinary skimmed milk.

Evaporated cultured skimmed milk, concentrated cultured skimmed milk, or *condensed cultured skimmed milk*—the product resulting from the removal of a considerable portion of the water from clean, sound skimmed milk which has been cultured by a suitable culture of lactic bacteria. It shall contain not less than 27 percent total solids.

Dried cultured skimmed milk—the product resulting from the removal of water from clean, sound skimmed milk which has been cultured by a suitable culture of lactic bacteria. It shall contain not more than 8 percent moisture.

Evaporated buttermilk, concentrated buttermilk, or *condensed buttermilk*—the product resulting from the removal of a considerable portion of water from clean, sound buttermilk derived from natural cream to which no foreign substances have been added, except such as are permitted and necessary in the manufacture of butter. It shall contain not less than 27 percent total solids, and for each percent of solids not less than 0.055 percent butterfat and not more than 0.14 percent ash.

Like condensed skimmed milk, it is valued for its protein and water-soluble vitamins, especially its vitamin B_2 content.

Dried buttermilk—the product resulting from the removal of water from clean, sound buttermilk derived from natural cream to which no foreign substances have been added, except such as are necessary and permitted in the manufacture of butter. It shall contain not more than 8 percent moisture, not more than 13 percent mineral matter (ash), and not less than 5 percent butterfat as determined by the Roese-Gottlieb method.

Condensed whole whey—the product resulting from the removal of a considerable portion of water from clean, sound cheese or casein whey, singly or together. It shall contain not less than 6 percent total whey solids; otherwise it shall be designated *"condensed whey" . . . percent solids."*

It is a valuable feedstuff because of its high protein content and because it contains also water-soluble vitamins, especially vitamin B_2.

Dried whole whey is dried, clean, sound cheese and/or casein whey. It shall contain not less than 65 percent lactose.

Like all *dried whey* products, it is an important ingredient in poultry feeding.

Condensed whey-solubles—the product resulting from the removal of albumin and the partial removal of milk sugar from clean, sound whey to which no foreign substances have been added, except such as are necessary in the manufacture of milk sugar.

Dried whey-solubles—the dried product resulting from the removal of albumin and the partial removal of milk sugar from clean, sound whey to which no foreign substances have been added, except such as are necessary in the manufacture of milk sugar.

Condensed whey-product results from the removal of a considerable portion of water and the partial removal of milk sugar from clean, sound whey to which no foreign substances have been added. It shall contain not less than 50 percent total whey-product solids; otherwise it shall be designated *"condensed whey-product . . . percent solids."*

Dried whey-product . . . percent lactose is the dried product resulting from the partial removal of milk sugar from clean, sound whey to which no foreign substances have been added. When listed as an ingredient in manufactured feeds, the term *"dried whey-product"* may be used.

Condensed cultured whey is partially dehydrated, clean, sound cheese and/or casein whey that has been cultured by a suitable ferment. It shall contain not less than 62 percent total cultured whey solids; otherwise it shall be designated *"condensed culture whey . . . percent solids."*

Casein is the product resulting from acid or rennet precipitation of skimmed milk. It must contain at least 80 percent crude protein.

Cheese rind is cooked, partially defatted cheese rind.

Dried milk albumin is the dried, coagulated protein fraction separated from whey. It shall contain not less than 75 percent protein (N x 6.25) on a moisture-free basis.

Dried whole milk results from the removal of water from milk and contains not more than 8 percent moisture, and not less than 26 percent butterfat as determined by the Roese-Gottlieb method.

NOTE: Tentatively, the following definitions of milk products have been accepted:

Condensed whey-product . . . percent solids—the product resulting from the removal of a considerable portion of water and the partial removal of milk sugar from clean, sound whey to which no foreign substances have been added. The solids of this product shall contain at least 55 percent lactose.

Dried hydrolyzed whey is clean, sound cheese whey in which at least 50 percent of the lactose has been hydrolyzed by a lactase enzyme.

Condensed hydrolyzed whey — resulting from the removal of a considerable portion of water from clean, sound cheese whey which has been hydrolyzed by a lactase enzyme so as to convert at least 50 percent of the lactose into glucose and galactose. It shall contain at least 50 percent total solids.

Caution: All milk products, not produced under known sanitary conditions, should be pasteurized before they are fed to farm animals.

References: F.6; E.12; H. 31.

See also BY-PRODUCT FEEDSTUFFS; VITAMIN A; VITAMIN D; PASTEURIZED DAIRY BY-PRODUCTS.

MILK PROTEINS. *See* MILK PRODUCTS.
MILK REPLACER PLAN.
                    See DAIRY RATIONS.
MILK SUGAR
            = LACTOSE.—*See also* GALACTOSE.
MILK-VETCHES = POISONVETCHES.
MILKWEEDS (*Asclepias* spp.) are perennial plants having milky juice; many of them contain RESINOIDS and are poisonous; e.g., *broadleaf milkweed*, *whorled milkweed*, *Mexican whorled milkweed*, and the species *A. labriformis* and *A. latifolia*.

Common milkweed plant with tufted seed and opening seed-pod. (H.48.)

Whorled milkweed with seed-pod. (H.48.)

When hungry sheep and cattle, especially in fall and winter, eat as little as 1½ to 3 oz. green plants in a day—particularly leaves—they develop some of the following symptoms of poisoning (varying with the milkweed species digested): weakness, shallow respiration, severe spasms, violent struggling, depression, inco-ordination; inflammation of rumen and intestine.—*See also* POISONOUS PLANTS.

MILL DUST is the fine powder that blows from milling machinery. Mill dust should be controlled as it is a fire and explosion hazard.—*See also* SCREENINGS.

MILLERS = MOTHS.

MILLER WHEAT = *Mediterranean wheat. See* COMMON WHEAT.

MILLET. *Proso millet* (also called *broomcorn millet, hershey millet,* or *hog millet*) = PROSO.

Cattail millet = PEARLMILLET.

Japanese millet is a variety of BARNYARDGRASS. — *See also* FOXTAIL MILLET; PASTURES.

MILL FEEDS are feed products obtained in flour mills; e.g., BRAN, MIDDLINGS, SHORTS, SCREENINGS, feed flour, etc.

"MILL RUN BRAN." *See* LABELS.

"MILL RUN SHORTS." *See* LABELS.

MILLS. *See* GRINDING OF FEED MATERIAL.

MILO is a classification of the grain SORGHUMS which includes many varieties, selections, and hybrids. Among the varieties are the following: *Colby, Day, Double Dwarf milo No. 38, Double Dwarf White Sooner, Double Dwarf Yellow Sooner, Dwarf White, Dwarf Yellow, Early White* (or *Sugar milo*), *Finney, Miloca, Sooner* (or *Sixty Day*), *Standard White, Standard Yellow, Texas,* and *Texas Double Dwarf.*

Of great importance are many KAFIR-milo derivatives.—*See also* COWPEA.

MILOCA, a *Double Dwarf White Sooner* (milo), is one of the grain SORGHUMS.

MILO DISEASE, or *milo root rot*—originally known as *"Pythium root rot of milo"* —is now called PERICONIA ROOT ROT.

MIMOSOIDEAE form one of the three subfamilies of the LEGUMES.

MINDUM. *See* DURUM WHEAT (variety).

MINERAL. *See* MINERALS.

MINERAL DEFICIENCIES of plants may cause symptoms that are likely to be mistaken for the symptoms of diseases of parasitic origin.

Mineral deficiency diseases of animals: Co, cobalt deficiency; Cu, copper deficiency; Fe, iron deficiency. Bone diseases: Ca, calcium deficiency; P, phosphorus deficiency. Other troubles: Se, selenium toxicity; GT, grass tetany; shaded area, goiter belt; X, unknown causes. (U.S.D.A.)

When soil is deficient in *potash*, a yellow mottling appears around the edges of the leaflet (e.g., of SOYBEAN), and tissues at the margin become brown and brittle. The breaking up of the dead tissues gives the leaflets a ragged look. The central area of the leaflets remains green, but from a distance the affected fields have a distinctively yellow appearance.

When plants suffer *iron* deficiency, the area between the veins fades from normal green to yellow, with the tissues around the veins remaining green. Later, the entire leaflet turns yellow.

The leaves of plants suffering from *manganese* deficiency look very much like those of plants with iron deficiency, a similarity which makes it difficult to distinguish between the two conditions. (A chemical test of the live plants may be necessary for accurate diagnosis.)

Phosphorus deficiency is characterized by prominent, reddish-purple veins, and a pale lower side of the undersized leaves.

Boron deficiency in ALSIKE CLOVER and other plants, is recognized by the brownish-red discoloration of the leaflets which later become ragged.

Chlorosis is a symptom of a number of mineral deficiencies, such as iron, manganese, *sulfur*, and *magnesium*. In the case of magnesium deficiency, the chlorosis is characterized by loss of (green-colored) chlorophyll between the leaves' veins. Similarly, leaves of sulfur-deficient plants turn light green.

In *nitrogen* deficiency, the lower leaves dry up.

Plants not receiving enough *copper* show wilting of the upper parts, inrolling of the leaf margins, and a deep color of the leaves.

Zinc deficiency is not uncommon. Its symptoms are small, crinkling leaves, chlorosis, and defoliation.

References: C.9; S.27.

MINERAL FEEDS are used to supplement RATIONS that are deficient in essential MINERALS. Some operators use them in all feeds or allow the animals free access to them. This practice may allow some animals to eat too much of some minerals.

The mineral composition of feeds will depend on many things—the acidity and fertility of soil, the temperature and rainfall, the species and age of plant, the methods of harvesting and storing, and many other factors. Plants, like animals, require minerals to grow. A plant will not be devoid of the minerals it requires for growth (of the minerals required by animals, plants do not seem to need sodium, chlorine, iodine and cobalt), but on soils of low fertility a shortage of some minerals may develop. The exact nature and degree of variation from year to year is not well understood, but some general ideas have developed.

Some plants are naturally rich in certain minerals. Legumes normally contain 1 percent CALCIUM, which is several times the amount a ruminant needs in its ration. Oil seeds and the hulls of cereal grains supply good amounts of PHOSPHORUS, but cereal phosphorus is not well utilized by poultry. Since hogs normally eat large amounts of grain and oil meals, they are not likely to suffer from a phosphorus deficiency.

Young plants usually contain more minerals than mature plants. The leaves contain more minerals that are valuable to the animal than do stems.

All animals need SODIUM CHLORIDE (salt); 0.5 percent is usually included in mixed rations and salt may be kept before hogs, sheep, and cattle at all times. Farmers and ranchers may mix salt with supplement to limit feed intake. The ratio of salt to concentrate is varied until the animals eat the desired amounts. 1 lb. salt to 3 to 5 lb. supplement will usually keep cattle from eating more than 1 lb. per day. When this method is used, the cattleman should be prepared to accept slightly lower gains than when the supplement is hand-fed, and plenty of water should be in front of the animals at all times.

Calcium and phosphorus supplements may be needed during some time of the year on most farms. Ground LIMESTONE contains some 36 to 38 percent calcium and is usually the cheapest source of calcium. Hogs on all vegetable rations can be self-fed a mixture of ground limestone and salt. 1 to 2 percent ground limestone is usually added to poultry rations and ground oyster shell is self-fed. Cattle and sheep on good fertilized pastures, or receiving legume hay, will not require additional calcium. High producing dairy cows should have access to a mineral mixture of limestone, BONE MEAL, and salt.

Phosphorus is the element most often deficient when animals are living on a roughage ration. Pastures that have been heavily fertilized with a complete fertilizer will contain adequate phosphorus, but, on unsupplemented pastures and during winter feeding, additional phosphorus may be needed. (If animals are not getting adequate phosphorus in their feeds, they will often chew bones or sticks.) Steamed bone meal, defluorinated DICALCIUM PHOSPHATE and defluorinated ROCK PHOSPHATE are good sources of phosphorus and, when mixed with salt (1:1), will be readily eaten if kept before the animals at all times.

It is important that the total calcium and phosphorus in the ration be present in a ratio of between 1:1 and 3:1. (1.3:1 is satisfactory for swine and 1.6:1 seems desirable for ruminants.)

Low grade roughages, such as straws and cobs, are often deficient in COBALT and may be deficient in IODINE. The grains are also deficient in cobalt. Legumes grown in some parts of the Great Lake areas and of the Northwest are low in cobalt. Usually, however, 5 lb. of good legume hay will supply all the TRACE ELEMENTS that cattle need; ½ lb. is sufficient for sheep, and 10 percent alfalfa leaf meal in the ration of swine and poultry will supply them with adequate trace minerals.

If legume hays are not available, an all-purpose mineral mix, to be self-fed in areas where trace element deficiencies are suspected, can be prepared according to the formula in the next column.

When trace element deficiencies are not likely, supply limestone, bone meal or dicalcium phosphate, and salt (1:1:1).

Minerals should not be "force-fed"— that is, included in the grain mixtures at high levels. The feed may already contain enough mineral and the addition may increase the intake to toxic levels. For example, if 1 percent calcium is added to

ALL PURPOSE MINERAL MIX

Ground limestone, lb.	300
Steamed bone meal or dicalcium phosphate, lb.	300
Salt, lb.	350
Iron sulfate, lb.	15
Copper sulfate, lb.*	4
Manganese sulfate, lb.†	4
Cobalt sulfate, lb.	½
Potassium iodide (stabilized), lb.	¼

* In areas where molybdenum (in excess of 8-10 ppm.) is known or suspected to be present, increase to 30 lb.

† If used in mixing poultry rations (3 lb. mineral mix per 100 lb. feed), increase to 8 lb.

swine rations, a condition known as *parakeratosis* develops. This can be cured by taking out the calcium, or by adding ZINC to the ration. In some areas the forages contain more COPPER than animals need. Extra copper could cause *jaundice* and other upsets.

The table on page 296 lists the minerals known to be needed by farm animals, indicates the requirements for the most important minerals, and shows some simple ways to supply the minerals. An oversupply of trace minerals may be more harmful than too little mineral. Too much IRON or copper reduces the activity of rumen bacteria. Some minerals are toxic. Excess MOLYBDENUM may be present if the land has been subjected to extensive leveling operations and limed to near neutrality or on soils that were derived from molybdenum-containing minerals.

Raw rock phosphate and colloidal phosphate contain enough FLUORINE to kill the animals and should not be fed to animals.

The following officially accepted regulations and resolutions regarding mineral feeds are now in force:

1. Substances which are intended, as feed for animals, other than man, and are to supply primarily *mineral elements* or inorganic nutrients, shall be classified as mineral feeds. A guaranty for protein, fat, and fiber is not required.

2. Feeds, mixed or unmixed—including mixtures containing both organic ingredients and 5 percent or more of mineral ingredients—shall be labeled to show (in addition to the other information required

by law) the minimum percent *calcium* (Ca), *phosphorus* (P), *iodine* (I), and the maximum percent *salt* (NaCl). The ingredients shall be stated in the form in which they are used. When feed mixtures contain relatively large amounts of salt, such as trace mineralized salt, the guaranty on the label shall show the minimum percentage of salt (NaCl).

3. The use of the word *"mineralized"* in the name of a feed is condemned, except for use in the brand name of salt such as *"trace mineralized salt," "trace mineral salt,"* or *"salt with trace minerals"* if the product contains significant amounts of trace minerals which are recognized as essential for the nutrition of farm animals. The ingredients shall be stated in the form in which they are used.

4. In the case of feeds containing inert *grit*, other added inert mineral matter, or *charcoal*, the words "with grit," "with charcoal," etc., must appear in the brand name; and in the statement of ingredients,

MINERAL REQUIREMENTS AND SOURCES

Mineral	Requirement					When is deficiency likely?	How to supply mineral
	Cattle	Sheep	Hogs	Growing chicks	Hens		
Calcium, % ration	0.30	0.25	0.60	0.7	1.85	Cattle and sheep on low-grade grass hay. Hogs in dry lot eating vegetable rations. Poultry unless supplemented.	Oyster shell, limed pastures, legume hay, ground limestone, bone meal.
Phosphorus, % ration	0.20	0.16	0.40	0.4	0.6	Cattle on low-grade roughage or on unfertilized pasture. Poultry on plant rations. Hogs on pasture with no protein supplement.	Fertilized pasture, steamed bone meal, dicalcium phosphate. Meat and bone scraps, oil meals.
Salt, % ration	0.50	0.50	0.25	0.4	0.5	In all animals, if not allowed salt.	Feeding grade salt.
Potassium, % ration			0.25	0.2		Never.	
Magnesium, % ration	0.03	0.03	0.06	0.04	0.06	On lush spring pastures following high N applications.	Magnesium carbonate.
Sulfur						Low-grade roughage supplemented with urea.	Add sodium sulfate to urea ration.
Iodine*						Pregnant animals drinking rain water or ground water in some areas.	1 oz. stabilized potassium iodide in 300 lb. salt.
Manganese*						Poultry rations. Never for other farm animals.	Add ¼ lb. manganese sulfate/ton poultry ration.
Iron, mg./day*						Baby pigs on board or concrete floors.	Sod in pen. Paint sow's udder with iron sulfate.
Copper, mg./day*	50	5–10				Not likely.	
Cobalt, mg./day*	0.70	0.10				When corn cobs or low-grade roughage are fed with corn.	1 oz. cobalt sulfate to 100 lb. salt.
Zinc, ppm.	?	?	50	?	?	When large amounts of Ca are added to swine rations.	Add 1 oz. zinc carbonate to 500 lb. feed or 100 lb. protein supplement, or reduce Ca in ration and self-feed.†

* 5 lb. good quality legume hay will often meet the needs of cattle for these trace elements.
† Feeds supply generally adequate amounts of zinc unless excess Ca is present in the ration.

their kind and amount must be given; provided that no more than 5 percent grit, charcoal, or other inert mineral ingredients, separately or together, is present in any feed other than the so-called mineral feed mixtures.

Inert materials are materials that do not contribute dietary factors.

5. When the word "iodized" is used in connection with a feed ingredient, the ingredient shall contain not less than 0.007 percent *iodine* uniformly distributed.

6. The use of the term *bone phosphate of lime* (BPL) is to be discouraged in connection with the labeling of feed ingredients.

In addition the following (at present) tentative regulations have been suggested by the Association of American Feed Control Officials:

1. The *fluorine* (F) content of any mineral or mineral mixture which is to be used directly for the feeding of domestic animals shall not exceed 0.30 percent for cattle, 0.35 percent for sheep, 0.45 percent for swine, and 0.60 percent for poultry.

Rock phosphate or other fluorine-bearing ingredients may be used only in such limited amounts in feeding stuffs as not to raise the fluorine content of the total concentration of the (grain) ration above the following amounts: for cattle, 0.009 percent; for sheep, 0.010 percent; for swine, 0.014 percent; and for poultry, 0.035 percent.

2. Formula feeds containing *drug* ingredients intended or represented for the cure, mitigation, treatment, or prevention of disease of animals other than man, and substances other than feeds, intended to affect the structure or any function of the body of animals other than man, shall be labeled to show, in addition to other information required by feed law: (a) the name of each ingredient including the name of each therapeutically active ingredient or agent stated as such, (b) adequate directions for use, and (c) adequate warnings against use under conditions which may be dangerous to health—provided that the term "drug" as used herein does not apply to vitamin, mineral, or other substances used solely for nutritional purposes, and (d) the use of medicaments

in proprietary mixed feeds for combating diseases of non-nutritional origin is not to imply an (official) endorsement by the above mentioned association.

3. *Phosphatic materials* for feeding purposes shall be labeled with a guaranty for the minimum percentages of calcium and phosphorus, and the maximum percentage of fluorine.

CLASSIFICATION OF MINERAL FEEDS

1. Packing house by-products:

Bone ash is the ash obtained by burning bones with free access to air, and containing a minimum of 15.3 percent phosphorus. The label shall show a guaranty for calcium and phosphorus.

Bone charcoal, or *bone black*, is obtained by charring bones in closed retorts. It shall contain not less than 14 percent phosphorus. The label shall show a guaranty for calcium and phosphorus.

Spent bone black, also called *spent bone*, is the product that results from repeated charring of bone charcoal, or bone black, after use in clarifying sugar solutions. It shall contain not less than 11.5 percent phosphorus. The label shall show a guaranty for calcium and phosphorus. Spent bone black contains about 30 percent carbon which has no nutritive value.

Steamed bone meal is the dried, ground product suitable for animal feeding, and obtained by cooking bones with steam under pressure.

> NOTE: A more specific definition, proposed in place of the above, is not official as yet; it reads: Steamed bone meal is the dried, ground product, obtained by cooking bones with steam under a minimum of 20 lb. pressure for at least one hour at a temperature of not less than 250° F. (121° C.). It shall contain not less than 12 percent phosphorus. The label shall include guaranties for calcium and phosphorus.

Special steamed bone meal—the dried, ground product suitable for animal feeding, obtained by cooking dried bones (after removal of grease and meat fiber) with steam under pressure, in the process of obtaining gelatin or glue.

NOTE: Not officially recognized as yet are the following definitions:

Cooked bone meal—the dried, ground product, obtained by cooking undecomposed bone in water at atmospheric pressure just enough to remove excess fat and meat. It shall contain not more than 25 percent protein and not less than 10 percent phosphorus. The label shall include a guaranty for protein, calcium and phosphorus.

Dicalcium phosphate from bone, or *precipitated bone phosphate*, is the residue of bones that have first been treated in a caustic solution, then in a hydrochloric acid solution, and then have been precipitated with lime and dried. It shall contain not less than 17 percent phosphorus, and the label shall include a guaranty for calcium and phosphorus.

2. Natural and prepared minerals:

Calcite, ground limestone, chalk rock, precipitated chalk, oyster shell flour, shell flour, and *precipitated calcium carbonate* are acceptable sources for calcium carbonate ($CaCO_3$). These minerals shall in each instance be true to name and shall contain not less than 33 percent calcium. They shall be designated as ingredients only by the names listed above.

Calcium carbonate is a product containing not less than 38 percent calcium.

Magnesium limestone, or *dolomite limestone*, is acceptable as a source of magnesium and calcium carbonates. The terms are synonymous and designate a native mineral, composed of mixtures of magnesium carbonate ($MgCO_3$) with calcium carbonate. Magnesium limestone shall contain not less than 10 percent magnesium (Mg) and shall be declared as an ingredient only by one of the names listed above.

Iodized salt is common salt containing not less than 0.007 percent iodine, uniformly distributed.

Low fluorine rock phosphate is ground phosphate rock containing not more than 0.5 percent fluorine. The minimum percentages of calcium and phosphorus, and the maximum percentage of fluorine, shall be stated on the label.

Defluorinated phosphate includes either calcined, fused, or precipitated calcium phosphate. It shall contain not more than 1 part fluorine to 100 parts phosphorus. The minimum percentage of calcium, and the maximum percentage of fluorine, shall be stated on the label. The term "defluorinated" shall not be used as a part of the name of any product containing more than 1 part fluorine to 100 parts phosphorus.

NOTE: *Partially defluorinated phosphate*—not as yet an officially accepted feedstuff—includes either calcined, fused, or precipitated calcium phosphate that contains not more than 0.30 percent fluorine. The minimum percentages of calcium and phosphorus, and the maximum percentage of fluorine, shall be stated on the label.

3. Additional officially recognized mineral ingredients (or inorganic chemicals):

The following inorganic (mineral) *chemicals* are recognized as suitable ingredients in animal feeds under controlled conditions, and when so used they shall be true to name and of purity equal to the commonly recognized "commercial grade" for the article; the chemicals shall be declared as ingredients only by the names herein used:

COBALT ACETATE
COBALT CARBONATE
COBALT CHLORIDE
COBALT SULFATE
COPPER CARBONATE
COPPER HYDROXIDE
COPPER SULFATE
IRON CARBONATE
IRON PHOSPHATE
IRON SULFATE
RED IRON OXIDE
MAGNESIUM SULFATE
MANGANESE CARBONATE
MANGANESE OXIDE
MANGANESE PHOSPHATE
MANGANESE SULFATE
DICALCIUM PHOSPHATE
MONOCALCIUM PHOSPHATE
ROCK PHOSPHATE
SOFT PHOSPHATE WITH COLLOIDAL CLAY
TRICALCIUM PHOSPHATE

SODIUM BICARBONATE
SODIUM CHLORIDE (salt)
SODIUM SULFATE (anhydrous)
GLAUBER'S SALT
SULFUR
ZINC CARBONATE
ZINC SULFATE

References: F.6; M.38.

See also BONE MEAL; RAW BONE MEAL; ANIMAL FEEDSTUFFS; WOOD ASHES; MINERAL DEFICIENCIES; LABELS; F.D.A.

"MINERALIZED." *See* MINERAL FEEDS.

MINERAL MATTER = ASH.

MINERAL MIXTURES = *mineral supplements.* *See* MINERAL FEEDS; COMMERCIAL MIXED FEEDS; GRAZING; HYDROCARBONS; TRACE ELEMENTS.

MINERAL OIL is a term often used for crude PETROLEUM and various petroleum products. Its viscosity (internal friction) can be determined with the help of the *Saybolt* viscosimeter which measures the rate of flow through a given orifice.

Mineral oil having Saybolt viscosity of 100 seconds is recommended for improving the stability of DINITRO-DUST-BARRIERS.—*See also* HYDROCARBONS; WHITE MINERAL OIL.

MINERALS, often referred to as *inorganic materials* or *ash* content of feedstuffs, are important nutrients. The animal body needs minerals since it contains approximately 3 to 5 percent of them. PHOSPHORUS is found in every cell of the body; CALCIUM, IRON, chlorine, phosphorus, and other minerals, occur in blood; and calcium and phosphorus make up a large part of the bones. Therefore, minerals must be included in all rations—lack of minerals causes a variety of livestock diseases, such as rickets, goiter, milk fever, lameness, swollen joints, etc.

Minerals contained in feed are absorbed unchanged into the blood stream from the small intestine.—*See also* MINERAL FEEDS; ABSORPTION; TRACE ELEMENTS; WOOD ASHES; NUTRIENTS.

MINGO. *See* SOYBEAN (variety).

MINNESOTA AMBER SORGO, a *Black Amber*, is one of the Amber-type sorgos which belong to the group of forage SORGHUMS.

MINNESOTA No. 1. *See* RYE (variety).

MINNESOTA No. 2 = *Swedish rye.*
                                   See RYE.

MINNESOTA RELIABLE
    = *Turkey wheat. See* COMMON WHEAT.

MINSOY. *See* SOYBEAN (variety).

MIRACLE WHEAT = *Fulcaster.*
                          See COMMON WHEAT.

MISCELLANEOUS FEEDS.
            See COMMERCIAL MIXED FEEDS.

MISCELLANEOUS PRODUCTS is an officially recognized group of feedstuffs comprising the following: DRIED APPLE POMACE; DRIED APPLE PECTIN PULP; DRIED BEET PULP; DRIED TOMATO POMACE; BABASSU OIL MEAL; COCONUT OIL MEAL; PALM KERNEL OIL MEAL; SESAME OIL MEAL; WHOLE PRESSED SAFFLOWER SEED; IVORY NUT MEAL; STRAW MEAL; GRASS MEAL; RAMIE LEAF MEAL; DEHYDRATED GARBAGE; YEAST DRIED GRAINS; also feedstuffs obtained from BUCKWHEAT and VELVETBEANS, and such FEED INGREDIENTS as ANISE SEED; CAPSICUM; CARAWAY SEED; DEXTROSE; FENNEL; FENUGREEK SEED; GENTIAN; GINGER; GLUCOSE; KELP; KAMALA; LACTOSE; and LOCUST BEAN.

Tentatively recognized feedstuffs of this miscellaneous group are GROUND ALMOND HULLS, CONDENSED BEET SOLUBLE PRODUCTS, AMMONIATED PLANT PRODUCTS, as well as a variety of OIL CAKES and OIL MEALS.

Reference: F.6.

MISSION OAT.
                See COMMON OAT (variety).

MISSOURI 0-205. *See* RED OAT (variety).

MISSOURI 8 is a CORN hybrid.

MISSOURI BLUESTEM = *Mediterranean wheat. See* COMMON WHEAT.

MISSOURI EARLY BEARDLESS.
            See SIX-ROWED BARLEY (variety).

MISSOY. *See* SOYBEAN (variety).

MITES belong to the SPIDER (and not to the INSECT) group. They are useful because they destroy such insect pests as the POTATO LEAFHOPPER and the WHEAT STRAWWORM.—*See also* CLOVER MITES; RED SPIDER.

"MIXED BRAN." *See* LABELS.

MIXED CONCENTRATES.
                        See CONCENTRATES.

MIXED FEEDS.

See COMMERCIAL MIXED FEEDS.

MIXED FERTILIZERS.

See FERTILIZERS.

MIXED GRAINS. See GRAIN GRADING.

MIXED HAY. See HAY GRADING.

MIXED SCREENINGS. See SCREENINGS.

MIXERS. See FEED MIXING.

MIXING FEED. See FEED MIXING.

Mn = MANGANESE.

MODOC is one of the intermediate-type grain SORGHUMS.

MOHAWK OAT.

See COMMON OAT (variety).

MOISTURE DETERMINATION.

See SILAGE.

MOLASSES is a feedstuff which has become increasingly important in recent years.

Feed molasses is a by-product of the cane sugar, beet sugar, corn products, and citrus industries. It is rich in energy value, highly palatable, and possesses other qualities that make it an excellent feed for animals.

The four types of feed molasses are very rich in carbohydrates and also contain such essential minerals as IRON and CALCIUM; however, they are poor in true protein. Molasses also is a good source of NIACIN and PANTOTHENIC ACID which is needed in swine and poultry rations.

A gallon of molasses weighs 11.7 lb., with the exception of citrus molasses, which weighs 11.3 lb. per gallon.

Molasses makes roughage more palatable and thus induces cattle to eat feed they would normally refuse. It is best handled by diluting it with 1 or 2 parts water and pouring this mixture on roughages, spreading it evenly over the roughage.

It has been determined that 6½ gal. molasses has approximately the same feeding value as 1 bu. yellow CORN. Therefore, molasses is worth about ¾ as much as corn, and if the molasses can be bought for less than ¾ of the price of corn, it pays to feed molasses. There are additional reasons for feeding molasses; e.g., it serves as a binding agent in mixed feeds and also reduces the dust in them. Furthermore, molasses may well replace other carbohydrates in mixed feeds. Such replacement by cane molasses may be as high as the following amounts:

	Molasses per total ration
Beef cattle	20%
Dairy cattle	20%
Work stock	15%
Sheep	10%
Swine	20%
Poultry	6%

The high molasses rations must be supplemented with extra PROTEIN. Molasses is valuable in stimulating RUMEN BACTERIA. About 5 to 10 percent will usually give maximum stimulation. Molasses also is a good silage preservative, and at least 75 percent of its feeding value is recovered in the silage.

Most molasses is sold in *liquid* form, but there are other forms on the market. These are mostly *dried molasses* products and are derived chiefly from cane and corn molasses. Various carrying agents, such as corn oil cake meal and BAGASSE pith, are used. These products are more expensive than liquid molasses but are easier to handle and mix.

Wet BEET PULP, when combined with molasses and dried, forms *dried molasses-beet pulp* which is palatable, bulky, slightly laxative, and keeps well in storage. Up to 30 percent beet molasses may be satisfactorily dried with beet pulp. Dried molasses-beet pulp is rich in carbohydrates and the protein content is about 9 percent; it is used chiefly for feeding dairy cattle.

DRIED CITRUS PULP may be mixed with as much as 30 percent citrus molasses. This type of feed has about the same value as dried molasses beet pulp and is consumed mostly by dairy cattle.

Molasses is also *pelleted* with other feeds. The exact percentage of molasses to be used depends on the type of processing equipment and the type of carrier used. PELLETS may contain from 15 to 35 percent molasses; at the higher level, however, the pellets sometimes disintegrate rather quickly when wet.

In addition to dried and liquid molasses there is now available *ammoniated molasses.* This product has an increased nitrogen content and, therefore, may contain a higher protein equivalent for animals.

NOTE: Feeding ammoniated molasses to cattle may be dangerous. Animals may develop a condition called "going crazy" in which they first get a wild look and then start to run. They will run into any kind of fence or object, but will appear normal after the "run", if they have not injured themselves.

Molasses containing *urea* (and, therefore, an increased nitrogen content) is also marketed.

NOTE: Experiments conducted to get information on the value of UREA as a partial substitute for oil meals for growing and fattening lambs, mature ewes, and beef steers, indicate that the rate of gain does not differ significantly from that obtained when the entire protein content of the ration was supplied in natural form.

There are four officially recognized types of feed molasses:

Feeding cane molasses—commercially called *blackstrap (molasses)*—is a by-product of the manufacture of cane sugar and shall contain 48 percent or more of total sugars expressed as invert sugar. Its solution in an equal weight of water shall test not less than 39.75 degrees Brix.

Cane molasses accounts for the largest volume of molasses used in livestock feeding. It has a sweet taste and odor, and is readily eaten by all kinds of livestock. It has a laxative action.

When cane molasses is fed free choice, beef cattle normally consume about 4 to 6 lb. (⅓ to ½ gal.) daily per head.

Feeding corn sugar molasses—widely known as *hydrol* or *corn molasses*—is a brown, sirupy by-product from the manufacture of corn sugar (dextrose), and shall contain 48 percent or more of reducing sugars expressed as dextrose and 60 percent or more of total carbohydrates (dextrins, and reducing and nonreducing sugars). Its solution in an equal weight of water shall test not less than 39.0 degrees Brix.

Corn molasses is higher in sugar content but it does not have the sweet odor of cane molasses.

Feeding beet molasses is a by-product of the manufacture of beet sugar and shall contain 48 percent or more of total sugars expressed as invert sugar. Its solution in an equal weight of water shall test not less than 39.75 degrees Brix.

Beet molasses is similar to cane molasses; however, it is more laxative than cane molasses because of its higher mineral content.

Suggested amounts of beet molasses for animals accustomed to it are as follows:

	Daily ration per 1,000 lb. live weight
Feeder cattle	4 – 6 lb.
Dairy cattle	2½– 3 lb.
Sheep	3 – 5 lb.
Horses, heavy	4 lb.
Horses, light	2½ lb.
Feeder pigs	5 –10 lb.

It is recommended that animals be accustomed gradually to beet molasses. Breeding animals should be given smaller allowances than those being fattened. The amount of beet molasses should be materially reduced six weeks before the young are born.

Citrus molasses is the partially dehydrated juice of citrus fruit. It shall contain not less than 45 percent total sugars expressed as invert sugar. Its solution in an equal weight of water shall test not less than 35.5 degrees Brix.

Citrus molasses differs from cane molasses in odor and taste. However, animals quickly become accustomed to it. It is used at low levels in concentrate mixtures for cattle, but when fed free choice to cattle will normally be consumed at a rate of about 5 to 6 lb. (⅖ to ½ gal.) daily per head. When fed in recommended amounts, molasses of any type does not produce any harmful laxative effects. Molasses frequently is beneficial when other feeds are constipating.

References: W.24; E.12; F.6; B.17; R.10.

See also WOOD MOLASSES; CORN-SUGAR MOLASSES; PLANT BY-PRODUCTS; BY-PRODUCT FEEDSTUFFS; DISTILLERS' PRODUCTS; YEAST; VITAMINS; SILAGE; SILAGE CROPS; ALFALFA-MOLASSES FEEDS; POISONED BRAN MASH.

MOLASSES DISTILLERS' CONDENSED SOLUBLES.

See DISTILLERS' PRODUCTS.

MOLASSES DISTILLERS' DRIED SOLUBLES. *See* DISTILLERS' PRODUCTS.

MOLASSES DISTILLERS' DRIED YEAST. *See* YEAST.

MOLASSES GRASS (*Melinis minutiflora*), also called *honeygrass* and *Gordura grass*, is a perennial reaching 4 ft. high. The plant is covered with hairs which secrete a sticky substance that has an odor like molasses.

This GRASS grows on sandy land in southern Florida and in Central and South America. Cattle have to get used to its characteristic odor before they graze this abundantly and rapidly growing pasture crop or its hay.

MOLASSES MIXING.

See FEED MIXING.

MOLD is (1) any FUNGUS that produces a superficial growth on various types of organic matter; (2) the growth itself. Molds occur most often on damp and decaying matter.—*See also* GRAY MOLD; SILAGE; STREPTOMYCES.

MOLD INHIBITORS are PRESERVATIVES used to prevent MOLDS from spoiling foods, beverages, and feedstuffs.

Officially accepted mold inhibitors for feeding stuffs are the following: SODIUM BENZOATE, SODIUM PROPIONATE; CALCIUM PROPIONATE; METHYL PARA-HYDROXYL BENZOATE; PROPYL PARA-HYDROXYL BENZOATE; and SORBIC ACID.

MOLDY FEED does not have to be injurious to livestock—many MOLDS are non-toxic, but simultaneous with mold formation other processes often take place in the feed (and particularly in SILAGE), which cause it to become *spoiled* and poisonous.

Most moldy HAYS and SILAGES can be safely fed to cattle. Exceptions are moldy sweetclover and LESPEDEZA hays or silages which contain DICOUMAROL. Horses and sheep should not be fed moldy hays.—*See also* FORAGE POISONING.

MOLDY HAY. *See* HAY.

MOLDY SILAGE. *See* SILAGE.

MOLE. *See* MOLES.

MOLE CRICKETS (*Gryllotalpa* spp.) are 1½ in. long brown INSECTS which, similar to moles, make small burrows in the soil. They may cause much damage to BERMUDA-GRASS.

MOLECULAR means: characteristic of, or pertaining to, MOLECULES.

MOLECULE is a unit of matter: the smallest portion of a chemical ELEMENT or compound that retains chemical identity with the substance in mass.

MOLES often do considerable damage to crops by raising the soil when making their burrows. Young plants usually die along these burrows.

Control: Poisons have proved unsuccessful in controlling moles. The use of traps, especially constructed to catch moles, is more reliable than any other method of control.

Reference: G.7.

See also RODENTS.

MOLYBDENUM, a heavy metal, occurs in some rare minerals. The more alkaline the soil, the greater is the amount of molybdenum taken up by plants. Some plants take up enough of this CHEMICAL ELEMENT to be poisonous to animals.

Molybdenum poisoning in cattle is characterized by the following symptoms: severe diarrhea, loss of weight, anemia, and change in the color of the coat—*Herefords* change to rusty orange, *Guernseys* and *Red Devons* to muddy yellow, *Hollstein-Friesians* to gray, and *black cattle* become rusty colored. The death rate of affected animals is 80 percent. Only cattle seem to be susceptible to molybdenum POISONING.

Addition of COPPER to the ration will overcome the toxic effects of molybdenum. —*See also* TRACE ELEMENTS.

MONANTHA VETCH (*Vicia articulata*), sometimes called *one-flower vetch*, is one of the not very winter-hardy VETCHES. The plant is weak-stemmed and viny, and in this respect resembles HAIRY VETCH. It has very fine stems and leaflets and matures early.

Monantha vetch is adapted to the extreme southern part of the Cotton Belt and to the milder parts of Washington, Oregon, and California. In orchards of Florida and southern Georgia it is used for winter green manure.

Wherever it can be grown, monantha vetch will make a good green-manure and forage crop. It is resistant to VETCH ANTHRACNOSE.

Reference: M.18.

See also LEGUMES; INOCULATION.

MONETTA. *See* SOYBEAN (variety).

MONGOL SOYBEAN
= *Midwest soybean. See* SOYBEAN.

MONGREL is any mixed breed.
—*See also* HYBRID.

MONO- is a prefix signifying *one*.

MONOCALCIUM PHOSPHATE, also called *monobasic calcium phosphate* or *calcium biphosphate*, forms colorless, deliquescent granules, or powder, of strong acid taste. It is partially soluble in water, forming free PHOSPHORIC ACID and various CALCIUM PHOSPHATES.—*See also* MINERAL FEEDS; SUPERPHOSPHATE.

MONOECIOUS (pronounced: mo-*nee*-shus) is any plant with flowers differentiated as to sex: their STAMINATE (male) and PISTILLATE (female) flowers are in separate INFLORESCENCES, but borne on one and the same plant, as distinct from DIOECIOUS plants.

Pertaining to FUNGUS, particularly a rust fungus, monoecious means: having all stages of its life cycle on a single species of plant.—*See also* HETEROECIOUS; RUSTS.

MONOSACCHARIDES.
See REDUCING SUGARS.

MONROE SOYBEAN.
See SOYBEAN (variety).

MONTANA COMMON ALFALFA is one of the COMMON ALFALFAS.

MONTREAL MANCHU.
See SOYBEAN (variety).

MORMON CRICKET (*Anabrus simplex*), a wingless form of GRASSHOPPER, is a bad pest in the western states. Its range extends from the Missouri River west to the Cascade and Sierra Nevada Mountains and from the Canadian line south to northern California, Nevada, Utah, and Colorado. Most of the crops grown there are susceptible, but the greatest damage has been done to range GRASSES as well as to dry-land WHEAT, OATS and ALFALFA.

The eggs are laid during late summer and autumn in well-drained, light sandy-loam soil. They are inserted just under the soil surface in bare spots between clumps of grass or sagebrush. Unlike grasshopper eggs which are in a pod, cricket eggs are laid singly. Each female lays about 150 eggs. The young crickets start hatching early in April and reach maturity about six to eight weeks later. There is one generation a year.

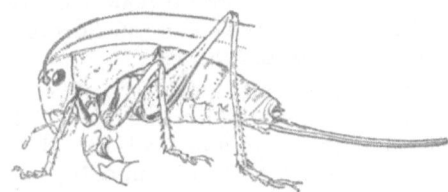

Mormon cricket. The female is wingless. (U.13.)

Mormon crickets persist in small numbers year after year in rough foothills and mountainous terrain. When weather conditions, over a period of years, are favorable to maximum reproduction, they increase rapidly and start migrating.

Control: Mormon crickets can be easily killed in all stages by application of wet bait, i.e., POISONED BRAN MASH (containing SODIUM FLUOSILICATE), or with DRY BAIT (containing TOXAPHENE or CHLORDANE).

Reference: U.3.

MORMON NEEDLEGRASS (*Stipa arida*) grows on rocky slopes up to 2½ ft. tall. This NEEDLEGRASS often makes good forage on ranges.—*See also* RANGE PLANTS.

MORPHOLOGICAL means: relating to MORPHOLOGY.

MORPHOLOGY is the science of form and structure. *Plant morphology* (or *structural botany*) deals with the forms of plants and their organs, their anatomy (or *gross morphology*), relationships, and development. TAXONOMY (or *systematic botany*) is based on morphology. *Cytology* (or *cell science*) and *histology* (*microscopic anatomy*) deal with *microscopic morphology* of plants (and animals).

Reference: D.9.

MORRISON FEEDING STANDARDS are based on pounds of total digestible NUTRIENTS and digestible protein. The more recent standards include values for Ca, P, and carotene, and are a modification of the old *Wolff-Lehmann Standards.* Because of variations between animals and climatic conditions, the standards are given as ranges with minimum and maximum values.

MORSE MUNGBEAN is a variety of the *green* MUNGBEAN.

MORSE SOYBEAN.
See SOYBEAN (variety).

MORTGAGE LIFTER = *Mediterranean wheat. See* COMMON WHEAT.

MOSAICS are VIRUS DISEASES of plants, characterized by the intermixing of light- and dark-green areas in the leaves. These mottled areas are irregular in outline and may follow the small veinlets. Mosaics attack WHEAT as well as SWEETCLOVER, SOYBEAN, FIELD PEA, and other legumes.

No control is known.—*See also* WHEAT MOSAIC; SOYBEAN MOSAIC; YELLOW MOSAIC.

MOTHS, or *"millers,"* are nightflying INSECTS with a large abdomen and often two "eye-spots" on the wings. Some moths are useful; others are destructive in the larval stage.—*See also* PEA MOTH; ARMYWORM; CORN EARWORM; GARDEN WEBWORM; EUROPEAN CORN BORER.

MOUNT. *See* MT.

MOUNTAIN BROME (*Bromus marginatus*) is a short-lived grass that is native to the Rocky Mountain and Pacific Coast regions. It is closely related to CALIFORNIA BROME and similar to it in soil and climatic adaption.

Plants usually grow to a height of 3 to 4 ft., with leaves 6 to 12 in. long and about ¼ in. wide. The flat leaf blades are usually hairy on their lower parts. Leaf sheaths are closed and form tubes.

Growth starts in early spring and continues with the production of large amounts of leafy forage that is liked by livestock. Rapid growth, vigor of seedlings, and a well-branched and deeply penetrating root system make mountain brome one of the most valuable grasses in cases where vegetative cover is required immediately in order to protect erodible, sloping land.

The forage that remains after removal of the seed crop is of fair quality and may be grazed, or cut and cured for roughage.

One of the important uses of this grass

Mountain brome. Panicle. (M.47.)

in the agriculture of the Pacific Northwest is that of seeding it in mixture with ALFALFA and SWEETCLOVER. In such a mixture, the grass adds materially to both, root and top growth. The combination offers greater soil protection than LEGUMES alone because of the greater density of plants and greater volume of roots. It is said also that the mixture of grass with the legumes, which have a high protein content, lessens the danger of BLOAT and provides a better-balanced forage.

Dangers: Among the diseases which attack this brome species are the GRASS SMUTS.

Variety: *Bromar*, an improved variety of mountain brome, is produced as certified seed.

Reference: H.1.

See also BROME GRASSES; GRASSES.

MOUNTAIN BUCKWHEAT
= TARTARY BUCKWHEAT.

MOUNTAIN-LAUREL. *See* KALMIAS.

MOUNTAIN-MAHOGANY.
See LITTLELEAF MAHOGANY.

MOUNTAIN RYE is a winter RYE variety.

MOUNT CARMEL SOYBEAN.
See SOYBEAN (variety).

MOUSE. *See* FIELD MICE.

MOUTH. *See* DIGESTIVE TRACT.

MOW means different things: (1) to cut with a scythe or mower; (2) to place hay in a mow; (3) a place to store hay inside. —*See also* HAY DRYING.

MOW-BURNT HAY. *See* HAY.

MOW DRYING. *See* HAY DRYING.

MT. BARKER is a SUBCLOVER strain.

MUCOUS MEMBRANE, or *mucosa,* is the fine lining of the hollow organs, such as the gastrointestinal tract. It plays an important role in the ABSORPTION of NUTRIENTS.—*See also* MUCUS.

MUCUNA. VELVETBEANS are *Mucuna* spp. However, their scientific nomenclature is sometimes confusing because authorities often disagree about it; in fact, many of them continue to use the (old) generic name *Stizolobium* instead of *Mucuna.*

M. deeringiana = DEERING VELVETBEAN (which includes *Florida, Arlington, Alabama, Bush,* and the hybrid *Osceola*); *M. cochinchinensis* = LYON VELVETBEAN; *M. hassjoo* = YOKOHAMA VELVETBEAN; *M. capitata* = TRACY BLACK VELVETBEAN; and *M. nivea* = CHINESE VELVETBEAN.— *See also* STIZOLOBIUM.

MUCUS is the sticky gland secretion which keeps the MUCOUS MEMBRANE moist. —*See also* BODY WASTES.

MUHLENBERGIA. *Muhlenbergia* spp. = MUHLY GRASSES.

MUHLY GRASSES (*Muhlenbergia* spp.), or *muhlies,* occur in the Southwest and Mexico. They are usually scattered over a wide variety of soils, from low deserts to high timbered areas. The palatability of the various species of this genus varies greatly. Some muhlies make very good forage, others are only fair. The perennial types, especially, are valuable for spring and summer use on ranges.

The so-called "seeds" of these GRASSES are closely enfolded by the firm outer flower SCALES; and the LEMMAS have AWNS or are sharp-pointed.

Reference: U.6.

See also RANGE PLANTS.

MUKDEN SOYBEAN.
See SOYBEAN (variety).

MULCH is the material which covers the ground around plants to protect them from heat and to conserve moisture. Leaves, hay, straw, or dust are often used as mulch.

MULE RATIONS.
See HORSE AND MULE RATIONS.

MULTI- is a prefix signifying many or numerous.

MULTIPLIERS PEA.
See FIELD PEA (variety).

MUNGBEAN (*Phaseolus aureus*), also known as *Chickasaw pea, Jerusalem pea, Oregon pea, Neuman pea,* or *chop suey bean,* is a summer annual LEGUME; it belongs to the FIELD BEANS. The mungbean somewhat resembles the COWPEA.

The plants, which grow 1 to 3 ft. tall, are used for soil building, hay, and pasture. The seed is used in livestock feed, and for growing bean sprouts (ingredients of Chinese dishes).

No other crop can be grown with so little trouble, and few other crops can be planted following SMALL GRAINS and give as profitable a return.

The mungbean has the advantage over other legumes of not being susceptible to many of the diseases which attack cowpeas, SOYBEANS, ALFALFA, and others. It is also less subject to the attack of field insects than are cowpeas and other large-seeded legumes.

The mungbean is adapted to the same general area now occupied by the cowpea, particularly to the central section, from north to south, of Oklahoma. This legume is also produced in Indiana, Illinois, California, northern Texas, and other states. It grows well on most any type of soil but is best suited to the warm, sandy loams.

Mungbeans may be planted from May to July. They will germinate in a soil having a low moisture content. The plant is fairly drought-resistant.

Mungbeans are usually planted in 3- to 3½-ft. rows, but drilling is becoming more popular; drilled mungbeans make a better hay and there is less waste in feeding. If the drill method of planting is used, a well prepared and firm seedbed is essential so the mungbeans can get a quick start and thereby control the growth of weeds and GRASS.

If the crop is to be used for forage, it may be harvested either with a mower or with a binder. If a mower is used, the plants should be allowed to partially cure in the swath, then be windrowed or cocked to complete the curing process. They should be put into the stack or barn just as soon as safety permits. The best time to harvest for forage is when the plants have reached their maximum growth, and only a few pods have turned brown in ripening.

Varieties: Mungbean varieties are of two major types, called "golden" and "green", because of the color of the seed.

1. *Golden mungbean*, or *golden gram*, is an erect, bushy legume, which has greenish-yellow flowers. The plant will reach a height of 3 to 4 ft. on average soil. It produces an abundance of short branches and leaves. The pods are about 3 in. long and turn black when ripe. There are 8 to 10 seeds per pod. The seed pods mature irregularly during the season and the seed shatters easily at maturity.

The golden variety is more satisfactory than the green-seeded type for hay, silage, forage pasture, and soil-building, because of the large amount of plant material produced. In general, the golden variety produces twice the weight of green forage or air-dry hay produced by the green variety, but the green variety produces about four to eight times more seed. Hay yield of the golden variety is 1 to 2 tons per acre. Golden mungbean hay averages 96 to 100 percent as high in feeding value as good quality alfalfa hay.

When used for pasture, the crop may be grazed when the plants have completed most of their vegetative growth.

The palatability of mungbean forage is improved by ensiling. Good quality silage can be made without the use of preservatives and with no special handling. On the average, 2.85 lb. golden mungbean silage is equivalent to 1 lb. good quality alfalfa hay for milk production.

2. *Green mungbeans*, or *green grams*, produce only half as much forage as the golden variety. They vary considerably in their growth habit. Some of the selections are viny and dwarfy, growing much like cowpeas; others are semierect, or bushy. The seed pods are very much like those of the golden variety, but the seed of some selections is dark green and slightly mottled; of others it is bright green and very hard in texture.

The feeding value of the green mungbean hay is 83 to 88 percent as much as that of alfalfa hay.

Ground green mungbean seed is a satisfactory ingredient for dairy cow concentrate mixtures. Liberal use of ground mungbeans in concentrate mixtures is considered economical if a source of cracked beans, or beans otherwise unsuitable for sprouting purposes, is available.

The two major types of green mungbean are *"natives"* (which are large, dull-colored beans) and *"orientals"* (shiny, hard-textured beans). Best known among the latter are "Selection No. 12" (characterized by its upright growth), *Chinese* (mungbean), and *Morse* (mungbean).

References: L.6; R.5.

See also INOCULATION.

Illustration: See FIELD BEANS.

MURIATE OF POTASH
= POTASSIUM CHLORIDE.

MURIATIC ACID.
See HYDROCHLORIC ACID.

MURUMURU OIL MEAL is produced from a Brazilian nut; the OIL MEAL contains about 10 percent fiber, 8 percent protein, and 6 percent fat and is occasionally used in cattle feeds

MUSTY HAY. *See* HAY.

MUTATION is a sudden variation in one or more characteristics of an organism which is later passed on through inheritance and marks the offspring as different from its parent stock. The offspring is called a *mutant*.

MUTTON BLUEGRASS (*Poa fendleriana*), also known as *muttongrass*, *Fendler bluegrass*, *winter bluegrass*, or (in some areas) *"wintergrass,"* occurs from northern Michigan and northern Wisconsin to south-eastern British Columbia, eastern Washington, California, east to the Chisos Mountains of western Texas and south into the mountains of Sonora, Chihuahua, and Baja California.

It is a perennial bunchgrass, tillering from the base, characteristically without rootstocks although, very rarely, small rootstocks (RHIZOMES) are produced. The erect, tufted stems, varying in height from 6 to 24 in., are roughened below the PANICLE. The tufts range up to about 1 ft. in diameter. The leaves are mostly basal. The firm and rather stiff blades are folded or inrolled, rarely flat. The leaf sheaths are somewhat roughened. The LIGULE is very small and scarcely noticeable. The oblong, contracted, usually green or pale bluish-purple flower head is protruding (EX-SERTED) from the top leaf sheath, and up to about 3 in. long. The individual SPIKE-LETS are 4- to 7-flowered, flattened, and about ⅛ in. long; the two GLUMES are broad and ⅛ in. long. The LEMMAS lack the cobwebby hairs at the base, so often found in the bluegrass genus, but have fine hairs on the lower part of the midrib and on the marginal nerves.

Mutton bluegrass occurs on all slopes, but especially on dry, southern exposures, chiefly inhabiting rich, well-drained clay loams; it is also found in sandy and gravelly soils. In the northern parts of its range, this grass often occurs in the foot-hills and on lower slopes, but in the Southwest it grows at higher elevations.

It is resistant to drought and, to a considerable extent, fire—often it is found on burns. The flowering period varies from March to early June; seed dissemination varies from July to November. Frequent associates are BROME GRASS, FESCUE, NEEDLEGRASS, REDTOP, and WILD-RYE GRASS species; also LUPINE, PENTSTEMON, SENECIO, YARROW, SEDGE, SAGEBRUSH, CINQUEFOIL, and SHRUBBY CINQUEFOIL.

Mutton bluegrass is one of the more important native range GRASSES, due to its high palatability, nutritiousness, wide distribution, fair abundance, and the fact that it starts growth very early in the spring and consequently is available when there is little other forage.

It rates as excellent forage for cattle and horses and very good or good for sheep and lambs, elk, and deer, especially in early spring. The foliage becomes rather harsh and dry with increasing maturity, and the palatability decreases somewhat as the season advances, although it is grazed well throughout the summer. In fall, when more tender and succulent forage is scarce, cattle and horses readily eat the air-cured herbage.

Reference: H.1.

See also BLUEGRASSES.

Mutton bluegrass. (H.47.)

MUTTONGRASS
= MUTTON BLUEGRASS.

MYCELIUM (pronounced: my-*see*-le-um; plural: *mycelia*) is the vegetative body of a FUNGUS. It is an aggregate of many threadlike strands, or *hyphae;* commonly, these "threads" are interwoven into a feltlike mass; from the mycelium the reproductive parts (SPOROPHORES) are produced. — *See also* FERMENTATION PRODUCTS.

MYCOSPHAERELLA. The FUNGUS *M. pinodes*, the sexual stage of *Ascochyta pinodella*, is one of the causes of the ASCOCHYTA BLIGHT OF PEA, a phase of the ASCOCHYTA BLIGHT COMPLEX. *M.*

arachidicola and *M. berkeleyii* cause LEAF SPOT OF PEANUT. *M. puераricola* (which occurs also in the imperfect stage) causes ANGULAR LEAF SPOT OF KUDZU.—*See also* CERCOSPORA.

N

NaCl, or (*common*) *salt*, is SODIUM CHLORIDE.

NANGEELA is a SUBCLOVER strain.

NANKING. *See* SOYBEAN (variety).

NAPHTHALENE DRAIN OIL, obtained when *naphthalene* is produced from the middle fraction of distilled tar, is used for making an OIL BARRIER which destroys CHINCH BUGS.

NAPIERGRASS (*Pennisetum purpureum*), *Napier* or *Nappier fodder*, also called *Carter grass* or *elephantgrass*, is important for forage. It is a perennial which grows in thick clumps. Its stalks are 6 to 10 ft. tall, the leaves ½ to 1 in. broad and 2 to 3 ft. long. The flowers are in erect, 6 to 12 in. long SPIKES.

In the South, where the grass is used for grazing and silage, it can be cut several times a season. Rich soil is required for good growth.

When Napiergrass is grown from seed, the seeding is made in greenhouse flats or in a nursery and the plants are transplanted to the field. The more common practice with Napiergrass, however, is to grow the crop from planted canes (mature stocks). The canes are cut into short lengths that contain one to three nodes, which are planted in rows and covered as one would cover potatoes. Plantings of Napiergrass will continue to produce for a number of years, but old stands decline in productive value and should occasionally be renewed. Improved strains of Napiergrass that are more disease-resistant and furnish superior grazing and production have been developed in *Florida* and *Georgia*.

Variety: A fine-stemmed horticultural variety of common Napiergrass is called *Merker grass*.

Reference: H.1.

See also GRASSES; SILAGE CROPS.

NARRAGANSETT is one of the VARIEGATED ALFALFAS.

NARROW BROWN LEAF SPOT on RICE is caused by the FUNGUS *Cercospora oryzae*. It occurs in Arkansas, Louisiana, and Texas. The spots which are found on the leaves are long and narrow and, therefore, are easily distinguished from those produced by other fungi. Usually the narrow brown spots do not appear in great abundance on plants in the field until late August or September. The bleached dead center, common to other leaf spots of rice, occurs only in very susceptible varieties. Injury to the affected plants is confined mainly to the reduction of leaf areas.

Control: Marked differences in varietal susceptibility indicate that this disease may be controlled by the selection and use of resistant varieties.

Reference: T.2.

See also BROWN-BORDERED LEAF SPOT; BROWN SPOT OF RICE.

NARROWLEAF VETCH (*Vicia angustifolia*), commonly known as *wild vetch* or *wild pea*, belongs to the VETCHES; it occurs in the United States mostly as a weed. Closely related to COMMON VETCH, it is very much like it but is usually distinguishable by its black pods, narrow leaflets, and smaller flowers. In the grainfields of the Spring Wheat Belt it is found in abundance, and in the Cotton Belt it is found along roadsides and in waste places.

A few orchardists of the South have found this species valuable for volunteering as a winter-cover and green-manure crop. The crop seems to succeed, however, only where there is good soil or weed growth, and an accumulation of organic matter. Narrowleaf vetch volunteers in pasture lands and makes excellent pasture; under cultivation it has seldom succeeded.

Reference: M.18.

See also LEGUMES; INOCULATION.

NARROW-LEAVED TREFOIL is a type of BIRDSFOOT TREFOIL.

NATALGRASS (*Rhyncelytrum roseum*), found in Florida, Texas, and Arizona, is well adapted to poor, dry, sandy soils and is especially valuable as pasture, hay, or cover crop on such soils in the extreme Southeast where few other GRASSES succeed. In Florida, Natalgrass yields two to four cuttings of highly nutritious hay, but there is considerable shrinkage in drying. The plants are free of nematodes.

Reference: W.16.

Natalgrass. Plant, spikelet, and floret. (H.26.)

NATIVE CORN = *squaw corn*.
See CORN.
NATIVE FORAGE. See GRAZING.
NATIVE GRASSES. See WILD GRASSES.
NATIVE PASTURE. See PASTURE.
NATIVE RYEGRASS
= COMMON RYEGRASS.
"NATIVES" is the name given to one of the two variety-types of the *green* MUNGBEAN.
NATOB is a *lespedeza* variety of the BICOLOR species.
NATURAL AND PREPARED MIN-ERALS. See MINERAL FEEDS.
NATURAL PASTURES
= *native pastures*. See PASTURES.
NAVY BEAN, or *White pea bean*, a COMMON BEAN variety, is one of the most important FIELD BEANS. It is a small, semi-trailing plant, characterized by its white flowers and seeds.
NEBRASKA No. 50
= *Cheyenne wheat*. See COMMON WHEAT.
NEBRED. See COMMON WHEAT (variety).
NECROSIS is the death of a tissue.
NECTAR is the sugary exudation of certain flower glands (NECTARIES); it is attractive to insects.
NECTARY, a flower gland which secretes NECTAR, is usually situated at, or near, the base of the COROLLA or PERIANT (floral envelope).
NEEDLE-AND-THREAD GRASS (*Stipa comata*) derives its name from the appearance of the seed, which is sharp-pointed; it has a bent, twisted, threadlike awn (usually 6 in. or more in length) that looks like a threaded sewing needle. It is a deep-rooted, long-lived, native bunchgrass that is often found on the western ranges and most abundantly on the sandy soils of the northern Great Plains. This grass occurs in almost pure stands as an invader on some of the abandoned croplands.

Seedstalks grow 1 to 4 ft. high, with leaves less than $\frac{1}{8}$ in. wide and 8 to 12 in. long. The LIGULE is membranous, notched, and prominent.

Growth starts in early spring, usually before associated native grasses green up, and continues throughout the summer if enough moisture is available; growth is resumed after a drought, if favorable moisture and temperatures are present in the fall.

Needle-and-thread grass. (H.1.)

Flowering of this species usually begins early in June, and the seed matures and is shed in July. Livestock graze the plants sparingly during this period because the sharp points of the seeds injure animals by working into parts of the mouth and the hide. Except for the period when seeds are present, livestock eat the forage readily and make good use of the standing cured forage for winter grazing.

References: H.1.

See also NEEDLEGRASSES; GRASSES; RANGE PLANTS; SAGE BRUSHES.

NEEDLEGRASSES (*Stipa* spp.), also often called *stipas* and *stipa grasses*, are distributed through the temperate zones. About 30 species grow in the western states. They rank fairly high as forage grasses because of their abundance, wide distribution, long growing period, and capacity to cure well on the ground. Each SPIKELET of the needlegrasses has one flower and terminates in a prominent awn. The injuries that the long, sharp awns cause on grazing animals are a serious objection to these grasses, regardless of their other virtues.

Dangers: Among the insect pests attacking the needlegrass is the WHEAT STEM SAWFLY.

Species: The most common needle-grasses found on western ranges and used as forage are NEEDLE-AND-THREAD GRASS and GREEN NEEDLEGRASS. Of some importance also are CALIFORNIA NEEDLE-GRASS and MORMON NEEDLEGRASS.

Reference: H.1.

See also GRASSES; RANGE PLANTS.

Illustration: See GREEN NEEDLEGRASS.

"NEMA" = NEMATODE.

NEMAHA. *See* COMMON OAT (variety).

NEMASTAN belongs to the Turkestan ALFALFAS.

NEMATODES, also called *"nemas"* or *eel-worms*, are microscopic worms which live in water, moist soil, and decaying or living organisms. Some are beneficial, others cause diseases, especially in plants; e.g., WHITE TIP of rice.

The fertilized female nematode may be seen with the naked eye as a small, white, pear-shaped worm. Each lays hundreds of eggs which hatch in five days and complete their live cycle in a month.

Among the more important nematodes are the following: ROOT-KNOT NEMATODES, STEM NEMATODES, STING NEMATODE, and MEADOW NEMATODE; they belong to different genera.

Control: Rotation with nonsusceptible crops, and growing susceptible crops only during the coolest part of the year, will help control this pest.

Genera: The more important nematode species are *Aphelenchoides oryzae*, *Belonolaimus gracilis*, *Ditylenchus dipsaci*, *Heterodera* spp., *Meloidogyne* spp., and *Pratylenchus* spp.

Reference: M.24.

See also CRIMSON CLOVER; FIELD PEA;

SOYBEAN; KUDZU; VETCHES; WHEATS; VEL-
VETBEANS; CROTALARIAS; RED LEAF.

NEOMYCIN is an important ANTIBIOTIC
produced by *Streptomyces fradiae*.

NEOSHO OAT. *See* RED OAT (variety).

NEOVOSIA. The FUNGUS *N. horrida*
causes KERNEL SMUT OF RICE.

NEPAL BARLEY.
          See SIX-ROWED BARLEY (variety).

NERIUM. *N. oleander*
                    = COMMON OLEANDER.

NERSCHINSK is a spring RYE variety.

NERVE is a term used for unbranched
and approximately parallel VEINS, applied
especially in the case of leaves, GLUMES,
LEMMAS, and PALEAS of GRASS flowers.

NET ENERGY. *See* ENERGY.

NET WEIGHT. *See* LABELS.

NEUMAN PEA = MUNGBEAN.

NEUTER are FLORETS or SPIKELETS
without STAMENS or PISTILS.

NEUTRAL means: neither ACID nor
alkaline; having a pH of (about) 7.—*See
also* ALKALI; ACID-BASE BALANCE; ABSORP-
TION.

NEVADA BLUEGRASS (*Poa nevadensis*)
is a perennial which grows in small bunches.
It is widely distributed throughout most
of the range countries of North America,
except in the Southwest; it occurs throughout an unusually wide elevation range:
from several hundred feet above sea level
to as high as 11,000 ft. Although this
BLUEGRASS species grows luxuriantly on
the rich soils of moist situations, it is most
common on relatively infertile, loose,
sandy, or loamy soils, often in association
with other GRASSES and weeds.

Nevada bluegrass, although seldom
abundant, is plentiful enough to furnish
considerable forage throughout its range.

This grass, one of the first to resume
growth in the spring, is very palatable and
highly relished by both game animals and
livestock during spring and early summer,
but cattle and horses continue to eat it
through the summer when the stalks and
leaves become slightly tough. In the fall,
the air-dried foliage—which is equal to
TIMOTHY in feeding value—is grazed
eagerly by all classes of livestock.

Reference: U.6.

See also RANGE PLANTS.

NEVADA EPHEDRA (*Ephedra nevadensis*), or *Nevada jointfir*, is characterized by
jointed branches. It is a shrub confined to
the more arid areas of the West; it grows
from Nevada to Old Mexico in gravelly or
sandy soils or, occasionally, in clay.

Nevada ephedra is an important forage
plant in such emergency periods as prolonged drought, or during scarcity of better
forage. It is only moderately palatable to
livestock as well as to deer; however, on
the winter range, the younger stems are
eaten with relish.

This ephedra species is olive or brownish
green, attains heights from 20 to 40 in.,
and has numerous stiff branches which
typically grow in pairs. The branches have
scales (modified leaves) about ⅛ to ³⁄₁₆ in.
long which occur in pairs jointed at their
base.

Reference: U.16.

See also RANGE PLANTS.

NEWAL. *See* SIX-ROWED BARLEY (variety).

NEWCHIEF = *Chiefkan.*
                    See COMMON WHEAT.

NEW ERA. *See* COWPEA (variety).

NEW MEXICAN FLAX is one of the
poisonous FLAXES.

NEW NORTEX. *See* RED OAT (variety).

NEW YORK WILD WHITE.
                    See WHITE CLOVER (variety).

NIACIN (pronounced: nye-a-sin), or *nicotinic acid*, or *antipellagra vitamin*, is one
of the water-soluble B-COMPLEX VITAMINS.
It forms white crystals and is essential for
the prevention of *pellagra* in man, swine,
and dogs. *Perosis* in poultry may be associated with a niacin deficiency in the
diet. Ruminants, and probably horses, do
not need dietary niacin for normal growth
and health. Other species require little, if
any, niacin if the ration contains large
amounts of TRYPTOPHAN.

The forage crops vary in niacin content;
it is much higher in grain SORGHUMS,
WHEAT, and BARLEY than in corn.—*See
also* VITAMINS; ANTIBIOTICS; MOLASSES.

NIACINAMIDE, or *nicotinamide*, forms
white, water-soluble needles. It has the
same biological function as NIACIN.—*See
also* VITAMINS.

NICARAGUA DURUM = *Arnautka.*
                    See DURUM WHEAT.

NICARBAZIN, a complex salt of *4,4-dinitrocarbanilide* and *2-hydroxy-4,6-dimethylpyrimidine*, is used in feed rations for the prevention of *coccidiosis* outbreaks in chickens. It is fed continuously to the birds during the danger period at the rate of 0.0125 percent.

NICOTIANA. *N. tabacum* = TOBACCO.

NICOTINE, an ALKALOID occuring in TOBACCO, is a colorless, oily liquid which turns brown when exposed to air. It is mixible with water and organic solvents.

Nicotine is poisonous and is used in two forms as INSECTICIDE for the control of CHINCH BUGS:

1. Nicotine-soap spray—gives very good results. It is prepared by dissolving ½ oz. of 40-percent NICOTINE SULFATE and 1 oz. soap in 1 gal. water.

2. Nicotine mineral-oil emulsion spray—is not used as extensively as the above. It is made as follows: Dissolve 1 lb. potash laundry soap in ½ gal. hot, soft water, add slowly 1 gal. highly refined white mineral oil, and beat until emulsification is complete. To 1 part (by volume) of this stock solution add 30 parts water; then add ⅛ oz. (approximately 1 teasp.) nicotinic sulfate to each gallon of the finished spray solution. The spray may be applied with any hand sprayer at the rate of 70 gal. per acre.

NICOTINE SULFATE forms water-soluble crystals. Commercially, it is mostly available as *nicotine sulfate solution* which is marketed as a dark-brown, heavy liquid. The latter is poisonous and used as disinfectant and INSECTICIDE.

NOTE: A 40-percent nicotine sulfate solution contains only a little more than 30 percent NICOTINE.

NICOTINIC ACID = NIACIN.

NIGGER WHEAT is a soft winter variety of COMMON WHEAT.

NIGHTBLINDNESS. *See* VITAMIN A.

NIGHTSHADE. *See* BLACK NIGHTSHADE.

NIGRA LOOSE SMUT is one of the two forms of LOOSE SMUTS OF BARLEY.

NINETY-DAY SPECKLED, better known as *Alabama*, is a name which was originally used to identify the *Georgia* variety of the DEERING VELVETBEAN.

NIRA. *See* RICE (variety).

NITRATE POISONING
= OAT HAY POISONING.

NITROFURAZONE is a patent-protected drug, used in some medicated feeds for protection against, and treatment of, *coccidiosis*. It forms yellow needles which are only slightly soluble in water.

NITROGEN is a gas which occurs in AIR to the extent of 78 percent by volume. *Compounds* of nitrogen, such as nitrates and ammonium salts, are used in FERTILIZERS to provide growing plants with nitrogen in usable form. Plants contain many nitrogen compounds, especially the PROTEINS (which include the ENZYMES). Plants cannot use nitrogen gas as such, but certain BACTERIA which grow on the roots of LEGUMES have the ability to convert the atmospheric nitrogen (which occurs in the soil) to nitrogen compounds.

AMINO ACIDS are converted into proteins in the animal body, but when there is a surplus of amino acids, they are converted largely into UREA and other nitrogenous waste products which are excreted in the *urine*. A part of the nitrogenous waste products can be recovered and utilized again in the form of MANURE.—*See also* AMMONIA; INOCULATION; NITROGEN FIXATION; NONPROTEIN NITROGEN.

NITROGEN FIXATION is the conversion of atmospheric NITROGEN to nitrogen compounds—either by soil organisms, or by organisms living in the roots of legumes, or by chemical means.—*See also* INOCULATION; BACTERIA.

NITROGEN-FIXING BACTERIA
= LEGUME BACTERIA.

NITROGEN-FREE EXTRACT is a term used for the substance obtained by subtracting from 100 percent the sum of the analytically determined percentages of ASH, PROTEIN, ETHER EXTRACT (fat), FIBER and MOISTURE (water). It is supposed to consist largely of CARBOHYDRATES (starches or sugars), but may contain many other substances, plus all of the errors in chemical determinations of ash, water, etc.—*See also* HAY.

NITROGENOUS means: NITROGEN-containing.

NITROGENOUS FERTILIZERS are AMMONIUM SULFATE, URAMON, UREA, CYAN-

AMIDE, etc.—*See also* NITROGENOUS; CRUDE PROTEIN; BLOAT.

3 - NITRO - 4 - HYDROXYPHENYL-ARSONIC ACID, or *4-hydroxy-3-nitro-benzenearsonic acid*, forms yellow, slightly water-soluble needles. It contains 28.5 percent ARSENIC and is used as a drug for the control of *coccidiosis* and to promote GROWTH of young animals and poultry. Like other ARSONIC ACID derivatives, it may be mixed in feed.

4 - NITRO - PHENYLARSONIC ACID, available under the trade name *Histostat*, is closely related to 3-NITRO-4-HYDROXY-PHENYLARSONIC ACID. It is used in concentration of 0.025 to 0.03 percent in poultry feed for the prevention of *blackhead* in turkeys and chickens. If given continuously, this drug exerts a GROWTH stimulating action on poults.—*See also* ARSONIC ACID.

NITTANY. *See* COMMON WHEAT (variety).

NODAL means: pertaining to a NODE.

NODE (sometimes called *knot*) is a closed and solid *joint;* the term refers especially to stems whose joints are enlarged and sometimes dark colored. The nodes are the points from which the leaves often spring.—*See also* GRASSES.

NODULE is a lump, knot, or "tubercle" on the root of many types of LEGUMES; it contains nitrogen-fixing bacteria (LEGUME BACTERIA).—*See also* INOCULATION; BLUE LUPINES; NITROGEN FIXATION.

NONALKALOID LUPINES
= SWEET LUPINES.—*See also* LUPINES.

NONGLUTINOUS RICES. *See* RICE.

NONHARDY ALFALFAS include the *African, Chilean, Indian,* and *Peruvian* varieties.—*See also* ALFALFAS.

NONPARASITIC SORGHUM LEAF DISCOLORATIONS, due to certain environmental conditions or hereditary factors, occasionally produce—on leaves of forage and grain SORGHUMS—symptoms that are frequently confused with those produced by fungus or bacterial diseases. A very common condition in sorghums is the presence of intensely colored leaf spots or stripes which are not covered with (bacterial) exudate or scales, have no dead areas in or around them, and show no evidence of the presence of fungus or

fruiting bodies. Much of this nonparasitic spotting may be due to mechanical injuries from insect punctures, wind, or sand particles. When grasshopper bait with ARSENIC or a CHLORATE-containing weed-killer falls on the leaves of sorghum, it causes a burning effect in irregular spots. Often, however, the cause of this leaf spotting is a physiological breakdown of the leaf tissues. Occasional plants have leaves so badly discolored that most of their leaf area is involved. The spots may be solid, or may follow various patterns. Certain of the latter types are known as hereditary.

One of the most common types of nonparasitic leaf discolorations is the yellow or yellow-striped (CHLOROTIC) appearance of the leaves of second-growth sorghums; nutritional deficiencies are believed to be the cause of this condition. A similar (chlorotic) condition in first-growth sorghum has been observed in certain varieties, particularly milos, grown on soils rich in calcium carbonate. Other chlorotic disturbances are definitely hereditary in nature.

Control: The remedy for the appearance of hereditary defects, and for those nutritional deficiencies which affect different varieties differently, is to choose seed stocks that produce only normal, healthy plants.

Reference: L.1.

NON POP and *Tennessee Non Pop* are SOYBEAN varieties.

NONPROTEIN NITROGEN in form of UREA, AMMONIUM CARBONATES, or AMMONIUM PHOSPHATES, is an acceptable ingredient in proprietary cattle, sheep, and goat feeds only; these ingredients are considered adulterants in proprietary feeds for other animals and birds. The following statement of guaranty of CRUDE PROTEIN for feeds containing these materials is to be used:

Crude protein, not less than . . . percent. (This includes not more than . . . percent *equivalent protein* from nonprotein nitrogen).

NOTE: The conversion factor of pure urea to equivalent crude protein is 2.915, but commercial sources usually contain a conditioner that reduces the factor to 2.62.

If feed contains more than 3 percent urea, or if the equivalent protein contributed by urea exceeds one-third of the total crude protein, the label shall bear a statement of proper usage and the following text:

Warning: This feed should be used in accordance with directions furnished on the label.

NONREDUCING SUGARS differ in their chemical structure from REDUCING SUGARS. The most important of the nonreducing sugars is the disaccharide *saccharose* (SUCROSE).—*See also* MOLASSES.

NONSACCHARINE SORGHUMS, or *grain sorghums*, are among the SORGHUMS which have value as forage plants.

NORDIHYDROGUAIARETIC ACID, which forms slightly water-soluble crystals, is one of the best ANTIOXIDANTS for fats and oils. It is used in amounts of 0.01 to 0.03 percent.

NORGHUM is one of the intermediate-type grain SORGHUMS.

NORKAN belongs to the forage SORGHUMS.

NORREDO. *See* SOYBEAN (variety).

NORSOY. *See* SOYBEAN (variety).

NORTEX and *New Nortex* are RED OAT varieties.

NORTHERN ANTHRACNOSE is caused by the FUNGUS *Kabatiella caulivora*. This disease is restricted largely to RED CLOVER and CRIMSON CLOVER in the cooler sections of North America, and damage is severe during cool, wet periods. Long, brown, sunken lesions cause much harm to stalks and leaf stems, and there is frequent girdling of stems.

Control: Only resistant varieties should be used. However, if a severe epidemic develops in a field, it may be practical to harvest early to salvage as much of the crop as possible.

Reference: T.1.

See also SOUTHERN ANTHRACNOSE; ANTHRACNOSE.

NORTHERN COMMON ALFALFAS include the *Dakota Common, Kansas Common,* and *Montana Common* ALFALFAS.

NORTHERN CORN LEAF BLIGHT, caused by the FUNGUS *Helminthosporium turcicum*, is a disease found in most of the humid areas where CORN is grown. In the United States it is frequently most abundant in the eastern half of the Corn Belt, extending to the Atlantic seaboard and southward. When favorable conditions prevail, the disease may become locally important in other parts of the Corn Belt. In some years symptoms may be found before silking, whereas in less favorable seasons symptoms may not appear until a month or more after silking. In extreme southern areas, such as Florida, an infection often becomes established on young seedlings.

Reductions in grain yield, resulting from this leaf blight, may exceed 30 percent; the later the disease appears, the less is the yield reduced. The nutritional value of fodder of severely blighted corn, though, is considerably lowered. Where much leaf area has been killed there is almost invariably a marked increase in STALK ROT.

Northern corn leaf blight is recognized by long, elliptical, grayish-green or tan spots on the leaves, ranging in size up to 6 in. by 1½ in. These lesions appear first on the lower leaves; they increase until—under severe conditions—all leaves are nearly covered. Under those circumstances plants have a dead, gray-green appearance, resembling frost injury. Many spores are produced on the lower surface of the lesions in damp weather. The spores are often arranged in concentric zones so that a faint target-like design is evident. Ears are not attacked, so that the possibility of seeds carrying the disease is extremely remote.

The fungus overwinters in infected corn leaves, and in the following summer spores are formed which are carried by the wind to growing corn leaves where, if moisture is present, they germinate and penetrate. Lesions become recognizable seven to ten days after penetration. Spores are formed on the lesions and these, in turn, are spread to healthy leaves to continue the disease cycle.

Besides corn, the fungus is known to

attack SUDANGRASS, SORGHUM and TEOSINTE.

Control: Northern corn leaf blight is most effectively controlled by planting resistant hybrids. Dusting or spraying with DITHANE is effective when the fungicide is applied to the plants soon after silking; in addition, the plants should be kept covered with the fungicide until corn is mature. Seed treatment and rotation have no effect in controlling this disease.

Reference: U.7.

NORTHERN CORN ROOTWORM
= CORN ROOTWORM.

NORTHWESTERN DENT.
See CORN (variety).

NOTCH-SEEDED BUCKWHEAT
= WINGED BUCKWHEAT.

NOXIOUS means: harmful, injurious, unwholesome; as, noxious weed seeds.— *See also* SCREENINGS.

N.R.C. = National Research Council.— *See also* NUTRIENT REQUIREMENTS; FEED-STUFF COMPOSITION.

NUCLEOPROTEINS. *See* PROTEINS.

NUCLEUS is the organ of the cell of a plant or animal which is essential in METABOLISM, GROWTH, reproduction, and heredity.

NUDA LOOSE SMUT is one of the LOOSE SMUTS OF BARLEY.

NUGGET DURUM is a white DURUM WHEAT variety.

NURSE COW METHOD.
See DAIRY RATIONS.

NURSE CROP is a temporary crop, sown or planted with a permanent crop, e.g., OATS sown in ALFALFA, or LESPEDEZA sown in oats.

NUT. *See* NUTS.

NUTRIENT REQUIREMENTS. The N.R.C. and its subcommittees, in preparing tables on nutrient requirements (also called *recommended nutrient allowances*) for farm animals and poultry, have developed standards of the highest possible degree of reliability. The daily nutrient requirements for domesticated animals are listed in a set of tables which have been compiled and published by the experts of the N.R.C. *See* pages 316-322.

NUTRIENTS are food ingredients which are essential to the GROWTH, development, or reproduction of plants and animals.

Some forty *simple* substances are *essential* to the body. For nonruminants these nutrients include at least thirteen MINERALS, ten to fifteen VITAMINS or vitamin-like substances, eight to ten AMINO ACIDS, and three FATTY ACIDS, besides WATER, SUGARS and some other substances. Certain nutrients can be synthesized from the diet or derived from other *complex* substances in the body, such as FATS and PROTEINS. There are variations with animal species as to essential nutrients and the ability of synthesizing nutrients in the body.

Digestible nutrients is a term referring to the portion of each nutrient that can be utilized by being taken into the body.

Total digestible nutrients, abbreviated T.D.N., is the sum of all digestible organic nutrients; i.e., protein + nitrogen-free extract + fiber + fat (the latter revised by multiplying it by 2.25 to bring its energy value in line with that of the other mentioned nutrients).—*See also* ABSORPTION; BALANCED RATION; METABOLISM; ENERGY; CAROB BEAN; FEED EFFICIENCY.

> NOTE: For livestock, the energy value of fat is approximately 2.25 times higher than that of carbohydrates or protein.

NUTRITION. *See* NUTRIENTS; RATIONS; FEED EFFICIENCY.

NUTRITIVE RATIO is the proportion between digestible protein and non-nitrogenous NUTRIENTS (whereby the latter term includes the fat content multiplied by 2.25 to bring its energy value up to that of the other nutrients).

NUTS are any kind of fruit whose seed is enclosed in a hard shell; e.g., ACORNS.— *See also* ALMOND HULLS; PASTURE PLANTS.

NUT-SIZE COTTONSEED CAKE.
See COTTONSEED.

NYMPH is a very young BUG.—*See also* MEADOW SPITTLEBUG; GRASSHOPPERS; CHINCH BUG; POTATO LEAFHOPPER.

DAILY NUTRIENT REQUIREMENTS FOR SWINE

Description of Pigs

	Market Stock						Breeding Stock			
							Pregnant females and breeding boars		Lactating females	
							Young stock	Adults	Gilts	Adults
Liveweight, lb	25	50	100	150	200	250	300	500	350	450
Expected daily gain, lb	0.8	1.2	1.6	1.8	1.8	1.8	0.75	0.5	....	....
Total feed (air dry) lb	2.0	3.2	5.3	6.8	7.5	8.3	6.0	7.5	11.0	12.5
Total digestible nutrients (75% TDN) lb	1.6*	2.4	4.0	5.1	5.6	6.2	4.5	5.6	8.3	9.4
Crude protein, lb	0.36	0.51	0.74	0.88	0.90	1.00	0.90	1.05	1.65	1.75
Inorganic nutrients:										
Calcium, gm	7.3	9.4	15.6	17.0	18.7	20.7	16.3	20.4	30.0	34.0
Phosphorus, gm	5.4	6.5	10.8	10.2	11.2	12.4	10.9	13.6	20.0	22.7
Salt (NaCl), gm	4.5	7.3	12.0	15.4	17.0	18.8	13.6	17.0	25.0	28.4
Vitamins:										
Carotene, mg	0.5	1.0	2.0	3.0	4.0	5.0	15.0	18.7	27.5	31.2
Vitamin D, I. U.	180.0	288.0	477.0	612.0	675.0	747.0	540.0	675.0	990.0	1,125.0
Thiamine, mg	1.0	1.6	2.6	3.4	3.8	4.2	3.0	3.8	5.5	6.2
Riboflavin, mg	2.4	3.2	5.3	6.8	7.5	8.3	7.2	9.0	13.2	15.0
Niacin, mg	16.0	19.2	26.5	34.0	37.5	41.5	30.0	37.5	55.0	62.5
Pantothenic acid, mg	10.0	16.0	23.8	30.6	33.8	37.4	27.0	33.8	49.5	56.2
Pyridoxine, mg	1.2	1.9	....	....	....	....	....	....	....	....
Choline, mg	800.0	....	....	....	....	....	....	....	....	....
Vitamin B₁₂, mcg	20.0	16.0	26.5	....	....	....	....	....	....	....

* For young pigs a high energy diet (80% T.D.N.) is recommended.

DAILY NUTRIENT REQUIREMENTS FOR DAIRY CATTLE
Based on air-dry feed containing 90 percent dry matter

Body Weight	Expected Gain		Daily Allowances per Animal[1]						
	Small Breeds	Large Breeds	Total Feed	Digestible Protein	T.D.N.	Calcium	Phosphorus	Carotene	Vitamin D
lbs.	lbs.	lbs.	lbs.	lbs.	lbs.	gm.	gm.	mg.	I.U.
			Normal	growth	of	dairy	heifers		
50	0.5	—	0.9	.20	1.0	4	3	6[2]	200
100	1.0	0.8	2.0	.40	2.0	8	6	6	400
150	1.3	1.4	4.0	.50	3.0	12	8	9	600
200	1.4	1.6	6.0	.60	4.0	16	11	12	800
400	1.2	1.8	11	.80	6.5	20	15	24	[3]
600	0.8	1.4	15	.85	8.5	18	15	36	
800	1.1	1.2	19	.90	10.0	16	15	48	
1000	—	1.3	22	.95	11.0	15	15	60	
1200	—	1.2	24	1.00	12.0	15	15	72	
			Maintenance	of	mature	cows[4]			
800	—	—	14	.50	6.8	8	8	48	[3]
1000	—	—	16	.60	8.0	10	10	60	
1200	—	—	18	.70	9.2	12	12	72	
1400	—	—	21	.80	10.5	14	14	84	
1600	—	—	23	.87	11.4	16	16	96	
		Reproduction (Add to maintenance during last 2 to 3 months)							
	2.0	2.0	8.0	.60	6.0	12	7	30	[3]
		Lactation (Add to maintenance for each pound of milk)							
	3.0% fat	—		.040	0.28	1	0.7	[5]	[5]
	4.0% fat	—		.045	0.32	1	0.7		
	5.0% fat	—		.050	0.37	1	0.7		
	6.0% fat	—		.055	0.42	1	0.7		
			Maintenance	of	breeding	bulls			
1200	—	—	18	1.00	10.3	12	12	72	[3]
1600	—	—	22	1.20	12.9	16	16	96	
2000	—	—	27	1.45	15.6	20	20	120	
2400	—	—	31	1.60	18.2	24	24	144	

[1] Thiamine, riboflavin, niacin, pyridoxine, pantothenic acid, and vitamin K are synthesized by bacteria in the rumen, and it appears that adequate amounts of these vitamins are furnished by a combination of rumen synthesis and natural feedstuffs. Manganese, iron, copper, and cobalt are clearly essential, but the amounts needed are not known. For growth, 0.6 gm. magnesium is needed per 100 lb. of body weight.

[2] Calves should receive colostrum the first few days after birth, as a source of vitamin A and other essential factors.

[3] While vitamin D is known to be required, the data are inadequate to warrant specific figures for older growing animals and for maintenance, reproduction, and lactation.

[4] When calculating the allowances for lactating heifers that are still growing, it is recommended that the figure for *growth* rather than *maintenance* be used.

[5] When adequate amounts of vitamins A and D are fed for normal reproduction, extra amounts will probably not stimulate milk production but will increase the vitamin content of the milk.

DAILY NUTRIENT REQUIREMENTS FOR BEEF CATTLE
Based on air-dry feed containing 90 percent dry matter

Body Weight, Pounds	Expected Daily Gain, Pounds	Daily Allowances per Animal						
		Total Feed		Digestible Protein, Pounds	Total Digestible Nutrients, Pounds	Calcium, grams	Phosphorus, grams	Carotene,* mg.
		Per cent of Live weight	Per Animal, Pounds					
Normal Growth, Heifers and Steers								
400	1.6	3.0	12	0.9	7.0	20	15	24
600	1.4	2.7	16	0.9	8.5	18	15	36
800	1.2	2.4	19	0.9	9.5	16	15	48
1000	1.0	2.1	21	0.9	10.5	15	15	60
Bulls, Growth and Maintenance (Moderate Activity)								
600	2.3	2.7	16	1.3†	10.0	24	18	36
800	1.7	2.1	17	1.4	11.0	23	18	48
1000	1.6	2.0	20	1.4	12.0	22	18	60
1200	1.4	1.8	22	1.4	13.0	21	18	72
1400	1.0	1.7	24	1.4	14.0	20	18	84
1600	...	1.6	26	1.4	14.0	18	18	96
1800	...	1.4	26	1.4	14.0	18	18	108
Wintering Weanling Calves								
400	1.0	2.8	11	0.7	6.0	16	12	24
500	1.0	2.6	13	0.8	7.0	16	12	30
600	1.0	2.5	15	0.8	8.0	16	12	36
Wintering Yearling Cattle								
600	1.0	2.7	16	0.8	8.0	16	12	36
700	1.0	2.4	17	0.8	8.5	16	12	42
800	0.7	2.3	18	0.8	9.0	16	12	48
900	0.5	2.0	18	0.8	9.0	16	12	54
Wintering Pregnant Heifers (Weights are for beginning of winter period; gains average for period)								
700	1.5	2.9	20	0.9	10.0	18	16	42
800	1.3	2.3	20	0.9	10.0	18	16	48
900	0.8	2.0	18	0.8	9.0	16	15	54
1000	0.5	1.8	18	0.8	9.0	16	15	60
Wintering Mature Pregnant Cows (Weights are for beginning of winter period; gains, average for period)								
800	1.5	2.8	22	1.0	11.0	22	18	48
900	1.0	2.2	20	0.9	10.0	18	16	54
1000	0.4	1.8	18	0.9	9.0	16	15	60
1100	0.2	1.6	18	0.8	9.0	16	15	66
1200	0.0	1.5	18	0.8	9.0	16	15	72
Cows Nursing Calves, 1st 3 to 4 Months After Parturition								
900–1100	None	...	28	1.4	14.0	30	24	300

DAILY NUTRIENT REQUIREMENTS FOR BEEF CATTLE—Continued

Body Weight, Pounds	Expected Daily Gain, Pounds	Total Feed		Digestible Protein, Pounds	Total Digestible Nutrients, Pounds	Calcium, grams	Phosphorus, grams	Carotene,* mg.
		Per cent of Live Weight	Per Animal, Pounds					

Daily Allowances per Animal

Fattening Calves Finished as Short Yearlings

400	Average for	3.0	12	1.1	8.0	20	15	24
500	period, 2.0	2.8	14	1.2	9.5	20	16	30
600	pounds	2.7	16	1.3	11.0	20	17	36
700	daily	2.6	18	1.4	12.0	20	18	42
800		2.5	20	1.5	13.5	20	18	48
900		2.3	21	1.5	14.5	20	18	54

Fattening Yearling Cattle

600	Average for	3.0	18	1.3	11.5	20	17	36
700	period, 2.2	3.0	21	1.4	13.5	20	18	42
800	pounds	2.8	22	1.5	14.0	20	19	48
900	daily	2.7	24	1.6	15.5	20	20	54
1000		2.6	26	1.7	17.0	20	20	60
1100		2.4	27	1.7	17.5	20	20	66

Fattening Two-Year-Old Cattle

800	Average for	3.0	24	1.5	15.0	20	18	48
900	period, 2.4	2.9	26	1.6	16.0	20	20	54
1000	pounds	2.7	27	1.7	17.0	20	20	60
1100	daily	2.6	29	1.8	18.0	20	20	66
1200		2.4	29	1.8	18.0	20	20	72

* The recommended carotene allowance for fattening animals is at the same rate as for cattle in other classifications because this is about the minimum rate that will result in significant storage and thus assure contribution of vitamin-A value for human use from the beef liver and fat. For optimum growth for feed-lot gains and freedom from clinical symptoms, 1.5 mg. carotene for each 100 lb. of body weight suffices for cattle previously depleted of body stores, and this level may be used when economically necessary except for pregnant or lactating cows. The vitamin-A value of the liver and the body fat of animals so fed, however, would be practically nil. Actually, no dietary carotene or vitamin-A is needed so long as the animals have sufficient storage reserve to meet physiological needs.

† During periods of moderate to heavy service, a level of about 2.0 lb. daily of digestible protein is tentatively suggested.

DAILY NUTRIENT REQUIREMENTS FOR SHEEP

Live Weight	Expected Daily Gain or Loss	Total Feed, Air Dry Basis	Total Digestible Protein	Total Digestible Nutrients	Calcium	Phosphorus	Salt	Carotene
Lb.	Lb.	Lb.	Lb.	Lb.	Gm.	Gm.	Lb.	Mg.
BRED EWES FIRST 100 DAYS OF GESTATION								
100	0.12	3.5	0.17	1.7	3.2	2.5	0.03	6.0
110	0.12	3.6	0.18	1.8	3.2	2.6	0.03	6.6
120	0.12	3.7	0.19	1.9	3.3	2.7	0.03	7.2
130	0.12	3.8	0.20	2.0	3.4	2.7	0.03	7.8

DAILY NUTRIENT REQUIREMENTS FOR SHEEP—Continued

Live Weight	Expected Daily Gain or Loss	Total Feed, Air Dry Basis	Total Digestible Protein	Total Digestible Nutrients	Calcium	Phosphorus	Salt	Carotene
Lb.	Lb.	Lb.	Lb.	Lb.	Gm.	Gm.	Lb.	Mg.
			BRED EWES LAST 6 WEEKS BEFORE LAMBING					
110	0.25	4.0	0.21	2.1	4.3	3.2	0.03	6.6
120	0.25	4.1	0.22	2.2	4.4	3.3	0.03	7.2
130	0.25	4.2	0.23	2.3	4.5	3.4	0.03	7.8
140	0.25	4.3	0.24	2.4	4.7	3.5	0.03	8.4
150	0.25	4.4	0.25	2.5	4.8	3.6	0.03	9.0
			EWES IN LACTATION					
100	−0.10	4.5	0.27	2.5	6.1	4.5	0.03	6.0
110	−0.10	4.6	0.28	2.6	6.2	4.6	0.03	7.1
120	−0.10	4.7	0.28	2.7	6.4	4.7	0.03	7.8
130	−0.10	4.8	0.30	2.8	6.5	4.8	0.03	8.4
140	−0.10	4.9	0.30	2.9	6.6	4.9	0.03	9.1
150	−0.10	5.0	0.31	3.0	6.8	5.0	0.03	9.7
			EWES—LAMBS AND YEARLINGS					
70	0.35	3.0	0.22	1.8	3.0	2.7	0.02	3.8
90	0.30	3.2	0.22	1.9	3.0	2.7	0.02	5.0
110	0.20	3.5	0.20	1.9	3.2	2.8	0.03	6.0
130	0.10	3.8	0.20	2.0	3.1	2.7	0.03	7.1
			RAMS—LAMBS AND YEARLINGS					
75	0.45	3.5	0.24	2.1	3.8	3.2	0.02	4.1
100	0.40	4.0	0.24	2.3	4.0	3.4	0.03	5.5
125	0.35	4.0	0.24	2.4	3.6	3.3	0.03	6.9
150	0.30	4.3	0.23	2.6	3.7	3.3	0.03	8.2
175	0.20	4.5	0.23	2.6	3.7	3.3	0.03	9.6
			FATTENING LAMBS					
50	0.25	2.1	0.17	1.2	2.5	2.1	0.02	3.0
60	0.30	2.3	0.18	1.4	2.6	2.2	0.02	3.6
70	0.35	2.7	0.19	1.7	2.9	2.4	0.02	4.2
80	0.35	2.9	0.20	1.9	2.9	2.4	0.02	4.2
90	0.25	3.0	0.20	2.0	2.7	2.3	0.02	5.4

DAILY NUTRIENT REQUIREMENTS FOR DUCKS [1]
In percentage or amount per pound of feed

	Starting and growing ducks
Total protein, per cent	17
Vitamins	
Vitamin D (I.C.U.) [2]	100
Riboflavin, mg.	1.8
Pantothenic acid, mg.	5.0
Niacin, mg.	25
Pyridoxine, mg.	1.2

[1] These figures are estimates of requirements and include no margins of safety.
[2] Birds generally use vitamin D_3, but do not use vitamin D_2 as do mammals.

DAILY NUTRIENT REQUIREMENTS FOR MATURE HORSES

Body Weight	Digestible Protein				Total Digestible Nutrients				Dry Matter—90% basis				Calcium*	Phosphorus*
	Maintenance	Work			Maintenance	Work			Maintenance	Work				
		Light	Medium	Hard		Light	Medium	Hard		Light	Medium	Hard		
lb.	lb.	lb.	lb.	lb.	lb.	lb.	lb.	lb.	lb.	lb.	lb.	lb.	gm.	gm.
400	0.31	0.38	0.46	0.52	3.9	5.1	5.9	7.3	6.2	8.2	9.4	11.0	7.3	8.5
600	0.42	0.52	0.62	0.70	5.3	6.9	8.0	9.9	8.5	11.0	12.8	14.8	9.9	11.0
800	0.53	0.65	0.77	0.87	6.6	8.6	10.0	12.3	10.6	13.8	15.9	18.4	12.2	13.7
1000	0.62	0.76	0.91	1.03	7.9	10.1	11.8	14.6	12.6	16.2	18.9	21.9	13.7	15.4
1200	0.71	0.88	1.04	1.18	9.0	11.6	13.5	16.7	14.4	18.6	21.6	25.1	15.7	17.7
1400	0.80	0.98	1.17	1.33	10.1	13.0	15.1	18.8	16.2	20.8	24.2	28.2	17.6	19.8
1600	0.88	1.09	1.29	1.47	11.1	14.4	16.7	20.7	17.8	23.0	26.7	31.0	19.4	21.8
1800	0.97	1.19	1.40	1.60	12.2	15.8	18.2	22.7	19.5	25.3	29.1	34.0	21.1	23.8

Mares—Last Quarter of Pregnancy

Body Weight	Digestible Protein	Total Digestible Nutrients	Dry Matter	Calcium*	Phosphorus*
400	0.46	5.4	8.6	7.8	7.8
600	0.62	7.3	11.7	10.6	10.6
800	0.77	9.0	14.4	13.1	13.1
1000	0.91	10.6	17.0	15.4	15.4
1200	1.04	12.2	19.5	17.7	17.7
1400	1.17	13.7	21.9	19.9	19.9
1600	1.29	15.2	24.3	22.1	22.1
1800	1.41	16.2	26.6	24.2	24.2

Mares—Lactating

Body Weight	Digestible Protein	Total Digestible Nutrients	Dry Matter	Calcium*	Phosphorus*
400	1.01	8.8	12.3	12.3	11.2
600	1.37	11.9	16.7	16.7	15.2
800	1.70	14.7	20.6	20.6	18.7
1000	2.01	17.4	24.4	24.4	22.2
1200	2.30	20.0	28.0	28.0	25.4
1400	2.59	22.4	31.4	31.4	28.5
1600	2.87	24.9	34.9	34.9	31.7
1800	3.12	27.1	37.9	37.9	34.4

* Calculated for horses at medium work.

DAILY NUTRIENT REQUIREMENTS FOR CHICKENS [1]
In percentage or amount per pound of feed

	Starting chickens 0-8 weeks	Growing chickens 8-18 weeks	Laying hens	Breeding hens
Total protein, per cent	20	16	15	15
Vitamins				
Vitamin A activity (U.S.P. Units) [2]	1200	1200	2000	2000
Vitamin D (I.C.U.) [3]	90	90	225	225
Thiamine, mg.	0.8	?	?	?
Riboflavin, mg.	1.3	0.8	1.0	1.7
Pantothenic acid, mg.	4.2	4.2	2.1	4.2
Niacin, mg.	12	?	?	?
Pyridoxine, mg.	1.3	?	1.3	1.3
Biotin, mg.	0.04	?	?	?
Choline, mg. [4]	600	?	?	?
Folacin, mg.	0.25	?	0.11	0.16

For footnotes 1-4, see next page at end of table.

DAILY NUTRIENT REQUIREMENTS FOR CHICKENS[1]—Continued

	Starting chickens 0-8 weeks	Growing chickens 8-18 weeks	Laying hens	Breeding hens
Minerals				
Calcium, per cent	1.0	1.0	2.25[5]	2.25[5]
Phosphorus, per cent[6]	0.6	0.6	0.6	0.6
Salt, per cent[7]	0.5	0.5	0.5	0.5
Potassium, per cent	0.2	0.16	?	?
Manganese, mg.	25	?	?	15
Iodine, mg.	0.5	0.2	0.2	0.5
Magnesium, mg.	220	?	?	?

[1] These figures are estimates of requirements and include no margins of safety.
[2] May be vitamin A or pro-vitamin A.
[3] Birds generally use vitamin D_3, but do not use vitamin D_2 as do mammals.
[4] Betaine can be used interchangeably with choline for methylating purposes but not for other functions such as prevention of perosis.
[5] This amount of calcium need not be incorporated in the mixed feed, inasmuch as calcium supplements fed free choice are considered as part of the ration.
[6] At least 0.45% of the total feed of starting chickens should be inorganic phosphorus. All of the phosphorus of non-plant feed ingredients is considered to be inorganic. Approx. 30% of the phosphorus of plant products may be considered as part of the inorganic phosphorus required. A portion of the phosphorus requirement of growing chickens and laying and breeding hens must also be supplied in inorganic form. For birds in these categories the requirement for inorganic phosphorus is lower and not as well defined as for starting chickens.
[7] This figure represents salt, or sodium chloride, added as such or in marine or fermentation products of high sodium chloride content.

DAILY NUTRIENT REQUIREMENTS FOR TURKEYS[1]
In percentage or amount per pound of feed

	Starting poults 0-8 weeks	Growing turkeys 8-16 weeks	Breeding turkeys
Total protein, per cent[2]	28	20	15
Vitamins			
Vitamin A activity (U.S.P.)[3]	2400	2400	2400
Vitamin D (I.C.U.)[4]	400	400	400
Riboflavin, mg.	1.7	?	1.5
Pantothenic acid, mg.	5.0	?	?
Choline, mg.	750	?	?
Folacin, mg.	0.4	?	?
Minerals			
Calcium, per cent	2.0	2.0	2.25[5]
Phosphorus, per cent[6]	1.0	1.0	0.75
Manganese, mg.	25	?	15
Salt, per cent[7]	0.5	0.5	0.5

[1] These figures are estimates of requirements and include no margins of safety.
[2] The protein content of rations for growing turkeys from 16 weeks to market weight may be reduced to 16 percent.
[3] May be vitamin A or pro-vitamin A.
[4] Birds generally use vitamin D_3, but do not use vitamin D_2 as do mammals.
[5] This amount of calcium need not be incorporated in the mixed feed, inasmuch as calcium supplements fed free choice are considered as part of the ration.
[6] At least 0.50% of the total feed of starting poults should be inorganic phosphorus. Approx. 30% of the phosphorus of plant products may be considered as part of the inorganic phosphorus required. Presumably a portion of the requirement of growing and breeding turkeys must also be furnished in inorganic form.
[7] This figure represents salt, or sodium chloride, added as such or in marine or fermentation products of high sodium chloride content.

O

O.A.C. 21.

    See SIX-ROWED BARLEY (variety).

OAKS (*Quercus* spp.) vary in size from low bushes to tall trees with small or large, evergreen or deciduous (dropping-off) leaves, and fruits which are ACORNS. Many oaks are important RANGE PLANTS and the leaves of a few species furnish fair forage; the leaves and leaf buds of others (e.g., of *Gamble oak*) contain TANNIN (which is not toxic but lowers the digestibility of forage) and a toxic principle as yet unknown.

    Eating too many acorns by the dam may contribute indirectly to the deformities found in the *acorn calf*.

Turkey oak. Twig with leaves and acorn. (W.35.)

    Acorn poisoning occurs occasionally in horses, but *oak leaf poisoning* is common in cattle and also in sheep when they feed largely on oak for two weeks or longer, especially in the spring. Emaciation, scabby nose, and constipation, followed by diarrhea and weakness, are symptoms of *oak poisoning.*—*See also* POISONOUS PLANTS.

OAT = COMMON OAT.—*See also* OATS; COMMON WILD OAT.

OAT CHOP. *See* OATS.

OAT DUST = *Oat shorts*. *See* OATS.

OATGRASS. *See* CALIFORNIA OATGRASS.

OAT GROATS. *See* OATS.

OAT HAY. *See* OATS.

OAT HAY POISONING is more correctly called *nitrate poisoning*, because not only oat hay but also CORN stalks, certain WEEDS, water, etc. may contain enough nitrate to cause heavy losses in cattle. Under certain soil and climatic conditions, immature oats cut for hay contain a considerable amount of POTASSIUM NITRATE. This in itself is not particularly poisonous, but the nitrate, when acted upon by bacterial fermentation, may be converted to nitrite which is very poisonous. POTASSIUM NITRITE combines with the hemoglobin of the blood to form methemoglobin. Hemoglobin carries the oxygen of the air to the tissues of the animals, but methemoglobin cannot unite with oxygen. The animal, therefore, dies of lack of oxygen, just as it would if all air were cut off by strangulation. Lips and membranes become a deep purple, and death follows in three to five hours after eating the hay. Some cows may recover, but if they are pregnant, they may abort their calves.

    Probably most of the potassium nitrate is converted into potassium nitrite in the cow's stomach but the nitrite may already be present in oat hay which is wet as the result of rain or snow.

    Oat hay containing more than 1.0 to 1.5 percent potassium nitrate should be considered dangerous for cattle and possibly also for sheep. More potassium nitrate can be safely fed if the ration is high in GRAINS or MOLASSES.

    A solution of METHYLENE BLUE, injected intravenously, will bring about recoveries in a very high percentage of cases.

    NOTE: Stockmen with a considerable supply of oat hay should have a sample analyzed for its content of potassium nitrate, or feed it for a day or two to one or two cattle.

    Reference: W.34.

OAT HULLER FEED. *See* OATS.

OAT HULLS. *See* OATS.

OAT LEAF SPOT, sometimes called *"leaf blotch,"* is caused by the FUNGUS *Helminthosporium avenae* and is one of the most prevalent OAT diseases in the southern states. Amounts of damage are difficult to determine, for oats often yield quite well even when the leaves show large numbers of spots. However, this disease is affected to a great extent by climatic conditions, and when humid conditions and mild temperatures prevail for a considerable period of time, serious losses often occur.

Primary infections appear on the first leaf soon after emergence, and result from either seed- or soil-borne fungi. The first spots on the new leaves resemble small (1/16 in.) dark-purple doughnuts. Later, the spots grow together and become oblong or linear in shape and reddish-brown in color. The infection may become so severe that the leaves turn yellow and die. Secondary infections, spreading from one plant to the other, occur rapidly under humid, shaded conditions where moisture stands on the oat plants for a considerable length of time.

Secondary infection spores that become lodged in or on the developing oat seed, are carried over to next year's crop.

Control: SEED TREATMENT with CERESAN M will kill the seed-borne spores and give some protection to the germinating seed. Crop rotation is beneficial in reducing the amount of fungus in the soil. Differences in susceptibility of oat varieties to leaf spot exist.

Reference: M.30.

OAT MIDDLINGS. *See* OATS.

OAT MILL FEED. *See* OATS.

OATS (*Avena* spp.) are annual GRASSES of the true cereal group. They constitute one of the most important small grain crops, particularly in Iowa, Minnesota, Illinois, Wisconsin, and Nebraska, but are also grown successfully in all other regions of the United States.

The oat plant produces three to five culms which grow 2 to 5 ft. high; the numerous roots penetrate the soil several feet deep. The leaves average 5/8 in. width and 10 in. length. Each PANICLE branch ends in a slender, drooping PEDICEL on which is borne the SPIKELET; the latter usually contains two perfect FLORETS and sometimes a third (imperfect) floret. Only in the *naked*, or *hull-less*, oats is the LEMMA longer than the two glumes. The color of the lemma is white, gray, yellow, red, or black. The awn, if present, may be straight, twisted, or bent into a knee. The kernel is spindle-shaped, furrowed, and covered with fine hairs; it is enclosed in the hulls, except in the naked varieties. Self-pollination is normal in most oat varieties, but in some there occurs natural crossing. Oat varieties differ widely in the degree of dormancy—some seeds germinate immediately after harvesting, others only after four to ten weeks of storage.

The United States has been divided into seven *oat regions:* (1) *North Central oat region,* where early and midseason common and red oat varieties are grown; (2) *Central* (or spring-sown) red oat region; (3) *Northeastern* oat region—where medium-early and midseason varieties are grown; (4) *Southern* (or fall-sown) red oat region; (5) *Great Plains* oat region—it grows midseason and early midseason common and red oat varieties; (6) *Rocky Mountain* oat region, where midseason common oats are mostly grown under irrigation; and (7) *Pacific Coast* oat region—where both common and red oats are cultivated.

Oats are not very profitable in actual cash returns. Nevertheless, they fit well into existing rotations and are a good nurse crop. They are a valuable feed for dairy cows and horses, young stock, the breeding herd, and as a scratch grain for poultry.

Because of their greater water-holding capacity, clay and loam soils are better than sandy soils for oats. Oats have a high water requirement, but should not be grown on very rich or on low, undrained land. Oats grown on poorly drained, wet soils may be injured by plant diseases.

Although oats respond well to liberal fertilization, it is more profitable to apply FERTILIZERS or barnyard manure to other crops used in rotation. Under some conditions, however, it may be profitable to apply fertilizers directly to the oat crop.

It is safe to apply well-rooted stable manure at the rate of 10 to 15 tons per acre on the poor soils a few months previous to sowing oats.

Many soils are deficient in phosphorus, and it must be supplied by the application of manure or commercial fertilizer. Where neither stable manure nor green manure is available, nitrogenous fertilizer should be applied to oats at seeding time. Potash ordinarily can be applied more profitably to some other crop in the rotation. Most experiment stations now suggest that fertilizer recommendations be made on the basis of a soil test. The needed nutrients are then applied for each crop in the rotation.

Frequently a complete fertilizer—one containing nitrogen, phosphorus, and potash—is used on the oat crop. A good combination on average soils is one containing 50 to 100 lb. sodium nitrate, 200 to 250 lb. superphosphate, and 40 to 80 lb. potassium chloride, applied at the rate of 200 to 300 lb. per acre. In any region where the nitrogen and potash content of the soil is taken care of by proper crop rotation and good cultural methods, the application of 150 to 200 lb. superphosphate per acre usually is one of the most satisfactory fertilizer treatments for oats.

Where oats are desired for forage rather than for grain production, the use of barnyard manure, or fertilizers high in nitrogen is sometimes advisable. The nitrogenous fertilizers induce a rank growth of straw which is necessary for the production of a large yield of forage.

On very acid soils, or when green-manure crops or stable manure are used, the application of lime may have a beneficial effect on oats.

Any rotation for oats should include both a cultivated and a LEGUME crop. In the northeastern states, one of the most common rotations consists of CORN, oats, and grass or CLOVER. This usually is a 4-year rotation, the clover and grass remaining as a meadow for pasture for two or more years. Oats also frequently follow POTATOES in a similar 4-year sequence, the oats in turn being followed by grass and clover. Where winter WHEAT is grown, the

Oats. Plant with spikelet and floret. (H.26.)

rotation may consist of corn, oats, winter wheat, and grass. In general, oats follow corn more successfully than any other crop; wheat can be grown after oats with good results. In the central part of the Great Plains area a popular rotation is the growing of corn and small grains in alternate years. In the southern part of the Great Plains area grain SORGHUM replaces corn.

A rotation which is often followed in irrigated areas in Colorado and Idaho includes the following: ALFALFA, three years or more; SUGAR BEETS or potatoes two or three years; and cereal crops two or three years, the cereal being used as a nurse crop for alfalfa in the last year.

In the Pacific area, oats frequently are sown as an intermediate crop between row crops and clover; a common rotation for this area is a row crop, such as corn or potatoes, followed by oats and clover, each one year.

In the South the best place in the rotation for oats is after a row crop, such as corn or cotton. Probably as good a rotation as has been devised for that part of the fall-sown oat area in which cotton is grown, is the following:

First year—cotton with CRIMSON CLOVER, VETCH, or Austrian winter FIELD PEAS, sown as a winter-cover crop. Second year—corn, following plowing-under of the winter legume. COWPEAS or SOYBEANS may be sown in the corn at the last cultivation, the corn to be cut as fodder and the cowpeas or soybeans and corn stubble turned under in time to seed fall oats. Third year —fall-sown oats, followed by cowpeas or soybeans sown as a hay crop, the cowpea or soybean stubble turned under and the soil sown to RYE as a winter covering, which is turned under in the spring as a green manure for cotton.

Outside the cotton-growing section of the South a satisfactory rotation is as follows: First year—corn, with cowpeas or soybeans sown at the last cultivation. Second year—fall-sown oats with clover or grass seeded in the oats. Third year— meadow or pasture.

Oats occasionally are grown in combination with other crops for either hay or grain. A common combination in many areas is oats and CANADA FIELD PEAS, which may increase both the yield and the feeding value of the crop, in addition to increasing the nitrogen content of the soil. BARLEY and an early oats variety combination also are grown.

Where RAPE will endure summer heat, it is sometimes sown with oats to be used as a pasture for hogs or sheep after the oats are harvested.

Frequently, too little attention is paid to the proper preparation of the seedbed for oats. The land usually is plowed and fitted for seeding by disking and harrowing. When fall plowing is not feasible, the land should be plowed as early as possible in the spring to give the soil time to settle and become firm and compact. The best seedbed for oats consists of from 2 to 3 in. of loose, mellow soil on the surface with a rather firm layer of soil beneath.

NOTE: Studies suggest that extra cost of good seed bed preparation will not always be returned in increased oat yield.

Seed oats should be fanned and screened. Fanning removes the light oats and trash, while the small oats and most of the weed seeds are removed by the screens. Wherever necessary, SEED TREATMENT is recommended to control SMUT.

Spring-sown oats should be sown early. This practice often permits the crop to escape injury from the effects of hot weather when the crop is nearing maturity. A good rule to follow is to sow oats in the spring, just as soon as the ground is in the condition to work. This can usually be facilitated by fall plowing. *Fall-sown oats* may be seeded in the South after October 1, in California after November 1.

Drilling is the best method of sowing oats; however, a considerable portion of the crop is still sown broadcast. With abundant moisture, shallow seeding from 1 to 1½ in. is sufficient; in loose, sandy loams, which loose their moisture easily, the seed should be placed at a depth of from 1½ to 2½ in. When oats are sown broadcast, the seed should be covered by a shallow disking or by harrowing the soil with a spike-tooth harrow.

Oats usually are cut with a grain binder or—especially if the crop is badly lodged or the straw is very short—sometimes with a mower.

Oat hay is obtained from a considerable acreage of oats, particularly in the Pacific area. Because of seasonal developments of weather or markets, oats sown for seed purposes are often harvested for HAY. When sown specifically for hay purposes, oats are frequently combined with peas, HUNGARIAN VETCH, or other vetches, which usually increase the yield of hay and improve its quality. When cut in late-milk or soft-dough stage and properly cured, oats make a very palatable and highly nutritious hay which is relished by all classes of livestock.

Hay from oats alone or from oats grown in combination with other crops is cut and cured similarly to other hay. Curing in the wind-row or cock is a common practice.

The proper time to cut oats for other uses than hay is when the grain is in the hard-dough stage and the straw still a little green. Earlier cutting usually results in shriveled grain and reduced bushel weight. If oats are allowed to become too ripe before binding, there is danger of loss by shattering.

Oats generally are shocked immediately after binding. The shock should be capped to prevent damage from rain and dew. If the oats are a little green or weedy, it may be advisable to let them cure in the bundle for a few hours or a day before shocking.

Oats are often stored in barns until threshed, but sometimes, because of lack of barn space, they are stacked. When they are stored in barns, threshing may be done when most convenient and usually after the oats have gone through the sweat and are cured completely in the mow.

When a combine (harvester-thresher) is used, the oats must be fully ripe. Ordinarily, combining should begin when all greenness has left the straw and the glumes turn a dull white. If combined earlier or when too wet, the grain may spoil in the bin.

Oat straw is of considerable value as forage and also as bedding for livestock. Usually, all oat straw is saved. It is protected from weathering either by being stored in barns or sheds, or by being carefully stacked. Due to its greater palatability, oat straw is preferable to other cereal straws as roughage˙ for livestock. Oat straw not needed for roughage should be converted into manure and returned to the land.

Clipped oats have a high bushel weight and lowered fiber content because the pointed ends of their hulls have been clipped off by an oat clipper.

Sprouted oats are sometimes fed to poultry.

A relatively small percentage of the oat crop is used for seed and for food in form of *rolled oats* (and *quick oats*) or *oatmeal*. These are manufactured from cleaned, hulled, polished, steamed, and finally rolled or ground kernels (groats). By far the largest portion of the oat crop, however, is used for feed.

Whole oats are often fed to horses, sheep, and poultry.

Oats in general have a better distribution of essential amino acids and B-complex vitamins than any other cereal. They are deficient in carotene, though.

Officially recognized are the following oat products:

Oat hulls—consisting of the outer covering of the oat.

Oat middlings—consisting of the floury portions of the oat groat obtained in the milling of rolled oats.

Oat shorts (or *oat dust*)—consisting of the covering of the oat grain lying immediately inside the hull, and portions of the floury part of the groat obtained in the milling of rolled oats.

Oat chop, ground oats, pulverized oats, crushed oats, and *crimped oats* consist of the entire product made by chopping, grinding, cutting, crushing, or crimping whole oats; it must not contain more than 10 percent of other grains, weed seeds, and other foreign materials. It is often preferred as feed for dairy cattle, breeding stock, and young stock.

Oat groats—the kernels produced from cleaned and dried oats in the process of manufacturing oat meal.

Hulled oats, or *undried oat groats*—the kernels produced from the undried grain in the process of hulling oats. (Considered equivalent to corn as hog feed).

NOTE: The oat groats and hulled oats are a very valuable feed for young animals. They are used in creep and starter rations for pigs and lambs.

Feeding oat meal—the broken rolled oat groats, oat groat chips, and floury portion of the oat groats, with only such quantity of finely ground oat hulls as is unavoidable in the process of milling; it must not contain more than 4 percent crude fiber.

Cut oat groats, cracked oat groats, and *ground oat groats*—produced by cutting, cracking, or grinding oat groats.

Clipped oat by-product—the light chaffy materials from the end of the hulls, empty hulls, light immature oats, and dust; it must not contain an excessive amount of oat hulls.

Oat mill feed—consisting of oat hulls, oat shorts, and oat middlings; it must not contain more than 30 percent crude fiber.

Oat mill feed is the most important oat by-product. It is high in fiber.

The principal outlet for oat mill feed is as a feed for cattle, sheep, and horses as a substitute for part of the grain allowance; it is generally worth somewhat less than half as much as corn or other cereals. In horses doing moderate work, oat mill feed has been used to replace both the hay and grain portion of the ration.

Oat huller feed—consisting primarily of oat hulls; it must not contain more than 34 percent crude fiber.

Dangers: Among the oat diseases are BLACK LOOSE SMUT, COVERED SMUT OF OATS, STEM RUST OF OATS, CROWN RUST, CULM ROT, VICTORIA BLIGHT, HALO BLIGHT, OAT LEAF SPOT, RED LEAF, DOWNY MILDEW, POWDERY MILDEW, BACTERIAL STRIPE OF OATS, and BLACKSTEM OF OATS.

Only some insect pests attack oats; e.g., LEAFHOPPERS, GRASSHOPPERS, the MORMON CRICKET, GREENBUG, CHINCH BUG, SOUTHERN CORN ROOTWORM, RED HARVESTER ANT, WESTERN HARVESTER ANT, the ARMYWORM, and STINK BUGS.

Immature oats cut for hay may cause OAT HAY POISONING. Oats may also cause GRASS TETANY.

FIELD BINDWEED is a WEED which is often found in oat fields.

Species: The most important oat species cultured in the United States are COMMON OAT and RED OAT.

Some oat species grow wild; e.g., COMMON WILD OAT, WILD RED OAT, SHORT OAT, SLENDER OAT, and SAND OAT.

References: S.16; S.17; S.18; S.19; S.20; S.21; T.6; M.30; E.12; F.6.

See also OAT PRODUCTS; CEREAL GRAIN BY-PRODUCTS; SHEAF OATS; VICTORIA OAT; HAIRY VETCH; KUDZU; RUSSIAN WILD-RYE; PEPPERGRASS; PROSO; RICE; SPELT; WHEATS; BUCKWHEATS; VELVETBEANS; ANNUAL LESPEDEZAS; CROTALARIAS; FIELD BEANS; DRIED CEREAL GRASSES; DODDERS; WHEAT JOINTWORM; WHEAT STEM SAWFLY; WHITEFRINGED BEETLES; PASTURES; SCREENINGS; SILAGE; SILAGE CROPS.

OAT SHORTS. *See* OATS.

OAT SILAGE. *See* SILAGE CROPS.

OAT SMUTS include BLACK LOOSE SMUT and COVERED SMUT OF OATS.

OAT STRAW. *See* OATS.

OBLONG is a figure having greater length than breadth and with the sides, sometimes gently rounded, approximately parallel.

ODERBRUCK BARLEY.
　　　　See SIX-ROWED BARLEY (variety).
ODESSA BARLEY.
　　　　See SIX-ROWED BARLEY (variety).
ODOR OF FEED is often of importance since the smell of feed influences its palatability. Cows may refuse to eat a certain hay or they may not utilize it sufficiently.

OFFAL is a term used to indicate animal or vegetable products not fit for human consumption, particularly (1) waste meat and (2) refuse or by-products of flour milling, such as BRAN, SHORTS, and MIDDLINGS.—*See also* ANIMAL PRODUCTS; WHEATS; BARLEYS; SOYBEAN.

OFFICIAL GRAIN STANDARDS.
　　　　See GRAIN GRADING.
OFFICIAL STANDARDS FOR HAY.
　　　　See HAY GRADING.
OGDEN. *See* SOYBEAN (variety).

OHIO 1. *See* SIX-ROWED BARLEY (variety).

OHIO C 38 is a CORN hybrid.

OHIO NO. 127 = *Fulhio. See* COMMON WHEAT.

OIL. *See* FATS.

OIL BARRIERS. NAPHTHALENE DRAIN OIL, PINE-TAR OIL, and other TAR OILS are used for the destruction of CHINCH BUGS. These *chemical barriers* are prepared and employed in the same manner as dirt-ridge CREOSOTE BARRIERS.

Reference: P.5.

... OIL CAKE, ... *oil chips, or* ... *oil flakes,* are tentatively recognized terms for the product remaining after removal of most of the oil from ground seed, cleaned of foreign materials. The name must include a term descriptive of the process of manufacture (hydraulic, expeller, or solvent extracted) and of its kind. It shall be designated according to its protein content.

Reference: F.6.

See also OILSEED MEALS; OIL MEAL; MISCELLANEOUS PRODUCTS; SORGHUMS.

OIL CHIPS. *See* OIL CAKE.

OIL FLAKES. *See* OIL CAKE.

... OIL MEAL (a tentatively accepted feedstuff term) is ... OIL CAKE, chips, or flakes ground to meal. The name must include a term descriptive of the process of manufacture (hydraulic, expeller, or solvent extracted) and of its kind. It shall be designated according to its protein content.

Reference: F.6.

See also OILSEED MEALS; MISCELLANEOUS PRODUCTS; SORGHUMS; MURUMURU OIL MEAL; PERILLA OIL MEAL; POPPY-SEED OIL MEAL; TUCUM-NUT OIL MEAL; UCUHUBA OIL MEAL; RUBBER SEED OIL MEAL.

OILS. *See* FATS.

OILSEED MEALS are BY-PRODUCT FEEDSTUFFS obtained as residues after removal of the oil particularly from SOYBEANS, PEANUTS, COTTONSEED and, to some degree, LINSEED. Small quantities of SESAME OIL MEAL, WHOLE PRESSED SAFFLOWER SEED, SUNFLOWER-SEED MEAL, RAPESEED MEAL, HEMPSEED MEAL, PALM KERNEL OIL MEAL, BABASSU OIL MEAL, and COCONUT OIL MEAL are produced. Other oilseed meals are of only limited local interest.

The oilseed meals are most important sources of PROTEIN for livestock. The oil of the seeds is removed either by hydraulic or screw presses or by extraction with organic solvents. All three methods are in general use and produce excellent feed.

The proteins of the soybeans and some other LEGUMES must be heated for maximum value to nonruminants; the heat may destroy antienzymes known to be present or may improve the availability of the protein, or may do both.

References: E.12; F.6.

See also SORGHUMS; OIL CAKE; OIL MEAL; MURUMURU OIL MEAL; POPPY-SEED OIL MEAL; AMINO ACIDS.

OKLA 44-14. *Combine White kafir Okla 44-14,* better known as *Dwarf kafir 44-14,* is one of the grain SORGHUMS.

OKLAHOMA 311 is a CORN hybrid.

OKLAHOMA COMMON ALFALFA belongs to the COMMON ALFALFAS.

OKLAHOMA No. 1 DARSO is one of the grain SORGHUMS.

OLD WORLD FLAX = FLAX.

OLEANDER. *See* COMMON OLEANDER.

OLIVE PULP, the by-product of the manufacture of olive oil from whole olives, has value as feedstuff only when it is first freed from the pits and *dried* before being fed to livestock.

OMASUM is the third STOMACH of a ruminant.

ONE-FLOWER VETCH

= MONANTHA VETCH.

ONE-WAY DISK PLOW is a useful plow in weed- or stubble-covered soils for seedbed preparation. It has a gang of disks that throw the soil to one side; it leaves the stubbles, weeds, straw, etc. well mixed with the surface soil, forming a rough surface which checks wind erosion.—*See also* SAGEBRUSHES.

ONION. *See* WILD ONION.

ONTARIO SOYBEAN.

See SOYBEAN (variety).

ONTARIO VARIEGATED.

See VARIEGATED ALFALFAS (variety).

OÖSPORE (pronounced: *oh*-oh-spore) is usually a *resting cell* which develops in nonflowering plants (CRYPTOGAMS) after sexual fertilization of a large, passive female cell by a small, active male cell.

The oöspore, whose hardened outer wall consists of cellulose, (usually) goes into a resting stage.

From the oöspores germinate the nonflowering plants, such as DOWNY MILDEW or related FUNGI and ALGAE which produce asexual SPORES.

OPEN FORESTS. *See* RANGE PLANTS.

OPHIOBOLUS. The FUNGUS *Ophiobolus*

oryzinus causes BLACK SHEATH ROT of rice; *O. graminis* causes TAKE-ALL.

OPIUM POPPY (*Papaver somniferum*) is an annual poppy which grows to 4 ft. with grayish-green foliage and large flowers of varying colors. It is used for the production of poppy-seed oil and its by-product, POPPY-SEED OIL MEAL.

OPPOSITE means: arranged in pairs, but on opposite sides (of a stem).

OPUNTIA. *Opuntia* spp. = PRICKLY PEAR CACTI.

ORANGE PULP, obtained as by-product in citrus-fruit canning factories, can be fed to livestock; the *dried orange pulp* is slightly laxative and less palatable to livestock than fresh pulp and should therefore be mixed with other feedstuffs.—*See also* DRIED CITRUS PULP.

ORANGE SORGO is a term used for various forage SORGHUMS (sorgos). *Red Orange* is a name erroneously used for *Colman sorgo,* and *White Orange* is erroneously used for *Sourless sorgo.* True Orange varieties are *Kansas Orange* and *Waconia Orange. See also* COWPEA.

ORANGE WHEAT = *Red May. See* COMMON WHEAT.

ORBIGNYA. *O. speciosa* = BABASSU.

ORCHARDGRASS (*Dactylis glomerata*), also known as *cocksfoot,* is a long-lived perennial, a distinctly bunch-type grass with folded leaf blades and compressed sheaths. Its tussock-forming habit is lessened somewhat by careful grazing management and by seeding with a legume, such as LADINO CLOVER, ALFALFA, or LESPEDEZA, particularly KOREAN LESPEDEZA. Both, orchardgrass and Korean lespedeza, thrive in summer and therefore extend the PASTURE season on land that would not maintain a growth of clover or alfalfa. It does not produce STOLONS or underground rootstocks (rhizomes) and, therefore, never makes a dense sod. The peculiar cluster formation of the INFLORESCENCE is characteristic and cannot be mistaken for that of any other cultivated grass.

Orchardgrass is widely distributed. Its persistency, leafiness, and ability to withstand relatively adverse soil and climatic conditions in the humid temperate regions of the United States make it a desirable pasture grass. The most extensive acreage stretches from southern New York State to southern Virginia and westward from the Atlantic coast to eastern Kansas and southeastern Nebraska.

Orchardgrass. Plant with spikelet and floret. (H.26.)

The grass flourishes on rich soil, but it also succeeds on light soil of medium fertility and on moist, heavy land. Orchardgrass is one of the best cultivated GRASSES for shade; when it is adapted, it is generally dominant in orchards, woodland pastures, and similar areas. It is quite cold-resistant and continues growth until the first severe frost. It can be seeded in early spring and is one of the first grasses to start growth. As a rule, it is necessary that the animals be turned on orchardgrass before seed heads form; otherwise the early spring growth becomes unpalatable for pasturage.

The practice of making grass silage has further demonstrated the value and usefulness of orchardgrass. When it is grown in combination with LEGUMES, such as RED CLOVER and Ladino clover, orchardgrass is able to produce the maximum tonnage of high-quality silage early in the season. If this early growth is removed, the orchardgrass will not crowd the legume, and its rapid recovery will produce an abundant, high-quality summer pasturage at a time when permanent pastures are dormant.

Grown alone, orchardgrass will average 1 to 2 tons of field-cured hay to the acre. From a combination of orchardgrass and clover or alfalfa, yields of 2 to 3 tons can be expected. In the seed-producing areas the growth remaining after seed harvest is used to good advantage for pasture or hay.

Nitrogen is probably the most effective fertilizer in increasing seed yields and quality. On soil of medium to good fertility, nitrogen appears to give the most effective results. Approximately 20 to 30 lb. of nitrogen are usually sufficient, except on poorer soil, or on old established fields of orchardgrass in which 40 to 50 lb. nitrogen may be necessary. Heavier applications often cause the grass to lodge and sometimes prevent the heads from filling out.

Dangers: Orchardgrass is not seriously troubled by insect pests. Diseases have been less severe in grazed fields than when the grass is allowed to grow to hay stage. The most serious diseases are ANTHRACNOSE, BROWN STRIPE, and STRIPE SMUTS.

Reference: H.1.

See also CHINCH BUG; TALL OATGRASS; ALSIKE CLOVER; SUBCLOVER; BIRDSFOOT TREFOIL; PASTURE PLANTS; SILAGE CROPS; HAY; SILAGE.

ORDER, in biological classification, is the division just above the FAMILY.

OREGON GRAY = *Winter Turf. See* COMMON OAT.

OREGON MARIOUT = *Club Mariout. See* SIX-ROWED BARLEY.

OREGON PEA = MUNGBEAN.

OREGON RYEGRASS
= COMMON RYEGRASS.

ORESTAN is one of the Turkestan ALFALFAS.

ORFED. *See* COMMON WHEAT (variety).

ORGANIC, as applied to chemistry, is that branch of science which deals with the CARBON compounds produced in plants and animals or, synthetically, in laboratories and industry.—*See also* ORGANIC COMPOUNDS.

ORGANIC ACIDS are ORGANIC COMPOUNDS showing the characteristics of ACIDS; e.g., CITRIC ACID.—*See also* FERMENTATION PRODUCTS; BACTERIA; ALCOHOL.

ORGANIC COMPOUNDS are substances formed by the union of CARBON and HYDROGEN, or carbon, hydrogen, and OXYGEN, or these CHEMICAL ELEMENTS together with others in definite proportions by weight.

ORGANIC MATTER in the soil is derived from vegetable material turned under.—*See also* GREEN MANURE.

ORGANIC SOLVENTS are ORGANIC COMPOUNDS which have the power to dissolve other organic substances; e.g., vegetable oil, gasoline, ALCOHOL, KEROSENE, CHLORINATED HYDROCARBONS, etc.—*See also* ALDRIN.

ORGANISM is a living thing that is organized in a definite pattern with a definite function and consists of an individual CELL or a complexity of cells.

"ORIENTALS" is the name given to one of the two types of the GREEN MUNGBEAN.

ORIUS. A blackish bug, *Orius insidiosus*, only $\frac{1}{16}$ in. long, is one of the most important INSECT enemies of the CORN EARWORM. It destroys eggs and small larvae of the earworm by puncturing them with its beak and sucking out the contents. Sometimes as many as 25 of these useful bugs may be found on a single corn plant.

Reference: B.9.

ORYZA. *O. sativa* = RICE.

ORYZOPSIS. *O. hymenoides* = INDIAN
RICEGRASS.

OS 420 is one of the inbred lines of CORN
hybrids.

OSAGE. *See* COMMON WHEAT (variety).

OSCEOLA VELVETBEAN is an intermediate-type variety of the DEERING
VELVETBEAN.

OSMOSIS is the passage of liquids, such
as blood, through porous substances, such
as membranes.—*See also* ABSORPTION.

OSTEOMALACIA is a bone disease characterized by softening and bending of the
bones due to mineral and/or vitamin
deficiencies.—*See also* CALCIUM DEFI-
CIENCY.

OTOE. *See* RED OAT (variety).

OTOOTAN. *See* SOYBEAN (variety).

OTTAWA. *Mandarin (Ottawa)* is the name
of a SOYBEAN variety.

OVAL means: broadly elliptical; it is not
a synonym for OVATE or OVOID.

OVARY is usually the basal cavity in the
PISTIL (female floral organ), in which the
rudimentary seeds are borne.—*See also*
GRASSES; PEG; PERICARP.

OVATE means: having the outline of a
hen's egg, with the broader end downward.
Ovate is a term used to describe *plane*
surfaces, such as a leaf; it is not to be
confused with OVOID or OVAL.

OVERGRAZING.

See RANGE MANAGEMENT.

OVERLAND OAT.

See COMMON OAT (variety).

OVOID means: shaped like a hen's egg
and with the broader end downward; this
term differs from OVATE in that it is used
in describing *solid* parts, such as a fruit.

OVULE is a rudimentary seed occurring
in the OVARY.

OWEN = *Goens.* See COMMON WHEAT.

OXALATE is any salt of OXALIC ACID.—
See also HALOGETON.

OXALIC ACID is an organic acid contained in notable quantities in the leaves
of BEETS and particularly in some POISON-
OUS PLANTS. Among them are HALOGETON
and GREASEWOOD; the latter's dried leaves
yield 9.4 percent of the toxic acid in form
of a salt (*oxalate*).—*See also* SILAGE CROPS.

OXIDATION is a chemical process involving the addition of OXYGEN or the
increase in the VALENCY of an ELEMENT.
The substance causing the oxidation is the
oxidizing agent.

OXYGEN is a gas. It exists in the earth's
atmosphere in free form diluted with nearly
four times its volume of NITROGEN, a variable and very small quantity of CARBON
DIOXIDE, about one-twentieth of its volume
of ARGON and traces of other inert gases.
In combined form, oxygen is very abundant in water, minerals, and the products
of plant and animal life. Oxygen is essential to the existence of animal life.—*See also*
AIR; SILICONE.

OXYTEMIA. *O. acerosa* = COPPERWEED.

OXYTROPIS. The *Oxytropis* spp. belong
to the LOCOWEEDS; *O. lambertii,* or *white
locoweed,* is one of the species which cause
true locoism.

OYSTER SHELL FLOUR is a CALCIUM
source. It consists of approximately 94
percent CALCIUM CARBONATE (and a small
percentage of a PHOSPHORUS compound).
The flour is widely used as a mixture for
poultry mash and MINERAL FEEDS for
livestock.

P

P = PHOSPHORUS.—*See also* MINERAL
FEEDS.

p = PARA.

P-762 = *Kanred.* See COMMON WHEAT.

PABA or *para-aminobenzoic acid,* is one of
the B-COMPLEX VITAMINS. It forms white,
water-soluble crystals. PABA is needed by
many organisms for GROWTH and counteracts the bacteriostatic action of SULFONA-
MIDES. It is widely distributed in nature;
a good natural source is YEAST. It is not
likely to be needed as a supplement for
animal feeds.

PACIFIC BUNCHGRASS RANGE is
of importance in eastern Washington and
Oregon, Idaho, and western Montana. It
consists of BLUEBUNCH WHEATGRASS, certain BLUEGRASSES, perennial FESCUES,
GIANT WILD-RYE, NEEDLE-AND-THREAD
GRASS, etc. It is grazed by both cattle and
sheep in the fall, winter, and spring.—*See
also* BUNCHGRASSES; RANGE PLANTS.

PACIFIC RYEGRASS

= COMMON RYEGRASS.

PACKING HOUSE BY-PRODUCTS are among the most important BY-PRODUCT FEEDSTUFFS. They are officially classified as ANIMAL PRODUCTS which include also those obtained from *poultry dressing plants, rendering plants*, and *whale fisheries*. Some of these by-products—those consisting chiefly of *bones*—are considered MINERAL FEEDS.

PADDOCK is a small, enclosed pasture lot near the stable.

PADDY is *rough rice.—See also* RICE.

PAGODA. *See* SOYBEAN (variety).

PALATABLE means: agreeable to the taste.

PALATABILITY is an expression of the relative relish with which food is consumed; specifically, the degree to which the herbage within easy reach of livestock is grazed when a range is properly utilized and managed. The percentage of the herbage grazed determines the palatability of the species.

Dairy cattle and animals being fattened should be fed only *palatable feeds*. Unpalatable feed should be mixed with sufficient amounts of palatable feed so as to induce consumption; or it may be tried as feed on another class of farm animal, since taste for certain feedstuffs often varies with animal species.

RYE is not palatable to dairy cows. It is often ground and mixed with CORN to get animals to eat it readily.

PALEA, or *palet*, is a thin, chaffy BRACT; it occurs, for instance, in a grass FLORET opposite the LEMMA, forming the upper of two leaves which enclose the flower.—*See also* GRASSES.

PALE WESTERN CUTWORM (*Agrotis orthogonia*) is a serious pest in the small-grain areas of the southern Great Plains, and of longstanding importance in the spring-WHEAT region farther north.

Control: The pale western cutworm works mostly underground and, therefore, cannot be satisfactorily controlled with the poison baits. In the spring-wheat region it is controlled by early spring starvation of the newly hatched worms; that is, by thorough cultivation of the wheat-stubble fields (to destroy all green vegetation as soon as the weeds and volunteer

grain show 1 to 2 in. of growth) and, after a delay of ten days, seeding a spring grain crop. In the southern Great Plains, winter wheat sown on land that has been cleanly fallowed during the preceding summer and that had been planted during the previous year to a row crop such as SORGHUM, almost always escapes serious injury.

Reference: P.16.

PALM. *See* BABASSU.

PALMATE = DIGITATE.

PALMETTO. *See* SOYBEAN (variety).

PALM KERNEL OIL MEAL is the ground residue obtained after the extraction of part of the PALM OIL by pressure or solvent from the kernel of the fruit of *Elaeis guineensis* or *E. melanococco*.

This officially recognized OILSEED MEAL is similar in composition to COCONUT OIL MEAL. It is fed mainly to dairy cattle but also to pigs; it is not palatable to the latter and, therefore, must be well mixed into their rations.

Reference: F.6.

See also ELAEIS.

PALM OIL is obtained from the fruit or the seed of two *Elaeis* spp.—*See also* PALM KERNEL OIL MEAL; PALMO MIDDLINGS.

PALMO MIDDLINGS, or *"palmo midds"*, consists of wheat middlings and 7 to 10 percent palm oil (obtained as industrial by-product of the scouring process of tin plates). Palmo middlings may be fed to swine or other livestock; it rates about the same or slightly less than standard wheat middlings.—*See also* PALM KERNEL OIL MEAL.

PANCREAS, or *sweetbread*, is a gland located below the stomach; it produces insulin and secretes PANCREATIC JUICE which plays an important role in the digestion taking place in the small intestine.

PANCREATIC JUICE, secreted from the PANCREAS, contains a number of ENZYMES, among them TRYPSIN and LIPASE.—*See also* DIGESTIVE TRACT.

PANGOLAGRASS (*Digitaria decumbens*) is adapted to Georgia, Florida, and California. It is a rapid-growing, vigorous, leafy perennial which prefers sandy soil, requires fertilization, and must be propagated vegetatively.

Pangolagrass is drought-resistant but

not frost-resistant; it is liked by cattle, and withstands grazing. It gives high yields of hay that cures rapidly.

Reference: W.16.

See also GRASSES.

PANICGRASSES (*Panicum* spp.) grow chiefly in the warm regions of the world. The species native to the United States occur primarily in the Southeast but are well represented also in the warmer parts of the West.

Species: VINE-MESQUITE and SWITCH-GRASS (with its improved variety *Blackwell*) are more important GRASSES than GUINEA-GRASS, PARAGRASS, BLUE PANICGRASS, and PROSO.

Reference: H.1.

Illustrations: See PARAGRASS; VINE-MESQUITE.

PANICLE is a compound INFLORESCENCE in which the lower branches are typically longer and blossom earlier than the upper branches.—*See also* GRASSES.

A panicle. (D.9.)

PANICUM. Among the 500 *Panicum* spp. are *P. obtusum* = VINE-MESQUITE; *P. virgatum* = SWITCHGRASS (and its improved variety called *Blackwell switchgrass*); *P. maximum* = GUINEAGRASS; *P. miliaceum* = PROSO; *P. purpurascens* = PARAGRASS; and *P. antidotale* = BLUE PANICGRASS (also called *giant panicum.*)

Reference: R.1.

See also GRASSES; PANICGRASSES.

PANOGEN contains 2.1 percent *methyl mercury dicyan diamide* in a liquid carrier.

This volatile mercurial is used for SEED TREATMENT of the SORGHUMS by the quick-wet method. When applied properly to sorghum seed at the rate of ¾ oz. per bushel, it improves emergence and stand, and controls SMUTS, particularly GRASS SMUTS.

Caution: Panogen is poisonous. Avoid breathing the fumes or getting the fungicide on the skin. Treated seed must not be used for feed or food.

Reference: L.1.

PANTOTHENIC ACID, one of the B-COMPLEX VITAMINS, occurs in animal tissues, liver and kidneys. It also occurs in plants; for instance, ALFALFA, WHEAT bran and wheat shorts are good sources, and grain SORGHUMS contain more of it than does corn.

Pantothenic acid is a pale-yellow oil which is soluble in water. Ruminants are able to synthesize this vitamin with the help of the rumen bacteria. A deficiency in pigs causes "goose stepping" in young animals and poor reproduction in mature animals. Deficient chicks have a dermatitis on the top of the feet and on the beak.

—*See also* MOLASSES; CALCIUM PANTOTHE-NATE.

PAPAVER. *P. somniferum* = OPIUM POPPY.

PAPILIONACEOUS is a term used to describe the butterfly-like shape of the petals of leguminous plants.

PAPILIONOIDEAE, more often called Faboideae, are a subfamily of LEGUMES.

PARA-, or *p*-, is a prefix indicating the 1,4- position in ORGANIC COMPOUNDS consisting of the atoms arranged in cyclic (ring) form.

PARA-AMINOBENZENEARSONIC ACID = ARSANILIC ACID.

PARA-AMINOBENZOIC ACID = PABA.

PARACASEIN is the CASEIN product prepared from fresh skim milk by the use of RENNIN.

PARAFFIN is a mixture of solid HYDRO-CARBONS; it is obtained from PETROLEUM.

PARAGRASS (*Panicum purpuracens*) is one of the PANICGRASSES. It grows in Florida, Texas, and the Gulf Coast region.

Paragrass is found on damp, heavy soils and is not injured by overflowing water; it is killed by temperatures below 18° F.

This GRASS grows 4 to 5 ft. high and is propagated vegetatively. It provides good grazing and withstands trampling. It gives good yields of high-quality hay if cut before it becomes woody. Paragrass can also be used as soiling crop. Sod-binding can be prevented by plowing under, in early summer, every few years.

Reference: W.16.

Paragrass. Plant, two views of spikelet, and floret. (H.26.)

PARAGUAY BAHIAGRASS.
                    See BAHIAGRASS (variety).

PARAKERATOSIS is a pig disease caused by ZINC deficiency.—*See* MINERAL FEEDS.

PARA RUBBER MEAL
                    = RUBBER SEED OIL MEAL.

PARA RUBBER TREE (*Hevea brasiliensis*) is the most important source of latex (the raw material of the rubber industry). The trees grow in South America and in South Asia; they may reach a height of 45 ft. An oil is obtained from their seeds, with RUBBER SEED OIL MEAL as a by-product.

PARASITE is any plant or animal which obtains its sustenance from another organism.—*See also* DODDERS; ECTOPARASITES.

PARASITIC means: living in or on another living organism. LIGHTING INJURY and MINERAL DEFICIENCY may be mistaken for *parasitic diseases* of plants.—*See also* PARASITE.

PARATHION is an organic INSECTICIDE containing SULFUR and PHOSPHORUS. The pure compound is a pale-yellow liquid very slightly soluble in water. The technical grades of parathion contain 95 to 97 percent active ingredient, are dark brown in color, and have a garlic-like odor. Formulations of parathion are commercially available as wettable powders, emulsifiable concentrates, dusts, and AEROSOLS.

For the control of the YELLOW CLOVER APHID, a spray delivering 3.2 oz. parathion to the acre is very effective. This spray may be used on hay but not on seed of alfalfa.

A dust containing 2 percent parathion applied at the rate of 25 lb. per acre kills STINK BUGS.

To kill larvae of the ALFALFA WEEVIL, dust or spray the crop with 4 oz. parathion per acre.

Caution: Parathion is very toxic to animals. Because it can be absorbed through the skin, lungs, or digestive tract, it is necessary to observe all possible precautions.—*See also* GRASSHOPPERS.

PARIS GREEN is a complex compound containing COPPER ACETATE and COPPER META-ARSENITE at the rate of 1:2 to 1:3; modifications of Paris green are also available. To obtain best results, the amount of ARSENIC (from the copper meta-arsenite contained in the product) must be known.

Paris green is valuable as INSECTICIDE and FUNGICIDE. It is used in POISONED BRAN MASH for the control of GRASSHOPPERS, ARMYWORMS, and CUTWORMS.

Caution: Paris green is a violent poison which must be handled with care.

PARRY ASTER (*Aster parryi*) is a POISONOUS PLANT found on ranges. Hungry sheep eating about 1¼ lb. green leaves and stems in a day may show these symptoms of poisoning: weakness, prostration, rapid weak pulse, increased urination, cyanosis.—*See also* SELENIUM POISONING; ASTER.

Parry aster. (B.29.)

PARSNIP (*Pastinaca sativa*) is a root crop. The biennial plant grows best in rich, moist soil. Its long, thick, white roots are sweet; they are rarely used as feed for cattle in this country.

PARTIALLY DEFLUORINATED PHOSPHATE. *See* MINERAL FEEDS.

PASPALUM. Among the *Paspalum* spp., or PASPALUM GRASSES, are *P. notatum* = BAHIAGRASS (or *bahia*), *P. dilatatum* = DALLISGRASS, *P. malacophyllum* = RIBBED PASPALUM, *P. urvillei* = VASEYGRASS, and *P. distichum* = KNOTGRASS.

PASPALUM GRASSES (*Paspalum* spp.) are primarily pasture GRASSES. Some species are short-lived, but Dallisgrass and

Bahiagrass maintain good stands and remain productive for a long time if properly fertilized and managed. To establish new stands, seedings should be made in the spring after corn-planting time (the latter part of April or early May). A good seedbed is essential. Lime and fertilizer should be applied before planting and also in later years. Dallisgrass requires greater amounts of plant food for growth than does Bahiagrass.

Species: Of the 400-odd species of *Paspalum*, the few that have economic importance in the United States are BAHIAGRASS and DALLISGRASS; VASEYGRASS, KNOTGRASS, and RIBBED PASPALUM are of minor value.

Reference: H.1.

PASTEURIZED DAIRY BY-PRODUCTS, such as skimmed milk, buttermilk, whey, are preferred to unpasteurized by-products (of creameries or cheese factories) when used for feeding purposes. Pasteurization (at 180° F.) kills disease-producing bacteria and reduces the danger of causing scours or spreading infections.

PASTINACA. *P. sativa* = PARSNIP.

PASTURE. *See* PASTURES.

PASTURE EVALUATION.

                    See PASTURE MANAGEMENT.

PASTURE MANAGEMENT is the planning for the use of PASTURE PLANTS when they are most palatable and nutritious; if that cannot be done by GRAZING, then a part of the PASTURE should be cut for HAY or SILAGE.

Provision should be made for adequate shade, water and rest areas, and for maintenance of good fences and cross-fences.

For highest returns from pastures, the best soil available should be selected. Planting pasture on droughty or poorly drained land is usually poor economy. Location of the water supplies, drainage ditches, and fences will influence the choice of land. Convenience of management should influence the picking of a suitable area.

After deciding on the species and varieties of pasture plants to be grown, it becomes a matter of establishing them and maintaining them in a productive condition. Experience supports the idea of using companionable components in simple mixtures (a GRASS or two, with one or two LEGUMES). On most farms there is probably a need for several simple mixtures, each grown primarily for a rather specific purpose; i.e., permanent pasture, hay, silage, aftermath grazing, late fall grazing, etc.

Feed is the largest cost item in the production of animal products. The proportion of the feed cost to the total cost of production ranges from about 50 percent for dairy cattle to 85 percent for beef cattle. A good pasture, properly managed, is recognized as the cheapest source of feed, largely because the animals themselves harvest the pasture crop. Other feeds must be harvested, stored, processed, and fed to the animals.

In *pasture evaluation* the amount of grazing consumed is calculated at market price; i.e., the price at which the feed could be sold or purchased or the cost of buying an equal amount of feed, whichever is cheaper.

Permanent pastures should be fully utilized in spring when the grasses are most nutritious.

NOTE: The *Hohenheim system* of management of permanent pastures, so called because it was first introduced at Hohenheim, Germany, utilizes two main practices; (1) rotating the cattle over separate parts of the same pasture, and (2) liberal application of fertilizers, particularly those carrying nitrogen. Division of the herd, in accordance with the quantity of milk produced, and giving the highest-producing cows the first chance at the fresh pasturage, and harvesting hay from some of the pasture fields early in the season are integral parts of the system. It is primarily intended for dairy cattle, but other livestock may also benefit. However, the differences in soil, climate, and economic conditions in Germany and in the United States have to be considered. On most farms in the United States an increase of 10 percent in the yield of nutrients (which is all that can be obtained from pasture by rotation grazing according to the Hohenheim

system) would not be enough to justify constructing permanent division fences and providing the necessary shade and water in each pasture. It does, however, justify the construction of a cheaper type of fence.

Renovation, rotation, and *irrigated pastures* must become well established. This means only moderate grazing the first year. Thereafter, care should be taken to prevent too intensive utilization, and grazing too closely, especially during dry periods.

The *grazing season* depends largely upon the length of the growing season and the type of pasture. Too often the season is extended at the expense of both pasture and livestock production. Permanent pastures of turf-forming grasses, such as KENTUCKY BLUEGRASS or BERMUDA-GRASS-WHITE CLOVER, can withstand heavy and rather intensive use without injury to the stand, if plant food requirements are maintained, but rotation or renovated pastures must be protected from grazing early in the spring or late in the fall. For this reason an improved permanent pasture, together with an *annual* or *supplementary pasture*, can be used to good advantage in protecting the rotation or renovated pastures.

Although forage should be utilized in a relatively immature stage in order to obtain the most nutritious feed, the pasture *plants* should be given due consideration at all times. Most species, particularly of the taller growing types, must be given periodic opportunities to recover. Too heavy utilization may reduce yields, weaken many of the more desirable species, and permit weeds and undesirable species to enter. In general, the competitive ability of a plant in mixture depends upon its height and density, and no two species are equal in their competitive abilities at all times of the year. Thus, grazing must be managed to subdue the more aggressive species by grazing or clipping at the time of strongest growth.

Mowing to control brush and weeds has increased forage yields as much as 50 to 60 percent, and the quality of the forage is materially benefited. The surplus growth can be utilized to good advantage by dry stock and light feeder cattle that would otherwise require harvested roughage.

Nearly all *soils* in the humid sections of the United States are deficient in calcium, phosphorus, potassium, and nitrogen. Adequate *fertilization* should, therefore, be given high priority in the pasture program. Most irrigated pastures of the drier sections are highly productive, but this means that the drain upon soil resources is heavy.

Commercial fertilizers are generally relied upon to furnish plant food to pastures. Although barnyard manure is ordinarily used on cropland, its value for pastures should not be overlooked.

Heavy applications of nitrogen help the growth of grasses but at the expense of the legumes in mixtures, whereas phosphorus and potassium help the growth of legumes. Nitrogen fertilizers are seldom used on mixtures of grasses and legumes except in small amounts on new seedlings. However, it is recommended to inoculate all CLOVER seeds, and to seed them immediately after the inoculum is applied.

Fertilizers are not fully effective unless they are mixed with the surface soil. Too often the sod is not torn up enough to form a fine, firm seedbed. Considerable improvement can be made by top dressing many permanent pastures. This will encourage the growth of the more desirable pasture plants and increase the nutritive value of the forage. In southern states, each pound of plant food, properly used, increases daily production equal to 4 lb. milk or ½ lb. in live-weight gain.

Pastures may be utilized in the following manner: A 5-year rotation plan for CORN, WHEAT, and pasture is employed to effectively furnish grazing throughout the pasture season and to provide reserve feeds in the form of silage and hay. Five fields are maintained, so that in each year one of the fields is in corn, one in wheat, and three in first, second, and third-year pasture. The first-year pasture is grazed rotationally throughout the season. From the second-year pasture, the first crop and sometimes the first two crops (depending upon the need for pasture) are taken for hay or silage and the aftermath is grazed. The third-year pasture is grazed early for a few

days before the permanent pasture is ready, followed by the removal of a crop of hay; it is then rotationally grazed during the rest of the season.

The order for grazing is: A few days of early spring grazing on the third-year pasture, from there to the permanent pasture and first-year rotation pasture. The first crops of the renovated pastures are either grazed or removed for hay or silage. By the time the permanent pasture drops in production, the second and third-year rotation pastures and the renovated pastures are usually ready for grazing.

The need for *supplemental feeding* of dairy cows on pasture should be kept at a minimum by a program that provides a continuous and adequate supply of nutrients through high-quality pasturage. Cows on good pasture will eat about as much roughage as they can handle well.

It may be advisable to feed the high-producing cows some supplemental grain, or dry hay may be kept before the cows at all times; when, however, it is necessary to feed supplemental hay because of pasture shortages, costs of production rise rapidly.

NOTE: Feeding hay to cattle on pasture can be a wasteful procedure. In experiments, the cows ate less grass and replaced it with hay.

In many sections of the RANGE country pasturage can be held in reserve for maintaining livestock during the fall and winter seasons. The same is true to a lesser extent in the bluegrass section of the Middle West and in the Appalachian regions. Many of our native grasses, such as BUFFALOGRASS and BLUE GRAMA, will, under deferred grazing, provide winter maintenance with a minimum of supplemental concentrates. Successful livestock men have such *reserve pastures* for feeder or light cattle.

References: W.26; W.30; W.31; W.32; B.23; S.29; H.38.

See also POISONOUS PLANTS; WATER.

PASTURE PLANTS, when short, young, and low in fiber content, are preferred by livestock. A dense sward of about 4 in. in height is eaten readily.

The grasslands, hay lands, and forested RANGE lands of the entire United States cover more than a billion acres, nearly 60 percent of the total land area. Two-thirds of this land is privately owned. The rest, mainly in the dry and mountainous parts of the western states, is publicly owned. More than half of the farms and ranges of the country depend largely on grassland for feed.

PASTURE PLANTS

Most abundant grasses and legumes comprising the forage in the grazing regions of the United States

Grasses	Legumes
North Central and Northeastern	
Kentucky bluegrass	White clover
Timothy	Korean lespedeza
Redtop	Sweetclover
Orchardgrass	Alfalfa
Canada- bluegrass	Common lespedeza
Tall oatgrass	Red clover
Meadow fescue	Alsike clover
Smooth brome	Hop clover
Sudangrass	Black medic
Ryegrasses	Crimson clover
Bentgrasses	
Reed canarygrass	
Bluestems	
Southern	
Bermuda-grass	Common lespedeza
Carpetgrass	Hop clover
Johnsongrass	White clover
Dallisgrass	Persian clover
Vaseygrass	Black medic
Redtop	Spotted bur-clover
Bahiagrass	
Rescuegrass	
Rhodesgrass	
Great Plains	
Gramas	Sweetclover
Wheatgrasses	Alfalfa
Buffalograss	
Brome grasses	
Galleta, tobosa, and	
curly mesquite	
Bluestems	
Sudangrass	
Bluegrasses	
Timothy	
Redtop	
Rescue	
Rhodesgrass	
Intermountain	
Wheatgrasses	Alfalfa
Gramas	White clover
Brome grasses	Sweetclover
Galleta, tobosa, and	Alsike clover
curly mesquite	Red clover
Bluegrasses	Black medic
Timothy	California bur-clover
Redtop	Strawberry clover
Bermuda-grass	
Bluestems	

Grasses	Legumes
North Pacific Coast	
Ryegrasses	Red clover
Kentucky bluegrass	White clover
Bentgrasses	Hop clover
Reed canarygrass	Alsike clover
Orchardgrass	
Meadow fescue	
Redtop	
Timothy	
Tall oatgrass	
Meadow foxtail	
South Pacific Coast	
Fescues	California bur-clover
Brome grasses	Alfalfa
Common wild oats	White clover
Bermuda grass	Black medic
Sudangrass	

The type of grazing land, its carrying capacity, and seasonal use vary according to topography, soil, and climate, all of which govern the number and kind of species constituting the forage.

The same physical and biological factors influence, though perhaps in varying degrees, the nutritional value of plants in farm pastures and on the range. In some areas and in certain seasons weeds constitute an important part of pasturage. Range forage is usually a mixture of herbaceous species, mainly native GRASSES and LEGUMES, but often includes an admixture of SEDGES, RUSHES, and other grasslike plants. The forage gathered from shrubs or trees is called *browse;* nuts that have fallen from trees and are used as feed are called *mast.* Mast and browse are important chiefly on ranges.

There are six definite *grazing regions* in the United States; the pasture plants in each of these regions are listed in the table.

It is seldom advisable to seed pasture land to a single plant species. Mixtures of grasses and legumes provide better balanced feed, more uniform seasonal production, higher yields, and savings on nitrogen fertilizer.

Besides providing such cheap succulent feed, pasturage is of great importance as a source of proteins, minerals, and vitamins. When the plants are immature they are much richer in protein than at a later stage of their growth. The young plants are softer, more tender, less fibrous, and more easily digestible. The stage of maturity is, therefore, the most important factor determining the value of the pasture.

Most, if not all, of the minerals of importance in animal nutrition are found in pasture herbage—to what extent depends largely on the differences in the soil, environment, and the inherent ability of the species to extract nutrients from the soil. Other factors are rain, growth period, time of grazing (or cutting), light intensity, temperatures, and type of fertilizer used.

Available moisture is probably the greatest single factor influencing the yield. Most species become more palatable, digestible, and nutritious under favorable moisture conditions. Many of the grasses of the dry Great Plains region, however, are able to retain their palatability and nutritive value at mature stages of growth when their moisture content is low.

Temperature is also an important influence on the nutritive value of many perennial pasture grasses. As the temperature rises during the summer months, production decreases and the ratio of stem to leaves increases, with an increase in crude fiber and a decrease in crude protein. High temperature reduces the quality to the greatest extent in the north humid regions.

Although grasses grown in the South are adapted to and withstand high temperatures, they too must have ample moisture and fertile soil.

References: W.28; C.15; S.27.

See also PASTURES; PASTURE MANAGEMENT; POISONOUS PLANTS; SAGEBRUSH; WEED.

PASTURES are areas of land covered with GRASSES, LEGUMES, or other forage plants used for GRAZING animals or fowls. The pastures provide about one-third of the nutrients used by dairy cattle, and about half those used by beef cattle and sheep. Livestock on the range get much more than half their feed from range forage.

Good pastures have other advantages—they provide more digestible protein per acre than most other feeds; labor is saved; wind and water erosion loss is reduced to a minimum; manure is returned to the land. Pastures, if properly managed, promote the health of animals; sanitation and control of parasites are easier than on dirt lots.

Farmers and ranchers are learning that grass is a crop to be grown with the same consideration given to any other crop. It will produce a profit equal to or exceeding

Principal source of grazed forage. Although native forage plants may furnish most of the feed in Region (*Ba*), tame grasses and grazed crops have increasing importance and may furnish more feed in some areas. (U.S.D.A.)

that of any other crop we can grow on dry or irrigated land.

In most regions of the United States pasture grasses do not grow at a uniform rate through the growing season. Usually the most rapid growth is during the spring. During this period most farms have more grass than can be consumed by the livestock, and a considerable amount of the grass matures. It is therefore essential to preserve the surplus growth of grass when it is at its best stage of growth, so that it may be fed at other seasons.

The term *grassland farming* refers to the use of grasses and legumes in a cropping system devised to fit the needs of the individual farm or ranch. Grassland farming makes full use of grasses and legumes in crop rotations and pasture improvement, in the production of more livestock products, and in achieving higher and more profitable crop yields. It is the most practical means of rebuilding soils while producing a profit at the same time. In fact, it provides a year-around farm income without much farm labor.

Another source of farm income connected with grassland farming is the production of grass seed and legume seed for sale.

Pastures have as many names as they have uses:

1. Tame pastures are lands covered with cultivated plants. The chief types of tame pastures are the following:

(a) *Permanent pastures*, also called IM-PROVED PASTURES or PLOWABLE PASTURES, are covered with perennial or self-seeding plants, and are kept for grazing indefinitely. They rarely require plowing or cultivation. They comprise a rather large percentage of pasture area on farms. Nearly every farm has some land not suitable for cultivation which is better left in permanent pastures. Much of such turf is thin, weedy, low in productivity, and otherwise run-down. Top dressing with phosphate, potash, and sometimes with nitrogen is one method of improvement; however, most of the increased production would result in increased spring grazing, with little improvement in midsummer production.

Ordinarily, the permanent pastures should be used only for grazing, because the plant species usually present are not easily harvested.

(b) *Rotation pastures* are established on the better cultivated croplands and are a part of the regular crop rotation. They are planted to quick-growing grasses and legumes for hay or grazing. Production in these pastures throughout the growing season is usually higher than it is in permanent pastures.

Rotation pastures are recommended for most dairy farms. A suggested rotation is one year of row crops, one year of SMALL GRAIN, and two or three years of pasture combined with hay and silage. Rotation pastures are especially productive in summer if they are seeded with nutritious, tall-growing, drought-tolerant species. These mixtures, for the North, usually contain such combinations as BROMEGRASS-AL-FALFA, ORCHARDGRASS-LADINO CLOVER, and (on poorly drained soils) REED CANARY-GRASS-Ladino clover.

Rotation grazing (where the animals are allowed to graze for a time and then moved to another area) is a desirable practice if the pasture just grazed is permitted to recover. Such rotation grazing is necessary with rotation pastures planted to the rapid-growing grasses and legumes such as brome-alfalfa, or orchardgrass-Ladino clover; a greater production is obtained and the stands of the desirable pasture plants are maintained over a longer period. Since more pastures of higher quality are available in summer, rotation pastures go a long way toward easing the midsummer feed shortage; if they are properly handled, considerable reserve feed of high quality, in the form of hay or silage, is obtained besides.

(c) *Irrigated pastures* are figuring more and more prominently in the dairy-pasture programs of the West. These pastures not only increase summer production but they are generally more economical than harvested feed or feed purchased. With an adequate water supply, proper fertilization, and adapted mixtures, herbage can be kept growing vigorously, and hence palatable and nutritious. Furthermore, irrigation can help maintain the more desirable species in the pasture sward.

Under proper management, 100 acres of irrigated pastures will carry as many animals, during the six to seven months of the summer, as 1,000 acres of range will support during the other five to six months.

(*d*) *Renovated pastures* have come into use more recently. Renovation increases the production of run-down permanent pastures by disking or other cultivation, by application of lime and other fertilizers, and by seeding rapid-growing and good grasses and legumes without subjecting the land to cropping.

The advantages of renovated pastures are several. Total production is greatly increased, the distribution of the seasonal production is superior, and the quality of the forage is improved. Renovation allows for an ideal seedbed, with an abundance of organic matter in the surface—an aid in establishing the seeded species and preventing erosion.

Renovation keeps the area involved out of production only in the spring, during peak pasture growth, when the farmer ordinarily has more pasture than he can handle efficiently anyway. The area usually will furnish grazing when it is most needed later in the season after permanent pastures start their decline in production. Renovated pastures continue to show improvement for at least two or three years.

Some of the more important tall-growing species commonly seeded in renovated pastures include Ladino clover, red clover, alfalfa, brome grass, and orchardgrass. To meet specific conditions, BIRDSFOOT TREFOIL, SWEETCLOVER, ALSIKE CLOVER, LESPEDEZA, TALL FESCUE, MEADOW FESCUE, reed canarygrass, and TALL OATGRASS are sometimes used.

(*e*) *Supplemental pastures*, also called *temporary pastures*, include several types. The second growth of a harvested hay meadow, if grazed, may be called *aftermath pasture*. Harvested fields of small grains and corn are often used as supplementary pastures and referred to as *stubble pastures;* perennial legumes, such as KUDZU and others in the South, are used to supplement the permanent and annual pastures.

Some dairymen depend upon annuals for supplementary pastures; SUDANGRASS is one of the best and most common crops used for the purpose. Small grains, soybean, sweetclover, and various millets are also often used as supplementary pastures.

In the South, supplementary pastures have become an integral part of the year-round grazing program. Small grains, particularly OATS, or RYEGRASS in combination with such legumes as CRIMSON CLOVER, VETCH, and WINTER PEAS, are becoming increasingly popular as winter pasture. A popular winter-summer combination is oats and ANNUAL LESPEDEZA; the oats may be grazed off completely or harvested for grain. Sudangrass, JOHNSONGRASS, and kudzu are used for summer grazing. The use of these supplementary crops, with permanent pastures of DALLISGRASS or CARPETGRASS on the lowlands and BERMUDA-GRASS with WHITE CLOVER, PERSIAN CLOVER, and HOP CLOVER on the uplands, will go far toward meeting the grazing needs of Southern dairymen.

(*f*) *Annual pastures* are used for grazing during a short period, not more than one crop season. The cereal crops, such as WHEAT, oats, and RYE, or Sudangrass, SOYBEANS, and PEARLMILLET are rather widely used on these pastures.

2. Native pastures, also called *wild* or *natural pastures*, are areas covered with native plants useful for grazing. When they cover an extensive area they are generally referred to as *range pastures*. Also included in this classification are the *shrub* or *brush pastures*, covered mainly with shrubs or browse plants; *woodland pastures*, on which grass or other forage grows among trees; and the *stump* or *cut-over pastures*, on land where the forest has been removed.

References: W.26; C.15; S.26; R.12; W.27; G.15.

See also PASTURE PLANTS; PASTURE MANAGEMENT; RANGES; POISONOUS PLANTS; SAGEBRUSH; WEED; DROUGHT; HAY; SILAGE.

PASTURE SAGEBRUSH
= FRINGED SAGEBRUSH.

PATHOGENS are micro-organisms which produce diseases.

Pathogenic, or *pathogenetic*, means: causing disease.

PATHOLOGY is the science which deals with diseased conditions. *Plant pathology* is concerned with plant diseases, their causes, effects, and treatment.

Pathological means: relating to pathology; due to disease.

PATNA. *Texas Patna* is a RICE variety.

PATOKA. *See* SOYBEAN (variety).

PAUNCH = *rumen*. *See* STOMACH.

PAWNEE. *See* COMMON WHEAT (variety).

PEA is a term used to indicate the FIELD PEA or other *Pisum* spp. or other crops not related to this genus, e.g., the ROUGHPEA and the COWPEA.—*See also* SOYBEAN; RYE; WHEAT; LEGUME BACTERIA; INOCULATION; PEA PODS; PEA VINES; CHARCOAL ROT; PLANT BY-PRODUCTS; SILAGE CROPS.

Wild pea = NARROWLEAF VETCH.

PEA ANTHRACNOSE, caused by the FUNGUS *Colletotrichum pisi*, attacks all above-ground parts of the FIELD PEAS. On the leaves the spots are irregular in outline and brownish in color, with somewhat darker borders. The affected areas on the pods are circular and sunken, and on the stems they are elongate. Pea anthracnose is seldom of much importance.

No control measures have been worked out, but crop rotation is helpful.

Reference: M.24.

See also ANTHRACNOSE.

PEA APHID (*Macrosiphum pisi*) injures garden peas as well as forage LEGUMES, such as FIELD PEA, CRIMSON CLOVER, RED CLOVER, other CLOVERS and ALFALFAS, by sucking the sap from the leaves, stems, blossoms, and pods. Even a few aphids may kill small plants and stunt larger ones. The pea aphid may also spread VIRUS DISEASES— eg., RED CLOVER VEIN MOSAIC— thus causing further damage to the plants.

The adult is a light-green, soft-bodied insect; it may or may not have wings. The *winged aphids* fly into the fields early in the spring and produce living young, which look like the *wingless* adult APHIDS. A single adult produces 10 to 14 young each day, and the young themselves begin to produce in one to two weeks. When food is plentiful, most of the adult aphids are wingless. When food conditions are un-

favorable, winged forms develop and fly to other fields. In the South this cycle continues throughout the year. In the North egg-laying adults develop in the fall and lay their black, shiny eggs on alfalfa or clover. In some climates only the eggs survive the winter.

Control: Dust or spray with ROTENONE or DDT.

Reference: U.3.

See also VIRUS.

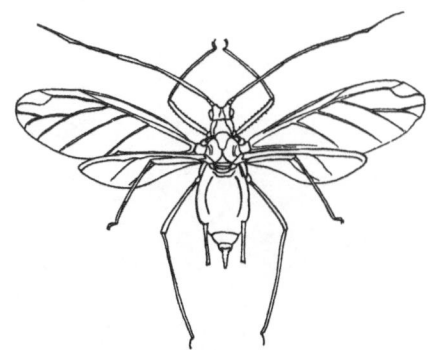

Pea aphid. (D.20.)

PEACHES. *See* FRUITS.

PEA HAY. *See* FIELD PEA.

PEA MOTH (*Laspeyresia nigricana*) is found often in the Great Lake regions and the Pacific Northwest. The moth appears soon after the FIELD PEA vines begin to bloom, usually about the middle of July. It lays its eggs on the pods, leaves, and stems. The eggs hatch in seven to ten days and the larvae enter the pods and feed on the seeds. Growth is completed in from sixteen to twenty-six days when the larvae emerge from the pods; they pass the winter in the soil.

Control: The most effective remedy is to grow early varieties of field pea, seed early, thresh early, and then burn the rubbish left on the field, or plow it under deeply, preferably in the fall. Whenever possible, new crops should be planted at a distance of at least 2 miles from fields that are known to have been infested during the previous season.

Reference: M.24.

PEANUT (*Arachis hypogaea*)—also called *earthnut, goober, pindar,* and *groundnut*— differs widely from other LEGUMES because of its habit of developing the pods underground.

For high yield and superior quality, peanuts require a moderately long growing period of four to five months, with a steady, rather high, temperature and a moderate, uniformly distributed supply of moisture (especially during the period that the peanuts are forming), followed by dry conditions during harvesting and curing.

Peanut. (M.3.)

Climatic conditions suitable for peanut growing are found along the Atlantic seaboard from southeastern Virginia southward and in the Gulf Coast region westward to southern California. Much of this region, however, consists of heavy-textured soils, mountainous areas, swamp, and river bottoms all unsuitable for the crop. A light-textured soil is needed that will not harden or bake; it should be free from stones, gravel, iron concretions, and other similar material. Peanuts grown for harvesting by animals, chiefly hogs, are adaptable to less favorable soils. The plants thrive best on soils having a slightly acid reaction (a pH of 6.0 to 6.5).

To be excellent for peanuts, soils must have good drainage and sandy loam surface layers, with friable sandy clay loam subsoils beginning 10 to 20 in. below the surface and extending to a depth of about 3 ft. or more. Many soils in the Coastal Plains have unfavorable deep, sandy subsurface layers.

Peanuts should not be grown on the same land year after year or even for two years in succession. They are able to use fertilizer residues left in the soil from the previous season's crop, and therefore, should follow some heavily fertilized crop such as CORN. Because land from which a crop of peanuts has been harvested is left practically bare and exposed to the full effects of water and wind erosion, the immediate sowing of a cover crop is extremely desirable. CRIMSON CLOVER, OATS, RYE, or any other crop that will form a quick ground cover can be used.

Deep, thorough plowing that completely buries all cover crops, or other material on the surface, is an important factor in the growing of peanuts. The use of fertilizers containing low nitrogen, fairly high phosphorus, and medium to low potash is generally recommended, but the mixture and method of application can only be determined by local conditions. For instance, in Alabama fertilizer recommendations for good soils are 300 to 400 lb. per acre of an 0-12-20 mixture, and on medium producing soils, 1,000 lb. basic slag and 100 to 150 lb. potassium chloride per acre. In the Southwest additional fertilizer is seldom used when the peanuts follow (heavily fertilized) COTTON. When peanuts are grown as a cash crop, and the vines are removed from the soil, 200 lb. per acre of a 2-12-6 fertilizer is often recommended. Lime or ground limestone is added to the soil (1) to change its reaction and make it less acid, and (2) to supply calcium for the nutrition of the peanut plant. Gypsum also supplies calcium, but without materially affecting the reaction of the soil. Gypsum is usually applied at 400 to 500 lb. per acre directly to the plants at full bloom; it is washed onto the soil around the plants by rain.

Planting peanuts in the shell is wasteful. Machine-shelled seed, when treated with an acceptable fungicide—CERESAN, ARASAN, or SPERGON—will usually give a higher germination than untreated hand-shelled seed. It is advisable to inoculate peanut seed when it is planted in soils that have never grown peanuts.

> NOTE: Care should be exercised in inoculating chemically treated peanut seed, because chemicals in contact with the nodule bacteria may make the inoculant ineffective.

In sections where crows, pigeons, squirrels, or other animals destroy the peanut seed after it is planted, it should first be spread on the floor or on a tarpaulin, sprinkled with a mixture of equal parts of PINE TAR and KEROSENE, and stirred to distribute the mixture uniformly. It is neither necessary nor desirable that the material cover the entire seed, because a little will be effective. A formulation of DINITRO - (ORTHO-SECONDARY) - BUTYL-PHENOL used as a pre-emergence spray on the soil, at a rate of 9 lb. per acre, gives effective control of most weeds, including crabgrass, for several weeks with no reduction in stand or yield of peanuts.

> NOTE: Pre-emergence treatments can be made to a narrow strip over the row, instead of to the entire surface, thereby greatly reducing the amount of the chemical required. The spaces between the treated strips are cultivated in the usual manner.

Cultivation, to keep the soil mellow around the plants as they spread and develop, provides favorable conditions for the penetration of the pegs, and for the enlargement and development of the pods.

Variety, moisture supply, soil fertility, and the tools used for handling peanuts determine the width of peanut rows and, to some extent, the spacing of the plants within the rows. Early planting, consistent with settled weather and a well-warmed soil, is usually desirable. A good plan is to plant early-, medium-, and late-maturing varieties at one time. Thus, differences of as much as three weeks may be obtained in the harvesting dates between the early and late sorts.

As peanuts approach harvest time they should be inspected every day or two to determine the right date for the harvest. Peanuts should be harvested during clear weather when the soil is sufficiently dry so it will not stick to the stems and pods. Peanuts should be loosened from the soil with a sharp implement (digger) that will cut the main root below the surface.

The removal of the peanuts from the vines with the help of portable pickers or combines is one of the most important phases of peanut production.

When harvested, peanuts usually contain about 40 percent moisture. This must be reduced to less than 10 percent before the peanut can be safely stored. Drying, or curing, is usually accomplished either by windrowing on the surface of the soil or by stacking; peanut stacks (arranged around 7 to 8 ft. long poles) should be slender, not more than 36 in. in diameter.

It usually pays to sow the land to a cover crop as soon as the peanuts have been harvested. However, if considerable quantities of peanuts are lost in the soil when harvesting, it is a common custom to turn hogs into the fields to feed on the peanuts remaining in the soil.

Peanut hay, also called *peanut straw*—the dry material which comes from the picker—is a valuable roughage for feeding to farm animals. The yield is from 1,500 to 4,000 lb. per acre. When properly cured, so that it retains most of the leaves and has a bright-green color, its feeding value is practically the same as that of SOYBEAN, COWPEA, RED CLOVER, and some of the other forms of legume hay.

The principal objection to peanut hay is the considerable quantity of dirt and dust it contains. While baled peanut hay is a standard roughage on the market, a large proportion of the hay is used on the farms where it is produced. It is often stacked in the field and animals are allowed to feed on the stacks, or it is hauled to the barn and fed as desired. Badly weather-damaged, moldy, or otherwise spoiled peanut hay is not suitable for feeding.

Peanut by-products of value as feed for domestic animals are the following:

Partly filled pods (lightweight), blown out during the picking operations.

Shrivels and *damaged peanuts* graded out at the shelling plant.

'*Splits, germs, skins*, and *sortings* from processing plants.

Presscake—the residue in the hydraulic oil press.

> NOTE: *Peanut shells*, or *peanut hulls*—obtained in the shakers of the shelling plant—are not recommended for feeding livestock. They do more harm than good; containing over 60 percent crude fiber, they interfere with proper digestion, especially of protein.

Officially recognized peanut products are:

Peanut skins—the outer covering of the peanut kernel, exclusive of the hulls as obtained in the usual process of commercial milling. This product may contain broken peanut kernels.

Peanut vine meal—the ground products obtained after separation of peanuts from peanut vines; it contains all or a large portion of the leaves and vines.

Peanut stem meal—the ground product obtained after the separation of peanuts from peanut vines from which practically all the leaves have been removed so that only vines and stems remain.

Unhulled peanut feed—the ground residue obtained after the extraction of part of the oil from whole peanuts. The ingredients must be designated as "*peanut meal and hulls.*"

Peanut feed—a product which contains at least 41 percent protein and not more than 20 percent crude fiber. It must be sold according to its minimum crude protein and maximum crude fiber content.

The recommended guaranties of crude fat and crude fiber, when used with the protein guaranties indicated, are:

Protein guaranty	Crude fat (minimum)	Crude fiber (maximum)
41%	4 %	20%
43%	4.25%	18%
45%	4.5 %	16%

NOTE: For all grades of *solvent extracted peanut feed* 0.50 percent crude fat is recommended.

Peanut cake—obtained after the extraction of part of the oil by pressure or solvents from peanut kernels as produced under reasonable milling conditions. It must be sold according to its protein content.

Peanut meal—the ground peanut cake; nothing shall be so recognized if it contains more than 11 percent crude fiber. It must be sold according to its protein content.

The recommended guaranties of crude fat and crude fiber, when used with the protein guaranties indicated, are:

Protein guaranty	Crude fat (minimum)	Crude fiber (maximum)
48%	5%	11%
50%	5%	9%

NOTE: For all grades of *solvent extracted peanut feed or meal* 0.50 percent crude fat is recommended.

Peanut meal contains no toxic compounds or antienzymes and therefore requires no heating or water treatment. It contains digestible nutrients and is one of the best protein supplements for livestock feeding, well liked by the stock. Its wide use for different classes of livestock, in areas where it is available, is comparable to that of soybean meal in the North. In the South it is widely used as the principal protein supplement to grain feeds in swine feeding.

Peanut meal contains VITAMIN E and many *B-complex vitamins;* it is rich in NIACIN, PANTOTHENIC ACID, and CHOLINE, but poor in vitamin B_{12}. Four of the AMINO ACIDS needed for promoting growth in fowls are found in fair proportions in peanut meal, namely cystine, histidine, lysine, and tryptophan. Peanut meal of the best grade is superior to the best grades of linseed oil meal or cottonseed meal, and nearly equal to soybean meal in feeding value for swine and poultry. It can replace soybean meal in ruminant rations.

Growing peanuts (especially the true runner peanuts) for *harvesting* by animals, mostly *hogs*, is very attractive. The cultural methods are the same as those followed with the crop grown for harvest. Usually the peanuts are planted alone, but they may be interplanted with another crop such as corn.

Succession plantings a few days apart, are often made. The animals are moved from field to field, or from one part of the field to another, as rapidly as they clean up each planting.

Dangers: Diseases to be considered by the grower are ROOT ROT, SOUTHERN BLIGHT, BACTERIAL WILT, HEAT CANKER, LEAF SPOT OF PEANUT, PEG ROTS, POD AND SEED DECAY, BLACK POD, and CONCEALED DAMAGE IN SEED.

Serious pests of the peanut are ROOT-KNOT NEMATODES, MEADOW NEMATODES, and the STING NEMATODE.

The most important insect enemies of the peanut are the CORN EARWORM, the FALL ARMYWORM, the POTATO LEAF HOPPER, CORN EARWORM, the SOUTHERN CORN ROOTWORM, the TOBACCO THRIPS, the VELVETBEAN CATERPILLAR, WHITE-FRINGED BEETLES, and BILLBUGS.

Classification: For commercial purposes the peanut varieties are divided into four groups; but intermediate forms are also found.

VIRGINIA TYPE

These large-seeded varieties are grown in southeastern Virginia, northeastern North Carolina, in Tennessee, parts of Georgia, and a few other localities where conditions are suitable. Supplying the CALCIUM requirements of the plants is necessary.

Virginia-type peanuts are either bunch or runner varieties. The bunch varieties are easier to cultivate and harvest, but the yields are usually lower than those of the runner group. Runner varieties also cover the ground more completely and thus give a greater degree of weed control.

(a) *Virginia bunch* is upright, but on fertile soil with ample moisture and good culture it assumes a somewhat spreading habit of growth late in the season. Typical varieties, e.g., *Virginia Bunch 67*, attain a height of 18 to 22 in., a spread of 28 to 30 in. The pods are borne within a few inches of the base of the plant.

(b) *Virginia runners*—which include *Holland Station Runner* and *Holland Jumbo*—have a spreading habit of growth, often completely covering the space between the 33- to 36-in. rows. The pods are borne on pegs arising on the side branches, often as much as 15 in. from the base of the plant.

(c) *Intermediate forms*, having an upright spreading habit, are also found.

RUNNER TYPE

Commonly known as *African Runner, Alabama Runner, Georgia Runner, Carolina Runner, Wilmington Runner*, or merely *Runner*, the plants are similar in general character to the runner varieties of Virginia-type peanuts. They are extremely vigorous, usually completely closing the spaces between the rows; however, on thin land, especially if dry weather occurs during the crop season, the growth may be very much restricted. The pods and seeds of runners are intermediate in size.

Because of the ability of the runner varieties to yield a heavier crop than the Spanish type and because of their widespread adaptability to the soil and climatic conditions of the Southeast, they are widely grown in that region.

SPANISH TYPE

The *Spanish peanuts* include forms commonly known as *White Spanish, Small Spanish, Improved Spanish*, and *Spanish*. The plants have an upright habit of growth, but the branches often spread until the rows are completely closed. Under favorable conditions the vines reach a height of 26 in. or more and a spread of 36 in. When grown on shallow soils with a scant supply of moisture, the plants are small, not over 1 ft. high with about the same spread.

The Spanish type peanut is grown throughout the Peanut Belt, but chiefly in the Southeast and Southwest. Its yield is lower than that of other types, but this can be partly overcome by planting in closer spaced rows and in closer spacing of the plants within the rows.

UNCLASSIFIED VARIETIES

Most of these peanuts, though distinct, resemble the Spanish in appearance. They include *Valencia, Tennessee Red*, and *Tennessee White*.

Various *wild peanut* species offer promise as legumes in the permanent pastures in the Southeast. Some species of the wild peanuts are perennials that are highly disease-resistant and furnish good forage.

References: B.10; E.10; S.14; E.12; F.6; S.15.

See also LEGUME FEED PRODUCTS; OIL-SEED MEALS; INOCULATION; VELVETBEANS; FIELD PEA; COMMON LESPEDEZA.

PEANUT CAKE. *See* PEANUT.

PEANUT FEED. *See* PEANUT.

PEANUT HAY.

        See PEANUT; HAY GRADING.

PEANUT HULLS. *See* PEANUT.

PEANUT MEAL. *See* PEANUT.

PEANUT MEAL AND HULLS.

        See PEANUT.

PEANUT SHELLS = *peanut hulls.* *See* PEANUT.

PEANUT SKINS. *See* PEANUT.

PEANUT STEM MEAL. *See* PEANUT.

PEANUT STRAW = *peanut hay.* *See* PEANUT.

PEANUT VINE MEAL. *See* PEANUT.

PEANUT VINES. *See* HAY.

PEA PODS, when empty, are a by-product of the modern process of quick-freezing the edible PEAS (*Pisum* spp.) in the green stage.—*See also* SILAGE CROPS; PLANT BY-PRODUCTS.

PEARL BARLEY, or *pearled barley*, is a pellet obtained by the removal of the outer parts of the BARLEY kernel by pressure against a whirling, rough pearling stone.

PEARL CORN is a term referring to *popcorn* with rounded kernels.—*See also* CORN.

PEARL KAFIR is one of the grain SORGHUMS.

PEARLMILLET (*Pennisetum glaucum*), sometimes known as *cattail millet*, is a summer annual that grows 6 to 10 ft. or more tall. The coarse stems form thick clumps. The leaves are about 1 in. wide, 2 to 3 ft. long, and quite numerous. The flowers appear in SPIKES 6 to 12 in. long.

Pearlmillet will mature seed as far north as Maryland, but it can be used economically only farther south, where it is used for grazing and silage.

Pearlmillet. (H.1.)

This grass requires a rich soil for best growth. Under favorable conditions it produces enormous amounts of green fodder and can be cut several times a season. It is a warm-climate crop and grows only in the warmer part of the year. Land used for growing this grass during the summer may be planted to OATS in the fall, to produce fall, winter, and spring grazing.

References: H.1; M.5.

See also BACTERIAL SPOT; ZONATE LEAF SPOT; GRASSES; VELVET BEANS; PASTURES.

PEARL VETCH.

        See COMMON VETCH (variety).

PEARS. *See* FRUITS.

PEAS. *See* PEA.

PEA-SIZE COTTONSEED CAKE.

        See COTTONSEED.

PEAT is slightly decomposed organic matter formed in the presence of excessive moisture.

PEATLAND is a SIX-ROWED BARLEY variety.

PEA-VINE CLOVER = *American single-cut clover*. See RED CLOVER.

PEA-VINE MEAL. See PEA VINES.

PEA-VINES are a by-product of the quick-freezing of PEAS in the green stage.

Pea vine silage is valued as cattle feed.

Pea vine meal can be used as poultry feed in place of alfalfa meal because it contains protein as well as vitamins A and B₂.—*See also* SILAGE CROPS; SILAGE; PLANT BY-PRODUCTS.

PEA VINE SILAGE. See PEA VINES.

PEA WEEVIL (*Bruchus pisorum*) is the most serious enemy of the FIELD PEA. It is a small grayish or brownish-gray BEETLE, marked with lighter spots. It lays its eggs on the young pod; the egg hatches and produces a larva which bores through the wall of the pod and enters the young seed which feeds the growing embryo. The pupa may emerge soon after harvest, or remain in the seed until the following spring.

Control: The most effective method of preventing damage to the seed harvested is to fumigate with CARBON DISULFIDE as soon as the seed is threshed.

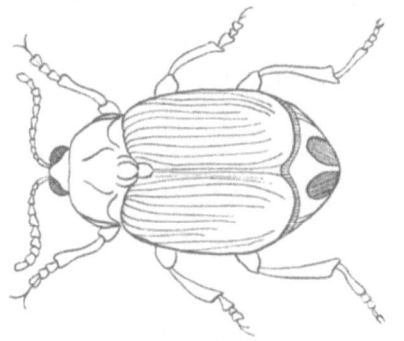

Pea weevil. (D.20.)

Plant only weevil-free seed. Dust weevil-infested parts of the fields with DDT, ROTENONE, or METHOXYCHLOR during the early-bloom stage before the eggs of the insect are laid. Destroy harvest refuse. Do not allow rubbish to accumulate around the pea fields and near farm buildings or the weevils might overwinter in it.

References: M.24; U.3.

PEBBLE-SIZE COTTONSEED CAKE.
See COTTONSEED.

PECK = *Fulcaster*. See COMMON WHEAT.

"PECKY RICE" are rice kernels, discolored by the *Helminthosporium* fungus which causes BROWN SPOT OF RICE, by other FUNGI, or by INSECT injury.

PECTINS are CARBOHYDRATES of the hemicellulose group which occur in many plants, especially in ripe fruits. They are water-soluble, but form jellies easily, e.g., by the addition of sugars or acids.—*See also* DRIED APPLE PECTIN PULP.

PEDICEL, or *pedicle*, is the footstalk or stem of an individual flower or fruit in an INFLORESCENCE which consists of *more than one* flower.—*See also* PEDUNCLE.

PEDICULOIDES. *P. ventricosus* is a small MITE which destroys the WHEAT STRAW-WORM.

PEDUNCLE is the part of the upper stalk of a flower cluster or of an INFLORESCENCE which consists of but *one* flower.—*See also* STALK ROTS; WEAK NECK; PEDICEL.

PEG is the stalk which bears the PISTILS of a flower.

After pollination, the peg of the PEANUT elongates as soon as the petals wither and grows into the soil, where the fertilized OVARY at the tip of the peg enlarges to form the POD, known as the peanut fruit.

PEGANUM HARMALA
= HARMEL PEGANUM.

PEG ROTS are caused by the FUNGI which also cause SOUTHERN BLIGHT and SOIL ROT of peanuts. They attack the immature pegs. As the PEANUTS approach maturity, the foliage of plants bearing a good crop of nuts fades to a lighter green color. Plants that retain the full green color and vigor have few or no nuts attached—only frayed stubs of pegs remain.

After the pod and seed are mature, the peg quickly dies; numerous nonparasitic, soil-inhabiting fungi and bacteria may then cause its decay. If the soil becomes moist and remains so for a week after the nuts are mature, the pegs become so weakened that they break when the plant is lifted, and the nuts are left in the soil; all fully matured nuts may be lost in harvesting.

Control: So far no satisfactory means have been developed for protecting peanuts from peg rots. Dusting with SULFUR for the control of LEAF SPOT OF PEANUTS helps the plant to retain its leaves and to continue its growth and vigor. Under these circumstances the pegs and pods are less susceptible to nonparasitic organisms of decay.

Reference: B.10.

PEKING SOYBEAN.

See SOYBEAN (variety).

PELICAN. *See* SOYBEAN (variety).

PELISS. *See* DURUM WHEAT (variety).

PELLETED SOYBEAN MEALS.

See SOYBEAN.

PELLETING is done in feed mills; they use special mixing equipment and *pellet mills* for compressing. The finished pellets must be cooled before sacking to avoid spoilage.—*See also* MOLASSES; LINSEED; SORGHUMS; GRANULATED FEEDS; IMPLANTATION; COMMERCIAL MIXED FEEDS; CRUMBLES.

PELLETS are used as poultry and livestock feed because they are handled easily and consumed without leaving much waste: the animals cannot pick out the most palatable feed ingredients.

PELLICULARIA. *P. filamentosa*, a FUNGUS, causes BROWN PATCH.

PENICILLIN, an ANTIBIOTIC, is a product of the mold PENICILLIUM. It is water-soluble and widely used for the treatment of infectious diseases and as GROWTH factor in feeds.

Procaine-penicillin, or *penicillin-procaine*, is used like *crystalline penicillin;* it is oil-soluble and more slowly absorbed by the body tissues than the latter.

PENICILLIUM. The *Penicillium* spp. are MOLD-type FUNGI. Some of them, e.g., *P. notatum* and *P. chrysogenum*, are used for the production of PENICILLIN and its salts. Other species attack sorghums, causing SEED ROT; they are also among the causes of CONCEALED DAMAGE OF SEED.

P. oxalicum causes SEEDLING BLIGHT.— *See also* FERMENTATION PRODUCTS.

PENN NO. 44 = *Nittany. See* COMMON WHEAT.

PENNISETUM. Of the many *Pennisetum* spp. only two are of importance for forage, namely *P. glaucum* = PEARLMILLET and *P. purpureum* = NAPIERGRASS. Of minor importance is *P. ciliare* = BUFFELGRASS.

PENNSCOTT RED CLOVER is a double-cut variety of RED CLOVER.

PENNSOY. *See* SOYBEAN (variety).

PENNYCRESS (*Thlaspi arvense*), also called *Frenchweed, fanweed,* or *stinkweed*, grows 1 to 3 ft. high, is slightly branched at the top and has alternate leaves, most of which are on the lower part of the plant. The WEED gets started in sparsely vegetated areas in overgrazed pastures and grows in thick patches, crowding out other vegetation. The brown, ridged seeds are borne in flattened, broadly-winged pods that are notched at the top. The infested patches may be easily seen after maturity because the mature yellow color contrasts with the surrounding green vegetation.

Pennycress. (B.28.)

Pennycress causes the most trouble to the dairy industry. Even small amounts of seed consumed by cows will impart a garlic-like flavor to milk and butterfat. To reduce the flavor, the cows should be removed from the pasture about four hours before milking time.

Control: Pennycress may be controlled or eradicated from pasture with the methods recommended for PEPPERGRASS. Infested areas should be cultivated or plowed in the fall or early spring; where the entire pasture is infested with the weed, the whole area should be revegetated.

Reference: H.25.

PENSACOLA BAHIAGRASS.
See BAHIAGRASS (variety).

PENTAD is a red DURUM WHEAT.

PENTOSANS are CARBOHYDRATES formed from *pentoses*, i.e. SUGARS with only five carbon atoms in the molecule (*hexose*-sugars, such as glucose and levulose, contain six). The pentosans occur in cereal brans and straws, and other woody plant tissues.—*See also* BACTERIA.

PENTOSES are SUGARS.—*See also* PENTOSANS.

PENTSTEMONS (*Pentstemon* spp.), also called *penstemons*, are perennial plants which occur chiefly in the western United States. Their approximately 150 species are considered range WEEDS; they vary in palatability from worthless to fairly good. The large shrubby species, from 1 to 8 ft. high, have browse value for sheep; the herbaceous species, with smooth, green foliage, generally rate as fair for cattle, and also occasionally for sheep.

Usually the pentstemons reproduce rather vigorously by means of rootstocks; they tend to increase on overgrazed ranges.

Reference: U.6.

See also MUTTON BLUEGRASS.

PEPPER. *Bird's pepper* = PEPPERGRASS.

PEPPERGRASS (*Lepidium apetalum*), also known as *green-flowered peppergrass, wild tonguegrass, large peppergrass,* and *bird's pepper,* is a weed which grows from 6 in. to 2 ft. tall. The plants are much branched at the top and are usually scattered rather than growing in patches. The flowers are small and the white petals drop off soon after the blossoms open. The seed pods are flattened on one side and are slightly winged with a notch at the top.

Peppergrass is most troublesome early in the spring when eaten by milk cows, because it imparts a disagreeable odor and flavor to milk that persists throughout all the processes of dairy production. The disagreeable sweetish taste becomes more noticeable with age of cream and butter. Removal of cows from the pasture about four hours before milking time helps to reduce the weedy flavor. However, lower milk production of the herd may result unless some feed is provided in the dry lot.

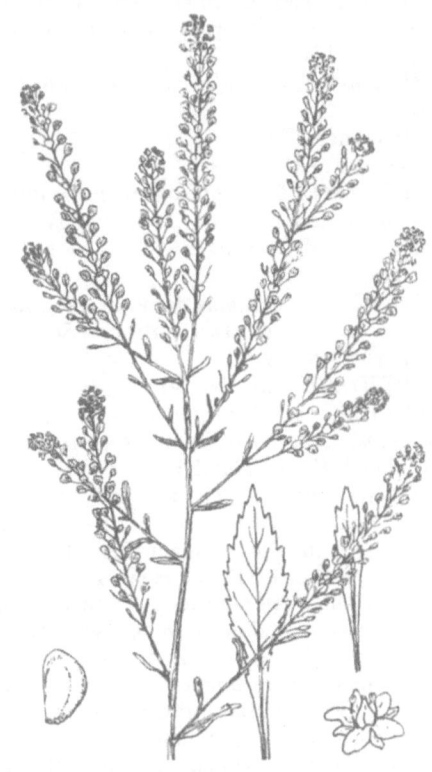

Peppergrass. (B.28.)

Control: If the infested areas are partially covered with grass, the livestock should be removed long enough for the grass to recover.

Where the infested areas are devoid of grass, they can be cultivated or plowed in the fall or early spring so as to kill the winter rosettes of these weeds. RYE may be seeded in the fall or OATS and SWEET-CLOVER may be planted in the spring for pasturing. If possible, large infested areas should be revegetated with a perennial GRASS.

Reference: H.25.

See also PENNYCRESS.

PEPPER SPOT, or *Pseudoplea leaf spot,* is caused by the FUNGUS *Pseudoplea trifolii.* It is generally distributed on TRUE CLOVERS, especially WHITE CLOVER, but occurs also on ALFALFA. Pepper spot is characterized by numerous black sunken spots on the leaves. As the spots grow older, they develop a reddish-brown margin. If the disease becomes severe it will cause browning of the leaves.

No control measures are known for this disease.

Reference: T.1.

See also LEAF SPOTS.

PEPPERWEED (*Lepidium scopulorum*) is one of the fair forbs (WEEDS); it has some value as feed on ranges.—*See also* RANGE PLANTS.

PEPSIN is a proteolytic ENZYME which occurs in the GASTRIC JUICE.—*See also* DIGESTIVE TRACT.

PEPTIDASES are ENZYMES.

PEPTIDES are combinations of two or more AMINO ACIDS.—*See also* ENZYMES.

PEPTONES are simpler in composition than the PROTEINS (milk, meat, gelatin, etc.) from which they are obtained during DIGESTION in the stomach by the proteolytic action of the enzyme PEPSIN. In the intestine, the peptones are further digested by the action of the enzyme TRYPSIN; thus, the AMINO ACIDS are formed which in turn are much simpler in chemical composition than the peptones. The latter form a white, or pale-yellow, water-soluble powder which is valuable as NUTRIENT.

PERENNIAL means: lasting for three or more years; said especially of herbaceous plants that are neither ANNUAL nor BIENNIAL.

PERENNIAL RYEGRASS (*Lolium perenne*), also known as *English ryegrass*, is quite similar in general appearance to ITALIAN RYEGRASS, except for its slightly flattened stem and the leaves which are folded in the bud. There are usually no awns on the seeds; the plants are commonly reddish at the base.

Plantings from commercial seed of perennial ryegrass often disappear in three to four years, probably because such plantings are from a seed mixture of perennial ryegrass with short-lived hybrid strains of the perennial and annual species. Perennial ryegrass plants become very tough as they grow older.

Truly perennial ryegrass is best adapted to regions with cool temperatures and ample moisture during the summer. Hot, dry weather in July and August affects the growth; and if the drought continues very long in September, the recovery is slow. Best results have been had on the Pacific Coast and in the central and southern parts of the Atlantic States. The grass has a wide range of adaptability to soils, but in general it prefers medium to high fertility.

Perennial ryegrass. (H.1.)

Perennial ryegrass is mainly used for permanent pasture seedings. It starts quickly and furnishes early grazing while other longer lived GRASSES are becoming established.

Dangers: Among the fungus diseases attacking this grass is the BLIND SEED DISEASE.

References: S.2; H.1.

See also COMMON RYEGRASS; RYEGRASSES; GRAZING.

PERENNIAL SOW THISTLE
= SOW THISTLE.

PERENNIAL VETCH (*Vicia tenuifolia*), one of the newer VETCHES, is a highly variable, sod-forming plant. It has narrow, smooth leaves and fine stems; it spreads by RHIZOMES. This species grows to 3 ft. in height and has profuse purple blooms. It is vigorous, productive, and early maturing. The seeds are hard and must be scarified. Plants are slow to establish, reaching maturity in the third year. This species appears useful on cut-over and burned-over timber areas of the Pacific Northwest, as well as for a cover on road cuts and fills.

Reference: H.22.

See also LEGUMES; INOCULATION.

PERERODKA = *Kubanka. See* DURUM WHEAT.

PERFECT is a *flower* which has both PISTILS and STAMENS.

Perfect (or *sexual*) *stage* of a FUNGUS is that period of life during which SPORES are produced sexually.—*See also* IMPERFECT.

PERIANTH is the floral envelope consisting of the CALYX and the COROLLA, both so modified that they cannot readily be distinguished.

PERICARP is matured OVARY wall; e.g., the outer covering, or bran layer, of a grain.

PERICONIA. The FUNGUS *P. circinata* causes PERICONIA ROOT ROT (or *milo disease*); this disease was first attributed to one of the *Pythium* spp. and erroneously called "Pythium milo disease" or "Pythium root rot of milo."—*See also* PYTHIUM.

PERICONIA ROOT ROT, or *milo disease,* or *milo root rot,* is sometimes—erroneously—called *Pythium root rot of milo.*

It is not caused, as was first thought, by one of the PYTHIUM species, but by the FUNGUS *Periconia circinata.* The disease has been observed in many fields in the western parts of Texas and Kansas and in Oklahoma, New Mexico, Nebraska, Arizona, and California.

Periconia root rot does not damage SORGHUMS on any land not cropped previously to milo, milo derivatives, or darso. It is the most serious disease known on milo, darso, and most of their hybrids.

In heavily infested soil, the disease may appear three to four weeks after planting, when the plants are only 6 to 9 in. high. Its first indication is a stunting of the plants and a slight rolling of the leaves, with the older leaves turning light yellow at the tips and margins. The disease progresses until all the leaves are affected and the plants die, usually without heading. In less heavily infested soil the disease may not appear until the plants are about ready to head. In such cases it progresses less rapidly; the plants may grow weakly until late in the season and may form small, poorly filled heads.

The disease attacks the roots before the above-ground parts of the plants show any symptoms. Examination reveals a water-soaked brown or reddish discoloration of the outer parts of the roots. Later, a soft rot destroys the root system and the central part of the large roots turns dark red or brown. The tissue of the base of the crown also turns dark red, and this discoloration extends up into the base of the stalk. The disease is commonly recognized by splitting the base of the diseased plant and finding the dark-red area.

After this root rot appears in a field, it usually becomes more severe each successive year if milo or other susceptible varieties are grown.

The disease may be spread by soil or carried in runoff or irrigation water and by farm implements. Although small areas of badly infested soil can be sterilized effectively by steam, FORMALDEHYDE, or CHLOROPICRIN, this is not practicable for field use.

Certain abnormal conditions of sorghums have been mistaken for symptoms of milo disease; e.g., stunted sorghum plants have been observed in low spots in irrigated fields where the soil had become partly waterlogged after being submerged for some time. The hereditary HEADLESS DEFECT, which prevents the formation of heads and sometimes side branches on most of the stalks of the plants, also has been mistaken for periconia root rot.

Control: Effective control measures are limited to the planting of resistant varieties. Finney milo, Texas milo, Double dwarf milo No. 38, Oklahoma No. 1 darso, Westland, darset, and selections from Sooner milo, are some of the highly resistant grain sorghums.

The kafirs, feteritas, sorgos, broomcorn, Early Kalo, Martin, Midland, Improved Wheatland 288, Dwarf kafir 44-14, and a number of the less common varieties of sorghums are, on the whole, resistant to the disease and may be safely grown in soil known to be infested. Colby and Wheatland are highly susceptible to this root rot; Extra Early Sumac is the only sorgo known to be particularly susceptible, but a resistant strain is now available.

Crop rotation offers small hope for controlling periconia root rot because the causal fungus persists in the soil for several years.

Reference: L.1.

See also QUADROON.

PERILLA OIL MEAL is obtained as by-product of the manufacture of perilla oil from the seed of the Chinese perilla plants (*Perilla* spp.). This oil meal contains approximately 38 percent protein, 8 percent fat, 18 percent nitrogen-free extract, 18 percent fiber, and 9 percent ash. It is used for feeding dairy cattle and sheep.

PERMANENT PASTURES.

See PASTURES; PASTURE MANAGEMENT.

PERONOSPORA. *Peronospora* spp. are FUNGI which cause DOWNY MILDEW; *P. trifoliorum* causes DOWNY MILDEW OF ALFALFA; *P. manshurica*, DOWNY MILDEW OF SOYBEAN; and *P. viciae*, DOWNY MILDEW OF VETCH. (The fungus formerly called *P. pisi* is now considered as possibly a specialized race of *P. viciae*.)

PEROSIS. *See* NIACIN.

PERSIAN CLOVER (*Trifolium resupi-*

natum) has found a useful place in the agriculture of the southern states. By producing feed in late winter and early spring, when southern GRASSES are dormant, it extends the grazing season. The forage is nutritious and is relished by all kinds of livestock and poultry. When grown with grass, Persian clover supplies nitrogen, as do other LEGUMES, thus improving the quality and the quantity of the grass. Once it is established with grasses and properly fertilized and managed, reseeding is not necessary because ample seed is produced for volunteer stands. Although primarily a pasture and hay legume, Persian clover is used in some places as a green-manure crop.

This TRUE CLOVER adapts best to low-lying, heavy, moist soils. It has been successfully grown as far north as Tennessee and also makes good growth in sections of the coastal region of the Pacific Coast states.

Persian clover is a true winter annual growing throughout the winter months in the form of a low rosette; with the advent of spring rapid growth occurs; seed is produced in late spring or early summer after which the plants die. Persian clover grows from 8 in. to 2 ft. high, depending upon the conditions. When grazed heavily or when the stand is thin, the stems bend down, giving the appearance of a low, spreading plant.

The light-purple flowers, forming a head somewhat flat in appearance during the early bloom stage, are self-pollinating and self-fertile and are borne in the leaf axils on stems from ½ to 2 in. in length.

Persian clover can be successfully grown on medium to slightly acid soils. On strongly acid soils 1 to 2 tons finely ground limestone is recommended per acre. This should be applied in midsummer for best results. In rotations the available residual minerals from applications to cultivated crops are sometimes sufficient to produce satisfactory stands and good yields. If they are not sufficient, phosphate and/or potash fertilizers should be applied by drilling, or broadcasting, immediately before seeding.

Lack of INOCULATION is one of the principal reasons for failures to obtain productive stands. Seed may be inoculated by using commercial cultures, by using inoculated soil obtained from a field that has grown a productive Persian clover crop the previous year, or by combining both methods. When the weather is hot and dry at seeding time, even inoculated seed sometimes fails to produce inoculated plants. It is not necessary to inoculate after Persian clover has been successfully grown in the same field for two years.

Persian clover is an excellent GRAZING plant. It produces a high quality, nutritious, protein feed from late winter to late spring. The grazing of Persian clover should be started early. Moderate grazing is desirable. The combination of Persian clover and lespedeza is not recommended, because the Persian clover makes its most rapid growth when the lespedeza is starting and thus crowds out the latter.

Properly cured Persian clover *hay* is relished by all kinds of livestock and has a high nutritive value. Yields average from 1 to 2 tons per acre, depending upon the rate of fertilizing, method of handling, and seasonal rainfall. Persian clover should be cut for hay when the plants are in from one-fourth to full bloom. The largest yield is obtained in the full-bloom stage, but the quality is higher when the hay is cut earlier.

On heavy, low-lying soil of the southern states the use of Persian clover as a *green-manure* crop is increasing, with satisfactory results. When used for this purpose, it is frequently lightly grazed during the late winter and early spring months. If it is allowed to approach maturity before being turned under, yields of as much as 15 tons per acre of green material have been obtained, and sufficient seed is placed in the soil for volunteer stands for two years.

Danger: Persian clover should be grazed carefully in order to avoid BLOAT of cattle and sheep. The danger from bloat may be reduced by early and continued grazing, by having a mixture of grass and clover, and by giving the animals free access to strawstacks or haystacks.

Reference: H.10.

See also PASTURES; PASTURE PLANTS; RICE; LAPPA CLOVER.

PERUVIAN ALFALFA belongs to the nonhardy ALFALFAS.

PESTICIDE is any agent that destroys pests; e.g., any FUNGICIDE, INSECTICIDE, etc.

PETAL is one of the (usually colored) separate, modified leaves of a COROLLA.

PETERSEN'S RELATIVE FEED VALUES are based on the ton price of the carbohydrate-rich CORN and protein-rich COTTONSEED meal. In much of the United States, SYOBEAN meal has replaced cottonseed meal as the basic protein and is used for comparison. To evaluate any feed, its "constant for corn" and "constant for cottonseed meal" must be determined: they show to what extent both base feeds affect any feed's value.

To calculate the relative value of 1 ton wheat middlings, the current ton price of corn (dent corn grade No. 2) is multiplied by the middlings constant for corn, i.e. 0.714, and then added to the product obtained by multiplying the current ton price of cottonseed meal (43 percent protein grade) by the middlings constant, for cottonseed meal, i.e., 0.283.

The constants have been computed for many feedstuffs; they are based on their percentage-contents of (1) total digestible nutrients or, in some cases, net energy content and (2) digestible protein. These values vary somewhat for different species of animals.

PETIOLE is a *leafstalk;* it attaches the blade of a leaf to the plant stem.

PETKUS. *See* RYE (variety).

PETROLATUM, or *petroleum jelly,* is a semisolid, purified mixture of HYDROCARBONS obtained from crude PETROLEUM. It is of white or amber color and soluble in organic solvents. Petrolatum can be used to protect the hands when handling poisons, such as POISONED BRAN MASH.—*See also* COAL-TAR CREOSOTE.

Liquid petrolatum = WHITE MINERAL OIL.

PETROLEUM in its natural, *crude* form is known as CRUDE OIL; when *processed,* it yields various HYDROCARBON fractions, such as benzine, gasoline, KEROSENE, PETROLATUM, and paraffin. It and many of its fractions are useful as INSECTICIDES or solvents for insecticides.—*See also* MINERAL OIL; PETROLEUM OILS.

PETROLEUM JELLY = PETROLATUM.

PETROLEUM OILS are the oily fractions obtained by the distillation of crude PETROLEUM.

Light petroleum oils, such as *stove oil*, *diesel oil*, and *kerosene*, are used for killing MESQUITE. The oil must completely envelop the main sprouting bud zone underground.

The oil quantity required varies with the size of the main stem, the degree of branching at the soil surface, the texture of the soil, and the depth to which the underground stem is buried; 1 qt. to 1 gal. per plant is adequate in most cases.

pH, an abbreviation of *potential hydrogen,* or HYDROGEN ion concentration, is the symbol of a scale designating the relative *acidity* or *alkalinity* of a solution. The scale ranges from 1 to 14; pH 7.0 represents the *neutral* solution. Numbers less than 7 indicate increasing acidity; numbers above 7 indicate increasing alkalinity.

The pH of *soils* is very important because most plants grow best between relatively narrow pH-limits. The soils may be classified as follows, according to pH values:
Extremely acid—pH below 4.5;
Very strongly acid—pH 4.5 to 5.0
Strongly acid—pH 5.1 to 5.5;
Medium acid—pH 5.6 to 6.0;
Slightly acid—pH 6.1 to 6.5;
Neutral—pH 7.0 (normally 6.6 to 7.3);
Mildly alkaline—pH 7.4 to 8.0;
Strongly alkaline—pH 8.1 to 9.0;
Very strongly alkaline—pH 9.1 and over.

See also ALKALI; ACID.

PHALARIS. Among the *Phalaris* spp., or CANARYGRASSES, are two which are cultivated for forage: *P. arundinacea* = REED CANARYGRASS and *P. tuberosa* var. *stenoptera* = HARDINGGRASS.

PHASEOLUS. The most important *Phaseolus* spp., or FIELD BEANS, are *P. vulgaris* = COMMON BEAN (with its more than 50 varieties, including *Navy bean*, *Pinto*, *Great Northern bean*, and *Red kidney bean*); *P. acutifolius* var. *latifolius* = TEPARY BEAN; *P. aureus* = MUNGBEAN, and

the two LIMA BEAN species: *P. limensis*, or *large lima bean*, and *P. lunatus*, or *small lima bean.*

PHASEOLUS VIRUS 2 causes YELLOW MOSAIC.

PHENOL, or *carbolic acid*, forms needle-shaped, water-soluble crystals. It is caustic, poisonous, and widely used as DISINFECTANT.

Crude phenol is recommended for the eradication of DODDERS.

PHENOTHIAZINE, a greenish-yellow powder, is one of the most effective water-insoluble *anthelmintics* (remedies against gastrointestinal worms). It is sometimes given mixed in feed to livestock and poultry; the drug is often administered in small, safe dosages over a prolonged period of time, mixed in feed or common salt.

PHENYLALANINE is one of the essential AMINO ACIDS. It forms white crystals which are only slightly soluble in water. Part of an animal's phenylalanine needs can be supplied by TYROSINE.

PHIALEA TEMULENTA, a FUNGUS, causes BLIND SEED DISEASE.

PHILEAENUS. *P. leucophtalmus* = MEADOW SPITTLEBUG.

PHLEUM. The genus *Phleum* contains about 10 grass species; of these only *P. pratense*, or TIMOTHY, is cultivated in the United States.

PHOMA. The FUNGUS *B. trifolii* causes BLACKSTEM OF CLOVER.

PHOROCERA. The fly *P. claripennis* is a natural enemy of the ALFALFA CATERPILLAR.

PHOSPHATE ROCK.

See ROCK PHOSPHATE.

PHOSPHATES are the salts of any PHOSPHORIC ACID.

PHOSPHATIC materials such as FERTILIZERS and certain MINERAL FEEDS, contain PHOSPHATES.

PHOSPHATIDES are fatty substances that occur in cellular structures of animals and plants and contain esters of PHOSPHORIC ACID.

PHOSPHOLIPIDS are substituted FATS containing PHOSPHORIC ACID and usually a nitrogen compound; most of them are good emulsifiers (which aid in the formation of stable EMULSIONS) and, thus, play important roles in life processes.

PHOSPHORIC ACID exists in various forms, all of which are derivatives of P_2O_5 (PHOSPHORUS PENTOXIDE). The most important is the common *orthophosphoric acid*, H_3PO_4, which occurs in colorless crystals (but is commercially available as aqueous solution), and the *pyrophosphoric acid* which forms a water-soluble crystalline powder.—*See also* ANTIOXIDANTS; PHOSPHATIDES: SUPERPHOSPHATE.

PHOSPHORIC ANHYDRIDE
= PHOSPHORUS PENTOXIDE.

PHOSPHORUS—the chemical element *P* —is not used as such, but its salts are widely used in MINERAL FEEDS, FERTILIZERS, and drugs; among them the PHOSPHATES are most important. Phosphorus-rich feeds of animal or vegetable origin are BONE MEAL, meat scraps, TANKAGE, fish meal, oil meals, mill feeds (especially bran and shorts), etc.

Phosphorus is necessary for life; it is found in many living tissues, blood, bones, teeth, etc. Some 80 percent of the body's total phosphorus content is in bones; the remainder is distributed throughout the body. Phosphorus deficiency in feed causes such conditions in livestock as rickets in young, and soft bones in adults.

Dry RANGE PLANTS and HAYS from phosphorus-low soils are low in phosphorus; cattle are often poor breeders under these conditions. Such animals sometimes eat bones or dirt, probably in an effort to overcome the deficiency.—*See also* MINERALS; PARATHION; HAY; GRAZING; VITAMIN D; VITAMIN B_{12}; SUPERPHOSPHATE.

PHOSPHORUS PENTOXIDE—P_2O_5— or *phosphoric anhydride*, is a white powder; when dissolved in water, and depending on the amount of water absorbed, it forms different PHOSPHORIC ACIDS.—*See also* FERTILIZERS.

PHOTOSENSITIZATION is characterized by the development of skin reactions in light-skinned animals exposed to sunlight, following consumption of certain plants and drugs.—*See also* RAPE.

PHOTOSYNTHESIS. *See* CHLOROPHYLL.

PHYGON, which contains 50 percent

dichloronaphthoquinone, generally improves germination and stand under unfavorable soil conditions; it also controls the KERNEL SMUTS. In SEED TREATMENT of SORGHUMS, apply at the rate of 2 oz. per bushel, using the same thoroughness and precautions as for COPPER CARBONATE.

If applied to sorghum seed by the slurry method, 1 lb. Phygon in 1 gal. water is sufficient to treat 16 bu. seed.

Reference: L.1.

PHYLLOSTICTA. The FUNGUS *P. glumarum* causes BROWN-BORDERED LEAF SPOT.

PHYSIOLOGY is the branch of biology which deals with normal life processes and functions.

Plant physiology is the study of plant organs and functions and is particularly concerned with growth and reproduction of plants.

Physiological means: relating to physiology.

PHYTELEPHAS. *P. marcrocarpa*
= IVORY PALM.

PHYTIN is a salt of *phytic acid*, an organic compound which contains PHOSPHORIC ACID.

PHYTOMONAS. Many pathogenic BACTERIA of the genus *Phytomonas* are now grouped in the genus *Pseudomonas*; e.g., *P. pisis.*—*See also* PSEUDOMONAS.

PHYTONOMUS. *P. meles* = CLOVER-HEAD WEEVIL; *P. nigrirostris* = LESSER CLOVER-LEAF WEEVIL.

PHYTOPHAGA. *P. destructor*
= HESSIAN FLY.

PHYTOPHTHORA. The FUNGUS *P. drechsleri* causes ROOT ROT OF SAFFLOWER.

PICKLEWEED (*Allenrolfea occidentalis*) is a shrub which is of very limited value as forage for sheep on winter ranges.—*See also* RANGE PLANTS.

PIERRE RYE. *See* RYE (variety).

PIERSON DURUM = *Arnautka*.
See DURUM WHEAT.

PIGEON RATIONS.
See POULTRY RATIONS.

PIGMENT is a coloring matter, i.e., an inorganic or organic substance which imparts a color to other materials. Pigments occur in animal or plant cells or tissues.

PIGWEEDS (*Amaranthus* spp.), e.g., *rough pigweed* (*A. retoflexus*), are WEEDS occurring in warm regions; they are often attacked by the GARDEN WEBWORM and STINK BUG.—*See also* ALFALFAS.

PILCHARD = SARDINE.

See also MARINE PRODUCTS.

PILCHARD OIL = SARDINE OIL.

PILOT WHEAT.

See COMMON WHEAT (variety).

PIMPINELLA. The dried ripe fruit of *P. anisum* is known as ANISE SEED.

PINCLOVER = ALFILERIA.

PINDAR = PEANUT.

PINEAPPLE PULP, when *dried*, is also called *pineapple bran*. This by-product of the canning industry consists of the outer shells and sometimes also of the cores of pineapples; occasionally, MOLASSES is added to this product. It is used for feeding purposes in Hawaii and in the western part of the United States. Its composition is similar to that of DRIED BEET PULP, but because it contains only 4 percent protein it replaces only a part of the concentrate mixtures in rations for dairy cows, horses, mules, or swine.

PINE TAR is a viscous, sticky, blackish-brown liquid obtained as residue from distillation of pine wood. It is used medicinally and as an INSECTICIDE and insect repellent.

PINE-TAR OIL is a *tar oil* obtained from PINE TAR. It is used for preparing OIL BARRIERS which are recommended for destroying CHINCH BUGS.

PINGUE = *Colorado rubberweed.*

See RUBBERWEEDS.

PINION = PIÑON.

PINK KAFIR is one of the grain SORGHUMS.

PIÑON (*Pinus edulis*), also spelled *pinion* or *pinyon*, is an evergreen tree which, with the JUNIPER tree, is typical of a large area —the piñon-juniper woodland formation— of winter ranges of the Intermountain region. Most of these trees grow in open stands, usually with SAGEBRUSHES in the understory.—*See also* RANGE PLANTS.

PINTO, one of the COMMON BEAN varieties, belongs to the FIELD BEANS. The plants, which are semitrailing, have white flowers and buff-colored, brown-speckled seeds.

PINUS. *P. edulis* = PIÑON; *P. ponderosa* = PONDEROSA PINE.

PINYON = PIÑON.

PIPERAZINE and several of its addition compounds, such as *piperazine citrate* and *piperazine phosphate*, are used increasingly in feed or drinking water for the removal of roundworms and some of the less common gastrointestinal parasites from poultry, swine, horses and (to a limited degree) from cattle.

The advantage of many piperazine compounds over other anthelmintics is that they are water-soluble and relatively safe. The drug may be administered as a single dose or over a period of two or three days.

NOTE: Since the composition of piperazine salts often varies as to water content, it is advisable to consider the content of *anhydrous piperazine* when buying or calculating the dose of the drug. Commercial *piperazine hexahydrate*, for instance, yields 40 to 44 percent anhydrous piperazine; piperazine citrate only 36 to 40 percent.

PIPER SUDANGRASS.

See SUDANGRASS (variety).

PIRICULARIA. The FUNGUS *P. oryzae* causes BLAST OF RICE.

PISTIL is the female floral organ. It consists of OVARY, STYLE, and STIGMA; the style is sometimes wanting.—*See also* GRASSES.

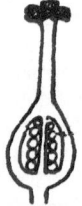

Pistil or female floral portion showing lobed stigma, style, and ovules in ovary. (D.9.)

PISTILLATE means: female; bearing PISTILS (or seed-producing organs) only.

PISUM. *P. arvense* = FIELD PEA.

PITH is the soft, spongy, central substance in the stems of plants.

PIT SILOS. See SILOS.

PLAINS LOVEGRASS (*Eragrostis intermedia*) grows on dry or sandy prairies in the southern states. It has some value in New Mexico and Arizona as range forage and ground cover.—*See also* GRASSES.

Reference: H.1.

PLAINSMAN MILO is not a true milo, but a kafir-milo derivative and, as such, belongs to the intermediate-type grain SORGHUMS.

PLANT is any organism belonging to the vegetable (or plant) kingdom.

PLANTAINS (*Plantago* spp.) are WEEDS with numerous basal leaves and small, greenish flowers. They are widely distributed throughout the United States at low elevations.—*See also* ALFALFAS.

PLANT BUG.

See RED-SHOULDERED PLANT BUG.

PLANT BY-PRODUCTS include a great variety of BY-PRODUCT FEEDSTUFFS obtained from vegetables or fruits; e.g., MOLASSES; BEET PULP, DRIED BEET PULP, BEET TOPS; and CONDENSED BEET SOLUBLE PRODUCTS; TOMATO POMACE and DRIED TOMATO POMACE; POTATO; SWEET POTATO and SWEET POTATO PULP; LIMA BEANS and LIMA BEAN VINES; SWEET CORN REFUSE; DRIED GRAPEFRUIT REFUSE, ORANGE PULP, DRIED CITRUS PULP, and CITRUS SEED MEAL; APPLE POMACE, DRIED APPLE POMACE, and DRIED APPLE PECTIN PULP; GRAPE POMACE; RAISIN PULP; ALMOND HULLS and GROUND ALMOND HULLS; OLIVE PULP.—*See also* AMMONIATED PLANT PRODUCTS; BROCCOLI; PALMO MIDDLINGS; PEAS.

PLANT DISEASES.

See DISEASES; ANTIBIOTICS.

PLANTER is a machine used for opening the soil and dropping (at certain intervals) seeds, seedlings, cuttings, or tubers.

PLANTER SORGO, or *Planter's Friend*, is one of the forage SORGHUMS.

PLANT LICE are known as APHIDS.—

See also GREENBUG.

PLANT MORPHOLOGY.

See MORPHOLOGY.

PLASMA. *See* BLOOD.

PLATHYPENA. *P. scabra*

= GREEN CLOVERWORM.

PLEOSPORA. The FUNGUS *P. herbarum* causes PLEOSPORA LEAF SPOT.

PLEOSPORA LEAF SPOT, occurring in warm weather, is caused by the FUNGUS *Pleospora herbarum*. It is one of the minor

diseases of the ALFALFAS. The irregular brown spots may destroy a large part of the affected leaves.

Reference: J.3.

PLOW is an implement for the preparation of seedbeds by breaking loose furrow slices of the soil. Among the many types of plows are the widely used DISK PLOWS; e.g., the ONE-WAY DISK PLOW.

PLOWABLE PASTURES

= *Permanent pastures. See* PASTURES.

PLUMOSE means: feathery or feather-like; also, having fine hairs on each side, like the plume of a feather.

See SIX-ROWED BARLEY (variety).

PLUMS. *See* FRUITS.

PLUSH BARLEY. *See* SIX-ROWED BARLEY (variety).

P₂O₅ = PHOSPHORUS PENTOXIDE.—

See also SUPERPHOSPHATE.

POA. Of the approx. 200 *Poa* spp. distributed throughout the world, 65 are native to the United States. Among those of agricultural importance are the following: *P. pratensis* = KENTUCKY BLUEGRASS; *P. compressa* = CANADA BLUEGRASS; *P. arachnifera* = TEXAS BLUEGRASS; *P. trivialis* = ROUGHSTALK BLUEGRASS; *P. bulbosa* = BULBOUS BLUEGRASS; *P. annua* = ANNUAL BLUEGRASS; *P. secunda* = SANDBERG BLUEGRASS; *P. ampla* = BIG BLUEGRASS; *P. fendleriana* = MUTTON BLUEGRASS; *P. nevadensis* = NEVADA BLUEGRASS; *P. palustris* = FOWL BLUEGRASS.

POACEAE = GRASSES.

POCKET GOPHERS are a serious pest in ALFALFA fields. If uncontrolled, these ground-living RODENTS may become so numerous that the crop is no longer profitable. The mounds interfere with mowing, and a large number of plants are smothered.

Control: Trapping is less effective than poisoning. The pocket gophers may be poisoned at any time they are active, but especially in late fall when burrows are being extended and in early spring when new mounds are being thrown up. Before placing the bait, the main runway must be found. This is done by probing with a wagon rod 8 to 12 in. from the mound on the side where the plug to the hole is seen. (The plug is a circular area on the indented side of the mounds). When the rod has been pushed into the ground a few inches and suddenly sinks about 3 in. more without increased pressure, the runway has been located. A sharpened broomstick is inserted in the hole made by the rod. Its top end is rotated in a circle to make the hole larger—but not deeper—and its wall firm. The stick is then drawn out and a tablespoon of STRYCHNINE-POISONED GRAIN is dropped in, and the hole covered, care being taken not to let dirt fall in and cover the bait. Every fourth or fifth mound should be baited and marked.

Reference: G.7.

POD, or *legume,* is a capsule, particularly a dry-seed-vessel; originally, the term was used of LEGUMES only; e.g., a pea pod.— *See also* COTTON; PEG; BLACK POD; LOMENT.

POD AND SEED DECAY affects PEA-NUTS. Like the pegs, the pods are quite resistant to attack by most soil-inhabiting FUNGI as long as the shell tissue is living, but they are susceptible to the fungi that cause SOUTHERN BLIGHT and ROOT ROT.

As the seeds mature, most of the nutrients in the shell tissues are moved into the seed, and the shell finally becomes a porous, corky mass of dead tissue protecting the seed. In this condition the shells are readily penetrated by the soil organisms which obtain nourishment from the stem, pass down through the pegs, and gain entrance to the interior or the pod. After the plants die from disease or overmaturity, and masses of dead leaves accumulate on the soil, growth of fungi becomes more accelerated and pod and seed decay more rapidly.

Reference: B.10.

POD AND STEM BLIGHT of SOYBEANS is caused by two varieties of the FUNGUS *Diaporthe phaseolorum;* the variety *batatatis* is more virulent and attacks vigorous plants; the variety *sojae* appears to be confined to older plants near maturity. The first indication of the disease is the appearance of dead plants with leaves still attached. Upon close examination, such plants show a brown, girdling area slightly sunken on the stem, usually located at the base of a branch or leaf stalk (PETIOLE)

close to the soil. The disease also affects the pods.

Pod and stem blight is likely to be worse in very rainy seasons, but it does not usually kill all the plants over a large area.

Among varieties of soybeans no differences in susceptibility are known.

Control: Since the disease organism overwinters on infected stems and seed, rotation and clean seed are the recommended control measures.

Reference: C.9.

POD CORN. *See* CORN.

POD SPOT. *See* LEAF AND POD SPOT.

POGONOMYRMEX. *P. barbatus* = RED HARVESTER ANT; *P. occidentalis* = WESTERN HARVESTER ANT.

POISON BAIT.
       See BAIT; POISONED BRAN MASH.

POISONBEAN (*Daubentonia drummondii*) is very poisonous to cattle, sheep, and goats. When they eat even small quantities of the seed of this POISONOUS PLANT, they soon become depressed, develop diarrhea, and the pulse becomes rapid.

POISONED BRAN MASH, a *poison bait* for the control of GRASSHOPPERS, FALL ARMYWORMS and other ARMYWORMS, and CUTWORMS is made up as follows:

Bran (preferably wheat bran)	25	lb.
Paris green (or arsenic trioxide)	1	lb.
Molasses (preferably crude cane molasses)	2	qt.
Oranges or lemons (if available)	6	(in No.)
Water up to	3½	gal.

The PARIS GREEN or ARSENIC TRIOXIDE and the BRAN are thoroughly mixed in a washtub or similar container. The juice of the oranges or lemons is squeezed into the water, the pulp and peeling chopped fine and added, after which some strong-smelling MOLASSES is dissolved in the water. The bran is then moistened with this solution until, after mixing, a crumbly mass is obtained.

NOTE: When bran is not available, coarse ALFALFA meal or coarse COT-TONSEED hulls may be used.

The early morning is the best time to scatter the damp mash about the fields where the *grasshoppers* are troublesome. The quantity of the foregoing formula is sufficient for 4 to 5 acres.

When the bait is used for the control of *cutworms* or *armyworms*, it should be spread late in the afternoon and at the rate of 20 to 25 lb. per acre. For each 1 lb. Paris green, one can substitute 1 lb. SODIUM FLUOSILICATE, or 4 oz. TOXAPHENE, or 1½ lb. CALCIUM ARSENATE.

Caution: To avoid soreness of the hands as a result of mixing and handling the bait, thoroughly grease hands with PETROLATUM or grease, working it well under and around the fingernails.

References: W.12; U.4; V.1.

POISONHEMLOCK (*Conium maculatum*) is a deadly "hemlock" now naturalized in western range country. When other feed is not available, sheep and cattle may eat the fruits and leaves of this POISONOUS PLANT. Symptoms of poisoning are nervous tremors, weakness, respiratory paralysis.

POISONING. *See* PRUSSIC ACID POISONING; FORAGE POISONING; OAT HAY POISONING; SWEETCLOVER HAY POISONING; ACORN; BITTER LUPINES; CROTALARIAS; ALKALOIDS; MOLYBDENUM; FLUORINE; SELENIUM POISONING; POISONOUS PLANTS.

POISONOUS PLANTS have caused extensive losses to the livestock industry.

Known poisonous plants are listed here. The information on many of these plants is still incomplete, however, and

Poisonhemlock. Leaf, details of fruit, portion of the inflorescence, details of flower, and root system. (G.19.)

other plants are also proving to be toxic under present GRAZING conditions.

Toxic substances found in poisonous plants include GLUCOSIDES—i.e., SAPONINS

POISONOUS PLANTS
Essentials about the principal poisonous plants in the United States

Common name of plant	Location	Animals most commonly poisoned
Arrow podgrass	Marshes and wet places throughout U. S.	Cattle and sheep
Baccharis	Hillsides of western Texas and southern New Mexico and Arizona	Cattle
Blacklaurel	Springy ground in mountains of California	Sheep
Black nightshade	Waste grounds from Maine to California	Cattle, sheep, goats, chickens, ducks, and geese
Bracken	Thickets and rich woods throughout U. S.	Horses and cattle
Cocklebur	Fields and waste lands of the eastern half and low wet places of the western half of U. S.	Pigs and cattle
Common oleander	Fields, edges of woods, roadsides in southern part of U. S.	All animals
Copperweed	Colorado Basin to southern California	Cattle and sheep

POISONOUS PLANTS—Continued

Common name of plant	Location	Animals most commonly poisoned
Deathcamasses...........	Gravelly hills, depressions, and meadows in western half of U. S.	Sheep and cattle
Dutchman's breeches.....	Woods in eastern half of U.S. north of Georgia	Cattle
Greasewood..............	Alkaline fields in western part of U. S.	Sheep
Greenstem paperflower....	Northern Arizona and southern Utah	Sheep
Groundsels—see Ragwort		
Harmel peganum........	Texas and New Mexico	Sheep and cattle
Horsebrushes............	Utah, Nevada, and eastern California	Sheep
Horsetails..............	Wet meadows throughout U. S.	Horses
Kalmias................	Hillsides, woods, and swamps, Maine and New York to Georgia	Cattle, sheep, and goats
Larkspurs..............	Mountains and plains throughout U. S.	Cattle
Locoweeds..............	Plains and mountain valleys, western half of U. S.	Cattle, horses, sheep, and goats
Lupines................	Throughout U. S.	Sheep and cattle
Milkweeds..............	Washington, Idaho, Utah, California, Arizona, Colorado to Mexico, and Kansas to Texas	Cattle and sheep
Oaks..................	Lower mountains of Colorado, Utah, and Southwest	Cattle
Parry aster.............	Dry flats of Wyoming	Sheep
Poisonbean.............	Coastal plains of Florida and Texas	Cattle, sheep, and goats
Poisonhemlock..........	Widely distributed	Sheep and cattle
Poisonvetches..........	Mountains, foothills, and valleys of Inter-mountain States	Cattle and sheep
Ragwort and groundsels..	Throughout U. S.	Cattle and horses
Rayless goldenrod.......	Fields along ditches in western Texas, New Mexico, and Arizona	Cattle, sheep, and horses
Rubberweeds...........	Western Texas and southeastern California	Sheep
St. Johnswort..........	Fields, waste places, and hills across northern half of U. S.	Animals with areas of white skin and hair
Tarweed...............	Northwest of U. S.	Horses, cattle, and swine
Thickleaf drymary.......	Denuded areas in western Texas and southern New Mexico	Cattle
Waterhemlocks.........	Wet places throughout the U. S.	Sheep and cattle
Western azalea..........	Moist places in mountains of California	Sheep
Western sneezeweed......	Mountains, meadows, and valleys from Montana to Arizona	Sheep and cattle
White snakeroot.........	Rich woods and ravines in eastern half of U. S.	Cattle and sheep
Wild cherry............	Hillsides, along streams, and woods throughout U. S.	Sheep and cattle

and CYANOGENETIC GLUCOSIDES—RESINOIDS, OXALIC ACID, TREMETOL, ALKALOIDS, and other chemical compounds, as yet undetermined.

References: H.43; S.10.

See also SORGHUMS; PRUSSIC ACID POISONING; SELENIUM POISONING; OAT HAY POISONING; SWEETCLOVER POISONING; DALLISGRASS POISONING; WILD WINTER PEA POISONING; ACORNS; BITTER LUPINES; GRASS TETANY; CROTALARIAS; FLAXES; HALOGETON; RANGE MANAGEMENT, FESCUE LAMENESS.

POISONVETCHES (*Astralgalus* spp.), or *milk-vetches*, belong to the same genus as some of the LOCOWEEDS, but they do not cause locoism; they produce entirely different symptoms, namely difficult breathing, nausea, and weakness. These symptoms appear when cattle and sheep eat large quantities of leaves and stems of such species as *two-grooved poisonvetch*, *four-winged poisonvetch*, *timber poisonvetch*, or *straight-stem poisonvetch.*—*See also* POISONOUS PLANTS; SELENIUM POISONING.

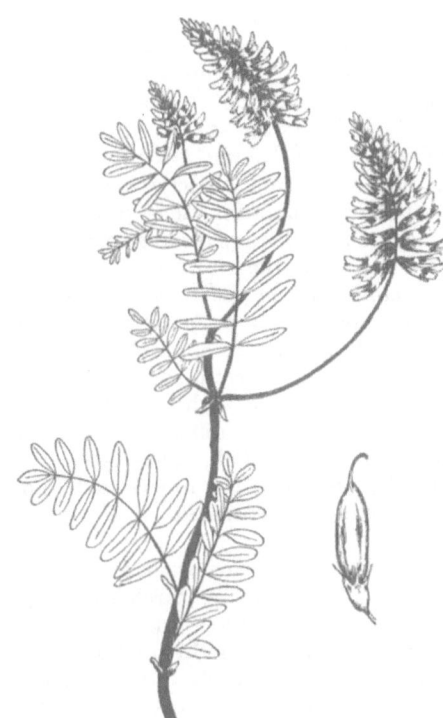

Two-groved poisonvetch. Twig and seed pod. (D.19.)

POLISHED RICE. *See* RICE.

POLISH WHEAT (*Triticum polonicum*) is a WHEAT species which is seldom grown in the United States. Polish wheat has a spring habit, the stem is tall, the SPIKE awned, large, and lax. The GLUMES are papery, very long, and narrow. The kernel is narrow and sometimes nearly ½ in. long; it is hard and its shape is somewhat similar to a kernel of RYE.

Polish wheat usually yields less than other adapted wheats.

White Polish is one of the varieties of this wheat species.

Reference: B.15.

POLLEN, or *pollen grains*, the fertilizing floral dust or powder, consists of the male germ cells of flowering plants.—*See also* POLLINATION.

POLLINATION is the transfer of POLLEN from the ANTHERS of the STAMENS to the STIGMA of a PISTIL, usually with the help of *pollinators* (beneficial insects) or wind.

The flowers of some plants are *self-fertile* and *self-pollinating*; others are self-fertile, but must be tripped or shaken to insure pollination. The flowers of some species of TRUE CLOVERS are *self-sterile*, and therefore require *cross-pollination* (the pollen grains must be transferred from plant to plant rather than from one FLORET to another on the same plant.)

Selfing is a synonym for self-pollination. —*See also* BEES; SELF-POLLINATED.

POLLINATORS. *See* POLLINATION.

POLY- is a prefix meaning many.

POLYGONUM. *Polygonum* spp. = KNOTWEEDS; *P. convolvulus* = WILD BUCKWHEAT.

POLYMYXIN is a generic term for ANTIBIOTICS produced by strains of the *Bacillus polymyxa*.

POLYSACCHARIDES are complex CARBOHYDRATES containing more than three molecules of simple sugars (hexoses). They include such natural products as STARCH, CELLULOSE, HEMICELLULOSES, and INULIN.

POMACE is the substance of fruits crushed by grinding, and used as food or feed; e.g., APPLE POMACE; DRIED APPLE POMACE; DRIED TOMATO POMACE.—*See also* PULP.

PONDEROSA PINE (*Pinus ponderosa*) is a dominating tree species in open forests on range land.—*See also* RANGE PLANTS.

POOR HAY. *See* HAY GRADING.

"POOR MAN'S ALFALFA" = SERICEA.

POPCORN. *See* CORN.

POPILLIA. *P. japonica* = JAPANESE BEETLE.

POPPY-SEED OIL MEAL, obtained as a by-product of the poppy-seed oil manufacture from OPIUM POPPY, contains some opium ALKALOIDS and therefore may exert a narcotic (sleep inducing) action when fed to young animals or to mature animals in large amounts; 2 lb. per dairy cow daily is considered safe.—*See also* OIL MEAL.

PORCUPINE GRASS (*Stipa spartea*) grows to a height of 3½ ft. It occurs on prairies of Canada and from Pennsylvania to Montana as well as in Missouri and New Mexico. When mature, this GRASS is injurious, especially to sheep, because of its sharp points which penetrate the skin.

Reference: U.6.

See also RANGE PLANTS.

POSEY = *Fultz*. See COMMON WHEAT.

POSO 48. See CLUB WHEAT (variety).

POSSUM-EARS = POUTS.

POTASH is a loosely used term to indicate many different POTASSIUM compounds, their *potash values* being expressed as the equivalent amount of K_2O (POTASSIUM OXIDE).—*See also* FERTILIZERS.

POTASH DEFICIENCY
= *Potassium deficiency*. See POTASSIUM.

POTASSIUM is a CHEMICAL ELEMENT. Some of its salts are found in blood, others are useful as FERTILIZERS. In small amounts, potassium is essential to normal nutrition, but this chemical element is so widely distributed in most crops, that *potassium deficiency* in livestock rarely develops.—*See also* SORGHUMS; WOOD ASHES.

POTASSIUM CHLORIDE, or *muriate of potash*, contains 42.5 percent POTASSIUM. It forms white, water-soluble crystals or powder and is used as FERTILIZER and as drug.

POTASSIUM IODIDE, a white, granular, water-soluble powder, contains 23.5 percent POTASSIUM and 76.5 percent IODINE and is widely used in the form of *iodized salt*.—*See also* MINERAL FEEDS.

POTASSIUM NITRATE, or *saltpeter*, contains 38.6 percent POTASSIUM and 13.8 percent NITROGEN. The white, granular powder has a cooling, saline taste and is soluble in water. It is valued as FERTILIZER and as an oxidizing agent. When reduced to POTASSIUM NITRITE, it becomes the cause of the so-called OAT HAY POISONING.

POTASSIUM NITRITE, the commercial grade of which usually contains 15 percent POTASSIUM NITRATE, is toxic. It has no place on the farm and is often the cause of so-called OAT HAY POISONING of farm animals.

POTASSIUM OXIDE—K_2O—is often misnamed "POTASH." It forms a gray powder or colorless crystals and is water-soluble. It is not used on the farm, but the term *"potash value,"* referring to K_2O—which is 2.4 times (or 140 percent) larger than the respective elemental POTASSIUM value—appears on tags, especially of FERTILIZERS.

POTASSIUM SULFATE contains 44.8 percent POTASSIUM. It occurs in white crystals or as granular powder and is water-soluble. Potassium sulfate is used as FERTILIZER and for eradication of such weeds as the DODDERS.

POTATO (*Solanum tuberosum*) is an annual with many varieties which grows best in cool climates. Tuber production is retarded at soil temperatures above 68° F.; freezing is injurious to the tubers. They are cultivated in different types of soils from sandy loam to peat; well-drained, uniformly moist soils are preferred.

The plant has erect branched stems 1 to 2 ft. long, with slightly hairy leaves. Flowers may have different colors; they develop into smooth, berry-type fruits, called *potato balls*. The tubers are found in the soil as enlargements of the STOLON ends.

The average tuber contains about 78 percent water and, chiefly, starch, some sugar, and other carbohydrates; the tuber is poor in protein, fiber, fat, and vitamins A and D.

Potatoes, particularly *surplus* and *cull potatoes*, make satisfactory animal feeds because they are high in starch and low in fiber; their protein content, on the basis of dry matter, is comparable to that of whole grains. Potatoes should be *cooked* before being fed to poultry or swine. They may be fed *raw* to ruminants, preferably after chopping, or *ensiled* with hay or dry corn fodder.

Sprouted potatoes, when fed in large quantities, are injurious to livestock; hogs and other animals die after eating potato sprouts. The toxic substance in the sprouts is SOLANINE.

Dehydrated potatoes—also called *potato flakes* or (if powdered) *potato meal*—may be a partial substitute for grain in livestock feeds, including those for swine and poultry (if the drying temperature is high enough for thorough cooking).

Dangers: Among the many potato diseases are MOSAICS, BLIGHTS, CHARCOAL ROT, RHIZOCTONIA STALK ROT; and insect pests such as BEETLES, APHIDS, WIREWORMS, the POTATO LEAFHOPPER, and the STEM NEMATODE.

Reference: E.12.

See also PLANT BY-PRODUCTS; SILAGE

CROPS; SWEET POTATOES; ALFALFAS; WHEATS; OATS; SOYBEAN; BUCKWHEAT; CORN; DISTILLERS' PRODUCTS; LIGHTNING INJURY.

POTATO DISTILLERS' DRIED RESIDUE.
See DISTILLERS' PRODUCTS.

POTATO FLAKES. *See* POTATO.

POTATO LEAFHOPPER (*Empoasca fabae*)
is a serious pest of ALFALFA in the eastern half of the United States. It also occurs throughout the country, with the possible exception of the Northwest, and has been found living on nearly 200 kinds of plants. It not only causes heavy losses to alfalfa, but damages RED CLOVER, Ladino WHITE CLOVER, PEANUT, and other forage LEGUMES, as well as POTATOES, BEANS and deciduous nursery stock.

The potato leafhopper pierces the leaves and leaf stalks, and sucks the juices of the plants, causing a yellowing and dwarfing of the foliage and, in heavy attacks, severe wilting. Various shades of pink, red, and purple, beginning at the midrib of the leaf, are typical of leafhopper injury.

Much of the loss to alfalfa is in young stands, where the injury to the alfalfa permits WEEDS and GRASSES to crowd it out. Alfalfa may be weakened so much that it cannot survive the winter. Usually the greatest damage is done to the second crop, but the third crop also is often severely injured in August and September.

Hay from alfalfa injured by the potato leafhopper, contains less protein and more carbohydrates and is only half as rich in carotene as hay from normal, green alfalfa.

The adult leafhopper is a pale-green insect about ⅛ in. long. It is very active, jumping or flying about when disturbed. Young and adult forms can run backwards and sideways as rapidly as forward.

The female leafhopper lays slender white eggs within the stems and larger veins of the leaves. In the summer the eggs hatch in six to nine days. At first the young leafhoppers are nearly white, but as they develop they become pale green. They shed their skins four times before they become full-grown and acquire wings. Soon after the adults appear they mate, and within a few days eggs are laid, and a

new life cycle begins. The period from egg to adult is about three weeks in warm weather.

The potato leafhopper has not been found overwintering north of the Gulf States. Early in the spring it moves northward with the warm winds. Near the Atlantic seaboard the insect feeds mainly on young OAK and hickory foliage to build up its migrating population. It can increase to enormous numbers within a short time. There are several generations a year.

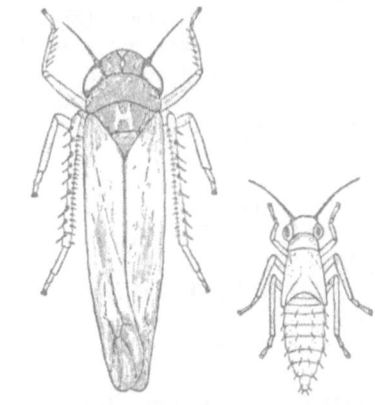

Potato leafhopper and fourth instar nymph. (Q.2.)

If there is any doubt as to whether this pest is present in injurious numbers, it is best to sweep the alfalfa with an insect-collecting net having a 15-in. opening. On a warm day when it is not windy, 20 sweeps from 5 different parts of the field should be taken. If, on an average, one potato leafhopper per sweep is captured, the infestation may damage the crop within three weeks.

Control: SPIDERS, MITES, certain insects and an infectious FUNGUS disease sometimes destroy both the nymphs and the adults of the potato leafhopper, but none of these natural enemies can be relied upon for satisfactory control. There are several ways to reduce potato leafhopper infestations: (1) Grow hardy varieties of alfalfa that have been found best adapted to your locality. (2) Plant alfalfa as far as possible from large plantings of beans, potatoes or other host plants. (3) Plant a 30-ft. strip of a grain or grass crop between alfalfa and other host crops to prevent immature leafhoppers from migrating. (4) Do not cut oftener than at approx. 45-day intervals, even if the crop begins to discolor. If leafhoppers are present, delay harvesting at the first crop about ten days. By that time large numbers of adults will have laid their eggs and will be taken away with the hay; any nymphs present will die of starvation, because they are too small to migrate to other food plants. (5) Do not interrupt cutting the field, or the leafhopper will migrate from the cut to the uncut area and cause severe damage. (6) If these cropping practices have not kept the potato leafhopper under control, spray with METHOXYCHLOR when the alfalfa is about half grown; spray earlier if the insect becomes abundant.

Reference: P.9.

See also LEAFHOPPERS.

POTATO MEAL. *See* POTATO.

POTATO SILAGE. *See* SILAGE CROPS.

POULARD WHEAT (*Triticum turgidum*), also known as *rivet-wheat*, is a WHEAT species with either winter or spring habit. The plants usually are tall with broad leaves. The culms are thick, the SPIKES long, and the GLUMES short. The kernels are thick, humped, and mostly hard, but they are usually very starchy (yellow berry).

The poulards are most closely related to DURUM WHEAT, and many varieties of poulard and durum look nearly alike. Only a very limited acreage of poulard wheat is cultivated in the United States, and the grain is of no commercial value except as feed for stock.

Alaska is probably the best-known poulard variety.

Reference: B.15.

See also COMMON WHEAT.

POULTRY BY-PRODUCT MEAL.
See ANIMAL PRODUCTS.

POULTRY CONCENTRATES.
See CONCENTRATES.

POULTRY DRESSING PLANT BY-PRODUCTS. *See* ANIMAL PRODUCTS.

POULTRY MANURE is richer in PHOSPHORUS and poorer in water than manure from livestock; it is a close second to sheep manure in NITROGEN content.—*See also* MANURE; FERTILIZERS; GRAZING.

POULTRY PASTURES AND RANGES.
See GRAZING.

POULTRY RATIONS. The nutritionist probably understands more about the nutrient needs of *chickens* than of any other farm animal. The efficiency with which the broiler industry operates today is in no small measure due to the recent nutritional advances. The nutrients that are of particular concern to the poultry industry are good quality PROTEINS in large amounts, high levels of ENERGY to maintain maximum production, large levels of CALCIUM and PHOSPHORUS, supplemental MANGANESE, and a very liberal supply of the VITAMINS, particularly fat-soluble vitamins A, D, E and, of the water-soluble groups, RIBOFLAVIN (B_2), CHOLINE, NIACIN, PANTOTHENIC ACID, and VITAMIN B_{12}.

In general, ordinary farm feeds are deficient in most of these important nutrients, with the exception of energy. For this reason, the poultry man normally buys large amounts of supplemental feed for successful operations. It is possible to provide satisfactory feeds based on low fiber farm grains and milk, fish, or meat products. These are rather expensive. The needs of the birds can be equally well supplied by replacing most of the animal protein with SOYBEAN meal, supplementing the ration with pure MINERALS, vitamines, and AMINO ACIDS. Extra FAT can be added to increase energy levels. These fortified RATIONS are particularly advantageous for starting poults and for broilers, but they can also be used for other birds.

BROILER RATIONS

Broiler producers usually depend on fast turnover of capital, a low mortality rate, and a well finished bird to insure them a reasonable profit at low margins. Well fortified rations are demanded to support this program.

Broiler rations are usually high energy rations with added fat, fortified with good quality proteins (and sometimes amino acids), high levels of vitamins and minerals. ANTIBIOTICS are recommended and should be included at the level of 2 to 8 gm. per ton of feed. Broiler rations usually contain one or more *medicaments*. During the finishing period, many producers dilute the standard broiler ration with CORN to give desired fat and color to the carcass. See table for sample broiler rations.

STARTER AND GROWER RATIONS

The development of replacement pullets is usually divided into two phases; from 0 to 8 weeks is the normal starting period, and from 8 to 20 weeks, or the beginning of lay, is the growing period.

The *starting period* is nearly always indoors and requires a complete ration that will encourage rapid growth but not excess fattening. The rations can contain limited amounts of fibrous grains such as OATS or

BROILER RATIONS

Ration:	1	2	3	4	5
Ground yellow corn, lb.	575	435	500	585	585
Milo or wheat, lb.	...	100	...	...	...
Barley or oats, lb.	50	...	...	...	...
Wheat middlings or shorts, lb.	...	100	50	...	...
Soybean oil meal, 44%, lb.	280	200	300	300	250
Stabilized animal fat, lb.	...	...	25	...	30
Meat and bone scraps, lb.	...	50	50	...	...
Fish meal, lb.	50	50	...	...	50
Fish solubles, condensed, lb.	...	...	30	25	...
Dried milk or whey, lb.	...	...	...	25	10
Distillers' solubles, lb.	...	...	...	...	20
Alfalfa meal, 17%, lb.	...	50	30	...	25
Alfalfa meal, 20%, lb.	...	...	...	20	...
Riboflavin supplement (500 mg./lb.), lb.	4	3	4	2	2
Methionine, lb.	...	½	½	½	½
Limestone, lb.	10	5	...	15	7
Steamed bone meal or CaHPO₄, lb.	20	3	...	18	12
Salt, lb.	5	3	5	5	3
Manganese sulfate, oz.	4	4	4	4	4
Vitamin A source (5,000 I.U./gm.), gm.	250	150	200	...	150
Calcium pantothenate, gm.	2	2	2	1½	2
Niacin, gm.	8	4	8	5	10
Choline, gm.	150	100	50	100	50
Vitamin B₁₂ supplement (6 mg./lb.), lb.	½	½	½	½	½
Dry vitamin D₃ (3000 I.U./gm.), gm.	4	4	4	8	4
Sulfaquinoxaline, oz.	...	2	...	...	+
Antibiotic, if desired	+	+	+	+	+
Arsanilic acid	+	...	+	+	...

STARTER RATIONS
Approx. 1000 lb.

Ration:	1	2	3	4	5
Yellow corn meal, lb.	360	580	650	360	250
Ground corn, milo, or wheat, lb.	200	...	...	100	...
Ground oats or barley, lb.	...	...	...	100	100
Wheat middlings or shorts, lb.	100	100	...	80	100
Soybean meal, 44%, lb.	200	200	250	150	...
Oil meals other than soybean, lb.	...	...	...	100	50
Fish meal or condensed fish solubles, lb.	20	25	25	...	50
Dried skim milk or whey, lb.	25	...	...	...	...
Meat scraps, 50%, lb.	50	50	25	40	50
Alfalfa meal, 17% or better, lb.	25	25	25	50	25
Ground limestone, lb.	5	5	7	7	...
Salt, lb.	5	5	5	5	...
Manganese sulfate, oz.	2	2	2	2	2
Riboflavin supplement, lb.	2	2	2	2	2
Feeding oil (vitamins A and D), lb.	2	2	2	2	2
Calcium pantothenate, gm.	...	...	2	...	...
Antibiotic source, if desired	+	+	+	+	+
Steamed bone meal, lb.	5	5	10	8	...

BARLEY and some mill by-products. The bulk of the protein usually comes from soybean meal, but limited amounts of other oil meals or corn gluten meal may be used. The ration should contain some ALFALFA meal or its equivalent, and one or two animal proteins. Unless fermentation or milk by-products are used, a riboflavin (B_2) supplement will be needed; either FISH LIVER OIL, or a dry source of vitamins A and D_3 will also be needed.

Examples of adequate starter rations are shown in the table on page 367.

During the *growing period*, chicks can be maintained on good quality range and a simple mixture of grains and oil meals, provided that adequate minerals are available. On this program, date of first lay may be later than on a heavy grain feeding program; but a healthy pullet will usually be produced if predators are controlled. Satisfactory feeds can be prepared that will enable the poultryman to conduct the growing phase indoors if he desires.

The growing bird does not require feeds as highly fortified as does the starting chick, but good quality ingredients must be used.

LAYING RATIONS

A 5-lb. hen needs about 90 lb. balanced feed each year to maintain body weight and produce 200 eggs.

Laying rations that will support good levels of production can be made reasonably well from farm grains and meat scraps and soybean meal. The required B-vitamins can be supplied in concentrate form or can be picked up from natural sources such as green feeds, insects, and cow MANURE. If eggs are being used for hatching, use more good quality protein and 50 percent more B-vitamins—riboflavin (B_2), pantothenic acid, folic acid, and vitamin B_{12}—in the breeding ration than is present in the normal laying ration.

The use of high energy laying rations with added fat will increase feed efficiency and help maintain winter egg production at high levels.

LAYING OR BREEDING RATIONS

	All mash			Mash to be fed with farm grains
Ration:	1	2	3	4
Ground yellow corn, lb.	630	510	612	400
Ground barley or oats, lb.	...	100	...	...
Wheat bran, lb.	...	...	...	70
Middlings or shorts, lb.	100	100	125	140
Soybean oil meal, lb.	100	100	100	205
Meat scraps, lb.	25	25	50	50
Fish meal, lb.	25	25	20	25
Dried milk or whey, lb.	25	25	...	25
Alfalfa meal, lb.	40	30	25	50
Dried distillers' solubles, lb.	...	25	...	...
Feeding oil, lb.	2	2	2	...
Limestone, lb.	30	30	30	...
Salt, lb.	5	5	5	10
Manganese sulfate, oz.	3	4	2	5
Steamed bone meal, lb.	10	10	8	30
Calcium pantothenate, gm.	...	...	1	...
Vitamin B_{12} supplement *	1	1	1	...
Niacin, gm.	...	...	10	...
Riboflavin supplement (250 mcg./lb.) *	6	6	10	...
Grain mixture to be fed with equal parts of mash in the scratch grain:mash system				
Corn and/or milo.	...	...	...	50
Heavy weight oats.	...	...	...	25
Wheat and/or barley.	...	...	...	25

* Vitamin B_{12} and riboflavin may be omitted from layer rations.

Rations for *caged layers* must be more carefully balanced than those for conventionally housed birds. It is known that built-up litter is a source of vitamin B_{12} and possibly other NUTRIENTS. In addition, the cage condition appears to increase the requirement of the birds perhaps through the "stress" mechanism.

Satisfactory rations are shown in the table. The rations can be formulated to suit the needs of the poultry keeper depending upon the feeds that are available to him at a reasonable price.

TURKEY RATIONS

Rations for three different stages of turkey development are most often prepared: a starting ration; a growing ration; and a breeding ration.

The first two rations are similar to the corresponding chick rations but contain more protein, vitamins, and minerals because the poult grows at a very rapid rate. Antibiotics are added to the starting ration as they are in broiler rations. The breeding ration should be similar to a hen ration but it may require additional fat-soluble vitamins.

The table contains an example of a starting and a growing ration.

TURKEY RATIONS

	Starter	Grower
Yellow corn, lb..............	312	650
Wheat middlings or shorts, lb................	50	...
Soybean oil meal, lb........	400	150
Meat scraps, lb.............	50	30
Alfalfa meal, 17%, lb.......	60	80
Fish meal or solubles, lb.....	40	30
Steamed bone meal or $CaHPO_4$, lb..............	10	10
Limestone, lb...............	12	20
Iodized salt, lb.............	4	5
Manganese sulfate, oz.......	4	4
Dried whey or buttermilk, lb.	25	20
Stabilized animal fat, lb.....	25	...
Riboflavin suppl. (250 mg./lb.), lb.........	4	4
Calcium pantothenate, gm...	2	1
Choline chloride, gm........	250	...
Niacin, gm................	15	5
Antibiotic.................	+	...
Vitamins A & D (2250 I.U. of A, and 400 I.U. of D/ gm.), lb................	2	2

DUCK RATIONS

Ducks appear to need slightly less protein and considerably more B-COMPLEX VITAMINS in their feed than do chicks of the same age. It would seem to be impossible to provide an economical feed without using supplements of riboflavin (B_2), pantothenic acid, and niacin.

PIGEON RATIONS

The feeding of pigeons is still an art rather than a science. Most pigeon rations are prepared from whole grains. A typical ration for squab production might contain sweet CORN, 40 percent; the SORGHUMS, kafir and milo, 30 percent; wrinkled PEAS, 20 percent; and WHEAT, 10 percent. MINERALS would be fed free choice. Such a ration will be improved by the addition of the vitamin B_{12} and riboflavin (B_2).—*See also* DRUGS.

POUTS, or *possum ears*, is a plant injury caused by TOBACCO THRIPS.

POWDERY MILDEW is caused by the FUNGUS *Erysiphe polygoni*. In the southern part of the Lupine Belt, it attacks the leaves and stems of LUPINES, producing irregular whitish blotches of varying size that give the plant the appearance of having been dusted with flour. The lower leaves are infected first, but most or all of the others are affected eventually. If the disease is severe, the leaves may drop off, depriving the plants of the foliage necessary to provide food for maturing their seed. Badly affected plants may produce little or no seed, and the plants may be killed.

Powdery mildew is also one of the most common and widespread diseases of CLOVER, especially of RED CLOVER, and to some degree of SUBCLOVER. Severe epidemics reduce forage yield and quality of hay. When the infection is severe, the leaves will turn from powdery gray to yellow and brown and fall off. The fungus can be controlled by using resistant clover varieties.

Other forage plants affected by powdery mildew are BARLEY, WHEATS, OATS, FIELD PEA and COW PEA.

References: T.1; W.8.

See also TRUE CLOVERS; RUSSIAN WILD-RYE; COMMON WHEAT.

Pr is one of the inbred lines of CORN hybrids.

PRAIRIE HAY, or *wild hay*, is obtained from a variety of native and wild GRASSES; its value depends chiefly on local conditions, stage of cutting, and the grass species available. Good sources of prairie hay are WHEATGRASSES, GRAMA GRASSES, and BUFFALOGRASS. It is important to maintain clean meadows for the production of high-quality prairie hay. If foreign material such as brush, weeds, and dead grass is present in prairie hay, animals will eat only small amounts of it.

The protein content of prairie hay is low. The carotene and calcium contents are relatively low, as compared to LEGUME hays, and under conditions of advanced maturity or weathering damage, in all probability, are critically low.

Satisfactory milk production can be obtained with good quality prairie hay, provided it is fed to cows with a concentrate mixture supplying the required amount of protein. A ration containing prairie hay and a 20-percent protein concentrate compares favorably with one made up of alfalfa hay and a 15-percent protein concentrate. To insure an adequate amount of carotene (provitamin A), prairie hay may require a supplement of it. Unless the hay is grown on high calcium soils, a calcium supplement should also be added to the ration.

It is advisable to offer prairie hay in combination with a legume roughage when possible. The protein thus furnished reduces the percentage of protein needed in the concentrate. A legume also increases the carotene and calcium in the ration.

In general, more pounds of weight-gain per acre of prairie grasses can be obtained from early-cut hay than from late-cut hays. References: R.11; B.25.

See also HAY; HAY GRADING; HAY MEASURING.

PRATYLENCHUS. *Pratylenchus* spp. (e.g., *P. leiocephalus*) are MEADOW NEMATODES.

PRECIPITATE is a fine deposit, obtained by its being *precipitated* (thrown out) from a solution (in which it was dissolved) by becoming insoluble, due to temperature change or action of a *precipitant* (reagent).

PRECIPITATED BONE PHOSPHATE, often labeled as *dicalcium phosphate from bone*, is one of the MINERAL FEEDS.

PRECIPITATED CALCIUM CARBONATE. *See* CALCIUM CARBONATE; MINERAL FEEDS.

PRECIPITATED CALCIUM PHOSPHATE. *See* CALCIUM PHOSPHATE.

PRECIPITATED CHALK. *See* CALCIUM CARBONATE; MINERAL FEEDS.

PRECIPITATED SULFUR. *See* SULFUR.

PRECURSOR is anything that precedes another physiologically active substance; or a physiologically inactive substance from which a VITAMIN, HORMONE, or ENZYME is derived, e.g., carotene is the precursor of vitamin A; tryptophan that of niacin.

PREDACEOUS means: living by prey.— *See also* PREDATORS.

PREDATORS are insects, birds, and other animals living by prey. Among the beneficial predatory insects are the STINK BUGS, LYGUS BUGS, and GARDEN WEBWORMS.

PRE-DIGESTING FEEDS for the purpose of saving on feeding cost, or improving milk or meat production, has proved to be a waste of time and money. In controlled feeding experiments, it has been repeatedly shown that pre-digested grains are often inferior to the same feeds when fed undigested. However, low-grade, high-fibrous, UREA-containing feeds have been experimentally fermented by rumen organisms and fed to hogs and poultry with good results.

PREGNANT COW RATION. *See* BEEF CATTLE RATIONS.

PREMIER WHEAT. *See* COMMON WHEAT (variety).

PREMIUM WHEAT = TRIUMPH. *See* COMMON WHEAT.

PREMO is one of the intermediate-type grain SORGHUMS.

PRESERVATIVES are used as *silage conditioners* and as MOLD INHIBITORS for feedstuffs.—*See also* SILAGE.

PRESSCAKE is the residue obtained as by-product of the remainder found in hydraulic presses used in oil mills or fermentation plants.—*See also* PEANUT; EXTRACTED PRESSCAKE.

PRICKLY COMFREY (*Symphytum asperrimum*), a perennial, is an inferior forage plant with large, hairy foliage.

PRICKLY PEAR CACTI (*Opuntia* spp.) are valuable as forage on ranges during drought periods, especially in southern Texas. It is necessary to remove the spines of the cacti before feeding to livestock; the gathered cacti are run through a special chopping machine, or the spines may be singed off with a torch.

The pears of this very slow-growing plant are watery and contain much crude fiber.—*See also* RANGE PLANTS.

PRIDESOY. *See* SOYBEAN (variety).

PROCAINE-PENICILLIN.
                                                     See PENICILLIN.

PROCESSED GARBAGE.
            See DEHYDRATED GARBAGE; GARBAGE.

PROCUMBENT = PROSTRATE.

PROLIFERATION is the production of new parts—occurring rapidly and repeatedly—from buds, offsets, etc.; also, the rapid succession of cell division or a new growth so formed.

PROLIFIC CORN is a name given various southern CORN varieties.

Early Prolific is a RICE variety.

PROPAGATION of plants, i.e., their multiplication, takes place according to one of two methods: (1) by *sexual reproduction* from seeds (which were previously produced through interaction between STAMENS and PISTILS), or (2) by *vegetative reproduction*, i.e., *asexually*, by grafting, budding, cuttings, or roots.—*See also* SPORES; SEED.

PROPRIETARY MIXED FEEDS are those manufactured under patent protection or under a trade-name or trade-mark which may or may not be registered.

PROPYL GALLATE and its esters are used in small amounts as ANTIOXIDANTS of fats. They should not be used in the presence of water and iron salts in feedstuffs to prevent their turning purple.

PROPYL PARA-HYDROXYL BENZO-ATE forms white crystals which are very slightly soluble in water. It is used as MOLD INHIBITOR.

PROSO (*Panicum miliaceum*), or *proso millet*—also called *broomcorn millet, hershey millet,* and *hog millet*—belongs to the PANICGRASSES.

It is distinguished from FOXTAIL MILLET chiefly by its head, which is a large open-branching PANICLE (like OATS). Proso has coarse, woody, hollow stems from 12 to 40 in. high. The stems are round or flattened and generally about as thick at the base as a lead pencil. The stems and leaves are covered with hairs. When threshed, most of the seed remains enclosed in the inner chaff or hull. The seed of proso is larger and not so tightly held in the hull as that of the millets of the foxtail group. The hulls of proso are of various shades and colors, including white, cream, yellow, red, brown, gray, and black. The bran, or seedcoat, of all varieties is a creamy white.

Proso. Black Voronezh proso showing the loose one-sided type of head. Red Lump proso showing the compact type of head.

Proso is cultivated from southern Carolina to the central Great Plains, and especially in eastern Colorado. Except in northeastern Colorado, proso should be grown only as a catch crop and not as a part of a regular rotation. Under average conditions other cereals will outyield proso if they are sown at the proper time. In rather dry seasons, however, proso often outyields other grains. Under extremely favorable conditions proso yields from 60 to 70 bu. per acre, but yields of 10 to 30 bu. per acre are the general rule.

Proso is sown in the spring and is adapted only to regions where spring grain is fairly successful. It is easily injured by frost either in the spring or fall and is not adapted to high altitudes or to localities where summer frosts occur.

Proso has the lowest water requirement of any grain crop. It is, however, less resistant to severe drought than well-adapted varieties of other grains mainly because of its shallow rooting habit. Hot winds occasionally dry and kill the plant before it forms seed, even when there is available moisture in the soil.

The plant is best adapted to a rich loam soil, but grows satisfactorily on nearly all types except coarse sandy soils. Because of its shallow feeding habits, a crop of proso leaves the upper layers of the soil somewhat impoverished.

Most varieties of proso require about sixty to eighty days from seeding to maturity. Seeding should be early enough to avoid early fall frosts. A fair rule is to sow proso from two to four weeks after CORN is planted.

Proso should be harvested when the seeds in the upper half of the heads are ripe, and the plant is still green. The seed is easily shattered when ripe, and the crop should be handled carefully to prevent loss. The binder is usually the best machine for cutting proso. The bundles left by the binder should be placed in rather small, or long and narrow shocks to prevent molding and to permit an early drying. The crop should be removed from the field as promptly as possible in order to prevent loss and damage from birds and field mice, which seem to prefer proso to any other grain. The seed of proso is separated from the straw very easily with an ordinary threshing machine.

Proso is not a hay crop and should not be grown for that purpose; *proso hay* is of poor quality. While the hay may be eaten, it is not relished, and considerable waste results. Proso can be made into *silage*, but the yields are too small to make it a profitable silage crop.

The unhulled proso *seed* has a composition very similar to that of oats but is slightly higher in feeding value than oats when fed to cattle. It is about equal in feeding value to corn and BARLEY for pigs and lambs. Additions of VITAMIN B_{12} significantly increase the rate of gain in swine. Proso is eaten very readily by all kinds of livestock but it should be ground before being fed. It is relished by poultry, and can be fed to them in unthreshed bundles, as the seed is shelled out easily by scratching. Proso is also a common ingredient in commercial birdseed.

Dangers: The only disease of any consequence that attacks proso is SMUT. This can be controlled by the standard formaldehyde SEED TREATMENT.

CHINCH BUGS attack proso.

Birds, gophers, and field mice are very destructive to proso after the seed is ripe. Harvesting as early as possible and prompt threshing or stacking will prevent much of the loss.

Varieties: There are three main groups of proso, according to the shape of the head of the panicle. These groups are further subdivided according to color of seed, color of chaff, height, and time of maturity. The leading varieties now grown in the United States are the *Yellow Manitoba* and *Early Fortune*. The *Hansen White Siberian*, *Red Russian*, *Turghai*, *Tambov*, *White French*, *Red Lump*, and *Black Voronezh* are grown to a lesser extent; a considerable amount of proso without a name is grown and merely referred to as "hog millet."

References: W.6; D.13; M.39.

PROSO HAY. *See* PROSO.

PROSO MILLET = PROSO.

PROSOPIS. *P. juliflora* = MESQUITE; *P. odorata* = TORNILLO.

PROSO SILAGE. *See* PROSO.

PROSTRATE, or *procumbent*, means: lying flat on the surface of the ground; said especially of stems.

PROTEID is a rather obsolete term for PROTEIN; its meaning is varyingly restricted by different authors to nitrogenous material of plant cells, ALBUMINS, etc.— *See also* PROTOPLASMA.

PROTEIN. *See* TRUE PROTEIN; CRUDE PROTEIN; PROTEINS; THYROPROTEIN.

PROTEIN ALFALFA MEAL.
                            See ALFALFAS.

PROTEIN COTTONSEED CAKE.
                            See COTTONSEED.

PROTEIN COTTONSEED MEAL.
                            See COTTONSEED.

PROTEIN PEANUT FEED.
                            See PEANUT.

PROTEIN PEANUT MEAL.
                            See PEANUT.

PROTEINS (formerly also called *proteids*) are composed of extremely complex combinations of AMINO ACIDS; they occur naturally and are essential constituents of all living cells, both animal and vegetable.

These nitrogenous substances contain (in addition to NITROGEN) CARBON, HYDROGEN, and OXYGEN; many contain SULFUR and some also PHOSPHORUS. More than fifty fairly definite proteins have been studied, among them the ALBUMINS, *nucleoproteins* (found in all cell-nuclear material) and such protein derivatives as PROTEOSES, PEPTONES, etc.

Most proteins are insoluble in water, but often somewhat soluble in salt solutions. When heated, they coagulate. Plants can produce their own proteins from simple nitrogen sources in the soil or air, but animals cannot and, therefore, must obtain them directly or indirectly from plant tissues. ENZYMES in the digestive tract break these food (feed) proteins down during DIGESTION, transforming them into amino acids which are recombined, after their ABSORPTION into the blood, into body proteins by other enzymes. The blood carries the proteins to the parts in the body where they are needed. Amino acids that are not used are destroyed in the liver, thus forming ENERGY and waste products, such as urea, which are excreted by the kidneys.

The proteins vary in their DIGESTIBILITY; some are *indigestible* and useless as food or feed; others are *incomplete*, lacking one or more of the essential acids. A chemical analysis alone does not give complete information on the real value of the "protein" of a given feedstuff for use in the different species and age groups of animals. In particular, many *vegetable proteins* have lower nutritional value than *animal proteins* which often are also cheaper.

Proteins are most important for animal health—they supply the building material for GROWTH, tissue repair, and HORMONE formation. Proteins can also be used for heat (body temperature) and ENERGY, but since proteins are usually the most expensive feedstuffs, this is a wasteful practice.—*See also* CRUDE PROTEIN; TRUE PROTEIN; METABOLISM; NUTRIENTS; LEGUME BACTERIA; OILSEED MEALS; ALEURONE; NITROGEN-FREE EXTRACT; LABEL; BROWNING REACTION.

PROTEIN SUPPLEMENTS. *See* COMMERCIAL MIXED FEEDS; SWINE RATIONS.

PROTEOLYSIS is the conversion of PROTEINS by decomposition or HYDROLYSIS to water-soluble products, such as PEPTONES and/or AMINO ACIDS.—*See also* SILAGE; PROTEOSES.

PROTEOLYTIC ENZYMES are those which are capable of splitting the PROTEINS; e.g., PEPSIN.—*See also* ENZYMES; DIGESTIVE TRACT.

PROTEOSES are water-soluble substances obtained from PROTEINS by splitting with ENZYMES (such as PEPSIN), acids, or alkalies. They differ from other protein decomposition products, such as PEPTONES, in that they can be precipitated by the addition of AMMONIUM SULFATE, while peptones remain soluble.

PROTHROMBIN occurs in the blood plasma where it plays an important role in the formation of blood clots.—*See also* SWEETCLOVER HAY POISONING.

PROTOPLASM is the colorless semifluid or almost jelly-like nitrogenous ("PROTEID") material found within the cell cavities of all living tissues.—*See also* ALBUMIN.

PROTOZOON (plural: *protozoa*) is a microscopic, single-cell form of animal life. *Protozoan parasites* occur in the blood, gastrointestinal tract, and other parts of the body; many of them cause diseases.—*See also* GASTRIC JUICE.

PROVITAMIN A = CAROTENE.

PROVITAMIN D₂ = ERGOSTEROL. *See also* VITAMIN D₃.

PROVITAMIN D₃
= 7-DEHYDROCHOLESTEROL.

PRUNUS. The genus *Prunus* includes these WILD CHERRY species: *P. demissa* = *western chokecherry; P. melanocarpa* = *black chokecherry;* and *P. virginiana* = *common chokecherry.—See also* POISONOUS PLANTS.

PRUSSIC ACID, or *diluted hydrocyanic acid*, is a water solution of the colorless gas HYDROGEN CYANIDE with a characteristic odor of bitter almonds. It is extremely poisonous.—*See also* SORGHUMS; JOHNSONGRASS; SUDANGRASS; FLAX; STRAW; SILAGE CROPS.

PRUSSIC ACID POISONING, also called *cyanide poisoning* or *sorghum poisoning*, causes a considerable loss of cattle each year. This danger is a serious disadvantage in the use of all SORGHUM species as a pasture or soiling crop. The danger is, however, not as great with certified sweet SUDANGRASS and some especially selected sorghum species.

To allow cattle, sheep, and goats to consume even a small quantity of sorghum grass before it has matured may cause death. Horses are less, and hogs least, susceptible. Farmers sometimes test the sorghum plants for poison by turning some animal of little value into the field first. If the poison is present in dangerous proportions it may very soon become apparent. Cattle, however, are not equally susceptible to poisoning by sorghums.

Symptoms of prussic acid poisoning are depression, paralysis, stupor, and difficult breathing. Death may occur in minutes.

The following facts about prussic acid poisoning must be kept in mind: (1) As *maturity* of the plants is approached, the PRUSSIC ACID content of sorghum decreases. Small plants (including those wilted from hot winds or retarded by drought or frost), early second growth, and young branches and suckers are high in prussic acid. Mature plants with ripe seeds are seldom dangerous, especially if the growth has been normal and few suckers and branches are present. Most of the prussic acid is found in the leaves, particularly the younger ones. (2) *Well-cured* sorghum fodder has lost much of its prussic acid and ordinarily is safe to feed to animals. Partly cured sorghums may be dangerous, as is hay produced by the drying of plants cut while in a toxic state. (3) Sorghum *silage* can be fed with safety. (4) *Hungry* cattle straying into a sorghum field usually suffer losses. Even if the sorghum appears to be safe to pasture, the herd should not be turned into a field with empty stomachs. A light feeding of grain, given prior to turning the animals on the sorghum, will do much to prevent injury. (5) Less trouble is experienced in the *southern* states than in those farther north.

Control: The following are remedies recommended for prussic acid poisoning, when they can be administered immediately after symptoms of poisoning appear: injections of METHYLENE BLUE, SODIUM NITRITE, or SODIUM THIOSULFATE, preferably given intravenously; especially promising is the injection of a combination of sodium nitrite and sodium thiosulfate. This treatment may be supplemented by other measures, such as injections of ATROPINE or GLUCOSE, the inhalation of AMMONIA, etc. The injection of these drugs is attended with some danger to the sick animal; the treatment, therefore, should be given by a veterinarian.

Drenching affected animals with a sugar or glucose solution may prove helpful in an emergency (it represses the prussic acid formation). The feeding of CORN or other grain supplement increases the amount of glucose in the paunch, thus decreasing the danger of poisoning. Similarly, the feeding of *alfalfa hay* and *cottonseed cake* supplements has been found to counteract the effects of prussic acid poisoning.

References: M.1; V.1.

See also POISONOUS PLANTS; CYANOGENETIC GLUCOSIDES; SOYBEAN; GRAZING.

PSEUDO- is a prefix signifying false.

PSEUDOMONAS. The BACTERIUM *P. syringae* causes BACTERIAL SPOT; BACTERIAL STRIPE is caused by *P. andropogoni*. *P. alfalfae* causes BACTERIAL LEAF AND STEM SPOT; *P. glycinea* causes BACTERIAL BLIGHT OF SOYBEAN; *P. tabaci* causes WILDFIRE; *P. pisi*, BACTERIAL BLIGHT OF PEA; *P. alboprecipitans* causes the BACTERIAL LEAF BLIGHT; *P. striafaciens*, BACTERIAL STRIPE OF OATS; *P. coronafaciens*, HALO BLIGHT OF OATS, and *P. phaseolicola* causes the same disease in KUDZU.

PSEUDOPEZIZA. The FUNGUS *P. jonesii* causes YELLOW LEAF BLOTCH; COMMON LEAF SPOT is caused by *P. medicaginis*.

PSEUDOPEZIZA LEAF SPOT (*Pseudopeziza trifolii*) is a LEAF DISEASE which seldom is severe enough to cause defoliation. The spot infections are restricted in size. This disease, although abundant, is not a major factor in either hay or seed production of RED CLOVER.

Reference: E.5.

See also FUNGUS LEAF DISEASES.

PSEUDOPLEA. The FUNGUS *P. trifolii* is the cause of *Pseudoplea leaf spot* which is better known as PEPPER SPOT.

PSILOSTROPHE. *P. sparsiflora* = GREENSTEM PAPERFLOWER.

PTERIDIUM. *P. aquilinum* = BRACKEN.

PTEROYLGLUTAMIC ACID is the modern term for FOLIC ACID.

PUBESCENCE means: (1) hairiness or (2) the downy substance which covers parts of some plants.

PUBESCENT means: hairy or downy haired; specifically, covered with fine, soft, short hairs.

PUBESCENT WHEATGRASS (*Agropyron trichophorum*), also called *stiffhair wheatgrass*, is a moderately coarse grass with short, thick rootstocks (RHIZOMES) which may not be apparent until the second or third year. It is easily established and matures in late August.

This WHEATGRASS species is able to survive dry, hot, and windy summers; it is suitable for pastures on burned-over land, but is not adapted to wet, poorly drained soils. Forage yields are high, especially in

early spring and late fall; only the new growth is grazed.

Pubescent wheatgrass is cultivated for seed in the Intermountain region, Washington, and at higher elevations of the Southwest. It can be used for range reseeding, pasture mixtures, or for conservation of eroded or wind-blown areas. This grass also furnishes palatable grazing for sheep in Oregon during the summer.

Reference: W.16.

See also REE WHEATGRASS; GRASSES.

PUCCINIA. The FUNGUS *P. purpurea* causes SORGHUM RUST; *P. carthami*, SAFFLOWER RUST; *P. graminis tritici*, WHEAT STEM RUST; *P. graminis avenae*, STEM RUST OF OATS; *P. coronata avenae*, CROWN RUST; *P. rubigo-vera*, LEAF RUST of grasses; *P. glumarum* = STRIPE RUST.

PUERARIA. *P. thunbergiana* = KUDZU.

PUFFED CEREALS, by-products of the breakfast cereal manufacture, are used, *ground* or *unground*, in some special feeds.

PULL-DOWN = FODDER.

PULLED FODDER. *See* FODDER.

PULP is the "flesh" of fruits—a soft, solid mass of organic matter (of vegetable or animal origin). It may be used in rations as SUCCULENT feed (if moist) or as dried pulp; e.g., DRIED APPLE PECTIN PULP; DRIED BEET PULP; DRIED CITRUS PULP.— *See also* POMACE.

PULVERIZED OATS. *See* OATS.

PUMPKIN (*Curcubito pepo*) has a very low feeding value because the water content of the fruit is nearly 90 percent. Pumpkins are often planted in cornfields since most farm animals relish them; the seeds of pumpkins, which are rich in fat, should not be fed alone because they may cause indigestion.

PUPA (plural: *pupae*) is the *resting stage* in the development of an INSECT. The LARVA changes in the PUPARIUM to the pupa, and emerges, after the intermediate (*pupal*) period has passed (mostly in a few weeks), as the adult insect, called *imago*.

PUPARIUM (plural: *puparia*) is the tough, leathery incasement, or *pupal case*, in which a LARVA *pupates*—i.e., changes to a PUPA, and finally to an adult insect.

PURDUE 21.

See SIX-ROWED BARLEY (variety).

PURDUE 31 is a *popcorn* hybrid.—
See also CORN.
PURDUE No. 4 = *Red May.*
See COMMON WHEAT.
PURPLE ALFALFAS
= *common alfalfas. See* ALFALFAS.
PURPLE SEED STAIN, or *purple stain,*
is caused by the FUNGUS *Cercosporina kikuchii.* The disease occurs in most states
where SOYBEANS are grown. Weather conditions, during the time the seeds mature,
have a pronounced influence on the percentage of discolored seeds.

The symptoms are most evident on the
seeds, but the fungus also attacks the
leaves, stems, and pods. On the seed, the
discoloration varies from pink to light
purple to dark purple and ranges from a
small spot to the entire area of the seed
coat. Cracks often occur in the discolored
areas, giving the seed coats a rough, dull
appearance. When diseased seeds are
planted, the fungus grows from the seed
coats into the seedling leaves (cotyledons)
and from them into the seedling stem of a
small percentage of the seedlings. The
fungus produces spores abundantly on the
diseased seedling, and the wind-blown and
rain-splashed spores lodge on the leaves of
nearby plants. The leaf spots caused by
the spores soon produce a secondary crop
of spores which infect other leaves, stems,
and pods. The fungus probably survives
the winter in diseased leaves and stems as
well as in infected seeds. Varietal differences in susceptibility have been observed.

Control: SEED TREATMENT aids in preventing loss of seedlings. Dusting with
COPPER CARBONATE or COPPER SUBSULFATE
during the growing season has also given
partial control of purple seed stain.
Reference: J.4.
PURPLE STAIN = PURPLE SEED STAIN.
PURPLESTRAW.
See COMMON WHEAT (variety).
PURPLE VETCH (*Vicia atropurpurea*),
one of the least winter-hardy of the commercial VETCHES, is a viny plant which resembles the growth habits of HAIRY VETCH
but differs in pod and seed characteristics.

In the milder parts of California it is
winter-hardy, but in western Oregon and
Washington it occasionally winterkills. In

the Cotton Belt, purple vetch has proved
entirely unsuited; even in the extreme
southern part of Georgia and Alabama, it
occasionally is severely injured.

Purple vetch has been grown as a seed
crop and has been used as a green-manure
crop and for hay, particularly in California.
Reference: M.18.
See also LEGUMES; INOCULATION.
PUSA No. 7. *See* SAFFLOWER (variety).
PUSH SWATHER is a windrower used
for harvesting RICE.—*See also* WINDROW.
PUSTULE is a pimple-like or blister-like
area, raised above the surrounding surface
and sharply circumscribed.—*See also* BAC-
TERIAL PUSTULE.
PYCNIOSPORES. *See* RUSTS.
PYRAUSTA. *P. nubilalis*
= EUROPEAN CORN BORER.
PYRETHRINS (pronounced: pye-*ree*-
thrins) are organic compounds present in
the flowers of the *Pyrethrum* spp. They are
quick-acting INSECTICIDES and have very
little toxicity to warm-blooded animals.
PYRIDOXINE HYDROCHLORIDE
= VITAMIN B_6.
PYTHIUM. Many *Pythium* spp. belong
to the FUNGI that cause SEEDLING BLIGHT
as well as SEED ROT. *P. arrhenomanes* does
not cause, as was first assumed, milo disease; this disease, now known as PERICONIA
ROOT ROT, was formerly called "*Pythium
root rot of milo.*" True PYTHIUM ROOT ROT is
caused by various *Pythium* spp.
PYTHIUM ROOT ROT is a term applied
to ROOT ROT caused by various species of
the genus *Pythium.* It attacks LUPINES
and other plants.
"*Pythium root rot of milo*" = PERICONIA
ROOT ROT.—*See also* PYTHIUM.

Q

QUACKGRASS (*Agropyron repens*) is one
of the WHEATGRASS species. It is an efficient
soil binder, but in cultivated fields this
perennial can become a WEED. However,
in pastures and meadows, because of its
nutritive value and permanency, quackgrass is of importance and well-liked by
both cattle and horses.

Control: To remove scattered patches
of quackgrass, spray with SODIUM CHLO-

RATE. A 4-year rotation consisting of grain, hay, and two years of cultivated crop will usually control quackgrass without loss of use of the land.

Reference: B.1.

See also GRASSES; ERGOT; RYE; ALFALFAS.

QUADROON COWPEA, better known as *Wonderful,* is a variety belonging to the "Clay" group of the COWPEA.

QUADROON SORGHUM is one of the intermediate-type grain SORGHUMS.

QUALITY. *See* MEAT QUALITY.

QUERCUS. The *Quercus* spp. are known as OAKS; e.g., *Q. gambelli* = GAMBEL OAK.

QUICKLIME = LIME.

QUICK-WET TREATMENT OF SEEDS. *See* SEED TREATMENT.

R

R 4 is one of the inbred lines of the CORN hybrids.

RABBITBRUSH is a name applied to approximately 70 *Chrysothamnus* spp. and a dozen subspecies; e.g., RUBBER RABBITBRUSH and SMALL RABBITBRUSH.

Spring rabbitbrush = *littleleaf horsebrush.* See HORSEBRUSHES.

RACE is a STRAIN of a domesticated SPECIES, produced by artificial SELECTION.

RACEME is a simple, elongated flower cluster, the RACHIS of which bears a series of one-flowered PEDICELS.—*See also* GRASSES.

Raceme. (D.9.)

RACHILLA, or *little rachis,* is the central axis which carries the SPIKELETS of GRASSES.—*See also* RACHIS.

RACHIS (pronounced: *ray*-kiss), or *rhachis,* is the axis of a SPIKE, RACEME, or branch of a PANICLE.

Little rachis = RACHILLA.

RADIATION. *See* IRRADIATION.

RADICAL. *See* VALENCE.

RAGWEED (*Ambrosia artemisiaefolia*), more exactly named *common ragweed,* is a troublesome WEED. It is an annual with rough, hairy, branching stems 1 to 4½ ft. tall; the deeply cut leaves are smooth. It possesses pollen-producing flowers borne in small clusters at the tips of the branches and a few seed-producing heads borne at the base of the leaves and in the forks of the upper branches. This weed is particularly bad in LESPEDEZA fields.

Control: Ragweed can be controlled by preventing seeding and by clean cultivation. The use of 2,4-D gives fair results. The weed is also attacked by COMMON DODDER and FIELD DODDER.

References: C.19; E.7.

See also DODDERS.

"RAGWORM INJURY."

See CORN EARWORM.

RAGWORT (*Senecio jacobaea*) and many of the closely related GROUNDSELS are widely distributed POISONOUS PLANTS which contain ALKALOIDS. Cattle and horses feeding for several days on leaves and stems of any of the poisonous *Senecio* spp. develop such symptoms as jaundice, scabby nose, loss of appetite, uneasiness, and loss of flesh.

RAINBOW OAT.

See COMMON OAT. (variety).

RAISIN PULP, a by-product from the manufacture of raisin sirup and other raisin foods, is not as valuable as CULL RAISINS are because it contains a higher fiber content. Nevertheless, when available, it can be fed to livestock in an otherwise well-balanced ration.

RAISINS. *See* CULL RAISINS.

RALSOY. *See* SOYBEAN (variety).

RAMIE (*Boehmeria nivea*), also called *rhea* or *China grass,* is a fiber crop. It is sometimes grown in the Southeast and in irrigated valleys of California. This perennial makes heavy vegetative growth.

RAMIE LEAF MEAL—an officially recognized feedstuff—is the ground product consisting chiefly of leaves from the RAMIE plant. It must not contain more than 18 percent crude fiber.

Reference: F.6.

See also MISCELLANEOUS PRODUCTS.

RAMONA 44.
            See COMMON WHEAT (variety).

RAMULISPORA. The FUNGUS *R. sorghi* causes SOOTY STRIPE.

RANCHER SORGO is one of the most valuable forage SORGHUMS because it has a very low PRUSSIC ACID content.

RANGE is an extensive area of natural pasture land.

Open range is unfenced land.—*See also* PASTURE; PASTURE MANAGEMENT; RANGE PLANTS; DROUGHT; GRAZING.

RANGE FORAGE consists of four classes: BROWSE, GRASSES, GRASSLIKE PLANTS, and WEEDS.—*See also* GRAZING.

RANGE INDICATORS.
            See RANGE MANAGEMENT.

RANGE MANAGEMENT regulates GRAZING in order to safeguard resources and to get sustained, maximum production of livestock and the best forage species.

Maximum livestock production requires adequate forage or other feed at all times. It is important to keep the RANGE in best possible condition at all times, year after year.

Under careful management, *excellent ranges* can be maintained. A considerable residue of the abundant plant material should remain unused on the ground at the end of the grazing season. Overuse, particularly in dry years, will result in the deterioration of dense sod, disappearance of important secondary species (including both taller GRASSES and perennial weeds), and consequent loss in grazing value. Any disintegration of the sod into distinct bunches, and a decrease in number of tall grasses are early signs of deterioration of the forage cover.

The prime management objective for range in the *good* class should be to improve it to an excellent condition. On very good soils and in favorable locations where moisture is plentiful, due to underground drainage or to run offs from higher ground, good range may be changed to excellent range in very few seasons of conservative use. Improvement of good range is indicated by an increase in the size of short-grass clumps and a decrease in the width of spaces between them. Deterioration is indicated by an increase in the amount of bare ground between clumps and the more or less complete disappearance of taller grasses and perennial weeds of fair or good forage value.

Range is ordinarily brought down to *fair* condition as the result of DROUGHT, of conditional overgrazing, or a combination of both. Insofar as overgrazing is the cause, the cure consists obviously in the reduction of the numbers of livestock to safe grazing capacity. On drought-reduced range every effort should be made, by adjusting numbers of animals and other desirable management, to keep the range from further deterioration, so that the full restorative effect of better moisture conditions may be realized when a good year comes.

Poor range in most cases has been reduced to this condition by severe drought, long-continued overgrazing, or both. The immediate objective should be to improve this condition to fair or good. To attain this, grazing must be carefully restricted. Natural recovery may be a slow process if most of the top soil has been removed by wind and sheet erosion. If depletion has progressed so far that recovery under use would be a very slow process, removal of livestock and artificial reseeding may be the most economical step.

Many factors besides drought and overgrazing have contributed to widespread range depletion. Reduced forage density, vigor, and production are important indicators of depleted ranges. Excessive trampling and grazing result in increased floods, siltation, and runoff, which reduce soil moisture available for forage growth. An enormous population of rabbits and mice destroys and consumes untold quantities of forage. Thousands of acres are utterly devastated by prairie dogs that congregate in dog towns.

A stable range program requires constructive action on many fronts.

Land use practices must be designed to safeguard topsoil and avoid unnecessary erosion, siltation, and waste of water or soil moisture. More adequate reserves of range forage, HAY, or concentrates, and also financial reserves, need to be built up against the inevitable droughts, GRASS-HOPPER invasions, or other emergencies.

Artificial reseeding is necessary where nearly all desirable native RANGE PLANTS have been destroyed by plowing, over-grazing, or other means. Reseeding is widely recognized as a profitable practice leading directly to range improvement. With the increased interest in reseeding has come a lessened demand for all-purpose grasses (formerly usually CRESTED WHEAT-GRASS or TIMOTHY) and a quick acceptance of specialized crops; e.g., TALL WHEATGRASS for alkaline soil, INTERMEDIATE WHEAT-GRASS for late spring grazing, RUSSIAN WILDRYE for fall grazing, etc.

Little or no soil preparation is necessary where seed is drilled in relatively clean stubble the first or second season after a grain crop, and before competing weeds take over to increase risks and costs.

Watering places are needed to avoid unnecessary penalties in reduced livestock gains, in wasted forage, and in range damage that occurs when livestock is not evenly distributed in relation to the forage supply. Even when the range as a whole is not overstocked, a third or more of it that is around watering places may be damaged; another third may be utilized properly; but on more remote areas forage is wasted when livestock is poorly distributed. *Salt* placed at strategic locations away from water helps to insure more uniform distribution.

Grazing should be as uniform as possible over the entire range. Adequate *fencing* is especially helpful in getting uniform use. At best, however, there will be variations in the degree of grazing on large areas within a pasture. The swales and ravine bottoms will invariably be grazed more closely than the adjoining hillsides because these usually remain green longer and have better forage than the open hillsides.

The class of livestock and season of use also have a bearing on proper use. Certain varieties and combinations of forage permit grazing of both cattle and sheep if skillful management is practiced. Unless very carefully regulated, however, grazing two classes of stock may result in slow but serious range deterioration.

Sheep ranges deteriorate more rapidly than cattle ranges under heavy stocking, closer repeated cropping, and trampling.

Segregating *winter* from *summer range* provides a change to fresh forage late in the fall and again in the spring. Fresh range available in December may save the costlier winter feed that otherwise would be required. Moving from winter to summer range about midway in the active spring growing season helps to minimize damage when soil is loose and plants are easily destroyed.

Winter ranges are especially important for profitable sheep production in Utah, Nevada, southern Idaho, and southwestern Wyoming. Knowledge of important species, annual measurements of the forage crop, and periodic evaluation of range condition are all essential in applying good management.

On many high mountain ranges weather limits their use to the summer. Proper management of such seasonal ranges is based on growth requirements of the forage plants and soil moisture conditions. When soils are wet, grazing should not begin until the key forage grasses are 4 to 6 in. high. If the soil is dry, grazing may begin when the grasses are 2 in. high, but should be light.

Plants that are poisonous or noxious only at certain seasons sometimes dictate the best season of use on certain ranges.

Grazing management is concerned primarily with the maintenance, production, and utilization of desirable species of range plants. *Deferred grazing*—i.e., excluding livestock from range pastures during the entire growing season—has proved to be extremely beneficial to the vegetation. Less successful is *deferred-and-rotation grazing* which can be arranged by dividing a summer range into three or more parts and grazing each at successively later dates in rotation so that one is grazed after seed maturity each year.

Ample feed reserves are essential for emergencies. Old cured forage that accumulates under conservative stocking will sustain livestock for a considerable period, but HAY, by-products of COTTONSEED, or SOYBEAN, or other reserves should be available.

Any temporary increase in stocking in favorable periods to use up accumulated reserves should be made with caution; production costs are increased by the use of supplemental feed for excess numbers.

Bad marketing practices also burden the range. Late-fall marketing often presents a problem. On matured forage, weight gains tend to level off or decline. Even when late-marketed cattle do make a slight weight gain, it may not compensate for the extra forage used and the interim risk.

Various types of brush control offer possibilities for improving many native ranges. Mowing and burning have given outstanding results in controlling some SAGEBRUSH species, and increasing grass production. Formulations of 2,4-D, when properly applied, give satisfactory control of sagebrush, SKUNKBRUSH, and many range weeds.

Better control is needed of range fires that rob wildlife of food and shelter and destroy forage, humus, watershed values, and sometimes lives.

References: H.39; C.17; B.24; H.41; S.30; P.22; H.45.

RANGE PASTURES.

See PASTURES; RANGES.

RANGE PLANTS, which form the natural wild vegetation of the RANGES, are often important as forage for grazing animals. Because of major differences in soil, temperature, and rainfall, range vegetation naturally falls into 10 readily recognizable types listed in the table (in descending order of extent in the United States).

In general, out of a total of at least 10,000 species growing on ranges, probably only about 1,000 are of major or secondary importance. Each range type is a complex of species, but only a comparatively few furnish the bulk of the forage. Generally speaking, perennial species—especially

RANGE PLANTS

Type of vegetation	Important forage species
Short grass	Western wheatgrass
	Blue grama
	Buffalograss
	Curly mesquite
Open forest	Ponderosa pine
	Fescues
	Brome grasses
Sagebrush-grass	Bluebunch wheatgrass
	Sagebrush
	Indian ricegrass
	Needlegrasses
Semidesert grass	Gramas
	Mesquites
	Sacaton
Piñon-juniper	Gramas
	Junipers
	Muhly grasses
	Piñon
Salt-desert shrub	Black sagebrush
	Shadescale
	Winterfat
Pacific bunchgrass	Bluebunch wheatgrass
	Sandberg bluegrass
	California needlegrass
	Giant wild-rye
Southern desert shrub	Saltbush
	Yuccas
	Various cacti
Tall grass	Slender wheatgrass
	Bluestems
	Porcupine grass
Woodland chaparral	Chamise
	California oatgrass
	Alfileria
	Oaks

GRASSES—are the backbone of a range. With the exception of certain LEGUMES, CALIFORNIA OATGRASS, some BROME GRASSES, and FINGERGRASSES, range annuals are, on the whole, of inferior palatability and forage value.

Many palatable range plants, i.e., some of the WHEATGRASSES, saltbushes, and native CLOVERS, compare favorably with ALFALFA and other cultivated feeds in their chemical composition, an indication of their nutritive value; some even have a higher ratio of minerals or protein. Chemical values of plants are, however, not constant; there is a higher protein and phosphorus content in the early stages of growth, which gradually decreases as the plants approach maturity. The chemical content of individual plants of some species also varies with soil, exposure, altitude, and other ecological factors. In general, range plants of higher altitudes—grasses or weeds—are higher in crude protein and lower in fiber than plants of lower altitudes.

Of special interest are the several hundred different plants found on *winter ranges*; only about 50 are considered important species. Some are good forage species, particularly for sheep on the ranges of the West, while others, like RUSSIAN-THISTLE, CHEATGRASS, HALOGETON, and BROOM SNAKEWEED, have little or no forage value on winter range, but are important because they provide plant cover on deteriorated ranges and serve as indicators of range condition.

On most winter ranges in good condition, desirable forage species are relatively abundant and furnish most of the forage. Species on the winter range, classified as to forage quality, together with recommended use (for sheep) in percent, are listed in the table.

WINTER RANGE PLANTS

Quality and recommended use (for sheep) in per cent of annual growth

Good forage species	
Shrubs:	%
Spiny hopsage	80
Black sagebrush	70
Winterfat	60
Bud sagebrush	50
Fourwing saltbush	50
Gardner saltbrush	40
Grasses:	
Indian ricegrass	75
Alkali sacaton	75
Squirreltails	75
Needle-and-thread grass	50
Galleta	45
Salina wild-rye	45
Mormon needlegrass	45
Forbs (weeds):	
Globemallows	80
Encelia	75
Fair forage species	
Shrubs:	%
Fringed sagebrush	40
Gray summer-cypress	35
Shadscale	25
Littleleaf mountain-mahogany	25
Big sagebrush	20
Nevada ephedra	20
Grasses:	
Bluebunch wheatgrass	30
Nevada bluegrass	30
Sand dropseed	25
Saltgrasses	25
Blue grama	20

Forbs (weeds):	%
Bassia	35
Lambsquarters	30
Pepperweed	25
Poor forage species	
Shrubs:	%
Broom snakeweed	15
Small rabbitbrush	10
Rubber rabbitbrush	10
Greasewood	5
Seepweeds	5
Littleleaf horsebrush	0
Cacti	0
Pickleweed	0
Grasses:	
Cheatgrass	10
Fendler three-awn	5
Sixweeks grama	5
Foxtail brome	5
Forbs (weeds):	
Locoweeds	10
Russian-thistle*	10
Halogeton	0
Other annuals	0

* Readily grazed on spring and early summer ranges, when the plants are green.

References: C.15; H.41; S.27; H.35.

See also SHRUBBY CINQUEFOIL.

RANGER ALFALFA.
    See VARIEGATED ALFALFAS (variety).

RANGES. *See* RANGE.

RAPE (*Brassica napus*) occurs in two types—annual, grown for seed, and biennial, valuable as pasture and forage crop. Rape reaches a height of 1½ to 2½ ft. The plant is leafy and rich in vitamins, proteins and minerals.

The seed of the *annual type* is fed to cage birds or used to produce oil (known as *colza oil* when refined); as a by-product, RAPESEED MEAL is obtained.

When the crop of the *biennial type* is 6 to 8 in. high, it can be used for sheep or hog pasture; when cattle are to be pastured on rape, it is better to wait for a larger leaf growth.

Dangers: There exists some danger of BLOAT when cattle or sheep are turned onto rape pasture without first being filled up on other feed.

Some rape pastures irritate light skinned breeds of swine. This condition, often falsely called "*scalding*," is the result of PHOTOSENSITIZATION.

Reference: H.29.

See also CHINCH BUG; OATS; GRAZING.

RAPESEED MEAL, or *rapeseed oil meal*, also known as *colza oil meal*, is obtained from the seed of RAPE as a by-product of the oil production. It has a pungent flavor (like all parts of the rape plant) and is often disliked by animals, except if mixed with other feeds.

Dangers: If fed in too large amounts, especially to young or pregnant animals, rapeseed meal may seriously irritate the digestive tract.—*See also* OILSEED MEALS.

RATION is the fixed, daily allowance of feed given an animal in one or more portions during a period of twenty-four hours.

A *balanced ration* contains all the NU-TRIENTS needed to properly nourish an animal during a 24-hour period.—*See also* BEEF CATTLE RATIONS; BULL RATIONS; DAIRY RATIONS; DRY COW RATIONS; SHEEP RATIONS; GOAT RATIONS; SWINE RATIONS; HORSE AND MULE RATIONS; POULTRY RA-TIONS; FLEXIBLE RATIONS; SELF-FEEDING; RATION FORMULATION; MINERAL FEEDS; COMMERCIAL MIXED FEEDS; WATER RE-QUIREMENTS; NUTRIENT REQUIREMENTS; VITAMINS; UREA; U.G.F.

RATION FORMULATION. To formulate FLEXIBLE RATIONS, the following steps must be taken (most of them shown in the *Ration Formulation Sheet*, pages 384, 385):

1. Fill in headings as to RATION number, date, and animal.
2. List critical NUTRIENTS across top of page.
3. At bottom of page list the requirement for critical nutrients.
4. Select cheapest sources of suitable GRAINS and PROTEIN sources in your locality and enter under ingredient column.
5. List amounts of ANTIBIOTICS, unidentified growth factors, and any other fixed ingredients that you desire; e.g., manganese sulfate.
6. Calculate the amounts of protein furnished by the fixed ingredients of the ration.
7. Add necessary amounts of the major protein and grain source to bring protein to desired level.
8. Calculate nutrient composition ot ration.
9. Add any needed ENERGY, AMINO ACID, VITAMIN or MINERAL sources to ration.
10. Adjust major grain and protein source so that ration has desired total weight.
11. Total all nutrients.
12. If chemical laboratories are available, have analyses of ration made.
13. If possible make a trial batch of ration and feed it experimentally.
14. Register ration with proper authorities.

RATTLE BOXES = CROTALARIAS.

RATTLEWORT (*Crotalaria sagittalis*) is one of the poisonous CROTALARIAS. It extends as far north as the New England States; there it behaves as an annual, but in the South it is a short-lived perennial.

RAW BONE MEAL is not an officially recognized feedstuff. If produced for use as a fertilizer, it is clearly unsafe for feeding purposes. Generally, the raw bone now sold for feed has been cooked in water at atmospheric pressure to remove excess fat and meat and to destroy any disease germs, but most of it has an odor that makes it unpalatable to certain classes of livestock, and many products are not ground fine enough for feed.

> NOTE: For animals other than poultry this or any other MINERAL FEED should be ground fine enough so that 60 percent of it will pass through a 40-mesh sieve. Too fine grinding results in a dusty product that is undesirable.

Reference: M.38.

RAYLESS GOLDENROD (*Aplopappus heterophyllus*), also known as *Jimmyweed*, is a bushy halfshrub growing 1 to 2, and occasionally 4 ft. high. Because of its extensive root system, it is a drought-resistant plant. The leaves and stems contain TREMETOL and are therefore poisonous to cattle, sheep, and horses; feeding on the plant frequently for several days, results in marked weakness and trembling of the animals, especially after exercise.—*See also* POISONOUS PLANTS.

REACTION of a soil is the degree of its acidity or alkalinity.—*See also* pH.

Rayless goldenrod. (N.6.)

RECESSIVE is a GENE that controls a character with less biochemical activity than a DOMINANT. In first generation HYBRIDS true recessives do not appear because this character is masked due to the presence of the contrasted (dominant) character from the other parent. Some genes are incompletely dominant.

RECOMMENDED NUTRIENT ALLOWANCES. *See* NUTRIENT REQUIREMENTS.

RECTUM. *See* DIGESTIVE TRACT.

RED AMBER SORGO is one of the forage SORGHUMS.

REDBINE. *Redbine-60* and *Redbine-66* are intermediate-type grain SORGHUMS.

RED CHAFF is a synonym of (1) *Goens*, (2) *Mediterranean wheat*, and (3) other varieties of COMMON WHEAT.

RED CHIEF.
                See COMMON WHEAT (variety).

RED CLOVER (*Trifolium pratense*) is by far the most important of the TRUE

CLOVERS. Wild, it is an extremely variable plant; there are known to be early, late, smooth, hairy, prostrate, erect, and semi-erect forms. The cultivated red clovers fall into two classes: (1) the important *early*, or *double-cut* giving two hay crops in a season; and (2) *late*, or *single-cut* giving but one hay crop in a season.

The plant is composed of numerous leafy stems arising from a thick crown; it usually lives only two years. The roots are much branched and the stems somewhat hairy.

Red clover. (M.3.)

Red clover heads are composed of from 50 to 150 FLORETS, which in themselves are complete units of reproduction. The florets develop and open in an ascending order from the base to the top of the head. The ANTHERS shed their pollen in the bud stage shortly before the floret (individual flower) of the head opens. Fertilization occurs from eighteen to fifty hours after pollination, the time depending on the

RATION FORMULATION SHEET

① | **56 B** | **December 1, 1956** |
| RATION NO. | DATE |

Ingredients	⑦ Amt.	Prot.	Therms	Ca lb.	P lb.	Vit. D I.U.
		② Lbs. Major Nutrients				
Milo	345	40.0	383	0.10	1.03	
Soybean Meal	245	108.3	139.6	0.73	1.71	
Yellow Corn	250	22.5	287.5	0.05	0.75	
Meat Scraps	50	25.3	36.0	4.85	2.10	
Animal Fat	10	—	29.0	—	—	
Alfalfa Meal	25	4.2	7.5	0.4	0.05	
Fish Solubles	25	8.75	10.0	0.05	0.20	
Dried Whey	25	3.05	12.5	0.22	0.20	
Salt	5	—	—			
MnSO$_4$	1/4	—	—			
Fixed Ingredients	90¼	16.0 ←⑥				
A + D, 2250, 400	1					1810
Choline Cl, 25 %	2/3					
B$_2$, 3.6 gm./lb.	2/5					
Methionine	1	1				
Niacin, gm.	4					
Ca pant., gm.	1.5					
Bone meal	7.0	0.9		2.05	1.06*	
Limestone	4.5			1.71		
Antibiotics 5gm./lb.	1					
Coccidiostat	+					
SUB TOTAL ⑧→		211.8	905.1	6.41	6.04	
TOTAL ⑪		213.7	905.1	10.17	7.10	1810
REQUIREMENT ③→		21.0	.900	1.0	0.6	180
ANALYSIS, CALC.		21.4	.905	1.0	0.7	181

*At least 0.45% of feed must be present as inorganic phosphorus. All nonvegetable sources supply inorganic phosphorus and 30% of vegetable phosphorus is considered to be inorganic.

Broiler
DESCRIPTION OF ANIMAL

Amino Acids, Micro Nutrients, Additives

Vit A 1000 I.U.	B$_2$ mg.	Niacin mg.	P. A. mg.	Choline mg.	Arg. lb.	Meth. lb.	Cyst lb.	Lys. lb.	Tryp. lb.
	138	4519	1725	69	1.03	0.35	0.69	1.03	0.31
	490	4091	1494	306	7.00	1.71	1.71	7.00	1.45
375	125	2450	650	50	1.00	0.75	0.25	0.50	0.20
	105	1070	75	45	1.50	0.50	0.30	1.75	0.20
—	—	—	—	—	—	—	—	—	—
1250	175	209	295	11	0.20	0.08	0.08	0.22	0.07
	200	3750	450	25	0.30	0.12	0.10	0.35	0.08
	325	127	560	18.7	0.07	0.04	0.08	0.22	0.03
1020									
				75					
	1440								
						1.0			
		4000							
			1380						
1625	1558	16,216	5245	525	11.10	3.55	3.21	11.07	2.34
2645	*2998*	*20,216*	*6625*	*600*	*11.10*	*4.55*	*3.21*	*11.07*	*2.34*
2.0	3.0	20	6.5	0.6	1.2	0.45	0.35	0.9	0.2
2.6	3.0	20	6.6	0.6	1.1	0.46	0.32	1.1	0.2

The ration formulation sheet is adapted from a form that is property of U. Mo. Dept. Animal Husbandry and Mo. AESt. Circled figures refer to steps in the *Ration Formulation* article.

temperature; when the atmospheric temperature is high, fertilization occurs much earlier than when it is low. The degree of success of pollination can be approximately determined two to three days after fertilization.

Upon self-fertilization many red clover plants do produce a few seeds, but the line cannot be maintained beyond the second or third generation. This phenomenon may be called *pseudo-self-fertility* as distinguished from *true self-fertility*. The number of true self-fertile plants in red clover is relatively small.

The structure of the red clover floret prevents cross-pollination by wind, since the anthers and PISTIL remain enclosed in the keel unless artificially tripped. Under natural conditions pollination is done by insects, nearly all of them bees; bumblebees and honeybees are the principal pollinators.

Red clover is the most important leguminous forage and soil-improving crop in the north-eastern quarter of the United States. It is also grown in Idaho, Washington, Oregon, and in many of the valleys of the Mountain States where the growing season is too short for three crops of alfalfa, and where a leguminous crop is desired in connection with the customary GRAIN crop. In the South red clover is grown locally, usually as a winter crop, though there are places with rich moist soil where it does well the year around.

This LEGUME will grow on any well-drained, fairly rich soil that has plenty of lime in it. It will not thrive on hard run-down land in which the organic matter has been exhausted by bad cropping practices.

When seeding clover upon land for the first time, it is well to provide for artificial INOCULATION, but after clover is established on a farm this procedure is usually unnecessary. The most common method of seeding is on winter grain, but it also is often seeded with spring grain. Late summer seeding is successful in much of the southern and eastern part of the clover area. Red clover is most often seeded with TIMOTHY, though sometimes with other GRASSES. With timothy the hay of the first crop year is mostly clover; the second year the timothy is predominant, and after that the clover largely disappears. Red clover is most used with a cultivated crop and a SMALL GRAIN crop in three, four, and five year rotations.

In order to obtain the best *hay*, the clover crop should be cut as a general rule at the half to full bloom stage.

Silage can be made from red clover as it can from other legumes, but the process is somewhat risky because the relatively low sugar content of clover results in poor fermentation. It is best to cut the clover fine, to mix it with grasses, and to trample carefully. A good preservative will help.

Red clover is a most excellent *pasture* for all livestock, especially while they are growing. For pigs it should be supplemented with a small grain ration, as this will induce much more rapid gains. Close early pasturing is injurious to the stand of clover.

Where pasturing is impracticable, red clover is often used as a *soiling crop;* that is, cut and fed green to livestock. Feeding in this way reduces or eliminates the danger from BLOATING. It makes a good early feed, is palatable, and from 6 to 10 tons of green feed per acre is not an unusual yield.

Dangers: The loss of stands because of continued growth of red clover has been called *clover sickness*. This supposedly toxic condition has repeatedly been studied, and been found to involve no mystery. The name "clover sickness" should, therefore, be discarded in favor of the name *clover failure*. This failure may result from one or a combination of the following causes: unfavorable soil conditions; unadapted or poor seed; poor methods of seeding; wrong treatment; and—last but not least—diseases and insects.

Damage caused to red clover by diseases includes various fungus diseases confined to the foliage. Among these FUNGUS LEAF DISEASES are STEMPHYLIUM LEAF SPOT, PSEUDOPEZIZA LEAF SPOT, CLOVER RUST, POWDERY MILDEW, BLACK STEM, and SOOTY BLOTCH. In addition to these LEAF DISEASES, there are other fungus diseases, including CHARCOAL ROT,

NORTHERN ANTHRACNOSE, and SOUTHERN ANTHRACNOSE found on stems; BLACK-PATCH which attacks the stems and the blossom heads; and two diseases which have proved to be major factors in the thinning of clover stands: CROWN ROTS and FUSARIUM ROOT ROT.

A complex of VIRUS DISEASES that are transmitted by aphids and leafhoppers occurs on many legumes; *red clover virus diseases*—particularly RED CLOVER VEIN MOSAIC—are widespread and may cause extensive damage in some areas.

A large number of insect species cause extensive injury to both the hay and seed crop. The most important of these are MEADOW SPITTLEBUG, various LYGUS BUGS (particularly the TARNISHED PLANT BUG), CLOVER ROOT-BORER, CLOVER ROOT CURCULIO, CLOVER-SEED-CHALCID, CLOVER-FLOWER-MIDGE, CLOVER MITE, several species of WEEVILS—e.g., CLOVER-LEAF WEEVIL, LESSER CLOVER-LEAF WEEVIL, CLOVER-HEAD WEEVIL, and CLOVER TYCHIUS LEAFHOPPERS (including the POTATO LEAF-HOPPER), YELLOW CLOVER APHID, PEA APHID, and other APHIDS, and GRASS-HOPPERS.

Varieties: *Mammoth red clover* is also known as *sapling clover*, *pea-vine clover*, *English clover*, *bull clover*, and *late red clover*; this multiplicity of names leads to confusion and it is desirable to use the more descriptive name of *American single-cut clover*. This clover makes a greater growth than the double-cut types and has a blooming period from ten days to three weeks later than the other clovers. Its general use is not recommended, except in the northern part of the Clover Belt, because the yield of its one cutting will not equal the yield of the two cuttings of the double-cut clovers. Because of the early loss of leaves in mammoth red clover, the hay crop should be cut at the bud to early bloom stage.

Of *double-cut clover*—also called *common red clover*, *medium red clover*, or *June red clover*—the following varieties are of importance in North America:

Kenland red clover, well adapted to that part of the Clover Belt where southern anthracnose is quite destructive. It has some resistance to CROWN ROT, and aside from superior yielding ability, Kenland has a longer life than other common red clover varieties. It is recommended in all, or parts, of 29 states, though its main area of adaptation centers in the Kentucky, Virginia, Maryland, Delaware, New Jersey, southern Ohio, Indiana, and Illinois area. At some locations, the yield advantage for Kenland has been over 40 percent. Average annual yields reach up to 3 tons of hay per acre.

Midland red clover, formerly called *Central Corn Belt blend*, is somewhat resistant to northern anthracnose, is winter-hardy, and has good growth characteristics. It is adapted to the middle section of the Red Clover Belt and should not be used where southern anthracnose is a factor in red clover survival.

Cumberland red clover is a blend of three highly productive disease-resistant strains of red clover that were developed in Kentucky, Tennessee, and Virginia. It averages up to 32 percent higher yield than other strains. Cumberland clover is preferred to ALSIKE CLOVER as a substitute for ALFALFA since its yields exceed those of alsike by at least 50 percent. The feeding value of good Cumberland clover hay is equal to that of alfalfa.

Pennscott red clover, originally designated *Lancaster 9* and later as *Scott*, produces high yields and has early seedling vigor and a high degree of persistency. It appears to have wide adaptation in states adjacent to Pennsylvania.

References: P.3; P.4; W.4; G.3; H.16; U.1; W.5; H.15; E.5; H.14; H.17.

See also WHEAT JOINTWORM; INSECTICIDES; LESPEDEZAS; ALFALFAS; PEANUT; CRIMSON CLOVER; ALSIKE CLOVER; FIELD DODDER; COWPEA; TALL OATGRASS; ORCHARDGRASS; WHEATS; OATS; BUCKWHEATS; LEGUME BACTERIA; SOIL CROPS; GRAZING; PASTURES; PASTURE PLANTS; HAY; NORTHERN ANTHRACNOSE; SOUTHERN ANTHRACNOSE.

RED CLOVER VEIN MOSAIC is one of the *red clover virus diseases*. This VIRUS is transmitted by the PEA APHID and causes a yellow discoloration along the veins; leaf mottling and dwarfing are the symptoms commonly observed. The infected plants produce only a few stunted flower heads.

Reference: E.5.

See also VIRUS DISEASES; RED CLOVER.

RED CLOVER VIRUS DISEASES affect RED CLOVER.—*See also* VIRUS DISEASES; RED CLOVER VEIN MOSAIC.

Red **fescue.** Plant, spikelet, and floret. (H.26.)

REDDISH-BROWN SHEATH ROT is a RICE disease caused by the FUNGUS *Helicoceras oryzae.* It is somewhat similar to BLACK SHEATH ROT but is not very common. Late in the season the lower leaf sheaths become reddish-brown and somewhat rotten, although rotting is not so severe as in the black sheath rot. Irregular, dark, seedlike fungus bodies are embedded in the sheaths and characteristic spores are found on the surface of the spots. It is not a serious disease on rice.

Reference: T.2.

RED DURUM.

> *See* WHEATS; DURUM WHEAT.

RED FESCUE (*Festuca rubra*), also called *creeping red fescue,* is one of the fine-leaf FESCUES. It resembles SHEEP FESCUE, but its leaves are bright green and it does not grow in tufts. As its name indicates, it is, in general, a creeping GRASS, but some of its varieties are more creeping than others, and at least one of them—known as *chewings fescue*—is not creeping.

Red fescue is occasionally used in grass mixtures for pastures in the northern states.

Varieties: The most important of the red fescue varieties is the (above mentioned) *chewings fescue.*

Reference: H.26.

See also SUBCLOVER.

REDFIELD.

See DAGHESTAN SWEETCLOVER (variety).

REDHART.

> *See* COMMON WHEAT (variety).

RED HARVESTER ANT (*Pogonomyrmex barbatus*) causes heavy losses in cultivated fields and orchards in the southwestern states. Bare areas ranging from 3 to 35 ft. in diameter, cleared of vegetation by colonies of these ANTS, stand out prominently in ALFALFA or GRAIN fields in this part of the country. A 20-acre field heavily infested with ants will often contain as much as a quarter-acre of land wasted in this way.

The ants also cause a direct loss to seed crops by collecting the seeds for storage. On 20 acres of alfalfa land, 100 colonies may consume 100 lb. seed. In newly seeded alfalfa fields the gathering of freshly sown seeds results in a thin stand for several seasons. Nests have been found with chambers containing large quantities of OATS collected from recently planted fields.

Colonies of the red harvester ant are found at the lower altitudes in Oklahoma, Texas, New Mexico, Arizona, and California, as well as in Mexico.

A closely related species, known as the WESTERN HARVESTER ANT, occurs in the higher altitudes and colder districts of approximately the same territory as the red harvester ant.

Running promiscuously about, harvester ants get on people and bite or sting them, sometimes causing severe pain and swellings. Livestock are also greatly annoyed by these pests; dairy cows may suffer numerous swellings, especially on the udder. Such attacks often reduce milk production.

The eggs, which are laid by the queen in the chambers of the colony, are about half the size of a pinhead. They are iridescent, milk-white, and usually clustered together. The larvae, which hatch from the eggs, are cream-colored and are shaped like crook-necked squashes. The full-grown larvae are about ¼ in. long.

The pupae are ¼ to ½ in long, of pale cream color, and have their legs and feelers, or antennae, folded on the underside.

There are four forms of the adult ant: (1) The sterile females, or *workers*—the reddish-brown ants seen in large numbers hurrying to and from the colony in search of food; (2) the winged *males*, and (3) the larger, *winged females*—the sexually mature individuals; (4) the *queen*—a mated female that has discarded her wings and established a colony.

The barren, circular area, commonly called an ant hill, which surrounds the entrance hole of the nest, is the result of the ants' cutting down any vegetation that attempts to grow there. The size of this area depends on the tunneled area underneath the surface. There are one or more pathways 1 to 4 in. wide and up to 200 ft. in length; they eventually disappear in the vegetation.

The entrance hole into the nest of the red harvester ant is usually located at about the center of the circle bared of vegetation. It joins the network of tunnels and chambers underground and is normally ¼ to ½ in. in diameter.

The nest is a series of subterranean tunnels and chambers. The tunnels, about ¼ in. in diameter, lead in and out of the nest and from one chamber to another. They extend downward at different locations, with series of chambers branching off from them. The chambers are flat-bottomed, with dome-shaped ceilings. They range in size from ¼ in. wide and ½ in. long to 10 in. wide and 1 ft. long. Stores may be found in any of the chambers.

The type of soil affects the general arrangement of the tunneling. If the soil is porous—such as sandy loam—for a considerable depth so that the nest can be extended downward without interruption, its general shape is that of a cone, usually from 8 to 10 ft. in depth.

The colony consists of one queen and innumerable workers, together with eggs, larvae, and pupae. The winged males and females appear only from April to October. Swarming occurs when the winged males and females emerge in large numbers and mate. The males die soon after mating.

The mated female establishes the new colony. She removes her wings and begins digging into the ground. She first digs a hole about 10 in. deep, and constructs from one to four small, dome-shaped chambers branching off from it. She next plugs up the entrance hole and deposits a cluster of about 50 eggs. When the eggs hatch, the young queen feeds the larvae a secretion from the fatty tissue of her body. These larvae develop into workers. The queen spends the rest of her life within the tunnels and chambers, continuing to lay eggs. She mates only once during her lifetime.

When the first of the workers issue in the newly formed nest, the queen again starts laying eggs, and thereafter limits her activity solely to this function. The colony increases in size for a few years.

Only a very small percentage of the

females live to establish colonies. Many are eaten by birds and toads. Some fail to mate in their hasty flight. Thousands are, no doubt, lost in their attempts to find suitable places in which to start colonies, and many females lack sufficient vitality to establish a colony that can survive the winter. The flooding of fields in irrigated districts or the occurrence of heavy rains also greatly hinders their attempts to colonize. Although the females may fly for considerable distances, most new colonies are established within a few hundred feet of the old colony.

The red harvester ant apparently harvests any seed that can be carried. Alfalfa, BUR-CLOVER, JOHNSONGRASS, oats, WHEAT, BERMUDA-GRASS, wild SUNFLOWER, and mesquite beans have been found in the chambers. Such products as BRAN or rolled oats are often carried in.

Rubber boots are an excellent protection against the ants, provided they have a really smooth surface. Knee-high shoes, with the trouser legs tucked inside, also afford considerable protection. Any method of fastening the trousers tightly about the ankle, such as the wearing of leggings, or tying the bottom of the trousers, is helpful. Regardless of the kind of protective measures taken, one should stamp one's feet on the ground frequently to dislodge any ants that may have started to climb.

Control: Only a treatment that destroys the queen and most of the workers, at a time of the year when there are no young in the colony, will provide complete control as the few remaining workers will eventually die.

When many of the ants are active on the surface, a dust containing DIELDRIN or CHLORDANE should be used. When nearly all the ants are in the nest use CARBON DISULFIDE or METHYL BROMIDE as fumigant. Two or more treatments, two to three weeks apart, are frequently necessary before a colony is destroyed or its activity suspended for a long period. All treated areas should be inspected from time to time and, if necessary, re-treated so that surviving ants can be killed.

Reference: B.8.

RED IRON OXIDE, consisting chiefly of *ferric oxide*, is a purplish-brown, heavy, water-insoluble powder. It is widely used as pigment and occasionally as ingredient of MINERAL FEEDS.

RED KAFIR is one of the grain SORGHUMS.

RED KENTUCKY SUGAR-CANE is the original name of *Leoti*, one of the forage SORGHUMS.

RED KIDNEY BEAN, a variety of the COMMON BEAN, is one of the important FIELD BEANS. It is bushy in appearance, the flowers are lilac-colored, and the seeds are red.

REDLAN is one of the intermediate-type grain SORGHUMS.

RED LEAF, or *yellow dwarf*, is caused by a soil-borne virus which has no scientific name. However, it is possible that more than one organism or condition causes OAT leaves to turn red, such as poor nutritional balance, aphid injury, root pruning, or possible nematode injury.

The typical symptoms of red leaf are reddening of the leaves and stems, a general decline in vigor, and a stunting of plants, with gradual yellowing and death of the plants. In many cases small areas in oat fields have been observed to become red and spread as the season advances.

Control: No control is known as yet, but rotation, adequate fertilization, and general sanitation measures are generally recommended.

Reference: M.30.

See also VIRUS DISEASES.

RED LEAF SPOT is a cowpea disease caused by the FUNGUS *Cercospora cruenta*. It is fairly common and widespread, but does not cause appreciable damage to the crop.

Reference: M.23.

RED-LEGGED GRASSHOPPER.

See GRASSHOPPERS.

RED LOCOWEED (*Astralgus drummondii*) is a nontoxic species of the LOCOWEEDS.

RED LUMP. *See* PROSO (variety).

REDMAN. *See* COMMON WHEAT (variety).

RED MAY, also known by many other names, is a COMMON WHEAT variety.

RED OAT (*Avena byzantina*) is an important OAT species, widely grown in certain parts of the United States. The kernels are large and plump and of a reddish color, characteristic of all red oats.

Varieties: In the following alphabetical listing the reference to "region" indicates the part of the country in which the variety is grown: (1) *North Central* oat region; (2) *Central* red oat region; (3) *Northeastern* oat region; (4) *Southern* red oat region; (5) *Great Plains* oat region; (6) *Rocky Mountain* oat region; (7) *Pacific Coast* oat region.

Appler (region 4) fall-sown.

Brunker (regions 1, 2, and 5) spring-sown; medium-early maturing; resistant to VICTORIA BLIGHT and SMUTS.

Burt (region 5) spring-sown; early maturing; grows very tall.

California Red (region 7) fall-sown.

Camellia (region 4) fall-sown; disease-resistant.

Coker's Victorgrain (region 4) fall-sown.

Columbia (regions 1 and 2) spring-sown; early maturing; brownish-gray rather than red grain.

Ferguson 922 (region 2) spring-sown; and (region 4) fall-sown—high yielding.

Forkedeer (region 4) fall-sown.

Fulghum (regions 1 and 2) spring-sown; and (region 7) spring- and fall-sown—one of the most important American oat varieties.

Fulgrain (region 4) fall-sown.

Fultex (regions 2 and 5) spring-sown; and (region 4) fall-sown—disease-resistant.

Fulwin (region 4) fall-sown.

Kanota (regions 1, 2, and 5) spring-sown; and (region 7) fall-sown—resistant to VICTORIA BLIGHT.

Missouri 0-205 (region 2) spring-sown—resistant to VICTORIA BLIGHT.

Neosho (region 2) spring-sown.

New Nortex (regions 2 and 5) spring-sown; and (region 4) fall-sown.

Nortex (region 2) spring-sown; and (region 4) fall-sown—dormant for sixty-six days; high yielding.

Otoe (regions 1 and 2) spring-sown—resistant to SMUTS and VICTORIA BLIGHT.

Red Rustproof, or *Texas Red* (regions 2 and 5) spring-sown—early maturing; relatively free from SMUTS.

Stanton (region 4) fall-sown.

Tennex (region 4) fall-sown.

References: R.7; S.21.

See also COMMON OAT.

RED ORANGE is a name sometimes—erroneously—used for *Colman sorgo.—See also* SORGHUMS.

RED PEPPER = CAPSICUM.

RED REPUBLIC = *Red May.*

See COMMON WHEAT.

"RED RICE" is a RICE variety which occurs as WEED in cultivated rice fields.

"RED RIPPER" is a name applied to a group of COWPEA varieties.

"RED ROT"

= SUGARCANE STALK ROT.

RED RUDY = *Goens.*

See COMMON WHEAT.

RED RUSSIAN is (1) the name of a PROSO variety and (2) a synonym of the more common name *Turkey wheat*, one of the COMMON WHEAT varieties.

RED RUSTPROOF.

See RED OAT (variety).

RED SEA WHEAT = *Mediterranean wheat. See* COMMON WHEAT.

RED-SHOULDERED PLANT BUG (*Thyanta custator*) is well distributed over the United States. It is one of the smallest species of STINK BUGS. The adults range from 9 to 11 mm. in length. They are green, but the shade varies greatly in different individuals. The antennae are usually reddish-brown. On some specimens there are narrow reddish bands and pale, light-yellowish, or reddish markings. The ventral surface of the body is pale.

Reference: R.6.

RED SPELT = *Red Winter. See* SPELT.

RED SPIDER (*Tetranychus telarius*) is a tiny, suckling MITE which is common on vegetables, shrubs, and trees. It does much damage to RUSSIAN WILD-RYE. Affected leaves become blotches with reddish-brown and pale-yellow spots; they dry gradually and drop off.

RED SPOT.

See BACTERIAL LEAF DISEASES.

RED STEM (*Ammannia coccinea*) is a WEED found in RICE fields. When mature, the entire plant above the ground is red in color.

Control is possible with 2,4-D.

RED TOP is a name used synonymously for (1) SUMAC SORGO, one of the forage SORGHUMS, and (2) *Mediterranean wheat*, a COMMON WHEAT variety.—*See also* REDTOP.

REDTOP (*Agrostis alba*)—not to be confused with RED TOP, possesses many common names, e.g., *whitetop, fiorin, white bent*, and *Herd's grass;* because all these names belong more properly to other GRASSES, they should not be used for redtop. It is a perennial with a creeping habit. Its stems are slender, the PANICLES

Redtop. (H.26.)

loose and usually reddish. Redtop is widespread in the United States: it grows from Canada to the Gulf of Mexico and from New York to California.

Of the many grasses of the genus *Agrostos*, redtop is the only one of much prominence for *hay*. It is used in pasture mixtures under humid conditions, as a soil binder, and as a winter lawn and golf-green grass in the Southeast. It ranks among the lowest of the standard northern *pasture* grasses in palatability, but it is valued in pasture mixtures because it comes quickly and vigorously and helps to form a compact turf that protects the soil until the slower growing grasses become established.

Redtop has the outstanding ability to grow under a variety of conditions. It is one of the best wet-land grasses, but it also resists drought and will grow on soils so low in lime that most other grasses fail. The strength of the creeping (rhizomatous) roots make it useful for holding banks to prevent erosion. It adds to the yield of TIMOTHY and CLOVER hay.

Best results are obtained from planting on a compact, well-prepared seedbed. The crop will persist for several years, depending upon the fertility of the soil and the management. It is ideal for use in lawn mixtures where quick establishment is desired because it germinates rapidly but does not offer extreme competition to the slower growing, more permanent grasses. Redtop is seldom seeded alone.

Reference: H.1.

See also AGROSTIS; RHIZOME; MUTTON BLUEGRASS; ALSIKE CLOVER; GRAZING; SILAGE CROPS; PASTURE PLANTS; VITAMIN A.

REDUCING SUGARS are able, because of their chemical structure (free aldehyde or ketone groups), to reduce in alkaline solution the ions of certain metals, such as copper, iron, and mercury. This ability of reducing sugars is used (1) to determine the presence of reducing SUGARS (causing color changes of the sugar-containing solution) and (2) to distinguish them from the NONREDUCING SUGARS. All *monosaccharides* are reducing sugars; e.g., GLUCOSE (DEXTROSE), GALACTOSE, and fructose (LEVULOSE). Also some of the *disaccharides* (formed from two monosaccharide molecules) belong to this group; e.g., maltose (MALT SUGAR), SUCROSE, and LACTOSE.— *See also* MOLASSES.

RED WINTER SPELT.

See SPELT (variety).

RED WONDER = *Fulcaster.*

See COMMON WHEAT.

RED X = *Rex Sorgo. See* SORGHUMS.

REED CANARYGRASS (*Phalaris arundinacea*) is a leafy perennial of wide agricultural importance as a wet-land grass.

It is adapted to the northern half of the United States; the largest acreages now are in Oregon, Washington, northern California, Minnesota, Wisconsin, and Iowa.

This grass species grows in clumps that often measure 3 ft. across. The dense heads are 2 to 8 in. long and become whitish as the seed matures. This grass makes its best growth in fertile and moist, or wet, soils; it is one of the best grasses for waterways and swamplands of a muck and peat nature. Contrary to earlier opinions, it is adapted to, and makes excellent growth on, upland soils that are frequently dry for long periods in the summer. It is winter-hardy and grows most rapidly during the cool spring months. Its long life, long grazing season, and large yields of nutritious, palatable forage make it a valuable PASTURE PLANT.

Reed canarygrass will not survive under continuous, close grazing. Its use as a silage crop is increasing. Yields of 3 to 4 tons of hay in a season are not at all uncommon. Early cutting of the hay crop improves the quality; otherwise it may be somewhat coarse.

Seed is harvested in all the regions where the grass is grown.

Reference: H.1.

See also CANARYGRASSES; GRASSES; PASTURES.

REED FESCUE = TALL FESCUE.

REED FOXTAIL (*Alopecurus arundinaceus*), also called *creeping foxtail*, is cultivated to some extent in the northern United States, particularly in North Dakota and in the Pacific Northwest, where it is best adapted. It is similar to MEADOW FOXTAIL in growth habits, except for more vigorous rhizomes and blackish seeds. This long-lived GRASS gives high forage yield, but it is weak in the seedling stage. Although the heads ripen rather early, thirty days before thimothy, the forage remains green throughout the summer except under severe drought. Reed foxtail is particularly well adapted to brackish conditions of tideland pasture and to river bottoms subject to overflow.

Reference: W.16.

REED KAFIR is one of the grain SORGHUMS.

Reed canarygrass. (S.32.)

REE WHEATGRASS, developed as a hybrid between INTERMEDIATE WHEATGRASS and PUBESCENT WHEATGRASS, shows no superiority over either parent. It occurs in South Dakota. Growth of established plants starts early in spring if moisture is favorable, continues during the heat of summer, and progresses rapidly during cool, moist weather of the late summer. Ree wheatgrass is drought- and frost-resistant, practically free from ergot, and adaptable to a wide range of soils. It is fairly readily established and recovers rapidly after cutting. The grass is highly nutritious and palatable; best hay is obtained when harvested in early bloom stage.

Reference: W.16.

See also GRASSES; WHEATGRASSES.

REGENT WHEAT.
   See COMMON WHEAT (variety).

REID YELLOW DENT.
    See CORN (variety).

REINDEER MEAT.
   See ANIMAL PRODUCTS.

REINDEER MEAT BY-PRODUCTS.
   See ANIMAL PRODUCTS.

RELATIVE FEED VALUES.
See PETERSEN'S RELATIVE FEED VALUES.

RELIANCE is one of the intermediate-type grain SORGHUMS.

RENDERING PLANT.
   See ANIMAL PRODUCTS.

RENNET, a white, RENNIN-containing, water-soluble powder, occurs in the inner lining of the (true) stomach of animals and (as *vegetable rennet*) in numerous plants.—*See also* MILK PRODUCTS.

RENNIN is an ENZYME found in the GASTRIC JUICES of many animal species; it changes milk into solid curd by transforming CASEIN to *paracasein*. The latter combines with CALCIUM and precipitates, making it possible for the proteolytic enzyme PEPSIN to digest the milk which, otherwise, passes undigested into the small intestine. Mature animals, which do not obtain milk or milk by-products in their feed, often lack rennin in their stomach juices.

RENO. *See* SIX-ROWED BARLEY (variety).

RENOVATED PASTURES.

See PASTURES; PASTURE MANAGEMENT.

REPRODUCTIVE VITAMIN.
    See VITAMIN E.

REQUA. *See* COMMON WHEAT (variety).

RESCUEGRASS (*Bromus catharticus*) appears spontaneously in many places in the southern states. It is a short-lived perennial adapted to humid regions with mild winters.

Plants grow to a height of 2 to 3 ft. Leaf blades are 8 to 12 in. long and about ¼ in. wide. Young plants are usually covered with short, fine hairs (PUBESCENT), but mature plants are only slightly so. PANICLES are branched with two to five SPIKELETS at the extremity of each branch.

Growth starts in the fall and continues through the winter; the plants mature in early summer. On poor land they make little growth but on rich soil they produce a good amount of forage which is relished by livestock. The growing vegetation also protects the soil against erosion in winter.

Dangers: Among the enemies of this grass species is the SOUTHERN CORN ROOTWORM.

Reference: H.1.

See also BROME GRASSES; GRASSES; PASTURE PLANTS.

RESCUE WHEAT.
   See COMMON WHEAT (variety).

RESERVE PASTURES.
   See PASTURE MANAGEMENT.

RESIN GUAIAC, also called *guaiac*, is obtained from the wood of *Guaiacum* spp.; the water-insoluble RESIN contains various guaiac compounds and is particularly rich in *guaiaconic acids*. It is used as ANTIOXIDANT.

RESINOIDS are somewhat indefinite, complex substances of the nature of a RESIN (a secretion product of plants). Resinoids are the active principle of such POISONOUS PLANTS as WESTERN AZALEA, WATERHEMLOCKS, and MILKWEEDS.

RESINS are (1) *natural gums* derived from trees and some other plants, or (2) *synthetic* organic products. All resins are water-insoluble, but soluble in organic solvents.—*See also* RESINOIDS.

RESISTANT WHEATLAND 288, like *Wheatland*, is one of the intermediate-type grain SORGHUMS.

RESTING SPORE. *See* OÖSPORE; SPORE.

RETICULATED means: net-veined; this term refers to leaf VEINS which look like a network.

RETICULUM is the *second* STOMACH of a ruminant.

REWARD WHEAT.
>   *See* COMMON WHEAT (variety).

REXORO. *See* RICE (variety).

REX SORGO is one of the forage SORGHUMS.

REX WHEAT.
>   *See* COMMON WHEAT (variety).

RHACHIS = RACHIS.

RHEA = RAMIE.

RHIZOBIUM (plural: *Rhizobia*). The *Rhyzobium* spp. are SYMBIOTIC, nitrogen-fixing BACTERIA which form NODULES on the roots of LEGUMES.

RHIZOCTONIA. Some of the *Rhizoctonia* spp. are among the FUNGI which cause SEED ROT, ROOT ROTS, and CONCEALED DAMAGE IN SEED; some species—especially *R. oryzae*, *R. zeae*, and *R. solani*—cause BORDERED SHEATH SPOT. Different strains of the mentioned *R. solani* are also involved in producing disease symptoms on different parts of practically all crop plants; they cause, e.g., RHIZOCTONIA STALK ROT, ROOT AND STEM ROT, DAMPING-OFF, as well as STEM CANKER OF CROTALARIA, SOIL ROT, BLIGHTS, etc.

R. crocorum is the cause of VIOLET ROOT ROT.—*See also* BERMUDA-GRASS; SORGHUMS; COTTON; VETCHES; POTATO; RICE.

RHIZOCTONIA STALK ROT is caused by the soil-borne FUNGUS *Rhizoctonia solani* that attacks POTATOES, COTTON, SORGHUMS, SOYBEAN, BIG TREFOIL, and other crops. This STALK ROT occurs in northern Texas. It differs from CHARCOAL ROT in that it first attacks the pith and produces in it a reddish discoloration, while the fibers remain as light streaks in the discolored pith. Later, compact masses (SCLEROTIA) formed by the fungus, and brown in color, may be found on the outside, under the leaf sheath.

Control: Little is known about varietal resistance to this stalk rot.

Reference: L.1.

RHIZOMA ALFALFA.
>   *See* VARIEGATED ALFALFAS (variety).

RHIZOMATOUS means: of the nature of a RHIZOME.

RHIZOME is an underground ROOTSTOCK; its simplest form is merely a creeping, usually thickened stem or branch growing entirely or partly beneath the surface of the ground. The fact that a rhizome is really a stem and not a root is evident from its manner of growth, its consisting of a succession of joints, and from the scales which (as true, though degenerated leaves) are borne at these joints and which often have buds in their axils.

Reference: D.9.

See also BENTGRASSES; STEM.

RHIZOPUS. The *Rhizopus* spp. are among the FUNGI causing SEED ROT and CONCEALED DAMAGE OF SEED.

RHODE ISLAND BENT
>   = COLONIAL BENT.

RHODESGRASS (*Chloris gayana*), one of the FINGERGRASSES, is a fine-stemmed and very leafy perennial which grows approx. 3 ft. high. The spreading, clustered SPIKES of the flowering head number from 10 to 15 and the seed is produced in abundance. The plant also spreads by running branches, or stolons, that are 2 to 6 ft. long and root and produce a plant at every node.

Rhodesgrass rarely withstands temperatures below 15° or 18° F. Its adaptation is therefore limited. However, it is winter-hardy in a narrow strip along the Gulf Coast from Florida to southern Texas and in southern Arizona and California.

It does best on fairly moist soil, although it will make growth during several months of drought. It grows well on sandy soils, well-drained peaty soils, and soils too alkaline for alfalfa, cotton, and other crops.

Rhodesgrass can be sown any time during warm weather, but early spring usually is preferred. The seed is usually broadcast. A well-prepared seedbed will help insure a good stand, although stands have been established on rather loose, rough ground.

Cultivated for pasture, this grass withstands trampling, recovers quickly, and is relished by livestock. Rotation grazing is the best method of management to insure greater production and maintenance of stand. It will also yield a leafy *hay* of high quality. The production for *pasture* or hay varies greatly, depending on soil fertility and the season. Yields of 5 to 7 tons of hay an acre have been reported.

Rhodesgrass. Plant and florets. (H.26.)

Rhodesgrass will produce three or four crops of seed a year.

Reference: H.1.

See also GRASSES; PASTURE PLANTS.

RHODOGYNE. *R. fuliginosa* is an insect parasite of the adult STINK BUGS.

RHUS. *R. trilobata* = SKUNKBRUSH.

RHYNCELYTRUM. *R. roseum*

= NATALGRASS.

RHYNCHOSPORIUM. The *Rhynchosporium* spp. are FUNGI which cause SCALD.

RIBBED PASPALUM (*Paspalum malacophyllum*) is a fine-stemmed, erect species of PASPALUM GRASS. It requires careful management because it is badly injured by heavy grazing.

Reference: H.1.

See also GRASSES.

RIBOFLAVIN = VITAMIN B_2.

RIBOFLAVIN SUPPLEMENT.

See VITAMINS.

RICE (*Oryza sativa*) is an annual GRASS; it belongs to the true cereal crops. There is probably more variation in growth habits of rice than of any other crop. For instance, hill rice is short and grows without flooding; other rices grow under water and are harvested from boats. Most commercially grown rice plants reach from 2 to 6 ft. in height. They mature in 120 to 139 days, depending on variety and climatic conditions. The hulled rice kernels vary in size —from 3.5 to 8.0 mm. in length and from 1.3 to 3.0 mm. in thickness.

There are three rice-producing regions in the United States: (1) the broad, level prairie region of southwestern Louisiana and southeastern Texas, the most important rice-producing region in the United States; (2) the prairie region of eastern Arkansas; (3) the interior valleys of California.

Successful rice culture is dependent on high temperatures during the growing season, a dependable water supply during the period of irrigation, soils that are comparatively flat and underlain with an impervious subsoil, and good surface drainage.

The heavy soils on which rice is grown and the high water table during the submergence season make it difficult to find crops that can be grown profitably in a rotation with rice. Although good yields of WHEAT and BARLEY have been reported on fallowed riceland, some growers do not think that these crops are profitable under such conditions; they prefer to alternate rice and fallow or to have two rice crops and one fallow. Other growers do not fallow, but leave the land idle and uncultivated until it is prepared for rice the following spring. Such idle lands are often *pastured*.

In the rice-pasture system of farming of the Gulf Coast, rice is grown from one to three consecutive years. Then for the next two to several years these fields are grazed by beef cattle as unimproved *rice-stubble* pasture. The vegetation consists of volunteer grasses and other plants. Improved pastures following rice are several times more productive than unimproved rice-stubble pastures.

Broadcast seeding of CLOVERS and grasses, without seedbed preparation, in standing rice (at last draining about ten days before harvest) or in stubble after harvest gives satisfactory stands of adapted grasses. In the Gulf Coast areas, OATS and RYEGRASS should be seeded from mid-September to December; DALLISGRASS from mid-September to November; KENTUCKY 31 TALL FESCUE, ALTA FESCUE, or clovers—especially LOUISIANA WHITE CLOVER, PERSIAN CLOVER, ALSIKE CLOVER and the HOPCLOVERS—should be seeded between October 15 and December 15.

NOTE: Phosphate fertilizer is needed to establish the clovers. From 150 to 200 lb. per acre of 0-45-0 is recommended for most rice lands. In some cases, lime and potash are required. INOCULATION of clover seed is essential.

The drains and levee ditches should be opened after the rice harvest to provide a free flow of water from the fields.

New ricelands are very productive, but yields are usually materially lower with each successive crop. The reduced yields may be due to lack of plant food elements, especially nitrogen which can be added to the soil as commercial fertilizer, or by growing a leguminous crop to be plowed

under. The second method should be followed wherever possible, as it adds organic matter and thus improves the physical condition of the soil as well as supplies nitrogen. Nitrogenous fertilizers, such as ammonium sulfate, uramon, urea, and cyanamide, may be applied at the rate of 100 to 200 lb. per acre just before seeding.

In preparing the seedbed the land should be spring-plowed 4 to 6 in. deep as early as it can be worked to advantage. After plowing, the land may be disked, har-

Rice. Plant and spikelet. (H.26.)

rowed, or dragged with a heavy float. A light rain, followed by a few warm days, helps to reduce clods. Where riceland is fall-plowed, the land is easily reduced to a good seedbed by the following spring through the action of rain and sun during the winter. It pays to kill all weed growth before sowing rice. If weed growth is killed on fall-plowed land before sowing, the cost of preparing a seedbed on spring-plowed and on fall-plowed land is about the same.

When the rice is sown broadcast and then submerged continuously, the seed-bed should be somewhat finer and more mellow than is necessary when sowing in the water.

Good seed rice should be free from red, immature, hulled, or broken rice, or the seeds of other rice varieties or weeds; and it should be well filled and uniform in size.

When rice is sown broadcast on the water, only shallow water (2 to 4 in. deep) is held until seeding is completed; then the depth of water is increased to 4 to 8 in. When the rice is sown on the surface of the soil, the fields are immediately submerged from 4 to 8 in. deep. The water is held continuously at this depth until the crop is nearly ready for harvest.

Relatively clear water is helpful in obtaining vigorous seedlings and good stands. Scum, which consists of various kinds of algal growth and partly decomposed organic particles, often prevents emergence of the rice seedlings to the surface of the water.

When the land is submerged, water flows from one check to another, and normally there is little danger that any part of the field will become stagnant if the water is maintained at a constant depth.

Good judgment is essential in deciding when the rice crop is ready to drain for harvest. The water should be held longer on soils that dry quickly than on those that dry slowly. It usually takes from ten to eighteen days after draining for the land to be dry enough to support harvesting machinery.

Rice is ready to harvest when the lower grains on the PANICLES are in the hard-dough stage, at which the moisture content of the standing grain ranges from 20 to 27 percent. The rice crop is largely harvested with push swathers and self-propelled, or tractor-drawn, combines. The windrow, placed on the stubble by the swather, is immediately threshed with a combine equipped with a pick-up attachment. The moisture content of the combined rice is reduced gradually to between 14 and 15 percent in artificial driers, and the dried rice is stored in bulk or sacks.

Rice straw, also called *paddy straw*, contains carbohydrates (32.2 percent), protein (4.2 percent), and fat (1.9 percent); it is often used as feed for farm animals.

Rough rice, or *paddy*, is the name of the rice kernel enclosed by the hull; if damaged or of low quality, it is used for feed. Otherwise, it is milled or used as seed rice.

In milling, rough rice yields (1) hulled rice kernels, known as *brown rice* (from which, after further treatment, *polished rice* is obtained); (2) *rice bran* (13 percent); (3) *rice polishings* (3 to 4 percent); and (4) *rice hulls* (20 percent). *Rice flour* is a by-product of the milling process.

Whole grains of polished rice are known as *head rice;* finely broken rice is called *brewers' rice;* it is used in breweries and distilleries, or for feed.

Brown rice is richer in vitamin B_1 than polished rice. Ground brown rice, called *rice meal*, makes a good stock feed.

These are the officially recognized rice products:

Rice bran—the pericarp (outer covering) or bran layer of the rice, with only such quantity of hull fragments as is unavoidable in the regular milling of rice.

Rice bran shows a high content of several of the B-complex vitamins, especially B_1, and niacin; and it contains more B_2 than the whole grain. It is not as high in protein as are the wheat by-products, but it is high in fat and becomes rancid rather readily.

The fiber content of rice bran varies according to the method of milling, the variety of rice, and as to whether rice hulls have been added to the bran. These variations in the fiber content of rice bran greatly influence its feeding values. Only high quality, rich bran, which is a highly nutritious feed—i.e., bran with low fiber and rice hull content—should be fed to farm animals.

Bran is fed to dairy and beef cattle, sheep, and swine. It may be fed to swine only in limited amounts; otherwise the body fats, or lard, become soft. Large proportions in the diet also tend to produce scours.

Rice polishings, or *rice polish*—obtained in the milling operation of brushing the grain to polish the kernel.

Rice polishings is very rich in niacin, richer in many other B-complex vitamins than rice bran, and higher in total digestible nutrients, a fact that is worthy of note in feeding cattle and sheep as well as swine. Like bran, it should be fed to swine only in limited amounts so as to avoid softening of the body fats. Because of the high fat content, polishings tends to become rancid on storage and may cause scouring, especially in small pigs. Therefore, polishings (and bran) should be fed as soon after milling as possible.

Rice meal—ground brown rice or ground rice from which the hull has been removed.

Ground rough rice—ground rice from which the hull has not been removed or ground paddy rice.

Rice stone bran—the siftings from the materials secured in removing hulls from rice; it contains rice germ, broken rice, and some rice hulls.

Rice huller bran—secured by the huller and cones from brown rice; consists mostly of the bran and germ.

Solvent extracted rice bran—obtained by removing part of the oil from rice bran or rice mill bran by the use of solvents; it must contain not less than 14 percent protein and not more than 14 percent fiber.

NOTE: Most of the rice by-products are low in protein content and should be supplemented with protein-rich feeds of good quality for the most satisfactory results.

Dangers: Weeds are a menace to rice production; the worst weeds in rice fields are BARNYARDGRASS, KNOTGRASS, ARROW-HEAD, WATERPLANTAIN, REDSTEM, UM-BRELLA PLANTS, CATTAILS, BULRUSH, SPIKE RUSH, WATER HYSSOP, and red rice (discussed below with other varieties of rice).

Insects that injure rice include GRASS-HOPPERS, the RICE WATER WEEVIL, FALL ARMYWORM, CHINCH BUG, RICE LEAF MINER, BILLBUGS, and LEAFHOPPERS.

Blackbirds are troublesome in rice fields in the spring and fall. In rice fields with open water, wild ducks and mud hens (coots) often cause serious damage.

Species of the TADPOLE SHRIMP may injure rice seedlings.

Among rice diseases are SEEDLING BLIGHT, BLAST OF RICE, BROWN SPOT OF RICE (causing "PECKY RICE"), NARROW BROWN LEAF SPOT, BROWN BORDERED LEAF SPOT, STRAIGHTHEAD, WHITE TIP, ROOT KNOT, LEAF SMUT OF RICE, BORDERED SHEATH SPOT, BLACK SHEATH ROT, REDDISH-BROWN SHEATH ROT, KERNEL SMUT, and STEM ROT OF RICE.

Varieties: There are two main groups of cultivated rice—the *nonglutinous rices* (which are grown most extensively) and the *glutinous rices*. Based on method of production, there are two general types of rice—*lowland* (irrigated or rain-fed), and *upland*. Based on kernel shape and size, the most important rice varieties cultivated in this country can be classed as follows:

1. *Short-grain rice*—grown almost exclusively in California; *Asahi* (also grown in the southern states), *Caloro* (with yellowish awns, widely grown in California), *Colusa* (awnless, earlier maturing than Caloro), *Conway* (awnless, earlier than Caloro, but later than Colusa).

2. *Medium-grain rice*—grown most extensively in Texas, Arkansas, and Louisiana; *Blue Rose* (late maturing), *Calrose* (grown in California), *Early Prolific* (early maturing), *Magnolia* (grown in Louisiana), *Zenith* (a high yielding variety).

3. *Long-grain rice*—grown mainly in the southern states: *Bluebonnet* (resistant to BLAST OF RICE), *Fortuna, Nira, Rexoro* (the most important variety in the South), *Texas Patna* (earlier than Rexoro).

NOTE: *Red rice*, a weed, is a variety of the rice species *Oryza sativa*. It is probably the worst pest in the rice fields of the United States. Red rice has a red seed coat when the hull is removed, whereas the other rice varieties have a brownish or whitish seed coat. Seed containing red rice should not be sown. The presence of red grains in milled rice reduces its value.

References: J.5; M.28; M.29; J.6; L.7; T.2; E.12; F.6.

See also CEREAL GRAIN BY-PRODUCTS; WILDRICES.

"RICE" is a term which refers to *popcorn* varieties with pointed kernels. It is not to be confused with RICE or RICE CORN.—*See also* CORN; WILDRICE.

RICE BRAN. *See* RICE.

RICE CORN, commonly called *durra*, is one of the grain SORGHUMS, as is *Jerusalem rice corn* (which is better known as *shallu*).

RICE FLOUR. *See* RICE.

RICEGRASS. *See* INDIAN RICEGRASS.

RICE HULLER BRAN. *See* RICE.

RICE HULLS. *See* RICE.

RICE KAFIR is one of the grain SORGHUMS.

RICE LEAF MINER (*Hydrellia griseola* var. *scapularis*) is one of the water-loving flies. The elongate, ribbed, white eggs are about $\frac{1}{40}$ in. long and are laid singly on the leaf blades of rice. From 1 to 15 eggs may be laid on the same blade. The eggs hatch in about four days into small maggots which almost immediately start mining inside the leaves. The maggots feed on the green cells causing the leaf blades to turn transparent. The leaves subsequently shrivel and lie prostrate on the surface of the water. The maggots reach a length of about $\frac{1}{8}$ in. when mature, and pupate inside the leaves. The usual length of time from laying of the eggs to emergence of the adult is from fifteen to twenty-four days. The adult flies, emerging from the puparia, are grayish in color and have a distinct metallic-greenish sheen. They range in size from $\frac{1}{7}$ to $\frac{1}{10}$ in.

The overlapping of the generations makes it possible to obtain all stages of the insect from any particular rice field, except newly emerged rice.

Control: The leaf miner can be controlled by lowering the water in the field to about 2 in., spraying with DIELDRIN, or HEPTACHLOR, and leaving the water low for forty-eight hours. The water is then raised and the checks are blocked off so that no water is spilled from the fields for two weeks to protect fish, domestic animals, and wildlife from the effects of the insecticide.

Reference: L.7.

RICE MEAL. *See* RICE.

RICE POLISHINGS is also known as *rice polish*. *See* RICE.

RICE STONE BRAN. *See* RICE.

RICE STRAW. *See* RICE.

RICE WATER WEEVIL (*Lissorhoptrus simplexi*) sometimes damages RICE in the fields.—*See also* WEEVILS.

RICHLAND SOYBEAN.

See SOYBEAN (variety).

RICINUS. *R. communis* = CASTOR BEAN.

RICKETS is a bone disease due to CALCIUM DEFICIENCY or lack of PHOSPHORUS or VITAMIN D.

RIDDELL GROUNDSEL. *See* GROUNDSELS; SENECIO; POISONOUS PLANTS.

RIDGE BUSTER is a disk implement for breaking down ridges and filling furrows of listed land.

RIND is the outer coat (covering) of trees, fruits, stalks, etc.; it may be a bark, skin, peel, or HUSK.

RINGLET. *Devil's ringlet* = DODDER.

RIO. *See* COMMON WHEAT (variety).

RIPLEY = *Purplestraw*.

See COMMON WHEAT.

RIVAL. *See* COMMON WHEAT (variety).

RIVET-WHEAT = POULARD WHEAT.

ROANOKE. *See* SOYBEAN (variety).

ROBBER FLIES, or *assassin flies*, the "hawks of the insect world," are predators which belong to 97 genera with over 500 species. They feed on VELVETBEAN CATERPILLARS and on many other larvae as well as eggs of INSECTS.—*See also* FLIES.

ROBINA. The fruit of the legume tree *R. pseudacacia* is known as CAROB BEAN.

ROCK PHOSPHATE is ground *phosphate rock*, a naturally occurring CALCIUM PHOSPHATE used as fertilizer, and as raw material for phosphorus and its salt.

Low fluorine rock phosphate, a CALCIUM and PHOSPHORUS source, is one of the officially recognized MINERAL FEEDS. However, it is not as well assimilated as other calcium phosphates.—*See also* FLUORINE POISONING.

RODENTICIDES, such as STRYCHNINE and its salts, are used to kill RODENTS.—*See also* STRYCHNINE-POISONED GRAIN.

RODENTS are gnawing mammals; many rodents are pests in fields; e.g., FIELD MICE, MOLES, POCKET GOPHERS, and rabbits.—*See also* STRYCHNINE-POISONED GRAIN.

ROGUE means: to weed out diseased or inferior plants from a seed-producing plot.

ROJO. *See* SIX-ROWED BARLEY (variety).

ROLLED BARLEY. *See* BARLEYS.

ROLLED GRAIN SORGHUMS.
                              See SORGHUMS.

ROMANELLA = *Turkey wheat.*
                       See COMMON WHEAT.

ROOSEVELT = *Midwest soybean.*
                    See SOYBEAN (variety).

ROOT is the (usually subterranean) part of a plant which lacks nodes.

ROOT AND STEM ROT, a combination of ROOT ROT and STEM ROT, is caused by the FUNGUS *Rhizoctonia solani.* It affects especially the VETCHES.

ROOT APHID. *See* CORN ROOT APHID.

ROOT BORER. *See* CLOVER ROOT-BORER.

ROOT CROPS, used occasionally as forage, include the SUGAR BEET, MANGEL, SWISS CHARD (most often grown for its leaves), CARROT, POTATO, SWEET POTATO, and TURNIP. They are very high in water and therefore low in dry matter which consists chiefly of highly digestible CARBOHYRDATES.

In general, growing, harvesting, and storing root crops is expensive. The preparation of SILAGE from the water-rich roots is also in most cases uneconomical.

ROOT DISEASES include ROOTS and basal portions of the CULMS; e.g., ASCOCHYTA FOOT ROT, PERICONIA ROOT ROT, and ROOT KNOT.

ROOT KNOT is a trouble most familiar to the farmer because of the effects it has on the roots of the diseased plants. Instead of the normal, tapering roots found on healthy plants, those attacked by root knot have numerous irregular swellings, so-called *knots* or *galls*, over the entire root system. They vary from small, inconspicuous swellings to rough, knotty enlargements 1 in. or more in diameter. At first they are the same color as the healthy roots, but they soon turn brown and decay, causing serious injury or death of the roots. They are quite different from the small, roundish nodules of the beneficial nitrogen-gathering organism, which rarely exceed ¼ in. in diameter. These nodules are attached loosely to the roots, while the root-knot galls are enlargements of the roots themselves and cannot be removed without breaking the roots.

This disease affects many crops, particularly ALFALFAS, LUPINES, KUDZU, CLOVERS, and other LEGUMES, RICE, etc.

Root knot is caused by a tiny ROOT-KNOT NEMATODE which lives for the most part in the roots of cultivated crops. It bores its way into the young roots, obtains its food from them, and so irritates the tissues that galls are produced. The formation of the galls not only uses up the food supply, but also interferes with its passage to the plant above, and results in stunting or death. If one of the galls is broken open, the enlarged female nematodes can be seen with the naked eye as pearly-white roundish bodies about the size of a pinhead.

Control: The most effective and practical method of controlling root knot is based on the use of resistant or immune varieties of forage crops. The control of the root knot is of vital importance in the South, where the COWPEA is the principal leguminous crop used in rotations to improve inferior soil areas. For intelligent planning of suitable crop rotations for such root knot infested fields, it is essential to know the crops which may be safely planted, as well as those which should be avoided. WINTER GRAINS, CORN, SORGHUM, VELVETBEANS, three varieties of cowpeas (*Iron, Brabham,* and *Victor*), and one of SOYBEANS (*Palmetto*), are highly resistant to root knot, while other farm crops for the most part are quite susceptible.

Reference: M.23.

ROOT-KNOT NEMATODES (*Meloidogyne* spp.) cause ROOT KNOT. They are very prevalent in fields of ALFALFA and ALYCECLOVER, but the other LEGUMES, except the CROTALARIAS, are also susceptible to them, as is BERMUDA-GRASS. The NEMA-TODES live in the soil and attack young fine roots, causing *galls*, or *knots*, which appear as small, scattered, tubercle-like growths. Scattered galls may be confused with nodules formed by nitrogen-fixing bacteria. Severely affected seedling plants are stunted in growth. Infection, usually, is more severe on sandy soils than on heavy soils, and is favored by high soil moisture and warm weather.

PEANUT fruits and above-ground parts of the peanut plants are attacked by the nematodes *M. hapla* and *M. arenaria* (the latter attacks also several other plants). The species usually found in Florida, Alabama, and Georgia, produces massive knots not only on roots, but also on pegs and pods of peanuts. The plants become yellow and stunted. In severely infected fields, the crop may be so reduced as to make harvesting unprofitable. The knots produced by the species in Virginia and North Carolina are not massive and the crop loss is less severe.

Control: Soil fumigation with DOWFUME W-40, or other soil FUMIGANTS, is occasionally used to kill root-knot nematodes.

In areas where root-knot disease prevails, only resistant legume selections should be planted, if available. Because the root-knot nematode infests the peanut pegs and pods, the shells from infested lots should be burned; fields known to be infested with these nematodes should not be planted to peanuts. The same precaution is necessary for Bermuda-grass.

References: B.10; W.18; S.12; W.15.

See also HETERODERA; CLUSTER CLOVER; HAIRY INDIGO.

ROOTLET is a small ROOT.

ROOT PLANTS. *See* ROOT CROPS.

"ROOT ROT" = SOIL ROT.

—*See also* ROOT ROTS.

ROOT ROT OF SAFFLOWER, caused by the FUNGUS *Phytophthora drechsleri*, often causes damage to the SAFFLOWER crop. The organism is soil-borne and widely distributed. High moisture favors the disease. Some varieties of safflower are very susceptible, but others are resistant.

Reference: T.3.

ROOT ROTS is a term applied to a number of FUNGUS DISEASES, such as FUSARIUM ROOT ROT, PYTHIUM ROOT ROT, PERICONIA ROOT ROT, VIOLET ROOT ROT, COTTON ROOT ROT, ROOT AND STEM ROT, and ROOT ROT OF SAFFLOWER. Root rots are related to the CROWN ROTS, but are caused by different FUNGI, all of which attack the underground parts of the plants where they cause varying degrees of decay. The decayed tissue may be reddish, gray, brown, or black. As a result of the death of the roots, the tops are stunted and often yellowish, and the plant may die.

Control: Rotation is recommended as a good practice, but it will be only partly successful as a control measure since some of the root rot fungi attack a great variety of crops; e.g., FIELD PEA, LUPINES, TRUE CLOVERS, and SWEET CLOVERS.

Reference: M.24.

See also SOUTHERN BLIGHT; VICTORIA BLIGHT; WHEAT SCAB; FOOT ROTS; STRIPE SMUTS; SOUTHERN ANTHRACNOSE; SEEDLING DISEASES.

ROOTSTOCK is a rootlike STEM, or branch, growing under the ground—i.e., a RHIZOME—or sometimes on the ground. On hard soil, a rootstock may become a RUNNER.

ROOTWORM. *See* CORN ROOTWORM; SOUTHERN CORN ROOTWORM.

ROSE CLOVER (*Trifolium hirtum*) is a winter annual LEGUME which will grow in soils where practically no other plants survive and will even provide some forage under such unfavorable conditions. Cattle, sheep, and deer appear to graze this clover well even when it is completely dried up. Close grazing does not seem to force rose clover back.

The nutritive value of rose clover is comparable to that of other legumes. Rose clover samples have a protein content of 25 percent when lush and of 12 percent when at the flowering stage. Dead, dry plants have 8 percent protein.

Rose clover is a many-branched legume that grows from 3 to 18 in. high. The spreading branches are densely covered with short, coarse hairs. As with most TRUE CLOVERS, each leaf has three leaflets at equal distances from the end of the leaf stalk, the center one somewhat larger than the other two. The leaflets have a scattering of short hairs over both surfaces and usually a small, reddish mark a little above the center. The leaf stalks are from ½ to 2 in. long, and have hairs like those on the leaflets. At the base of the leaf stalks are hairy STIPULES ¼ to ½ in. long. The flower heads are rose-colored, spherical, about ¾ in. across, and profusely covered with stiff, white hairs. A leaf with somewhat smaller leaflets grows directly from the base of the flower head.

For best results treat the seed with a general *Trifolium* inoculant just before planting in the fall. Apply 100 to 300 lb. superphosphate per acre; disk the field before the fall rains. After seeding roll the area with a ringroller or cultipacker.

Rose clover. (L.4.)

If the soil is extremely infertile and supporting practically no growth of native plants, use rose clover alone. If there is a fairly good cover of native, weedy annual and other type grasses, use a mixture of 50 percent rose clover, 25 percent SUB-CLOVER, and 25 percent CRIMSON CLOVER. The field may be grazed throughout the winter, or heavily for a shorter period in the spring.

No other range legume has proved to be so well adapted to such a great variety of soil types and climatic conditions as rose clover. When it is grown on GRAIN land it will provide a good aftermath feed to supplement the cereal stubble, will volunteer in succeeding years, and will add nitrogen to the soil.

Rose clover is the best winter legume to use to reclaim abandoned grain land. Gradually it builds up the soil to a point where it will support more and more growth of desirable forage plants.

Reference: L.4.

ROSEN. *See* RYE (variety).

ROSE SOYBEAN.

See SOYBEAN (variety).

ROSETTE is a dense basal cluster of leaves, caused by dwarfing of the true (leafy) stem. It resembles the PETALS of a double rose.

Rosettes are common among winter annuals.

ROT. *See* ROTS.

ROTARY HOE is an intertillage implement consisting of numerous hoe wheels fitted with teeth which penetrate and stir the soil. It is often used for uprooting weeds or for breaking soil crusts.

ROTATION is a systematic plan of changing the crops from field to field or on one field from year to year.

ROTATION GRAZING. *See* PASTURES.

ROTATION PASTURES.

See PASTURES; PASTURE MANAGEMENT.

ROTENONE is a white, crystalline compound occurring in DERRIS, CUBE, and other plant roots. It is a very effective INSECTICIDE.

Dust containing 0.75 percent rotenone applied at the rate of 20 lb. per acre, controls PEA WEEVILS. Repeat the treatment in three or four days, if necessary.

PEA APHIDS can be destroyed by application, per acre, of 35 to 40 lb. dust containing 1 percent rotenone, or of a spray containing 4 percent rotenone in 125 gal. water.

A *rotenone mineral-oil emulsion spray* is prepared in the same manner as the NICOTINE mineral-oil emulsion spray, but ⅛ oz. derris or cube extract containing 5 percent rotenone (in place of ⅛ oz. nicotine sulfate) is added to each gallon of finished spray solution. This spray may be applied at the rate of 70 gal. per acre for the control of CHINCH BUGS.

ROTS are FUNGUS DISEASES which attack thousands of plants; they affect roots, stalks, corn ears, and other parts of grasses, flowers, vegetables, shrubs, trees, etc. Among the more important rots are the following: ROOT ROTS (e.g., PERICONIA ROOT ROT and COTTON ROOT ROT); STEM ROTS, such as VICTORIA BLIGHT, BROWN STEM ROT, SCLEROTINIA STEM ROT, and STEM ROT OF RICE; STALK ROTS (particularly CHARCOAL ROT, FUSARIUM STALK ROT, COLLETOTRICHUM STALK ROT, and RHIZOCTONIA STALK ROT); SHEATH ROTS, e.g., BLACK SHEATH ROT and REDDISH-BROWN SHEATH ROT; FOOT ROTS (e.g., ASCOCHYTA FOOT ROT, CULM ROT, and LEAF AND POD SPOT); CROWN ROTS; PEG ROTS; BUD ROTS, and SEED ROT.

"ROTTEN NECK," a condition of RICE, is caused by the same *Helminthosporus* FUNGUS which is the cause of BROWN SPOT OF RICE.

ROUGHAGES include bulky and coarse feedstuffs, particularly ALFALFA stems, dry sorgo, CORN fodder, HAY, STRAW, COTTONSEED hulls, OAT hulls, and SILAGE. They are rich in fiber content and therefore of low digestibility. The best quality roughages contain at least 50 percent T.D.N. and poor ones as little as 10 percent. LEGUME roughages are valuable sources of vitamins, minerals, and proteins. Nonlegume roughages, unless cut rather early, are poor sources. Some roughage is needed in the rations of ruminants and horses to supplement and dilute feed concentrates.—*See also* CHOPPING ROUGHAGE; GRAZING; BACTERIA; VITAMIN D; CRIMSON CLOVER; AL-FALFA PELLETS; ALFALFA MOLASSES FEED; LEGUMES; FORAGE; SORGHUMS.

ROUGHPEA (*Lathyrus hirsutus*)—also known as *wild winter pea, Caley-pea,* and *Singletary pea*—is a winter annual with weak stems and decumbent growth, except in thick stands when it is ascending. In general this LEGUME species has the appearance of sweet peas (a climbing plant with beautiful flowers), the leaves having one pair of long, narrow leaflets and terminating with a coiled TENDRIL. The lavender flowers are usually borne in pairs on a long stem and are fairly conspicuous. The seed pods are rough, or HIRSUTE, and the seed is round and characteristically TUBERCULATE.

Roughpea is adapted to the southern third of the United States wherever moisture conditions are favorable. It prefers lime soil but will grow on the average acid soil of the South. In most places, however, the use of 200 lb. lime is beneficial. The use of 200 lb. or more of 20-percent superphosphate will give greatly increased yields in most soils of the South; in some cases the use of potash is also beneficial. Although the roughpea does best on well-drained soils, it will grow on soils too wet for clover or small grain.

The principal uses of roughpea are for pasturage, winter cover crop, and hay. It is a weed in GRAIN fields. In most parts of the South it makes much less growth than AUSTRIAN WINTER FIELD PEAS or HAIRY VETCH, and is, accordingly, inferior to these crops as green manure to precede COTTON or early planted CORN. When it can be allowed to stand until late in spring it makes a good soil-improving crop.

Roughpea should be seeded in the fall. Once it is established further seeding is not necessary since plants can be volunteered almost indefinitely.

INOCULATION has not been found necessary in any section.

Dangers: Injury to livestock from grazing roughpea nearing maturity has been observed in Alabama and Louisiana and, therefore, caution is advised when grazing maturing plants, or feeding hay containing plants with well-developed seed. It can be pastured safely through

April, however, and later, if the plants are grazed rather closely and kept from forming seed.

Among the plant diseases attacking roughpeas is the BROWN SPOT.

Reference: M.3.

See also WILD WINTER PEA POISONING.

ROUGH PIGWEED.
See PIGWEEDS; WEEDS.

ROUGH RICE. *See* RICE.

ROUGH-SEED BULRUSH.
See BULRUSH.

ROUGH SPOT, caused by *Ascochyta sorghina*, is rather widespread in the southeastern states.

This FUNGUS LEAF DISEASE attacks forage and grain SORGHUMS and JOHNSONGRASS. It is first observed as circular to oblong, light-colored spots. Then the red or tan pigment, depending on the variety, becomes apparent as the FUNGUS spreads and injures the leaf tissue. Soon small, black specks (the fruiting bodies of the fungus) develop abundantly in the spots. On older leaves the spots are grayish to yellowish-brown or purplish-red, usually $\frac{1}{8}$ to 1 in. long, and $\frac{1}{16}$ to $\frac{1}{4}$ in. wide, running lengthwise on the leaf. As the spots enlarge, they grow together so that the size of diseased areas is extremely variable. In some cases the areas are surrounded by a reddish or tan border, while in others no color develops; sometimes the pigment is distributed as small specks throughout the infected areas.

When the areas affected with rough spot are rubbed between the fingertips, the sandpaper-like roughness, caused by the hard, raised fruiting bodies of the fungus, can be detected readily. By the time the leaves die and become dry, the fruiting bodies often cover most of the leaf surface. Similar lesions occur on the leaf sheaths and occasionally on the stalks.

The disease decreases the forage value of the crop materially and also the production of sugar in the stalk. Rough spot is heaviest where grain sorghum or Sudangrass is grown on the same land for several seasons.

The following varieties are relatively free of this disease: Schrock (a grain sorghum), and the Straightneck, Silvertop, and the McLean sorgos.

Control: Grain sorghums and Sudangrass should not be grown on land where rough spot occurred the preceding season. SEED TREATMENT is advisable; the use of available resistant varieties is recommended.

Reference: L.1.

See also LEAF DISEASES.

ROUGHSTALK
= ROUGHSTALK BLUEGRASS.

ROUGHSTALK BLUEGRASS (*Poa trivialis*), known also as *roughstalk*, resembles KENTUCKY BLUEGRASS but differs from it in that it has no creeping rootstocks (RHIZOMES). The branches of the PANICLE are more slender and spreading. It is used to some extent in this country as a wet-pasture grass, but more often as a grass for shady lawns.

Reference: H.1.

See also BLUEGRASSES; GRASSES.

ROX ORANGE, more often called *Waconia Orange*, is one of the forage SORGHUMS.

RUBBER RABBITBRUSH (*Chrysothamnus nauseosus*) is a perennial shrub growing 20 to 40 in. high. Its range extends from Canada south to Wyoming, Utah, Nevada, and eastern California. Under normal conditions the forage value of this plant is very low, but its flowertops, herbage, and tender stems are occasionally eaten by all classes of farm animals.

Reference: U.6.

See also RANGE PLANTS.

RUBBER SEED OIL MEAL, or *para rubber meal*, is obtained as a by-product from the manufacture of oil from the seed of the PARA RUBBER TREE. It is a feedstuff of low palatability; however, it has some feeding value when mixed with other feedstuffs since it contains about 28 percent protein, 9 percent fat, and 37 percent nitrogen-free extract.

RUBBERWEEDS (*Actina* spp.), particularly *bitter rubberweed* and *Colorado rubberweed*, or *pingue*, are SAPONIN-containing POISONOUS PLANTS. Eating small quantities of the leaves, stems, or flowers for several days causes vomiting and weakness in sheep.—*See also* ACTINEA.

Colorado rubberweed. (N.6.)

RUDIMENTARY means: undeveloped.
RUDY. *See* COMMON WHEAT (variety).
RUFFLYN.

    See SIX-ROWED BARLEY (variety).
RUMEN is the *first* STOMACH of a ruminant.
RUMEN BACTERIA play an important role in the decomposition and utilization of feed. Many kinds of BACTERIA are present in large numbers. The total may reach 2 billion in a drop of rumen fluid. At birth, ruminants have no rumen bacteria but they can be transferred to young calves by direct *rumen inoculation* with *cud* material from healthy cows; or calves, as well as mature animals, may be fed small amounts of carefully dried and enriched rumen bacteria mixed with the ration.

Rumen bacteria are indicated as after-treatment for animals having received medication that has reduced their natural rumen bacteria content to a large degree. It has also been established that properly treated rumen bacteria, collected from healthy animals of the right kind, will help feeder cattle and sheep adjust to a grain ration and produce more rapid gains, especially during the first month of the feeding period.—*See also* GROWTH.
RUMINANTS are animals which have a compound STOMACH (the first section of which is called *rumen*); cattle, sheep, and goats are the domesticated ruminants.
RUMINATION is the process of "chewing the cud." Ruminants swallow dry feed, moistened with saliva, in the form of boluses; the latter absorb fluid in the rumen; after some time, the solid parts of the feed, mixed with liquid, are forced into the mouth where they are chewed, again swallowed and then moved through the four compartments of the complex STOMACH of the animal. Coarse roughages may remain in the rumen many hours before they are returned to the mouth.
RUNNER is a long, slender, leafless form of creeping branch, prostrate on the ground. Each runner, after having grown to its full length, strikes root from the tip, then forming a bud at that point; this develops later into a tuft of leaves and so gives rise to a new plant. Sometimes the runner roots at the joints also, in which case it may merge into a STOLON.

    Reference: D.9.

    See also ROOTSTOCK; PEANUT.

RUSH. *See* SPIKE RUSH; BULRUSH; RUSHES.
RUSHES form the Juncaceae, or rush family, which includes the genera *Juncoides* and *Juncus*. The rushes belong to the GRASSLIKE PLANTS which represent one of the four classes of RANGE FORAGE.

    The INFLORESCENCE of rushes has a six-parted flower, showing a close relationship to lilies.—*See also* PASTURE PLANTS; BILLBUGS.
RUSSIAN GIANT. *See* RYE (variety).
RUSSIAN RED = *Galgalos*.

    See COMMON WHEAT.
RUSSIAN-THISTLE (*Salsola pestifer*) is closely related to LAMBSQUARTERS; it is a

WEED widely distributed over the western United States and Canada. This plant is salt-resistant and hence grows well, though not exclusively, on alkaline soils. It is an annual which reaches a height of 4 ft., forming a dense, bushlike plant from 2 to 6 ft. in diameter. The Russian-thistle contains more protein and carbohydrates than CLOVER and as much or more ash than ALFALFA, but is less palatable and digestible than the latter.

On early spring ranges this species rates a fair forage for all classes of livestock. However, after the plant matures and forms sharp spines, it is worthless. In many of the drought-striken areas, the quite drought-resistant Russian-thistle has been used successfully as emergency feed to prevent livestock from starving. If cut when in bloom, before the sharp spines form, this plant makes good emergency hay; and it is eaten readily when ensiled, especially if mixed with ALFALFA or grain.

The Russian-thistle is often attacked by GARDEN WEBWORMS.

Reference: U.6.

See also SILAGE CROPS; RANGE PLANTS.

RUSSIAN WILD-RYE (*Elymus junceus*) is an introduced, long-lived perennial bunchgrass adapted to the conditions of the northern part of the Great Plains and Intermountain region. It is especially useful as a pasture grass for spring and summer grazing. Russian wild-rye remains palatable during the entire grazing period and is readily eaten by sheep and cattle even after the forage has turned brown in the fall.

This wild-rye species has erect, naked stems about 3 ft. tall, arising from an abundance of soft, dense, basal leaves which are from 6 to 18 in. in length and up to $\frac{1}{4}$ in. in width. Plant color varies from light to dark green, with many shades of blue-green. The flowering head is a dense, erect SPIKE; the seed has short, stiff hairs.

The roots are fibrous and many penetrate to a depth of 8 to 10 ft.; they have a wide horizontal spread and draw heavily on moisture for a distance of 4 to 5 ft.

Russian wild-rye grows best during the cool seasons, but its period of summer growth makes this species an excellent competitor with weeds and other GRASSES or LEGUMES in mixed plantings.

Russian wild-rye is adapted to a fairly wide range of soil types, but it is most productive on fertile loams. It may also be of value for use on alkaline soils, but it does poorly on soils of low fertility.

In areas where it is adapted, Russian wild-rye is most useful as a pasture grass. The hay is of excellent quality and highly palatable and nutritious. Good-quality hay can often be cut from seed-production plots after seed harvest. Despite its rather low forage yield, livestock gains on Russian wild-rye pasture are generally high because of the great digestibility and superior nutritive qualities of this grass.

Because of a shortage of available nitrogen, forage yields of Russian wild-rye may drop rapidly as stands become older. Fertilizing with nitrogen greatly increases the yield.

In the northern Great Plains early fall, seeding on fallow has met with great success. A light seeding of OATS with the grass will prevent soil blowing on fallow; the oats will also aid in protecting the seedlings from wind whipping. Other nurse crops should not be used. Late fall-seedings, in October and November, can be made in clean grain stubble. Spring seedings are to be made as early as possible.

A grain drill should be used for seeding; the depth of seeding should extend only $\frac{3}{4}$ in. Drill spacings of 6 to 12 in. are satisfactory for pasture; wider spacing between rows may be better in the drier areas.

The rate of seeding is approximately 8 lb. an acre. If ALFALFA or YELLOW SWEET-CLOVER is to be grown with the grass, the legume seeding should be at the rate of about 2 lb. an acre; when the grass is early-fall-seeded, the legume is not to be planted until the following spring.

Newly established stands should not be grazed the first year. Weed growth is generally rather heavy, but it should be clipped only when the growth is so heavy that the grass or legume seedlings are seriously hampered. If clipping is necessary, it should be done only in cool weather and at a height of about 6 in. Stands are usually so well established by the second year that weeds cease to be a problem.

Dangers: SEEDLING BLIGHTS that cause damping-off are probably the most damaging of all the diseases of Russian wild-rye; early-fall planting is a good method of avoiding the loss. FOOT ROT and ROOT ROTS of mature plants tend to thin older stands. POWDERY MILDEW occasionally does some damage in a cool, damp spring. Diseases that may cause loss in seed-production plots are HEAD SMUT, STEM RUST, and a FUNGUS affecting the lower internodes and causing lodging.

GRASSHOPPERS do more damage than other insects to young plants. The larva of the WHEAT STEM SAWFLY is damaging by causing blasted heads. In some years the RED SPIDER does some damage to Russian wild-rye.

References: R.2.

See also WILD-RYE GRASSES; RANGE MANAGEMENT.

RUST. *See* RUSTS.

RUST FUNGUS. *See* RUSTS.

RUSTPROOF. *Red Rustproof* is a RED OAT variety.

RUSTS are FUNGUS DISEASES or FUNGI themselves. The life cycle of a *rust fungus* may involve up to five types of SPORES. Some rusts parasitize only one species of plant during their lives (MONOECIOUS) or two species (HETEROECIOUS). The WHEAT STEM RUST is heteroecious and has five types of spores, namely: the red *uredospores* which spread the rust from grain plant to grain plant; the dark *teliospores* which remain on straw or stubble, resisting winter temperatures, and germinating in spring to produce *basidiospores;* the latter carry rust to barberry, infect it, germinate, and produce *pycniospores;* these fuse sexually to produce *aeciospores* which infect the grain plant.

Other important rusts are STEM RUST OF OATS, SORGHUM RUST, STRIPE RUST, CLOVER RUST, ALFALFA RUST, LEAF RUST, CROWN RUST, and SAFFLOWER RUST.—*See also* TRUE CLOVERS; EMMER.

RUTABAGA (*Brassica napobrassica*), or *Swedish turnip*, is a biennial, cultivated as an annual vegetable. It needs cool weather and moist soil to develop. The tops of these plants are a VEGETABLE WASTE PRODUCT, used as feedstuff.

Reference: E.12.

RYANIAS (*Ryania* spp.) are tropical shrubs and trees; the wood of some of them possesses insecticidal action. A dust containing 40 percent ryania is recommended for use against the SUGARCANE BORER. Four applications are made at weekly intervals while the first- (or spring-) generation borers are hatching from the eggs laid on the leaves by the moths. The INSECTICIDE is applied with airplanes or ground equipment at the rate of 10 lb. per acre, very early in the morning while the plants are wet from dew. Where ryania is used to control the second (or midsummer) generation, the same number of weekly applications is made by airplane. First-generation dusting is preferable.

RYE (*Secale cereale*) belongs to the GRASSES and is one of the important true cereals. It is an annual or winter annual plant which in many respects is similar to WHEAT. However, its stems are larger and longer, usually 5 ft., and up to 10 ft., high; its leaves are coarser and more bluish, its roots, which may extend 5 to 6 ft. into the soil, branch near the soil surface more profusely than those of wheat.

Rye is a plant which is naturally cross-pollinated. The SPIKE of the common type, which is the rye cultivated in the United States, consists of a SPIKELET at each RACHIS joint. The spikelet is composed of two fertile and one abortive FLORET. The broad LEMMA is awned and barbed. The rye kernel is narrow and brownish-olive, bluish-green, or yellow in color.

Rye is grown very largely as a cash grain crop in the western half of the United States. It is a valuable green-manure crop when green manuring is a profitable practice, chiefly because of its rank growth and its adaptation to low temperatures, which enable it to grow in late fall and early spring. Rye can be sown on poor soil or on a poorly prepared seed-bed without much expense, but it responds well to more favorable conditions. It is better adapted to sandy soils than any other grain crop. Plowing under rye to bind the soil while another crop is becoming established is a common practice on newly irrigated sandy lands.

Rye, especially when mixed with legumes, such as VETCHES or PEAS, is a valuable soiling crop or green feed in some sections. It makes a good growth in early spring before other crops are well started, and often produces high yields of forage.

In many sections of the United States rye is of chief importance as a *pasture crop*. In nonirrigated lands in Kansas, Nebraska, Colorado, Montana, Idaho, and parts of other states, it is especially useful for this purpose. Its vigorous fall growth allows it to be more heavily pastured in the autumn than wheat. It makes a good growth in the spring and can be pastured until the crop is nearly mature. If enough of the crop remains it can be harvested for grain. Where soil moisture and climatic conditions are favorable, spring rye sown in late summer makes abundant fall pasture.

Rye is often successfully used as a nurse crop for LEGUMES, notably ALFALFA and SWEETCLOVER. Its earliness enables it to be harvested in time to avoid smothering the legume crop.

The rank of growth, early vigor, and heavy tillering of rye make it useful as a smother crop to keep down COMMON WILD OATS and many other weeds. Rye ripens early and usually can be harvested before wild oats mature. It thus is a valuable aid in the control of this noxious annual weed pest.

If rye is drilled into small-grain stubble in the fall without previous preparation, no expense is incurred except for seed and seeding. It matures early and is harvested before any of the spring grains are ready to cut. The little labor required to produce rye by the above method often makes the crop a profitable one, although the GRASS-HOPPER menace may be increased.

There are some disadvantages to growing rye, the principal one being its tendency to volunteer. The seed shatters rather easily, so that some is always scattered on the land on which this crop has been grown. It is therefore very difficult to

eradicate rye in a system of continuous small-grain farming.

Summer fallowing is not a profitable seed-bed preparation for spring grains or for winter rye in the Great Plains and prairie regions.

Rye. Plant, spikelet, and floret. (H.26.)

In northeastern North Dakota summer fallowing often is necessary to control SOW THISTLE, QUACKGRASS, and CANADA

THISTLE. Winter rye sown on this fallow makes a rapid growth and serves to check the weeds which have survived the fallowing operation, and it is the best crop to grow on such land.

Fall plowing for rye, although commonly practiced in the East, usually is not profitable in the western half of the United States because of the additional expense involved.

Winter rye can be sown at any time during the fall. If sown early, it will produce more fall pasture than if sown later. As a rule, rye should be sown at about the time winter wheat is sown, but the time of sowing is of less importance with rye than with wheat. Because of the ability of rye to germinate at low temperatures and with limited moisture, it can be sown safely at a later date than wheat. Winter rye sown very late sometimes does not emerge until spring, but even then may produce fair yields, though less than from earlier seeding. Sowing winter rye in the spring usually results in failure.

Spring rye should be sown as soon as the soil conditions permit. Early seeding almost always results in the highest yields of grain.

For best results rye should be sown with a grain drill at a depth of 1½ to 2½ in. On stubble land a single-disk drill will penetrate best, while on a well-prepared seed bed other types of drills are equally satisfactory. On fallowed land the hoe drills give excellent results.

The rate of seeding best suited to the growing of rye is about the same as for wheat. Late seeding should be thicker than early seeding, and thick seeding is preferable to medium or thin seeding when the crop is grown for pasture, green manure, soiling, or hay.

The same fertilizer recommended to be applied in the fall for wheat should be used for rye and the crop grown on sandy soils and light-colored silt loams is equal to wheat in its response to top dressing with nitrogen in the spring. On light-colored fine sand, fertilizing with 300 lb. 3-12-12 at seeding time and top dressing in early April with 125 lb. of 33-percent ammonium nitrate per acre is recommended.

Rye should be harvested with a binder when the first kernels are ripe. If harvested in the stiff-dough stage, threshing is more difficult. If the crop is left standing until it is dead-ripe, there will be some loss from the shattering of the grain.

Usually rye is threshed with the ordinary threshing machine. If unbroken straw is wanted, a special threshing machine is used.

Most of the rye consumed in this country is ground into flour. A good deal of the crop is used for feed.

Rye is similar to wheat in composition but has a slightly lower feeding value. It is fed most satisfactorily in mixtures with other grains because it is less palatable than other grains. Most of the trouble that has been experienced in feeding rye probably has been due to overfeeding, or using grain that was spoiled or contained ERGOT. Rye is nearly equal in feeding value to corn and is equal to barley for feeding hogs. Mixtures of rye and BARLEY produce more rapid gains than rye alone when fed to hogs getting alfalfa hay.

In general, rye is a less palatable *pasture* crop than wheat and most grasses but is consumed readily when other feed is not available.

Rye hay, only when cut sufficiently early, is satisfactory for feeding purposes, but it is less palatable than the hays made from legumes and the better grasses and grains, and is inferior to them. The rye stems are hairy and frequently coarse and tough and are not greatly relished by animals which have access to other forage. The value of rye as a hay crop lies in its ability to grow where other crops are not productive and to furnish an early hay crop.

Rye straw is useful for the bedding of livestock but has little value as feed, being the least desirable of all cereal straws.

These are the rye products officially recognized as feedstuffs:

Rye bran—the coarse outer covering of the rye kernel as separated from cleaned and scoured rye.

Rye feed—obtained in the usual process of milling rye flour, consisting principally of the mill-run of the outer covering of the rye kernel and the rye germ with small quantities of rye flour and aleurone; it must not contain more than 9.5 percent crude fiber.

Rye feed is the most important rye by-product. It can replace part of the cereal or other cereal grain by-products in cattle, sheep, and swine rations. However, the supply of rye feed is limited.

Rye red dog—consisting principally of aleurone with small quantities of rye flour and fine rye bran particles; it must not contain more than 3.5 percent crude fiber.

Rye low-grade feed flour—consisting principally of rye flour and small quantities of aleurone and fine rye bran particles; it must not contain more than 1.5 percent crude fiber.

Rye middlings—consisting principally of rye feed and rye red dog combined in the proportion obtained in the usual process of milling rye flour; it must not contain more than 8.5 percent crude fiber.

Rye flour middlings—consisting of rye feed, rye red dog, and rye flour combined in the proportions obtained in the milling of rye flour; it must not contain more than 5 percent crude fiber.

Dangers: Many of the insects, plant diseases, and rodent pests attacking rye are the same as those which attack other small grains, and the same methods of control are applicable. Except in the cases of ERGOT, STEM SMUT, and ANTHRACNOSE OF GRASSES, diseases cause relatively little injury to the rye crop. WHEAT STEM RUST, WHEAT SCAB, TAKE-ALL, and LEAF RUST attack rye, but usually the injury is not great, owing to the early maturity of the crop. LOOSE SMUT OF RYE is of rare occurrence.

Among the insects attacking rye are GRASSHOPPERS, the HESSIAN FLY, CHINCH BUG, WHEAT STEM SAWFLY, ARMYWORM, and SOUTHERN ROOT WORM.

Varieties: Four groups of rye can be distinguished; one has composite spikes and three have simple spikes of brown, red, or white to yellowish color; only the last mentioned group, called *common rye*, is of practical importance. It includes two distinct types of rye, known as *winter rye* and

spring rye. It is of utmost importance to know whether one is obtaining winter or spring rye for seeding purposes.

(1) WINTER RYE

The hardiest and earliest of all cereals is winter rye; it successfully survives the winters, even in the extremely cold sections of northern North Dakota and Montana. Its earliness frequently enables it to escape injury from drought and rust. Rye sprouts more quickly and grows more vigorously than wheat at low temperatures, and consequently can be sown successfully at a much later date in the fall. It is better adapted to light sandy soils, acid soils, poor thin land, and poorly prepared seed beds than the other small grains. It produces a fair crop under conditions where the other small grains would fail completely.

Rye can be grown in practically all parts of the United States, although its culture is limited in most regions because other crops are more profitable. It is most profitable in the northern and eastern states. North Dakota, Minnesota, Nebraska, South Dakota, and Wisconsin lead in the acreage of rye. Hardy winter rye is successfully grown in all parts of the Dakotas and Minnesota, where winter wheat usually will not survive the winters. Rye is the best small-grain crop for the sandy lands of northeastern Colorado and the adjacent states.

In the arid and frosty portions of the Great Basin and Intermountain regions, particularly in the higher and drier sections of eastern Washington and Oregon, in northern California, and in Wyoming, rye is more certain than other grain crops. It is grown there for both grain and hay. The hay serves as a winter feed for the range stock and is the only feed crop which can be grown with safety in the driest sections. In other parts of the United States rye has a place as a pasture or hay crop or for soiling or green manure.

These are the leading varieties of winter rye: *Abruzzes* (adapted to the Cotton Belt, grown principally as a pasture or cover crop); *Aland; Advance* (this has been the most productive variety in South Dakota but has lost through contamination with other varieties); *Azor; Balbo* (a newer variety used for pasture and winter cover in Kansas, Nebraska, Indiana, and other eastern states); *Dakold* (once the most productive variety in North Dakota, it has now lost through contamination); *Dean*, or *Minnesota No. 1* (similar to Swedish); *Emerald* (developed in Minnesota); *Excelsior; Heinrich; Imperial* (vigorous, hardy, adapted to Wisconsin); *Ivanov; Johannes; Mammoth White* (grown in the Northeast); *Mountain*, or *Alpine; Petkus* (very similar to Rosen); *Pierre* (of superior winter-hardiness, developed in North Dakota); *Rosen* (late, adapted to the Corn Belt region, Colorado, and Michigan); *Russian Giant; Spanish Double; Schlanstedt*, or *Wisconsin Pedigree No. 2; Swedish* or *Minnesota No. 2* (similar to Dakold); *Thousandfold* (adapted to the northeastern states); *Zeeland*.

(2) SPRING RYE

This type is of minor importance in comparison with winter rye. It sometimes outyields winter rye on clay land where the latter is injured by heaving of the soil from freezing and thawing. It matures quickly and will produce seed from late-spring sowing, although early seeding gives the highest yields. Spring rye is well suited to the dry lands in the high plateau districts of the Great Basin region in central Oregon, and the adjacent sections of Idaho and California where late-spring frosts frequently injure winter rye at blossoming time. Spring rye also has produced fair returns on the sandy soils in northeastern Colorado and a few other localities.

These are the most important varieties of spring rye; *Cataluna; Irkutsk; Magellan; Nerschinsk; Palermo; Saxon Spring;* and *Vern* (adapted to Oregon).

References: M.31; G.10; P.13; E.12; F.6.

See also RYE PRODUCTS; DRIED CEREAL; DISTILLERS' PRODUCTS; BREWERS' PRODUCTS; YEAST DRIED GRAIN OR VINEGAR DRIED GRAIN; DODDERS; DUST STORMS; WHEAT JOINTWORM; BUCKWHEATS; FIELD PEA; CORN; POLISH WHEAT; PEPPERGRASS; PEANUT; CRIMSON CLOVER; KUDZU; CROTA-

LARIAS; LESPEDEZAS; OATS; SILAGE CROPS; PASTURES; GRAZING.

RYE BRAN. *See* RYE.

"RYE BUCKWHEAT"
= TARTARY BUCKWHEAT.

RYE DISTILLERS' DRIED GRAINS. *See* DISTILLERS' PRODUCTS.

RYE DISTILLERS' DRIED GRAINS WITH SOLUBLES. *See* DISTILLERS' PRODUCTS.

RYE DISTILLERS' DRIED SOLUBLES. *See* DISTILLERS' PRODUCTS.

RYE FEED. *See* RYE.

RYE FLOUR MIDDLINGS. *See* RYE.

RYEGRASSES (*Lolium* spp.)—which should not be confused with the WILD-RYE GRASSES—are forage grasses which have a wide range of adaptability to soils, but prefer medium to high fertility. In some sections the GRASSES are considered wetland grasses, although production usually declines as the drainage gets poorer. Ryegrasses can be seeded in the fall or early spring. Spring seedings are preferable where winters are severe. Fall seedings are more successful in more temperate regions. The seed should be covered with approximately ½ in. soil.

Ryegrasses make very rapid winter and spring growth, and new seedings often are ready to pasture in three months. Unless pastured too heavily, the grass can be used continuously until summer in the West and until late in the spring in the South.

The ryegrasses are generally cut for *hay* when the seed is in the soft-dough stage. The hay cures rapidly and, when handled properly, has a bright-green color. Because of its leafiness and its medium fine stems, it makes a high-quality hay that is considered excellent for horses and is fed successfully to cattle and sheep. Under favorable soil and moisture conditions pasturage is produced after the hay crop is removed.

When seeded alone for forage 20 to 25 lb. seed per acre is sufficient. When seeded with SMALL GRAIN or a LEGUME for annual pasture, 8 to 10 lb. per acre will give a satisfactory stand.

Dangers: The WHEAT STEM SAWFLY is one of the more persistent enemies of the ryegrasses.

Species: The common name ryegrass is applied to a group of plants comprising two species: ITALIAN RYEGRASS and PERENNIAL RYEGRASS. Selections and hybrids of these two species have received special names; to designate these mechanical or hybrid mixtures grown domestically, the term COMMON RYEGRASS is used. Certain annual species (other than Italian ryegrass) are WEEDS.

References: H.1; S.2.

See also CRIMSON CLOVER; ALSIKE CLOVER; SUBCLOVER; ANNUAL LESPEDEZAS; RICE; PASTURES; PASTURE PLANTS.

Illustration: *See* ITALIAN RYEGRASS.

RYE HAY. *See* RYE.

RYE LOW-GRADE FEED FLOUR.
See RYE.

RYE MALT SPROUTS.
See BREWERS' PRODUCTS.

RYE MIDDLINGS. *See* RYE.

RYE PASTURE. *See* RYE\

RYE RED DOG. *See* RYE.

RYE SILAGE. *See* SILAGE CROPS.

RYE STRAW. *See* RYE.

S

39-20-S = DACOTA AMBER 39-20-S.

S 100. *See* SOYBEAN (variety).

S-1 SWEET SUDANGRASS is a strain of *Sweet Sudangrass.—See also* SUDANGRASS.

SABADILLA POWDER, the powdered dried seed of *Schoenocaulon* spp. which grow in Mexico and South America, is very poisonous. It contains ALKALOIDS and is used as drug and as INSECTICIDE.

Sabadilla powder is very effective against CHINCH BUGS when used in 4 percent concentration at the rate of 50 lb. to the acre. It may be applied with a duster; all bugs receiving a coating of the dust die within an hour.

SACATON (*Sporobolus wrightii*) is a robust perennial bunchgrass, occurring from Arizona to western Texas and south into Mexico. The young shots of this semi-desert GRASS are highly relished early in the season by cattle and horses. The herbage cures well and constitutes fairly good winter forage.

The leafy stalks grow 2 to 8 ft. tall.

Sacaton must not be confused with the related, but smaller and less coarse, ALKALI SACATON.

Reference: U.6.

See also RANGE PLANTS.

SACCHARINE SORGHUM = *forage sorghum. See* SORGHUMS.

SACCHAROMYCES. The *Saccharomyces* spp. are higher FUNGI commonly known as YEASTS.

SACCHAROSE = SUCROSE.

SACCHARUM. The *Saccharum* spp., or SUGARCANES, include *S. sinense* = CHINESE SUGARCANE and its variety, the JAPANESE SUGARCANE.

SACKED FEED. *See* STORING FEED.

SAFFLOWER (*Carthamus tinctorius*), also known as *false saffron*, is an annual plant grown in the western part of the Great Plains and California. It reaches 1 to 3 ft. in height, has an extensive root system, light-colored stems and branches, white, yellow, orange, or red flowers, and small light-colored, four-angled seeds.

It is a relatively new oil crop in the United States and gives best yields when the atmosphere is hot and dry during flowering time, in conjunction with a favorable supply of stored subsoil moisture throughout the growing season. Unless a field is known to be relatively free from weeds, planting in cultured rows is preferred to solid drilling.

The seeds of the safflower are used in the manufacture of WHOLE PRESSED SAFFLOWER SEED.

Dangers: A leaf spot disease and a bud rot (neither of which has yet been identified definitely) have been observed, mostly under the conditions of high rainfall or high humidity. Other diseases are SAFFLOWER RUST and ROOT ROT OF SAFFLOWER. GRASSHOPPERS sometimes cause severe damage to the crop; other enemies of safflower are the CLOVER LEAFHOPPER, CUTWORMS, and RODENTS.

Varieties: The safflower varieties do not differ markedly in seed yields, but do in oil content. *Pusa No. 7, Ahmednager No. 1,* and *Simla* are among the best varieties now available for production.

References: C.13; T.3.

SAFFLOWER MEAL

= WHOLE PRESSED SAFFLOWER SEEDS.

SAFFLOWER OIL FEED

= WHOLE PRESSED SAFFLOWER FEED.

SAFFLOWER RUST, caused by the FUNGUS *Puccinia carthami*, is a serious and widely distributed disease of SAFFLOWER. The organism is unique among the rusts in that it is seed-borne. The rust is both a seedling disease and a leaf trouble; it is best controlled by the use of resistant varieties.

Reference: T.3.

SAFFLOWER SEED MEAL

= WHOLE PRESSED SAFFLOWER SEED.

SAFFMEAL

= WHOLE PRESSED SAFFLOWER SEED.

SAGE. *See* BIG SAGEBRUSH.

SAGEBRUSH CONTROL.

See SAGEBRUSHES; RANGE MANAGEMENT.

SAGEBRUSHES (*Artemisia* spp.) are often a desirable forage on ranges used by livestock in winter. Likewise, on ranges used by big game in winter, sagebrush growth is an important source of browse.

In tall, dense stands sagebrush—regardless of species—is undesirable. It is relatively unpalatable to sheep and cattle. It uses moisture and nutrients that should be producing good grasses.

Sagebrush prevents grazing of grasses hidden under its woody stems and crown. It hampers movement of livestock, especially sheep. The brush snags wool from fleeces and causes lambs and calves to stray and become lost. Heavy brush makes conditions ideal for predators, such as coyotes.

Control of sagebrush results in major increases in grass production on millions of acres of western range. Establishing or restoring a good stand of forage plants (through natural or artificial seeding) enables ranges to supply forage for more sheep and cattle, and is helpful in improving watersheds.

Overly dense sagebrush is largely the result of overgrazing, together with drought. Sagebrush control is recommended primarily for ranges used by livestock in the spring, fall, and summer but not for those grazed chiefly in the winter.

Methods to kill sagebrush vary with density, height, and age of the sagebrush stand, associated shrub species, amount of grass understory, topography, amount of rock in the area, type of soil and its susceptibility to erosion, facilities available for doing the work, size of the area to be treated, and other factors.

The methods most often used for control of sagebrush on rangelands are as follows:

1. *Planned burning* is most useful on fairly level tracts of 1,000 acres or more, either to permit the increase of perennial grasses already present or to prepare the land for seeding. Many stands of sagebrush, however, can be burned only under conditions of extreme fire hazard. On ranges where the erosion hazard is high or the dominant perennial grasses would be seriously damaged, burning should not be used.

Burning is one of the most difficult and dangerous methods. It should first be tried out on a small scale to test its effectiveness and to determine the safeguards that should be followed to meet the hazards for a given set of conditions. The use of fire to remove sagebrush for range improvement should be attempted only with the approval of local fire-control authorities.

2. *Plowing or disking* is primarily used on rock-free to slightly rocky ranges where seeding is to be done afterwards. The heavy offset disk and one-way disk plow have proved effective in eradicating all types and ages of sagebrush, except silver sagebrush, on soils that are almost rock-free. An extra heavy offset disk has proved the best implement for use on heavy, crusted soils. The brushland plow is effective on lands with scattered large rocks.

3. *Railing* has proved satisfactory as a low-cost method for removal of old, mature brittle stands of big sagebrush on large acreages, especially when the areas support a fair stand of native grass. It is not a desirable method where there are considerable numbers of young sagebrush plants, where associated undesirable plants, such as RABBITBRUSH, SPINELESS HORSE-BRUSH or CHEATGRASS are abundant, or where seeding is necessary.

4. *Harrowing* with the pipe harrow is recommended for soil disturbance, seed covering, and control of open stands of mature sagebrush on moderately or extremely rocky grounds. The ripping and gouging action caused by the toothed pipes, bouncing along among the rocks, tears out some of the sagebrush, loosens rocks, and disturbs the soil enough for seed coverage. The harrow is useful for covering broadcast seed on rocky alpine areas, sagebrush burns, or on areas where sagebrush has been killed by herbicides.

5. *Beating* with beaters and cutters can be especially useful for control of stands of big sagebrush that are uniformly old and large, and where rocks are absent or protrude less than 3 in. above the soil surface. Beating is useful for the release of native perennial grasses growing beneath the sagebrush or of seeded stands being suppressed by sagebrush, or where hazard of wind or water erosion is high.

6. *Grubbing* with mechanical grubbers (blades, cultivators, root planes) is useful primarily where control of rabbitbrush as well as sagebrush is necessary. Their use is limited to rock-free, deep soils where the potential production of seeded grass will justify the cost, or where conversion of the sagebrush range to cropland is the objective.

7. *Spraying with herbicides* is most useful either on ranges that have a good understory of native grass to thicken up and replace sagebrush, or on seeded areas that have been invaded by sagebrush. It may also be useful for killing sagebrush preparatory to seeding on areas where sagebrush is short or sparse.

Lack of knowledge regarding effects of herbicides (especially 2,4-D and 2,4,5-T) on the associated desirable broadleaved forage plants and shrubs, and on animal life that depends upon this vegetation, gives reason for caution in using herbicides for sagebrush control. County and state laws must be observed.

Common native grasses, e.g., the FES-CUES, BLUEGRASSES, and WHEATGRASSES, increase production from 20 to 85 percent the first year following chemical control. Such species as NEEDLE-AND-THREAD GRASS

often become an important component. Native forbs, however, such as dandeloin, lupine, arnica and many others may be killed or damaged by 2,4-D compounds; they may remove as much as 75 percent of the susceptible plants.

Black sagebrush. (H.41.)

8. *Other methods* have been tried in clearing sagebrush for farming or for range improvement but they are useful only if the necessary equipment is readily available and local conditions are suitable. Among these methods are (a) *mowing* with an ordinary power-takeoff mower, having a heavy cutter bar, snub-nosed guards,

heavy-smooth sections, and a double set of clips to hold the sections snugly against the blunt guards; (b) *ripping* with a 2-ton self-clearing road clipper with teeth placed 14 to 16 in. apart, pulled by a crawler tractor; (c) *rolling* with the heavy type of rolling brush cutter; (d) *clearing* tracts with *road graders* or *bulldozers;* and (e) the *flooding* of areas where high spring runoff waters are available.

Regrassing: Eradicating sagebrush for purposes of range improvement is of little avail unless a good stand of desirable forage plants promptly reoccupies the area to protect the soil from erosion, prevent the early return of sagebrush in large quantity, prevent invasion by other undesirable plants, and repay the cost of sagebrush control through a rapid increase in grazing capacity.

Some ranges have enough desirable perennial GRASSES and other forage plants growing beneath the sagebrush to revegetate the area quickly after the sagebrush is killed. If these are not destroyed by control operations, and if good grazing management is used afterwards, there is little need for range reseeding. As a general rule seeding is not needed where more than one fifth of the total plant cover is made up of desirable plants, provided they are fairly well distributed.

Successful seeding to perennial grasses may not entirely prevent establishment of undesirable annuals, but these are likely to be so sparse as to provide little competition to perennials. By the second or third year native or seeded perennials should fully occupy the site.

Even where initial eradication is complete and good stands of grass develop, sagebrush will sometimes come in again. Usually the re-established sagebrush stands are sparse and not objectionable, but in about one year out of five, conditions are favorable for a dense stand. Eradication of nearly established sparse or dense sagebrush stands is a major problem. In some cases it pays to get rid of even the sparse seedling stands within three to seven years, before the young sagebrush plants have produced a crop of seed.

In grazing ranges after sagebrush control the following three recommendations should be observed: (1) Avoid trailing livestock the first fall and winter across areas where sagebrush has been eradicated. (2) Delay grazing until native grasses are vigorous and seeded forage plants are well established. (3) Practice good grazing management after the new forage stand is well established. To obtain proper grazing intensity, to insure grasses being grazed at the proper season, and to help secure proper distribution of livestock, it may be necessary to fence the seeded and improved areas and to provide better stockwater facilities.

Species: There are many shrubby species of sagebrush, all of which are often called just "sagebrush." BIG SAGEBRUSH is by far the most common. Associated species are SILVER SAGEBRUSH, THREETIP SAGEBRUSH, and BLACK SAGEBRUSH (which is often useful as winter forage on ranges), BUD SAGEBRUSH, and FRINGED SAGEBRUSH.

References: P.19; P.20; B.20 ; P.18.

See also SLENDER WHEATGRASS; MUTTON BLUEGRASS; MORMON CRICKET; RANGE PLANTS; RANGE MANAGEMENT.

SAGEBRUSH-GRASS RANGE.
See RANGE PLANTS.

SAGITTARIA. *S. latifolia* = ARROWHEAD.

SAGRAIN, better known as *Schrock,* is one of the grain SORGHUMS.

SAINT. *See* ST.

SALINA WILD-RYE (*Elymus salinus*) grows on rocky slopes and sagebrush hills from Wyoming and Colorado to Idaho, Nevada, and Southern California. This GRASS is useful as forage on many ranges. —*See also* RANGE PLANTS; WILD-RYE GRASSES.

SALINE SOIL contains more than 0.2 percent water-soluble salts, but has a pH less than 8.5.

SALIVARY FERMENTS.
See DIGESTIVE TRACT.

SALMON (*Salmo salar*), a fish with orange-pink flesh, is often used in fish meal.—*See also* MARINE PRODUCTS.

SALMON LIVER OIL. *See* VITAMINS.

SALMON OIL. *See* VITAMINS.

SALSOLA. *S. pestifer* = RUSSIAN-THISTLE.

SALT is the reaction product of a base and an ACID.

Common salt = SODIUM CHLORIDE.—*See also* ALKALI; RANGE MANAGEMENT; GRAZING; MINERALS.

SALTBUSH. *See* SHADSCALE SALTBUSH; GARDNER SALTBUSH; FOURWING SALTBUSH; RANGE PLANTS.

SALT-DESERT SHRUB.

See RANGE PLANTS.

SALTGRASSES (*Distichlis* spp.) are low, perennial GRASSES with creeping RHIZOMES, sometimes STOLONS. They have some value for forage in the interior basins, such as the vicinity of Great Salt Lake.—*See also* RANGE PLANTS; STRAWBERRY CLOVER.

SALTING RANGES.

See RANGE MANAGEMENT.

SALTPETER. POTASSIUM NITRATE.

SALT POISONING.

See SODIUM CHLORIDE.

SALTSAGE = GARDNER SALTBUSH.

SALT WITH TRACE MINERALS

= TRACE MINERAL SALT.

SAMPLE GRADE OF HAY.

See HAY GRADING.

SANALTA.

See TWO-ROWED BARLEY (variety).

SAND is small rock or mineral fragments with a diameter of 0.05 to 1 mm.—*See also* SCREENINGS.

SANDBERG BLUEGRASS (*Poa secunda*) is the most common native bluegrass species. It is a hairless (GLABROUS), tufted perennial that occurs generally throughout the northern Great Plains and the western states.

Plants may grow to a height of 8 to 24 in. The INFLORESCENCE is a PANICLE only about ½ in. wide. The forage is scanty but palatable. Since Sandberg bluegrass begins growth early in the spring, it supplies green, succulent forage at a time when it is most beneficial to grazing animals. It usually matures and dries by the first part of July. This bunchgrass is highly drought-resistant.

When used for revegetation, Sandberg bluegrass is ordinarily seeded in mixtures with other adapted GRASSES, being dominant in spring but yielding dominance to other species of the mixture as the season advances.

Reference: H.1.

See also BLUEGRASSES; RANGE PLANTS.

SAND DROPSEED (*Sporobolus cryptandrus*) is a tufted, widely distributed, native grass. It occurs most abundantly in the southern Great Plains and the Southwest. It is an invader species on raw, denuded soil and is most prevalent on sandy soil.

Plants grow about 2 to 3 ft. tall, with solid stems and fairly numerous leaves about ¼ in. wide and 4 to 12 in. long. Seed heads are open, the finely branched

Sand dropseed. Plant, glumes, and floret. (H.26.)

PANICLES average 8 to 12 in. in length and terminate in single SPIKELETS. Characteristically, many of the seed heads remain within the upper portion of the surrounding sheath, so that the plants tend to retain many of their very small seeds. Roots are coarse, fibrous, and penetrating, a characteristic that accounts partly for the wide adaption of this species.

Sand dropseed produces a fairly large amount of foliage, which is taken readily by livestock while green but only sparingly after the plants reach maturity. It is a prolific seeder and, when protected and properly grazed, tends to increase in density on the depleted range.

When the grass is grown in rows and cultivated or grown under irrigation, exceptionally high yields—exceeding 1,000 lb. an acre—may be obtained.

Sand dropseed is valuable for revegetation use because it is widely adapted as to soils and climate, and forage production is satisfactory.

Reference: H.1.

See also DROPSEEDS; GRASSES; RANGE PLANTS.

SAND LOVEGRASS (*Eragrostis trichodes*), a vigorous, long-lived, native bunchgrass, occurs on sandy soils of the central and southern parts of the Great Plains. Plants normally grow to a height of 3 to 6 ft. The PANICLES are sometimes half as tall as the plant and have a distinctive purple color. The slightly hairy leaf blades are $\frac{1}{4}$ in. wide and about 12 in. long. Leafy foliage, primarily basal, is abundant. Roots are vigorous, spreading, and deeply penetrating, and therefore of value in conservation.

Plants begin growth very early in the spring and remain green until late fall. Sand lovegrass is generally considered one of the most palatable and nutritious of the range GRASSES; frequently it suffers from continuous overuse.

Sand lovegras is easily established from seed and volunteers aggressively. It makes excellent growth when seeded either alone or in mixture on sandy soils but does not thrive on heavy soils, except in pure stands.

Reference: H.1.

See also LOVEGRASSES.

SAND OAT (*Avena strigosa*) belongs to the wild growing OATS.

SAND VETCH = HAIRY VETCH.

SANETT = *Sanford. See* COMMON WHEAT.

SANFORD is a soft, intermediate spring variety of COMMON WHEAT.

SAPLING. *Sapling clover*, more correctly called *American single-cut clover*, is one of the RED CLOVER varieties.

Sapling sorgo is one of the forage SORGHUMS.

SAPONINS are GLUCOSIDES which, when taken with water, form foam; they occur in POISONOUS PLANTS, e.g., in RUBBERWEEDS and BLACK NIGHTSHADE. Some ALFALFA contains 1 to 3 percent saponin. When this material is extracted and given to cattle or sheep, severe cases of BLOAT are soon produced.

SARCOBATUS. *S. vermiculatus* = GREASEWOOD.

SARDINE (*Clupea pilchardus*), or *pilchard*, is often used in fish meal.—*See also* MARINE PRODUCTS.

SARDINE OIL, or *pilchard oil*, is a vitamin-rich fish oil. *See also* VITAMINS.

"SAUSAGE SMUT" is one of the STRIPE SMUTS.

SAWDUST. *See* HYDROLYZED SAWDUST.

SAWFLY. *See* WHEAT STEM SAWFLY.

SAXON SPRING. *See* RYE (variety).

SAYBOLT VISCOSITY.

See MINERAL OILS.

SAY STINK BUG (*Chlorochroa sayi*), formerly known as the *grain bug*, is a STINK BUG species widely distributed in the western states from Mexico to Canada. The adults vary considerably in size and color; the females average 13.3 mm. in length and the males slightly less. The color ranges from dark green in spring, to light green in summer, and to olive and reddish-brown in fall and winter.

Reference: R.6.

SCAB. *See* WHEAT SCAB.

SCALD, caused by *Rhynchosporium* spp., is a FUNGUS DISEASE that attacks MEADOW FOXTAIL and other grasses. It occurs primarily on leaf blades and less extensively on sheaths and is most destructive in spring and autumn.

The symptoms consist of small, water-soaked, bluish-gray spots that enlarge to form irregular, scaldlike blotches which usually have brown margins. Scalded leaves are often killed, and almost complete defoliation results.

The FUNGUS overwinters on dead leaves in old crowns and in lesions in perennial grasses. Abundant spores are produced and the wind carries them to infect new leaves during cool, wet periods.

Control: Crop rotation, elimination of old plant residues, and spring burning help control the disease.

Reference: K.11.

SCALDING is a term which is often used, although wrongly, for an irritation observed on light-skinned swine exposed to sunlight.—*See also* PHOTOSENSITIZATION; RAPE.

SCALE is a plant organ or part reminiscent of the scale of a fish; e.g., a modified leaf forming part of the protective covering of a flower bud, or the rudimentary leaf on a RHIZOME.

There is a growing tendency in botany to confine the use of the term scale to the basal and underground portions of the plant and to use the term BRACT for analogous parts in the INFLORESCENCE.

Reference: D.9.

See also SHIVE.

SCANDINAVIAN FEED-UNIT SYSTEM compares the productive value of different feeds with that of 1 lb. of a standard grain feed, such as barley. The *feed-unit value* of wheat bran for dairy cows is 1.1 lb., i.e., it takes 1.1 lb. of it to equal 1 lb. barley; the feed-unit value of cottonseed meal for dairy cows is only 0.8 lb., that of corn silage is 6 lb., etc. The feed-unit values are somewhat different for the various species of farm animals.

> NOTE: It is to be emphasized that these values do not express true net ENERGY.

Too high net energy values have been assigned to proteins in this system for conditions prevalent in the United States and it is, therefore, rarely used in this country. In Scandivanian and some other European countries, where roughages are mostly low in proteins, this feed-unit system is of practical value.

SCARBOROUGH DWARF No. 7 is a dwarf-type of *broomcorn.*—*See also* SORGHUMS.

SCARIFIED SEED is seed that has been treated so that it may germinate more quickly.

Seed can be scarified (1) mechanically—e.g., by rubbing with an abrasive, such as sandpaper or emery cloth or (2) by acids, such as commercial SULFURIC ACID. The former method is the more practical, but the latter can sometimes be used on small lots of seed. The seed is put into the acid for about thirty minutes and then thoroughly washed with water. If not immediately planted, the seed must be dried.

Reference: M.9.

See also SCARIFYING MACHINES; KUDZU; LESPEDEZAS.

SCARIFYING MACHINES. The abrasives used mostly are sandpaper, emery cloth, or sandstone. A small concrete mixer can be operated as a *scarifier* by first adding to the seed coarse stones, the size of large marbles or larger. If the concrete mixer is rotated for about one hour with stones and seed in equal bulk, reasonably good *scarification* should be affected. The scarified seed can be easily separated from stones by screens.

Reference: M.14.

SCARLET CLOVER = CRIMSON CLOVER.

SCHLANSTEDT is a winter RYE variety.

SCHOENOCAULON. From the seeds of various *Schoenocaulon* spp., which resemble barley, SABADILLA POWDER is obtained.

SCHROCK, or *Schrock kafir,* is one of the grain SORGHUMS.

SCIOTO. *See* SOYBEAN (variety).

SCIRPUS. *S. mucronatus* = BULRUSH.

SCLEROTIA (sing.: *sclerotium*) are the hard, rounded masses of the *fungus filaments* (HYPHAE) which usually serve as resting bodies to carry the FUNGI through unfavorable weather. Some fungi can survive for many years in soil, plant refuse, or seed by means of sclerotia which vary in size from microscopic to several inches in diameter.—*See also* CHARCOAL ROT; RHIZOCTONIA STALK ROT; SCLEROTINIA STEM ROT.

SCLEROTINIA. The FUNGI *S. scleroti-*

orum and *S. trifoliorum* cause SCLEROTINIA STEM ROT.

SCLEROTINIA STEM ROT, a FUNGUS disease, is caused by *Sclerotinia* spp.

S. sclerotiorum most commonly attacks and destroys the stems of LUPINES, especially of BLUE LUPINE plants, but any part of the plant above ground may be attacked and killed. The part above the dead area then wilts and dies, but usually the losses are not serious. This disease has been found largely in northern Florida.

Sclerotinia stem rot may attack the pods, and small SCLEROTIA—i.e., black fungus masses—may replace the decayed seeds. These sclerotia, which often resemble seeds, may be carried to the field and initiate the disease. Therefore, care should be exercised to remove all sclerotia when cleaning the seed.

The disease can be distinguished from BOTRYTIS STEM CANKER, which it resembles closely, by the irregularly shaped sclerotia in the pith of the stem; botrytis also produces sclerotia, but these are commonly formed on the surface of the diseased plant rather than within the pith.

S. trifoliorum, which is widely distributed in the southern United States, attacks VETCHES, SOYBEANS, and many other legumes, non-leguminous forage plants, and weeds.

Control: No satisfactory control for this soilborne fungus has been found as yet.

References: W.8; M.18.

SCLEROTIUM. Among the *Sclerotium* spp. are the FUNGI *S. rolfsii* which cause SOUTHERN BLIGHT and, together with *S. bataticola*, CONCEALED DAMAGE IN SEED; with other organisms, they also cause SEEDLING BLIGHT.—*See also* SCLEROTIA.

SCOLECOTRICHUM. The FUNGUS *S. graminis* causes BROWN STRIPE.

SCOTCH PEA. *See* FIELD PEA (variety).

SCOTT = *Pennscott red clover. See* RED CLOVER.

SCOURING-RUSH (*Equisetum hyemale*) is a slender HORSETAIL species with 4 ft. tall, evergreen stems. It is one of the POISONOUS PLANTS.

SCOURINGS consist of such portions of the cuticle, brush, white caps, dust, smut, and other materials as are separated from the grain in the usual commercial process of *scouring.—See also* LABELS.

SCREENED CORN CHOP. *See* CORN.

SCREENED CRACKED CORN.
                                          See CORN.

SCREENED GROUND CORN.
                                          See CORN.

SCREENINGS is the by-product obtained in the cleaning of grains that are included in the United States Grain Standard Act, and of other agricultural seeds (e.g., clovers). It may include light and broken grains and agricultural seeds, weed seeds, hulls, chaff, joints, straw, elevator or mill dust, sand, and dirt.

No grade of screenings shall contain any seeds or other material in amount that is either injurious to animals or will impart an objectionable odor or flavor to their milk or flesh. The screenings shall contain not more than four viable primary noxious weed seeds per pound and not more than 100 whole secondary noxious weed seeds per pound. The primary and secondary noxious weed seeds shall be those named as such by the individual state seed control laws.

NOTE: *Weed seeds*, that have been screened from various seeds, can be used for feed if they are finely ground and mixed with other feeds. The protein content of weed seeds is extremely variable, and only limited information on their digestibility is available.

In general, finely ground weed seeds may substitute for as much as 25 percent of the grain in rations. The fine grinding required to kill seeds makes them a dusty product, and many seeds have an objectionable odor, or are unpalatable. The addition of 10 to 15 lb. MOLASSES to each 100 lb. of ground weed seeds is, therefore, recommended.

All grades of screenings must bear minimum guarantees of protein and fat and maximum guarantees of fiber and ash.

When used in mixed feeds the portion of the description appearing in parenthesis in these (official) definitions shall appear in the list of ingredients:

Grain screenings (from WHEAT, CORN, OATS, etc.) shall consist of 70 percent or more of grain (light and broken), including WILD BUCKWHEAT and COMMON WILD OATS. It shall contain not more than 6.5 percent ash.

Mixed screenings (grains, seeds, hulls, chaff) is screenings excluded from the preceding grade. It shall not contain more than 27 percent fiber. If it contains more than 13 percent ash, the words *"sand"* and *"dirt"* shall be included in a parenthetical statement.

Chaff and/or dust is material that is separated from grains or seeds in the usual commercial cleaning processes; it may include hulls, joints, straw, mill or elevator dust, sweepings, sand, dirt, grains, and seeds. If it contains more than 15 percent ash, the words *"sand"* and *"dirt"* shall appear on the label.

Reference: F.6.

See also CEREAL GRAIN BY-PRODUCTS; BY-PRODUCT FEEDSTUFFS; LABELS.

SCREWBEAN = TORMILLO.

SEA ISLAND COTTON. *See* GOSSYPIUM.

SEASIDE CLOVER (*Trifolium wormskjoldii*), a productive native species of the West, belongs to the TRUE CLOVERS. It is of only minor importance as forage plant.

Reference: M.3.

SEAWEEDS of the families Fucaceae and Laminariaceae are sources of KELP, one of the officially recognized FEED INGREDIENTS.

SECALE. *S. cereale* = RYE.

SECTION is a natural division of a botanical group, especially of a GENUS and, hence, usually synonymous with SUBGENUS. In very large genera the section is often a division of the subgenus.

Reference: D.9.

SEDAN KAFIR is one of the grain SORGHUMS.

SEDGES (*Carex* spp.), or *umbrella plants*, are members of the Cyperaceae, or sedge family. They occur in perennial or annual form and belong to the GRASSLIKE PLANTS.

Sedges have three-ranked leaves and usually solid stems which are often triangular in cross section. The floral organs are in the AXILS of scales.

The small, dry, one-seeded and one-celled fruits are called ACHENES.

Sedges are frequently important to wildlife and some moist- and wet-meadow sedges comprise a large percentage of the hay crop in western mountain valleys, where livestock are winter-fed. The sedges tend to produce less crude fiber and ash and more crude protein and nitrogen-free extract than grasses, especially with increasing altitude. They are also very effective in soil protection. Nevertheless, sedges—especially the annual forms—are often considered WEEDS; for instance in rice fields.

Reference: U.6.

See also RANGE FORAGE; PASTURE PLANTS; MUTTON BLUEGRASS; BILLBUGS; CHINCH BUG.

SEED is a fertilized and matured OVULE; it encloses a rudimentary plant and food necessary for its germination.—*See also* EMBRYO; PASTURE; SCREENINGS; PROPAGATION.

SEEDBED is the land prepared for seed to be sown in it.

SEED COAT = BRAN.

SEED DECAY. *See* POD AND SEED DECAY.

SEEDER is an implement used for sowing seed, either by broadcast or by drill method.

SEEDLING is a young plant grown from a seed.

SEEDLING BLIGHT is a broad term that includes anything causing seedlings to die. Like other BLIGHTS and the related WILTS, it is caused by a variety of soilborne micro-organisms, particularly FUNGI, and affects the seedlings of various crops, including SORGHUMS, TRUE CLOVERS, RICE, RUSSIAN WILD-RYE, etc. Some of these fungi also produce SEED ROTS or DAMPING-OFF.

In *sorghums*, seedling blight is caused by *Pythium* spp., particularly *P. arrhenomanes*, that attack the young sprout in its early development and prevent its emergence; they also attack and rot the primary roots of the young seedlings. The species *Fusarium moniliforme*, besides rotting the seed, frequently attacks sorghum seedlings at the surface of the soil after they have emerged and causes them to rot or damp off and fall over; this fungus may destroy the primary roots of young seedlings also.

The species *Fusarium culmorum* is capable of completely inhibiting germination in cold soil. *Penicillium oxalicum*, in addition to arresting germination, may also kill the seedlings, even after they have reached the fourth-leaf stage. The injury by this fungus is characterized at the start by a grayish or silvery-green color of the leaves, followed by a gradual yellowing. The leaves become limp and curled, and finally die, but the plants remain upright even after they are dead. Seedlings attacked and killed by *Pythium* spp. and *Helminthosporium* spp. display similar symptoms and also remain erect.

Control: Seedling blight is controlled in the same manner as SEED ROT: by careful selection and SEED TREATMENT along with proper cultural practices.

References: L.1; M.2.

See also LEAF BLIGHT; BROWN SPOT OF RICE; WHEAT SCAB; SCLEROTIUM.

SEEDLING DISEASES. Loss of seedlings from disease occasionally causes serious depletion of stands. Among the more important seedling diseases are SEEDLING BLIGHT, ROOT ROTS, SOUTHERN BLIGHT, HEAT CANKER, SOIL ROT, and BACTERIAL WILT.—*See also* ROTS.

SEED POD. *See* POD.

SEED ROT of SORGHUMS is most severe when the soil is cold and wet after planting. Under such conditions, much of the seed of sorghum fails to germinate and rots because it is attacked by various seed-borne and soil-inhabiting FUNGI, chiefly species of *Fusarium, Aspergillus, Rhizopus, Rhizoctonia, Penicillium, Pythium* and *Helminthosporium*.

Control: Seed rot (and SEEDLING BLIGHT) may be controlled to a considerable extent by careful selection and SEED TREATMENT. Seed should be well matured and properly cured, and the seed coat should be as free as possible from cracks and nicks. Before being planted, the seed should be treated with a good disinfectant that will protect it not only from seed-borne fungi but also from the harmful fungi present in the soil. The seed should not be planted until the soil is warm enough for prompt germination. The fields should be thoroughly tilled. Seed rotting is most

common in feterita, hegari, and similar soft-seeded types of sorghum.

Reference: L.1.

See also LEAF BLIGHT; SEED TREATMENT.

SEED STAIN. *See* PURPLE SEED STAIN.

SEED TREATMENT. Seed may be treated with FUNGICIDES and/or BACTERICIDES in five ways:

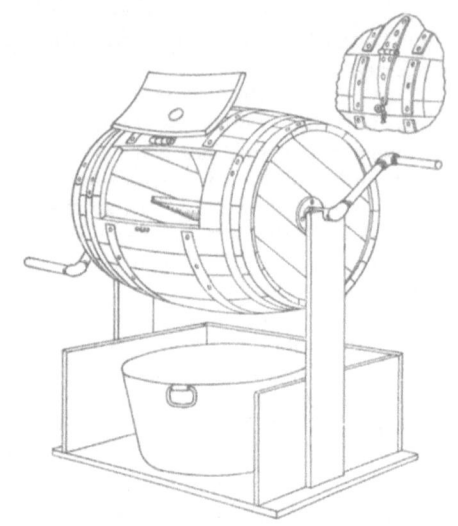

Seed treatment. Barrel mixer for treating seed with dust fungicides. (L.1.)

1. *Dust treatment:* The seed is mixed with a dust fungicide and well coated with it. Such fungicides are COPPER CARBONATE, COPPER SUBSULFATE, CERESAN M, ARASAN, SPERGON, PHYGON, CERESAN, and SULFUR.

2. *Slurry treatment:* To overcome the discomfort and health hazard caused by chemical dusts flying in the air, they may be applied as a water suspension, or slurry, of about the consistency of buttermilk. A specially constructed slurry treater is required to coat the seed with the slurry so that undue wetting is avoided. CERESAN M, PHYGON, or Arasan SF may be used as fungicide in the slurry method.

3. *Liquid treatment:* The seed is soaked in a fungicidal liquid—e.g., FORMALDEHYDE SOLUTION—for a definite period of time and then spread out to dry.

4. *Quick-wet treatment:* Seed is thoroughly mixed with a small quantity of a concentrated liquid fungicide, usually a volatile mercurial, such as PANOGEN.

5. *Hot-water seed treatment* is effective for the control of LOOSE SMUT OF WHEAT and of the LOOSE SMUTS OF BARLEY. The treatment does not give prolonged protection.

Reference: L.1.

See also SEED ROT; SEEDLING BLIGHT; BACTERIAL LEAF DISEASES (namely BACTERIAL STRIPE OF SORGHUM, BACTERIAL STRIPE OF OATS, BACTERIAL STREAK, and BACTERIAL SPOT); FUNGUS LEAF DISEASES (especially ROUGH SPOT, LEAF BLIGHT, ZONATE LEAF SPOT, and TARGET SPOT); LOOSE KERNEL SMUT, and HEAD SMUT; SEED ROT; PURPLE LEAF SPOT; BROWN SPOT OF RICE; GRASS SMUTS; FLAG SMUT OF WHEAT; BLACKPATCH; HALO BLIGHT; WHEAT SCAB; OATS; ANTIBIOTICS.

SEEPWEEDS (*Suaeda* spp.) are shrubs that have only little forage value.—*See also* RANGE PLANTS.

SELECTION is a term usually applied to selected plant progenies. It is used by plant breeders to indicate the isolation of an individual plant, a *line*—i.e., related plants which originated from the same progenitor (ancestor)— or a type derived from a VARIETY or HYBRID.

SELECTION No. 12 MUNGBEAN is a variety of the *green* MUNGBEAN.

SELENIUM is a CHEMICAL ELEMENT similar to SULFUR with which it occurs in ores and soils. It is insoluble in water; its toxicity resembles that of arsenic.—*See also* SELENIUM POISONING; LOCOWEEDS.

SELENIUM POISONING of livestock, variously called "*alkali disease*" or "*blind staggers*", is caused by daily ingestion of small amounts of SELENIUM, an element found only in certain soils and taken up in amounts toxic to animals by a limited number of plants. The symptoms and the duration of the poisoning depend upon the amount of selenium ingested daily.

Definite areas in South Dakota, Nebraska, Wyoming, New Mexico, Montana, and Utah contain sufficient selenium to cause poisoning mainly of cattle and horses, although sheep are occasionally affected.

Among the plants that have a tendency to take up selenium if it is present in the soil are LOCOWEEDS, POISONVETCHES, and the PARRY ASTER. The amount of selenium found in these plants indicates the extent of selenium in that particular area. The selenium-charged plants grow up and die, year after year, until the surface soil becomes high in selenium. After this, GRASSES and other plants also may contain selenium, creating a definite "poison area" for livestock.

Chronic selenium poisoning in cattle causes emaciation, abnormal growth of hoofs, stiffness, loss of hair from the tails of cattle and horses, aimless wandering (blind staggers), and the depraved appetite resembling that caused by phosphorus deficiency.

A diagnosis, in most cases, must be based on the selenium content of the forage. Amounts as small as 10 ppm. may produce chronic poisoning if fed over a long period.

Reference: W.34.

SELFED = SELF-POLLINATED.

SELF-FEEDING is also called "*cafeteria system*". Rations or ration ingredients are put before the animals who are allowed to eat as much as they want. The biggest advantage of this system is in labor saved to put out feed several times each day. The disadvantages are: (1) the animals may waste feed; (2) some animals do not balance their rations properly—e.g., a sheep may eat too much grain and founder, or a hog may eat too much supplement and increase the cost of production; (3) self-fed hogs may eat too much concentrate and become too fat to produce a desirable carcass.

Self-feeding is usually thought to be more successful if a complete mixed ration is fed. This is especially true for ruminants in dry lot. Hogs, however, may make more economical gains if they are fed shelled CORN, PROTEIN supplement, and MINERALS in separate hoppers. Poultry also do well on the cafeteria system.—*See also* RATION.

SELF-FEEDING TRENCH SILO.

See SILOS.

SELF-FERTILITY. *See* FERTILITY.

"SELFING" = SELF POLLINATION.

SELF-POLLINATED, or *selfed*, means: pollinated in the bud or by pollen from the same flower.

SELF-POLLINATION is often called "*selfing*".—*See also* POLLINATION.

SELF-STERILE. *See* STERILE.

SEMI- is a prefix meaning half.

SEMIARID is the climate of regions usually having an annual precipitation of 10 to 20 in.

SEMIDESERT GRASS RANGE.
See RANGE PLANTS.

SEMINOLE. *See* SOYBEAN (variety).

SENECA. *See* SOYBEAN (variety).

SENECIO. The genus *Senecio* belongs to the COMPOSITE family and probably comprises a larger number of species—about 2,600—than any other genus of flowering plants. Some of these species are poisonous; among them are *S. spartioides = broom groundsel; S. riddellii = Riddell groundsel; S. longilobus = threadleaf groundsel; S. integerrimus = lambstongue groundsel;* and other *groundsels* as well as *S. jacobaea = ragwort.—See also* MUTTON BLUESTEM.

SEPTORIA. The FUNGI *S. tritici* and *S. nodorum* cause SEPTORIA BLIGHT OF WHEAT; *S. glycines* causes BROWN SPOT OF SOYBEAN; *S. pisi* is the cause of LEAF BLOTCH.

SEPTORIA BLIGHT OF WHEAT is caused by (1) *Septoria tritici*, a widely distributed species that attacks mainly leaves (the disease, therefore, is often called *septoria leaf blotch* or *speckled leaf spot*); and (2) *S. nodorum*, a species that attacks leaves, culms, and heads (causing the disease also known as *septoria glume blotch*). The leaf blotches and spots frequently are indefinite in color, shape, and margin. Usually, the leaf and culm tissue surrounding the lesion bleaches slowly and turns straw-color or brown. As the leaves bleach or as the diseased plants reach maturity, the small, regular spore cases develop beneath the leaf surface of the diseased areas.

The damage the spores cause results largely from lodging and shriveled kernels which are common in susceptible varieties of COMMON WHEAT and other species of WHEAT.

Control is difficult as the spore cases are produced in abundance on the old straw and stubble. Crop rotation, plowing under old straw, and the use of resistant varieties offer the only means of control.

References: D.11; B.16.

See also FUNGUS.

SEPTORIA GLUME BLOTCH.
See SEPTORIA BLIGHT OF WHEAT.

SEPTORIA LEAF BLOTCH.
See SEPTORIA BLIGHT OF WHEAT.

SERICEA (*Lespedeza cuneata*, first introduced as *L. sericea*) is one of the upright growing, perennial summer LEGUMES. Its roots are woody and widely branched, penetrating the soil to a depth of 3 ft. or more, and making sericea very drought-resistant.

This lespedeza species is well adapted to the middle latitudes of the eastern United States but has survived winters in Michigan. In southern Georgia and in Florida fairly good growth has been made. It is usually considered a "poor land" crop and is sometimes called the *"poor man's alfalfa,"* because it will make a satisfactory growth on land not suited to alfalfa. Sericea will grow on soils ranging from strongly acid to alkaline, but it does best on soils with a reaction of from pH 5.5 to 6.5.

On poor soils, in addition to responding to lime, sericea has responded favorably to phosphate and potash. When not properly fertilized, BROOMSEDGE will often appear in a field and may crowd out the sericea.

INOCULATION of sericea is not always necessary. The organism that produces nodules on the roots of sericea is the same as that on the annual COMMON LESPEDEZA.

The seed may be sown behind the cultipacker and left without covering, except on slopes where heavy rain might wash the seeds down. Good stands cannot be expected if the seeds are covered deeper than $\frac{1}{4}$ to $\frac{1}{2}$ in.

Sericea is used for ground covering and soil improvement. It makes a very good hay crop, provided it is adequately fertilized and cut before the bloom stage; not only does the protein content decrease, but the tannin content increases rapidly as the plants grow taller than 10 to 15 in. from the ground. The sericea plant comes back after cutting by sprouting from the stubble; so it should not be mowed closer than 2 to 3 in. from the ground.

Sericea is a pasture plant that is grazed readily and profitably. It is best to start grazing early, when the plants are only 3 to 4 in. high. This legume is especially valuable during dry weather when other pasture plants fail.

Good silage can be made by using molasses.

NOTE: Sericea *hay* was not satisfactory for growing heifers in Arkansas.

Dangers: Sericea is not ordinarily troubled seriously with insects and diseases. ARMYWORMS sometimes damage a stand of sericea. The THREE-CORNERED ALFALFA HOPPER has damaged sericea in certain areas. WHITE GRUBS may prove a serious pest in old sericea stands. Sericea is suceptible to COTTON ROOT ROT, but this disease is confined largely to the blacklands of Texas.

References: W.7; E.6.

See also WEEPING LOVEGRASS; LESPEDEZAS; LUPINES.

SERRATE means: saw-toothed.

SESAME (*Sesamum indicum*) is one of the oldest oilseed crops. It is an erect annual plant, ranging in height from less than 2 to over 10 ft. The flowers vary in color from white to black. The sesame stalks have only little value as livestock roughage. The seed and the by-product of the sesame oil production—SESAME OIL MEAL—are valuable for feeding.

Reference: K.7.

SESAME OIL MEAL is the ground residue obtained after the extraction of part of the oil by pressure from SESAME seed. It must be designated and sold according to its protein content.

This officially recognized by-product is, like the whole seed, a rich source of fair-quality protein (it is especially high in the amino acid METHIONINE, but is very deficient in lysine). The digestibility is high. All species of farm animals like this feed-stuff.

References: F.6; K.7.

SESAMUM. *S. indicum* = SESAME.

SESSILE means: (literally) sitting, i.e., without a stem or stalk; a sessile leaf sits directly on the axis or stem of the plant. In a SPIKE, all the flowers are sessile.

SETA (plural: *setae*) is a bristle, or a stiff thick hair.

SETARIA. *S. italica* = FOXTAIL MILLET.

SEXUAL REPRODUCTION.
                        See PROPAGATION; SPORES.

SHADSCALE SALTBUSH (*Atriplex confertifolia*), *shadscale*, or *spiny saltbush*, is a compact, low shrub 8 to 24 in. high. It grows in individual clumps or bushes, but sometimes forms dense, almost solid clumps 8 to 10 ft. in diameter. The base of the plants and stems is woody and the stems are spine-tipped. The thick leaves and seed BRACTS densely cover the stems. Shadscale is grayish-green during the growing season but takes on a reddish-purple autumn coloration. In the wintertime the plants turn grayish-brown. Shadscale grows in pure stands on the lower valley areas where the soil is fine and rather heavily alkaline.

Shadscale saltbush. (U.6.)

Shadscale is one of the most important native species on the *winter range*. The leaves and seeds are readily eaten by sheep but the stems are not eaten much because they are protected by woody spines. This species, therefore, invades or increases on overgrazed range when the other more palatable vegetation is destroyed or injured. Prolonged drought causes heavy mortality of shadscale.

Reference: H.41.

See also RANGE PLANTS.

SHAK CROTALARIA (*Crotalaria incana*), one of the CROTALARIAS, is not poisonous. It is readily eaten by cattle and mules.

SHAKES, a nervous disorder in livestock, is caused by BERMUDA-GRASS affected with a variety of FUNGUS diseases.

SHALLU is one of the grain SORGHUMS.

SHANTUNG KAOLING. *Dwarf Shantung kaoling*, a variety of *kaoling*, is one of the grain SORGHUMS.

SHARK LIVER OIL. *See* VITAMINS.

SHARK MEAL is a FISHERY BY-PRODUCT obtained from sharks after the removal of liver, hide, and fins. The remaining material, after being rendered and ground, contains up to 90 percent protein, 10 percent of which is in form of UREA. Shark meal has been used successfully in poultry, calf, and swine rations.—*See also* MARINE PRODUCTS.

SHARON is one of the grain SORGHUMS.

SHEAF is a bundle of grain cut and tied; sheaves are piled up to form a SHOCK.

"SHEAF OATS" are obtained in the semiarid areas of the Northwest of the United States and Canada from OATS cut in the dough stage or later, and cured. The dry leaves and stems of "sheaf oats" make excellent forage for cattle, horses and sheep.

SHEATH is the basal portion of a leaf which envelops the stem. The edges of sheaths grow together to form a tube.—*See also* GRASSES; BOOTH; BLADE.

SHEATH ROTS. *See* BLACK SHEATH ROT; REDDISH-BROWN SHEATH ROT; ROTS.

SHEATH SPOT.

See BORDERED SHEATH SPOT.

SHEEP FESCUE (*Festuca ovina*) is a bunchgrass that forms dense tufts with numerous stiff, rather sharp, bluish-gray leaves. It is adapted to about the same climate as BLUEGRASSES and can be grown in the most northern agricultural areas. It succeeds better than most GRASSES on sandy or gravelly soils. Cattle and sheep will graze sheep fescue, but it is not generally recommended for pastures. Its

Sheep fescue. (H.1.)

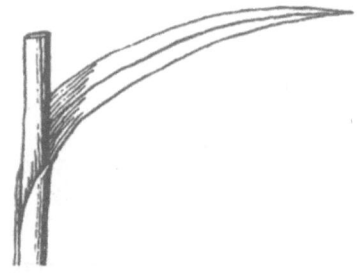

The sheath of a grass. It envelops the culm (or stem) and continues into the blade (or lamina). (D. 9.)

greatest use is for making a durable turf on sandy soils.

Reference: H.1.

See also FESCUES.

SHEEP GRASS = ST. AUGUSTINE GRASS.

SHEEP-LAUREL. *See* KALMIAS.

SHEEP RATIONS. Of all the animals that are kept on the farm, the sheep is the best adapted to the utilization of ROUGH-AGES. Over 90 percent of the total feed consumed by sheep in this country comes from roughage and PASTURE. The sheep is the only animal that can repeatedly produce choice products without concentrate feeding.

FLUSHING EWES

It is good practice to have ewes on a rising plane of NUTRITION at the time they are *bred* in the fall. This can be achieved by turning them to good quality pasture which has been saved for this purpose, or by adding a limited amount of a mixture of OATS and OIL MEAL to the ration about three weeks before the proposed beginning of the breeding season. This practice results in more ewes settling at first service, and may increase the number of twins.

WINTERING PREGNANT EWES

After the ewes have been successfully mated, they may be wintered on a low plane of nutrition until the last six weeks before gestation. Ewes have been found to winter well if one-half of their total roughage consists of either LEGUME HAY or SILAGE; the latter should be free of mold and spoilage. Under prevailing conditions in the south central and southern parts of the country, the ewes will winter satisfactorily on winter pasture and no supplemental feed will be needed. If all roughage is of poor quality, add ½ to ¾ lb. GRAIN and $\frac{1}{10}$ lb. PROTEIN supplement to the ewe's ration. Yearling and two-year old ewes especially need the extra feed, but mature ewes may produce satisfactorily without added concentrate. Ewes may also be wintered on cured RANGE grasses with ⅛ lb. protein supplement added.

During the last six weeks before *lambing*, ewes should be placed on a higher plane of nutrition; the bulk of the ration should be cut down since the developing fetus will take up a large portion of the body, and less room will be available to accommodate coarse fibrous feeds. This can usually be achieved by reducing the amount of silage or coarse HAY in the ration and replacing it with, perhaps, ½ lb. grain mixture. If the ewes are bred for late lambs, excellent results can be obtained by putting the ewes on lush spring pasture, such as small grain, during this period. The succulence of the pasture produces a desirable laxative condition which seems to make lambing easier, and also stimulates a high rate of milk flow.

If the ewes are on pasture, no difficulty is experienced at lambing time, and ewes can be expected to produce satisfactorily on pasture alone. If ewes have early lambs in dry lot, a bran mash or other slightly laxative feed can be used for a few days before lambing, and for a few days after. Early born lambs usually respond to extra feed provided in a creep after they are two

SHEEP RATIONS
Amounts are given for 10 head

	Dry lot conditions							
	Wintering ewes					Fattening lambs		
	Early winter			6 weeks before lambing				
Ration:	1	2	3	1	2	1	2	3
Legume hay, lb.	..	20	..	20	20	15	10	8
Grass hay, lb.	20	..	35	20	..	..	..	7
Corn or sorghum silage, lb.	..	50	..	..	..	..	20	..
Grass silage, lb.	50	..	..	..	40	..	..	..
Grain, lb.	..	..	..	2	10	15	15	15
Oil meal, lb.	..	..	2	..	2	1	1	1
Molasses, lb.	..	..	..	3	..	2	..	2
Minerals, free choice	+	+	+	+	+	+	+	+
Wheat bran, lb.	..	..	..	2	..	..	..	2

weeks old. One of the most satisfactory *creep feeds* is cracked CORN.

During *lactation* the ewe may be continued on the same ration that she received during the latter part of the gestation period. Once the ewes and lambs are turned to pasture, there is usually no need for additional feed.

FATTENING LAMBS

Lamb fattening rations must provide proper physical and chemical balance. Most experimental results show that rations providing 60 parts hay and 40 parts grain during the early part of the feeding period, and only 45 parts hay and 55 parts grain during the latter part of the fattening period, are satisfactory. However, if the ration is *pelleted*, higher levels of roughage can give excellent results. Any changes should be made gradually. Grains do not need to be ground for fattening lambs, because they do a satisfactory job themselves. Better results can probably be obtained if the animals are self-fed on a mixture of grains and hays. This mixture may be held together by addition of about 7 percent MOLASSES and will usually give very good rates of gain.

SHEEP-SIZE COTTONSEED CAKE.
See COTTONSEED.

SHELBY RED CHAFF = *Goens. See* COMMON WHEAT.

SHELLED CORN. *See* CORN.

SHELL FLOUR and *oyster shell flour* consist mainly of CALCIUM CARBONATE and sometimes also of a large amount of MAGNESIUM CARBONATE. They are used as MINERAL FEEDS.

SHEPHERD'S-PURSE (*Capsella bursa-pastoris*) is an annual or winter annual with a short, whitish taproot and smooth, branched stem 8 to 16 in. high. The alternate leaves are arrow-shaped. Very small white flowers are borne on the end of the branches. This plant is often a troublesome WEED.

Control: Shepherd's purse can be controlled by crop rotation and clean cultivation or with 2,4-D when spread early in spring.

Reference: C.19.

See also ALFALFAS.

SHERMAN BIG BLUEGRASS.
See BIG BLUEGRASS (variety).

SHERMAN'S CLOVER.
See STRAWBERRY CLOVER (variety).

SHINNEY = *Whippoorwill. See* COWPEA.

SHIPPING FEVER
= HEMORRHAGIC SEPTICEMIA.

SHIPPING PNEUMONIA
= HEMORRHAGIC SEPTICEMIA.

SHIVE is the external SCALE of a stalk or of the bark of FLAX.

SHOCK is a pile of usually six to ten crop sheaves or cut stalks set together to dry.

To shock means: to set into shocks.

SHOCK CORN. *See* CORN.

SHOOT is a stem with its attached parts.

SHORT GRASS RANGE.
See RANGE PLANTS.

SHORT-GRAIN RICES are RICE varieties.

SHORT OAT (*Avena brevis*) is a close relative of COMMON OAT, but it grows wild and is often considered a WEED.—*See also* OATS; AVENA.

SHORTS is livestock feed made from the by-product of wheat milling; it includes some flour, wheat germs, and bran.—*See also* OFFAL.

SHOWY CROTALARIA (*Crotalaria spectabilis*) is one of the CROTALARIAS grown extensively in the United States. It is poisonous to cattle, sheep, goats, swine, and poultry, but preferred for soil improvement because its stem is less woody than that of other crotalaria species, and therefore easy to turn under and handle with light plows. This species is the least palatable of the crotalarias. It is attacked by CROTALARIA ANTHRACNOSE, GRAY MOLD, STEM CANKER OF CROTALARIA, and other crotalaria diseases.

SHREDDING means: cutting or tearing into shreds. CORN or SORGHUM stover and other FODDERS are often shredded so as to induce animals to eat larger quantities of this roughage and to reduce the amount of refuse. It is important that the feed is dry before it is shredded, or it may mold or heat in storage.

SHRIMP MEAL. *See* MARINE PRODUCTS.

SHRIMPS. *See* TADPOLE SHRIMPS.

SHRUB is a woody, perennial plant, differing from a perennial HERB by its persistent and woody stems, and from a tree by its low stature and habit of branching from the base. There is, of course, no hard-and-fast line between herbs and shrubs or between shrubs and trees. Under very favorable growth conditions, species of shrubs frequently become trees and vice versa. Also, there are a few plants which are herbaceous in temperate climates but shrubby in tropical or subtropical regions.

Reference: D.9.

See also RANGE PLANTS.

SHRUBBY CINQUEFOIL (*Dasiphora fruticosa*) is a many-branched bush. It ranges from Labrador to Alaska and south to California, New Mexico, Minnesota, Illinois, and New Jersey. The plant is mostly from 10 to 24 in. high, but occasionally reaches a height of 5 ft. With its profusion of yellow flowers, its hairy, grayish-green leaves divided into five to seven small leaflets, and its reddish-brown, shreddy bark, this shrub is seldom confused with other RANGE PLANTS.

In general, shrubby cinquefoil has low palatability for livestock; it makes fair to good forage for cattle and sheep on closely cropped summer-range meadows. In New England, this species is sometimes considered an aggressive WEED in agricultural lands.

Reference: U.6.

See also MUTTON BLUEGRASS.

SHRUB PASTURES. *See* PASTURES.

SIBERIAN BUCKWHEAT
= TARTARIAN BUCKWHEAT.

SIBERIAN FOXTAIL MILLET.
See FOXTAIL MILLET (variety).

SIBS, or *siblings*, are offsprings of the same parenteral plant.

SIDE-DELIVERY RAKE is an implement used to gather HAY into loose windrows, so it may dry quickly.

SIDE-OATS GRAMA (*Bouteloua curtipendula*) is a long-lived, native grass with an exceptionally wide distribution; it is most abundant in the Great Plains. This grass has short, scaly rootstocks; usually it makes a bunch-type growth.

The flowering stalks are 2 to 3 ft. tall. The leaves are about 6 in. long and nearly ¼ in. wide. The seed heads consist of a large number of SPIKES set on a slender,

zigzag main axis (RACHIS). The root system is fairly deep and well branched.

Side-oats grama produces an abundance of leafy forage, which is well liked by all classes of livestock. Forage production ranges from ½ to 1½ tons per acre. Good hay can be obtained if the plants are mowed at the proper stage of growth. Ordinarily this species is seeded in mixtures with other adapted GRASSES.

Side-oats grama is considered excellent for conservation use. It is adapted to a

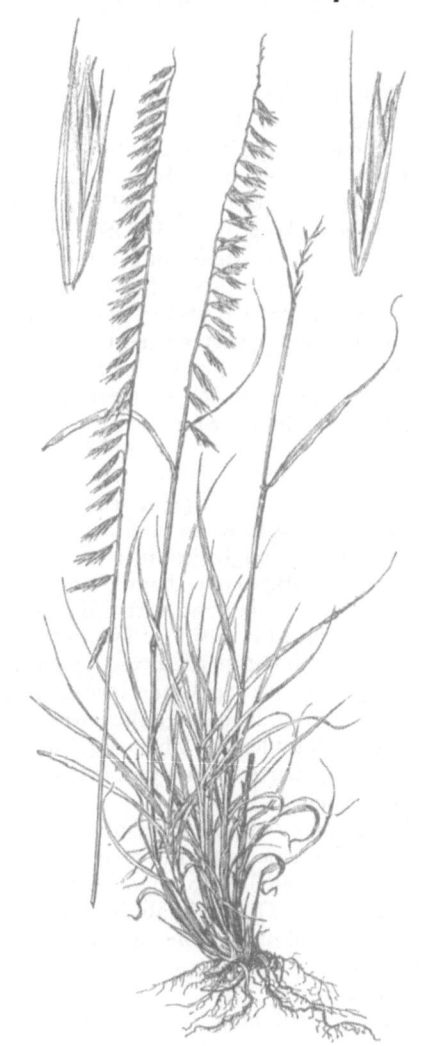

Side-oats grama. Plant, spikelet, and florets. (H.26.)

wide range of soil and climatic conditions. Seedling vigor is good, and failure to obtain a satisfactory stand seldom occurs if minimum care is used in seedbed preparation and drilling.

Variety: *Elreno side-oats grama* is a new variety of this grass developed in Kansas. *Tucson* is a variety widely used in Oklahoma.

References: H.1; H.23.

See also GRAMA GRASSES.

SIFTING FEED. *See* BOLTING.

SILAGE, or *ensilage,* is partially-fermented forage plants. When its quality is good, it is a succulent and nutritious feed that can be used for all kinds of farm animals when green forage is not available.

There are four methods of harvesting forage plants: pasturing, soiling, haying, and silage making. Pasturing is usually the most efficient method, but often forage plants grow faster in the spring and early summer than normal pasturing can remove them. The next most efficient method of harvesting forage plants is haymaking, but this requires drying weather. *Ensiling,* on the other hand, which can be done in periods unfavorable to the field-curing of hay, is an efficient method for the conserving and storing of green feed for use as supplemental feed when necessary.

Converting roughages into silage has certain advantages over haymaking: losses caused by bleaching and scattering and by rain are avoided in ensiling. The CAROTENE content is also better preserved by ensiling than by hay making. When coarse roughage is converted into silage the entire plant is made palatable, but when dried for hay the stemmy parts of many plants are not readily eaten.

The feeding value of good-quality silage is about the same as that of an equal quantity of the same crop made into good-quality hay. The palatability of silage encourages animals to consume more roughage and, as a result, they produce more milk and meat.

CORN and SORGHUMS usually make more satisfactory feeds in the form of silage. In fact, one or the other of these species will, if properly ensiled, produce more T.D.N.

per acre of land than any other crop handled in other ways.

Silage made with corn and sorghum is widely used for both dairy and beef cattle and for fattening lambs. Its slightly laxative effect often is an advantage where legume hay is not included in the ration.

Silage is an especially good way in which to store feed for an emergency period. A number of by-products can be stored by ensiling; e.g., PEA VINES from canneries and sugar BEET PULP from beet-sugar factories that cannot be readily and profitably dried. Silage requires less storage place than hay and creates no fire hazards. Still another advantage of the ensiling process is the killing of weed seed so that cattle feeding on silage do not scatter viable weed seed.

Standards by which the quality of silage may be judged were set up by the American Dairy Science Association committee on silage methods. These standards are:

1. *Very Good:* Clean, acid odor and taste, no butyric acid, no mold, sliminess, or proteolysis, acid pH of 3.5 to 4.2, ammonia nitrogen less than 10 percent of total nitrogen.

2. *Good:* Acid odor and taste, trace only of butyric acid, acid pH of 4.2 to 4.5, ammonia nitrogen 10 to 15 percent of total nitrogen.

3. *Fair:* Some butyric acid, slight proteolysis or some mold content, acid pH 4.5 to 4.8, ammonia nitrogen 15 to 20 percent of total nitrogen.

4. *Poor:* High butyric acid, high proteolysis, sliminess, or mold, acid pH 4.5 to 4.8, ammonia nitrogen about 20 percent of total nitrogen.

Silage has been found to be particularly well adapted as a feed for *dairy cows.* In many sections silage has come to be the dairy farmer's basic winter roughage.

The average dairy cow will eat about 3 lb. silage and 1 lb. hay a day for each 100 lb. live weight. If less hay is fed, the amount of silage will increase. Some large cows have eaten 75 to 100 lb. grass silage per day.

Good silage may be fed successfully to vigorous *calves* as soon as they will eat it, but it should be added to the ration gradually, and the uneaten silage removed from the feed trough each day. Many feeders prefer to omit it from the ration until the danger of serious digestive disturbances is past, say sixty or ninety days, after which it is fed safely in quantities up to the capacity of the calf. *Yearling heifers* consume about half as much as mature stock, that is, from 12 to 24 lb. a day. When the silage is supplemented with some good legume hay, no grain is required to keep the yearlings in a thrifty growing condition.

Beef cattle are perhaps capable of making a greater utilization of coarse roughages than most other kinds of domestic animals. Yet cattle frequently eat only the leaves and fine stems of many mature forages. Silage made of sorghums or corn from a given acreage will feed fully twice as many cattle as the fodder produced from a similar area.

Silage has its greatest usefulness as the main feed so far as wintering rations for beef cattle are concerned. In fattening rations, well-made silage may serve chiefly as a source of vitamins, minerals, and sometimes as a source of good-quality protein. The succulence of the silage is desirable to help keep the animals "on feed" over extended periods. One may feed 3 to 6 lb. corn silage per 100 lb. of live weight of beef cattle.

The use of silage in the winter ration of the *sheep* flock is increasing. Good silage is an economical as well as valuable part of the ration of both breeding and feeder lambs. As a rule, to breeding flock not more than 4 lb. silage per head for small ewes, and 6 lb. for large ewes per day should be fed, and some hay always should be in the ration. Part of the silage should be removed the last two or three weeks before lambing, and replaced with less bulky feeds.

Lambs weighing from 50 to 60 lb. should consume about 1½ lb. silage per head per day when receiving grain and hay in addition. Larger quantities of silage can be fed, but some protein supplement should be added to balance the ration.

Silage is not generally used in *horse* and *mule* feeding, but it is a safe feed for both if it is of good quality and is carefully fed.

This is particularly true of corn silage. Under no circumstance should silage be considered as the principal roughage for horses and mules, but it should serve as a partial substitute for hay in the ration. Because of its bulky nature, horses and mules doing hard work should not be fed large quantities of silage; but, owing to its tonic, laxative, and appetizing effects, it is well suited for the maintenance of idle horses and mules, brood mares, and growing stock. Silage should be introduced gradually into the ration and the amount fed should generally not exceed from 10 to 15 lb. daily per animal.

Swine can use considerable amounts of silage. LADINO or other fine-stemmed legume silage can be fed to pigs weighing over 60 lb. Such pigs can consume 1 lb. silage per meal per day. As they get older, and the capacity of the digestive tract increases, they can eat larger amounts.

Sows and gilts can be fed corn silage, grass silage, or legume silage, at the rate of 2 to 3 lb. per 100 lb. body weight.

In *poultry*, poor GRASS silage may cause hens to produce "grass eggs" with olive-colored yolks. This discoloration can be prevented by feeding high-grade silage or by restricting the use of silage to 2 or 3 lb. per 100 hens.

Biochemical changes taking place from the time green forage is put into the SILO until it becomes sour, aromatic silage are as follows:

1. *Respiration*—free oxygen trapped in the space around the silage oxidizes the carbohydrates to carbon dioxide and water, and releases energy. Respiration ceases when the plant cells die. The oxygen is used up to a considerable extent.

2. *Enzymes* present in the green feed begin to digest the carbohydrates and reduce them to sugars. Other enzymes digest proteins to amino acids.

3. *Molds, yeasts, and aerobic bacteria* begin to multiply and feed on the available sugars. Their activity diminishes as the oxygen supply becomes exhausted and is replaced by carbon dioxide, a gas, which settles like a blanket over the surface of the silage. In air pockets or around the outside where air is not excluded, these micro-

organisms continue to multiply and if not checked, will produce poor-quality, moldy feed.

The temperature of the material may rise as high as 100° F. within a week. This is caused by the activity of various micro-organisms. As the exhaustion of oxygen slows down biological activity, the temperature gradually decreases, and the whole mass settles to a surprising extent.

4. *Anaerobic bacteria*, especially *Lacto-bacillus spp.*, which thrive in the absence of air, break down the sugars first to alcohol, then to acids—mainly acetic acid and some lactic acid; the acid content rises until it is sufficiently strong to kill the bacteria that produce the acids.

In about a month, the ensiling process is complete. The feed, if left untouched in the silo, will remain unchanged; if exposed to air (with opening the silo or through air pockets) the mold organisms again will become active.

Silage-making is affected by various factors:

1. *Moisture content* is the most important factor. Forage plants used for silage should be harvested at the same stage of maturity that produces the best quality of hay. The best moisture content for successful silage production is between 60 and 70 percent. Most corn and grain sorghums will contain the right amount of moisture. Other crops may need wilting.

The length of time necessary to wilt silage to the desired moisture content varies with the moisture content of the green forage and with the weather. Highly succulent forage needs a half day to wilt on a partially cloudy day, while an hour's exposure on a sunny, breezy day may be sufficient.

In general, forage with a high moisture content makes a sour silage. A corrective measure is to mix with the silage 5 to 20 percent dry straw or hay or corn cobs.

Forage that is too low in moisture content does not pack well. Air pockets are trapped and around these the silage becomes moldy. Low moisture content can sometimes be corrected by adding water during the silo-filling process.

NOTE: Moisture content can be determined by a number of methods, but the cheapest and most direct is the oven method. The time required to complete a test will vary with the moisture content, but usually will be less than an hour.

Procedure: (1) Preheat kitchen oven to 275°F. (2) Weigh a wide, shallow metal container. (3) Spread cut forage in thin layer in container. (4) Weigh container and forage; hold total at less than 25 lb. (5) Subtract weight of container to obtain *green weight* of forage sample. (6) Place container and sample in preheated oven; if oven has no vent, leave door slightly open. (7) When plant material appears to be dry, weigh; then return it to oven. Repeat at 5- or 10-minute intervals until weight remains constant. (8) Subtract weight of the container to obtain *dry weight* of forage. (9) Obtain *water content* of sample by subtracting dry weight from green weight. (10) *Moisture percent* is weight of water lost (No. 9 above) divided by the green weight of forage (No. 5 above).

2. *Length of the cut sections* has an important bearing on the packing of the ensilage and, hence, the quality of the silage. The most efficient length varies with the crop that is being ensiled. Silage made of corn and sorghum is cut with the *silage cutter* into sections ½ to 1½ in. long. Grass and LEGUME silage material is cut into shorter sections; at 75 percent moisture, the feed should be cut into pieces ½ to ⅝ in. long; at less than 70 percent moisture, ¼ to ⅜ in. long; wilted grass silage material should be finely cut, not exceeding ¼ in.

3. *Silage conditioners*, or *preservatives*, should be used when immature grasses or freshly cut legumes are put into the silo without wilting. These plants contain lower percentages of fermentable sugars than either corn or sorghum. In the absence of the acetic- and lactic-acid formation, other reactions take place between bacteria and the media from which the butyric and other objectionable acids are formed. These make undesirable silage.

A solution to the problem is to add ground GRAINS, such as grain SORGHUMS, OATS, corn, BARLEY, or WHEAT, which are high in carbohydrates and let the natural forces break the carbohydrates down to acids. The ground grain can be spread over the silage with a scoop as it is put into the silo. It will require about 200 lb. ground grain to each ton of *green silage* for best results. In addition to preserving the silage, ground grain also increases its feeding value. About 90 percent of the feeding value of the ground grain is retained in the silage.

Blackstrap molasses is also often used for this purpose. It is high in fermentable sugars that are readily attacked by the bacteria to form the preserving lactic and acetic acids. The amount needed depends on the material. Usually 40 to 100 lb. MOLASSES is used per ton of silage. The higher rates are used if the silage is made of immature legume plants. Legume silage made with molasses does not need further treatment before feeding.

Molasses is mixed with water and spread over the silage as it goes into the silo. Many farmers prefer to use *dried molasses* since it can be applied either by hand or through the silage blower. Molasses preserves the silage in good condition and increases its feeding value.

SODIUM META-BISULFITE (anhydrous) is a satisfactory chemical preservative for silage. It preserves the silage in excellent condition and with a minimum loss of nutrients from seepage or drainage. The silage has good color, is very palatable, and objectionable odors are reduced to a minimum. The chemical is used at the rate of 8 to 10 lb. to each ton of green forage. In trench silos, it can be spread in dry form by hand as the chopped forage is being put into the silo; it should be applied in layers approximately 6 in. apart. In filling upright silos the sodium meta-bisulfite is usually applied through the silage blower.

SULFUR DIOXIDE, a gas, is injected carefully into the forage at 2-ft. intervals or closer—after storage in the silo—or into trailer loads before storage, at the rate of 5 to 6 lb. per ton of forage. It produces sulfurous acid upon coming in contact with crop moisture and reduces plant respiration and bacterial activity. The silage is preserved with a low fermentation rate and low acid content. More of the carotene is preserved than by any other method. Sulfur dioxide has no effect on silage seepage. It must be metered carefully into the forage, since too large a quantity (7 lb. or more per ton) will make the silage less palatable.

Inorganic acids, such as SULFURIC ACID or HYDROCHLORIC ACID, when added to silage as preservatives often cause very sour conditions, an unnatural odor of the silage and destruction to the masonry of the silo. Inorganic acids generally prove unsatisfactory in silage making.

4. *Exposure to air* causes spoilage of silage. Therefore, after a silo is opened for use, about 3 to 4 in. of the surface should be removed daily to prevent spoilage. Even a 2-day interval between the removal of feed gives molds opportunity to develop.

5. UREA can be added to increase the protein level of silage. Tests have been run that indicate that the proper amount, i.e., 10 to 15 lb. per ton, considerably improves the feeding value of sorghum and corn silage.

Dangers: *Moldy silage* should not be used for sheep and horses because it causes digestive disturbances. *Frozen silage* should be thawed before being used; it should then be fed immediately, before decomposition sets in. No harm will result from feeding it after it is thawed nor is the nutritive value known to be changed in any way.

References: S.23; W.25; P.20; S.24; C.14; F.9; S.25.; P.17.

See also SILAGE CROPS; ALFALFAS; SOYBEAN; SUNFLOWERS; BARNYARDGRASS; ROOT CROPS; ALMOND HULLS; PASTURES; PASTURE MANAGEMENT; GRAZING; SOILING CROPS.

SILAGE BLOWER. *See* SILAGE.

SILAGE CONDITIONERS are PRESERVATIVES used in SILAGE making; e.g., SODIUM META-BISULFITE.

SILAGE CROPS are numerous. In fact, any forage crop that is palatable when grazed, or when fed green, or as dry hay, will also make palatable SILAGE; and any crop that is unpalatable in the green stage or as hay will make unpalatable silage. In addition, certain materials other than forage crops can be ensiled; e.g., POTATOES and ALMOND HULLS.

"*Grass silage*" is a term used to designate silage made of various GRASSES (including CEREALS, especially SMALL GRAINS, such as WHEAT, BARLEY, RYE, and OATS) and/or LEGUMES that are ordinarily considered hay crops. If grasses and legumes are used, the term *grass-legume* (or *hay crop*) *silage* is sometimes applied; the term *legume silage* refers to silage made from legumes alone, e.g., from ALFALFA, SOYBEANS, VETCHES, LESPEDEZAS, COWPEAS, KUDZU, TRUE CLOVERS (such as SWEETCLOVER, LADINO CLOVER, ALSIKE CLOVER, CRIMSON CLOVER and RED CLOVER), etc.

Many conditions on the farm may make ensiling a more practical method of handling a crop than haying. The first cutting of weedy alfalfa may be chopped and ensiled to make palatable feed. Dry, hard, and piercing beards of many grasses in hay are dangerous to animals, but by the process of ensiling they become soft and harmless. Also, the curing of clippings for hay on a permanent pasture may not be convenient; under these conditions the wilted forage can be ensiled profitably.

In making grass silage certain principles should be kept in mind. First, the quality of the silage is best when the plants are cut at a stage that makes the best hay. Second, green forage as it is cut may contain as much as 85 percent moisture. To ensile this material, part of the moisture should be removed by allowing the cut forage to dry in the swath or windrow until it is wilted. When its moisture is reduced to 60 to 70 percent it can be ensiled. In satisfactory drying weather, two hours are sufficient time for drying. In poor drying weather, longer periods are required.

The greater cost involved in making silage instead of hay is in transportation of the wilted material from the field to the SILO and in chopping it in the silage cutter.

Loosely packed silage spoils. Forage material therefore should be firmly packed and the sunken places leveled out during the settling.

The feeding value of grass and grass-legume silages made of immature plants may be improved by adding MOLASSES at the rate of 40 to 100 lb. per ton of silage. If the silage is mostly grass the 40-lb. rate should be used. The amount of molasses should be increased in proportion to the amount of legumes in the ensiled material. The more important silages are as follows:

1. *Corn silage* is obtained from CORN; this widely grown crop produces good forage yields that are utilized most efficiently as silage. The yields are generally estimated at 1 ton of silage for each 5 bu. corn that would be produced; some silage corns produce more forage.

In general, if the corn crop is intended for silage, choose a variety maturing ten days to two weeks later than a variety grown for grain. The best variety for any locality is the latest to mature to the silage stage in the available growing season.

There is a high, positive correlation between the number of days to silking and yield: the earliest silking yields the least, and the latest the most.

The ear weights are about one-third of the total weight for all corn varieties. In total digestible nutrients, however, the ears contribute more than half the total. The planting rates for silage corn should be regulated to produce as many plants as possible with good ears.

Corn should be cut for silage when most of the grains are dented and three-fourth of them are past the soft-dough stage. At this time the lower leaves may be turning brown. The moisture content is 65 to 70 percent. If a long harvest period is expected, it is best to begin harvest a little earlier. If cut too soon, the corn will be too high in moisture, and the silage will become sour and soggy and will require drainage. If cut too late, the forage will be difficult to pack solidly enough in the silo to prevent mold. When corn is ensiled too dry, add water to aid in packing, but even then silage made from ripe corn lacks aroma and palatability.

2. *Cereal grain silages* are made from oats, wheat, barley, and rye. In feeding value they rank in the order listed, but wheat may produce more feed per acre. It is recommended that oats are ensiled when the dry-matter content has reached

30 to 35 percent. To harvest oats at an earlier stage is undesirable; but if necessary the forage should be wilted before it is ensiled, or some dry feed, such as ground corn, should be added to bring the dry-matter content to the needed level.

Other cereal grain silages are made like *oat silage*, but bearded barley and rye may be cut earlier to prevent the beards from becoming sharp and dangerous. Best results are usually obtained when 40 to 60 lb. MOLASSES per ton of silage is added.

3. *Sorghum silage* in the United States is second in importance to corn silage and replaces it in the more arid regions where corn is an unreliable crop. SORGHUM is dangerous to graze after a drought or frost because of the presence of prussic acid. But sorghum made into silage is free of this poison.

Sorghum is usually cut for silage when the seeds are hard. It may be cut earlier with little loss in feeding value, when it is allowed to wilt before ensiling. At younger stages, the water content is more than 80 percent, which is too high for successful ensiling. The grain content of sorghum is lower than that of corn, because the ratio of grain to total weight is less. Generally, sorghum silage is no more acid than corn silage. When, in case of drought, sorghum is cut before heading, the silage is more acid than corn silage unless the plants are allowed to lose some of their moisture before ensiling.

The *sorgo* (forage sorghum) varieties are most commonly used for sorghum silage; the sorghums with sweet, juicy stalks give a high yield.

In *grain sorghum* silage there is some loss of grain passing through the animal undigested. Grain loss in hard-seeded KAFIR has been found to rank from one-third to as high as one half. Feeding value of sorghum is utilized more efficiently as silage than as dry stover. The feeding value of forage sorghum is less than that of the kafirs, but in many areas the difference in yield more than balances the nutritional deficiency.

Sorghum silage is often stored in trench silos for feed in dry years.

Broomcorn silage is sometimes prepared from the green broomcorn plant after the brush is harvested. Molasses applied at a rate of not less than 50 to 70 lb. per ton is needed for palatability.

4. *Grass silage*, made from any of the common grass species—e.g., ORCHARDGRASS, KENTUCKY BLUEGRASS, BROME GRASS, NAPIERGRASS, TIMOTHY, REDTOP, BERMUDA-GRASS, TALL FESCUE, etc.—can replace legume hay in the dairy ration at the rate of about 3 lb. silage to 1 lb. hay. Dairy cattle have been fed grass silage as the only roughage for a considerable length of time without ill effects. Many dairymen feed a small amount of hay, and then give the cows all of the grass silage they will clean up. Dairymen with dryland pastures often start feeding silage as soon as their pastures start to dry up. Silage is the best substitute for good, lush pasture.

Beef cattle are commonly fattened on grain and grass silage as the sole roughage.

Grass silage is also used by the sheep men. They find it an excellent feed for ewes while carrying lambs, and during the period following lambing.

Grass silage put up with ground wheat or molasses is a supplement to the regular rations for chickens and turkeys. It can be used to advantage to replace fresh succulent green feed during the fall and winter months. The silage must be young, tender, and finely chopped if it is to be used for poultry.

Brome grass silage with satisfactory keeping qualities can be made from brome grass without a preservative; adding cane molasses increases its palatability.

Kentucky bluegrass-molasses silage made from very good forage in a prebloom stage of maturity contains approximately the same amount of total digestible nutrients as an excellent grade of corn silage but more than two and one-half times as much digestible crude protein.

Bluegrass-molasses silage made from early bloom bluegrass grown on fertile land that has previously been nitrated may equal or surpass in palatability, rate of gain, and economy of gain excellent alfalfa-molasses silage as a roughage for steers fattened in dry lot.

Nearly 4 tons of high quality bluegrass forage wilted and ready for ensiling can be produced per acre from ungrazed fertile land in one cutting. This makes a very high quality roughage from the surplus forage usually produced by permanent pastures during the spring, and thereby evens out forage supplies.

5. *Alfalfa silage*, if properly made, loses none of the ALFALFA leaves and only a small percentage of its carotene content. The interest in the use of alfalfa as a silage crop is growing. The making of alfalfa silage requires more care than the making of silage from most other crops. The carbohydrate content of alfalfa is relatively low and the protein content high. The reverse is true with crops like corn or sorghums that are considered ideal ensilage crops. The fermentation and bacterial action necessary to make good ensilage will not take place in the alfalfa unless the carbohydrates are supplied in the form of grain or molasses (or by adding acid to give the proper acidity); otherwise the bacteria may act on the proteins, breaking them down into vile-smelling substances that make the ensilage unfit for feed.

Alfalfa in the later stages of bloom will have more carbohydrates in the forage than when cut in the bud stage; also, a lower percentage of water, which is desirable in making alfalfa silage. A moisture content of from 60 to 70 percent is most satisfactory. When moisture is insufficient, water may be added. If the moisture is above 70 percent, dry roughage should be added; dry sorgo or corn fodder, threshed alfalfa stems, etc., are suitable. From 15 to 25 percent of the total silage mixture may consist of dry roughage. The amount of molasses to be added varies from 40 to 80 lb. per ton of ensilage. Ear corn chops also may be used successfully by mixing at the rate of 200 to 300 lb. per ton.

6. *Soybean silage*, if made from soybeans alone, usually is rather bitter and has a strong, disagreeable odor. If wilted to a proper moisture content—60 to 65 percent—and tightly packed, soybeans may produce a satisfactory high-protein feed. The addition of corn meal will give a well-preserved silage. About 2 or 3 parts corn forage to 1 part soybeans makes a well-balanced silage that keeps well, is readily eaten by stock, and has no bad effects on the quality of milk and its products.

7. *Ladino clover silage*, because of its exceptionally high content of digestible crude protein, is especially well adapted to the feeding of dairy cows, pregnant and lactating ewes, rapidly growing young stock of all species, and brood sows during the winter months.

8. *Pea vine silage* is prepared from PEA VINES and empty PEA PODS, two by-products of the modern development of quick-freezing of PEAS in the green stage. The moisture content of the green material as it comes from the viners is high, about 87 percent. Drying the material before ensiling is impractical. Pea vine silage is usually made by stacking the vines in straight-sided stacks 20 to 30 ft. wide and at least 20 ft. high. The high stacks require less artificial packing. Since the vines are not chopped, they should be distributed evenly to prevent the formation of air pockets.

Silage made from pea vines is eaten readily by animals. Feeding results generally have been good.

9. *Lima bean silage*—made from the vines and empty pods obtained as by-product of the quick-freezing of LIMA BEANS—has not given uniform results when fed to animals.

10. *Beet pulp silage* is prepared from BEET PULP, a by-product of the beet-sugar industry. Pulp is often ensiled at the factory. After fermentation it is also dried and sold for feed.

Sugar beet pulp has a water content of about 90 percent, considerably higher than the optimum required for silage. This probably accounts for the strong odor of beet pulp silage, but in spite of this, it is a palatable feed, used extensively to fatten steers and lambs. Feeding lots are often located near the sugar-beet factory to eliminate costs of long-distance hauling.

In beet pulp silage the losses from breakdown of the nutrients to acids continue longer than in silage with a lower moisture content. The total digestible nutrient content is about half that of good corn silage;

therefore it is important to supplement beet pulp silage with concentrate feeds.

11. *Beet top silage* is obtained from the tops of sugar beets. The tops are estimated to be about one-tenth of the beet yield on the dry basis. They are now mechanically separated from the beets with the help of modern machinery and are an important by-product of the beet-sugar industry.

The feeding quality of beet top silage is preserved through all kinds of weather. The silage process softens the beet crowns and reduces the danger to livestock from choking. There are, however, two drawbacks to the use of sugar beet tops as silage: dirt clinging to the crowns, which reduces the quality of the silage, and the high oxalic acid content of the leaves, a chemical uniting with calcium to form insoluble calcium oxalate crystals.

The tops are usually ensiled in trench silos or in stacks. The leaves should be allowed to wilt to reduce the moisture to 70 percent or lower. The general practice is not to chop them into short sections; if not chopped, beet top silage after curing will have to be removed with the aid of a hay knife or some other cutting tool. The average production of beet tops is 5 to 6 tons of green weight per acre, of which 40 percent is in the crowns and 60 percent in the leaves.

12. *Sunflower silage* is used in the northern and western states, where the growing season is short and cool; it is a poor substitute in areas where corn can be grown. Sunflower silage is less palatable than mature corn silage, with only about two-thirds of the feeding value, but its feeding value is favorable when compared with silage made from immature corn. Its feeding value for cattle is about the same as that of oats and vetch silage, but for sheep it is inferior. Sunflower silage is reputed to be somewhat constipating.

The crop is cut when one-half or two-thirds of the heads are in bloom. If cut too early the silage is watery and leaks badly. If cut too late the heavy stalks are too difficult to pack. The older plants are decidedly less palatable than the younger.

13. *Potato silage* has the advantage that it keeps the feeding value of the potatoes used, while the feeding value of the tubers deteriorates under ordinary storage conditions. Whole potatoes do not ensile satisfactorily and sprouted or frozen potatoes should not be used.

Potatoes contain 78 to 80 percent moisture and therefore cannot be ensiled alone. To take up excess moisture, dry forage is mixed with the potatoes, in the proportion of 20 to 25 lb. per 100 lb. potatoes. The potatoes and the dry feed can be run through the ensilage cutter together for mixing. To produce good silage 2 percent ground corn is then added to furnish the inoculum of the acid-forming bacteria.

Potatoes can also be added to corn silage and sorghum silage at the rate of 500 lb. per ton of silage.

Potato silage must be packed thoroughly. Trench silos with sloping walls are better than silos with perpendicular walls. As the silage settles in the trench silo, it slides against the wall. In the upright silo it settles away from the wall near the top, and its excessive weight increases the pressure on the walls at the bottom. Adequate drainage should be provided at the bottom of the silo.

14. *Almond hulls silage* has received some attention as a source of feed for cattle and sheep. If fed dry, the almond hulls should be ground. To keep ground hulls from spoiling, the moisture content must be about 12 percent. The moisture content ranges from 10 to 45 percent when the hulls come from the huller. The expense of drying and grinding may be avoided by dumping the hulls into a trench silo and adding water to bring the moisture content up to 65 percent. The hulls are rich in carbohydrates that form the organic acids needed to preserve silage.

15. *Other silage materials* are APPLE POMACE; WET BREWERS' GRAIN; RUSSIANTHISTLE; SUGARCANE; JOHNSONGRASS; WEEDS of various kind; SWEET CORN refuse from canneries (husks, cobs, shanks, etc.); small grains mixed with CANADA FIELD PEAS or vetches; mixtures of grasses with red clover; soybeans with Sudangrass; etc.

References: S.23; G.13; G.14; N.3; H.32; W.25.

See also SILAGE.

SILAGE CUTTER.

See SILAGE; SILAGE CROPS.

SILAGE MAKING. *See* SILAGE.

SILAGE PRESERVATIVES.

See PRESERVATIVES; MOLASSES; SILAGE.

SILAGE STANDARDS. *See* SILAGE.

SILICON is a CHEMICAL ELEMENT which is widely distributed in nature in quartz, sand, clay, granite, mica, and many other minerals.—*See also* SILICONES.

SILICONES are organic long-chain compounds containing alternate SILICON and OXYGEN atoms. Some of the compounds are known to act as anti-foaming agents and are also used in feed for the prevention of BLOAT.

SILK is the name of the long STYLE of CORN.

SILOS are light-walled structures used for making and keeping silage. Various kinds of silos are in use for forage crops and *stacks* are used for siloing cannery refuse and roughage, but rarely for siloing grasses. Success with the latter method depends upon plenty of moisture in the crop, thorough packing, and in the case of hay crops, upon a covering to exclude air as well as to provide weight. Large losses can occur from wind blowing the silage from stacks.

Common types of *silo* structures are as follows:

1. *Trench silos*—simple trenches dug into the ground—which predominate in the South, Southwest, and the Great Plains.

2. *Tower silos* built of wood, concrete, tile, brick, stone, or steel; they predominate in the North and in the East.

3. *Pit silos* built underground in semi-arid climates.

4. *Fence silos* built with successive rings of 2- by 4-in. welded steel, steel mesh, or wood slate fencing and lined with a fiber-reinforced paper.

5. *Box silos*, 20 ft. high and consisting of two boxes, each of 6 by 10 ft. in width and length.

All these silos preserve silage effectively if the air is excluded and adequate drainage provided for any excess liquid that may accumulate by seepage from the soil or by expression of juice.

The quantity of spoiled silage varies directly with the surface exposed to the air. This accounts for the heavy loss in stacks and in shallow silos. For the same reason, the *gas-tight steel tower silo* is most efficient. This type of silo aids in silage making by keeping out the air and keeping in the carbon dioxide that develops from natural fermentation of the silage. The fermentation process in the gas-tight silo

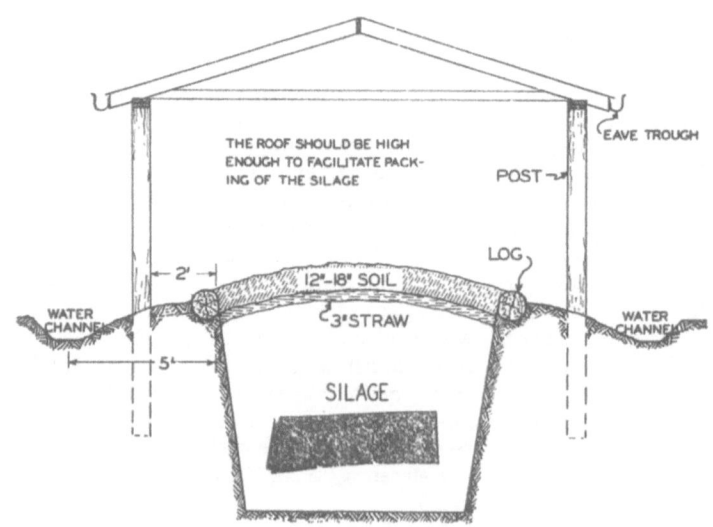

Trench silo. Cross section of unlined trench showing logs used to retain back-fill of soil, also roof construction. (C.18.)

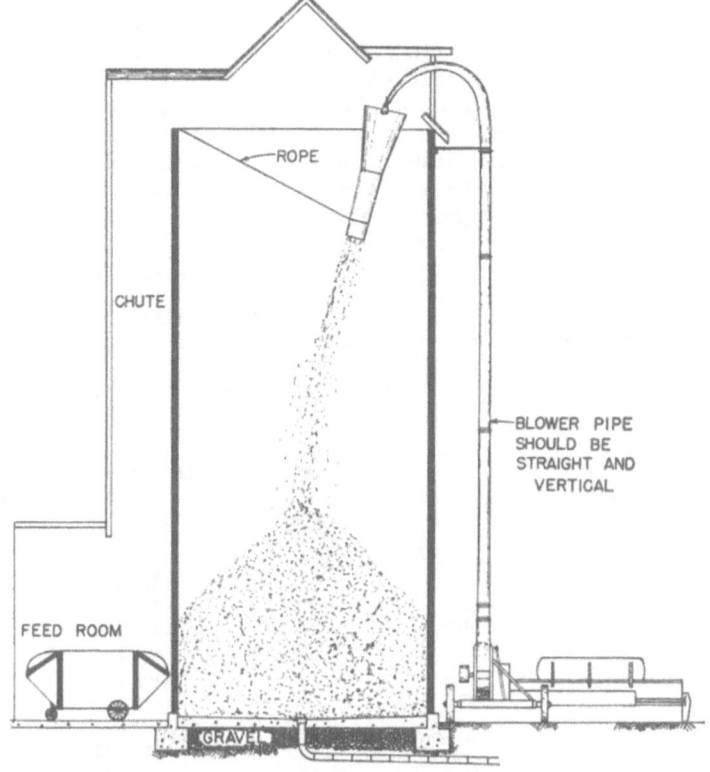

Silo filling. Distributor pipe should be adjusted to give a uniform cone in the center when filling without trampling. (H.32.)

is much the same as in a conventional silo except that top spoilage, side spoilage, and flake mold are prevented. The silo can be partially filled on widely separated dates and the silage can be fed out over a long period of time. This silo is equipped with a top filling hatch and an automatic bottom unloader. A vinyl plastic breather bag and a pressure relief valve, in the top part of the silo, compensate for differences in atmospheric pressure inside and outside the silo (which are due to temperature changes and to gas formation within the silo) and thus keep outside air from coming in contact with the silage.

The weight of material that can be put into the silo depends upon the moisture content of the crop, the depth and diameter of the silo, the rapidity of filling, the length of cut, and the character of the crop. Corn with a normal content of moisture, chopped in $\frac{1}{4}$- to $\frac{3}{8}$-in. lengths, and put into a silo 14 ft. in diameter by 40 ft.

in height at the rate of about 30 to 40 tons a day will contain on an average about 13 lb. dry matter to the cu. ft. and, when taken out after several months, will contain about 13.5 lb. dry matter to the cu. ft. The average weight, therefore, of 1 cu. ft. of such silage immediately after filling would be 46 to 47 lb. if the moisture content was 72 percent; and the average weight after several months, assuming the moisture content to be the same as in the freshly chopped material, would be about 48 lb. to 1 cu. ft. The reason that the silage after several months weighs only a little more per cu. ft. than it did when filled in, is that the increase in weight from the reduction in volume by settling is partly offset by the losses from evaporation and fermentation and also sometimes from expression of juice. The dry matter per 1 cu. ft. grass or legume silage is usually about the same as per 1 cu. ft. corn silage.

The sizes of silos suitable for herds of

different sizes and fed at different rates are shown in the tables:

A *forage blower* is often used for elevating the properly cut silage material into the silo. The pipe of the blower should be as nearly perpendicular as possible; horizontal pipes are liable to clog. *Forage elevators* have also been adapted for silo filling. Either a sloping or vertical type may be used. They are not apt to become clogged.

Trench silos can be filled without any elevating mechanism.

The usual practice of packing the silage is to have at least one man to tramp and *distribute* the silage. Unless the material is distributed, the leaves are blown to one side of the silo. Tramping can be dispensed with so far as the quality of the silage is concerned, but a silo filled rapidly without tramping does not hold quite so much. In view of the fact that the silage settles

TOWER SILOS
Sizes (inside measurements) required for different-sized herds when fed at various rates

No. of animals	Quantity fed per animal daily	For a winter feeding period of 200 days		For a summer feeding period of 100 days	
		Total amount needed	Diameter and height of silo	Total amount needed	Diameter and height of silo
	lb.	tons	ft.	tons*	ft.
5	30	15	8 by 18	....	....
5	40	20	8 by 22	....	....
5	50	25	8 by 26	....	....
10	20	20	8 by 22	10	....
10	30	30	10 by 22	15	8 by 18†
10	40	40	10 by 28	20	8 by 22†
10	50	50	{10 by 32}{12 by 24}	25	8 by 26
20	20	40	10 by 28	20	8 by 22†
20	30	60	12 by 28	30	10 by 22†
20	40	80	{12 by 36}{14 by 28}	40	10 by 28
20	50	100	14 by 34	50	{10 by 32}{12 by 24}
30	20	60	12 by 28	30	10 by 22†
30	30	90	14 by 30	45	10 by 30
30	40	120	{14 by 40}{16 by 32}	60	12 by 28
30	50	150	16 by 38	75	12 by 34
40	20	80	{12 by 36}{14 by 28}	40	10 by 28
40	30	120	{14 by 40}{16 by 32}	60	12 by 28
40	40	160	16 by 40	80	{12 by 36}{14 by 28}
40	50	200	{16 by 48}{18 by 40}	100	14 by 34
50	20	100	14 by 34	50	{10 by 32}{12 by 24}
50	30	150	16 by 38	75	12 by 34
50	40	200	{16 by 48}{18 by 40}	100	14 by 34
50	50	250	18 by 48	125	{14 by 40}{16 by 32}

* A silo that would hold only 10 tons or less would be too small to be practicable.

† Too shallow to permit 3 in. to be removed daily. Removal of less than 3 in. daily is not practicable for summer feeding.

TRENCH SILOS

Sizes required for different-sized herds when fed on the basis of the removal of 4 in. of silage daily to avoid spoilage

Depth	Width		Weight of silage per linear ft.	Animals that can be fed with a daily allowance per head of:			
	Bottom	Top		40 lb.	30 lb.	20 lb.	10 lb.
ft.	ft.	ft.	lb.	No.	No.	No.	No.
6	6	9	1575	13	17	26	53*
6	7	11	1890	16	21	32	63
6	8	13	2205	18	24	37	74
8	6	10	2240	19	25	37	75
8	7	12⅙	2695	22	30	45	90
8	8	14⅔	3185	27	35	53	106
10	6	11	2975	25	33	50	99
10	8	14⅔	3955	33	44	66	132
10	10	18⅔	5005	42	55	83	167
12	10	16	5460	45	61	91	182
12	12	20	6720	56	75	112	224
12	14	26	8400	70	93	140	280

* For example, a trench silo 6 ft. deep, 6 ft. wide at the bottom, and 9 ft. wide at the top will contain 1,575 lb. silage per linear ft. (sufficient for the removal of 4 in. silage for *three days*). Therefore, for a 60-day feeding period for 13 cows the silo should be 20 ft. long, and for a 120-day feeding period, 40 ft. long.

more in the middle than at the sides, it is thought that keeping the center higher than the sides while filling the upper part of the silo may lessen the extent to which the silage will draw away from the sides at the top. Of various contrivances used for distributing the cut material, metal pipes put together loosely in sections is commonly recommended.

In trench silos, livestock can be driven back and forth over the chopped material, or a tractor of either the wheel or crawler type can be used.

When a crop is placed in the silo the *temperature* rises until all the air in the interstices (small spaces) of the silage is exhausted. Coarse chopping and poor packing favor the development of high temperatures. Ordinarily silage should not reach a temperature above 100° F. The maximum temperature is usually reached within a week, after which the silage cools slowly and steadily. If a high temperature persists, it is evidence that air is penetrating from the top or sides of the silo. If the high temperatures continue long enough the silage will eventually become moldy.

The silage in tower silos should be removed in even layers and the entire top should be kept level. Any spoiled material found around the wall should be removed

every day or two rather than allowed to remain. The aim should be to expose as little of the silage as possible and to feed it rapidly enough to prevent heating and spoiling.

Trench silos are emptied by removing the silage in vertical layers from the end of the silo. In many cases the truck or wagon is backed into the trench for easy loading.

The self-feeding trench silo will reduce the labor involved in silage feeding. The silo is prepared with concrete floor and sides, and a movable feeding gate is fixed across the front. Cattle eat their way to the back. Some maintenance is required to make sure that the silage is being evenly consumed. Provide one opening for each three to four head of cattle.

Dangers: Suffocating gas from fermenting silage, mostly carbon dioxide, forms in all silos shortly after filling begins and continues until fermentation stops. The gas, being heavier than air, collects and remains in any depression or enclosed space.

If forage contains large amounts of nitrate, oxides of nitrogen will form; these are very dangerous yellow or reddish-brown gases.

The above-ground silo offers better ventilation through its doors, and large quantities of gas seldom accumulate unless the doors are put in too far above the silage level. Gas is a particular hazard in a below-ground silo. Many lives have been lost because of carelessness in entering a silo where there may be danger of gas. The following precautionary measures should be observed: (1) Keep open the silo's sectional doors level-by-level during filling; (2) maintain operation of the ensilage blower while workers are inside; turn it on ten to fifteen minutes before workers enter; (3) after the filling is completed, do not enter the silo for about ten days. By that time, carbon dioxide is reduced to a nontoxic level, but the status of the nitrogen dioxide is conjectural.

References: W.25; S.23; S.24; S.25; M.40; C.18.

See also SILAGE, SILAGE CROPS.

SILT is a term applied to small mineral soil particles of a diameter of 0.002 to 0.05 mm.; they are deposited in form of fine earth and mud from running or standing water.

SILVERHULL.
    See COMMON BUCKWHEAT (variety).

SILVER KING. *See* CORN (variety).

SILVERMINE.
    See COMMON OAT (variety).

SILVER SAGEBRUSH (*Artemisia cana*) is a plant which sprouts heavily from roots and stem base.—*See also* SAGEBRUSHES.

SILVERTOP is one of the forage SORGHUMS.

SIMLA. *See* SAFFLOWER (variety).

SIMPLE SUGARS. *See* SUGARS.

SINGLETARY PEA = ROUGHPEA.

SIRUP, often spelled *syrup*, is a concentrated, viscous, aqueous solution of SUCROSE (cane sugar or beet sugar); sirup may contain a small amount of other CARBOHYDRATES.

Glucose sirup is a GLUCOSE product obtained from the hydrolysis of corn or sorghum starch; *corn sirup* is more commonly used.—*See also* SORGHUMS.

SITANION. The *Sitanion* spp. are known as SQUIRRELTAILS.

SITONA. *S. hispidula* and *S. flavescens* are commonly known as CLOVER ROOT CURCULIOS.

SIX-ROWED BARLEY (*Hordeum vulgare*), the most important BARLEY species, is an annual with erect culms. It grows 2 to 4 ft. tall; the blades are flat and mostly ⅕ to ⅜ in. wide. The SPIKE is (nearly) erect and ⅝ to 4 in. long (excluding awns). All the spikelets produce large seeds.

> NOTE: If the lateral florets overlap, this species is sometimes called *four-rowed barley*.

Varieties: The six-rowed barley varieties may be divided according to their winter-hardiness into two groups:

WINTER BARLEYS

This group includes the following winter-hardy varieties:

Cordova (smooth-awned, resistant to POWDERY MILDEW);

Ferguson, or *Texas Winter* (rough-awned);

Iredell (hooded);

Jackson (smooth-awned);

Kentucky 1 (rough-awned);

Marett Hooded 4;

Marnobarb (smooth-awned);

Michigan Winter (rough-awned);

Missouri Early Beardless (hooded);

Ohio 1 (rough-anwed);

Purdue 21 (rough-awned);

Reno (rough-awned);

Sunrise (short-awned);

Tennessee Beardless 5 and *6*, or *Tennessee Hooded 5* and *6* (hooded);

Tennessee Winter (rough-awned);

Texan (smooth-awned);

Ward, or *Woodward* (rough-awned);

Wintex (rough-awned);

Wisconsin Barbless, or *Wisconsin Pedigree 38* (smooth-awned);

Wong (short-awned).

SPRING BARLEYS

The following are the most important nonwinter-hardy barley varieties:

Arivat (semismooth-awned);

Atlas (rough-awned);

Beecher (semismooth-awned);

Beldi Giant (rough-awned);

Blue (rough-awned);

Byng (semismooth-awned);

California Coast, also called *California Feed* or *Bay Brewing* (rough-awned);

Club Mariout, also called *Oregon Mariout* or *Golden Mariout* (rough-awned);

Coast (rough-awned, was for many years the common barley of the western states);

Colsess (hooded);

Custer (smooth-awned, an improved *Velvon*);

Feebar (semismooth-awned), resistant to WHEAT STEM RUST);

Flynn 1 (semismooth-awned);

Flynn 37 (smooth-awned);

Gem (semismooth-awned);

Glacier (smooth-awned, feed-type);

Hiland (semismooth-awned, SMUT-resistant);

Himalaya (rough-awned);

Horsford, or *Success Beardless*, also called *Great Beardless* (hooded);

Kindred, or "*L*" *barley* (rough-awned);

Lico (smooth-awned);

Manchuria (rough-awned);

Meloy 3 (hooded);

Nepal (hooded);

Newal (smooth-awned; grown in Canada);

O.A.C.21, or *Arctic*, (rough-awned, extensively grown in Canada);

Oderbruck, or *Wisconsin Pedigree 5* (rough-awned);

Odessa (rough-awned);

Peatland (rough-awned, for peatlands or where SCAB is a problem);

Plush (smooth-awned);

Rojo (smooth-awned);

Rufflyn (rough-awned);

Stravropol, also called *Hog barley* and *Feed barley* (rough-awned);

Trebi (rough-awned);

Vaughn (semismooth-awned);

Velvet (smooth-awned);

Velvon (smooth-awned).

References: H.26; H.27; W.20; W.21; A.6. *See also* TWO-ROWED BARLEY.

SIXTY DAY MILO, better known as *Sooner milo*, is one of the grain SORGHUMS. It is not to be confused with *Day milo*.

SIXWEEKS GRAMA (*Bouteloua barbata*) is an annual GRASS with 2½ ft. long CULMS. It is found on ranges where it is considered poor forage.—*See also* RANGE PLANTS; GRAMA GRASSES.

SKELETONWEED (*Lygodesmia juncea*), also called *wild asparagus*, is a WEED.—*See also* WHEATS.

SKIMMED MILK. *See* MILK PRODUCTS.

SKIM MILK = *skimmed milk.*

See MILK PRODUCTS.

SKUNKBRUSH (*Rhus trilobata*), also known as *skunkbush*, is a much-branched shrub from 2 to 7 ft. high, widely distributed from Alberta to Illinois, northern Mexico, California and southern Oregon. The plant has a disagreeable odor and acid fruits.

The palatability of this species to livestock is low to good, depending on the local supply of palatable herbaceous vegetation.

Control: Skunkbrush can be controlled with 2,4-D.

Reference: U.6.

See also RANGE MANAGEMENT.

SLAG is the fused material which separates in smelting and floats on the molten metal. It contains CALCIUM, SILICON, MAGNESIUM, PHOSPHORUS, and other compounds as well as the elements of the metal being produced. Slag is often used as fertilizer.—*See also* SLAG POISONING.

SLAG POISONING is a term applied to the poisoning of livestock on pastures treated with lumps of SLAG or SUPERPHOSPHATE. Small amounts probably are consumed without doing harm, but when an appreciable amount is eaten the effect may prove fatal.

These materials should be put out in a pulverized form and disked in, or the treated pasture should not be grazed until after sufficient rain has fallen to wash the material off the foliage and to dissolve any lumps. All empty bags should be burned.

Reference: S.28.

See also GRAZING.

SLENDERLEAF CROTALARIA (*Crotalaria intermedia*) is nonpoisonous and belongs to the CROTALARIAS that are grown extensively in this country, especially for soil improvement and also for feeding purposes in form of HAY or SILAGE. It has been computed that 107 lb. dry matter in slenderleaf crotalaria silage is equivalent to 100 lb. dry matter in No. 1 green alfalfa hay for milk production.

Reference: M.14.

SLENDER OAT (*Avena barbata*) is a wild growing plant that is often considered a

WEED. It resembles COMMON OAT. On the Pacific Coast it is often utilized for HAY.— *See also* OATS; AVENA.

SLENDER WHEATGRASS (*Agropyron trachycaulum*)—sometimes called *western ryegrass* (a name more correctly used for COMMON RYEGRASS)—is a native perennial bunchgrass distributed throughout the United States except in the Southeast and South Central States; it is most prevalent throughout the Northern Great Plains and the Rocky Mountain States. This grass commonly grows to a height of 3 ft., and

Slender wheatgrass with spikelets, glumes, and lemmas, the right one ending in an awn. (U.6.)

the dense, leafy bunches may reach a foot or more in diameter. The leaves are from 3 to 13 in. long and about $\frac{1}{4}$ to $\frac{1}{2}$ in. wide. Propagation is entirely by seed.

Although slender wheatgrass occurs on most soil types, it prefers the sandy loams. It is less drought-resistant than either crested wheatgrass or western wheatgrass.

Slender wheatgrass begins growth rather early in the spring and provides an abundance of palatable forage that is well liked by all classes of livestock. The forage cures well on the ground and furnishes considerable quantities of nutritious feed for winter grazing. The plants are not so resistant to close grazing as sod-forming wheatgrass species, and careful management is required for satisfactory grazing returns.

Good yields of high-quality hay are obtained if the plants are cut before the foliage becomes harsh and woody. The plants are relatively short-lived, and stand density decreases rapidly after the fourth production year. On the other hand, seedling vigor is exceptionally good and excellent vegetation cover is provided a few weeks after planting. These characteristics suggest using this grass in mixture with other GRASSES that are slow to become established; they make slender wheatgrass also well adapted to reseeding of sagebrush and burned areas.

Dangers: Among the diseases attacking this grass species are the GRASS SMUTS.

References: H.1; B.1.

See also WHEATGRASSES; ALFALFAS; RANGE PLANTS.

SLICKER is a weeding implement; its knifelike blade or blades run under the soil surface.

SLICKHEAD = *Fultz. See* COMMON WHEAT.

SLOBBERING of horses may be due to the presence of much ripe seed in hay prepared from overripe ALSIKE CLOVER and other plants.

Slobbering of cattle and sheep sometimes occurs when RED CLOVER hay is consumed. The substance causing slobbering has not been identified.

SLOP. *See* DISTILLERS' PRODUCTS.

SLOP FEEDING of hogs is often thought to produce better results than dry feeding.

A *slop* or *swill* is formed by mixing ground feeds with water and/or dairy products. Usually this turns out to be a time consuming and rather expensive procedure. Furthermore, slop-fed hogs may be discounted at the market because they are mostly less trim in the middle than nonslopfed hogs.

Slop feeding does insure that the animal gets adequate water, and this may be an advantage with hired laborers who might not otherwise give the hogs water when needed.—*See also* SWINE RATIONS.

SLOUGH is a (marshy) SWALE, a place of deep mud, or a whole full of mire.

SLUGS are mollusks. They hide under rubbish in day time and feed at night on leaves.

The common *garden slugs* in particular are pests of SUBCLOVER and other forage plants; they are most destructive to the young seedlings of fall plantings. These slugs leave a shiny trail wherever they crawl and lay jelly-like eggs in the soil, under stones, etc.

Control: The slugs can be controlled by poison bait pellets containing 3 percent METALDEHYDE and 5 percent CALCIUM ARSENATE applied at the rate of 5 lb. per acre.

Reference: R.3.

SLURRY TREATMENT OF SEEDS.
See SEED TREATMENT.

SMALL GRAINS include such CEREALS as RYE, OATS, WHEATS, BARLEYS, grain SORGHUMS, etc.—*See also* FOXTAIL MILLET; RED CLOVER; MUNGBEAN; FIELD PEA; CORN; LESPEDEZAS; RYEGRASSES; VETCHES; LUPINES; ALFALFAS; PASTURES; HAY; STRAW; DUST STORMS; EUROPEAN CORN BORER; CHINCH BUG; ARMY CUTWORM; STINK BUGS; BILLBUGS; ARMYWORMS; BACTERIAL LEAF BLIGHT.

SMALL HOP CLOVER (*Trifolium dubium*), also called *suckling clover*, is one of the TRUE CLOVERS. It is a winter annual widely scattered throughout the southern states and Pacific Northwest. This species tolerates unfavorable soil and climatic conditions, and like LARGE HOP CLOVER, is very valuable for early spring pasturage.—*See also* INOCULATION; LEGUMES.

SMALL LIMA BEAN. *See* LIMA BEANS.

SMALL RABBITBRUSH (*Chrysothamnus stenophyllus*) is a low, compact shrub 8 to 15 in. high. When the plant begins to grow, new shoots are produced from the woody base of the old stems. The new twigs are green but as they mature they become shiny white. The stems are fine and the leaves very narrow, $\frac{1}{16}$ in. wide, about 1 in. long, and often somewhat twisted. In August, the plant produces a mass of tiny yellow flowers at the ends of the branches.

Small rabbitbrush grows most abundantly on loose sandy soils. The presence of small rabbitbrush in noticeable quantities on WINTERFAT and BLACK SAGEBRUSH range probably indicates that these areas are in deteriorated condition.

Ordinarily, small rabbitbrush is not very palatable to sheep and very little of its herbage is utilized except on ranges that are heavily grazed.

Reference: H.41.

See also RANGE PLANTS.

SMALL-SEEDED ALFALFA DODDER (*Cuscata planiflora*), like the related LARGE-SEEDED ALFALFA DODDER, is a parasitic plant. It belongs to the DODDERS. This WEED is particularly troublesome in ALFALFA in the West but is not found in the East.

Reference: H.19.

SMALL-SEEDED VETCHES are varieties of the HORSEBEAN.

SMALL SPANISH is a Spanish-type PEANUT.

SMOOTH BROME (*Bromus inermis*), or *smooth brome grass*, a long-lived, perennial sod grass with strong creeping rootstocks, is grown widely throughout the country. It is adapted especially to regions of moderate rainfall and low to moderate summer temperatures.

Plants reach a height of 3 to 4 ft. and produce an abundance of basal and stem leaves. Leaf blades vary from 8 to 12 in. in length and from $\frac{1}{4}$ to $\frac{1}{2}$ in. in width. The leaf sheaths are smooth and closed, forming a tube. The root system is extensive, with strong rootstocks; the interlaced roots and rootstocks form a coarse but dense sod, which resists grazing and trampling and protects the soil against wind and water erosion. Smooth brome makes its best growth on moist, well-drained clay loam soils of relatively high fertility.

Smooth brome. (S.32.)

As a pasture or hay grass, smooth brome scarcely has an equal in the area of its best adaption. Growth begins early in the spring and continues through the summer if enough moisture is available. Smooth brome is well liked by all classes of livestock either as hay or as pasture. It should not be grazed closely, or the stand will soon thin out.

Satisfactory stands have been obtained in the Great Plains. Smooth brome fits well into a grass-LEGUME seeding, and many acres are now in brome-ALFALFA or brome-CLOVER mixture that formerly were devoted to legumes alone.

Such grass-legume seedings have special conservation and economic values. They give greater forage yield and protection to the soil than either the brome or the alfalfa would give if seeded alone, and the danger of BLOAT to grazing animals is less. The presence of a legume in the mixture also prolongs the useful life of the stand, presumably because it increases the amount of nitrogen available in the soil and thereby keeps the grass stand from becoming sod-bound.

Smooth brome is also recommended for inclusion in silage mixtures.

Varieties: Two distinct types of smooth brome, differing in growth behavior, are generally recognized:

The *southern* type is best adapted to the Corn Belt states and to the parts of the central Great Plains that have protracted dry periods and high summer temperatures. Improved varieties include *Achenbach*, *Lincoln*, *Elsberry*, and *Fisher* smooth bromes.

The *northern* type has been found well adapted to Canada and the northern Great Plains, where long periods of hot weather seldom occur. It includes the varieties *Kidder*, *Fargo*, *Martin*, and *Manchar*.

References: H.1; G.4; A.1; A.3.

See also BROME GRASSES; GRASSES; PASTURE PLANTS.

SMOOTH-LEAVED PERUVIAN is a strain of the nonhardy *Peruvian alfalfa.*— *See also* ALFALFAS.

SMOOTH-LEAVED TREFOIL is a type of BIG TREFOIL.

SMOOTH VETCH.

See HAIRY VETCH (variety).

SMUT. *See* SMUTS.

"SMUT BALLS" are smutted kernels.— *See also* COMMON BUNT.

SMUT GALL are galls (KNOTS) developed on overground parts of plants.—*See also* CORN SMUT; LOOSE KERNEL SMUT OF SORGHUM; COVERED KERNEL SMUT.

SMUTS are plant parasites. Nearly 140 different species of smut FUNGI attack approximately 300 species of GRASSES and many CEREALS.

The dusty black or brown smut masses are made up mostly of millions of tiny cells or groups of cells—spores that serve for reproduction and dissemination.

Some smuts destroy the flowering structure. Others are restricted to certain parts of it. Some are confined almost exclusively to the stems of the plants. Others produce galls or tumorlike structures in various parts of their host plants.

The smut fungi exhibit a remarkable degree of specialization not only to certain species of plants but also to certain varieties or strains within those host species. Likewise, there are often strains or races of the smut fungi to contend with. Some smuts persist for years in the soil or in old manure piles, especially if humidity is low; they remain capable, however, to attack when their host plants become available again.

Many smuts are seed-borne. Smut spores, carried by the wind in millions, become lodged in or on the developing seeds of healthy plants. When the seeds germinate, the smut spores germinate also and infect the young seedlings, which then develop into smutted plants.

Some of the smut fungi are not limited to seedling infection but apparently can infect any succulent or rapidly growing part of their host plants and make them smutty. The time between infection and the appearance of smut varies rather widely.

Some of the GRASS SMUTS are destructive to cereal crops also.

Control: SEED TREATMENT is effective only against those smuts in which infection occurs at the time of seed germination.

References: F.4; M.26.

See also STRIPE SMUTS; HEAD SMUT; STEMS SMUT; COVERED KERNEL SMUT; COVERED SMUT OF OATS; COVERED SMUT OF BARLEY; LOOSE KERNEL SMUT; LOOSE SMUTS OF BARLEY; LOOSE SMUT OF WHEAT; BLACK LOOSE SMUT; KERNEL SMUT OF RICE; LEAF SMUT OF RICE; OAT SMUTS; CORN SMUT; DWARF BUNT; COMMON BUNT; FLAG SMUT OF WHEAT; RED OAT; WHEATS.

SNAKEROOT.
See WHITE SNAKEROOT; POISONOUS PLANTS.

SNAKEWEED. *See* BROOM SNAKEWEED.

SNEEZEWEED. *Orange Sneezeweed* = WESTERN SNEEZEWEED.—*See also* POISONOUS PLANTS.

SNOW = *Fultz. See* COMMON WHEAT.

SOAKING FEED is advisable if grinding facilities are not available or when grain with hard or small kernels is to be fed to livestock; e.g., old CORN that hardened in summer (and therefore may cause sore mouths when fed untreated) should be ground or soaked. If soaked feed is left standing for a period of time, it may become stale.

SOAPS are compounds of FATTY ACIDS and ALKALIES (or metals). They are sometimes formed in the intestine when fatty acids (from FATS) are neutralized before being absorbed into the blood stream.

Laundry soap, which is a soft *potash soap*, is used in insecticidal sprays.—*See also* ABSORPTION; NICOTINE; LEAD ARSENATE.

"SOCIAL BEES." *See* BEES.

SOD is (1) a plowed meadow or pasture, or (2) the top few inches of soil filled with grass roots.—*See also* RANGE MANAGEMENT; BUNCHGRASS; GRAZING.

SOD CROPS spread laterally, often by means of RHIZOMES or STOLONS, to form sod. Many of these plants can easily be established by transplanting pieces of sod. Among the sod crops are KENTUCKY BLUEGRASS, FESCUES, REDTOP, LESPEDEZAS.

SODIUM ARSANILATE is the sodium salt of ARSANILIC ACID; the anhydrous form of this compound contains 31.3 percent ARSENIC (the hydrous form 23 percent). It occurs as a white, crystalline powder that is water-soluble and poisonous. Sodium arsanilate is used in poultry and hog feeds at the rate of 0.01 percent or less for the control of *infectious diarrhea* or to stimulate GROWTH.—*See also* ARSONIC ACIDS.

SODIUM ARSENITE, commercially available in form of *sodium meta-arsenite*, is an extremely poisonous, grayish-white powder that contains approximately 82 percent ARSENIC TRIOXIDE; it is water-soluble. An alkaline solution is marketed also and, like the powder, used as an INSECTICIDE.

For killing MESQUITE, a shallow basin is dug that encircles the base of the plant; a diluted sodium arsenite solution is poured into the earthen basin. Since this arsenical is a powerful soil sterilizer, all plant growth in the poisoned spot is prevented for some time (sometimes for years). The solution is most effective when applied in the winter months.

> NOTE: Concentrated sodium arsenite solution (with an arsenic trioxide content of 4 lb. per gal.) should be diluted at the rate of 1 part with 6 parts water; use 1 qt. of this diluted solution for mesquites up to 12 in. basal stem diameter; larger quantities for stems exceeding 12 in.

Caution: When sodium arsenite solution penetrates the clothes, the skin should be washed immediately with soap and water. Keep all equipment and materials out of reach of children and animals. Wood from poisoned trees must not be used as fuel since the smoke may be poisonous.

SODIUM BENZOATE is a white powder with a sweetish, astringent taste. It is very soluble in water and widely used as PRESERVATIVE.

To prevent mold formation, sodium benzoate may be added to feed ingredients at a concentration not exceeding 0.1 percent; the label must carry a notation to that effect.—*See also* MOLD INHIBITORS.

SODIUM BICARBONATE, or *baking soda*, is a white, water-soluble powder. It is sometimes added to feedstuffs as an *antacid* i.e., as an agent that counteracts acidity.—*See also* MINERAL FEEDS.

SODIUM BISULFITE.

> *See* SODIUM META-BISULFITE.

SODIUM CHLORATE forms colorless crystals that have a saline, cooling taste and are soluble in water. It is an oxidizing agent and sometimes used as weed killer. Care must be taken when handling the powder since it may become a fire hazard.

A solution containing at least 2.5 percent sodium chlorate, when sprayed on LARK-SPURS during their active growing period, is effective in eradicating this poisonous plant. A sodium chlorate spray is also recommended for the removal of scattered patches of QUACKGRASS and CRAB GRASS.

For the control of FIELD BINDWEED, sodium chlorate may be applied as spray or dry, preferably in the fall. An original application of 3 to 4 lb. per 1 sq. rod, followed by re-treatments in subsequent years, is an economical control method.

SODIUM CHLORIDE, known as *common salt* or *"salt"*—the chemical compound *NaCl*—forms colorless crystals or a white crystalline powder soluble in water.

Blood contains sodium chloride. If too large quantities of salt are mixed in slop and fed to hungry pigs, *salt poisoning* may develop. As much as 10 percent salt, however, has been fed in protein supplements without serious trouble.

Common salt applied at the rate of 20 lb. per 1 sq. ft. of crown area will kill the EUROPEAN BARBERRY; sprinkling the salt on the ground around the canes will kill the native species of the BARBERRY BUSHES. It is also used for the eradication of DODDERS.—*See also* MINERAL FEEDS; TRACE MINERAL SALT; GRAZING.

SODIUM FLUORIDE, a white, poisonous powder, which can easily be confused with sugar or salt, is soluble in 25 parts water.

> NOTE: The technical grade of sodium fluoride is usually tinted.

It is used as drug; mixed with dry feed at the rate of 1 percent, and fed in this mixture for one day, it is effective in killing the *large intestinal roundworms* of swine.

SODIUM FLUOSILICATE, or *sodium silicofluoride*, is a violent poison. The white, granular powder is slightly water-soluble. It is an INSECTICIDE sometimes used in POISONED BRAN MASH against ARMYWORMS and CUTWORMS. Sodium fluosilicate is very satisfactory for the control of the COWPEA CURCULIO. As a dust, it is applied in the evening at the rate of 8 lb. per acre when the first blossoms appear. The application should be repeated at weekly intervals throughout the fruiting season. If used as a spray, 6 lb. sodium fluosilicate is needed per 100 gal. water.

SODIUM HYDROXIDE is an ALKALI available in form of white flakes, pellets, or sticks. It is very soluble in water and must be handled with great care.—*See also* CAUSTIC.

SODIUM HYPOSULFITE
= SODIUM THIOSULFATE.

SODIUM META-BISULFITE forms water-soluble, white crystals and is used as a SILAGE CONDITIONER. The *sodium bisulfite* of commerce usually consists chiefly or almost entirely of sodium meta-bisulfite.

SODIUM NITRATE, or *Chile saltpeter,* available in form of colorless crystals or white, granular powder, has a bitter-saline taste. It is used as nitrogen fertilizer and for the eradication of DODDERS.

SODIUM NITRITE occurs in white or yellowish granules or powder. It is very soluble in water; in air, it slowly oxidizes to SODIUM NITRATE.

Sodium nitrite is very useful as an antidote to PRUSSIC ACID POISONING, especially if it is injected—in combination with SODIUM THIOSULFATE—into the vein immediately after symptoms of poisoning appear in the animal. The recommended dose for cattle is 2 to 3 gm. sodium nitrite in water, followed by 4 to 6 gm. sodium thiosulfate in water; for sheep, the dose is up to 1 gm. sodium nitrite and 2 to 3 gm. sodium thiosulfate. This treatment should be given by a veterinarian who may supplement it by injections of ATROPINE or GLUCOSE, inhalation of AMMONIA, etc.
Reference: M.1.

SODIUM PROPIONATE, which forms transparent, water-soluble crystals or granules, is used as a drug and a MOLD INHIBITOR.

SODIUM SILICOFLUORIDE
= SODIUM FLUOSILICATE.

SODIUM SULFATE is available in two forms, both of which are white, water-soluble powders or granules used as feed ingredients:

1. *Anhydrous,* also called *salt cake.*
2. *Hydrous* (56 percent water content), commonly called *Glauber's salt;* it is also

used as purgative for ruminants.—*See also* MINERAL FEEDS.

SODIUM TCA, or *sodium trichloroacetate,* is the sodium salt of TCA, or trichloroacetic acid. It is a noncorrosive, water-soluble powder used as a weed killer. Most commercial preparations contain 90 percent sodium trichloroacetate (equivalent to 80 percent trichloroacetic acid). For best results, weeds must be uniformly covered by a spray of proper strength, applied at a time of not too excessive rainfall (or irrigation) and when the weed is not too tall.

A spray containing 150 to 175 lb. sodium TCA (90 percent strength) in 125 gal. water gives adequate coverage for an acre and kills 90 to 95 percent of established stands of JOHNSONGRASS.

Caution: Sodium TCA is irritating to the skin and eyes. Any sodium TCA on the skin should be washed off as soon as possible with plenty of water. Contaminated clothes must be thoroughly washed.

Care should be taken in spraying near crop plants and ornamentals because sodium TCA may severely injure them.

Soil treated with sodium TCA is temporarily sterilized, but the effect generally disappears in thirty to sixty days depending on rainfall, irrigation, and soil texture.
Reference: D.2.

SODIUM THIOSULFATE, or *sodium hyposulfite,* forms colorless crystals or white granules. It contains 36 percent water and is water-soluble.

Sodium thiosulfate has mild purgative as well as antibloat action. It is also an antidote used in the treatment of animals affected with PRUSSIC ACID POISONING; this drug is particularly valuable if administered intravenously following an injection of SODIUM NITRITE.—*See also* BLOAT.

SODIUM TRICHLOROACETATE
= SODIUM TCA.

SOFT BONES. See PHOSPHORUS.
SOFT CORN. See CORN.
SOFT PHOSPHATE WITH COLLOIDAL CLAY is sometimes called "*colloidal phosphate.*" It is recognized as a PHOSPHORUS source.—*See also* MINERAL FEEDS.

SOFT RED WINTER WHEAT. *See* WHEATS; COMMON WHEAT; HESSIAN FLY.

SOIL, the upper layer of the solid land, consists of particles of decomposed rocks mixed with decaying vegetable and animal matter (humus).—*See also* pH.

SOIL CONSERVATION. *See* DROUGHT.

SOIL EROSION. *See* EROSION.

SOIL FUMIGATION gives good control of the STING NEMATODE. About three weeks before planting, a *fumigant*, such as DOWFUME W-40, should be applied (the latter at a rate of 7 gal. to the acre). After fumigation the soil should be left undisturbed until planting time. Care must be taken to plant the seed directly above the area in which the fumigant was applied.

SOILING CROPS are cut and fed green to livestock; e.g., SORGHUMS, RED CLOVER, or CORN.

If land produces an abundance of forage, a large number of animals must be used to harvest it. This often results in forage lost by trampling or MANURE damage. Soiling prevents these losses, thereby increasing the number of animals that can be fed from a given area. It is said to reduce danger from BLOAT (but for this purpose it is not 100 percent effective). The chief disadvantage is the increased requirement of labor and machinery to harvest and feed the forage, and to haul the manure from the lots.

SOIL ROT, also called *"root rot"* or *dry rot*, is a SEEDLING DISEASE of PEANUTS. It is caused by the FUNGUS *Rhizoctonia solani* which attacks the stem at or just below the soil line, in localized spots in a field. The stem usually decays completely, and the plant dies. The disease rarely kills a plant more than six weeks old.

Reference: B.10.

SOIL STERILANTS are agents that sterilize the soil by killing living matter in it; e.g., SODIUM CHLORATE.—*See also* DOWFUME W-40; FUMIGANTS; FIELD BINDWEED.

SOJA BEAN = SOYBEAN.

SOJA VIRUS 1 is the cause of SOYBEAN MOSAIC.

SOLANINE, an alkaloid, is a toxic substance occurring in *Solanum* spp.; e.g., in potato sprouts.

SOLANUM. *S. nigrum* = BLACK NIGHTSHADE; *S. tuberosum* = POTATO.

SOLITARY BEES include most of the *wild bees.*—*See also* BEES.

SOLVENT EXTRACTED COTTONSEED FLAKES. *See* COTTONSEED.

SOLVENT EXTRACTED COTTONSEED MEAL. *See* COTTONSEED.

SOLVENT EXTRACTED LINSEED OIL PRODUCTS. *See* LINSEED.

SOLVENT EXTRACTED PEANUT FEED. *See* PEANUT.

SOLVENT EXTRACTED PEANUT MEAL. *See* PEANUT.

SOLVENT EXTRACTED RICE BRAN. *See* RICE.

SOLVENT EXTRACTED SOYBEAN FEED. *See* SOYBEAN.

SOLVENT EXTRACTED SOYBEAN FLAKES. *See* SOYBEAN.

SOLVENT EXTRACTED SOYBEAN OIL MEAL. *See* SOYBEAN.

SONCHUS. *S. arvensis* = SOW THISTLE.

SOONER. *Sooner milo, Double Dwarf White Sooner,* and *Double Dwarf Yellow Sooner* are grain SORGHUMS.

SOOTY BLOTCH (*Cymadothea trifolii*) is a common FUNGUS LEAF DISEASE; ALSIKE CLOVER may practically be defoliated by the disease which also is quite common on WHITE CLOVER. It does little damage to RED CLOVER.

Most of the lesions develop on the underside of the leaflets, in some instances being so thick that the surface appears black.

References: E.5; H.8.

See also LEAF DISEASES.

SOOTY STRIPE, which is caused by *Ramulispora sorghi*, a FUNGUS, occurs on grain and forage SORGHUMS and JOHNSONGRASS. It has been found in Alabama, Arkansas, Florida, Georgia, Illinois, Louisiana, Mississippi, North Carolina, Oklahoma, and Texas.

The disease attacks the leaves and sheaths. On the leaves, the spots begin as small, oblong, reddish-purple areas that develop into conspicuous, elongated lesions with purplish borders and straw-colored dead centers. These dead centers usually are covered with black resting bodies of the fungus, which may impart a sooty appearance to the lesions; hence the name of this LEAF DISEASE, sooty stripe.

In some varieties, particularly Leoti sorgo, and Tift and Sweet Sudangrasses, the borders around the leaf stripes are tan instead of purple.

Control: No distinctly resistant sorghum varieties are known.

Reference: L.1.

See also FUNGUS LEAF DISEASES.

SORBIC ACID forms white crystals that are only slightly soluble in water. It is used as MOLD INHIBITOR.

SORGHASTRUM. *S. mutans*
      = INDIAN GRASS.

SORGHUM. *See* SORGHUMS.

SORGHUM BLIGHT.
    See BACTERIAL LEAF DISEASES.

SORGHUM HAY. *See* SORGHUMS.

SORGHUM-LEAF DISCOLORATION.
See NONPARASITIC SORGHUM-LEAF DISCOLORATION.

SORGHUM MIDGE (*Contarinia sorghicola*), a fly, costs American farmers many millions of dollars every year in the damage that it inflicts on the grain SORGHUMS. Losses are also caused in the seed crops of all other sorghums, including SUDANGRASS; the damage to these plants is only slight.

The larvae of this fly damage the crops by consuming the plant juices of the developing seeds. The affected grain or seed becomes shrunken and sometimes discolored, so that the infested SPIKELETS resemble sterile ones. The injuries popularly are known as *blast* or *"blight."*

If the heads of any of the sorghum plants are examined while in bloom, many small, reddish, gnatlike flies, or midges, may be found crawling on the spikelets. They are the adult females, which are laying their eggs within the spikelets. The eggs hatch in two days, resulting in larvae, or maggots, that develop into pupae in from seven to eleven days; the adults, or midges, emerge from the pupal stage at the end of three days. At that time the male midges are swarming about the same heads waiting for the emerging females. Mating occurs soon after emergence, and the females fly to the nearest suitable sorghum heads to lay their eggs.

The males live only a few hours; the females live more than one day during the summer, but they may live a little longer in cooler weather. Each female lays approx. 100 tiny eggs.

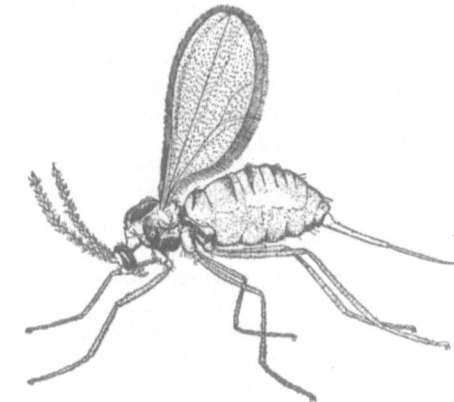

Sorghum midge. Adult female with ovipositor extended. (G.1.)

Successive life cycles occur throughout the season from the first emergence of midges (that hibernate as larvae within cocoons) in the spring until the host plants are killed by freezing temperatures in the fall. JOHNSONGRASS, when allowed to head, provides an excellent place for hibernation. It blooms very early, thus permitting the midgets to increase in number before the sorghum fields come into bloom.

Adult midges may fly a considerable distance from an infested source, especially with the help of the wind, which is an important factor in the dispersion over nearby fields.

Control: Because all stages of the midge, with the exception of the adult, are passed within the seed husks, there are at present no practical means of controlling it by the use of insecticides. Considerable loss can be prevented by putting into effect the following recommendations:

Use only pure seed of as uniformly blooming a strain as it is possible to obtain. Plant at the time of the season best suited for the variety selected. Prevent Johnsongrass from producing heads in or near the sorghum field previous to the blooming of the crop. Cultivate or burn over fields of

Johnsongrass early in the spring to destroy hibernating midges before they can emerge. Heads of sorghum that bloom much before the main crop should be destroyed. Where sorghum grain is threshed, plow under or destroy before the spring emergence of midges all refuse not eaten by livestock.

No variety of sorghum has yet been found to show any great resistance to the attacks of the sorghum midge.

References: G.1; V.1.

SORGHUM PASTURE. *See* SORGHUMS.

SORGHUM POISONING

= PRUSSIC ACID POISONING.

SORGHUM RUST, caused by the FUNGUS *Puccinia purpurea*, attacks most varieties of SORGHUM. It occurs frequently in the humid Gulf Coast region and occasionally during wet seasons in states as far north as Kansas and Indiana. Usually it does not become evident until the seed is well developed, so that it causes relatively slight losses to the grain sorghum crop. Abundant rust, however, causes the leaves to dry and break off so that the forage value of the crop may be lowered.

The rust appears on both the upper and lower surface of the leaves as blisters, or pustules, covered with a brownish coating that eventually breaks open and allows the chestnut-brown rust spores to escape. Before the blisters appear, small purple, red, or tan spots may be seen at the points where the infestation is developing. As the pustules develop, the colored regions around them become larger and considerable areas of the leaves may be destroyed. It is probable that JOHNSONGRASS serves as the principal overwintering host for the fungus.

Control: Growing resistant varieties is the only feasible method for controlling this LEAF DISEASE. Milos, certain hybrid strains involving milo, shallu, and a Leoti-Atlas derivative, appear to be highly resistant. Kafirs and sorgos tend to be moderately susceptible, and feteritas highly susceptible. Broomcorn and Sudangrass also are susceptible.

Reference: L.1.

See also FUNGUS LEAF DISEASES.

SORGHUMS (*Sorghum* spp.) consist of *S. halepense* = JOHNSONGRASS, *S. su-* *danense* = SUDANGRASS, and *S. vulgare*, or (common) sorghum.

Of *S. vulgare* (which will be discussed here) many VARIETIES, SELECTIONS, and STRAINS exist. The differences between most of the varieties are indistinct and unstable because of intercrossing. Therefore, no definite varietal Latin names are assigned to most of them.

The sorghums generally are grown like CORN; during the early stages of growth, corn and sorghum look somewhat alike, but the sorghum usually has finer stems and leaves. The leaves become broad and the coarse stems reach from 2 to 15 ft. in height, depending on variety and growing conditions.

Most of the sorghums are grown in the South Central States. At least half of the acreage of these GRASSES is harvested for fodder and silage. The sorghums are especially valuable for winter feeding in the Great Plains where other forage crops are not well adapted. They are used extensively to feed all species of farm animals. Though somewhat laxative, the sorghums are nutritious. For cured *hay* it is best to cut the plants by the time 10 to 20 percent of them have developed heads.

If the sorghums are grown as a *soiling* crop (i.e., cut and fed green), care must be taken to avoid BLOATING and PRUSSIC ACID POISONING. Because cattle, sheep, and goats are particularly susceptible to "*sorghum poisoning*," the use as *pasture*— as long as the crop is not nearly mature— must be discouraged.

The sorghums do best where the temperatures are uniformly high during the growing season. They are of most value in regions of uncertain rainfall, because they resist wilting and remain practically dormant during periods of drought, resuming growth as soon as there is sufficient rain to wet the soil. The plants have many secondary feeding roots and small leaf areas. The combination of an efficient moisture-absorbing system with a reduced evaporating surface accounts in part for their great ability to withstand drought.

The sorghums thrive on a variety of soils. Deep fertile sandy loams are best, but fair crops can be produced on heavy clays if they are well drained. In humid areas or on irrigated land, LEGUMES, like COWPEAS or CLOVER, should be substituted in the cropping system and the productive power of the soil restored by applications of commercial fertilizers or barnyard manure.

The sorghums are more tolerant of alkali in the soil than most crops. Because cotton, corn, oats, wheat, and other crops yield approximately 15 percent less on fields that have produced a crop of sorghum the previous year than on fields preceded by most other crops, it is said that sorghum is "hard on the land." The sorghum species, however, have no permanent injurious effect on the soil. It is best to follow sorghum with a spring-sown crop or summer fallow and not with a fall-sown grain.

In the Great Plains, where the limiting factor is moisture, the use of fertilizers does not pay. In the southeastern states any fertilizer known to be beneficial to corn may be expected to be equally beneficial to sorghum; where forage is the chief consideration, it is desirable to have the nitrogen content of the fertilizer rather high. Highest yield of sorghum silage may be expected when a complete fertilizer is used; phosphorus is most effective in increasing yields, followed by nitrogen, potassium, and lime.

> NOTE: If the levels of nitrate are too high, i.e., if too much nitrate fertilizer is used, *nitrate poisoning* may result; this condition is better known as OAT HAY POISONING.

Dangers: Losses in livestock from prussic acid poisoning are greatest from pasturing sorghum (1) in second growth following harvesting of the crop for silage, forage, or sirup, and (2) when the plants are stunted, especially during periods of drought or following light frost. Frosted sorghum can be salvaged if cut and stored immediately following frost. Cured into dried forage or made into silage, frosted forage is free from danger as a feed. Practically all of the PRUSSIC ACID disappears as the fodder is being cured or made into silage. (Varieties having a low, non-toxic level of prussic acid are being developed to overcome the hazard.)

Sorghum. Fruit cluster with leaf. (W.35.)

The sorghums are injured by SEED ROT, SEEDLING BLIGHT, and WHEAT SCAB, caused by fungi; BACTERIAL LEAF DISEASES, namely BACTERIAL STRIPE OF SORGHUM, BACTERIAL STREAK, and BACTERIAL SPOT; the fungus diseases of leaves, i.e., ZONATE LEAF SPOT, ROUGH SPOT, ANTHRACNOSE OF GRASS, LEAF BLIGHT, NORTHERN CORN LEAF BLIGHT, GRAY LEAF SPOT, TARGET SPOT, SOOTY STRIPE, and SORGHUM RUST; smuts, i.e., COVERED KERNEL SMUT,

LOOSE KERNEL SMUT OF SORGHUM, HEAD SMUT; and root and stalk diseases, namely WEAK NECK and the fungi-caused rots (PERICONIA ROOT ROT, CHARCOAL ROT, FUSARIUM STALK ROT, COLLETOTRICHUM STALK ROT, RHIZOCTONIA STALK ROT).

Among the many insect pests attacking the grain sorghums are the SORGHUM MIDGE, WEBWORMS, SUGARCANE BORER, CORN EARWORM, ARMYWORM, GREENBUG, CORN LEAF APHID, CHINCH BUG, LEAF-HOPPER, WHITE GRUB, GARDEN WEBWORM, and STINK BUGS.

Classification: Confusion in names, the overlapping of groups, and the hybrid derivatives which are constantly arising make exact classification of the sorghum varieties difficult. For practical purposes, classification of the sorghums on the basis of uses is preferable. The *forage sorghums*, the *grain sorghums*, and the *grass sorghums* are valuable for forage, silage, hay, or pasture; *broomcorn*, which forms the fourth group, is used chiefly for industrial purposes and occasionally as silage crop.

1. (A) FORAGE SORGHUMS

Sorghum grown for forage is often referred to as *sorgo*. It is also known as *sweet sorghum, saccharine sorghum,* or *cane;* the word "cane" properly belongs to SUGARCANE (an entirely different group of plants), and the use of this term for the forage sorghums leads to confusion, especially in regions where sugarcane is commercially grown.

The sorgos have sweet, juicy stems and usually are 6 to 14 ft. tall. The seeds in general are smaller than those of the grain sorghums and are not especially desirable for feed because their taste is bitter, and because a relatively large proportion passes through the animal undigested; only white-seeded varieties (such as Atlas, White African millet, and Tricker) are free from bitterness. The sorgos usually yield 25 to 50 percent more forage than the grain sorghums, except under very dry conditions.

The principal areas where the sorghums are harvested for forage are the southern Great Plains, western Texas, Oklahoma, Kansas, eastern New Mexico, and south-eastern Colorado. Except at high elevations, they can be grown as a secondary forage crop in all parts of the United States, almost up to the northern boundary.

The sorgos are rapidly replacing corn for *silage* in states where drought frequently injures the corn crop. In good seasons the better varieties of the forage sorghums frequently yield one-third to two-thirds more silage than corn, whereas under conditions of severe drought or grasshopper injury, when corn is nearly a failure, they may still produce 3 to 8 tons of silage per acre.

At least 30 distinct varieties of sorgo are grown in the United States. Among the most important are the following:

Amber sorgos

The Amber sorgos mature quite early. They are satisfactory for silage and forage and produce a better quality of cured forage than most other sorgo varieties, but rarely equal them in tonnage.

There are two main groups—Black and Red Ambers.

(a) The Black Amber group includes the following varieties:

Black Amber (formerly known as *Early Amber*) is the leading sorgo grown north of the latitude of northern Kansas. It is grown extensively in Nebraska, Colorado, South Dakota, Minnesota, and north-western Kansas, and to a limited extent in many other states. This important Amber variety is early in maturity and of medium height (usually 6 to 9 ft.) and has slender stalks with rather few leaves. The heads are loose, open, and nodding. The seeds are light brown and, after threshing, are nearly all enclosed in the black, shiny chaff. Black Amber is grown chiefly for forage, but is also the leading type for sirup. It is susceptible to HEAD SMUT.

Dakota Amber (*sorgo*) is shorter and earlier than other Black Ambers and is grown under the cool, dry, short-season conditions commonly prevailing in the Dakotas, Wyoming, and Montana.

Dakota Amber 39-30-S, also called *39-30-S*, is important because of its low PRUSSIC ACID content.

Minnesota Amber (sorgo) is of local importance.

Waconia Amber (sorgo) is superior to the Minnesota Amber from which it was developed.

(b) The Red Amber group consists of only one distinct variety:

Red Amber (sorgo) is similar to regular Black Amber except that it has dark-red instead of black chaff on the seed and is slightly later and heavier and slightly more compact. It is adapted to sections just south of where Black Amber is a leading variety and is grown most commonly in southern Nebraska and eastern Colorado. Red Amber yields more and better forage than Black Amber where both varieties reach maturity, but south of northern Kansas it is inferior to many other sorgo varieties that require a longer season. Red Amber is highly susceptible to LOOSE KERNEL SMUT, HEAD SMUT, and susceptible to BACTERIAL STRIPE OF SORGHUM.

Atlas (sorgo), a white-seeded sorgo, was originated by selection from a cross between Sourless sorgo and Blackhull kafir. It is now very popular, especially in the eastern half of Kansas and Nebraska and in western Missouri; it is gradually increasing in acreage in Oklahoma, Texas, Iowa, Kentucky, and other states. Atlas is taller, later growing, and yields a heavier forage than either parent from which it was produced. It grows 6 to 10 ft. high and usually does not reach maturity where the frost-free period is less than 170 days, but matures and produces high yields of forage in most sections south and east of east-central Nebraska. It has strong stalks that lodge less easily than those of many of the tall varieties of sorgo. The seeds are white, almost indistinguishable from those of kafir (a grain sorghum) and, unlike most other sorgos, can be used readily as a grain feed.

The plant exhibits a remarkable tenacity; it can remain green but dormant through long periods of drought and resume growing when rains occur.

Atlas has sweet, juicy stalks, but is rarely used for sirup making; it is excellent for fodder and silage production and also

for cured forage. It produces 20 to 30 percent more tonnage of both stalk and grain than corn, even on the very best of corn land.

Atlas is highly ANTHRACNOSE-resistant and somewhat resistant to COLLETOTRICHUM STALK ROT and LEAF BLIGHT.

Axtell is a mid-early, white-seeded sorgo variety. It requires from 106 to 116 days to make a crop. The grain yield of Axtell tends to be higher than for Atlas, but the forage yield is less. Otherwise the two varieties have much in common. The stalks are 6 to 8 ft. tall, juicy, sweet, and leafy. The variety is well adapted to the eastern half of Kansas and where there is danger that Atlas will not mature before frost.

Collier (sorgo) is one of the sorgos that, until recently, found only limited interest. Because of high sugar content, some of the Collier strains, however, have attracted attention as potential sources of food (sugar) and feeds.

Kansas Collier 704-D, a promising selection, has the highest sugar content of all Collier sorgos. The variety is grown to a limited extent near dehydrating plants as a source of pellets for prepared feeds. It is best adapted to a fertile soil with an abundance of rainfall and is capable of high tonnage. Its height ranges from $5\frac{1}{2}$ to $7\frac{1}{2}$ ft., depending on the season. The heads are slightly drooping, the seeds small and reddish-brown.

Ellis is an important sorgo selected from a cross between Leoti and Atlas. It has an open head with slightly drooping seed branches and medium-sized, palatable, waxy, white seed. The seed of Ellis will average about 15 percent more niacin than do Leoti and Atlas. Ellis ranges in height from 5 to $6\frac{1}{4}$ ft. and requires from 100 to 110 days to mature. The stems are sweet and juicy, with bright green leaves resistant to SORGHUM RUST and to the several BACTERIAL LEAF DISEASES that, in seasons of high humidity, often redden and destroy large areas of the leaf surface in some of the common forage types. Ellis, however, is susceptible to HEAD SMUT. Ellis shows good feeding qualities as silage and cured forage.

Fremont is an early variety of sorgo adapted to northeastern Colorado where it is grown for forage.

Gooseneck (sorgo), or *Texas-seeded Ribbon cane*, is the latest and largest sorgo grown in the United States. It is adapted only to the southern states, where long growing seasons and ample moisture prevail; there the variety produces high yields of forage and sirup. Gooseneck is distinguished by thick, compact heads (that bend over and hang down), black chaff, and rather large brown seeds. It is somewhat resistant to LEAF BLIGHT.

Honey (sorgo) is known by many names, including *Honey Drip*, *Japanese Ribbon cane*, *Japanese-seeded Ribbon cane*, and *Sprangle Top*. It is among the most popular sorgo varieties south of Kansas. Honey is late in maturity, has very tall, thick stalks with large, open, brushy heads and bright red chaffs that remain on the seeds after threshing. This sorgo is grown both for forage and sirup; it is one of the most productive varieties where the season is long enough and ample moisture is available to bring the crop to maturity. It is susceptible to COLLETOTRICHUM STALK ROT.

Leoti (sorgo), or *Leoti Red*, was found growing under the name of *"Red Kentucky Sugar-Cane"* on a farm in Leoti, Kansas. The variety had considerable distribution in western Kansas; currently it is grown in Nebraska. Leoti is attractive, juicy, and one of the sweetest forage sorgos; it grows from 6 to 7 ft. high and matures in about 110 days. It has an open head with lax upper branches which droop. The reddish-brown seed is of medium size, more than one-half covered by red glumes which are not easily separated from the grain in threshing. The rind of the stalk is rather hard which distracts to some extent from the feeding value of Leoti; also, it looses its leaves during high winds. Leoti is resistant to some of the BACTERIAL LEAF DISEASES and, therefore, has been used extensively in sorghum crosses; it is one of the parents of Ellis, Cody, Sweet Sudangrass, and Tift Sudangrass. Leoti is susceptible to SOOTY STRIPE, HEAD SMUT, TARGET SPOT, and COLLETOTRICHUM STALK ROT.

Norkan is the result of a cross between Atlas and Early Sumac. It possesses the earliness and tonnage of Early Sumac and the white but somewhat smaller grain of Atlas. The plants have 10 to 12 mid-wide leaves, are mid-early, mid-tall, the stems are slender, juicy, and a little less sweet than Early Sumac.

In the drier regions Norkan often produces an abundance of grain at the expense of stored sugars in the stems; this may limit the feeding value of the fodder. Norkan is somewhat resistant to LEAF BLIGHT.

Orange sorgos

Several varieties of Orange sorgo are grown; in addition, the name *"Orange"* often is applied erroneously to other sorgo varieties; e.g., to Sourless sorgo which is also called *White Orange*, and to the Colman sorgo, usually called *Red Orange*. The ripe heads of the Orange varieties are of medium height and mature later than the Amber sorgos, but earlier than Atlas and Sumac. The Orange sorgos are excellent for silage, sirup, and cured forage.

There are two Orange varieties of practical importance—Kansas and Waconia Orange sorgos.

Kansas Orange (sorgo) has rather small, oblong heads with dark reddish-brown chaffs, and rather small brown seeds. It is 8 to 11 ft. tall and usually produces a slightly larger yield of forage than Atlas. This Orange variety is grown extensively in eastern Kansas and somewhat in other states, but is being replaced rapidly by by Atlas and other white-seeded sorghum varieties.

Waconia Orange (sorgo), or *Rox Orange*, is shorter and earlier than Kansas Orange and has bright-red chaffs and medium-sized yellow-brown seeds. The seeds extend well out of the chaff and give the ripe heads a yellowish-red appearance. Under favorable conditions Waconia Orange grows to a height of 8 to 9 ft., but usually it is shorter. It is an important variety in Iowa, where it is popular for both forage and sirup, but is grown in other states as well.

Rancher is of growing importance because it has the lowest *prussic acid* content of any sorghum variety used as forage. (It was developed from Dakota Amber strains.) Rancher is early in maturity and produces a large tonnage of high quality fodder. Silage made from more mature plants has little acidity and does not spoil when properly ensiled.

Sourless (sorgo), or *African millet*, called also—erroneously—*White Orange*, is grown extensively in southern Kansas and western Oklahoma; it is usually badly mixed as to type. Growers of this sorgo claim that the fodder does not sour during winter but retains its juiciness and sweetness until spring; this accounts for the name "sourless." It is a popular variety for forage under conditions of limited rainfall. The seeds are mostly pale buff with some mixture of white seeds. The yellowish-brown chaff fades out to a straw color at maturity. Sourless is earlier, and shorter, than Kansas Orange.

Kansas Sourless 704-H is a promising selection. The heads are upright and will average 6 to 10 in. in length. The seed is fairly small and of light-buff to light-brown color. Kansas Sourless 704-H is best adapted to a fertile soil where the rainfall is high. Because of its good tonnage and high total sugar content, Kansas Sourless 704-H is dehydrated for the production of pellets which are used in prepared feeds.

Sugar Drip (sorgo), an important variety with tall, large stalks, is characterized by high lodging resistance.

Sumac sorgos

This group includes the following Sumac varieties:

Early Sumac (sorgo). In western Kansas and other localities, where the season is short, Early Sumac is preferred to Sumac. It grows from 5½ to 7 ft. tall and matures in about 100 days. Early Sumac appeals to farmers because of its uniformity, attractive appearance, earliness, ease of handling, and the fact that the stalks are juicy, sweet, and not too large and coarse.

Extra Early Sumac matures earlier than all other Sumac varieties; however, it is the only sorgo particularly susceptible to PERICONIA ROOT ROT.

Sumac (sorgo), often called *Red Top*—not to be confused with the grass REDTOP

—is the leading variety of the sorgos. It matures later than other important sorgos, e.g., Amber, Orange, and Atlas, but earlier than Honey. Sumac is grown extensively in Texas, Oklahoma, Kansas, and many other states. It has short, compact heads and very small, dark reddish-brown seeds that thresh free from the hulls. The stalks are easily handled and harvested. This, together with its usual leafiness and satisfactory yields, accounts for its great popularity.

Sumac 1712 has heavy, very juicy, sweet stalks; for this reason it is not only a forage plant but also valued for sirup-making. It is drought-resistant and characterized by wide adaptation and very reliable yielding ability.

Tricker is a white-seeded, relatively early, drought-resistant sorgo variety. It is grown in western Kansas, but it is so short and low in forage yields that kafir (a grain sorghum) is often grown in preference to it. The seed of Tricker is so similar to Atlas that it has frequently been sold, fraudulently or mistakenly, as Atlas.

Other sorgo varieties are grown on a limited ′acreage and are valued for sirup production as well as for forage by many farmers in the southern states. Among these sorgo varieties are the following:

Clubhead;

Colman (sorgo), erroneously called *Red Orange*, is susceptible to HEAD SMUT;

Cowper (resistant to LEAF BLIGHT);

Denton (sorgo) is somewhat resistant to LEAF BLIGHT;

Dwarf Ashburn, or *Dutch boy;*

Folger (sorgo), or *Folger's Early;*

McLean (sorgo) is relatively free of ROUGH SPOT and LEAF BLIGHT;

Planter (sorgo), or *Planter's Friend*, is resistant to ANTHRACNOSE and COLLETOTRICHUM STALK ROT;

Rex (sorgo), or *Red X;*

Sapling (sorgo);

Silvertop is relatively free of ROUGH SPOT;

Straightneck (sorgo) is practically free of ROUGH SPOT;

White African (sorgo), or *White Mammoth*, is white-seeded.

All of these varieties, with the exception

of Dwarf Ashburn, are rather tall and medium late in maturity. They produce a satisfactory yield and quality of forage, but most of them are of only local importance.

1. (B) FORAGE SORGHUM PRODUCTS

No sorgo products have been officially recognized as yet.

> NOTE: For *sweet sorghum meal*, the following definitions have been proposed:
>
> *Dehydrated sweet sorghum meal*—obtained by dehydrating the entire plant of selected sorghum varieties. It shall contain not less than 25 percent total sugars.
>
> *Dehydrated sorghum meal*—obtained by dehydrating the entire sorghum plant. If the name of a variety is indicated, the product must conform thereto.

2. (A) GRAIN SORGHUMS

The *nonsaccharine sorghums*, or grain sorghums, are of great importance in the southern Great Plains, the irrigated parts of New Mexico, Arizona, and California, and the hot, dry areas of Texas, Oklahoma, Kansas, and adjacent states. Some of them, such as the hegaris, kafirs, Freeds, darsos, and Grohoma, are of special interest; they may properly be considered *dual-purpose sorghums*, since they produce good yields of grain and also are satisfactory for fodder, bundle feed, or silage. Other grain sorghums, such as the milos, durras, many feteritas, and shallu, have stalks so dry and unpalatable that they are seldom used for forage; there is always considerable waste in feeding them unless they are chopped, and at best they are far inferior to the sorgos.

The dual-purpose grain sorghums usually yield about two-thirds as much forage as do well-adapted varieties of sorgo. Nevertheless, there are certain advantages in growing them for forage: (1) The bundles can be topped after harvest, and the heads threshed or ground if grain is desired, and the yield of grain is satisfactory. (2) When most of the stalks fail to produce heads, they are harvested for forage without sub-sequent topping of the stalks to remove the heads. (3) Grain can be utilized as a grain crop in seasons of favorable rainfall, or it can be utilized for forage if it does not promise a profitable grain yield. (4) The shorter stalks make harvesting, shocking, loading, stacking, and feeding much less laborious than the taller sorgos. (5) The feeding problem is simplified since a single bundle of grain sorghum, weighing 10 to 15 lb., furnishes a suitable quantity of both grain and forage for one animal's feed. (6) The best types of grain sorghums have about the same feeding value as corn (which they often replace in rations), analyze slightly higher in protein, vitamins niacin and biotin, and in pantothenic acid, but lower in fat, pyridoxine, and riboflavin, than corn.

Grain sorghum seed should be crushed or ground for dairy, beef cattle or hog feed, but the grain can be fed whole to sheep and poultry. There seems to be little difference in the feeding value of the various varieties, but there is a decided difference in palatability.

The grain sorghums have been used to make good *silage*, especially when mixed with a LEGUME (such as SOYBEANS), CORN, or forage sorghums. If used for silage, the grain sorghums should be cut when the seeds are in the hard dough stage; if they are allowed to mature before being put into the silo, a large portion of the small, hard seeds is not chewed or digested, thus causing an unnecessary waste of feed.

When used as fodder, the grain sorghums should be free of mold and fed while in a fairly moist condition. VELVETBEANS and grain sorghum, planted following SMALL GRAIN crops, furnish palatable and nutritious grazing in many areas.

More than 40 distinct varieties of grain sorghum are now grown in the United States, in addition to miscellaneous varieties of hybrid origin which are not easily classified.

> NOTE: In this survey, the true grain sorghum varieties are, probably for the first time, separated from those derived from crosses of different sorghum varieties, and are classified as separate intermediate types; e.g.,

Martin is neither kafir nor milo but one of the kafir-milo intermediates.

The grain sorghums differ in time of maturity, height, juiciness of stalk, size and color of grain, leafiness, color of chaff, beardiness, shape and compactness of heads, and many other characteristics. Some varieties reach heights of more than 6 ft., but extra-dwarf varieties, grown for harvesting with combines, usually do not exceed 2½ ft. in height.

The following are the most important grain sorghums used in one form or another to feed farm animals: As whole plants—either green, cured, or ensiled; as unthreshed sorghum heads—fed whole, or ground; or as grain.

Cody originated from the cross of Leoti sorgo and Club kafir; it ranges in height from 40 to 55 in. The stalk is somewhat juicy, slightly sweet, and has good standability. The white seed is tinged with sienna-brown. The grain from this relatively unimportant variety is rich in NIACIN and a source of glutinous starch. It is used as a feed grain on farms. Cody is somewhat resistant to BACTERIAL LEAF DISEASES.

Darsos

Three varieties belong to this group:

Darso, a dual-purpose sorghum, has juicy stalks; erect, loose, open heads; and reddish-brown seeds. The plants usually are slightly shorter than those of the kafir varieties and mature about the same time as the early kafirs. Darso has considerable resistance to injury from CHINCH BUGS, but is susceptible to PERICONIA ROOT ROT, STALK ROTS, and SMUTS, including COVERED KERNEL SMUT. Because of its bitter seeds, the darso crop is seldom injured by birds. This bitter taste is disadvantageous in that the grain is rather unpalatable to livestock and poultry, too.

Oklahoma No. 1 darso is grown to a considerable extent in central Oklahoma, where milo cannot be grown safely due to CHINCH BUG injury; this variety is also highly resistant to PERICONIA ROOT ROT.

Darset (pronounced: dar-*set*), or *little darso*, is the most important variety. It was only recently developed in Oklahoma as a combine-type grain sorghum resistant to damage by birds and weathering. Darset is essentially Oklahoma No. 1 darso but with an earlier maturity of between a week and ten days. This variety is of value as a feed grain but of little value as forage, or silage, because of its short growth. It is also resistant to PERICONIA ROOT ROT and CHINCH BUG, and will remain standing under conditions that produce lodging of regular darso.

Durras

The durras, formerly grown under the names of *Jerusalem corn* and *rice corn*, are characterized by dry stalks; compact, gooosenecked, bearded heads; and extremely flattened kernels. They mature early and are fairly well suited to conditions of limited moisture. The plants sucker rather freely. The stalks, being scarcely leaved and dry, are of little value for forage. The chief objections to durra are the frequent shattering of the grain from the heads, the excessive number of hairs on the chaff, and its susceptibility to COVERED KERNEL SMUT and LEAF SMUT. Practically all the durra now being grown in the United States is found in California.

The three varieties of durra do not rank as important sorghums:

White durra. The grain of the standard White durra, also known as *White Egyptian corn*, or "*Gyp*," has long been used for poultry feeding; poultrymen in certain sections of California pay a premium for it.

Dwarf White durra is now replacing the standard White durra because it is more easily harvested.

Brown durra, or *Brown "Gyp*," is a variety nearly identical with the standard White durra, except that is has brown, rather than white, seeds. Its grain is less palatable than that of other durras.

Feteritas

The feteritas are less leafy than kafirs and milos. They have dry stalks; erect, rather compact, oval heads; and very large, chalky-white seeds. The seed coats are covered with small lines or checks. The heads are beardless. The plants as a rule sucker less than those of milo but more than nearly all the varieties of kafir. Feteritas are generally resistant to PERICONIA ROOT ROT and SMUTS (e.g., LOOSE

KERNEL SMUT OF SORGHUM, COVERED KERNEL SMUT, and HEAD SMUT), but somewhat susceptible to CHINCH BUG injury and CHARCOAL ROT; they are very susceptible to SEED ROT (especially when planted early and without a seed treatment), and to SORGHUM RUST.

The feteritas mature early and are better able to produce a crop in seasons of limited moisture than the later maturing milos and kafirs. They are well adapted to planting as a catch crop in midsummer.

There are three varieties of feterita:

Dwarf feterita, the least important of the three; it is similar to Spur, except shorter.

Spur feterita, a dual-purpose variety, is grown extensively in Texas. It is somewhat resistant to LEAF BLIGHT, and highly resistant to all races of COVERED KERNEL SMUT.

Standard feterita, most frequently grown in Kansas; it usually outyields the other feteritas.

Freeds

Of some importance are two Freed varieties:

Dwarf Freed, a newer variety, is grown to some extent in western Kansas. It is considerably shorter than the standard Freed, and therefore lodges less and is more easily harvested.

Freed, a forage-type grain sorghum, has slender, sparcely leaved, sweet, juicy stalks; erect, loose, bearded heads; and small white kernels. It tillers rather freely and matures very early. Freed is adapted to conditions of extreme drought and short seasons which prevail in western Kansas and eastern Colorado.

Grohoma is a dual-purpose sorghum. It is medium late in maturity and has large, loose, beardless heads and brown kernels. The stalks are rather dry, but the juice in them is slightly sweet. Grohoma probably originated as a hybrid between feterita and some sorgo. It is fairly productive under favorable conditions, but less so than kafir and milo under conditions of severe drought; it does not belong to the important sorghums.

Hegaris

Some of the hegaris are of great importance as dual-purpose or as grain sorghum. Among the hegaris are these five varieties:

Bonita, or *Little hegari*, is an early dwarf variety. The stalks are slender, and, while pithy, are palatable to livestock in bundles, or for grazing. Bonita has a wide adaptation and is recommended for its high yields. It can be grown at higher altitudes and under lower rainfall than regular hegari.

Combine Bonita is a new variety which differs from regular Bonita chiefly in size: it can easily be harvested with the combine.

Early hegari, or *Combine hegari* (a kafir cross), is widely distributed in Texas, Oklahoma, and southern Kansas. The stalk of this variety is shorter and more slender than that of regular hegari, bears 11 or 12 leaves, and tillers freely. This dwarf combine type does not stand well but is valuable as forage. Seed is chalky white with a reddish undercoat. Early hegari is somewhat resistant to LEAF BLIGHT.

Hegari, originally called *Dwarf hegari*, and frequently known on farms as "*high gear*" or "*higeary*," is intermediate between kafir and feterita in general appearance and in characteristics. It has juicy and rather leafy stalks. The seeds resemble those of kafir but have a more chalky appearance, and a purplish-brown inner seedcoat. The plants of this variety usually sucker very freely. Hegari has a wide range with respect to maturity, being relatively early under some conditions, and later than any of the kafirs under others. It is also extremely variable with respect to yield, producing very high yields under favorable conditions and very low yields in other cases. This dual-purpose sorghum often fails to produce grain in dry seasons. Hegari is now grown extensively in the western and southern parts of Texas, in eastern New Mexico, and under irrigation in the Salt River Valley of Arizona. The best results usually are obtained from rather late planting, ordinarily in June. This sorghum is resistant to ANTHRACNOSE, COLLETOTRICHUM STALK ROT, and race 1 of the COVERED KERNEL SMUT, but is susceptible to SEED ROT and, to some extent, to CHARCOAL ROT.

Hegari is regarded rather highly as a bundle feed because of its leafy, palatable stalks.

Hi-hegari is a tall, non-sweet forage-type variety, developed from regular hegari primarily for forage and silage uses. It has the same maturity, adaptation, and grain yield as hegari, but it gives approx. 20 percent more forage which is relished by livestock.

Kafirs

The kafirs belong to the most important dual-purpose sorghums; they have juicy (but not sweet) stalks and for this reason are grown to a considerable extent for forage. The kafirs have long, slender, cylindrical heads; and small white, pink, or red seeds. The plants usually have few suckers. The heads, except those of Bishop kafir, are beardless.

The kafirs vary in average height from 4 to 7 ft. and usually do not lodge until after frost or late in the season. They are very susceptible to KERNEL SMUTS (especially COVERED KERNEL SMUT and LOOSE KERNEL SMUT OF SORGHUM), and moderately susceptible to BACTERIAL STRIPE OF SORGHUMS, HEAD SMUT, WEAK NECK, and SORGHUM RUST; they are fairly resistant to CHINCH BUG injury, and thus can be grown farther east than the milos. Kafirs are also resistant to LEAF BLIGHT, BACTERIAL STREAK, CHARCOAL ROT, PERICONIA ROOT ROT, and immune to LOOSE KERNEL SMUT OF JOHNSONGRASS.

The kafirs can be divided in two groups —Blackhull and other kafir varieties.

(a) The Blackhull kafirs include these three varieties:

Standard Blackhull (kafir), often simply called *Blackhull kafir*, produces seeds which are partly covered with black hulls (or chaffs). It matures in 115 to 130 days and is grown in the eastern half of Kansas, Oklahoma, and Texas, and in southwestern Missouri. Standard Blackhull grows from 5 to 7 ft. tall and is characterized by juicy, stout, short-jointed stems; its 12 to 16 broad, stiff leaves are set close together. The heads are 8 to 12 in. long and relatively larger in diameter than those of most other kafirs because of longer seed branches. The medium-sized, oval-shaped seeds are white.

with a splash of red or black on the tips,

Texas Blackhull (kafir) is earlier and shorter than Standard Blackhull. This variety is well suited to the western parts of Kansas, Texas, and Oklahoma.

Western Blackhull (kafir) is a variety very similar to Texas Blackhull in size and adaptation. It is resistant to ANTHRACNOSE and COLLETOTRICHUM STALK ROT.

(b) Other kafir varieties of practical importance are as follows:

Club kafir is resistant to COLLETOTRICHUM STALK ROT.

Combine kafir-60 has juicy stalks which stand up well in the field. This is an important variety which yields well.

Dawn kafir, sometimes erroneously called *Dwarf kafir* (a name that belongs to another variety), is early, and short, and best suited to conditions of limited moisture, or to short seasons under which other varieties are likely to fail to produce grain. It is somewhat resistant to LEAF BLIGHT.

Hydro kafir is similar to Reed, except it is later and has thicker heads.

Pink kafir, a medium early sorghum of local importance, is grown in western and southern Kansas chiefly for forage to be fed as bundle feed and for silage purposes. Its height is from 4 to 6 ft. The slightly juicy stalks are rather leafy. The heads are 10 to 14 in. long and cylindrical in shape, with very short branches well filled with small, pinkish-white seeds. Pink kafir is one of the parents of Early Kalo and Midland, and has appeared in other crosses.

Red kafir gives relatively high yields. Different strains of this variety grow from 4 to 8 ft. tall and mature in 100 to 130 days. The most common strain is tall and late, and is a good forage producer. The heads of Red kafir are typically long and cylindrical. The juicy stalks are slender and the medium-sized seeds are light red. Red kafir is immune to LOOSE KERNEL SMUT OF JOHNSONGRASS.

Reed kafir has long, black GLUMES; it produces good yields of grain in many sections of the Great Plains, but it is of limited value for forage because of its brittle leaves and infrequent tillering; it also lodges easily. Reed kafir is immune to LOOSE KERNEL SMUT OF JOHNSONGRASS.

Sunrise kafir is grown for bundle feed in the western parts of the Great Plains because of its tall, slender, sweet, and numerous stalks, and its early maturity.

White kafir is similar to Blackhull except that it has drier stalks and white chaffs on the seeds; it is earlier than the Standard Blackhull.

Other kafirs are *Pearl kafir* and *Rice kafir* (both not being grown in this country at the present time); *Sedan kafir;* and *Sharon kafir* (which is immune to LOOSE KERNEL SMUT OF JOHNSONGRASS).

Kaolings

Kaolings are characterized by dry, slender, sparsely leaved stalks; erect, loose, or particularly compact heads; and small, brown or white kernels. They mature early and can be grown farther north in this country than any of the other grain sorghums, but they cannot successfully compete with corn in the northern states and are unlikely to reach maturity in a cool season. The crop is of little value for forage because the plants have few leaves and the stalks are too dry and woody for good feed. Therefore, none of the kaolings is at present grown in the United States.

Two of the leading kaoling varieties may be mentioned here:

Dwarf Shantung kaoling is a variety resistant to LOOSE KERNEL SMUT OF SORGHUM.

Manchu Brown kaoling, another variety, has a compact head and small, brown seeds.

Milos

Milo—as a class in the grain sorghum group—is represented by a number of varieties which differ principally in height, seed color, and resistance to PERICONIA ROOT ROT and SEEDLING BLIGHT. They are extremely susceptible to CHINCH BUG injury, CHARCOAL ROT, and STALK ROT, and are often affected by NONPARASITIC SORGHUM-LEAF DISCOLORATIONS. They are highly resistant to SORGHUM RUST, LOOSE KERNEL SMUT OF SORGHUM, HEAD SMUT, and some races of COVERED KERNEL SMUT. Characteristics common to all true milos are the rather dry stalks and large, oval or egg-shaped, mostly pendent or gooseneeked

heads. The stalks have limited value for forage as compared with the kafirs and sorgos. The fully developed leaves show a distinctive yellow midvein (or midrib). The seeds of the milos are large, somewhat flattened, pale reddish-yellow, or pure white. They are about one-third enclosed in nearly black, hairy GLUMES which are transversely wrinkled. (The transverse wrinkle is an important mark of identification to distinguish true milo from Wheatland and other kafir-milo derivatives.)

The milos, in general, are somewhat earlier than the kafirs, requiring 100 to 130 days to reach maturity. All milos mature at approximately the same period, with the exception of Sooner and Early White, which are earlier.

The most important varieties of milo are as follows:

Colby, which is susceptible to PERICONIA ROOT ROT, CHINCH BUG, and WEAK NECK, is of limited local importance.

Double Dwarf milo No. 38 stands only 1½ to 3 ft. high and is resistant to PERICONIA ROOT ROT. It is of increasing local importance.

Double Dwarf White Sooner (milo) has dry stalks.

Double Dwarf Yellow Sooner (milo) is of early maturity and can be harvested with a combine. The slender stalks carry 10 to 12 leaves. This Sooner variety is among the earliest and most consistent varieties of grain sorghums.

Dwarf White milo is nearly identical with Dwarf Yellow milo except in seed color (which is white).

Dwarf Yellow milo usually grows to a height of 3½ to 4½ ft. and responds more favorably to irrigation than almost any of the other sorghums cultivated in the southern Great Plains.

Finney (*milo*) is resistant to PERICONIA ROOT ROT and SEEDLING BLIGHT; it originated from Dwarf Yellow milo and is very similar to it.

Miloca is Double Dwarf White Sooner whose starch is of a waxy type.

Sooner (*milo*), or *Sixty Day milo*, originated from a cross between Dwarf Yellow milo and Early White milo and is grown from southern Texas to central South

Dakota. While its yield is low in the dry years, it is one of the very few varieties that produce grain in the drought areas; and it is therefore favored by farmers in urgent need of feed grain. However, Sooner is highly susceptible to PERICONIA ROOT ROT, WEAK NECK, CHINCH BUG, and lodging. (Selections resistant to periconia root rot are now being grown in Texas and Oklahoma). It is one of the earliest varieties to mature, and therefore well adapted to regions with short growing seasons. Sooner is similar to Dwarf Yellow milo, except that it is earlier and its stalks are shorter, more slender, and less leafy.

Other milos of only local importance include *Standard Yellow milo, Standard White milo, Early White milo* (also called *Sugar milo*), *Day milo* (an early maturing combine which originated from a cross between Dwarf Yellow and Early White milos), *Texas milo* (resistant to PERICONIA ROOT ROT), and *Texas Double Dwarf milo*.

Schrock, also known as *Schrock kafir*, or *Sagrain*, is a dual-purpose sorghum of local importance. This variety grows in eastern Kansas, Oklahoma, and some of the southern states. It grows 4 to 5 ft. high, has a very dense foliage and stout, moderately juicy stalks. The forage rivals that of the kafirs in quality and tonnage, but falls well below the tonnage of the leading sorgos. The seeds are bitter, yellowish-brown, and of medium size, and contain the waxy type of starch. Schrock is resistant to LOOSE KERNEL SMUT OF SORGHUM and ROUGH SPOT, but susceptible to COVERED KERNEL SMUT.

Shallu, which was developed from a kafir cross, has been grown and exploited under many names, such as *Egyptian wheat, California wheat, Jerusalem rice corn,* and *Mexican desert wheat corn.* It has tall, slender stalks; long, open, leaning heads; and small white kernels. The stalks are dry, lacking in leafiness, and of little value for forage. The crop is resistant to LEAF BLIGHT, BACTERIAL LEAF DISEASES, and SORGHUM RUST, but susceptible to BACTERIAL SPOT, LOOSE KERNEL SMUT OF SORGHUM, and other SMUTS; it shatters and lodges easily, and matures late. It is

seldom grown, except in the Gulf region of Texas and Louisiana.

Intermediate grain sorghums

There are four groups of intermediate grain sorghum varieties:

Kafir—Feterita derivatives

This group includes the following five varieties.

Ajax is a medium-late dual-purpose grain sorghum that produces large heads and chalky-white seeds. The stalks are thick, leafy, and rather short. Ajax is a productive variety in Texas and Oklahoma when ample soil moisture is available.

> NOTE: *Imperial kafir,* originally selected from a plot of kafir in the Imperial Valley of California, is indistinguishable from Ajax in appearance and performance.

Chiltex is a grain sorghum of medium height, maturing in midseason. The heads are erect and similar in shape to those of kafir; and they have small, white seeds. The stalks are dry, like those of feterita. Chiltex is grown to some extent in northern Texas and southern Oklahoma, where it is more productive than kafir under the limited moisture conditions prevailing in that section.

Dwarf kafir 44-14, formerly called *Combine White kafir Okla. 44-14,* is very similar to Texas Blackhull kafir. This important variety was developed from a cross between Sharon kafir and Early feterita. It is resistant to PERICONIA ROOT ROT and CHINCH BUG injury.

Premo is similar to Chiltex, except that it has thicker and more compact heads and matures later; for this reason it is adapted only to conditions in the southern half of the sorghum region.

Wonder (kafir) is rather early and of striking appearance, but has dry stalks and is susceptible to SEED ROT.

Kafir—Freed derivatives

Six of the intermediate varieties of this group are as follows:

Cheyenne (*kafir*), or *Sweet Stalk kafir,* a dual-purpose sorghum, is grown widely in northwestern Kansas and the adjacent sections of Nebraska and Colorado. It has heavier, more compact heads than Freed and produces more grain.

Coes, isolated from a cross of Pink kafir and Freed, is a dual-purpose sorghum which has extreme earliness and is valued in short-season areas particularly as bundle feed. The plant is of medium height and has slender, semi-juicy (but not sweet) stalks. The seed is white, with splashes of red, and small to medium in size.

Highland Improved Coes, a variety developed from regular Coes, is a dual-purpose sorghum of local importance.

Greely, a dual-purpose sorghum, is a cross of Pink kafir and Freed.

Modoc, also a cross of Pink kafir and Freed, is used principally for plant breeding purposes.

Weskan, another cross of Pink kafir and Freed, is a late sorghum of only limited importance.

Kafir—Milo derivatives

The third group of intermediate sorghum varieties consists of the following:

Beaver (milo) is a cross of milo and kafir which grows in the Panhandle region and in Kansas. The heads are slightly pendent, the stalks dry, and not so resistant to lodging as Wheatland.

Bishop (kafir), also called *Algeria*, has large, bushy heads, is rather late in maturing, and is susceptible to CHINCH BUG injury. It is promising only under conditions of ample moisture.

Caprock is a double-dwarf type sorghum; it matures in 110 to 115 days. The stalks are thick and stout; the large and yellow to reddish-yellow seeds are resistant to PERICONIA ROOT ROT.

Combine-7078 is a double-dwarf combine variety, adapted to dry areas along the western edge of the Sorghum Belt.

Fargo (milo), or *Straightneck milo*, is no longer cultivated in this country.

Kalo resulted from a natural cross between Pink kafir and Dwarf Yellow milo. It is of little importance at present.

Early Kalo, a dual-purpose sorghum, grows to an average height of from 42 to 50 in. and will mature in 95 to 110 days. It can produce a high yield of grain. The slender stalks are somewhat dry and have from 8 to 12 leaves, producing cylindrical heads from 7 to 10 in. long. The seeds are similar to Pink kafir in shape and size, but of a pale reddish-yellow color. Early Kalo is subject to lodging if left until the stalks become fully dry before harvesting. The variety is susceptible to CHINCH BUG injury and to WEAK NECK, but resistant to PERICONIA ROOT ROT and COLLETOTRICHUM STALK ROT.

Early Kalo makes an excellent bundle feed when drilled at the rate of 10 lb. to the acre. The thick seeding limits its grain production. It is then readily cut with a grain binder and is easily handled. Such a crop makes a satisfactory roughage for beef and dairy cattle.

Martin, also called *Martin's Combine milo*, is a dwarf combine variety which resembles kafir in general appearance and behavior, has slender stalks, and bears 11 or 12 leaves. The grain can be harvested shortly after the seeds are mature, which is the chief merit of this variety. Seeds are brown to brownish-red in color, contain more tannin than do other combine varieties, and are less palatable, medium in size, and hard. Martin matures in about 100 to 105 days. It is susceptible to lodging from CHARCOAL ROT, but very resistant to PERICONIA ROOT ROT and ANTHRACNOSE.

Midland is a combine-type sorghum which originated from a cross between Pink-kafir and Dwarf Yellow milo. This double-dwarf variety ranges from 38 to 40 in. in height, has stout, somewhat juicy stalks, and a high resistance to lodging. It matures in 100 to 110 days. The round, reddish seed is of medium size, threshes free of GLUMES easily but does not shatter readily. Its best adaptation is in northern and central Kansas, where it is grown extensively, as well as in Nebraska. Since this variety is earlier than Westland, the two often supplement each other when the need arises for early and late planting. Midland is resistant to PERICONIA ROOT ROT, CHARCOAL ROT, to some degree to WEAK NECK, and (when planted early) to CHINCH BUG.

Since the stalks are slightly juicy, the standing stover has some value as fall pasture after the grain has been combined.

Plainsman (milo) is an important double-dwarf combine. The stalk is thick, stout, and bears 14 to 15 mature leaves. The variety is resistant to PERICONIA ROOT ROT.

Quadroon (sorghum) is a dwarf variety It is resistant to LEAF BLIGHT, PERICONIA ROOT ROT, and CHARCOAL ROT. Quadroon grows tall enough to be used as forage.

Redbine-60 is one of the widely used newer grain sorghums. This variety is a red-seeded combine kafir and blooms in sixty days.

Redbine-66 is very similar to Redbine-60, but it is awnless and matures in sixty-six days.

Redlan is suitable for combine harvesting. The stalk of Redlan is juicy; its upper part and the seed branches of the head dry as the seed ripens, while the remainder of this stalk remains green. It is somewhat resistant to CHINCH BUG injury, and Redlan grain is not readily eaten by birds.

Resistant Wheatland-288 is an improved, new variety which is not affected by PERICONIA ROOT ROT.

Westland is a double-dwarf combine sorghum, probably a natural outcross out of the Wheatland, but resistant to PERICONIA ROOT ROT and to some degree to WEAK NECK. The stalks are kafir-like but somewhat dry, range in height from 36 to 46 in., and bear from 12 to 14 mature leaves. The seeds are of medium size and hardness, with a bright, reddish-yellow color. Westland matures in from 105 to 115 days. It stands well if not attacked by CHARCOAL ROT, which may cause some lodging. This grain sorghum is especially recommended for southwestern Kansas, the irrigated region of the Arkansas River Valley, and areas where the "milo disease" (PERICONIA ROOT ROT) is known to be in the soil.

Wheatland was the first combine sorghum to become well established in Kansas and western Oklahoma, from where it spread rapidly into other states. It grows to a height of 31 in., depending upon seasonal conditions. The heads are elongated and vary in length from 6 to 9 in. Wheatland is highly susceptible to PERICONIA ROOT ROT.

Other intermediate sorghums

To the intermediate varieties which are derivatives of other sorghums than kafirs belong these three:

Gurno originated from a cross between Dwarf Freed and Dwarf feterita. It is a double-dwarf type sorghum, with slightly juicy stalks, which shows considerable standability. Gurno ripens early enough for western Kansas where earliness is important. The plants are not resistant to most races of COVERED KERNEL SMUT, other KERNEL SMUTS, and CHINCH BUG. The grain is palatable to poultry and livestock.

Norghum, as its name implies, is a "sorghum of the north"; this early-maturing grain sorghum was developed to fit the short growing season and climate of South Dakota, and resulted from a cross between a selection of Kalo and one developed from Dwarf feterita and Gurno. Norghum is a combine grain sorghum that is about equal to CORN in feeding value. It grows to a height of 36 to 46 in. The stalks are medium slender and moderately leafy. The seeds are of small to medium size, have a mottled, reddish-brown color, and germinate readily. This variety is early, but it may be used as a catch crop in many areas. A thorough preparation of the seedbed may increase the yield 50 percent.

Reliance is a cross of Modoc and Sooner milo. It has a larger range in planting time, and stands longer after killing time than Norghum. Reliance is grown chiefly in South Dakota, where it was developed.

2. (B) GRAIN SORGHUM PRODUCTS

Grain sorghum *by-products* are very much like similar corn products but do not contain carotene.

Grain sorghum products—especially those of milo, hegari, kafir, or feterita—are commercially listed among the CEREAL GRAIN BY-PRODUCTS. Officially recognized are the following:

Grain sorghum chop—a product made by grinding, cutting, or chopping the grains of grain sorghums.

> NOTE: If the name of the variety is given for this or any other grain sorghum product, it must be true to name.

Grain sorghum head chop—the chopped head of grain sorghums, including grain and stems.

Grain sorghum head stems—consisting of the stems of grain sorghums from which the grain has been removed.

Grain sorghum gluten feed—the part of the grain that remains after the extraction of the larger part of the starch and germ in the wet milling manufacture of starch or sirup.

Grain sorghum gluten meal—the part of the grain that remains after the extraction of the larger part of the starch and germ, and the separation of the bran, in the wet milling manufacture of starch or sirup.

Grain sorghum oil cake—consists of the germ of the grain from which part of the oil has been pressed; it is the product obtained in the wet milling manufacture of starch, sirup, and other grain sorghum products.

Grain sorghum oil meal is ground grain sorghum oil cake.

Grain sorghum mill feed is a mixture of bran, germ, and a part of the starchy portion of the grain as produced in the manufacture of grits from grain sorghums; it shall contain not more than 5 percent crude fiber.

Grain sorghum feed meal is the fine siftings obtained in the manufacture of rolled or flaked grain sorghum grains.

Rolled grain sorghums is the product obtained by running whole grain sorghums over smooth flaking rolls, properly tempering them, removing most of the fine particles, and subsequently drying and cooling them.

3. GRASS SORGHUMS

Both sorghums of this class—SUDANGRASS and JOHNSONGRASS—are now considered separate species (and for this reason are not discussed in this section which is only concerned with the many selections, hybrids, strains, and varieties of the species *Sorghum vulgaris*). They are valuable for forage and in many respects similar; however, Sudangrass is an annual without rootstocks, Johnsongrass is a perennial with vigorous rootstocks.

4. BROOMCORN

There exists only a limited demand for broomcorn; it is a cash crop grown for its brush, the fibers of which are used in the production of brooms. Only in years when all livestock forage is in very short supply, may broomcorn be grazed or even bound after the heads have been harvested.

The plants are seldom, if ever, attacked by HEAD SMUT or PERICONIA ROOT ROT, but are very susceptible to ANTHRACNOSE, COVERED KERNEL SMUT, and COLLETOTRICHUM STALK ROT.

Broomcorn grows well on the sandier soils; it is considered one of the best row-crops for preventing wind erosion.

These broomcorn varieties are most commonly grown:

Black Spanish (broomcorn), or *Black Jap*, is the leading standard type. It usually grows to a height of 6 to 11 ft. The GLUMES are dark brown to black. It is relatively early, and tends to produce a fairly straight brush.

Japanese Whisk Dwarf is the only whisk type grown in this country. It reaches a height of from 2½ to 4 ft. and produces a fine, slender brush 12 to 18 in. in length. Three-fourths of the brush is covered by an upper leaf sheath. In recent years Scarborough has largely replaced Japanese Whisk Dwarf.

Scarborough (Dwarf No. 7) is a dwarf type. The plants grow 4 to 6 ft. tall. The glumes are reddish-tan in color.

References: H.5; H.1; L.1; G.1; M.2; S.3; S.4; H.6; M.1; W.35; E.3; F.1; S.8; K.3; K.4; D.5; D.4; F.2; N.2; F.3; J.1; C.6; L.5; E.12; F.6.

See also CRIMSON CLOVER; DISTILLERS' PRODUCTS; SILAGE CROPS; SILAGE; HAY; DUST STORMS; FIELD BINDWEED; ROOT KNOT; SOUTHWESTERN CORN BORER; PALE WESTERN CUTWORM; CLUSTER CLOVER; SHREDDING; BAGASSE; CYANOGENETIC GLUCOSIDES; ROUGHAGE.

SORGHUM SILAGE. *See* SORGHUMS.

SORGO is also called *sweet sorghum* or *forage sorghum.*—*See also* SORGHUMS.

SORGO SILAGE.

See SILAGE CROPS; SORGHUMS.

SORT is synonymous with the word *"genotype"* which refers to types, different in one or more inherited characteristics.

The variability in varietal designations causes some agronomists to use the term "sort" in preference to "VARIETY."

SORTINGS. *See* PEANUT.

SOURCLOVER (*Melilotus indica*), sometimes called *bitter clover*, is a species of the SWEETCLOVERS. This winter annual is a yellow plant, has small, rough, dark-green seeds and makes its best growth along the Gulf Coast, in Southern New Mexico, Arizona, and California. In the North it has no value.

Reference: M.3; P.7.

See also INOCULATION.

SOURLESS SORGO (sometimes erroneously called *White Orange*) and its selection, *Kansas Sourless 704-H*, are forage SORGHUMS.

SOUR SILAGE. *See* SILAGE.

SOUTH AMERICAN POPCORN.
                                                    See CORN.

SOUTH AMERICAN TYPE OF ITALIAN RYEGRASS = COMMON RYEGRASS.

SOUTHERN ANTHRACNOSE, caused by the FUNGUS *Colletotrichum trifolii*, is a major disease of RED CLOVER in the southern Clover Belt of the United States. It has been recorded as far north as southern Canada, but is primarily a high-temperature disease that flourishes at about 82° F. This disease is of little economic importance in the northern clover areas. It occurs occasionally on ALFALFA, BIG TREFOIL, CRIMSON CLOVER, SUBCLOVER, BUR-CLOVER, and WHITE SWEETCLOVER, but not on white clover. Alsike clover is practically immune.

A resistant variety, Kenland RED CLOVER, is available.

Symptoms resemble those of NORTHERN ANTHRACNOSE; in fact, a positive identification in the field is frequently difficult and sometimes impossible. Dark tufts of setae (bristles) in the older lesions indicate that the disease is southern anthracnose. Southern anthracnose also commonly attacks the upper part of the taproot; that has not been observed for northern anthracnose. However, like northern anthracnose, it may occur on plants at any stage of development.

The disease appears on the leaves as dark-brown spots of irregular shape, which vary from pin-point lesions to a general infection over most of the surface. PETIOLES (leafstalks) are very susceptible, become dark brown, and the leaflets droop. Lesions near the base of a stem often cause death and browning of the entire stem. Dark lesions develop on the upper part of the taproot, gradually girdle it, and cause the plant to wilt and die. Diseased crowns become brittle so that the stems are readily broken off at the soil level. CROWN ROT or ROOT ROT, caused by southern anthracnose, kills some plants and weakens others so that they cannot survive long droughts, adverse winter conditions, and attacks of other diseases.

Reference: H.21.

See also ANTHRACNOSE.

SOUTHERN BLIGHT, also known as "*white mold*," is caused by the soilborne FUNGUS *Sclerotium rolfsii* and is largely a disease of the root and crown; it affects various kinds of plants, including LUPINES, PEANUTS, and CROTALARIAS. The plants are usually attacked near the surface of the soil, and a canker is formed that quickly involves the entire stem and often girdles it. Diseased plants may at first be stunted, but later wilt and die.

In dry soil the fungus may attack the roots 1 in. or more below the soil surface. The roots decay, and white fungus growth often covers the affected part. Typical small, round, white SCLEROTIA, which later turn brown, frequently form on the lesions at or near the surface of the soil, or on the adjacent soil.

Often, several neighboring plants, or consecutive plants in a row, are attacked. Young seedlings may be girdled and killed in a few days, but death comes much more slowly to older plants. The decayed tissue usually is lighter brown in color than that attacked by RHIZOCTONIA ROOT ROT. Southern blight is also distinguished from rhizoctonia root rot by the color of the sclerotia.

Control: No effective control measure is known for this fungus.

Reference: W.8.

See also SEEDLING DISEASES.

SOUTHERN BUR-CLOVER
                                        = SPOTTED BUR-CLOVER.

SOUTHERN COMMON ALFALFA includes *Arizona Common* and *Oklahoma Common.—See also* ALFALFAS.

SOUTHERN CORN LEAF BLIGHT is caused by the FUNGUS *Helminthosporium maydis*. This disease is found throughout the world, wherever CORN is grown under warm, humid conditions. In the United States, it occurs chiefly in the southeastern states; it is also found in the southern parts of Missouri, Illinois, Indiana, Ohio, and eastward to the Atlantic Coast.

Typical lesions of southern corn leaf blight range from ¼ in. to 1½ in. in size. A distinguishing characteristic of the lesions is their parallel sides. The lesions are tan or straw-colored, and may have a brownish-purple margin. Ears are not affected; seed transmission of the disease, therefore, is unlikely.

The asexual stage is the most abundant stage in the life cycle of the parasite.

Control of this disease may be effected through the use of resistant hybrids. Rotation or seed treatment does not control southern corn leaf blight.

Reference: U.7.

SOUTHERN CORN ROOTWORM (*Diabrotica undecimpunctata howardi*) is commonly known as the *budworm;* it is also sometimes termed *drillworm.* The adult is known as the *spotted cucumber beetle.* The rootworm is found from Maine to Florida, and from the Atlantic Coast to the Rocky Mountains, but only in the South is it a serious pest. During years of heavy worm infestation, it is necessary to replant CORN as many as three times, and even then only a poor stand is obtained.

Serious infestations of the rootworm may be expected during seasons of abundant rainfall and cool weather. Corn in lowlands is more seriously affected than that in uplands.

In addition to corn, the rootworm is known to feed on the roots of JOHNSONGRASS, RESCUEGRASS, WHEAT, RYE, young OATS, ALFALFA, and PEANUT plants.

The "worm" does not confine its injuries to the roots, but also feeds on the interior of the stem, bores out the crown and kills the bud. Plants affected in this way will break off at the injured point

Southern corn rootworm. Larva and adult. (L.9.)

when an attempt is made to pull the plants.

The bud-leaves of injured plants dry up and die, while the rest of the plant retains its original color for a time.

When the worms feed on the roots of older plants, they cause them to fall over, lodge, and appear sickly or stunted.

The southern corn rootworm has four distinct stages through which it passes in its development. The yellow egg is the size of a large pinhead; hundreds of eggs are deposited by the female beetle slightly below the surface of the ground among various herbaceous plants. The eggs hatch in about three weeks, early in spring, and in from six to eight days in midsummer.

The grub, or larva, when first hatched, is slender and yellowish-white with a dark-brown head, and a dark patch on the top of the last body segment. It is active at this stage of growth but so small that it is scarcely perceptible to the unaided eye. The full-grown worm usually is deep yellow. The mature larva tapers toward the head-end with the last few segments of the body much wider than the head, which makes them look swollen. The grub stage lasts from three to four weeks; occasionally, however, the insect may live in this stage from five to six weeks. As the larva increases in size, it sheds its outgrown skin twice. Usually only one larva is found at one plant.

After the larva is full-grown, it enters the ground to a depth ranging from a few inches to $\frac{1}{2}$ ft. or more. It then makes a cell and begins to shorten, sheds its skin, and passes into the pupal (or resting) stage.

The pupa is soft, yellowish, about $\frac{1}{4}$ in. long, and has two very conspicuous spines at the tip of the abdomen. The pupal stage lasts six to eight days in summer and from ten to thirteen days in spring and fall. At the end of this period the *spotted cucumber beetle*, or adult, comes forth and works its way through the soil to the surface of the ground.

The adult is about $\frac{1}{4}$ in. long, yellowish-green, with a black head, black legs, and 12 black spots on the back. The beetle often does considerable damage to young corn by cutting off the bud leaves; it is even more injurious to certain truck crops, especially to squashes, cucumbers, etc.

The southern corn rootworm passes the winter in the beetle stage (except possibly in southern Florida and Texas). In more northerly regions the beetles crawl under rubbish and other places that afford them protection. With the return of warm weather, they again become active.

This insect has two complete generations annually in South Carolina. Serious injury to young corn in the South is largely done by the larvae of the first generation of the year.

Control: The bobwhite and a number of other birds prey upon the adults of the southern corn rootworm. Among its insect enemies the most important is a two-winged fly, *Celatoria diabroticae*, which attacks the insect in the beetle stage.

Measures which have been recommended for the control of the southern corn rootworm include the following: (1) If possible, plow lowlands early in the season, at least one month before planting of corn. (2) Plant lowlands in southern Georgia and *western* Florida between April 20 and May 1; in central Georgia and the southern half of South Carolina, between May 1 and May 10; in northern Georgia, the northern half of South Carolina, and all of North Carolina, between May 10 and May 20. (3) Destroy beetles by burning over field borders and terraces in winter, or by applying DDT or CHLORDANE. (4) Plant low places in fields twice as thickly as the uplands; thin out later if necessary. (5) Enrich the soil by raising leguminous crops. (6) Plant profusely rooting hybrid corn varieties in areas where root damage is severe. (7) Apply ALDRIN, HEPTACHLOR, or TOXAPHENE to the soil.

References: L.9; B.10.

SOUTHERN DESERT SHRUB.
See RANGE PLANTS.

SOUTHERN SOYBEAN
= *Mammoth Yellow. See* SOYBEAN.

SOUTHERN TYPE SMOOTH BROME.
See SMOOTH BROME.

SOUTHLAND. *See* COMMON OAT (variety).

Sow thistle. (B.28.)

SOUTHWESTERN CORN BORER

(*Diatraea grandiosella*) often attacks CORN; farmers have tended to substitute SORGHUMS for corn when this pest is prevalent.

SOUTHERNWESTERN PORCUPINE GRASS = CALIFORNIA NEEDLEGRASS.

SOW RATIONS. *See* SWINE RATIONS.

SOW THISTLE (*Sonchus arvensis*), also called *perennial sow thistle*, is a WEED which spreads by seed and by creeping roots. The erect stems reach a 2 to 5 ft. height, the leaves are up to 1 ft. long, deeply cut, and of light green color. The yellow flower heads are dandelion-like. All parts of the plant are filled with a milky juice. It occurs in the northern parts of the United States.—*See also* RYE.

Illustration: See page 469.

SOYBEAN (*Glycine max.*), also called *soya bean*, *soja bean*, or *Manchurian bean*, is an annual summer legume. It is an erect, branching plant, resembling in its early growth the ordinary FIELD BEAN.

Nearly all varieties are pubescent; that is, the stems, leaves, and pods are covered with fine tawny (brown) or gray hairs. The leaves of the soybean vary widely in shape, size, and color; they nearly always begin to turn yellow as the pods ripen, and usually have fallen by the time the pods are mature. The small inconspicuous flowers are borne in the axil of the leaf and are either white or purple. The pods, containing two to three seeds, range in color from very light straw through numerous shades of gray, and brown, to nearly black. The seeds are usually straw yellow (sometimes very pale and then erroneously called white), greenish-yellow, green, brown, or black. Bicolored seeds occur in several varieties, the most common pattern being green or yellow, with a saddle of black or brown; the brown and black colors are concentrically arranged.

Acreage and production of soybeans in the Western Hemisphere are concentrated chiefly in the North Central States of the United States.

In general, the climatic adaption of soybeans may be said to be about the same as those for CORN. The crop is especially well adapted to the northern half of the Cotton Belt, and to central and southern parts of the Corn Belt. In these localities the large and later varieties, which give yields that make soybean cultivation profitable, can be grown.

After the soybean plant is well started, it withstands short periods of drought and is not seriously retarded in growth, nor reduced in yield, by a wet season, provided weed growth is controlled. It is less susceptible to frost than are corn, COWPEAS, or field beans.

The soybean will succeed on nearly all types of soil, but the best results are obtained on mellow, fertile loams or sandy loams. In general, the soil requirements are about the same as those of corn; however, the soybean will make a more satisfactory growth on soils low in fertility, provided inoculating organisms are present. The soybean will do better than CLOVERS or ALFALFAS on soils of low fertility or on acid soils, but for the best results acid soils must be limed, and soils low in fertility supplied with those mineral elements in which they are deficient.

A well-drained soil is not necessary. Excellent yields are procured on some muck soils; the crop grows well on drained swampland. However, being an intertilled

Soybean. (M.3.)

crop, the soybean should not be planted on land that is subject to severe erosion.

The successful production of soybeans is dependent, in part, on a well-prepared seedbed. Spring plowing is favored by most growers because soybeans frequently follow corn, which often is not harvested in time for fall or early winter plowing. A soil free from clods insures the best results especially when seeding is done in close drills.

When soybeans are grown on land giving good yields of corn, they should produce a good crop without any additional fertilizers. On sandy soils, or soils of low fertility, an application of 200 to 300 lb. mixed fertilizer is desirable. An adequate quantity of lime to sweeten acid soils is usually necessary before satisfactory response will be obtained from applications of potash or phosphate. On some infertile acid soils the application of 20 to 30 lb. manganese sulfate, in addition to lime, is necessary to prevent severe manganese deficiency.

Where the crop is grown for the first time, soybeans make a rather poor growth unless inoculated. The lack of INOCULATION is usually indicated by a pale or yellowish-green color of the plant. Natural inoculation now occurs throughout much of the region where soybeans are extensively grown. Inoculation is most easily accomplished when the soil is neutral or alkaline. When a soil once becomes well inoculated, no further attention to this feature is necessary, provided a crop of soybeans is grown occasionally (every three to four years) on the land.

The bacteria of the soybean nodules will not inoculate any of the other commonly cultivated legumes, nor will the bacteria found in the nodules of other legumes inoculate soybeans. Some varieties of soybeans are more difficult to inoculate than others.

Soybeans may be sown during a period extending from early spring until mid-summer, depending on the variety being sown, the latitude, and the use to be made of the crop.

In the northern states the soybean planting season extends from the first of April to the last of June. For pasture, green manure, or even for hay, the soybean may be sown as late as the first of August in the southern states, or the first of July in the northern states and still produce a fair crop.

The two principal methods of seeding soybeans are by drilling in close rows, and in rows wide enough apart for cultivation. One of the advantages of row planting is higher yield of high-quality seed. In some sections, especially in the Southeast, the row method is employed in the production of forage. In the southern states, where larger and later soybean varieties are grown, soybeans are usually planted in rows 3 to 4 ft. apart. The seeding of one or two drill-widths of soybeans about fields of corn has become a common practice in several areas of the Corn Belt.

The depth of seeding is of much importance, since poor stands frequently result from covering the seed too deeply. In clay and other heavy types of soils, shallow seeding—about 1 in.—is recommended; in light loams and sandy soils the seeding may be deeper, but it should not exceed 3 in.

Under favorable conditions soybeans germinate in a few days, and cultivation should begin as soon as the seedlings appear. If the soil is of a heavy type and forms a hard crust after a rain, a light cultivation with the rotary hoe, weeder, or harrow, should be given to break the crust. Cultivation of soybeans should be frequent enough to keep down weeds. Usually, two or three cultivations after the beans are up will be sufficient.

In the Corn Belt a common rotation is corn, soybeans, SMALL GRAIN, and a deep-rooted legume. In regions where cowpeas are grown, soybeans are adapted to practically the same place in the rotation as cowpeas. In general, soybeans should replace an intertilled crop, such as corn, rather than a small-grain or sod crop. A winter cover crop in the rotation following soybeans is important to reduce soil erosion and leaching.

In certain sections of the South the soybean is especially valuable as a crop after early POTATOES or canning peas.

The soybean can also be used as a catch crop where new seedings of GRASS and clover have failed; it furnishes an excellent emergency hay crop in the rotation. In Louisiana, soybeans are grown extensively for green manure in rotation with SUGARCANE.

Soybeans may be satisfactorily grown in combination with other crops. The chief advantage of the mixture is the production of better balanced feed, and the yields are often somewhat better than when the crops are grown separately. When grown for hay, the mixture is more easily cured.

Soybeans are more generally grown with CORN than with any other crop. The mixture is commonly used for pasture or silage. Pasturing sheep on soybeans and corn is practiced throughout the Corn Belt. This combination furnishes feed not only for lambs being prepared for market but also for breeding ewes. Sheep will eat weeds, soybeans, and corn leaves before touching much of the ear corn.

Soybeans and cowpeas in combination make a very satisfactory mixture for hay, pasture, or green manure. Varieties of these crops, that mature about the same time, should be used. In sowing a mixture of both, it is best to have more soybean plants than cowpeas so that the vining cowpeas may have support. The plants should be cut for hay when the soybean seed is half- to full-grown, and the first pods of the cowpea are mature.

SUDANGRASS is an excellent crop for growing in combination with soybeans for hay, soilage, or pasture. The best results of this mixture are obtained in regions where irrigation is possible; there, not only a better yield, but a better balanced forage, is furnished. The yields of hay range from 2 to 4 tons per acre. The mixture should be cut for hay about the time the Sudangrass is in full bloom.

Sudangrass and soybean mixture is a valuable annual pasture to supply extra feed for dairy cattle during the summer, when grass pastures are low in yield. The crop is ready for pasture when from 18 to 24 in. high; it should not be pastured earlier because of the danger of PRUSSIC

ACID POISONING. Close pasturing also should be avoided.

Good results have also been obtained by drilling in close rows soybeans and FOXTAIL MILLET.

Soybeans may be grown in combination with SORGHUM for hay, as a soilage crop, or for silage. The best results are obtained in cultivated rows.

The soybean has an important place among soiling crops because of its palatability at all stages of development and the ease of handling.

Alone, the soybean is grown in the United States principally for its seed which is utilized for the production of oil and meal. Most of the meal is used for livestock feed. The soybean is also used for forage—preserved either as hay or silage, cut and fed green as soilage, grown as a green manure crop (only under exceptional conditions), or as a summer crop in orchards, and also pastured with hogs and sheep.

Pasturage for livestock of all kinds and for poultry is especially desirable when harvesting is hindered by weather conditions and lack of labor, or when the crop is grown for soil improvement. Soybeans furnish a satisfactory pasturage in late summer and early fall when perennial pasturage may be short. If a longer grazing period is desirable, varieties of soybeans, differing in maturity may be sown, or the same variety may be sown at different dates. Soybeans may be grazed until most of the leaves are removed, allowed to grow again for about thirty days, and then regrazed.

Soybeans make excellent supplementary hog pastures and are one of the best forage crops for general use in fattening the hogs, especially in the southern states; it is more profitable to allow the beans to ripen before the crop is grazed. However, when grown alone, soybeans do not make a satisfactory pasturage for beef and dairy cattle; grazing materially reduces growth, and trampling causes a large waste of feed, and damages the growing plant.

Soybeans are a good summer pasture for poultry. They supply succulent green feed continuously through late summer and early fall, and, because of their luxuriant growth, provide shade for the growing birds.

Soybean silage has a somewhat bitter taste and strong odor. If ensiled with corn-meal, its palatability is greatly improved.

Soybean hay is fed profitably to all kinds of livestock and poultry. As a source of digestible protein, soybean hay reduces the quantity of high-priced concentrates. Feeding soybean hay alone is not advisable, as digestive troubles may result. The chief objection to the soybean for HAY is that it has rather coarse, woody stems. This objection may be overcome to some extent by using heavier rates of seeding, growing a good forage variety, and harvesting at the proper time.

Yields of hay harvested at full-bloom stage are approximately half of those obtained when seeds are one-half to three-fourths developed. The percentage of leaves in the hay decreases steadily with advancing stages of maturity.

A most common method for making hay is to cut the crop as soon as the dew is off the plants, and leave it in the swath until thoroughly wilted. After wilting, and before the leaves become dry and brittle, the hay is racked into windrows and left a day or two, depending on the weather, and then placed in tall, narrow cocks or bunches to complete the curing. After four or five days of fair weather, soybean hay is ready to be stacked or housed. The cocks should be opened a few hours before hauling to dry out thoroughly.

Curing frames or poles are still used in some areas, especially in the South. The plants, as a rule, should be well wilted when placed on a frame or pole.

Drying in barns has proved helpful in many areas for producing high quality soybean hay. Water loss from plants is greatest for the first few hours after cutting. To take advantage of this situation, soybean hay should be allowed to field-dry from four to six hours, during which time the moisture content will drop from approximately 75 percent to 40 percent. This incompletely dried hay can be placed in a barn, equipped with air ducts and a blower, and reduced to 20 percent-moisture containing hay in four days during clear weather. With cloudy weather, curing takes more time, but if adequate, forced ventilation is provided there is little danger from spoilage.

To test the fitness of the hay for storage, or baling, twist a handful of hay in the hands. If it breaks easily when twisted once or twice, it is well cured, but if it is hard to break and sap is squeezed out of the stems or pods, it is not in condition to be stored or baled.

Yields of approximately 2 tons of soybean hay per acre can be expected under favorable conditions. Yields of 4 to 5 tons per acre have been obtained from well-fertilized soils in the Coastal Plain areas of the southeastern states.

Soybean hay compares favorably with other important hay crops in its content of digestible nutrients. Artificially dried hay is more nutritious than field-cured hay, especially if the latter is weathered in curing. *Chopping* or *grinding* the hay reduces waste in feeding.

Soybean seeds, or *beans*, are of great value as a high protein feed for livestock and poultry. The beans contain from 30 to 48 percent protein and can be used as a protein supplement to replace, at least partially, the expensive commercial protein concentrates necessary for stock feeding and milk production. They are the cheapest and most sufficient source of vegetable protein available to the farmer. Soybean protein is superior to other vegetable protein in common use because it includes in well-balanced amounts the essential AMINO ACIDS.

Soybeans should not be stored until they are thoroughly dry. When first harvested, the seeds often contain too much moisture for satisfactory storage. For storage, soybean seed should not contain more than 15 percent moisture. If artificial drying equipment is not available, it is a good plan to spread the beans on a clean, dry floor (or canvas) immediately after threshing, and to shovel them over from time to time until they are thoroughly dry. They may then be put in sacks or bins, but the storage room should be well ventilated and dry.

The beans can be fed whole to sheep and hogs. It is better to *crack* or *grind* them, however. It is advisable first to mix the beans with corn, OATS, or PEAS, and then grind them together into meal; soybeans alone are difficult to grind because of their high fat content. Due to the high protein content, soybeans should always be fed in mixtures with a less concentrated feed.

Soybean straw left after the threshing of the beans from the plants, until recently was generally used for feeding. Since the advent of the combine, it has been used more extensively for spreading on the land because of its fertilizing value; it also effects lessening of soil erosion.

Soybean straw from mature plants is low in digestible protein and high in fiber. Although relatively poor as a roughage for high producing dairy cows, soybean straw does have value for wintering dry cows and heifers, provided the animals are fed liberally and not forced to eat the coarse portions, and provided also, that the straw is supplemented with a good legume hay. Soybean straw has been found satisfactory as roughage for wintering idle work horses and mules.

These are the officially recognized soybean products:

Ground soybeans—obtained by grinding whole soybeans without cooking or removing any of the oil.

Soybean mill feed—the by-product resulting from the manufacture of soybean flour or grits; it is composed of soybean hulls and the offal from the tail of the mill.

Soybean hay meal—obtained from the grinding of the entire soybean hay, without the addition of any stems, straw, or foreign material, or the abstraction of leaves or beans. It shall be reasonably free from other crop plants and weeds, and shall contain not more than 33 percent crude fiber.

Soybean hulls—consisting of the outer covering of the soybean.

Expeller soybean oil chips—obtained after expressing part of the oil from soybeans by crushing, cooking, and mechanical pressure, using an expeller, screw press, or any other mechanical press.

Expeller soybean oil meal—resulting from grinding expeller soybean oil chips.

Hydraulic soybean oil meal—resulting from grinding hydraulic soybean oil cake.

Hydraulic soybean oil cake—obtained after expressing part of the oil from soybeans by crushing, cooking, and hydraulic pressure.

Solvent extracted soybean feed—resulting from the partial removal of protein and nitrogen-free extract from solvent extracted soybean oil meal.

NOTE: All soybean by-products obtained in the manufacture of soybean oil must be designated and sold according to their protein content. The process used in the manufacture of these products must be indicated in their respective brand names.

Solvent extracted soybean flakes—obtained after extracting most of the oil from soybeans by cracking, heating, flaking, and the use of solvents. After extraction of the oil, the product is cooked.

Dehulled solvent extracted soybean flakes—obtained after extracting most of the oil from the dehulled soybeans by cracking, heating, flaking, and the use of solvents. After extraction of the oil, the product is cooked. It shall contain not more than 3 percent crude fiber.

Dehulled solvent extracted soybean oil meal—obtained from grinding and carefully grading dehulled solvent extracted soybean flakes. It shall not contain more than 3 percent crude fiber.

NOTE: TRICHLOROETHYLENE-extracted soybean meals are dangerous. They cause bleeding, digestive disturbances, and death. HEXANE and other organic solvents commonly used in the solvent process usually give satisfactory feeds.

Solvent extracted soybean oil meal—resulting from grinding solvent extracted soybean flakes.

Soybean cubes or soybean pellets consist of soybean oil meal which has been processed through a cubing or pelleting machine. The product shall be firm but not flinty, of sweet odor, and free from mold. Its name must include either the term "hydraulic," "expeller," or "solvent extracted" to specify the method of manufacture of the source material.

NOTE: On the range, *pelleted soybean meals* have met with favor because of convenience and economy in feeding.

Known primarily as a protein concentrate, *soybean meal* (of any type) is also a significant source of VITAMIN B₁ (equal to, or better than, such grains as corn and WHEAT), and supplies appreciable amounts of CHOLINE and NIACIN, but is not rich in other B-complex vitamins (especially riboflavin). Some meals contain appreciable amounts of VITAMIN E and VITAMIN K, but vitamin A is too low in soybean meal to be important in regular dietary practice.

In total digestible nutrients, soybean meal compares favorably with other meals. However, to avoid digestive troubles that may result from the high protein content, the meal should be fed with the same precautions observed with other highly concentrated feeds.

Soybean meal is the leading protein supplement for nearly all classes of livestock. It is used interchangeably with linseed meal and cottonseed meal in rations for cattle (dairy and beef), for sheep, and even for horses and mules. For dairy cattle, many feeders prefer expeller- or hydraulic-produced soybean meal to solvent-process meal in concentrate mixtures otherwise low in fat. Tests with fattening beef cattle generally have not shown a significant difference between meals produced by different processes.

As a protein supplement to corn in feeding swine, soybean meal has met with wide favor. Pigs on pasture grow and fatten efficiently and rapidly on rations of corn and soybean meal, supplemented with ground limestone and salt. In dry lot, inclusion of ground legume hay and an animal-protein concentrate is generally beneficial. Depending on availability, other OILSEED MEALS, such as cottonseed, linseed, and peanut, can be used as replacements for part and, if the pigs weigh more than 75 lb., sometimes all of the soybean meal.

The proteins of soybean meal were once thought to be much less effective than those of ANIMAL BY-PRODUCTS in supplementing the proteins of grains in the diets of farm animals other than ruminants.

We know now that diets, composed mainly of grains and soybean meal, are deficient in some previously unknown vitamins that are supplied by animal-protein supplements of good quality. Vitamin B₁₂ is believed to be the one most often deficient in practical diets for poultry. Its presence makes possible the successful rearing of chickens on diets in which all the protein is of plant origin, most of it being supplied by soybean meal and grains. Care must be taken to see that such diets contain adequate contents of vitamin B₂, calcium, and phosphorus. The quantities of these nutrients in soybean meal are much less than those in the animal-protein supplements. The other oilseed meals are similar to soybean meal in that respect.

Dangers: Soybeans usually are comparatively free from serious insect pests, but losses due to their attacks are gradually increasing. At present, the most important insect pests are GRASSHOPPERS, the VELVETBEAN CATERPILLAR, LEAFHOPPERS, GARDEN WEBWORM, CHINCH BUG, BLISTER BEETLES, BEAN LEAF BEETLES, MEXICAN BEAN BEETLE, JAPANESE BEETLE, FLEA BEETLES, GRAPE COLASPIS, ARMYWORM, FALL ARMYWORM, GREEN CLOVERWORM, CORN EARWORM, and CUTWORMS.

Serious diseases of the soybean, occurring quite generally over the major producing areas, are the following: (1) *Fungus diseases*—BROWN STEM ROT; BROWN SPOT; STALK ROT; FROG-EYE; SCLEROTINIA STEM ROT; ALTERNARIA LEAF SPOT; POD AND STEM BLIGHT; DOWNY MILDEW OF SOYBEANS; CHARCOAL ROT; PURPLE SEED STAIN; and RHIZOCTONIA STALK ROT. (2) *Virus diseases*—BUD BLIGHT; SOYBEAN MOSAIC; and YELLOW MOSAIC. (3) *Bacterial diseases*—BACTERIAL BLIGHT; BACTERIAL PUSTULE; WILDFIRE. (4) *Mineral deficiencies*. (5) *Lightning injury*.

Among other enemies of soybeans are rabbits, which are exceedingly fond of this plant. Where rabbits are numerous, soybean culture is practically impossible unless the field can be enclosed with rabbitproof fencing, or very large areas of the crop can be grown. The dusting of the plants in small plantings, or of the outer rows in larger plantings, with LIME, and the dusting or spraying with some arsenical poison (CALCIUM ARSENATE), have prevented serious damage from rabbits.

Pigeons and pheasants may cause considerable damage to soybean plantings by picking off and eating the cotyledons just as the seedlings are emerging, or picking out the planted seed from the rows.

In many sections deer and rabbits have done much damage, and in some localities in northern states, woodchucks.

Varieties: The United States Department of Agriculture has made more than 10,000 introductions from China, Manchuria, Japan, Korea, East Indies, and India, representing 2,500 distinct soybean types; more than 100 named varieties of soybeans are now handled by domestic growers and seedmen. Unfortunately, there is confusion in the names of varieties, the same variety frequently being known under different names.

Varieties may be divided into three commercial utilization groups; namely commercial, forage, and vegetable. (1) *Commercial* varieties for seed production are preferably yellow-seeded and are used for oil, oil meal, flour, and grits, but these varieties may also be used for forage purposes if heavier rates of seeding are used. (2) *Forage* varieties. (3) *Vegetable* varieties are those that have been found best for eating purposes.

Coming into general use among plant breeders is a *classification* according to relative maturity groups. The varieties being grown in the United States range in maturity from very early (about 75 days) to very late (200 days or more) and have been divided into nine maturity groups (0 through VIII), group 0 and group I being adapted to the northern part of the country. The succeeding groups are adapted farther south, group VIII being grown in the Gulf Coast region.

The varieties mentioned in the table are recommended either for forage or com-

SOYBEAN VARIETIES

Maturity group	Utilization	
	Commercial	Forage
0	Capital, Flambeau, Goldsoy, Kabott, Minsoy, Montreal Manchu, Norsoy, Pagoda, Pridesoy.	
1	Blackhawk, Cayuga, Habaro, Manchu 3, Manchu 606, Manchukota, Mandarin, Mandarin(-Ottawa), Mandarin 507, Monroe, Ontario, Wisconsin Black.	Cayuga, Wisconsin Black.
2	Bavender Special, Earlyana, Granger, Harman, Hawkeye, Mandell, Mingo, Mukden, Richland, Seneca.	
3	Adams, Chief, Dunfield, Illini, Lincoln (soybean), Manchu, Pennsoy, Scioto, Viking.	
4	Boone, Gibson, Hongkong, Macoupin, Mansoy, Midwest (also called Roosevelt, Mongol, or McClave), Morse (soybean), Mount Carmel, Patoka, Wabash.	Ebony (or Black Beauty), Kingwa, Norredo (or Vanderburg Black), Peking, Virginia, Wilson.
5	Haberlandt, Herman, Hollybrook, S 100.	
6	Arkan, Arksoy (or Early Woods Yellow), Arksoy 2913, Armredo, Delsoy (formerly called Edsoy), Dorman, Dortchsoy 2, Magnolia, Mamredo, Ogden, Ralsoy, Rose Non Pop.	Laredo.
7	Charlee, Clemson, C.N.S. (or Clemson Nonshattering), Georgian, Hayseed, Mammoth Brown (or Giant Brown), Mammoth Yellow (also called Southern or Late), Missoy, Monetta, Palmetto, Roanoke, Tennesse Non Pop, Tokyo, Volstate, Woods Yellow, Yelredo.	Barchet, Charlee, Clemson, Georgian, Hayseed, Missoy, Monetta, Palmetto, Tanner, Yelredo.
8	Acadian, Arisoy, Creole, Delsta, LZ, Mamloxi, Mamotan, Nanking, Pelican, Seminole, Yelnando.	Avoyelles, Biloxi, Creole, Gatan, Otootan.

mercial (and forage) uses or for forage use (exclusively), and are listed according to maturity-classification groups.

References: M.19; C.9; W.14; M.20; M.21; M.22; F.6.

See also LEGUME FEED-PRODUCTS; HAY DRYING; STRAW; BUCKWHEAT; COMMON LESPEDEZA; CLUSTER CLOVER; FIELD PEA; MUNGBEAN; PEANUT; DODDERS; WHEAT JOINTWORM; LIGHTNING INJURY; SILAGE CROPS; PASTURES; GRAZING; RANGE MANAGEMENT; MINERAL DEFICIENCIES; LEGUME BACTERIA; TRICHLOROETHYLENE; VITAMIN E; LEGUME.

SOYBEAN CUBES. *See* SOYBEAN.

SOYBEAN HAY.
                   See HAY GRADING; SOYBEAN.

SOYBEAN HAY MEAL. *See* SOYBEAN.

SOYBEAN HULLS. *See* SOYBEAN.

SOYBEAN MEAL.
                   See SOYBEAN; HAY GRADING.

SOYBEAN MILL FEED. *See* SOYBEAN.

SOYBEAN MOSAIC is caused by *Soja virus 1.* The leaves of SOYBEAN plants infected with MOSAIC are distorted, narrower than normal, and their margins turn downward. They may have a yellowish cast and usually show a dark green, blister-like puckering along the veins. Under certain conditions, the leaves may be severely distorted. The infected plants usually show some stunting.

As the summer advances and the weather gets warmer, oil-type soybean varieties show progressively less evidence of the VIRUS DISEASE, and the new leaves show practically none of the distortion observed earlier in the season. Certain vegetable varieties of soybean continue to show leaf distortion throughout the season.

Mosaic is seed-born and transmitted by insects.

Control: Infected plants should be rogued from the field where soybeans are grown for seed.

NOTE: The chemical 2,4-D, which has become popular for weed control, sometimes produces a leaf distortion on soybeans almost identical with that produced by soybean mosaic.

Reference: C.9.

SOYBEAN OIL MEALS. *See* SOYBEAN.

SOYBEAN PASTURE. *See* SOYBEAN.

SOYBEAN PELLETS = *soybean cubes.*
                                  See SOYBEAN.

SOYBEAN SEED. *See* SOYBEAN.

SOYBEAN SILAGE.
                   See SOYBEAN; SILAGE CROPS.

SOYBEAN STRAW. *See* SOYBEAN.

SPANCROSS. *See* CORN (variety).

SPANISH. *Black Spanish* is a *broomcorn.*
                         —*See also* SORGHUMS.

SPANISH DOUBLE. *See* RYE (variety).

SPANISH PEANUT.
                   See SPANISH-TYPE PEANUTS.

SPANISH SWEETCLOVER, formerly called *Madrid White,* is a WHITE SWEETCLOVER variety.

SPANISH-TYPE PEANUTS include *White Spanish, Small Spanish, Improved Spanish,* and *Spanish.*—*See also* PEANUT.

SPARING ACTION. Under certain conditions an animal seems to need less of a given NUTRIENT if another nutrient is present. Thus, the AMINO ACID cystine spares methionine, tryptophan spares NIACIN, FAT spares VITAMIN B_1, and ANTIBIOTICS, while not classed as nutrients, may spare certain B-COMPLEX VITAMINS or protein feedstuffs.

SPARTAN.
            See TWO-ROWED BARLEY (variety).

SPECIAL STEAMED BONE MEAL.
                            See MINERAL FEEDS.

SPECIES. The concept of species has undergone numerous changes in the past hundred years. One of the more satisfactory definitions states that species is a group of individuals with so many characteristics in common as to indicate a very high degree of relationship, and a common descent. Thus, a species is a closed system; that is to say, it will not cross with another species and produce fertile offspring.

NOTE: Therefore, from the genetic standpoint, SUDANGRASS probably should not be separated from the SORGHUMS since it readily hybridizes with them. But there are always exceptions to the rules, and Sudangrass recently has been given species rank by the Botanical Classification group;

for this reason it is now referred to as *Sorghum sudanense* and no longer as *S. vulgare.* var. *sudanense.* Another exception is TALL WHEATGRASS which hybridizes with certain WHEAT species.

GENERA are composed of species which in turn often comprise various SUBSPECIES and VARIETIES.—*See also* HYBRID; SUBGENUS; TYPE.

SPECKLED. "NINETY-DAY SPECKLED," "HUNDRED-DAY SPECKLED," and "EARLY SPECKLED" are DEERING VELVETBEAN varieties.

SPECKLED BLOTCH
= BLACKSTEM OF OATS.

SPECKLED COWPEA = *Whippoorwill.*
See COWPEA.

SPECKLED JAVA = *Taylor.*
See COWPEA.

SPECKLED LEAF SPOT.
See SEPTORIA BLIGHT OF WHEAT.

SPELT (*Triticum spelta*) is a WHEAT species. It may be of either winter or spring habit, and awnless or awned. It is similar to EMMER. The kernels, which remain enclosed in the GLUMES after threshing, are pale red, long, and laterally compressed, with an acute tip and a crease.

Spelt is grown commercially only to a slight extent, mostly in the eastern half of the United States. It is used principally as feed for livestock; its feeding value about equals that of OATS but is somewhat less than that of BARLEY and CORN.

Varieties: Only two varieties are grown commercially in this country; namely, *Alstroum* (beardless) and *Red Winter* (or *red spelt* bearded). Neither is resistant to STEM RUST, LEAF RUST, LOOSE KERNEL SMUT, or COVERED KERNEL SMUT.

References: B.15; M.33.

"SPELTZ" is a name incorrectly used for EMMER.

SPENT BONE BLACK, or *spent bone,* is one of the MINERAL FEEDS.

SPENT HOPS. *Dry Spent hops* is one of the BREWERS' PRODUCTS.

SPERGON, a fungicide containing 98 percent *chloranil,* or *tetrachloro-para-benzoquinone,* controls the KERNEL SMUTS of SORGHUM and, to some extent, improves

emergence. In SEED TREATMENT, apply it at the rate of 2 oz. per bushel in the same way as COPPER CARBONATE, and take the same precautions, even though Spergon is not highly poisonous.

For treating PEANUTS, Spergon is applied as dust at the rate of 3 to 6 oz. per 100 lb. shelled seed.

Reference: L.1.

SPHACELOTHECA. The *Sphacelotheca* spp. are FUNGI which cause diseases that attack the heads of the SORGHUMS: *S. holci* = LOOSE KERNEL SMUT OF JOHNSONGRASS; *S. cruenta* = LOOSE KERNEL SMUT OF SORGHUM; *S. sorghi* = COVERED KERNEL SMUT; *S. reiliana* = HEAD SMUT.

SPHAERALCEA. *Sphaeralcea* spp.
= GLOBEMALLOWS.

SPHEX. One of the *Sphex* spp. is an insect enemy of the ARMYWORM.

SPIDER is the name for a large order which does *not* belong to the true INSECTS. Most of the spiders secrete a thick fluid which they use to form webs. They have eight legs and eight eyes.

The MITES belong to the spiders. Many species are useful because they destroy such insect pests as the POTATO LEAFHOPPER.—*See also* RED SPIDER.

SPIKE is an elongated flower cluster in which the flowers are SESSILE. WHEATS and WHEATGRASSES are examples of plants whose flowers are borne in spikes.—*See also* GRASSES.

SPIKELET is a small cluster. The name is applied to each of the main components of a grass or sedge INFLORESCENCE. A grass spikelet usually consists of two GLUMES and one or more FLORETS.

GRASSES are distinguished primarily by differences of spikelet construction.

Reference: D.9.

SPIKE RUSH (*Eleocharis palustris*), also known as *wiregrass,* grows on poorly drained land and is a troublesome WEED in RICE fields. It is a leafless perennial plant with underground rootstocks.

Reference: J.5.

SPIKE-TOOTH HARROWS with their wide steel teeth are used to break clods, level the ground, and kill weeds.—*See also* HARROW.

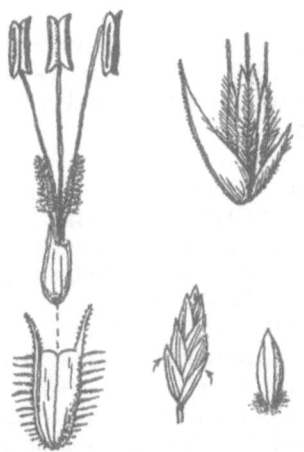

Spikelets. Left, timothy showing anthers, filaments, stigmas (plumose), and lemma which parts together form the floret; glumes are underneath. Right top, blue grama; an incomplete spikelet showing lemma within glumes (large second glume and small first glume). Right bottom, Kentucky bluegrass showing glumes (see arrows) and six lemmas, and separately, lemma with hairs at base.

SPINACH (*Spinacea oleracea*) is a widely used herb. Its leaves and stems are vegetable by-products which, when available, are valued as feedstuffs.

Reference: E.12.

See also PLANT BY-PRODUCTS.

SPINE is a sharp-pointed, rigid, deep-seated emergence from a plant.

SPINELESS BUR-CLOVERS. Experience has shown that the forms of BUR-CLOVER with large spineless burs—e.g. BUTTONCLOVER—cannot be maintained in pastures, except when given special attention and protection. Spineless varieties of CALIFORNIA BUR-CLOVER with small burs have seeds which escape grazing animals more readily and consequently are more persistent.

Reference: M.6.

See also CLOVERS.

SPINELESS HORSEBRUSH (*Tetradymia canescens* var. *inermis*) is a shrub which sprouts quite heavily from stem base and roots. It is a POISONOUS PLANT.—*See also* SAGEBRUSH; HORSEBRUSHES.

SPINY HOPSAGE (*Grayia spinosa*) is a shrub which is considered good forage for

sheep on winter ranges.—*See also* RANGE PLANTS.

SPINY SALTBUSH

= SHADSCALE SALTBUSH.

SPISSISTILUS. *S. festinus*

= THREE-CORNERED ALFALFA HOPPER.

SPITTLE is a whitish, froathy mass found on RED CLOVER and other plants infested with the MEADOW SPITTLEBUG.

SPITTLEBUG. *See* MEADOW SPITTLEBUG;

DWARF DISEASE.

SPLITS. *See* PEANUT.

SPOILED FEED.

See FORAGE POISONING; HAY; SILAGE.

SPOILED SILAGE. *See* SILOS; SILAGE.

SPORANGIUM = SPORE CASE.

SPORE is a microscopic body from a FUNGUS, BACTERIUM, protozoon, or fern, that has entered the *resting state* and is capable of growth and reproduction when the conditions become favorable. Spore formation often provides a plant or an animal organism with a means of surviving unfavorable seasons, or extremes of temperatures.

The one- to many-celled FRUITING BODY (reproductive unit) of a CRYPTOGAM (a seedless plant, such as a fungus) functions like the seed of a higher plant.

Fungus spores may be formed by sexual reproduction or vegetatively. Some have thick walls and are called *resting spores*. Many spores are very light in weight so that they can be blown about by the wind; others fall to the ground and are moved about by water, animals, insects, or machinery. When conditions are favorable, the spore is capable of germinating and producing a new fungus body.—*See also* RUSTS.

SPORE CASE, or *sporangium*, is the minute, usually globular case in which the SPORES of CRYPTOGAMS (seedless plants) are produced; it is somewhat analogous to the ovary of a flowering (seed) plant.

SPORIDIUM (plural: *sporidia*) is a small SPORE produced on a FRUITING BODY of RUSTS, SMUTS, and some other FUNGI.

SPOROBOLUS. The genus *Sporobolus* includes the following DROPSEEDS: *S. airoides* = ALKALI SACATON; *S. cryptandrus* = SAND DROPSEED; *S. wrightii* = SACATON.

SPOROPHORE is that part or organ of a plant which produces the SPORES.

SPORT is an abrupt deviation from the TYPE of a plant or animal. Sports usually breed true to the new type.

SPORULATION is the process of producing SPORES.

SPOTS are definite, diseased areas; they are mostly caused by FUNGI or BACTERIA; e.g., BROWN SPOT OF LUPINE; BROWN SPOT OF RICE; BROWN SPOT OF SOYBEAN; LEAF SPOT OF PEANUT; ALTERNARIA LEAF SPOT; OAT LEAF SPOT; ZONATE LEAF SPOT; GRAY LEAF SPOT; BROWN BORDERED LEAF SPOT; LEAF AND POD SPOT; BORDERED SHEATH SPOT; ROUGH SPOT; TARGET SPOT; and BACTERIAL SPOT.

SPOTTED BUR-CLOVER (*Medicago arabica*), or *southern bur-clover*, is one of the BUR-CLOVER species. It is widely cultivated in the United States and distinguishable by the reddish-brown spot in the center of each leaflet. Spotted bur-clover is adapted to the southern states and is becoming well established on the Pacific Coast, especially in California, along streams and in shady places; it is also preferred in the Cotton Belt.

Spotted bur-clover. (M.6.)

Varieties: *Manganese spotted bur-clover* and *Giant spotted bur-clover* are improved varieties which are being used in many new plantings.

References: M.6; M.10.

See also CLOVERS; PASTURE PLANTS.

SPOTTED CUCUMBER BEETLE is the adult of the SOUTHERN CORN ROOTWORM. This beetle is a subspecies of the WESTERN SPOTTED CUCUMBER BEETLE.

SPOTTED WATERHEMLOCK.

See WATERHEMLOCKS; POISONOUS PLANTS.

SPRANGLE TOP, commonly called *Honey sorgo*, is one of the forage SORGHUMS.

SPREADING OAT = COMMON OAT.

SPRING RABBITBRUSH

= LITTLELEAF HORSEBRUSH.

SPRING-SOWN OATS.

See OATS; COMMON OAT; RED OAT.

SPRING VETCH = COMMON VETCH.

SPROUTED OATS. *See* OATS.

SPROUTING GRAIN. The method of soaking and then sprouting grain by means of special "processes" and equipment has proved valueless, since any increase in weight is due to water absorption.

The sprouting process causes a loss in *carbohydrate* but may increase the VITAMIN content of some grain species.—*See also* PREDIGESTING FEEDS.

SPROUTS. *See* BREWERS' PRODUCTS.

SPUR FETERITA is one of the grain SORGHUMS.

SQUASHES (*Cucurbita* spp.), particularly the large, coarse varieties, may be used for feed. However, if used for fattening hogs, squash causes the fat to turn yellow. For feeding purposes in fall or early winter, the succulent squashes are preferably sliced or chopped.

SQUAW CORN. *See* CORN.

SQUIRRELTAILS (*Sitanion* spp.) are perennial bunchgrasses, varying from GLABROUS (hairless) to densely PUBESCENT (hairy). They are widespread in the western states and have good forage value when young, but at maturity the joints of the SPIKES, with their pointed RACHIS joints and long-awned SPIKELETS, often cause injury to stock.—*See also* RANGE PLANTS; GRASSES.

STACK is a large pile of hay, straw, or grain.—*See also* HAY MEASURING; SILAGE CROPS; SILOS.

STACK-BURNT HAY. *See* HAY.

STACK DRYING. *See* HAY DRYING.

STACKER is a device used for lifting a load of hay and dropping it on a STACK.

STALE BREAD. *See* BREAD.

STALK is a STEM or the support of such organs of leaves, flowers, or flower clusters as PETIOLE, PEDUNCLE, PEDICLE, the FILAMENT (of a STAMEN), etc.

STALK BORER

= COMMON STALK BORER.

STALK DISEASES affect particularly the SORGHUMS. They include WEAK NECK and such FUNGI-caused ROTS as CHARCOAL ROT, FUSARIUM STALK ROT, COLLETOTRI-CHUM STALK ROT, and RHIZOCTONIA STALK ROT.—*See also* NORTHERN ANTHRACNOSE.

STALK ROTS are attributed to FUNGI. Some of them invade SORGHUM plants (especially darso, milo and some of its selections) through openings caused by insects or mechanical injuries. Probably BACTERIA also invade the stalk, and thus help bring about a water-soaked and, later, a rotted condition.

If the upper stalk (PEDUNCLE) is invaded by fungi or bacteria, the injury usually is confined to the peduncle and RACHIS. In some varieties this may result in premature ripening of the head, drying of the rachis and peduncle, and a breaking-over of the upper stalk as in typical WEAK NECK, whereas the lower part of the stalk may remain healthy, and the side branches that grow from the lower nodes may produce good heads. Infections in the lower part of the stalk usually are more destructive. External symptoms of such infections may at first consist of a water-soaked appearance of the stalk, with or without red or purple discoloration, or streaks on the surface of the stalk and in the veins of the sheaths and leaves. Later, one may see poorly developed kernels, premature ripening, and frequently a softening of the base of the stalk, followed by lodging. The inside of the stalk may show water-soaked or discolored pith, or both. The inside of the roots of affected plants likewise appears water-soaked and discolored, and the tips of the diseased roots are frequently dead.

There are four fungi to which stalk rot has been attributed. While none of these is definitely known to be the sole cause, each may play its part. The diseases caused by them are known as CHARCOAL ROT, FUSARIUM STALK ROT, COLLETOTRI-CHUM STALK ROT, and RHIZOCTONIA STALK ROT.

Control: Definite methods for control of stalk rots are as yet unknown. Rotation and other cultural practices may prove helpful, as will also the control of the insects that attack the stalks of sorghum plants.

Reference: L.1.

See also ROTS; STEWART'S WILT; BAC-TERIAL LEAF BLIGHT; NORTHERN CORN LEAF BLIGHT; WHEAT SCAB.

STAMEN is the male floral organ which bears POLLEN grains.—*See also* GRASSES.

STAMINATE means: male; bearing STA-MENS (or pollen-producing) organs only.

STANDARD = BANNER.

STANDARD BLACKHULL KAFIR is one of the grain SORGHUMS.

STANDARD CRESTED WHEAT-GRASS, or *desert wheatgrass*, was for a long time thought to be a uniform WHEAT-GRASS species, namely *Agropyron desertum*. However, it is now recognized as a mixture of two species: primarily *A. desertum* with some *A. cristatum*, which is better known as FAIRWAY CRESTED WHEATGRASS. As compared to the latter, standard crested wheatgrass is much more variable as to size, leafiness, and head type. It has ascending SPIKELETS.

From a nomenclature standpoint, most of the "CRESTED WHEATGRASS" of seedsmen and agronomists in the United States belongs to this type.

Reference: W.16.

STANDARD FETERITA is one of the grain SORGHUMS.

STANDARD WHITE MILO belongs to the grain SORGHUMS.

STANDARD YELLOW MILO is one of the grain SORGHUMS.

STANDBY = *Mediterranean wheat.*

See COMMON WHEAT.

STANTON. *See* RED OAT (variety).

STARCH is a white, odorless, tasteless, granular or powdery, complex CARBOHYDRATE. It occurs widely in plant cells, especially in seeds, bulbs, and tubers.

Starch is insoluble in water, but forms a paste when soaked in it. During digestion, starchy feed hydrolyzes to GLUCOSE and other simple sugars.—*See also* ALCOHOL; AMYLOSE; AMYLOPECTIN; GLUTEN GLYCOGEN; BACTERIA; ENZYMES.

STARTER AND GROWER RATIONS.
See POULTRY RATIONS.

STAR-THISTLE.
See YELLOW STAR-THISTLE.

ST. AUGUSTINEGRASS (*Stenotaphrum secundatum*) is sometimes called *sheep grass* in British Guiana. It is naturally a seashore plant and will withstand salt spray. This grass is found along the southern Atlantic coastal regions as an extensively creeping, rather coarse and hairless (glabrous) perennial that produces STOLONS with long internodes and branches that are short, rather leafy, and flat.

The sheaths are flat and folded, the blades 4 to 6 in. long, the flowering culms 4 to 12 in. tall; the flower SPIKES are 2 to 4 in. long, both terminal and borne at an axil (AXILLARY).

Because it thrives in shaded areas, St. Augustinegrass is especially useful for lawns. It affords also good pasturage but has not been used extensively for that purpose, except for grazing on muck soils in the Everglades in Florida. The creeping, flat stems of St. Augustinegrass root to form dense sods which will stand trampling.

Because practically no seed is produced, vegetative material must be used in making new plantings. For this purpose rooted runners are planted in rows, or disked into the soil during moist periods, and packed subsequently. Establishment is not difficult, and good stands are usually obtained.

St. Augustinegrass should be well fertilized. Nitrogen is especially essential with annual applications on sandy land. Ample moisture is also necessary for best growth and development.

Dangers: St. Augustinegrass is subject to BROWN PATCH fungus, which does most damage in warm weather when there is undue moisture. But so far the disease has not been a serious factor in growing the grass.

CHINCH BUGS also may do damage.

Reference: H.1.

See also GRASSES.

ST. CHARLES WHITE.
See CORN (variety).

STEAMED BEANS. *See* FIELD BEANS.

STEAMED BONE MEAL.
See MINERAL FEEDS.

STEAMING FEED. *See* COOKING FEED.

STEARIN, or *tristearin*, consists of *stearic acid* and GLYCERIN. It is the colorless, solid material contained in many hard animal fats, or obtained after chilling oils. Stearin is insoluble in water. When used in feedstuffs, it must be labeled; i.e., the name stearin must be prefixed by the name of

St. Augustinegrass. (H.1.)

the vegetable or animal oil, or fat, from which it was obtained.—*See also* VITAMINS.

STEER RATIONS.
                    See BEEF CATTLE RATIONS.

STEIGUM.
          See TWO-ROWED BARLEY (variety).

STELLARIA. *S. media* = CHICKWEED.

STEM is the continuation of the root, i.e., the ascending axis of a plant from which such organs as the leaves, flowers, and fruits develop. Special forms of stems for specific functions are the RHIZOMES, ROOT-STOCKS, STOLONS, CROWNS, and TENDRILS.

A *jointed* or *solid* stem or STALK is known as CULM.—*See also* GRASSES.

STEM BLIGHT.
                See POD AND STEM BLIGHT.

STEM CANKER OF CROTALARIA, caused by the FUNGUS *Rhizoctonia solani*, destroys many young CROTALARIAS, particularly the SHOWY CROTALARIA varieties. This FUNGUS DISEASE spreads in hot, dry weather; it is reported to have killed as many as 30 percent of the plants in blooming time.

STEM CANKERS include ASCOCHYTA STEM CANKER, BOTRYTIS STEM CANKER, and STEM CANKER OF CROTALARIA.

STEM DISEASES are BLACKSTEM, SOUTHERN ANTHRACNOSE, and various STEM CANKERS.

STEM MEAL. *Alfalfa stem meal* is an official feedstuff.—*See also* ALFALFAS.

STEM NEMATODE (*Ditylenchus dipsaci*) causes injury in some of the important ALFALFA producing regions. Damage may result in a reduced production of hay, or a killing of the alfalfa stand in two or three years. Moist weather favors NEMATODE infection.

Nematode injury is located principally in the crown, affecting the young buds and stem bases. Infected buds become thickened, deformed, and usually do not elongate into stems. Later in the season, particularly where damage has not been severe, infected plants may be dwarfed and have a reduced number of stems, some of which may have become yellow or distorted. Such plants often have stems swollen at the base and dark brown in color. They are also brittle and easily broken off. Infected shoots sometimes

have swollen areas near the tip or at nodes along the stem.

The nematodes which cause this disease may be distributed from infected areas to neighboring fields through the movement of soil and irrigation water. They may also be transported in infected hay, or in poorly cleaned seed.

Control: Crop rotation is helpful as a means of control. Heavily infested fields should be plowed and seeded to other crops for at least two or three years with special care to eliminate all alfalfa and CLOVER plants. Crops suitable for rotation include WHEAT, CORN, BARLEY, POTATOES, and GRASSES.

At present, NEMASTAN is the only variety known to be resistant to stem nematode in alfalfa fields.

Reference: S.12.

STEMPHYLIUM. The FUNGUS *S. sarinaeforme* is the cause of STEMPHYLIUM LEAF SPOT.

STEMPHYLIUM LEAF SPOT (*Stemphylium sarinaeforme*) is quite common on RED CLOVER. Leaf infections gradually enlarge and eventually destroy the entire leaf. Many of the lower leaves often are destroyed, particularly in thick stands. The lesions on leaves are typically irregular with concentric zones of brown spots of various sizes.

This FUNGUS LEAF DISEASE is most injurious to the hay crop. It reduces the hay's quality because of the loss of lower leaves. Infections can be found on most plants during the entire growing season, but the fungus appears to be most injurious in early summer.

Reference: E.5.

See also LEAF DISEASES.

STEM ROT OF RICE is one of the more important diseases of RICE. It is caused by either of two species of FUNGI, *Leptosphaeria salvinii* and *Helminthosporium sigmoideum* var. *irregulare*, both of which may be found on the same plant. The disease is widespread in Arkansas and Louisiana, but occurs also in Texas and California.

The earliest signs of the disease may be seen late in July or early in August, when small, black, discolored areas appear on the leaf sheaths at the surface of the water, or slightly above. When the spots on the lower, outer leaf sheaths enlarge, the inner sheaths are also invaded and discolored. When the infection reaches the stalk, numerous small, shiny, black, appressed bodies are produced on it by the fungus. Dark masses of fungus threads then develop on the stalk, and longitudinal brown to black streaks appear in the stalk above and below these masses. Soon, the fungus threads may be found in the interior of the stalk as a cottony, white mass, and by the time the rice is nearly mature, numerous small, black, seedlike fungus bodies may be found in the sheaths and stalks. It is at this stage that the stalks break over and the plants lodge. Plants infected at an early stage of development produce only lightweight heads.

The fungus remains alive in the soil for six years. It lives through the winter as seedlike fungus bodies (SCLEROTIA) in stubble and soil, and these bodies floating on the water produce the primary infection on the sheaths of plants the following summer. Primary infection may also be produced from the fungus living-over on BARNYARDGRASS and on southern WILD-RICE growing in and near fields, and in irrigation ditches. Secondary infection by spores produced on the sheaths occurs after the fungus has been established on the rice plants.

Control: The most satisfactory method of control is to drain the water from infected fields before the infections have reached the rice stalks. Enough water should be added from time to time thereafter to keep the soil saturated, but not submerged. Such treatment results in a light reduction in yield. No highly resistant varieties of commercial value are available in the South.

Reference: T.2.

See also BLACK SHEATH ROT.

STEM ROTS are FUNGUS DISEASES which affect the stems of forage and other plants. —*See also* ROTS; CROWN ROTS; ROOT AND STEM ROT; STEM ROT OF RICE; STEM CANKERS; SUBCLOVER; VETCHES.

STEM RUST OF OATS, caused by *Puccinia graminis avenae*, is a rather mild FUNGUS disease of OATS. It usually occurs late in the season and is not noticed until the crop is nearly mature.

The *red-spore stage* of stem rust of oats primarily attacks the stems and leaf sheaths, but in some cases may attack the leaves. The pustules are long and irregular and rupture the stem, causing a scaly appearance. The *black-spore stage*, which follows the red-spore stage, occurs very late in the season on dead or dying plants, and is often overlooked at harvest time. Heavy infections of stem rust of oats can cause premature ripening of the grain, with a resulting decrease in yield and quality.

Control: Stem rust of oats can be controlled by planting resistant varieties. Growing early maturing varieties will decrease chances of severe damage.

Reference: M.30.

STEM RUSTS are FUNGUS DISEASES occurring on the stems of plants; e.g., STEM RUST OF OATS and WHEAT STEM RUST. —*See also* RUSTS; BARBERRY BUSHES; MEADOW FOXTAIL; RUSSIAN WILD-RYE; RYE; EMMER; DURUM WHEAT; SPELT; CLUB WHEAT; COMMON WHEAT; WHEATS; SWITCHGRASS.

STEM SAWFLY.

See WHEAT STEM SAWFLY.

STEM SMUT, caused by *Ustilago* spp., is a FUNGUS DISEASE that develops in significant abundance on economic grasses; e.g., *U. occulta* which affects RYE.

Stem smut is characterized by the development of conspicuous, dusty, brown or black layers of SMUT around the internodes of the stems. At first the smut is hidden by the leaf sheath that envelopes the stem, but as the stem grows, the smut is exposed.

Reference: F.4.

STENOTAPHRUM. *S. secundatum* = ST. AUGUSTINEGRASS.

STERILE means: barren. It is said of shoots that produce leaves but no flowers, and of ANTHERS (pollen sacs) that are rudimentary and do not function. The causes of *sterility*—i.e., the factors which hinder normal flowering, POLLINATION, FERTILIZATION, or fruit bearing—are manifold; they may be inherent, or due to environmental influences.

Self-sterile is a plant which will not set seed to its own pollen.—*See also* TRUE CLOVERS.

STEROID is a generic name for a variety of compounds which includes STEROLS, SAPONINS, and sex HORMONES.

STEROLS are complex, unsaturated ALCOHOLS which are widely distributed in animal and plant tissues; e.g., CHOLESTEROL and ERGOSTEROL.—*See also* VITAMIN D; VITAMINS.

STEWART DURUM is a white DURUM WHEAT variety.

STEWART'S WILT, or *bacterial wilt of corn,* is caused by *Bacterium stewartii;* it is more destructive to sweet corn than to dent corn. In this country Stewart's wilt is most abundant in the Central States from Iowa eastward to the Atlantic Coast.

This CORN disease is carried over the winter within the bodies of CORN FLEA BEETLES. Adult beetles begin to feed on young corn seedlings in late spring and early summer, and in so doing start infections on the leaves. During the growing season these beetles continue to spread the disease from infected to healthy plants. When winters are mild, a large number of corn flea beetles will usually survive. As a result, the disease is abundant the next summer.

Infected plants of *sweet corn* varieties wilt rapidly, and resemble plants with inadequate water supply. They die or are stunted, tassel prematurely, and may produce no ears. Infected plants frequently show irregular, pale-green to yellowish streaks in the leaves. The vascular bundles in the entire plant become filled with bacteria, so that when the stalks are cut, yellow masses of the organism ooze out as beads on the cut surface. In severely infected plants cavities may form in the pith of the stalk. Bacteria spread through the vascular system of the entire plant, passing through the cob into the kernels. Infected kernels are a means of spreading the disease to new localities.

In *dent corn,* the characteristic symptom is a number of long, irregular, pale-green streaks in the leaves which eventually turn yellow, die, and become straw-colored. The dead tissue often becomes covered with fungi, which may mistakenly be assumed to be the cause of the disease. Where infection is severe, much of the leaf area is destroyed. Susceptibility to STALK ROT increases, and stalk breaking becomes prevalent when this occurs. Dent corn kernels may become infected, but this is very rare.

Control: The most practical means of control of this disease is by use of resistant hybrids. There is some indication that well balanced soil fertility, with an optimum supply of potassium, tends to minimize the disease. High levels of nitrogen, however, may predispose plants toward susceptibility.

References: U.7; E.11.

STICK. *See* ANIMAL PRODUCTS.

STIFFHAIR WHEATGRASS

= PUBESCENT WHEATGRASS.

STIFF LAMB DISEASE. *See* FIELD BEANS; FISH-LIVER OILS; VITAMIN E.

STIFFSTEM FLAX belongs to the *poisonous* FLAXES.

STIGMA is the feathery (plumose) receptive organ of a PISTIL through which FERTILIZATION by the POLLEN grains is accomplished.—*See also* GRASSES.

STILBESTROL, or *diethylstilbestrol,* is a synthetic substance having the action of a female sex hormone. This drug forms a white, crystalline powder which is soluble in fatty oils.

Stilbestrol *pellets* can be implanted subcutaneously (under the skin) in the region of the shoulders, or near the ear; they increase growth rate and feed efficiency of beef cattle, sheep, and poultry. An alternate method to treat animals is to supply them with feed which contains stilbestrol in proper proportion, and well mixed in accordance with F.D.A. regulations. In general, a daily oral dose of 10 mg. stilbestrol per head is considered a safe, effective dose for beef steers weighing more than 600 lb.; it results in significant increase of GROWTH and feed efficiency, without noticeable effect on carcass quality.

NOTE: Encouraging experimental work is in progress concerning the use of stilbestrol-containing feed for dairy animals, lambs, and poultry, but no favorable effects have been seen with swine.

STILLAGE. *See* DISTILLERS' PRODUCTS.

STIMULANTS are drugs which produce stimulation in general—these are often called *"tonics"*—or of a particular organ.

Stomachic stimulants, or *stomachic tonics,* are used to stimulate appetite and promote digestion of food in the stomach. ANISE SEED, CARAWAY SEED, FENNEL, BITTERS, GINGER and GENTIAN are also used as stomachic stimulants.—*See also* FEED INGREDIENTS; TONICS.

STING NEMATODE (*Belonolaimus gracilis*) is a very serious pest of PEANUTS in Virginia. It is associated with severe stunting and loss of yield. This NEMATODE, in contrast to the ROOT-KNOT NEMATODE and the MEADOW NEMATODE, is an ectoparasite; that is, it feeds on the roots from the outside by puncturing them with a needlike "sting" and then withdrawing the juices. This nematode will not be found in the roots but in the surrounding soil.

Control: Good control of the sting nematode, and correction of associated stunting, has been obtained by SOIL FUMIGATION.

References: B.10; W.10; W.18.

See also DOWFUME W-40.

STINK BUGS are important pests of ALFALFA, but they feed also on BUR-CLOVER, BARLEY, COTTON, grain SORGHUM, OATS, WHEAT, PIGWEED, JOHNSONGRASS, BEANS, SUGAR BEETS, CORN, COWPEAS, VETCH, and other plants. Adults and nymphs injure seed by removing the liquid contents while the seeds are immature. The damaged seed collapses into a flattened shell, which soon shrivels, dries, turns brown, and is usually blown out with the chaff during threshing. A round, and sometimes slightly depressed, scar develops at the puncture.

About five weeks are required in summer, and seven to eight weeks in the spring and fall, for most stink bug species to complete their life cycle from egg to adult. Females lay on an average 60 to 150 eggs. Adults usually live forty to sixty days in spring and summer, but some overwintering adults live seven to nine months.

Stink bugs are seldom active below 60° F. In southern climates they pass the winter in the adult stage. They may be found near the soil surface under low-growing plants or plant debris. Weedy and grassy fence rows, roadsides, and ditchbanks are important winter habitats in irrigated areas.

It is usually early March before the adults resume normal activity on growing plants. Mating does not become common until the temperature reaches 75° or above. The females lay their eggs on weed growth, and the first generation nymphs, feeding mainly on the immature seed, grow to maturity on it. Most of the first generation move into fields of alfalfa, small grains, sugar beets, and other crops.

The adults fly readily from field to field or from crop to crop during spring, summer, and early fall. The nymphs do not fly, and usually travel shorter distances.

Control: A large percentage of stink bug eggs is destroyed by parasitic and predaceous insects. Adult stink bugs also are often parasitized.

Good kills of stink bugs are obtained with fairly high dosages of CHLORDANE, LINDANE, PARATHION, or TOXAPHENE (the latter preferably with sulfur dust), and with mixtures containing DDT plus lindane, chlordane, or toxaphene. Insecticidal treatment should be given before the blooming period is over.

The most important natural enemies of stink bugs are *Telenomus utahensis,* an egg parasite, and *Rhodogyne fuliginosa,* a parasite of the adults.

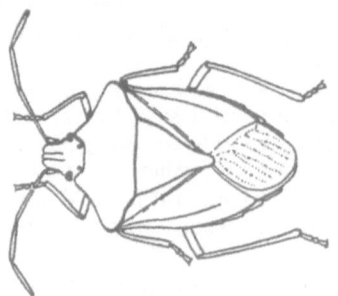

Stink bug. (U.13.)

Destructive populations of stink bugs, particularly in seed alfalfa, can usually be prevented by (1) destroying weeds or dead plant material that provide adult bugs with winter shelter, (2) controlling weeds late in the winter and in the spring, in and surrounding seed-alfalfa fields, (3) growing seed alfalfa as far removed as practicable from fields of sugar beets or SMALL GRAINS, (4) starting and harvesting the seed crop on uniform dates in all fields in a wide area, and (5) growing the seed crop as rapidly as good agronomic practices permit.

Species: Among the most important stink bug species are the SAY STINK BUG, the BROWN COTTON BUG, and the RED-SHOULDERED PLANT BUG.

Reference: R.6.

STINKING SMUT = COMMON BUNT.

STINKWEED = PENNYCRESS.

STIPA. The genus *Stipa* includes about 30 species of *stipa grasses*, or *stipas*, better known as NEEDLEGRASSES; among them are *S. comata* = NEEDLE-AND-THREAD GRASS; *S. viridula* = GREEN NEEDLEGRASS, and its variety *green stipa grass*; *S. spartea* = PORCUPINE GRASS; *S. pulchra* = CALIFORNIA NEEDLEGRASS; and *S. arida* = MORMON NEEDLEGRASS.

STIPULE is one of a pair of leaflike appendages borne at the base of certain PETIOLES (leafstalks).

STIZOLOBIUM. There is no real authority for retaining the generic name *Stizolobium* in place of the new official name *Mucuna*, which has been conserved for the VELVETBEANS.

ST. JOHN'S BREAD = CAROB BEAN.

ST. JOHNSWORT (*Hypericum perforatum*), more correctly called *common St. Johnswort* or *goatweed*, is a POISONOUS PLANT found as a very aggressive WEED pest on many ranges. Cattle, sheep, and goats with white skin and hair, when standing in bright sunlight and feeding on the leaves of the plant, soon show sore, scabby areas on their skin, followed by itching and rapid respiration.

STOCKSBILL = ALFILERIA.

STOCKYARD DISEASE

         = HEMORRHAGIC SEPTOCEMIA.

St. Johnswort. Flowering shoot and lower part of plant showing shoots of runners. (H.48.)

STOLON, similar to a RUNNER, is a trailing or reclining STEM or branch above ground; however, it strikes root wherever it touches the soil and at that point sends up new shoots which later become separate plants. Many plants multiply vegetatively in this way.

STOMA (plural: *stomata*) is one of the small openings—breathing pores—on the surface of a leaf, especially on its underside.

STOMACH—the "sack" between gullet and small intestine—contains HYDRO-

CHLORIC ACID, GASTRIC JUICE, and the ENZYMES which play an important role in proper DIGESTION.

The stomach of RUMINANTS consists of four "compartments": the first and largest called RUMEN, or *paunch;* the second, *reticulum,* or *honeycomb;* the third, *omasum, manifold,* or *manyplies;* and the fourth—the only one that produces gastric juice and enzymes—*abomasum,* or *true stomach.*

The simple stomach of a horse holds 3 to 5 gal., that of a pig over 2 gal.; the compound stomach of a mature cow, however, holds up to 65 gal., making it possible for the ruminant to consume and digest large quantities of roughage.—*See also* INTESTINES; ABSORPTION; RUMINATION.

STOMACHIC TONICS.
                    See STIMULANTS; TONICS.
STOMACH JUICE = GASTRIC JUICE.
STOMACH STIMULANTS.
                    See STIMULANTS.
STOMATA. *See* STOMA.
STONE BRAN. *Rice stone bran* is an officially recognized RICE feed.
STONER = *Fulcaster.*
                    See COMMON WHEAT.
STORING FEEDS differs for bulk and sacked materials. For storage of grains, *grain elevators* with properly designed hoppered bins, conveyors, and spouts are used. Large *steel tanks,* similar to silos, are also practical for feed storage, as are *portable grain bins.*

When *ground feed* is to be stored, its moisture content and its keeping qualities must be known. Ground feed is more perishable than whole grains, as it becomes rancid and loses its vitamins more rapidly. Loss of vitamins is more rapid, too, if the feed contains TRACE ELEMENTS.

Sacked feed can easily be stored in a warehouse provided it offers protection against rodents. Bags should be stacked so that the ones that have been in storage for the longest time, will be moved out first

For *unloading* bulk material from trucks, dump hoppers and truck lifts or hoists are used, while for sacked feed trucks, portable conveyors, and hand pilers are useful. Various types of scales are available for weighing the feed material.

STOVER is the remainder of the CORN or SORGHUM plant after the corn ear, or sorghum head, has been removed. It is rough feed of low nutritional value. Mature corn gives 50 to 65 percent stover, grain sorghums (and occasionally sorgos) yield from 45 to 75 percent stover. Only those sorghums which have juicy stalks, such as hegari and some kafirs, are satisfactory.

STOWELL'S EVERGREEN.
                    See CORN (variety).
STRAIGHTHEAD is a condition characterized by the failing of normal-appearing RICE plants to set seed in many of the flowers. As a result, the heads remain erect instead of turning down as they do when they fill normally. Straighthead is most prevalent on new land, and on land that has not been cropped to rice for several years and on which a heavy growth of weeds and grass has been plowed under. In fields where this disease does strike, large areas are often so seriously affected that the yield of grain is not sufficient to justify harvesting and threshing.

Control: Land on which straighthead is likely to occur should be drained at least once and allowed to dry until the surface cracks, before the rice is in the boot.

Reference: T.2.

STRAIGHTNECK. *Straightneck milo,* better known as *Fargo,* belongs to the kafir-milo derivatives and, as such, to the grain SORGHUMS.

Straightneck sorgo is one of the forage SORGHUMS.

STRAIGHT-STEM POISONVETCH.
                    See POISONVETCH.
STRAIN is a SELECTION which has progressed to the point where it is relatively true-breeding and ready for testing. A VARIETY is a former strain which has survived the testing program of the plant breeders, and has been released to the seed growers.—*See also* HYBRID; RACE; ECOTYPE; DERIVATIVE.

STRANGLEWEED = DODDER.
STRAVROPOL.
                    See SIX-ROWED BARLEY (variety).
STRAW is the dried residue of fine-stemmed plants from which the seeds have been removed. It is widely used as ROUGHAGE. Straw has a higher fiber content and, therefore, fewer digestible nu-

trients and lower nutritional value than HAY made from the same crop. Among the crops whose straw is used for stock feeding are ALFALFAS, various CLOVER species, FIELD PEAS, FIELD BEANS, LESPEDEZAS, SOYBEANS, and other LEGUMES; BUCK-WHEATS, SMALL GRAINS, and other CEREALS and GRASSES; FLAXES; etc.

Cattle make better use of straws than do other species, and even for cattle only little straw should be included in production rations. Straw has its greatest value when fed as the bulk of a maintenance ration for mature cows. If supplemented with 5 lb. legume hay, good grass hay, and OIL MEAL, or 15 to 20 lb. SILAGE and MINERALS, straw makes a suitable winter ration for mature cattle or horses.—*See also* SCREENINGS; BLOAT; HYDROLYZED STRAW.

STRAWBERRY CLOVER (*Trifolium fragiferum*) plays an important role in the reclamation of seepy, saline, and alkaline soils.

This true clover is a perennial, low-growing, pasture legume which spreads vegetatively by creeping stems that root at the nodes. It is difficult to distinguish—except when in bloom—from certain types of WHITE CLOVER. However, its flower heads are round or slightly pointed, and in color they are mostly pink to white, resembling a strawberry.

The flowers of strawberry clover are self-fertile. Strawberry clover is palatable and is relished by all classes of livestock and poultry, but seldom grows tall enough to harvest for hay.

Strawberry clover is adapted to a wide range of conditions, having been established successfully in most of the western states. It has been observed to make a good growth in association with SALT-GRASS; in established stands, plants have survived salt concentrations of over 3 percent. In general, strawberry clover will make its best growth in wet to moist soils, and its favorable adaptation to such conditions makes it extremely valuable for large areas in most irrigation projects where drainage is a limiting factor in crop production. Another valuable characteristic is its ability to survive flooding for one to

two months. However, the clover will not make sufficient growth to warrant its use on dry lands. Strawberry clover thrives under wide extremes of temperatures, ranging from 40° below zero to high summer temperatures.

Seeding should be done early in the spring on a firm, moist seedbed; the seed may either be broadcast or drilled in not very deep. This clover spreads rapidly by the creeping stems, and originally thin stands may become thick by the end of the second year. On nonsaline soils, strawberry clover may be spring-seeded with a companion GRAIN crop, if an ample supply of moisture is available throughout the season.

Strawberry clover is principally a pasture plant, though it may be used as a green-manure crop, particularly on soils where salinity prevents the growth of other LEGUMES. It is very palatable and is perhaps equal to white clover. When grown on saline soils, the composition of the vegetation is somewhat higher in minerals than when grown on salt-free soils, but all available reports indicate that animals have not been injuriously affected from grazing it. All types of animals and poultry have grazed strawberry clover with good results. It will survive under close grazing, and has been grazed continuously from early spring until late in the fall without affecting the stand.

Farmers who have used large acreages for grazing claim that on similar soils the carrying capacity of strawberry clover is far superior to that of other pastures. Many claim that one acre will carry one or two cattle through the entire growing season, provided growing conditions are favorable.

Dangers: Like other legumes, strawberry clover may cause animals to BLOAT, and necessary preventative measures should be taken.

Variety: *Shearman's clover* differs from common strawberry clover in that it is easily winter-killed.

Reference: H.12.

See also TRUE CLOVERS; INOCULATION; PASTURE PLANTS.

STRAW-COLORED CARBOLIC ACID
= CRESYLIC ACID.

STRAW MEAL, an officially recognized feedstuff, is the ground product that remains after separation of the seed from mature forage plants. The source of material shall constitute a part of the name of the product; for example, "bluegrass straw meal" or "alfalfa straw meal."

Reference: F.6.

See also MISCELLANEOUS PRODUCTS.

STRAWWORM.
See WHEAT STRAWWORM.

STREAK. *See* BACTERIAL STREAK.

STREPTOMYCES. Among the MOLDS which are important as producers of ANTIBIOTICS and VITAMIN B_{12}, are the *Streptomyces* spp.; particularly *S. aureofaciens*, used for the manufacture of AUREOMYCIN; *S. venezuelae* produces CHLOROMYCETIN, *S. rimosus*, TERRAMYCIN; and *S. griseus*, STREPTOMYCIN.—*See also* FERMENTATION PRODUCTS.

STREPTOMYCIN is one of the ANTIBIOTICS used not only as drug but also as a feed supplement.

It is produced by the mold *Streptomyces griseus* and marketed in form of various salts; e.g., *streptomycin sulfate* and *streptomycin hydrochloride*. These salts are water-soluble, white to tan powders.

> NOTE: 1,000,000 units streptomycin base (i.e., the active streptomycin itself which is combined with inert, inorganic material to form the streptomycin salts) is equivalent to 1 gm.

Dihydrostreptomycin is a derivative of streptomycin, used for the same purposes as the latter.—*See also* GROWTH.

STRIATA. *Giant Striata* is a variety of STRIPED CROTALARIA.

STRIATA CLOVER (*Trifolium striatum*), one of the TRUE CLOVERS, forms productive stands in only a very few areas.

Reference: M.3.

STRIATE means: marked with parallel, longitudinal grooves or ridges.

STRIP CROPPING is the growing of crops in long, narrow strips across a slope on a line of contour with dense-growing and intertilled crops alternating.—*See* DUST STORMS.

STRIPE. *See* BACTERIAL STRIPE OF SORGHUM; BACTERIAL STRIPE OF OATS; SOOTY STRIPE; BARLEY STRIPE; STRIPE RUST; STRIPE SMUTS.

STRIPED CROTALARIA (*Crotalaria mucronata*, or *C. striata*) is one of the non-poisonous CROTALARIA species which are grown extensively in the United States, especially for soil improvement and for pasture, particularly in open woodland. The plants are often attacked by CROTALARIA ANTHRACNOSE, GRAY MOLD, and other crotalaria diseases.

Variety: *Giant Striata* is the name of a large-growing type of the striped crotalaria.

Reference: M.14.

STRIPE RUST (*Puccinia glumarum*) is also known as *yellow rust* because of the color of its pustules in the summer stage: the pustules are arranged in rows, giving the appearance of narrow yellow stripes. The FUNGUS attacks WHEATS and other cultivated and wild GRASSES; however, this fungus disease has not become established in the grain-growing regions. It develops most abundantly on the GLUMES or CHAFFS, leaves and leaf sheaths, but may also attack the stems and the kernels. If the attack of stripe rust is heavy, considerable damage is likely to result; under such conditions the kernels may be shriveled and the yields considerably reduced.

The *black stage* of the rust is formed after the *red* (or summer) stage. The black pustules also are produced in rows that look like dark-brown or black stripes; they may appear on all above-ground parts of the plant. The *yellow* stage of the rust overwinters under a variety of climatic conditions.

Epidemics of stripe rust are most likely to occur when there has been abundant infection in late summer and fall of the previous year, and when spring and summer conditions—such as cool nights, and warm, sunny (but not hot) days—prevail during the growing season.

Badly rusted grain may germinate

poorly, but infection is not transmitted to plants grown from such diseased seed.

STRIPE SMUTS are caused by various *Urocystis* and *Ustilago* spp. These belong to the *leaf smuts* which are evident in the leaves of GRASSES as black stripes and contain the spores. The spores are shed and disposed into the wind. Afterwards, the affected leaves take on a shredded appearance and wither. Affected plants quite frequently are dwarfed and contorted, and produce abnormal, sterile heads, if any at all. Seedlings of grasses affected with stripe smut are predisposed to drought injury and ROOT ROT. Mature plants often are so weakened that they cannot survive severe winters. One of the most virulent and common stripe smuts is known as FLAG SMUT OF WHEAT.

Another of the common and destructive stripe smuts is the one that has long gone under the name of *(true) stripe smut*, or *"leaf smut."* Still another type of the leaf smut is the *"sausage" smut;* it produces small, but conspicuous, blister-like (often sausage-shaped), black pustules on the leaves of the GRAMA GRASSES.

Control: Some of the stripe smuts may be controlled by SEED TREATMENT with FORMALDEHYDE SOLUTION.

Reference: F.4.

See also ORCHARDGRASS; SMUTS.

STRONGER AMMONIA WATER.
See AMMONIA.

STRYCHNINE is an ALKALOID, available in form of water-insoluble crystals; however, many of its salts—e.g., *strychnine sulfate*—are water-soluble and, like the free alkaloid, used as RODENTICIDES.

Caution: Strychnine is a violent poison and must be handled with the greatest of care even by professional pest-control operators.—*See also* STRYCHNINE-POISONED GRAIN.

STRYCHNINE-POISONED GRAIN, i.e., grain impregnated with STRYCHNINE, is used as bait to control FIELD MICE, POCKET GOPHERS, moles, and other RODENTS, as well as birds, but it is not commonly used as rat poison. The bait is scattered in the areas frequented by the animals to be exterminated.

STUBBLE is the basal portion of the STEMS of plants remaining standing after cutting.

STUBBLE-MULCH FARMING.
See DUST STORMS.

STUBBLE PASTURES are *supplemental* PASTURES.

STUMP PASTURES. *See* PASTURES.

STYLE is the stalklike and often slender portion of the PISTIL (female floral organ); it connects the STIGMA with the OVARY.—*See also* GRASSES.

SUAEDA. *Suaeda* spp. = SEEPWEEDS.

SUB- is a prefix signifying "low" or "below."

SUBCLOVER (*Trifolium subterraneum*), also known as *subterranean clover*, is a winter annual LEGUME that is well adapted to use in pastures. Subclover is so named because of its habit of burying the developing seed heads in the soil, or beneath the vegetative mat that the plant forms on the surface of the soil. This legume is suited to drained, somewhat acid, summer-droughty soils, has a vigorous growth of nutritious forage, and possesses persistence in pastures in combination with GRASSES. The plant is of particular value on lands that are too dry, or too low in fertility, for WHITE CLOVER.

No other legume has been able to offer so much to stockmen in western Oregon as subclover; it is found also in other areas of the northwestern states.

The plant normally starts from seed in the fall. Development during the winter is rather slow. Growth during the spring is rapid, and by midsummer the seeds are matured and the plants die. Most plants from spring planting, however, live over into the second year. Well established seedlings can survive temperatures as low as 5° F.

Subclover, when it is to be used for forage, should always be planted in mixtures with adapted, companionable grasses. Growers often establish subclover on old grass sod by broadcasting the seed in the fall without seedbed preparation. A better method of planting is to scatter threshed subclover straw over the land, and trample it into the soil with livestock. The best method of establishing the clover in mixture with grasses is to plant at the time the grass is planted.

This clover is naturally adapted to early fall planting. Spring planting in April, May, or June, on well prepared seedbeds, is often practiced where soils heave badly with winter frosts.

Subclover develops root nodules with the same nodule bacteria that inoculate other TRUE CLOVERS. Where these bacteria are not present in the soil, they must be applied artificially to the seed, or distributed in inoculated soil at the time of planting. Threshed subclover straw and heads from the current year's crop will carry the bacteria to fields over which it may be scattered.

The seed may be planted by drilling or by broadcasting—followed by harrowing-in where possible—and should not be covered more than 1 in.

Subclover requires fertilization to maintain high production. A general recommendation is a yearly application of 200 to 300 lb. superphosphate (16 percent) per acre, or 100 to 150 lb. each, of gypsum and triple superphosphate (45 percent), per acre. Lime, applied at 1 to 2 tons per acre, may be beneficial on the more acid hill soils. Subclover is relished by all kinds of livestock. It is also suited as silage or hay.

Good management is essential to the maintenance of a balance between subclover and grasses in pastures. Tall-growing and vigorous grasses, such as ALTA FESCUE, if allowed to make considerable growth, may reduce excessive amounts of subclover in the pastures. Dense-sodded grasses, such as RED FESCUE and CHEWINGS FESCUE, tend to prevent domination by subclover. Other pasture plants, grown together with subclover, are RYE-GRASS, ORCHARDGRASS, and/or MEADOW FOXTAIL.

Dangers: When subclover grows on soil with an unbalanced MINERAL make-up, it may contain so much estrogen-like material that sheep become sterile.

STEM ROT causes rapid dying of infected plants and is most active on subclover in late winter and early spring. (A legume-free crop rotation is the only practical control measure). Less destructive diseases that may appear are POWDERY MILDEW, LEAF RUST, SOUTHERN ANTHRACNOSE, and CLOVER RUST.

VIRUS DISEASES, spread chiefly by APHIDS, are sometimes destructive. The effects on subclover are: crinkling, mottling, and yellow streaking of the leaves, with stunting and occasional killing of the plants. Spring plantings are most likely to be affected.

The principal pests of subclover are the common garden SLUG and the WESTERN SPOTTED CUCUMBER BEETLE.

GOPHERS destroy many plants. FIELD MICE inhabit the abandoned tunnels of gophers, moles, and digger squirrels, and eat the clover seeds and seedling plants.

Strains: Subclover is quite variable in plant types. More than 40 different strains have been recognized. These strains are grouped into three types—*early, midseason,* and *late*.

1. The various *early* strains are relatively low in production, and are useful in areas of low rainfall where later maturing strains cannot persist.

2. *Mt. Barker* is a midseason strain.

3. *Tallarook* is a medium-late strain; *Nangeela* is a vigorous midseason-to-late strain. In general, the later strains are the highest forage and seed yielders and require the most favorable growing conditions.

Reference: R.3.

See also INOCULATION.

SUBCUTANEOUS means: under the skin. Subcutaneous implantations of STILBESTROL pellets are made to accelerate meat production.—*See also* ANTIBIOTICS.

SUBFAMILY = TRIBE.

SUBGENUS is a division of a GENUS—a group of closely related SPECIES within a genus. There will never be complete unanimity of opinion among botanists as to what are genera and what are subgenera. The subgenera, especially of very large genera, are sometimes divided into SECTIONS.

Reference: D.9.

SUBHUMID CLIMATE is characterized by sufficient precipitation—20 to over 30 in. annually—to support a moderate to dense growth of GRASSES, but not one of dense forests.

SUBLIMED SULFUR
= FLOWERS OF SULFUR.—*See also* SULFUR.

SUBSOIL is below plow depth, i.e., below the surface and the subsurface soils which contain most of the soil organic matter. Usually it is lighter in color.

SUBSPECIES is a rank immediately below a SPECIES; some botanists regard this term as a synonym of VARIETY, but if subspecies and varieties are both recognized, then the variety ranks below subspecies.

Reference: D.9.

See also TYPE.

SUBTERRANEAN CLOVER
= SUBCLOVER.

SUBTILIN is an ANTIBIOTIC produced by the *Bacillus subtilis.*

SUCCESS BEARDLESS = *Horsford.*
See SIX-ROWED BARLEY.

SUCCULENT means: juicy, fleshy, or pulpy. Some succulent feed should be included in production rations, if possible. GRASS, ROOTS, SILAGE, PULPS, or MOLASSES are satisfactory.

SUCKER is a branch, or shoot, from a creeping underground STEM which ascends above ground and tends eventually to become a separate individual plant. Suckers are common in many woody plants.—*See also* TILLER.

SUCKLING CLOVER
= SMALL HOP CLOVER.

SUCRASE is an ENZYME.

SUCROSE, also known as *saccharose, cane sugar,* or *beet sugar,* is a sweet, crystalline CARBOHYDRATE; it is colorless when pure and water-soluble. By hydrolysis, sucrose splits into DEXTROSE and fructose (LEVULOSE).—*See also* SUGARS.

SUDANGRASS (*Sorghum sudanense*)—formerly classified as a variety, *S. vulgare* var. *sudanense*—is also called *Sudan.* It is one of the SORGHUMS. This GRASS was obtained from the Sudan as the result of a systematic search for a plant that resembled JOHNSONGRASS, except for its (RHIZOME-type) rootstock.

It is a fibrous rooted annual that grows 3 to 7 ft. high and has stems about ³⁄₁₆ in.

in diameter. The PANICLE is loose and open; the numerous leaves are long and narrow. The plants TILLER readily when planted in thin stands. The grain usually fades to a pale yellow or straw color with maturity but may also range from all shades of brown to black; the seeds of Sweet Sudangrass may approach red when grown under droughty conditions.

Sudangrass. (S.32.)

Sudangrass requires warm climate for its best development. It thrives on fertile, well-drained loamy soils, but can be grown successfully also on almost any class of soil from heavy clay to light sand.

Production on soils of low fertility can be greatly increased by application of commercial fertilizers.

Sudangrass tolerates droughty conditions. It is used chiefly as a temporary *pasture* grass to be grazed in the later part of the summer, particularly in the central and southern states, and in the irrigated districts of the Southwest. It can be grazed much earlier if planted early in the season, but early seedings are not so satisfactory for late-season GRAZING. For a long grazing season two plantings are best, one early and one late. Yields of forage range from 1 to 7 tons an acre, depending upon rainfall and soil fertility.

This sorghum species probably has its widest use by dairymen as a supplementary pasture crop. If grazed when 18 to 24 in. in height, the dry forage will contain about 10 to 13 percent protein. As the plants reach maturity the protein content drops rapidly to about 4 percent.

Sudangrass may also be harvested for *hay*; the best quality HAY is obtained if it is cut and properly cured as the first heads begin to appear. At this stage the dry forage will contain about 6 to 8 percent protein. This hay is an excellent roughage for work animals and stock cattle, and only slightly less valuable than alfalfa hay for milk cows. In irrigated sections of the Southwest, Sudangrass yields practically as much hay as alfalfa.

Sudangrass is a good soiling crop. It may also be ensiled and, when properly stored, makes good quality *silage* that has a feeding value similar to grass silage but lower than corn or Atlas (sorghum) silage.

If grazed when the plants are actively growing, Sudangrass is readily eaten by livestock. Sweet Sudangrass is much more readily grazed than other varieties of Sudangrass when in their advanced stages of growth. As the plants reach maturity, the sugar concentration tends to increase in the stems of the sweet type, whereas the other types tend to become somewhat dry and pithy.

When Sudan is grazed down, it should be rested long enough to recover. This may make it desirable to plant two or more separate fields and to use a rotation system of grazing.

Sudangrass seems to do best in pure stands. However, it is well adapted for planting in areas which are infested with JOHNSONGRASS, but are to be used for hay or pasture. Rotation studies have shown that Sudangrass, grown in rotation with SWEETCLOVER, may double yields over comparable plots planted continuously to Sudangrass. Planting of VETCH in the winter, followed by Sudan in the summer, has also increased the yields of Sudan. Where rainfall is adequate, Sudangrass may be grown with SOYBEANS.

Dangers: Precaution must be exercised against PRUSSIC ACID POISONING when the animals are first turned into the Sudan fields, particularly in the northern latitudes; however, the risk is not as great as it is with the other sorghums. The PRUSSIC ACID content varies in the different types of Sudan, with the sweet type tending to be more dangerous.

If humidity and temperatures are high, Sudangrass is attacked by leaf diseases, namely the three BACTERIAL LEAF DISEASES of sorghum, as well as CHARCOAL ROT, NORTHERN CORN LEAF BLIGHT, HEAD SMUT, particularly COVERED KERNEL SMUT, SOOTY STRIPE, and TARGET SPOT. Such varieties as Tift Sudangrass and Piper Sudangrass, however, have shown a high degree of resistance to the bacterial leaf diseases; Sweet Sudangrass, in addition, is somewhat resistant to charcoal rot.

The CHINCH BUG is more harmful to Sudangrass than any other insect. GRASSHOPPERS also may severely damage Sudangrass. Plant lice (APHIDS) feed upon the plants, and tend to infest the Sweet Sudangrass more than the other varieties. The SORGHUM MIDGE damages the plants only lightly, but may prevent seed formation.

Johnsongrass is often an adulterant in Sudangrass seed south of the 38th degree of latitude. Sudangrass hybridizes freely with other sorghum species, and care is necessary to keep it from becoming a mongrel crop.

Varieties: For many years after the introduction of Sudangrass, only one variety was grown in this country. Since then, in addition to the common type of Sudan (called *Common Sudangrass*), other varieties have been developed, of which the following are the most important:

California No. 23 Sudangrass, widely used in California, grows taller than Common Sudangrass, and produces more forage.

Ft. Collins Sudangrass, which originated in Colorado, is one of the newer varieties. It is low in prussic acid content.

Piper Sudangrass is a high-yielding type of Sudan which in productivity is second only to Tift Sudan. It also possesses some resistance to LEAF DISEASES.

Sweet Sudangrass was developed by hybridizing Leoti sorgo and Common Sudangrass, and then making appropriate backcrosses and selections. The seeds are non-shattering, the GLUMES have a distinctive sienna color. The sweetness and juiciness of the stalk make this variety more palatable to livestock than Common Sudangrass. Sweet Sudangrass is slightly later than Common Sudan; it grows and remains green over a longer season, yet retains its good grazing qualities. The palatability of this grass does not deteriorate with maturity as much as does that of Common Sudan. Sweet Sudangrass is resistant to several of the BACTERIAL LEAF DISEASES, and to CHINCH BUG. It is more resistant to CHARCOAL ROT than Common Sudangrass. However, it is attacked by SOOTY STRIPE.

S-1 Sweet Sudangrass, a single-strain type developed from Sweet Sudangrass No. 372, is superior as a silage crop and for grazing; it is characterized by a heavy glume of sienna color. This variety is grown in Texas, New Mexico, Arizona, and California.

Tift Sudangrass, one of the Leoti sorgo crosses, is probably the most productive of the commercially available Sudangrasses. It has a pithy stem, which is not sweet, and tends to tiller and develop side branches. Tift Sudan is later in maturity and more resistant to BACTERIAL LEAF DISEASES and LEAF BLIGHT than Common Sudangrass. However, it is attacked by TARGET SPOT and SOOTY STRIPE.

Wheeler Sudangrass is a fine-stemmed,

fine-leafed plant with small seeds, much superior to the ordinary coarse, cane-type Sudangrass. It is low in PRUSSIC ACID content; however, it is highly susceptible to the BACTERIAL LEAF DISEASES.

References: D.3; H.1; C.4; S.3; V.1; L.5.

See also PASTURES; PASTURE PLANTS; ALFALFAS; COWPEA; DUST STORMS.

SUDANGRASS HAY.
                          See SUDANGRASS; HAY.

SUDANGRASS PASTURE.
                                    See SUDANGRASS.

SUDANGRASS SILAGE.
                    See SUDANGRASS; SILAGE CROPS.

SUGAR. *See* SUGARS.

SUGAR BEET (*Beta vulgaris* var. *crassa*) is a biennial herb of great economic importance. It grows in humid regions or in semiarid and arid regions under irrigation and is used as raw material in the manufacture of *beet sugar* (SUCROSE). The best sugarbeet gives also a number of PLANT BY-PRODUCTS valued as feeds, e.g., BEET TOPS; BEET PULP, and CONDENSED BEET SOLUBLES PRODUCT.

Sugar beets are attacked by CHARCOAL ROT, DODDERS, STINK BUGS, CHINCH BUGS, and other enemies.—*See also* OATS; MOLASSES; BAGASSE.

SUGARCANE BAGASSE. *See* BAGASSE.

SUGARCANE BORER (*Diatraea saccharalis*) damages SUGARCANE, CORN, and SORGHUM in the Gulf States in much the same way as the EUROPEAN CORN BORER injures corn in the northern states. The yellowish-white, brown-spotted larva is about 1 in. long when full-grown, and turns into a small, straw-colored moth. It is able to survive only in tropical and semitropical regions, and is one of the worst insect pests. It produces several generations a year.

Control: A dust containing RYANIA is recommended for use against the sugarcane borer.

Reference: P.16.

SUGARCANES (*Saccharum* spp.), also called *canes*, are grown for sugar production in Louisiana and Florida.

NOTE: *Red Kentucky Sugar Cane* is the original name of *Leoti*, one of the forage SORGHUMS; today, the term

CANE is sometimes used for forage sorghums in general: this causes confusion and should therefore be stopped.

Sugarcane is a perennial GRASS; its stalks have a diameter of 1 to 2 in., and reach a height of more than 15 ft.; each of its numerous nodes produces a leaf.

Valuable by-products of the cane sugar manufacture are MOLASSES and BAGASSE. Sugarcane is also used as a SILAGE CROP.

Sugarcane. Plant, racemes, spikelet with pedicel, and rachis joint. (H.26.)

Dangers: The plants are attacked by such diseases as ZONATE LEAF SPOT and SUGARCANE STALK ROT ("red rot"), and by such insects as BILLBUGS and SUGARCANE BORER.

Species: Among the American sugarcanes are five species—such as the CHINESE SUGARCANE with its varieties and hybrids —some of which are important as forage plants.—*See also* SOYBEAN; COWPEA.

SUGARCANE STALK ROT, or *"red rot,"* is caused by the FUNGUS *Colletotrichum falcatum.*—*See also* COLLETOTRICHUM STALK ROT.

SUGAR DRIP SORGO is one of the forage SORGHUMS.

SUGAR MILO, better known as *Early White milo,* belongs to the grain SORGHUMS.

SUGARS are CARBOHYDRATES, characterized by sweet taste. They occur in nature, or can be obtained by hydrolysis from STARCH and other complex compounds. Sugars can be distinguished as REDUCING SUGARS or NONREDUCING SUGARS, depending on their chemical structures.

Weight for weight, sugars have slightly less net energy than starch, but often exert an appetite-stimulating action and favor the establishment of certain desirable BACTERIA in the digestive tract.

Simple sugars, like GLUCOSE, are contained in blood, and play an important role in life processes.

Complex sugars, e.g., SUCROSE, are formed by the union of two or more of the simple sugar molecules.—*See also* ABSORPTION; PRUSSIC ACID POISONING; ALCOHOL; HEMICELLULOSES; PENTOSANS.

SUITER'S GRASS, better known as *Kentucky 31 tall fescue,* is a strain of TALL FESCUE.

SULFADIAZINE forms a white or yellowish powder. It is toxic and almost insoluble in water, except as *sulfadiazine sodium.*

Added to regular mash at a level of 0.5 percent for two to four days, sulfadiazine is effective in reducing the severity of *coccidiosis* of poultry.—*See also* SULFONAMIDES.

SULFA DRUGS = SULFONAMIDES.

SULFAGUANIDINE, a white, crystalline powder, is practically insoluble in water. This drug is particularly effective for the prevention and treatment of *dysentery* and *coccidiosis* of calves, swine, and poultry. Mixed in feed or milk, it may be fed to very young animals over a period of one to four weeks (calves may receive 30 gr. daily) during the first month of their lives for the prevention of *white scours*.

For the control of coccidiosis, chicks may be fed a mash ration containing 1 percent sulfaguanidine for one to two weeks.—*See also* SULFONAMIDES.

SULFAMERAZINE occurs as white powder and is practically insoluble in water, but *sulfamerazine sodium* is water-soluble. A mash containing 0.5 percent sulfamerazine is helpful in controlling *cholera* as well as *cecal coccidiosis* in poultry; it is fed to the flock once or twice at 3- or 4-day intervals.—*See also* SULFONAMIDES.

SULFAMETHAZINE, a derivative of SULFAMERAZINE, is similar in properties and action to the latter. *Sulfamethazine sodium* is water-soluble and may be used at a rate of 0.2 percent in mash for two or three 2-day periods, with 3-day intervals, for the treatment of poultry affected with *cholera* or *cecal coccidiosis*.

SULFAQUINOXALINE occurs as a light-tan powder which is only slightly soluble in water, but *sulfaquinoxaline sodium* is water-soluble and widely used at the rate of 0.01 percent in drinking water for the prevention of *cholera* in poultry; more concentrated solutions—0.025 to 0.04 percent in drinking water—are helpful in the control of *coccidiosis* if the medication is given for 2-day periods; or this SULFONAMIDE may be fed at the rate of 0.0125 percent in mash to chicks for the first two months of their lives.

SULFATES are salts of SULFURIC ACID.

SULFATHIAZOLE, which forms white crystals or powder, is practically insoluble in water; *sulfathiazole sodium*, however, is water-soluble. This SULFONAMIDE is sometimes given to birds affected with *coryza* (cold); it is mixed in mash at the rate of 3.3 percent and fed to the flock for three days. Or the sodium salt is used at the rate of 1:1,000 in drinking water for a 5-day period.

SULFITES are the salts of SULFUROUS ACID.

SULFONAMIDES, or *sulfa drugs*, is a term which includes SULFANILAMIDE and its derivatives, e.g., SULFADIAZINE, SULFAGUANIDINE, SULFAMERAZINE, SULFAMETHAZINE, SULFAQUINOXALINE, and SULFATHIAZOLE. These drugs are systemic anti-infectives and must be administered carefully, since overdosing may cause toxic reactions. Some of them are mixed in feed and thus used for the prevention or control of diseases of livestock, and particularly of poultry.

Many sulfa drugs are also available as *sodium* salts which are water-soluble and can therefore be administered in drinking water.—*See also* PABA.

SULFUR occurs in nature in form of yellow lumps, or as SULFATES, or other sulfur compounds. This mineral element is necessary for normal nutrition; ruminants can use it in its elemental form, and poultry may have a requirement for inorganic (sulfate) sulfur which can be met by organic sulfur. Most of the requirement consists of the sulfur-containing AMINO ACIDS cystine and methionine.

In general, ruminant feeds contain adequate amounts of sulfur. If, however, very low grade ROUGHAGE is fed and the water supply contains no sulphur, high UREA feeds may require additional sulfur for good performance.

Commercially, finely powdered SULFUR DUST is available in a variety of grades: e.g., FLOWERS OF SULFUR, also called *sulfur flowers* or *sublimed sulfur* (obtained by cooling sulfur vapors); *precipitated sulfur*, or *milk of sulfur* (prepared by chemical reaction); *washed sulfur* (derived from flowers of sulfur by washing so as to remove its impurities); *sulfur flour* (i.e., ground sulfur); and *wettable sulfur* (very finely ground sulfur of 320 or less mesh, mixed with material which helps to form SUSPENSIONS when added to water).

Flowers of sulfur and other forms of sulfur dust (if fine enough to stick to the seed) are fairly effective in controlling the KERNEL SMUTS of the SORGHUMS when applied at the rate of not less than 2 oz. per bushel. They are not effective against soil fungi, and may even reduce emergence. The sulfur dusts have the advantage, however, of being cheap and nonpoisonous. With good seed and warm soil, they are fairly good substitutes for other SEED TREATMENTS. Apply them in the same way as COPPER CARBONATE.

LEAF SPOT OF PEANUT can be effectively controlled by dusting with sulfur dust; at the same time, this treatment helps to control PEG ROT and TOBACCO THRIPS. The first application should be made as soon as spots appear on older leaves at the base of the plant. Subsequent applications should be made at intervals of ten to fourteen days until two weeks before harvest.

Sulfur, when applied in the late afternoon at the rate of 15 lb. per acre, controls young VELVETBEAN CATERPILLARS.

Caution: Sulfur dust should not be inhaled; some persons are allergic to this chemical.

See also DILUENT; CUBE; DERRIS; CRYOLITE; TOXAPHENE; PARATHION; MINERAL FEEDS; FERTILIZER; GYPSUM; PROTEINS.

SULFUR DIOXIDE is a colorless gas (or, under pressure, a liquid); it has a suffocating, strong odor and is water-soluble. When SULFUR is burned, it becomes sulfur dioxide which is used as disinfectant, preservative, and particularly as *silage conditioner;* in the silo, sulfur dioxide forms SULFUROUS ACID. However, sulfur dioxide must not be added to any ANIMAL PRODUCT or any mixed feed containing animal products.—*See also* SILAGE.

SULFUR DUST is SULFUR in very fine form; e.g., FLOWERS OF SULFUR. Good sulfur dust must be finely ground, so that 97 percent pass a 320-mesh screen. It is used as DILUENT in INSECTICIDES.

SULFUR FLOUR. *See* SULFUR.

SULFUR FLOWERS

= FLOWERS OF SULFUR.—*See also* SULFUR.

SULFURIC ACID, an inorganic acid, is, when concentrated, a heavy, clear, very corrosive liquid. In dilution of 0.5 percent it is used as spray for the eradication of DODDERS.

Caution: Sulfuric acid is dangerous—handle it with great care. When diluting it, add it very slowly, under stirring, *to* the water.—*See also* SCARIFIED SEED; SILAGE.

SULFUROUS ACID, a colorless liquid of suffocating odor, is formed from SULFUR DIOXIDE and water; it occurs, for instance, when the former gas comes in contact with moisture in silos, where it acts as a *silage conditioner.*

The addition of sulfurous acid or of any of its salts, known as *sulfites,* to ANIMAL PRODUCTS or mixed feed containing animal products, is officially disapproved.—*See also* SILAGE.

SUMAC SORGO—also known as *Red Top* (not to be confused with the grass REDTOP) —as well as *Sumac 1712, Early Sumac,* and *Extra Early Sumac,* are forage SORGHUMS.

SUMMER BLACK STEM, or *cercospora leaf spot,* is caused by *Cercospora zebrina* which is sporadic in occurrence. It attacks ALFALFA. The FUNGUS, usually producing only an unimportant leaf spot, does produce considerable blackening of stems, especially in late summer. It attacks chiefly the upper part of stems—not stem bases. The leaf spots are dark brown or black at the center. In moist weather the spots and blackened stems may have a whitish sheen from the spores that develop at the surface.

Reference: J.3.

SUMMER-CYPRESS.

See GRAY SUMMER-CYPRESS.

SUMMER LEGUMES include SOYBEAN, KUDZU, PEANUT, VELVETBEANS, CROTALARIAS, ANNUAL LESPEDEZAS, SERICEA, HAIRY INDIGO, COWPEA, ALYCE CLOVER, FLORIDA BEGGARWEED, etc.

SUMMER RANGE.

See RANGE MANAGEMENT.

SUNFLOWERS (*Helianthus* spp.) are annual or perennial plants which reach a height up to 20 ft., with heads up to 1 ft. and more in diameter, and hairy stems of 1 to 3 in. in diameter.

The seeds of the cultivated sunflower species are used as poultry feed or in the manufacture of oil and its by-product, SUNFLOWER-SEED MEAL. In localities where summer temperatures are low, sunflowers are often used as SILAGE CROP of good quality; they can also be fed as green *soiling crop.*

NOTE: *Wild sunflower* (a variety of the common species *H. annuus*) is a WEED found particularly in the Great Plains. Its seeds are very small and therefore of no economic value.

Dangers: Sunflowers are attacked by CHARCOAL ROT, CORN EARWORMS, HARVESTER ANTS, etc.—*See also* COMPOSITE; GRAZING.

Sunflower. (B.28.)

SUNFLOWER-SEED MEAL, one of the OILSEED MEALS, is obtained as by-product in the manufacture of oil from the seeds of SUNFLOWERS. It is similar in composition to LINSEED OIL MEAL and well liked by all species of farm animals.

SUNFLOWER SILAGE.
　　　　　　　　　　　See SILAGE CROPS.
SUNN CROTALARIA (*Crotalaria juncea*), also called *Sunn hemp*, is a poisonous species of the CROTALARIAS, yet used to some extent for forage and soil improvement.

Reference: M.14.
SUNN HEMP = SUNN CROTALARIA.
SUNRISE BARLEY.
　　　　　　See SIX-ROWED BARLEY (variety).
SUNRISE KAFIR is one of the grain SORGHUMS.
SUNSHINE VITAMIN = VITAMIN D.
SUPER- is a prefix signifying "above," "beyond."
SUPERGOLD is a popcorn variety.
　　　　　　　　　　　—See also CORN.
SUPERHARD = *Blackhull wheat.*
　　　　　　　　　　See COMMON WHEAT.
SUPERPHOSPHATE, or *acid phosphate*, is the most important PHOSPHORUS fertilizer. It is a mixture of MONOCALCIUM PHOSPHATE and GYPSUM, and yields 14 to 20 percent available PHOSPHORIC ACID (as P_2O_5).

Superphosphate may cause so-called SLAG POISONING in animals having access to this fertilizer.—*See also* TRIPLE SUPERPHOSPHATE.
SUPERRED = *Red Chief.*
　　　　　　　　　　See COMMON WHEAT.
SUPPLEMENTAL PASTURES.
　　　See PASTURES; PASTURE MANAGEMENT.
SUPPLEMENT FEEDS. *See* CONCENTRATES; COMMERCIAL MIXED FEEDS.
SUPRA- is a prefix meaning "over," "above," or "beyond" (where growth or action is implied).
SURFACE SOIL, or *topsoil*, consists of the upper 5 to 8 in. of the soil, or the depth commonly stirred up by a plow.
SURFACTANTS, or *wetting agents*, are contained in DETERGENTS which are widely used in place of SOAP. Experimental work indicates that some of the surfactants have value as GROWTH stimulants, especially for chicks over eight weeks of age. Feeds, containing surfactants, must first be cleared through the F.D.A. which, at present, considers them as medicated feeds containing a new drug.
SURPLUS FRUITS. *See* FRUITS.

SURPLUS POTATOES. *See* POTATO.

SUSPENSION is a dispersion of finely divided solid particles in a liquid.—*See also* COLLOIDAL.

SUTURE is a seam, or seamlike ridged or furrowed, medial line; e.g., the sutures of a POD are the lines where the valves of the pod separate to discharge the seeds.—*See also* LEGUMES.

SUWANNEE BERMUDA.
　　　　　See BERMUDA-GRASS (variety).

SWALE is a tract of low, and often marshy, land.—*See also* RANGE MANAGEMENT.

SWAMP WHEAT = *Mediterranean wheat. See* COMMON WHEAT.

SWARD is a matlike, grassy piece of land. —*See also* PASTURE; PASTURE PLANTS; GRAZING.

SWATH is (1) a strip of cut herbage lying on the STUBBLE for curing, or (2) the width of the cutting bar when cutting HAY.—*See also* GRAZING; SWATHER.

SWATHER is a device used on a mower, or reaper, to straighten up uncut, fallen grain, and to mark the boundary of the SWATH.

SWAYBACKED is a grain kernel which has a depression on the back; e.g., the kernel of Chiefkan, a WHEAT variety.

SWEATING. *See* FERMENTATION.

SWEDISH CLOVER = ALSIKE CLOVER.

SWEDISH RYE is a winter variety of RYE.

SWEDISH SELECT.
　　　　　See COMMON OAT (variety).

SWEDISH TURNIP = RUTABAGA.

SWEEP is a double-bladed V-shaped knife on a cultivating implement.—*See also* SWEEPS.

SWEEPINGS. *See* SCREENINGS.

SWEEPRAKE is an implement with long teeth, used to pick up hay from the SWATH, COCK, or WINDROW and carry it to a STACK.

SWEEPS are taken with the *sweep net* to count the number of insects in the field.— *See also* LYGUS BUGS; POTATO LEAFHOPPER.

SWEETBREAD = PANCREAS.

SWEETCLOVER. *See* SWEETCLOVERS.

SWEETCLOVER HAY POISONING is due to feeding moldy SWEETCLOVER *hay* or *silage* for about three weeks. When green and growing, sweetclover is not poisonous to stock. Molds and mildews present on hay, while in themselves not poisonous, are the source of the trouble. Sweetclover contains a substance known as COUMARIN which gives the rather overpowering sweet odor to freshly-cut sweetclover. The action of molds upon this coumarin transforms it into DICOUMAROL. The latter replaces VITAMIN K, and thereby prevents the formation of a substance called PROTHROMBIN which is needed for blood clotting. So-called *sweetclover disease* is really a hemorrhage, or a series of hemorrhages, that may be fatal.

After feeding damaged sweetclover hay or silage, any wound (such as that of dehorning or castration or parturition) may be followed by fatal bleeding. Any severe bruise against a post or tree may be followed by a collection of large masses of semiclotted blood under the skin with great swelling over the hips, shoulders, or ribs.

Usually, sweetclover hay can be safely fed when it is mixed half and half with TIMOTHY or other hay; it can also be fed alternatingly with other hay, each type being fed for 10-day periods throughout the winter. At least thirty days should elapse after the last feeding of sweetclover before any surgery, such as dehorning, is attempted.

Injections of vitamin K, or the feeding of good quality ALFALFA, will reverse the effects of spoiled sweetclover.

Reference: W.34.

SWEETCLOVERS (*Melilotus* spp.) are LEGUMES but do not belong to the true clovers. They are important forage crops of the northern United States and Canada. Annual and biennial forms of sweetclover occur, permitting a wide use in rotations. The forage is high in protein, and makes an abundance of pasturage.

The habit of growth of sweetclover varies. The *annual types* make all their growth in one season, mature seed, and die. The *biennial types* grow from 10 to 24 in. during the season of seeding. Then buds are formed on the crown of the roots; these buds remain dormant over the winter and, allowed to grow, the plant attains a height of 4 to 5 ft., or 8 to 10 ft. on good soil.

However, when the second season's growth is cut in June or late May, the new growth comes from dormant buds on the lower portion of the stem; when completed, seed is produced, and the roots and tops die.

Sweetclover will grow almost anywhere, provided there is more than 17 in. of rain suitably distributed, and there is sufficient lime in the soil. Depending upon the degree of acidity, applications of from 300 lb. to 4 tons of ground limestone per acre are generally made. Sweetclover is very resistant to cold and drought, and is not troubled by heat. It is able to obtain phosphorus from relatively unavailable soil phosphates and will thrive in soils too high in alkali for most cultivated crops.

If properly inoculated, approximately two-thirds of the nitrogen of the plant is obtained from the air. The bacteria that cause the plants to produce these nodules are the same as those that inoculate ALFALFA, and hence, where alfalfa or sweetclover have not been grown before, INOCULATION should be provided. A firm seedbed is necessary to get a stand of sweetclover.

WHEAT and OATS are the common companion crops of sweetclover. BARLEY, because of its early maturity and open growth, is an ideal companion crop. FIELD PEAS are often successfully used in the northwestern states. In the Great Plains states, drilling the seed in wheat, SUDAN-GRASS, or SORGHUM stubbles has given the best stands.

No other plant is known that will furnish as much grazing, under as wide a range of conditions, as a good stand of sweetclover in its second season. The first season's growth may be grazed from the time it is 8 to 10 in. high until frost, and the second season's growth may be grazed from early spring until the plants are too woody to be palatable, which is usually about the end of July.

Hogs, cattle, horses, and sheep may all be pastured on sweetclover. Cattle should be turned on this (so-called) CLOVER early, when it is 6 to 8 in. high and tender.

The value of the deep-rooted sweetclover as a soil improver lies in its ability to take nitrogen from the air and to deliver it by rapid decay when turned under.

As a *hay* crop, sweetclover is not equal to alfalfa or red clover. Occasionally, when the first season's growth is 18 to 24 in. high and late fall weather is favorable for curing, good sweetclover HAY can be made. Except in dry regions, the sappy, coarse stems of the second year's growth are most difficult to cure. Because of these difficulties, many farmers are successfully ensiling sweetclover, using the proper amounts of molasses or acid. It is a particularly valuable feed when ensiled mixed with ground corn, or corn-and-cob meal.

Dangers: Losses of stands of sweetclover during the second year have been of increasing frequency in many sections of the Corn Belt, and particularly where sweetclover has been used regularly in rotations for ten to fifteen years. A rotting of the new shoots at the soil surface is characteristic of the early dying that may occur in small areas, or may spread until the whole field becomes valueless for grazing A blackening and distortion of the stems is usually related to dying at the blooming period. A combination of various diseases appears to be the cause of these troubles. The most common of such diseases are DOWNY MILDEW, LEAF SPOTS, MOSAIC, BLACKSTEM, and ROOT ROT.

Insects, having the most serious effect on this crop, are the GRASSHOPPERS, CUT-WORMS, and the weevils.

There is less danger from BLOAT with sweetclover than with alfalfa or red clover, but the danger does exist. When spoiled, sweetclover hay contains DICOUMAROL which may cause animals to bleed to death.

Species: There are approximately 20 species of sweetclover; the three species of agricultural importance are WHITE SWEET-CLOVER (and its variety *Madrid White*); YELLOW SWEETCLOVER (and its variety *Madrid*); and SOURCLOVER. White and yellow sweetclovers are principally biennial in growth habit, but annual forms do occur with some of them having considerable agricultural value, e.g. *Hubam* (*clover*), a white-flowering plant. The species DAGHESTAN SWEETCLOVER, which is yellow-flowering, has both annual and biennial forms, e.g. one of its varieties, *Redfield*.

References: P.7; M.3; G.5.

See also SWEETCLOVER HAY POISONING; FIELD BEANS; RYE; CRESTED WHEATGRASS; MOUNTAIN BROME; TALL OATGRASS; WEEPING LOVEGRASS; PEPPERGRASS; ALSIKE CLOVER; HAIRY VETCH; SILAGE CROPS; PASTURE PLANTS; PASTURES; LEGUME BACTERIA; WHITE GRUBS; WHEAT JOINTWORM; TAKE-ALL.

Illustration: See WHITE SWEETCLOVER.

SWEET CORN. *See* CORN.

SWEET CORN REFUSE from canneries is a PLANT BY-PRODUCT that can be used as silage material.—*See also* SILAGE CROP.

SWEET LUPINES, or *nonalkaloid lupines*, are free of ALKALOIDS.—*See also* BLUE LUPINES; WHITE LUPINE; YELLOW LUPINE; LUPINES.

SWEET POTATO (*Ipomea batatas*) has roots which are low in fat, protein, and fiber, but high in starch and VITAMIN A potency. They are often attacked by CHARCOAL ROT and other plant diseases.

Surplus and cull sweet potatoes can be dehydrated and fed to animals where they have a value equal to about 75 percent of an equal weight of grain.

Dehydrated sweet potatoes are valuable as a partial substitute for grain in rations for cattle, sheep, swine, and poultry.

Reference: E.12.

See also PLANT BY-PRODUCTS; SWEET POTATO PULP.

SWEET POTATO PULP. *Dried sweet-potato pulp*, the residue from the manufacture of starch from SWEET POTATOES, is occasionally used as feedstuff.

SWEET SORGHUM is a synonym of *forage* SORGHUM.

SWEET SORGHUM MEAL.
See SORGHUMS.

SWEET STALK KAFIR, better known as *Cheyenne*, is an intermediate-type grain SORGHUM.

SWEET SUDANGRASS is a SUDANGRASS variety.

SWEET VERNALGRASS (*Anthoxanthum odoratum*) is one of the GRASSES found in the northern states and as far south as the Cotton Belt. It starts growth early in the spring, even on poor soils. Because of its bitter taste, it is an inferior fodder grass. Sweet vernalgrass improves the odor of HAY, but is undesirable in more than a very small amount. The seeds are eaten by birds.

Reference: W.16.

SWILL. *See* SLOP FEEDING.

SWINE RATIONS are given in two tables on pages 503 and 504.

The first table gives the formula for a *flexible protein supplement* and the other shows how this supplement can be *combined with other feed* to give satisfactory rations.

On many farms the swine herd is kept as a means for marketing CORN or other farm GRAINS. In other areas, hogs convert GARBAGE and other wastes into food, and some farmers keep a sow to raise animals for their own meat. Many segments of the swine industry could profit by using more ROUGHAGE or other bulky feeds in the rations at certain times.

BRED SOWS AND GILTS

Bred sows and gilts should not gain over 1 lb. per day during the 114 day gestation period. Experimental work has shown that a litter of 10 pigs and the increased maternal tissues will weigh 60 lb. The other 50 lb. is growth of new tissues in gilts, or improved condition in the sows, so that they are better able to suckle their litters. This rate of gain can usually be obtained by feeding some 6 lb. CONCENTRATE per day. This is at the rate of about 2 lb. per 100 lb. of live weight for gilts, and less for a sow. However, these animals can be self-fed on bulky mixtures which contain 30 to 40 percent ground ALFALFA hay or corn cobs. The same effect can be obtained by using large amounts of SILAGE in the ration. Supplements for the silage should provide any NUTRIENTS which are missing from the silage. For example, if corn silage is fed, large amounts of PROTEIN, CALCIUM, and PHOSPHORUS are necessary to balance the deficiencies of the corn silage. If a high quality LEGUME silage, like Ladino clover, is fed, the major deficiency will be ENERGY and about 2 lb. grain will balance the ration. Pregnant sows and gilts can be satisfactorily carried on a ration of good quality PASTURE and limited amounts of grain.

At *farrowing time*, sows should be put on a bulky ration. The usual grain ration can be diluted with 20 percent wheat bran for three to five days. The BRAN is then removed gradually and the sows returned to the standard ration after one week. Sows with no more than 5 pigs should be hand-fed, but sows with larger litters may be self-fed. A satisfactory feeding plan allows the sow 6 lb. feed for maintenance and 1 lb. feed for each pig in the litter for milk production. As the pigs get older, they will gradually begin to consume some sow feed and the allowance may be slightly increased. However, a better management practice is to give the pigs a creep ration of their own.

CREEP-FEEDING PIGS

As soon as the pigs are a week old, they will usually begin nibbling at feed if they can reach it. Make small, comfortable *creeps* and provide palatable, highly nutritious rations and fresh water at all times. The sow's milk provides good quality PROTEIN, and if she is properly fed, will be reasonably rich in VITAMINS. The main thing that is needed in creep rations is a good supply of available energy, but proper levels of proteins, MINERALS, and vitamins, are also essential. Farm grains are better than no creep ration, and under some price conditions will produce the cheapest gains, though not the most rapid. If hogs are a good price, the use of a creep ration containing good quality protein, and fortified with minerals and vitamins, may be expected to pay large dividends. While there are a number of reports about the tastes of pigs for special feeds, a healthy, active pig will usually eat rations (similar to those shown in the table) without any trouble. If offered a choice, however, the pigs seem to prefer rations which contain SUGAR, although it is diffi-

A FLEXIBLE PROTEIN SUPPLEMENT

Tankage or meat scraps and/or fish meal, lb.	200-250
Soybean oil meal, lb.	200-300
Other oil meals, if available at low prices, lb.	50
Alfalfa or alfalfa leaf meal, lb.	100-200
Wheat shorts or middlings, lb.	100
Milk or whey, if available at economical prices, lb.	50
Fermentation by-products or fish solubles (creep ration), lb.	50
Antibiotic supplement, if desired	
Iodized salt, lb.	15
Limestone, lb.	20
Bone meal or equiv. phosphorus source, lb.	15
Total, lb.	750

The amounts of soybean oil meal, meat by-products, and dairy by-products should be varied with price. If soybean meal is very expensive, reduce to a minimum of 100 lb. for young pigs and omit after pigs weigh 125 lb.

Creep rations should contain: a milk by-product; fish meal or solubles; and a fermentation by-product.

The alfalfa can be omitted if hogs have access to good pasture, or if a vitamin supplement is used to provide riboflavin (B_2) and pantothenic acid; only minimum amounts should be included for small pigs.

An antibiotic may be included at recommended levels if desired. If pigs are on good legume pasture, feed only ½ lb. supplement per day until they weigh 125 lb. and then reduce to ¼ lb. or omit if the pasture is very lush.

The flexible protein supplement can be mixed with grains to provide a complete meal for each age group or it may be self-fed. Many groups of hogs might eat too much of the supplement; so many farmers will desire to hand-feed the supplement to limit consumption. The pigs should receive about 1 lb. supplement per day until they weigh 125 lb. The amount can be reduced to ½ or ¾ lb. per day.

COMPLETE SWINE RATIONS
Amounts given are for 100 lb. ration

Ration:	Creep ration		Growing fattening rations							Gestation				Lac-ta-tion
			Weaning to 75 lb.		75 lb. to 125 lb.		125 lb. to market			hand-fed		self-fed		
	1	2	1	2	1	2	1	2	3	1	2	1	2	1
Corn or milo and/or wheat, lb.	58	65	60	65	50	88	55	95	45	60	45	39	15	55
Oats, lb.	..	..	..	..	10	..	10	..	..	20	10	13	..	10
Barley, lb.	..	..	10	..	10	..	..	..	..	..	10	..	..	10
Wheat middlings or shorts, lb.	..	..	..	..	5	..	..	..	..	..	5	..	..	5
Molasses, lb.	5	..	..	..	5	..	5	..	..	..	10	..	..	5
Legume pasture, free choice..	..	+	..	..	..	+	..	+	..	..	..	..	..	..
Cereal pasture, free choice...	..	..	..	+	..	..	..	..	..	..	..	..	..	+
Legume silage, lb.	..	..	..	..	..	..	..	..	50	..	..	..	..	..
Corn silage, lb.	..	..	..	..	..	..	..	..	..	..	..	..	70	..
Wheat bran, lb.	..	..	..	..	..	..	20	..	..	..	..	..	..	..
Alfalfa hay, lb.	..	..	..	..	..	..	..	..	..	..	..	35	..	..
Protein supplement, lb.	37	35	30	25	20	12	10	5	5	20	20	13	15	15
Minerals, free choice	+	+	+	+	+	+	+	+	+	+	+	+	+	+

If swine are hand-fed allow the following amounts per animal per day:

Creep rations, ¾ lb.	100 lb. pig, 5½ lb.
25 lb. pig, 2 lb.	150 lb. pig, 7 lb.
50 lb. pig, 3¼ lb.	Bred sows and gilts, 6 lb.
75 lb. pig, 4 lb.	Nursing sows, 12 lb.

For growing fattening pigs allow ¾ lb. 40-percent protein supplement per day and all of the corn they will eat.

cult to demonstrate a growth-promoting effect of the additional sugar over and above the added energy.

MARKET HOGS

Rations for pigs from weaning up to 75 lb. of body weight should be especially well fortified with vitamins, and should contain proteins which are rich in these three AMINO ACIDS: LYSINE, TRYPTOPHAN, and METHIONINE. A mixture of proteins is usually more satisfactory than any one protein alone. It is recommended that good quality animal protein make up at least one third of the protein supplement for pigs of this size. The remainder of the protein supplement can be supplied by SOYBEAN oil meal, limited amounts of other OIL MEALS, wheat shorts, and alfalfa meal. The ration should not be high in fiber, and unless high quality ingredients are used, it will probably be necessary to add supplements of riboflavin (B_2), PANTOTHENIC ACID, and VITAMIN B_{12}, to increase these vitamin levels to those required by the pig for best results. There is little reason to believe that additional minerals other than salt (sodium chloride) are needed in a ration of this kind. If the supplement does not contain a good quality animal protein, it may be necessary to add limited amounts of LIMESTONE and BONE MEAL. Excesses should be avoided, because calcium will tie up TRACE ELEMENTS, particularly ZINC, causing skin disorders and lowering rates of gain. These conditions can be overcome by adding zinc to the ration. Trace-element additions, which have been shown to give the best response in swine, include IRON and COPPER to the rations of very young pig, COBALT and zinc to those of the growing and fattening pig. Most rations seem to contain an adequate amount of MANGANESE to meet the animals' need for growth. In areas that have produced goiterous livestock, IODINE must be added for satisfactory results.

From 75 lb. to 125 lb., a pig can receive a ration containing less protein, minerals, and vitamins, but large amounts of grains and bulky feeds. The amount of supplement can be kept constant, and the grain increased as the pig gets larger and eats more.

Between 125 lb. to 200 lb. is a critical time in producing a carcass of the desired type. Most lard-type hogs are bred to

fatten rapidly at this stage, and will produce a carcass too fat for today's market.
By feeding rations high in fiber, or by
limiting the energy intake to about 80 to
85 percent of the normal feed, a more
desirable carcass is obtained. This limited
feeding may be done by hand-feeding or
by diluting the ration with alfalfa meal,
corn cobs, or other fibrous feeds.—*See also*
SLOP FEEDING; CREEP FEEDING.

SWISS CHARD (*Beta vulgaris* var. *cicla*),
or *leaf beet*, is a small-rooted plant whose
large, thick-stemmed leaves are used as
green feed.—*See also* GRAZING.

SWITCHGRASS (*Panicum virgatum*), a
native perennial, is distributed throughout
most of the United States. It is abundant
and important as a forage and pasture
grass in the central and southern parts of
the Great Plains. Switchgrass usually
grows 3 to 5 ft. high. The flowering head
is a widely branching open PANICLE. The
leaves are usually from ¼ to ½ in. wide,
and 6 to 18 in. long, and green to bluish-
green in color. Switchgrass occurs on
nearly all soil types, but is most abundant
and thrives best on moist, low areas of
relatively high fertility.

Heavy, vigorous roots and underground
stems make the species excellent for conservation use. Seedling growth is aggressive. Usually, switchgrass (a bunchgrass)
is seeded with the grass species with which
it occurs naturally. Best seedling stands
have been obtained where plantings were
made on a clean, firm, well-prepared
seedbed.

Growth begins in late spring and continues through the summer if there is
enough moisture. Forage is produced in
abundance and—especially during the
period of early growth—is acceptable to
livestock. Hay of good quality can be had
by mowing the grass when seed heads
begin to form.

Dangers: STEM RUST is a disease injurious to most native switchgrass plants
except the Blackwell variety.

Variety: *Blackwell*, a new and improved
variety, developed through plant selection and breeding work, yields excellent
forage and shows considerable resistance
to stem rust.

Switchgrass. Plant, spikelet, and floret. (H.26.)

References: H.1; O.1.

See also PANICGRASSES; GRASSES; INDIAN
GRASS.

SYMBIOSIS is the living together of dis-
similar organisms, especially when the relation is naturally beneficial, as, for example,
the nitrogen-fixing bacteria on the roots
of LEGUMES or the BACTERIA in the RUMENS
of cattle, sheep, and goats.

Symbiotic means: of, or pertaining to, symbiosis; e.g., the symbiotic bacteria known as *Rhizobia.—See also* INOCULATION.

SYMPHYTUM. *S. asperrimum*

= PRICKLY COMFREY.

SYNTHETIC means: produced by artificial means, or *synthesis;* i.e., by a reaction which forms a complex compound from simpler compounds or from CHEMICAL ELEMENTS.

Synthesize means: to produce something by synthesis or to unite numerous elements into one.

SYRUP = SIRUP.

SYSTEMATIC BOTANY = TAXONOMY.

SYSTEMIC means: relating to the entire organism, or system; as, systemic disease. *—See also* ANTIBIOTICS.

T

2,4,5-T, or *2,4,5-trichlorophenoxyacetic acid*, is more effective than the related 2,4-D in killing woody plants, such as trees and BRUSHES. It occurs as white solid, which is practically insoluble in water; however, its *sodium salts, amine salts*, and *esters* are similar in properties and solubility to the corresponding compounds of 2,4-D; the low-volatile ester formulations have the preference of many users.

If sprayed at the rate of 1 lb. in 5 to 20 gal. diesel oil per acre, 2,4,5-T kills SAGE-BRUSH and many other kinds of brush.

TABLE SCRAP MEAL, or *garbage tankage*, is now called *degreased, dehydrated garbage.—See also* DEHYDRATED GARBAGE; GARBAGE.

TADPOLE SHRIMPS (*Apus* spp.) may injure RICE seedlings.

Control: These shrimps can be controlled with COPPER SULFATE.

TAGANROG = *Kubanka. See* DURUM WHEAT.

TAGS on feed bags are guaranties required in most states of this country. The intrastate trade with feeds is regulated by state laws. According to most state regulations, a tag must be attached to each bag of feed; the tag must give the composition of the feed, especially with respect to the guaranteed minimum percentages of pro-

tein and fat, and the maximum percentage of fiber.—*See also* LABELS; CARBOHYDRATES.

TAKE-ALL, sometimes called *whitehead*, is a disease caused by the FUNGUS *Ophiobolus graminis;* it attacks chiefly WHEAT, sometimes RYE, and BARLEY, and occurs in the eastern United States. Fortunatel,y infected fields are comparatively few; but because of possible heavy losses from this disease, it must be carefully watched.

Take-all may kill plants in the rosette stage; it may so dwarf them that only a few low culms with small heads are formed, or it may kill plants that have attained about normal size, as the heads are beginning to fill. Such plants turn almost white. Nearly all the plants in certain spots or scattered plants in a field may be killed. Stems of infected plants usually are black to a height of 1 to 2 in. above the soil; the plants pull easily because of the rotted, infected roots.

The take-all fungus is not carried with the seed but persists in the soil for several years.

Control: Much of the damage can be avoided by maintaining a proper balance of soil nutrients. Soil requirements seem to differ in different areas. Severe infection occurs when nitrogen, phosphorus, and potassium are inadequate, or phosphorus and potassium in short supply. Nitrogen may increase the incidence of root infection.

An adequate supply of readily available nitrogen during the fall and winter permits the causal fungus to prolong its existence on infested stubble. If nitrogen is available to young plants, the roots are more readily invaded. Adequate nitrogen, however, will also permit the rapid replacement of diseased roots with new ones if phosphorus is present.

The best control measure known is to keep wheat, barley, and ryè off infested land for about four years. In the meantime it is advantageous to plow under a green-manure crop, such as SWEETCLOVER. No resistant varieties are known.

References: B.16; M.37.

TALC, or *talcum*, a natural magnesium silicate, is a very fine, water-insoluble, white or light-gray powder with adhering qualities. It is therefore often used as an inert DILUENT in insecticidal dusts.—*See also* CUBE; DERRIS; CRYOLITE.

TALLAROOK is a SUBCLOVER strain.

TALL BANK FESCUE = TALL FESCUE.

TALL FESCUE (*Festuca arundinacea*), also called *reed fescue, king fescue, giant fescue, ditch fescue,* or *tall bank fescue,* is a deeply rooted and strongly tufted perennial with stems 3 to 4 ft. high, erect and smooth. The numerous dark-green basal leaves are broad and flat, the sheath is smooth, and the LIGULE short. The nodding PANICLE head is 4 to 12 in. long and has lance-shaped (lanceolate) SPIKELETS that are ½ in. long or more; there are many florets to each spikelet. This grass flowers in June and July.

Tall fescue is found growing in damp pastures and wet places throughout North America. It can be distinguished from the closely related MEADOW FESCUE by its greater height and its broad leaf that is deep green on the upper surface and ribbed and rough. Tall fescue is higher in forage yield and longer-lived than meadow fescue.

This grass is adapted to a variety of soils and in general to the same region as meadow fescue; i.e., throughout the TIMOTHY region and also farther south and west. In the South, the grass stays green enough during the winter to be eaten readily by livestock. Tall fescue that is to be used for both summer and winter pasture, should be rested in fall. It does best on heavy soils that have considerable humus and will grow as well in wet as in dry situations. In fact, it grows better on wet-land PASTURES than other grasses in common use.

Tall fescue does well with LEGUMES in soil-conserving crop rotations. Since it makes good pasture and stays green in winter, it cuts down on the acreage needed for feed crops. And it has the deepest and strongest root system of any grass now grown in the South. This root system holds up cattle on wet land where they would otherwise mire. It makes an excellent turf in waterways; and it controls erosion on steep slopes.

Danger: Under certain conditions tall fescue may cause FESCUE LAMENESS in cattle.

Strains: Two strains, *Alta fescue* and *Kentucky 31 tall fescue*—also called *Suiter's grass*—are receiving the most attention now.

1. *Alta fescue* is rather widely used for pasture for cattle and sheep in the Pacific Northwest, and may be desirable for winter pasture in the South. Farther north it serves well for pastures of somewhat low fertility. Alta is quite drought-resistant and aggressive; it competes strongly with the LEGUMES and sometimes wins out over them.

2. *Kentucky 31 fescue* is a good pasture grass for land which is too wet for cultivation, and cannot be economically drained. Planted along with a suitable legume it makes satisfactory early pasture. In mixtures with legumes this fescue grows more vigorously and needs less nitrogen fertilizer than pure stands. LADINO CLOVER and Louisiana WHITE CLOVER appear to be ideal for most lowlands and highly fertilized uplands. Tall fescue with ALFALFA controls erosion more effectively than alfalfa alone and also reduces *bloating* of livestock.

Kentucky 31 fescue, in combination with MADRID SWEETCLOVER or alfalfa, promises to increase soil productivity, reduces COTTON ROOT ROT, and supplies better grazing for livestock in southern areas.

References: H.1; H.3; B.3.

See also BLOAT; FESCUES; GRASSES; SILAGE CROPS.

TALL GRASS RANGE.

See RANGE PLANTS.

TALL MEADOW OATGRASS

= TALL OATGRASS.

TALL OATGRASS (*Arrhenatherum elatius*), also called *tall meadow oatgrass,* is grown quite generally in the Central and Northern States. It is a hardy, upright perennial which reaches a height of 30 to 60 in. and produces many leaves. Tall oatgrass does not propagate by rootstocks but tends to be bunchy. It produces seed in open heads resembling those of cultivated oats. Tall oatgrass prefers well-drained soil and seems to be especially adapted to *light* sandy or gravelly land; it does not grow well in shade.

This grass has many desirable forage qualities. It can be used for pasture or meadow and gives a heavy yield of hay which is quite palatable. Although it does not produce a very good sod, it stands pasturing well and furnishes abundant grazing.

Tall oatgrass. Plant, spikelet, and upper floret. (H.26.)

Tall oatgrass comes on early in spring and remains green until late in the autumn. Best results are obtained if it is grazed in rotation and given controlled GRAZING. Under continuous, close grazing the stand is easily reduced. It is becoming a common practice in the Pacific Northwest to use a mixture of tall oatgrass and SWEETCLOVER for a short-rotation hay or pasture combination; mixtures of tall oatgrass, OR-CHARDGRASS, and RED CLOVER have also been used. For the best hay the mixture should be cut at about the time it begins to bloom.

The seed of tall oatgrass is often of low viability and shatters before fully matured. In sections where there is a reasonable amount of moisture in autumn or late summer and where winters are not severe, best results will probably be obtained by seeding in September or early October; otherwise, spring seedings by broadcasting are best. A well-fitted seedbed is essential. After sowing, the seed should be covered by cultivating or harrowing lightly.

Dangers: The GRASS SMUTS are among the diseases which attack tall oatgrass.

Variety: An improved form of tall oatgrass, developed in Oregon, is being introduced under the name of *Tualatin oatgrass*. Its seed shatters less readily than that of tall oatgrass. When compared with the latter, it is finer stemmed, leafier, and later in maturity; many of the SPIKELETS bear two fertile FLORETS rather than just one.

Reference: H.1.

See also GRASSES; ALSIKE CLOVER; PASTURES; PASTURE PLANTS.

TALL WHEATGRASS (*Agropyron elongatum*), a SPECIES of the WHEATGRASSES, hybridizes with COMMON WHEAT and with DURUM WHEAT.

This perennial is hardy, drought-resistant, matures after most other GRASSES have dried up, and produces excellent forage on grounds too alkaline for any other useful crop.

Tall wheatgrass is a blue-green, tall, erect, perennial bunchgrass which tends to become coarse as it approaches maturity; it may get tough in late summer.

Reference: B.1.

See also RANGE MANAGEMENT.

TAMBOV. *See* PROSO (variety).

TAME HAYS. *See* HAYS.

TAME PASTURE. *See* PASTURE.

TANKAGE is a protein-rich by-product of packing houses.—*See also* ANIMAL PRODUCTS; WHEATS; DISTILLERS' PRODUCTS; DEHYDRATED GARBAGE; GARBAGE.

TANKING UNDER LIVE STEAM.
                         See ANIMAL PRODUCTS.

TANNER. *See* SOYBEAN (variety).

TANNIN, or *tannic acid*, occurs in many barks and fruits; particularly rich in tannin are the nutgalls of OAKS. The seeds of *Martin*, a grain SORGHUM, contain more tannin than do other combine varieties. Tannin is also contained in LESPEDEZAS, especially in the SERICEA species, in relatively large amounts.

Tannin is commercially available as a tan or brown powder or spongy mass which is water-soluble but incompatible with many chemicals. It is widely used as an astringent (i.e., tissue-contracting or secretion-stopping) drug.

TAPROOT is a single central root of a plant; it grows vertically downward, giving off small lateral roots.

TAR, a thick, dark-brown or black liquid or semisolid, is the residue of the destructive distillation of such organic materials as wood, coal, or crude oil.—*See also* PINE-TAR OIL; GAS TAR; COAL TAR; TAR BARRIER.

TAR BARRIER. This CHEMICAL BARRIER for CHINCH BUGS is made by pouring a narrow line of COAL TAR on a path prepared by packing the soil firmly along the margin of the field. The bugs are repelled by the odor of the tar, and while fresh, it also acts as a physical barrier because of its stickiness. However, a tar barrier has to be renewed more frequently than a CREOSOTE BARRIER.

NOTE: Only tars from which the creosote and cresylic acid have not been distilled should be used. Tars that are by-products of the manufacture of water-gas have little value against chinch bugs.

Reference: P.5.

TARES = COMMON VETCH.

TARGET SPOT, caused by the FUNGUS *Helminthosporium sorghicola*, occurs on forage and grain SORGHUMS, and JOHNSON-GRASS, and has been reported from Florida, Georgia, Louisiana, North Carolina, and Texas. It produces small, well-defined spots that are tan on Tift Sudangrass and certain other Leoti crosses, and reddish-purple on Common Sudangrass and other sorghum varieties. Older lesions have light centers surrounded by a series of alternate dark and light bands, so that they resemble a target. They range in size from tiny spots, barely visible, to more elongated areas limited somewhat in length by the leaf vein. The lesions may grow together so as to involve most of the leaf areas. Under ordinary field conditions the spread of this FUNGUS LEAF DISEASE is not rapid; however, in very moist atmosphere spores form in abundance.

Control: Sorghum varieties, resistant to target spot, have not yet been developed. SEED TREATMENT will prevent the spread of this LEAF DISEASE to new areas.

Reference: L.1.

TARNISHED PLANT BUG is one of the LYGUS BUGS.

TARTAR. *White Tartar* is a COMMON OAT variety.

TARTARY BUCKWHEAT (*Fagopyrum tataricum*) is also known by many other names, such as "*India Wheat*", "*Rye buckwheat*", "*Duck wheat*", *Bloomless, Hull-less, Marino, Mountain* (*buckwheat*), *Siberian* (*buckwheat*), *Wild Goose*, and *Calcutta* (*buckwheat*). Seeds of Tartary are much smaller than those of COMMON BUCKWHEAT, nearly round in cross section, and usually pointed. The color ranges from dark gray to black and the seed coat from smooth to decidedly rough and spiny. The leaves are narrow and arrow-shaped. The plants are inclined to approach a vine condition. The flowers are very small and have inconspicuous greenish-white petals.

Tartary buckwheat is grown in the mountains because it is less subject to injury from frost and seems a little better adapted to the rougher, less favorable lands than other BUCKWHEAT species.

Its principal use in this country is as feed, especially for poultry. It may be fed whole to chickens, but before being fed to horses and hogs, it should be ground and bolted to remove the hulls as the latter may cause dietary disturbances.

Reference: Q.1.

TARUANIAN = *Turkey*. See COMMON WHEAT.

TARWEED (*Amsinckia intermedia*) is a POISONOUS PLANT. It affects horses, cattle, and swine; the seeds of tarweed are sometimes mixed with wheat chaff or screenings. Characteristic effects of tarweed poisoning are loss of appetite, jaundice, emaciation, and, in horses, a tendency to walk continuously.

TAXONOMY, also called *systematic botany*, is the science of classification—the arrangement of plants according to their natural relationships.—*See also* MORPHOLOGY.

TAYLOR. See COWPEA (variety).

TCA, or *trichloroacetic acid*, is a herbicide, but SODIUM TCA is preferred. The free acid forms corrosive, water-soluble crystals.

TDE, also called *tetrachlorodiphenylethane*, *DDD*, or *dichlorodiphenyl dichloroethane*, is *1,1-dichloro-2,2-bis(p-chlorophenyl) ethane*. It is closely related to, but less toxic than, DDT. TDE forms water-insoluble crystals and is used as INSECTICIDE.

A dust containing 5 percent, if applied at the rate of 40 lb. per acre, controls FALL ARMYWORMS.

Caution: Hay or forage that has been treated with TDE should not be fed to dairy animals or to meat animals being finished for slaughter.

T.D.N. = *total digestible nutrients.* See NUTRIENTS.

TEDDER is an implement for loosening HAY in the swath or windrow so that it may cure more quickly.

TELENOMUS. *T. utahensis*, an egg parasite, is an important enemy of the STINK BUG. This WASP is only 1.2 mm. long, and relatively broad and oval in outline.

TELIOSPORES. *See* RUSTS.

TEMPORARY PASTURES = *supplemental pastures.* See PASTURES.

TENDERNESS. *See* MEAT QUALITY.

TENDRIL is a slender, modified leaflet, branch, or STEM, commonly spirally coiling at the tips; it serves as an organ of support, as in PEA VINES and VETCHES.

TENMARQ. *See* COMMON WHEAT (variety).

TENNESSEE 76 is a variety of the COMMON LESPEDEZA.—*See also* ANNUAL LESPEDEZAS; LESPEDEZAS.

TENNESSEE 101 is a CORN hybrid.

TENNESSEE BEARDLESS.
      See SIX-ROWED BARLEY (variety).

TENNESSEE NON POP.
      See SOYBEAN (variety).

TENNESSEE RED AND TENNESSEE WHITE are unclassified PEANUT varieties.

TENNESSEE WINTER.
      See SIX-ROWED BARLEY (variety).

TENNEX. *See* RED OAT (variety).

TEOSINTE (*Euchlaena mexicana*) is a close relative of CORN; it has broad, flat blades and reaches 6 to 16 ft. high. This GRASS is grown in the southeastern part of the United States as a forage crop.

TEPARY BEAN (*Phaseolus acutifolius* var. *latifolius*), one of the FIELD BEANS, has white seeds. The plant is drought-resistant; for HAY it outyields many other annual LEGUMES.

TERMINAL means: topmost.

TERRAMYCIN is now a trade-name for *oxytetracycline;* it is an ANTIBIOTIC produced by the MOLD *Streptomyces rimosus*. This drug and its salts are water-soluble and widely used (1) in drinking-water to combat infectious diseases and (2) in rations fed to various poultry, swine, and calves to increase weight gain.—*See also* GROWTH.

TERSAN is a FUNGICIDE which contains THIRAM and a MERCURIAL. Applied in dry form, or as a fluid, at the rate of 3 oz. per 1,000 sq. ft. area, it will control BROWN PATCH. It is also used in the same manner as ARASAN for SEED TREATMENT, e.g., in the control of GRASS SMUTS.

TEST WEIGHT PER BUSHEL.
      See GRAIN GRADING.

TETANY is a CALCIUM DEFICIENCY disease characterized by low calcium level of the blood. Symptoms of this disorder are convulsions, nervousness, rigid muscles, and stiff legs.—*See also* GRASS TETANY; VITAMINS.

TETRACHLORODIPHENYLETHANE
      = TDE.

TETRACHLORO-PARA-BENZO-QUINONE = CHLORANIL.—*See also* SPERGON.

TETRADYMA. The *Tetradyma* spp., or HORSEBRUSHES, include *T. glabrata* = LITTLELEAF HORSEBRUSH and *T. canescens* = SPINELESS HORSEBRUSH.

TETRAMETHYLTHIURAM DISUL-FIDE = THIRAM.—*See also* ARASAN; TERSAN.

TETRANYCHUS. *T. telarius* = RED SPIDER.

TEXAN BARLEY.
> *See* SIX-ROWED BARLEY (variety).

TEXAS 8 is a CORN hybrid.

TEXAS BLACKHULL, one of the *Blackhull kafirs*, belongs to the grain SORGHUMS.

TEXAS BLUEGRASS (*Poa arachnifera*) is a vigorous, sod-forming, native perennial that occurs in the southeastern states and the warmer parts of the southern Great Plains. This grass is one-sexed (DIOECIOUS) —a plant that produces pollen does not produce seed, and vice versa: The most characteristic difference in the appearance of the two kinds of plants is that the females have a mass of fine cobwebby hairs on their SPIKELETS; the spikelets of the males are smooth and hairless.

Plants grow to a height of 1 to 3 ft., with numerous leaves ¼ in. wide and from 6 to 12 in. long, light greenish in color, and dense. Long, webby hairs at the base of the LEMMA are very prominent.

Texas bluegrass grows through the winter, producing an abundance of leafy, nutritious forage at the season when most range forage is harsh and least palatable to livestock.

Because of the palatability and abundance of the forage it yields, this species is valuable for range and pasture in the area to which it is adapted.

Reference: H.1.

See also BLUEGRASSES; GRASSES.

TEXAS MILO and *Texas Double Dwarf milo* are grain SORGHUMS.

TEXAS PATNA. *See* RICE (variety).

TEXAS RED = *Red Rustproof. See* RED OAT.

TEXAS-SEEDED RIBBON CANE, better known as *Gooseneck*, is a forage SORGHUM.

TEXAS WINTER = *Ferguson. See* SIX-ROWED BARLEY.

TEXAS YELLOW BEARDGRASS = *King Ranch bluegrass. See* TURKESTAN BLUESTEM.

THATCHER is a COMMON WHEAT (variety).

THEISS = *Turkey. See* COMMON WHEAT.

THEOBROMA. *T. cacoa* = COCOA TREE.

THERAPEUTIC, or *therapeutical*, means: curative or relating to the remedial treatment of disease.—*See also* MINERAL FEEDS.

THERIOAPHIS. *T. ononidis* = YELLOW CLOVER APHID.

THERM. *See* CALORIE.

THIAMINASE is an ENZYME that destroys thiamine, or VITAMIN B₁.—*See also* BRACKEN.

THIAMINE CHLORIDE = VITAMIN B₁.

THIAMINE HYDROCHLORIDE
> = VITAMIN B₁.

THICKLEAF DRYMARY (*Drymaria holosteoides*) has leaves and stems which, when eaten by cattle at the rate of 8 oz. per 100 lb. body weight in a day, will cause poisoning. The consumption of this POISONOUS PLANT causes such symptoms as depression, weakness, and inflammation of stomach and intestines.

THIO-DIPROPIONIC ACID and its *esters* are used as ANTIOXIDANTS; e.g., DISTEARYL THIO-DIPROPIONATE.

2-THIOURACIL forms small, bitter-tasting crystals which are practically insoluble in water, but soluble in alkaline solutions. It reduces the activity of the THYROID gland. This drug has been used experimentally to affect GROWTH and fattening of farm animals, but the results, especially with heifers, have not been encouraging, particularly when compared with those obtained with STILBESTROL.

> NOTE: 2-thiouracil seems to do very well in increasing fat formation in hogs weighing more than 150 lb., but with consumer preference for lean pork the drug has not been important.

See also HORMONES.

THIRAM, which is *bis(dimethyl-thiocarbamyl)disulfide*, also known as *tetramethyl thiuram-disulfide*, forms crystals which are soluble in organic solvents but insoluble in water. It is a fungicide used in ARASAN (50 percent) and *Arasan SF* (75 percent). It is also contained in TERSAN.

THISTLES are prickly-leaved herbs; many of them are WEEDS, e.g., the SOW

THISTLE, RUSSIAN-THISTLE, YELLOW STAR-THISTLE, and CANADA-THISTLE.

THLASPI. *T. arevense* = PENNYCRESS.

THOMAS PHOSPHATE = BASIC SLAG.

THORNE is a winter variety of COMMON WHEAT.

THOUSANDFOLD. *See* RYE (variety).

THOUSAND-HEADED KALE. *See* KALE.

THREADLEAF GROUNDSEL. *See* GROUNDSELS; SENECIO; POISONOUS PLANTS.

THREE-CORNERED ALFALFA HOPPER (*Spissistilus festinus*) is one of the harmful insects which often cause ALFALFA during late summer to be unthrifty, short, and distinctly yellow in color. It sometimes also damages SERICEA plants.

The nymphs, as well as the adults of this insect pest, feed by thrusting their beaks into the plant and sucking out the plant juice. Feeding is often in a continuous line, forming a ring or girdle around the stem. The latter type of feeding weakens the stem with the usual loss of color and stunting of growth. In several cases most of the stems yellow or die before they reach a height of 6 in. Stand losses have been observed on heavily infested areas.

Control: No alfalfa strain with any degree of resistance to the three-cornered alfalfa hopper is known. In late season stands, when alfalfa is not cut for hay or used for grazing, satisfactory control of this insect has been obtained by dusting with LINDANE or with a mixture of DDT and lindane.

Reference: K.6.

THREETIP SAGEBRUSH (*Artemisia tripartica*) is one of the less troublesome SAGEBRUSHES; it can sometimes be found where the BIG SAGEBRUSH grows.—*See also* SAGEBRUSHES.

THREONINE is one of the essential AMINO ACIDS. It occurs in eggs, milk, casein, and other protein products, and forms water-soluble crystals. Most practical rations appear to contain sufficient threonine to support good GROWTH and reproduction.

THRIPS *See* TOBACCO THRIPS.

THUMPS is a baby-pig disease.—*See also* IRON.

THYANTA. *T. custator* = RED-SHOULDERED PLANT BUG.

THYROID GLAND. *See* IODINE.

THYROPROTEIN contains the hormone THYROXINE; it can be synthesized from IODINE and proteins of skimmed milk, casein, etc. It is used to supplement the *thyroid gland* in stimulating cows to produce more milk, richer in butterfat.

Daily feeding of 10 to 15 gm. thyroprotein mixed in 3 lb. additional grain ration may increase the milk production of healthy *cows* 5 to 20 percent and the fat content of milk 25 to 50 percent. It is recommended to limit thyroprotein feeding to the declining phase of lactation, or from the 70th day of lactation to the end of the sixth month of pregnancy. The drug should not be fed to young stock, nor should it be fed in the summer.

Overdosing may cause reduction of milk production and can also result in loss of weight, other symptoms of stepped up METABOLISM, and sometimes heart failure.

NOTE: Purebred breed associations have banned the practice of increasing milk production by the use of thyroprotein since the milk production records, after the administration of the drug, do not truly reflect the cow's inherited milk-producing ability.

Thyroprotein has also been fed successfully to lactating *goats* for increased milk production; for the promotion of GROWTH, it may be fed to *pigs* at the rate of 15 gm. per 100 lb. complete ration, from the time the animals weigh 40 to 50 lb. until they reach 200 to 300 lb. body weight. When fed to *poultry*, thyroprotein promotes feathering and growth, and has caused older hens to lay eggs with better-quality shells.

Improved sex drive has been observed in aging *bulls* and in *rams* who have been fed thyroprotein for a limited period of time.

THYROXINE is a HORMONE obtained synthetically, or from the *thyroid gland;* it contains 64 percent IODINE and is the active ingredient of THYROPROTEIN.

Thyroxine forms white, water-soluble crystals. It influences METABOLISM and GROWTH as well as reproduction.

TICK CLOVER = DESMODIUM.

TIFTON BUR-CLOVER (*Medicago rigidula*) is one of the newer species of burclover. It is the most winter-hardy of all BUR-CLOVERS and is distinguished by its hard, spiny, comparatively large bur.

Reference: M.6.

See also CLOVERS.

TIFT SUDANGRASS.

See SUDANGRASS (variety).

TILLAGE is the preparation of land for seed, keeping it from WEEDS, and thus improving the physical condition of the soil for the cultivation of field crops.—*See also* DUST STORMS.

TILLER is an erect shoot which arises from the crown of a GRASS.

To tiller means: to produce tillers.

TILLETIA. The *Tilletia* spp. are seed SMUTS. The FUNGUS *T. caries* causes DWARF BUNT and COMMON BUNT. The latter is also caused by *T. foetida*.

TILTH is the state of cultivation or the physical condition of soil.

TIMBER POISONVETCH.

See POISONVETCHES.

TIMOTHY (*Phleum pratense*) was once called *Herd's grass*. The stems, or culms, are 20 to 40 in. tall. They emerge from a bulblike base and form large clumps. Timothy differs from most other GRASSES in that one of the lower INTERNODES is swollen into an egg-shaped (OVOID) body, referred to as a *bulb*, or CORM. The corm is annual in duration, forming in early summer and dying the next year, when seed matures. The PANICLE is cylindrical and commonly 2 to 4 in. long and often longer.

Timothy grows better on clay loams than on light-textured, sandy soils. It is well adapted to the cool, humid climate of the northeastern area and North Central States, and also to the valleys of the Rocky Mountains, and to the coastal region of the Pacific Northwest.

Fall seedings alone or with WINTER WHEAT are best; seedings started then are less likely to be injured by dry weather in late spring or early summer than are those from spring seedings.

Timothy is commonly sown with CLOVER —especially medium red clover, mammoth red clover, or ALSIKE CLOVER—or with ALFALFA so as to get a hay with higher protein content and to maintain a better soil productivity.

A change in the quality of *hay* occurs as the season advances. The percentage of nitrogen-free extract, fat, and protein gradually decreases while the less digestible and less valuable crude fiber increases with maturity. Therefore, timothy should be cut when it is in early bloom in order to get the greatest value per acre of high-quality hay. The acreage cut for hay in the United States is estimated at more than 6 million acres.

Timothy. Plant, glumes, and floret. (H.26.)

Dangers: ARMYWORMS, BILLBUGS, and CHINCH BUGS often attack timothy.

Reference: H.1.

See also RED CLOVER; REDTOP; TALL FESCUE; MEADOW FESCUE; REED FOXTAIL; WHEATS; PASTURE PLANTS; SILAGE CROPS; RANGE MANAGEMENT; VITAMIN A; SWEET-CLOVER HAY POISONING.

TIMOTHY HAY.

See HAY GRADING; HAY MEASURING.

TIP. *See* WHITE TIP.

TIP CLOVER. *See* WHITE TIP CLOVER.

T.N.1006 = THORNE.—*See also* COMMON WHEAT.

TOASTED CORN FLAKES. *See* CORN.

TOBACCO (*Nicotiana tabacum*) contains the alkaloid NICOTINE. Finely ground tobacco can be used as DILUENT in insecticidal dusts.—*See also* CUBE; DERRIS; CRYOLITE: WHEATS.

TOBACCO RING SPOT VIRUS is the cause of BUD BLIGHT.—*See also* VIRUS DISEASES; VIRUS.

TOBACCO THRIPS (*Frankliniella fusca*), an insect pest, causes widespread disease-like injury to PEANUT leaves, which delays the growth of seedlings by two to three weeks. Growers often refer to the thrips injury, which is most severe in dry weather, as "*pouts*" or "*possum-ears*". Injured terminal buds of very small seedlings may become black as if they had been burned. Much of the feeding by this insect, however, is done on the upper surface of the leaflets before they unfold, and is not apparent until they open up later.

The tobacco thrips is dark brown or black, and so small that it is not easily seen. It moves rapidly. The young stages are yellow and hatch from eggs placed within the tissues of the peanut leaflets. Several generations develop each season.

Control: Peanuts grown on soils of low fertility should be treated with DDT as soon as thrips injury appears on the young plants, and again about ten days later. SULFUR dust or BORDEAUX-MIXTURE spray is also useful.

References: B.10; W.17.

TOBOSA (*Hilaria mutica*), or *tobosa grass*, grows 2 ft. high culms from a tough rhizomatous base. The SPIKELETS are bearded at the base. This GRASS is used as a forage in the Great Plains region.

Reference: H.26.

See also PASTURE PLANTS.

TOCOPHEROL. *See* VITAMIN E.

TOCOPHERYL ACETATE. The *dl*-type of *alpha*-tocopheryl acetate forms a pale-yellow, viscous, water-insoluble liquid which is not affected by air or light.—*See also* VITAMIN E.

TOKYO. *See* SOYBEAN (variety).

TOMATO (*Lycopersicon esculentum*) is a widely cultivated vegetable with many varieties.—*See also* TOMATO POMACE; VITAMIN A; LIGHTNING INJURY.

TOMATO FRUITWORM

= CORN EARWORM.

TOMATO POMACE consists of the skin, pulp, and crushed seeds that remain after manufacture of tomato juice. It is high in protein, fat, and fiber, and has been fed successfully in the wet form to swine.

Reference: E.12.

See also PLANT BY-PRODUCTS; DRIED TOMATO POMACE.

TONICS and other "*conditioners*" are often valueless or unnecessary. Healthy animals do not need agents to stimulate their appetite; sick animals should be treated expertly rather than fed "conditioners"; in no case will the latter prevent or cure infectious diseases.

Tonics should be termed more specifically as *bitter tonics*, or BITTERS, which stimulate appetite and digestion; *hematinic tonics*, which improve blood qualities (e.g., ARSENICALS; IRON); *arsenic tonics*, such as ARSONIC ACIDS, which have hematinic properties; *stomachic tonics*, or *stomachic* STIMULANTS, which act like bitters; etc.

TOOTHED BUR-CLOVER

= CALIFORNIA BUR-CLOVER.

TOPOGRAPHY is the description of the physical characteristics of a tract of land, or any other particular place, or region.

TOPPED CORN. *See* CORN.

TOPPED FODDER. *See* FODDER.

TOPSOIL = SURFACE SOIL.

TORNILLO (*Prosopis odorata*), or *screwbean*, is easily distinguished from the closely related MESQUITE by its tightly coiled seed pods. It is confined largely to valley bottom lands where it may form heavy thickets and is seldom, if ever, a noxious plant on the uplands.

Reference: P.11.

See also LEGUMES; INOCULATION.

TORULOPSIS. The *Torulopsis* spp. are FUNGI which are useful in the form of dried *torula* YEAST.

TOTAL DIGESTIBLE NUTRIENTS.

See NUTRIENTS; GRAZING.

TOWER SILOS. *See* SILOS.

TOXAPHENE, formerly called *chlorinated camphene*, is one of the CHLORINATED HYDROCARBONS; it contains 67 to 69 percent chlorine. This slow-acting INSECTICIDE is a yellow, waxy mass and has a piney odor. It is insoluble in water, but soluble in many organic solvents. Toxaphene formulations are available as emulsifiable concentrates, dusts, and wettable powders.

A 20-percent toxaphene dust, applied at the rate of 20 lb. per acre, controls ARMYWORMS and CUTWORMS; the same 20-percent dust used at 10 to 15 lb. per acre kills FALL ARMYWORMS. For the control of the VELVETBEAN CATERPILLAR, use a 10-percent toxaphene dust at the rate of 15 lb. per acre. A 10-percent granulated toxaphene applied at the rate of 250 lb. per acre controls the SOUTHERN CORN EARWORM.

For the control of GRASSHOPPERS, 1½ lb. toxaphene per acre is required; it has a residual action for two to four weeks. Toxaphene controls GARDEN WEBWORMS as long as the larvae are small and if applied—per acre—at a rate of 3 lb. dust, or as a spray containing 2 lb. toxaphene. Adult CHINCH BUGS can be controlled with sprays or dusts applied at the rate of 1½ lb. toxaphene per acre. For the control of the MEADOW SPITTLEBUG, toxaphene is sprayed at the rate of 1½ lb. per acre. Sprays of 3 lb. per acre control LYGUS BUGS and, at the same time, grasshoppers; however, for the simultaneous control of lygus bugs and STINK BUGS, 6 lb. toxaphene per acre should be used as spray, or 25 lb toxaphene-sulfur dust applied, the dust consisting of 20 percent toxaphene and 40 percent SULFUR. (Figures are based on undiluted toxaphene.)

Caution: Forage treated with toxaphene should not be fed to dairy cows, poultry, or animals that are being finished for slaughter. However, the crops treated with toxaphene may be fed to animals if at least forty days are allowed to elapse between application of the insecticide and cutting or pasturing the crop.

To protect *bees*, toxaphene should be applied to ALFALFA in bloom only before 7 a.m. or after 7 p.m.—*See also* POISONED BRAN MASH; DRY BAIT.

TOXAPHENE-SULFUR DUST.

See TOXAPHENE.

TOXIC means: poisonous.

Toxicity is the degree of virulence of a poison.

TOXINS are organic poisons secreted by PATHOGENS while growing.—*See also* ANTIBIOTICS.

TOXOPTERA. *T. graminum* = GREENBUG.

TR is one of the inbred lines of CORN.

TRACE ELEMENTS, also called *trace minerals*, are MINERALS required only in minute amounts by animals or plants. While most rations normally used contain sufficient quantities of all elements, they may be deficient in some areas. The deficient elements, then, must be added to the diet to prevent, or overcome, certain deficiency diseases.

The most important trace elements are MANGANESE, IRON, ZINC, MOLYBDENUM, COPPER, COBALT, and IODINE. The last three CHEMICAL ELEMENTS are more often lacking in rations than others.—*See also* MINERAL SALTS.

TRACE MINERAL SALT, or *trace mineralized salt*, may also be labeled *salt with trace minerals*.—*See also* TRACE ELEMENTS; MINERAL FEEDS.

TRACY BLACK VELVETBEAN (*Mucuna capitata*), an early, black-seeded hybrid, is more prolific than the DEERING VELVETBEAN. It belongs to the earliest VELVETBEANS and matures seed as far north as Washington, D.C. The beans are shiny black, flat, and of large size. Both seeds and pods are somewhat softer than those of other types of velvetbeans.

Reference: P.10.

See also LEGUMES; INOCULATION.

TRANSLUCENT means: semitransparent.

TREBI. *See* SIX-ROWED BARLEY (variety).

TREFOILS (*Lotus* spp.) are perennial, fine-stemmed, and leafy plants that are somewhat decumbent when grown as single plants but are fairly upright in thick stands.

The leaves are SESSILE along the stem and have five leaflets, which vary in shape from linear to oval. The yellow flowers are quite showy; they are borne on long flower stalks, which carry several characteristically spreading seed pods that resemble a bird's foot.

The primary use of trefoils is for pasturage, both alone and in mixtures, but they also make good hay.

The establishment of stands of trefoils has been somewhat difficult. Since the seed is very small, good seedbed preparation is essential. In the North seeding is done in the spring. Farther south seeding can be made either in the spring or fall. A very firm seedbed is one of the prime essentials.

The trefoils require a special inoculant, which must be supplied at time of planting.

As with most other LEGUMES, a fertilizer high in phosphate will increase yields.

The harvesting of trefoils for hay offers no special difficulties, but harvesting for seed is not so easily accomplished, because the seed is apt to shatter when ripe.

In comparison with other legumes, the trefoils have high feeding value both for hay and pasturage. They are especially useful for furnishing late summer feed.

Trefoils are of special value because they are deep-rooted, make growth in late summer, and will grow in situations where ALFALFA and CLOVER cannot be grown to advantage.

Species: The two species of practical importance are BIG TREFOIL and BIRDS-FOOT TREFOIL.

Reference: M.3.

See also INOCULATION.

Illustration: *See* BIRDSFOOT TREFOIL.

TREMETOL, an alcohol of oily consistency, is the toxic principle of WHITE SNAKEROOT and RAYLESS GOLDENROD, two POISONOUS PLANTS which cause *trembles* in livestock and *milksickness* in man drinking milk or eating butter made from milk of affected cows.

Trifoliate leaf, as in clover. (D.9.)

TRENCH SILOS. *See* SILOS.

TRI- is a prefix signifying the number three.

"TRI" = TRICHLOROETHYLENE.

TRIBE, or *subfamily*, is a group of related genera forming a natural division of a FAMILY; for example, the CLOVER tribe (or Trifolieae) of the LEGUME family (Fabaceae, or Leguminosae).

> NOTE: Occasionally the term subfamily is not used synonymously with tribe, but as a division between the family and the tribe.

TRICALCIUM PHOSPHATE, or *tribasic calcium phosphate*, is available as a white, water-insoluble powder. It occurs in nature as ROCK PHOSPHATE and is widely used as fertilizer and in MINERAL FEEDS.

TRICHLOROACETIC ACID = TCA.

TRICHLOROETHYLENE, or *"Tri"*, is a colorless, heavy, nonflammable, toxic liquid. It is practically insoluble in water, but mixable with the common organic solvents and is widely used as an extraction medium for many processes.

The distribution of *trichloroethylene solvent extracted* SOYBEAN oil meal, flakes, or pellets for animal feeding purposes is prohibited.

2,4,5-TRICHLOROPHENOXYACETIC ACID = 2,4,5-T.

TRICHOGRAMMA. The wasplike INSECT *Trichogramma minutum*, an important egg parasite, is an enemy of the CORN EARWORM. In some seasons in the southeastern United States fully 90 percent of the earworm eggs fail to hatch because of the work of this insect.

Reference: B.9.

TRICKER is one of the forage SORGHUMS

TRIFOLIEAE—i.e., the *clover tribe*—include not only the TRUE CLOVERS (*Trifolium* spp.), but also the other CLOVERS (*Melilotus* spp. and *Medicago* spp.).—*See also* TRIBE.

TRIFOLIOLATE means: having three leaflets; i.e., the leaves are three-divided, as in CLOVER and ALFALFA.

TRIFOLIOSIS is a disease of horses and mules which is caused chiefly, if not exclusively, by ALSIKE CLOVER. Seemingly this clover, under certain conditions, is the cause of sores on the animals. The trouble is very widespread, even though it has not been observed in all areas where alsike clover is grown. It is well for farmers pasturing stock on this clover to watch for signs of sore mouths or sore spots on the forelegs or body. When such are noted, the animals should at once be taken from alsike clover pastures and put on grass. Unless the disease has progressed too far, the animals will recover promptly.

Reference: P.2.

TRIFOLIUM. The *Trifolium* spp., or TRUE CLOVERS, include *T. repens* = WHITE CLOVER; *T. pratense* = RED CLOVER; *T. incarnatum* = CRIMSON CLOVER; *T. hybridum* = ALSIKE CLOVER; *T. dubium* = SMALL HOP CLOVER; *T. procumbens* = LARGE HOP CLOVER; *T. fragiferum* = STRAWBERRY CLOVER; *T. resupinatum* = PERSIAN CLOVER; *T. subterraneum* = SUB CLOVER; *T. fendleri* = FENDLER CLOVER; *T. wormskjoldii* = SEASIDE CLOVER; *T. variegatum* = WHITE TIP CLOVER; *T. ambiguum* = KURA CLOVER; *T. glomeratum* = CLUSTER CLOVER; *T. striatum* = STRIATA CLOVER; *T. hirtum* = ROSE CLOVER; *T. nigrescens* = BALL CLOVER; *T. lappaceum* = LAPPA CLOVER; *T. carolinianum* = CAROLINA CLOVER; *T. pannonicum* = HUNGARIAN CLOVER; *T. alexandrinum* = BERSEEM (or *Egyptian clover*); and *T. medium* = ZIGZAG CLOVER.

TRIGLOCHIN. *T. maritima* = ARROW PODGRASS.

TRIGONELLA. The dried fruit of *T. foenumgrascum* is known as FENUGREEK SEED.

TRIPLE SUPERPHOSPHATE contains no gypsum, but three times as much PHOSPHORUS as SUPERPHOSPHATE. It forms a dry, granular, gray product and yields about 50 percent available P_2O_5.

TRISTEARIN = STEARIN.

TRITICUM. The genus *Triticum* includes the following WHEATS: *T. compactum* = CLUB WHEAT; *T. dicoccum* = EMMER; *T. durum* = DURUM WHEAT; *T. polonicum* = POLISH WHEAT; *T. spelta* = SPELT; *T. turgidum* = POULARD WHEAT; and *T. vulgare* = COMMON WHEAT.

TRIUMPH is a winter variety of COMMON WHEAT.

TRUE ARMYWORM = ARMYWORM.

TRUE CEREAL CROPS. *See* CEREAL.

TRUE CLOVERS, or the *Trifolium* spp., are LEGUMES; however, the name CLOVER is incorrectly applied to many more plants which, in some respect, resemble the true clovers.

Throughout the world there are approximately 250 species of the genus *Trifolium;* of these, more than 80 species are native to the United States, but none of them has proved to be of agricultural value, although they contribute to GRAZING and to the wild hay crop.

The true clovers are perennial or annual. For the most part, except at high latitudes, the growth period of the annual species is confined to the fall, winter, and spring months; in general, they thrive in a cool, moist climate on soils where there is an available supply of phosphorus, potassium, and calcium. Many of the perennial species may behave as biennials and annuals because of unfavorable climatic conditions, attacks of diseases, and insect pests.

Most of the species are long-day plants, although many continue to flower into early fall. So-called winter clovers (e.g., some varieties of white clover and red clover) have frost resistance and high feed quality, making them valuable in the pastures of southern areas. Some of the nitrogen fixed by the clover remains in the soil to benefit the associated grasses for three to five months after the legume crop has completed growth, thus influencing pasture over a long season.

Wide differences exist in the ability of different species to tolerate unfavorable environments and in the habit of growth, flowering, and reproduction.

The flowers of all species are borne on heads, with the number of FLORETS from as low as 5, in sub clover, to as high as 200 per head, in red clover and white clover. The flowers of some species are self-sterile, requiring cross-pollination. Others are self-fertile but must be tripped or shaken to insure pollination; still others are self-fertile and self-pollinating. Seeds per pod vary from one to eight, depending upon the species.

Dangers: Diseases of the clovers cause heavy losses in many areas, especially when the crops are grown for forage or seed instead of being plowed under for green manure. The diseases of most species are the same or are similar, namely SEEDLING BLIGHT, CROWN ROT, ROOT ROT, BLACK STEM OF CLOVER, MILDEW, LEAF SPOTS, and RUST. In general they cannot be completely controlled, but severity can be reduced considerably by (1) use of adapted and resistant varieties; (2) seed treatment; (3) sanitation; and (4) good cultural practices.

Species: Nine species are of national or regional agricultural importance, namely RED CLOVER, WHITE CLOVER, CRIMSON CLOVER, ALSIKE CLOVER, SMALL HOP CLOVER, LARGE HOP CLOVER, STRAWBERRY CLOVER, PERSIAN CLOVER, and SUBCLOVER.

The following species are of local importance: CLUSTER CLOVER, SEASIDE CLOVER, STRIATA CLOVER, ROSE CLOVER, BALL CLOVER, LAPPA CLOVER, CAROLINA CLOVER, FENDLER CLOVER, WHITE TIP CLOVER, HUNGARIAN CLOVER, ZIGZAG CLOVER, and KURA CLOVER (an excellent nectar-producing plant).

References: M.3; H.7; T.1.

See also TRIFOLIUM; HAY; INOCULATION.

TRUE CRESTED WHEATGRASS = FAIRWAY CRESTED WHEATGRASS.—*See also* CRESTED WHEATGRASS; WHEATGRASSES.

TRUE FATS. *See* FATS.

TRUE PROTEIN is a term referring to substances which, chemically, are pure PROTEINS, not mixtures of nitrogenous compounds (including, among others, an amount of true protein), as contained in many feedstuffs.—*See also* CRUDE PROTEIN; ARMSBY FEEDING STANDARDS.

TRUE STOMACH = *abomasum. See* STOMACH.

TRUMBULL.

See COMMON WHEAT (variety).

TRYPSIN is an ENZYME contained in PANCREATIC JUICE.—*See also* PEPTONES.

TRYPTOPHAN is an essential AMINO ACID. It forms crystals which are slightly soluble in water. Tryptophan is needed for the GROWTH of birds and other animals.

CORN and most other cereal GRAINS are poor sources of tryptophan, and even such highly regarded animal PROTEINS as fish meal and meat scraps, may not contain enough tryptophan to adequately supplement corn for young pigs or chicks. SOYBEAN meal and MILK PRODUCTS are excellent sources and should therefore usually be used as part of the protein supplement for pig and chick starters.

TUALATIN OATGRASS, or *Tualatin (meadow) oatgrass,* is a variety of TALL OATGRASS.—*See also* GRASSES.

TUBER is a thickened, usually starchy swelling of a subterranean STEM serving for food storage and vegetative reproduction. POTATO and SWEET POTATO are examples of tubers.

Tubers sprout from buds known as *eyes.*

TUBERCULATE means: beset with small, pimple-like prominences (*tubercles*).

TUBER WATERHEMLOCK.

See WATERHEMLOCK; POISONOUS PLANTS.

TUCUM-NUT OIL MEAL, or *tucum oil meal,* is obtained from the seed (nut) of the TUCUM PALMS; it is occasionally used as cattle feed in place of cocoanut oil meal. An average analysis gives 11 percent protein, 6 percent fat, and 9 percent fiber.—*See also* OIL MEAL.

TUCUM PALMS (*Astrocaryum* spp.) are tropical plants.—*See also* TUCUM-NUT OIL MEAL.

TULE = CATTAIL.

TUNA OIL is a yellow to red-brown liquid of characteristic odor. It is obtained from cannery refuse of tuna fish, and is a good source for the VITAMINS A and D.

TURF. *Winter Turf* is a COMMON OAT variety.

TURGHAI. *See* PROSO (variety).

TURKESTAN ALFALFAS, also called *Turkistan alfalfas,* include (*Commercial*) *Turkestan, Hardistan, Nemastan,* and *Orestan.—See also* ALFALFAS.

TURKESTAN BLUESTEM (*Andropogon ischaemum*), also known as *East Indian bluestem* or *yellow bluestem,* is a perennial bunchgrass with leafy foliage. It is not widely used as yet.

Dangers: Turkestan bluestem, and particularly the variety King Ranch bluestem, is very susceptible to a LEAF RUST which may become serious during wet periods in the summer time and thus reduce production sharply. If wet periods occur during flowering time, ERGOT may become a serious problem.

Varieties: *King Ranch bluestem,* also known as *K. R. bluestem* (or *K-R bluestem*) and *Texas yellow beardgrass,* is the most important variety of Turkestan bluestem. It can be distinguished from the other varieties of this species by a ring of small, white hairs at each node. This perennial bunchgrass is much finer stemmed and leafier than the native American bluestems. It grows 3 to 5 ft. high and produces seed heads continuously from early summer to frost.

Like all Asiatic bluestems it prefers fine-textured soils. The grass has a very extensive root system and because it is aggressive and a heavy feeder, it tends to rapidly deplete the nitrogen supply in the soil.

King Ranch bluestem has been used chiefly in Oklahoma, Texas, Louisiana, and Arkansas for pasture, seed crop, and hay; also for overseeding in brush-infested or depleted rangeland, soil conservation, and erosion control. The forage is relished by livestock early in the season, but, as with all bluestems, the quality declines with maturity.

Care of seeding stands is similar to other warm-season GRASSES; i.e., weeds should be controlled by mowing, and GRASSHOPPERS curbed by spraying. Seedlings grow very slowly at first, and little growth can ordinarily be expected before the first fall after seeding. Once well established, little care is required. Production can usually be increased with nitrogen fertilizers and on some soils phosphate should be added. King Ranch bluestem is somewhat cold-sensitive, and some growth should be left on in the fall for cover; following severe winter injury, there may be little or no forage until mid-summer.

Elkan bluestem is more erect than King Ranch; it has the greatest cold-resistance of any Turkestan bluestem variety in use. References: H.2; W.16.

See also INDIAN GRASS.

TURKEY RATIONS.
                          See POULTRY RATIONS.

TURKEY WHEAT is a winter variety of COMMON WHEAT.

TURKISTAN = *Turkestan.*
                          See ALFALFAS.

TURNIP (*Brassica campestris* var. *rapa*) is a root crop sometimes grown for forage; the tops of turnips are among the vegetable waste products used as feedstuff.

Dehydrated turnip leaf meal can be used instead of ALFALFA meal in poultry diets. It contains protein and VITAMINS A and B$_2$.

Swedish turnip = RUTABAGA.—*See also* PLANT BY-PRODUCTS.

TUSSOCK is a tuft (e.g., a tuft of grass or sedge) often growing on wet ground.

TWIGS. *See* LEAVES AND TWIGS.

TWO-GROOVED POISONVETCH.
                          See POISONVETCHES.

TWO-ROWED BARLEY (*Hordeum distichum*) is an annual BARLEY very similar to the closely related SIX-ROWED BARLEY. The central florets are fertile; the lateral florets are fairly well developed but sterile. The species is not grown extensively in North America; it is mostly found in the semiarid region.

Varieties: Among the two-rowed barley varieties—all of which are spring barleys—are the following:

Alpha (rough-awned);

Charlottetown 80 (rough-awned, grown in Canada);

Compana (semismooth-awned, drought-resistant);

Hannchen (rough awned);

Horn (rough awned);

Sanalta (smooth-awned, grown in Canada);

Spartan (smooth-awned, feed type);

Steigum (smooth-awned);

White Smyrna (semismooth-awned, drought-resistant).

References: H.26; H.27; W.21; A.6.

TWO-STRIPED GRASSHOPPER.
       See GRASSHOPPERS.

TYCHIUS. *T. picirostris*, a WEEVIL, is commonly called CLOVER TYCHIUS.

TYMPANY = BLOAT.

TYPE is a term indicating the specimen or specimens on which a SPECIES (or SUB-SPECIES or VARIETY) is based, and from which the latter can be described.

Typical means: having the character-istics of, or well matching the type of, a species, variety, etc.

Reference: D.9.

See also ECTOTYPE; SPORT.

TYPHA. *T. latifolia* = CATTAIL.

TYROSINE is one of the dispensable AMINO ACIDS. It forms water-soluble crystals and is used by birds and other animals as GROWTH factor. It can replace part of the (amino acid) PHENYLALANINE in rations.

U

UCUHUBA (*Virola surinamensis*) is a tropical tree whose nuts are used for the production of ucuhuba fat and UCUHUBA OIL MEAL.

UCUHUBA OIL MEAL, obtained from the Brasilian UCUHUBA nuts, is sometimes used in dairy feeds. The OIL MEAL contains approximately 16 percent protein, 21 percent fat, and 25 percent fiber.

U.G.F. stands for *unidentified growth factor(s)*. Poultry nutritionists pay more attention to these factors than do others, though the U.G.F. may be of significance for baby pigs and perhaps other non-ruminants.

The term U.G.F. is applied to an un-known NUTRIENT or to a combination of unknown nutrients present in certain feeds that, if added to RATIONS containing all of the known nutrients, will cause a stimulation in GROWTH rate.

At least three such factors are postulated —the *fish factor,* the *whey factor,* and the *grass* (or *alfalfa*) *factor; distillers' solubles* are thought to contain a fourth factor (or it may be the same as the whey factor).

Sources of such factors are often added to broiler rations at the level of 1.5 to 3 percent.

NOTE: Some investigators believe that the factors will eventually be identified as water-soluble VITAMINS; most of the U.G.F. sources, however, supply good amounts of essential AMINO ACIDS and may have their favorable effect by improving amino acid balance. In some experiments, the ASH from U.G.F. sources has given growth stimulation; this may mean that the factors are inorganic compounds.

ULTA = *Turkey wheat.*
       See COMMON WHEAT.

ULTRAVIOLET RAYS have short wave length, are invisible, and able to exert chemical action; e.g., ultraviolet irradia-tion of ERGOSTEROL gives VITAMIN D_2, while radiation of certain CHOLESTEROLS results in the formation of VITAMIN D_3.—*See also* YEAST; VITAMIN D.

UMBRELLA PLANTS = SEDGES.

UNDRIED OAT GROATS
      = *hulled oats. See* OATS.

UNHULLED PEANUT FEED.
       See PEANUT.

UNI- is a prefix signifying one.

UNIDENTIFIED GROWTH FACTOR
       = U.G.F.

UNISEXUAL means: one-sexed; i.e., having the organs or flowers of one sex only—either STAMENS (male) or PISTILS (female), as, for instance the unisexual SPIKELETS of BUFFALOGRASS.

UNKNOWN, more often called *Wonderful,* is a "Clay" variety of COWPEA.

UNSATURATED FATTY ACIDS.
       See VITAMIN F.

UNTHRIFTINESS. *See* VITAMIN A.

UPLAND COTTON. *See* GOSSYPIUM.

UPLAND RICE. *See* RICE.

URAMON is a nitrogenous fertilizer which contains 42 percent UREA nitrogen.

URATES are salts of *uric acid* dissolved in urine of mammals, but occurring in solid form in the urine of birds.—*See* VITAMINS.

UREA, or *carbamide*, contains over 46 percent NITROGEN. It is formed in the body during the PROTEIN metabolism.

UREA IN A WINTERING RATION FOR SHEEP

Daily ration:	1	2	3	4	5	6	7	8
Corn lb....................	0.15	0.15	0.15	0.15	....	....	....	....
Molasses, lb................	....	....	....	....	0.20	0.20	0.20	0.20
Casein, lb..................	0.10	0.20	....	....	0.10	0.20	....	....
Soybean meal, lb...........	....	....	0.18	0.36	....	....	0.18	0.36
Urea-corn mixture [1]........	+		+		+		+	
Average daily gain, lb.......	0.07	0.04	0.20	0.11	0.06	0.06	0.14	0.12

[1] 7 parts corn, 1 part urea, fed to replace either 0.10 lb. casein or 0.18 lb. soybean meal.

Urea forms crystals which have a cooling, salty, and bitter taste and are very water-soluble. It is widely used as NITROGENOUS FERTILIZER, as a drug because of its DIURETIC action, and as a protein extender for ruminants.

When used as a protein extender, in a mixed ration, it may replace one-third of the total protein, or make up 3 percent of the grain ration, or 1 percent of the total ration; 5 percent urea is permissible in thoroughly mixed protein supplements.

About 1 lb. urea and 7 to 8 lb. corn are equal to 7 lb. oil meals. It is usually economical to use urea if oil meal sells for 1.3 times the price of corn.

Most feeds contain plenty of SULFUR, but it will be advisable to add some sulfur to rations containing urea if the base feed contains less than 0.15 percent. SODIUM SULFATE is satisfactory and may be added at the rate of 1 lb. per 10 to 20 lb. urea, depending upon the sulfur level present in the rest of the feed. In some areas the WATER supply contains enough sulfur to meet the animals' need.

Urea has a slightly bitter taste and animals may not like it when they are first given a urea-containing feed. The use of 5 to 10 percent MOLASSES in the ration will encourage them to eat more feed and may supply some TRACE ELEMENTS. Urea will probably give better results when used to supplement a ROUGHAGE-type ration than when used with a GRAIN ration which contains a readily available source of nitrogen to stimulate bacterial growth.

Avoid using raw soybean products as carriers for urea feeds. Raw soybeans, or improperly heated soybean oil meal, contain an enzyme, *urease*, that will cause the urea to be broken down into ammonia gas which will escape and ruin the feed. The effects of urea-containing rations on weight gains and carcass grade are shown in the tables.

Urea cannot be used to advantage in feeds for animals with simple stomachs, such as swine and chickens. It can be used by ruminants, such as cattle and sheep, through co-operation with the *rumen bac-*

UREA FOR FATTENING STEERS ON A RATION CONTAINING NO HAY

	Daily ration:	1	2
Average initial weight, lb....................................		537.5	538.1
Average final weight, lb......................................		747.2	740.5
Average daily feed:			
Ground ear corn, lb...		13.3	13.7
Ground cobs, lb...		2.4	2.0
Cottonseed meal, lb...		2.0	1.0
Vitamin A (2250 I.U./gm.), lb..............................		.1	.1
Urea (262), gm...		....	61
Molasses, lb...		1.45	1.45
Minerals, free choice.......................................		+	+
Crude fiber consumed, lb...................................		1.83	1.74
Gain in 112 days, lb...		209.7	202.4
Average daily gain, lb..		1.87	1.81
Average dressing, percent....................................		60.4	59.1
Average carcass, grade.......................................		Low choice	Low choice

teria that grow on the ruminants' feed and are then digested in the lower gastro-intestinal tract. In the fore-stomach CARBOHYDRATE-feed materials are broken down by fermentation into simpler compounds and re-combined into new ones by BACTERIA and other micro-organisms naturally residing there. The bacteria use many sources of nitrogen for growth.

Bacterial synthesis of protein from urea proceeds in two major steps: (1) the urea is broken down to AMMONIA, and (2) the ammonia is then combined with carbohydrate fragments to form protein in the bacterial cells. The second step must keep pace with the first to prevent the accumulation of ammonia; thus, rapid growth and multiplication of the rumen bacteria are necessary. This can best be assured by providing in the ration readily-available carbohydrates as contained in cereal grains and molasses, a relatively low level of natural protein supplements, and necessary MINERALS.

Caution: Overdosing with urea, or allowing animals to consume large amounts over a short period of time, may lead to disastrous results.

Reference: G.18.

See also SILAGE; SHARK MEAL.

UREASE is an ENZYME that causes UREA to decompose into AMMONIA (gas).

UREDOSPORES. *See* RUSTS.

URINE, the solution of waste products secreted from the kidneys, is rich in organic NITROGEN compounds such as UREA and uric acid and contains also mucus, HORMONES, coloring matter, etc.

UROCYSTIS. The *Urocystis* spp. are FUNGI which are among the causes of STRIPE SMUTS; *U. tritici* causes FLAG SMUT OF WHEAT.

UROMYCES. The FUNGUS *U. trifolii* is the cause of CLOVER RUST; *U. striatus* causes ALFALFA RUST.—*See also* LEAF RUSTS.

URONIC ACIDS are derived from GLUCOSE. *See also* HEMICELLULOSE.

UROPHLYCTIS. The FUNGUS *U. alfalfae* causes CROWN WART.

U.S.13 is a CORN hybrid.

U.S. GRADES OF HAY.

See HAY GRADING.

U.S. GRAIN STANDARDS ACT.

See GRAIN GRADING; SCREENINGS.

U.S.P. = U.S. PHARMACOPOEIA.

U.S.P. units are those defined in the U.S.P.; they apply mostly to vitamins, such as A and D, and are often, but not always, identical with the *international units.—See also* A.O.A.C.

U. S. PHARMACOPOEIA, usually abbreviated *U.S.P.*, is an official book describing the standards of a large number of officially accepted drugs.

USTILAGO. FUNGI belonging to the *Ustilago* spp. cause STRIPE SMUTS and STEM SMUTS; e.g., *U. oculta*, which affects RYE. *U. avenae* is the cause of BLACK LOOSE SMUT; *U. kolleri* causes COVERED SMUT OF OATS; *U. hordei* is the cause of COVERED SMUT OF BARLEY; *U. tritici* causes LOOSE SMUT OF WHEAT; *U. maydis* causes CORN SMUT; *U. nuda* and *U. nigra* cause the LOOSE SMUTS OF BARLEY.

UTON. *See* COMMON OAT (variety).

V

VALANCE is the capacity of an ATOM (or atom group, called *radical*) to combine directly with one or more atoms of HYDROGEN or other CHEMICAL ELEMENTS.

Some elements exhibit more than one valence; e.g., for COPPER, the valences are 1 and 2. Such elements can undergo changes of the valences; copper, for instance, may be reduced from its *cupric* form (having a valence of 2) to its *cuprous* form (having a valence of 1); an OXIDATION reaction would result in the converse.—*See also* VALENCY.

VALENCIA is an unclassified variety of PEANUT.

VALENCY is the numerical value of the VALENCE. HYDROGEN has a valency of 1.—*See also* OXIDATION.

VALINE is one of the essential AMINO ACIDS. It forms a white, water-soluble, crystalline powder and occurs in nature in fibrous proteins; it can also be synthesized. Valine is needed for maintaining GROWTH. Most practical rations supply adequate amounts.

VANDENBURG BLACK

= *Norredo.* See SOYBEAN.

VAR. (plural: vars.) is the abbreviation for VARIETY.

VARIANT is any individual that derivates from the usual, recognized characteristics of a TYPE within a VARIETY, SPECIES, or SUBSPECIES.—*See also* SPORT.

VARIEGATED means: marked by a diversity of coloration.

VARIEGATED ALFALFAS, are a species of ALFALFA (*Medicago media*) which are distinguished by their variegated flower color, ranging from yellow to light purple. They are the result of a natural cross between the COMMON ALFALFA group and the YELLOW-FLOWERED ALFALFAS.

Varieties: *Atlantic* (*alfalfa*) yields well and persists on the relatively shallow, infertile, disease-infested soils frequently found in the East. Throughout most of the area, Atlantic outyields the other alfalfa varieties by 10 percent or more.

The large number of alfalfa strains that entered into the ancestry of Atlantic shows up in its plant characteristics. Individual plants may vary from light to dark green in color and from erect to almost prostrate in growth. Most flowers are light purple, but occasionally other colors or shades are found. Differences occur in size and shape of leaves, and in other characteristics which have much to do with the ability of the plants to survive and grow vigorously. Atlantic will usually persist as long as, or longer than, any other variety when managed properly.

Atlantic has considerable resistance to the insects and diseases that attack alfalfa under eastern conditions. It is fairly tolerant of BACTERIAL WILT disease and gives better results under wilt-infested conditions than do any but the most resistant varieties (none of which have all of Atlantic's other good qualities). Atlantic also is among those varieties which show least injury as a result of infection with leaf spot diseases.

Baltic (*alfalfa*) takes its name from Baltic, South Dakota, near which it was grown for several years before being introduced into commercial production. In yield and adaptability this variety compares favorably with Grimm. Like Grimm it is susceptible to bacterial wilt.

Cossack (*alfalfa*) is superior in hardiness and yielding ability. It grows more erect than Ladak and has better recovery after cutting. Cossack usually is more resistant and slightly less susceptible to WILT than Grimm but in general characteristics it is similar.

Grimm (*alfalfa*) is the best known variety in the variegated alfalfa group. The winter-hardiness of Grimm makes it valuable in the northern half of the United States and in Canada. It is also known for its excellent root development, good recovery after cutting, good seed production, and ability to compete with weeds and grasses. Because of susceptability to BACTERIAL WILT, it is not long-lived, but still is highly useful in stands to be harvested for not more than two years.

Hardigan (*alfalfa*) traces to selections of Baltic made for high seed production and desirable forage characteristics. It shows little flower variegation, is adapted to the same general regions as Grimm and Baltic, and, like those alfalfas, is susceptible to BACTERIAL WILT.

Ladak (*alfalfa*) is largely purple flowered. It is extremely winter-hardy and somewhat drought-resistant. Other characteristics include its semiprostrate growth habit, slow recovery after cutting, early fall dormancy, and exceptionally heavy first crop. It is more resistant to BACTERIAL WILT than Grimm or common alfalfas, and has given excellent results in the northern Great Plains.

Meeker Baltic originates from the variety known as Baltic. It has been grown rather extensively in Colorado, but under WILT condition looses its stand very rapidly. It is very similar to Grimm.

Narragansett (*alfalfa*), developed for persistance and high forage yields, equals Ranger and Grimm in winter-hardiness and exhibits extreme decumbent growth during fall. It is susceptible to WILT, but is promising where the disease is not prevalent.

Ontario Variegated, also known as *Canadian Variegated*, is very similar to Grimm in color, growth habits, and yield. It is of local importance in the northeastern part of the United States and in eastern Canada.

Ranger (*alfalfa*) is a multiple strain variety, having been synthesized from five selections that originated from the three alfalfa varieties. It is distinctly variegated in flower color and varies in habit of growth from decumbent to upright. This variety recovers rapidly after cutting. It is slightly more susceptible to leaf spot diseases than Grimm.

Ranger is one of the high-yielding varieties, ranking slightly below Buffalo. It is highly resistant to BACTERIAL WILT and is also relatively cold-resistant. It is therefore adapted to the Northern, Central, and New England States.

Rhizoma (*alfalfa*) is characterized by having a deep-set crown and ability to spread by producing RHIZOMES, or underground stems. In eastern Canada Rhizoma has produced good pasture fields. In western Canada it is susceptible to BACTERIAL WILT. In the United States it has yielded somewhat less than Ranger or Buffalo.

Vernal, a BACTERIAL WILT-resistant variety, is equal or superior to Grimm in winter-survival. Hence, its greatest value is in those northern states where wilt is a problem and winter conditions are rather severe.

References: G.6; G.7; W.10; B.6.

VARIEGATED CUTWORMS, which belong to the CUTWORMS, are very damaging to ALFALFA.

VARIETY is a term for which a number of definitions has been established; e.g., an *agronomic variety*, also called *crop variety*, or *agricultural variety* (applied to grain or field crops), or a *horticular variety* (applied to VARIANTS of ornamental plants), is quite different from a *botanical variety* (e.g., a variant occurring in the wild state), and all attempts to bring the terminologies together and into agreement, have so far met with failure.

An *agronomic variety*, known by a common rather than a Latin name, may be defined as the kind of crop a farmer grows and which reproduces its kind and remains uniform (such as a certain SORGHUM variety).

NOTE: Because of the variability of varietal designations which change considerably from one part of the country to the other—as may be demonstrated by the fact that one and the same variety is often called by two, three, four, and more common names—some agronomists prefer to use the word "sort" instead of the word "variety."

In the early breeding program one starts with SELECTIONS, but once they have been found superior, they are named and then called a variety and released to growers.

Such an agronomic variety differs from other agronomic varieties in performance, or in origin, even when it cannot be distinguished on the basis of morphological characteristics.

A *botanical variety* has been described as a STRAIN (or group of strains) which can be differentiated from other groups by its structural or functional characters. It can also be described as a group of individuals within a species, or subspecies, that differs from the rest of either one in morphological or physiological characteristics or pathological reactions.

NOTE: How widely experts differ in their ideas of this term is demonstrated by the fact that Bailey recognized only 8 botanical varieties of sorghum, Stephens and Martin 72, and Snowden 165.

The scientific name of any botanical and many of the other varieties is expressed either by a *Latin trinominal* (three-word name, often stated with the authority) or by the species name followed by the varietal name preceded by var. (and followed by the varietal authority), e.g., *Phaseolus acutifolius* var. *latifolius.—See also* TYPE; ECOTYPE; VARIANT; SORT; CROSS; HYBRID.

VASCULAR means: containing vessels, or ducts, for the transportation of fluids.

Vascular tissue is tissue characterized by the presence of ducts.

VASEYGRASS (*Paspalum urvillei*) is a tall, erect PASPALUM GRASS. It is best adapted to fertile soils and common in the South along highways or railroads where it has not been heavily grazed. It can be easily eradicated by close continuous grazing.

Reference: H.1.

See also GRASSES; PASTURE PLANTS.

VAUGHN.

See SIX-ROWED BARLEY (variety).

V.E., or *vesicular exanthema,* is a swine disease.—*See also* GARBAGE.

VEGETABLE. *See* VEGETABLES.

VEGETABLE BY-PRODUCTS.

See PLANT BY-PRODUCTS.

VEGETABLE FATS. *See* FATS.

VEGETABLE IVORY MEAL

= IVORY NUT MEAL.

VEGETABLE PROTEIN.

See PROTEIN; SOYBEAN.

VEGETABLE RENNET. *See* RENNET.

VEGETABLES are plants cultivated for their edible parts; the latter may be *roots,* (e.g., CARROT), *tubers* (e.g., POTATO), *stems* (e.g., CELERY), *leaves* (e.g., SPINACH), *flower heads* (e.g., CAULIFLOWER), *fruits* (e.g., TOMATO), or *seeds* (e.g., CORN).—*See also* VEGETABLE WASTE PRODUCTS.

VEGETABLE WASTE PRODUCTS, resulting from the production and processing of VEGETABLES for canning, freezing, and dehydration, can be used as feedstuffs. Such PLANT BY-PRODUCTS include tops of CARROTS, TURNIPS, and RUTABAGAS; leaves and stems of BROCCOLI, SPINACH, and KALE; PEA VINES and LIMA BEAN VINES.—*See also* BY-PRODUCT FEEDSTUFFS.

VEGETATIVE REPRODUCTION is asexual reproduction; i.e., PROPAGATION without sexual cells or germs (seeds).—*See also* SPORES; ROOTSTOCK.

VEIN—so called because of a fancied resemblance to a human vein—is one of the fibrovascular bundles forming part of the framework (skeleton) of a leaf.

The term *nerve* is often used as a synonym, but it is preferable to confine the term NERVE to simple, parallel VENATIONS (as in a grass blade).—*See also* MIDVEIN; VASCULAR.

VEINLET is a small VEIN, usually one that branches off from a larger one.

VELVET BARLEY.

See SIX-ROWED BARLEY (variety).

VELVETBEAN. *See* VELVETBEANS.

VELVETBEAN CATERPILLAR (*Anticarsia gemmatilis*) is an insect pest which feeds on various crops in the southeastern states. It chews up VELVETBEANS, SOYBEANS, and PEANUTS. Sometimes the caterpillar infests fields of KUDZU, ALFALFA, HORSEBEANS, COWPEAS, and COTTON.

The insects produce three generations during a season. They die when winter comes. Female moths lay their eggs singly on the lower surfaces of leaves. The eggs are white when first laid and orange-colored when ready to hatch. They hatch in three to five days.

The young larvae are greenish and about $\frac{1}{10}$ in. long. After three weeks of continuous feeding, they are full grown and $1\frac{1}{2}$ in. long. They are black or green with several narrow light stripes along their backs and sides. The larvae are very active. When disturbed they spring into the air and wriggle rapidly. They spit a brownish liquid.

Their feeding done, the larvae burrow $\frac{1}{4}$ to 2 in. beneath the soil or crawl under trash, make earthen cells, and change to pupae. The pupae are dark brown and about $\frac{3}{4}$ in. long. The adults emerge in about ten days.

The adults, or moths, are not destructive but are flying danger signs. They are grayish-brown with a wingspread of about $1\frac{1}{2}$ in. They can be distinguished from other moths by the brown or black zigzag lines across their wings.

Control: Apply DDT, METHOXYCHLOR, TOXAPHENE, SULFUR, or CRYOLITE, in the morning or afternoon when the air is quiet.

Velvetbean caterpillars have many natural enemies. Red-winged blackbirds, killdees, upland plovers, grackles, sparrows, bobolinks, and mockingbirds feed on them. So do lizards, skunks, frogs, toads, and poultry.

WASPS, ROBBER FLIES, GROUND BEETLES and FIRE ANTS destroy eggs and caterpillars. Moles and ground beetles feed on the pupae in the soil.

A FUNGUS disease frequently attacks velvetbean caterpillars; the larvae killed by the fungus are greenish-white and hang on the stems or leaves.

References: U.5; U.3.

VELVETBEAN MEAL.

See VELVETBEANS.

VELVETBEANS (*Mucuna* spp.) are vigorous-growing annual LEGUMES, the vines of which (except for the bush varieties) usually attain a length of between 10 to 25 ft. The leaves are trifoliolate, with large, ovate, membranous leaflets shorter than the PETIOLE. The flowers of the different species and varieties vary in color from white to dark purple, are 1 to 1½ in. long, and are borne singly or in twos and threes, in long pendent clusters.

The pods are covered with hairs. The pubescence which sheds to a large extent soon after maturity, may cause skin irritation. The pods of some species are only 2 to 3 in. long, while those of others may reach a length of 5 to 6 in.

Velvetbeans have numerous rather fleshy surface roots, which are often 20 to 30 ft. long and are abundantly supplied with nodules.

The greatest acreage of velvetbeans is found in the well-drained, sandy Coastal Plain soils of the South Atlantic and Gulf States. The legume can be successfully grown on newly cleared land as well as on land that has been cultivated many years. It has been used extensively as a green-manure crop on cut-over pineland and sandy soils. It makes a good growth on clay soils in the northern portion of the Cotton Belt.

While continuous cropping of velvetbeans does not affect the vegetative growth from year to year, it results in decreased seed yields.

Although velvetbeans make a fair growth on poor soils, it is sometimes advisable to apply a fertilizer at the time of seeding, with kind and quantity depending on local conditions. For instance, in Mississippi, 100 to 200 lb. phosphatic fertilizers per acre may be necessary to obtain good yields; the addition of COTTONSEED MEAL at the rate of 200 lb. per acre further increases the yield of beans. However, in other localities, these fertilizers do not result in increased yields. The velvetbean is not sensitive to sour soils but is helped by lime. Velvetbeans, through their root nodules, are able to obtain nitrogen from the air, much of which is returned to the soil when only the pods are harvested, or when the crop is pastured and the roots and uneaten portions of the plants are allowed to decay.

INOCULATION is unnecessary, as apparently all of the velvetbean area is provided with the organism that forms nodules on the roots of the plants. No lack of root nodules occurs when they are planted on land for the first time, but instances have been noted where the growth of the vines has been materially increased by inoculation. The same strain of the organism that inoculates LIMA BEANS, COWPEAS, and LESPEDEZA also inoculates velvetbeans.

Although the velvetbean is easy to raise, the best results are obtained on well-prepared seedbeds. In general, the land should be prepared as for CORN. The ground should be plowed thoroughly to a depth of about 6 in. in December or January, and harrowed at intervals to kill the weeds.

Velvetbeans will not germinate well in cold or wet soils, and as the young plants are very susceptible to injury by frost, planting should be delayed until all danger of frost is past.

As most of the velvetbeans are grown with corn, it is better, in general, to grow late varieties that can be planted with the corn; but if early varieties are used, it is best to plant the beans by hand in the corn rows some time after planting the corn.

In addition to corn, PEARLMILLET, JAPANESE SUGARCANE, SORGHUM, and other strong-growing plants are also planted with velvetbeans as supporting crops.

A popular method is to plant corn and beans in seperate rows, and to plant every third row to beans. Where this method is used, the beans may be planted at the same time as the corn or at a later date, as the cultivation of corn will not interfere with the planting and cultivation of the velvetbeans. Many farmers, who feed hogs extensively, plant alternate rows of PEANUTS and corn, with velvetbeans in the corn rows, and pasture the entire crop. On land where peanuts do well, this combination yields an abundance of feed in a well-balanced form.

Cultivation of velvetbeans during the early growing period will increase the yield sufficiently to more than pay the cost; it should be continued until the plants begin to put out long runners, after which the ground will be covered so completely as to smother all weeds.

The time of harvesting velvetbeans depends largely on the variety, and the use to be made of the crop. Late varieties to be used for pasture may be left in the fields all winter. Early varieties should be pastured very soon after the beans mature, because mature pods are likely to shatter. When grown for seed, the entire crop should be harvested as soon after the pods mature as practicable.

The most important use of the velvetbean is as a GRAZING crop for cattle and hogs in the autumn and winter; it is never *grazed* readily by stock until well-matured or frosted. On sandy soils, the leaves, vines, and pods often furnish feed until early spring. It is usually better to delay grazing until the crop is well-matured or killed by frost, as the leaves will be off the plants at that time, and the corn may be gathered with less difficulty. Many cattlemen allow one-third to one-half acre per month for each steer or cow, the grazing period being about three months, but this may be shortened or lengthened as deemed advisable. Hogs should be allowed to follow the cattle to consume the beans which they have wasted; a common practice is to allow one or two hogs, in addition to the cattle, for each acre of beans. A good stand of velvetbeans should produce about 200 lb. beef and 100 lb. pork per acre.

Velvetbeans are seldom used for *hay* because of the difficulty in handling the long, tangled vines. When used for this purpose it is necessary to cut the vine before many of the pods mature in order to save the leaves, which shatter rapidly as they approach maturity. The hay is coarse and rough, at best, and is not relished by horses and mules. Yields of 2 to 3 tons per acre may be obtained.

Velvetbeans, particularly the early varieties, and corn have been used to some extent for *silage*. Most of the vine growth of the early varieties is wrapped about the cornstalks, and little trouble has been experienced in cutting the corn with corn knives and running it through the silage cutter. Silage made from this mixture turns black after it has been in the silo for a short time because of the juices in the velvetbean plants, but this condition apparently does not impair its keeping qualities or feeding value. Corn and velvetbean silage is as palatable as corn silage. Some dairymen, who have fed silage made from the mixture, prefer it to corn silage.

The velvetbean is one of the best soil-improving crops both for naturally poor soils and for those on which yields have decreased markedly. The ability of this plant to make a profitable growth on land so poor that most legumes do not thrive on it, places it among the most important crops for the South. Even though the crop is grazed, much of the nitrogen in the plants consumed by the stock will be returned to the soil in the manure.

In some sections velvetbeans have proved to be the most profitable crop to plant for one to two years on newly cleared land, as they not only supply considerable grazing or feed but also improve the soil for the crops that follow.

When late varieties of velvetbeans are planted without a supporting crop, they produce such a dense growth of vines that weeds, persistent GRASSES, and in many cases tree sprouts, are smothered.

It requires about 2½ lb. velvetbean meal or 1½ lb. ground beans to equal the feeding value of 1 lb. high-grade cottonseed meal.

When the beans of the velvetbean are to be fed to cattle, it is preferable to feed them together with the hulls.

Two types of velvetbean products have been recognized officially:

(1) *Ground velvetbean and pod*—derived by grinding velvetbeans with the pods and without additional pods or other materials.

It is manufactured by crushing together (with specially designed machinery) beans and the pods. Although no standard of fineness has been established, a meal of the fineness of corn meal is preferred, especially in the manufacture of mixed feeds. Velvetbeans may be ground with other feeds, e.g., with corn in the shuck.

(2) *Velvetbean meal*—ground velvetbeans containing only an unavoidable trace of hulls or pods.

Velvetbean meal is often used in mixed feeds. In horse feeds it seldom forms more than 25 percent of the mixture, whereas for dairy cows it may run as high as 70 percent.

A popular mixed feed for dairy purposes is composed of 15 percent cottonseed meal, 45 percent corn-and-cob meal, and 40 percent velvetbean meal, while a popular feed for horses contains, in addition to the velvetbean meals, corn, OATS, and ground hay or straw.

Dangers: As velvetbeans are very high in digestible protein, fat, and carbohydrates, great care should be exercised in feeding them to livestock, especially at first. After the stock becomes accustomed to the beans, they should be kept in the field for only a short period each day until the crop is somewhat reduced, as excessive consumption is a waste of concentrated feed. In addition, overfeeding sometimes has as LAXATIVE effect. For these reasons, and because better gain will be obtained, velvetbeans should be fed in combination with other feeds.

The velvetbean is notably free from disease or insect enemies. Only under very unusual conditions is it affected by a NEMATODE which causes ROOT KNOT.

The VELVETBEAN CATERPILLAR is the only insect which causes serious injury to the velvetbean.

Species: Nomenclature of velvetbeans is often confusing. DEERING VELVETBEAN (with its varieties *Florida, Georgia, Alabama, Arlington, Bush,* and *Osceola,* a hybrid), TRACY BLACK, LYON VELVETBEAN, CHINESE VELVETBEAN, and YOKOHAMA VELVETBEAN are the most important velvetbean species.

References: P.10; F.6.

See also LEGUME FEED PRODUCTS; DODDERS.

Illustration: See DEERING VELVETBEAN.

VELVET CHAFF

= *Galgalos. See* COMMON WHEAT.

VELVETGRASS (*Holcus lanatus*) is widely distributed in both eastern and western United States. Because of low palatability, this GRASS is regarded as a WEED on the Pacific Coast. However, difficulty of eradication has encouraged its use for HAY. Velvetgrass gives two cuttings a year under favorable conditions, but with much shrinkage. It is disliked by horses.

Reference: W.16.

VELVET MESQUITE.
                    See MESQUITE (variety).

VELVON.
                    See SIX-ROWED BARLEY (variety).

VENATION is the arrangement of VEINS in a leaf.

VENTILATION. *See* AIR.

VENTRAL—the opposite of DORSAL—means: of, or pertaining to, that side of a PISTIL or other organ which faces the AXIS, or center, of a flower.

VERN. *See* RYE (variety).

VERNAL is (1) a variety of EMMER or (2) one of the VARIEGATED ALFALFAS.

VERNALGRASS.
                    See SWEET VERNALGRASS.

VERNUM is a white DURUM WHEAT variety.

VETCH. *See* VETCHES.

VETCH ANTHRACNOSE—not to be confused with FALSE ANTHRACNOSE OF VETCH—is caused by the FUNGUS *Colletotrichum villosum.* On the leaves of many vetch species it produces small, round spots, which first are light green and later become light brown or gray, with a brown or red border. The stem lesions are linear and usually dark brown. On the pods the lesions are dark red, with a darker margin and lighter center. Severe defoliation and death of the entire plant may occur during wet weather.

Control: The disease can be controlled by rotation or by planting such resistant species as *monantha vetch* or *Hungarian vetch.*

Reference: W.15.

See also ANTHRACNOSE.

VETCHES belong to the *Vicia* spp., as does the HORSEBEAN. Most vetches are annuals. The common agricultural species are all viny and weak-stemmed. The stems attain a length of from 2 to 5 ft. In most cultivated species the leaves have many leaflets and are terminated with a TENDRIL. From few to many flowers are borne in a cluster, or RACEME. In general the seed is round or oval and the pods are elongated and compressed.

The vetches require cool temperatures for their best development. In the regions with mild winters, as in the southern regions and Pacific Coast areas, they make their growth during the fall, winter, and early spring months, maturing in late spring and early summer. In the North, where winters are severe and moderately cool, they start growth early in spring and mature late in summer or fall. Species vary with reference to winter hardiness and the minimum temperature at which they will make growth.

All of the commercial vetches make good SILAGE, hay, pasture, and green manure, and can be used for cover crops and feeding green. The seed is used as one of the ingredients in ground poultry feed.

The vetches are comparable in feeding value to CLOVER, ALFALFA, and other common LEGUME crops. The protein content of hay usually ranges from 12 to 20 percent, depending upon the stage of development when the crop is cut.

Vetches are not particular in regard to soil. All do well in rich loam. They are more tolerant of acid soil conditions than most legume crops, and, outside the lime-belt areas in the eastern part of the United States, succeed without the addition of lime.

A moderate moisture supply is necessary for vetches; none are drought-resistant.

In the South most of the vetch is seeded following COTTON; then, little or no preparation of the soil is needed. The same is true when vetch is planted following other cultivated crops. In the Pacific Northwest disked seedbeds are used when vetch follows cultivated crops, while plowing and subsequent preparation is practiced on fall grain stubble or uncultivated land.

North of latitude 40°, from the Rocky Mountains to the Atlantic Coast, all commercial vetches should be sown early in spring, except hairy vetch, which should be sown during August or early September. On the Pacific Coast, west of the Sierra Nevada and the Cascade mountains, vetches can be safely sown in the fall. In western Oregon, western Washington, and northwestern California, vetch should be seeded as early as the seasonal rains will permit. In other parts of California, where the climate is mild and where irrigation is practiced, and in the mild parts of Arizona, seeding should be made from the middle of August to the first of October. In the Northern part of the Cotton Belt the best time for seeding is the latter half of September and in the southern part early in October.

To sow vetch with GRAIN is the common practice where the crop is grown mainly for forage, as the grain furnishes a support for the weak stems of the vetch and to a considerable extent prevents lodging. Where OATS succeed, they are the favorite grain to use in combination with vetch, though WHEAT, RYE, and BARLEY may be used. When vetch is used mainly as a green manure crop, it is nearly always sown alone. Less seed is needed with the drill method than with broadcasting methods.

In the Pacific Coast states fertilizers usually are not necessary for a successful growth of vetch. In western Oregon, however, gypsum, commonly applied at the rate of 75 to 150 lb. per acre, is often used with beneficial results. In the southern states east of the Mississippi River it is almost universally necessary to use 300 to 400 lb. of 16% superphosphate per acre; on land that has not grown legumes before, sodium nitrate or ammonium sulfate, at the rate of 100 lb. per acre, should be used. A well-rotted barnyard manure, at the rate of 15 to 20 tons per acre, may be applied to insure a stand and good growth of vetch.

In the Pacific Coast states vetch is nearly always naturally inoculated. In the eastern part of the United States it is advisable to introduce the proper bacteria artificially unless it is known that they are already present in the soil. Many failures with vetch are directly attributed to the lack of INOCULATION. Inoculated plants are easily recognized by their greener color and more vigorous growth, and by the nodules on their roots.

Vetches make good *hay* either alone, or in mixture with the SMALL GRAINS, and are relished by all kinds of livestock. Vetch, planted with one of the small grains, is often cut green and fed to cattle or other livestock. Succulent late-winter and early-spring feed can be supplied in mild climates this way with little expense.

Vetch is ordinarily cut for hay when the first pods are well developed. After being cut the vetch should be windrowed or bunched, and then shocked. This handling should always be done before the leaves are dry. Vetch should be allowed to cure in the shocks several days. Without a swather, harvesting is considerably more difficult. From 1½ to 3½ tons of hay per acre is the usual harvest.

For *pasture* the vetches, alone or in mixture, extend the GRAZING season by supplying late-fall and early-spring feed. They stand trampling and are well suited for pasture.

As a general rule vetch is pastured only when the ground is dry, to avoid packing the soil and to reduce the possibilities of BLOAT in cattle and sheep.

Even when vetch is grown primarily for hay or for seed, a limited degree of pasturing is often desirable, especially where the growth is likely to be usually rank and where it is necessary to delay the harvest period. Hogs should not be used for this purpose, as they destroy many of the plants by biting them off below the crown. Sheep and calves do the least damage in pasturing vetch to be used primarily for hay or seed crop.

Probably the greatest use of vetch is for green manuring. In limited areas of the United States vetch is grown for seed.

Dangers: More than 20 fungus diseases have been reported that attack vetch species in the United States. The most common include the following: DOWNY MILDEW OF VETCH, VETCH ANTHRACNOSE, FALSE ANTHRACNOSE OF VETCH, LEAF AND POD SPOT, GRAY MOLD, BROWN SPOT, ROOT AND STEM ROT, and SCLEROTINIA STEM ROT.

If vetches are grown only as a cover crop, it has been generally observed that diseases do no extensive damage.

Many insect pests of alfalfa, clover, and other forage legumes also attack vetch. Among the more important of these are APHIDS, the CORN EARWORM, GRASSHOPPERS, STINK BUGS, CUTWORMS, the FALL ARMYWORM, various WEEVILS, and LEAFHOPPERS. All the varieties of vetch are subject to attack by NEMATODES, and at times serious damage may result.

Species: While HAIRY VETCH (and its variety *smooth vetch*), COMMON VETCH, and HUNGARIAN VETCH are the kinds most commonly used in the United States, others are used in limited areas, e.g., PERENNIAL VETCH, WOOLLYPOD VETCH, MONANTHA VETCH, PURPLE VETCH, BITTERVETCH, NARROWLEAF VETCH, and BARD VETCH; some of these offer possibilities of more extended use. One of the widely distributed wild species is BIRD VETCH.

Reference: M.18.

See also SUDANGRASS; CRIMSON CLOVER; KUDZU; HAIRY INDIGO; LUPINES; LEGUME BACTERIA; HAY; SILAGE CROPS.

Illustration: See HAIRY VETCH.

VETCH HAY. *See* HAY GRADING.

VETCH SILAGE. *See* SILAGE CROPS.

VIABILITY is the state of being VIABLE.

VIABLE means: capable of living, growing, and developing.

VICIA. Among the *Vicia* spp. are *V. tenuifolia* = PERENNIAL VETCH; *V. ervilia* = BITTERVETCH; *V. dasycarpa* = WOOLLYPOD VETCH; *V. sativa* = COMMON VETCH; *V. pannonica* = HUNGARIAN VETCH; *V. atropurpurea* = PURPLE VETCH; *V. angustifolia* = NARROWLEAF VETCH; *V. villosa* = HAIRY VETCH; *V. monantha* = BARD VETCH; *V. articulata* = MONANTHA VETCH; *V. faba* = HORSEBEAN; and *V. cracca* = BIRD VETCH.

VICTOR. *See* COWPEA (variety).

VICTORIA BLIGHT, a FUNGUS DISEASE of OATS, is caused by *Helminthosporium victoriae*. It attacks many oat varieties which derive their CROWN RUST resistance from VICTORIA, a South American oat variety. The Victoria type of resistance to crown rust is closely associated with susceptibility to Victoria blight.

Victoria blight is seed-borne and soil-borne. It is sometimes so destructive that seeds fail to germinate, and many seedlings die of root rot or stem rot at ground level. Infection on older plants advances upward and is favored by rather high temperatures. Leaves become half-striped yellow and then turn reddish-yellow before they wilt and die.

As the infected plants mature, the lower parts of the stems become black, due to external development and sporulation of the fungus. These plants are easily pulled up because the roots have rotted away. Infected plants which reach maturity tend to lodge, with stem breakage occurring just above the joints. They yield small amounts of chaffy seed.

Control: Grow resistant varieties. If susceptible varieties are to be grown, SEED TREATMENT will reduce seed-borne infections but will not control secondary or soil-borne infections. Late planting of susceptible varieties is desirable to avoid higher fall temperatures which favor this organism. Crop rotation also will help to keep down Victoria blight, as well as several other important small-grain diseases.

Reference: M.30.

See also CERESAN M; RED OAT.

VICTORIA OAT is an important South American OAT variety. It is resistant to CROWN RUST, but susceptible to VICTORIA BLIGHT.

VICTORY. *See* COMMON OAT (variety).

VIGNA. *V. sinensis* = COWPEA.

VIGO. *See* COMMON WHEAT (variety).

VIKING. *See* SOYBEAN (variety).

VINE. *Love-vine* = DODDER.

VINEGAR is weak (4- to 5-percent) ACETIC ACID, obtained by fermentation of apple cider, wine, or other alcoholic liquids.

The dried residues of vinegar production are called VINEGAR DRIED GRAINS.

VINEGAR DRIED GRAINS are used as feedstuffs.—*See also* YEAST DRIED GRAINS OR VINEGAR DRIED GRAINS.

VINE-MESQUITE (*Panicum obtusum*) does not belong to the MESQUITE genus. It is a vigorous, long-lived, native perennial of the southwestern states. The stiff, erect

culms are 1 to 2 ft. tall; leaves are 4 to 6 in. long and about ¼ in. wide. STOLONS are numerous, long, and have swollen nodes. The seed PANICLE is 2 to 5 in. long.

This grass grows where rainfall is scant, but is most abundant where additional water is received in occasional floods. It produces a fair amount of forage, which

Vine-mesquite. Plant, spikelet, and floret. (H.26.)

livestock relish when it is green and succulent; however, stock may not eat the mature plants if other feed is available.

Because the stolons may grow 15 ft. in a single season, the plants can well be propagated by transplanting sod pieces. Vine-mesquite hay, cut after the seed heads mature, can be used as another means of establishing new seedings.

Reference: H.1.

See also PANICGRASSES; GRASSES.

VIOLET ROOT ROT, caused by the FUNGUS *Rhizoctonia crocorum*, affects AL-FALFA. It is prevalent in years of heavy rainfall, and in fields where there has been inadequate drainage. The disease occurs in circular areas in the field. In each case the disease starts from a center of infection and the fungus spreads in all directions through the soil, killing most of the plants as it progresses. Diseased alfalfa plants are characterized by the reddish-brown or violet color of the roots and crowns. The fungus penetrates the roots and crowns and produces a decay. In advanced stages of the disease the bark sloughs off very easily.

Control: Violet root rot may be controlled by crop rotation and adequate drainage.

Reference: G.7.

VIOSTEROL = VITAMIN D_2.

VIRGINIA SOYBEAN.

See SOYBEAN (variety).

VIRGINIA-TYPE PEANUTS include *Virginia bunch*, *Virginia runners*, and intermediate forms.—*See also* PEANUT.

VIROLA. *V. surinamensis* = UCUHUBA.

VIRUS is an infectious PATHOGEN, too small to be seen with a compound microscope. This ultramicroscopic protein body is capable of multiplying. Many of these viruses, when present in tissue, cause VIRUS DISEASES.

A virus which has no scientific name causes RED LEAF; another one attacks the ALFALFA CATERPILLAR.

VIRUS DISEASES are caused by viruses and best controlled by breeding resistant plant varieties. These diseases are often spread by insects; APHIDS and LEAFHOP-PERS, for instance, are sometimes destructive to SUBCLOVER, RED CLOVER, and other

LEGUMES by spreading *red clover virus diseases*, particularly RED CLOVER VEIN MOSAIC. BUD BLIGHT is caused by the *tobacco ring spot virus;* SOYBEAN MOSAIC by the *Soja virus 1*; and YELLOW MOSAIC by *Phaseolus virus 2*. Other virus diseases are WITCHES'-BROOM; DWARF DISEASE, RED LEAF, YELLOWS, and various MOSAICS.— *See also* LUPINES; CROTALARIAS; ALFALFA CATERPILLAR; VIRUS.

VISCID means: sticky or gummy to the touch.

VISCOUS means: being of sirupy, slow flowing consistency.

VITAMIN A, formerly called *anti-infective vitamin*, or *growth-promoting* or *anti-xerophthalmia vitamin*, is a light-yellow oil which is soluble in fats and fat solvents. It is destroyed by exposure to air, high temperature, and sunlight. When stored in a full bottle, away from light and in a cool place, it is quite stable; but when spread out in a fine film, as when mixed with feed, some destruction of the vitamin is inevitable. The practice of storing feed in a warm hatching or brooder room is decidedly objectionable.

NOTE: 1 U.S.P. unit or 1 I.U. vitamin A is equivalent to 0.3 mcg. *vitamin A alcohol* or 0.344 mcg. *vitamin A acetate*. 1 I.U. *carotene*, or *provitamin A*, equals 0.6 mcg. pure *beta-carotene*.

Vitamin A, which can be produced synthetically, occurs in nature only in animals and animal products. Plants contain CAROTENES, sometimes called *provitamin A*, which are changed in the animal body to true vitamin A. Excess amounts are stored in the liver and other organs, and may be utilized in seasons of partial dietary deficiency. These carotenes are yellow pigments and give the yellow color to CARROTS and some other yellow plants. They are also found in all green-colored plants although the yellow there is masked by the green pigment of CHLOROPHYLL. The color of forage is usually a fairly accurate index of the amount of carotenes it contains. Disappearance of the green color indicates a considerable degree of destruction. Rations may slowly loose their vitamin-A activity during storage because the vitamin is destroyed by oxidation.

The best source naturally available to livestock is fresh, rapidly growing forage, such as BLUEGRASS or ALFALFA pasture. Good quality forage will supply an adequate amount of carotenes. Dried forage is also an excellent source, if cut at the proper stage and cured under favorable conditions. Alfalfa hay is especially reliable, and the grass hays, such as TIMOTHY and REDTOP, also contain a liberal amount if cut before they are too mature. Dehydrated hays have a higher content than sun-cured. Yellow carrots, green CLOVER and other LEGUMES, CORN gluten meal, and green KALE, are rich sources of vitamin A. Not quite so rich, but still good, are green CABBAGE, yellow CORN, HOMINY FEED, LIVER, SWEET POTATOES, TOMATOES, and whole MILK.

As a rule the vitamin-A requirement of poultry can be satisfied in the form of carotenes by including yellow corn and alfalfa meal in the rations.

Feeds over one year old are unreliable sources of carotene. Circumstances most conducive to a deficiency of vitamin A are long periods of drought, especially during the summer and early fall, when temperatures are high. Vegetation ceases to grow and the carotene already there is destroyed by exposure to light and oxygen.

As long as livestock have access to green vegetation, it can be assumed that they receive enough vitamin A. If, however, they are confined to drylot pens, the feeder must take pains to insure an adequate supply to prevent the usual deficiency symptoms of xerophthalmia (dryness of the conjunctiva, i.e., the mucous membrane covering the eyeball and lid), night blindness, swollen joints, unthriftiness, low fertility, weak young, weakness in the hindquarters, lowered resistance to infectious diseases, and scours.

References: R.4; M.15; S.10; H.20; S.11.

See also VITAMINS.

VITAMIN A ACETATE forms pale-yellow, water-insoluble crystals. This ester is more stable to oxidation than the free VITAMIN A (alcohol) and therefore widely used in feedstuffs.

NOTE: 1 U.S.P. unit, or 0.3 mcg.

vitamin A = 0.344 mcg. vitamin A acetate.

See also VITAMINS.

VITAMIN A & D FEEDING OIL.
See VITAMINS.

VITAMIN A CONCENTRATE.
See VITAMINS; VITAMIN A.

VITAMIN B₁—also known as *"thiamine,"* *thiamine hydrochloride,* or *thiamine chloride,* or *antineurotic vitamin*—is one of the B-COMPLEX VITAMINS which occurs in plant and animal tissues, especially in the hulls of cereal grains, rice polish, green leaves, roots, tubers, YEAST, liver, eggs, and milk. It is needed for carbohydrate oxidation; its lack causes slow GROWTH, loss of appetite, and nervous conditions.

Vitamin B₁ forms water-soluble, white, stable crystals.

NOTE: 1 gm. vitamin B₁ (hydrochloride) = 333,000 I.U.

See also VITAMINS; FIELD BEANS; THIAMINASE.

VITAMIN B₂, or *riboflavin,* also called *vitamin G,* is one of the B-COMPLEX VITAMINS. It forms an orange-yellow, crystalline powder that is stable in solid form, but is destroyed by light in acid or alkaline solution. Vitamin B₂ occurs in all plant and animal cells where it aids protein metabolism. Particularly rich in this vitamin are milk, eggs, YEAST, malted barley, liver, heart, and kidney. Green LEGUMES are good sources. No cereal contains enough for young chicks or pigs.

Vitamin B₂ is a nutritional factor for many animal species; healthy ruminants and horses, however, do not need additional amounts of it in their normal rations.

A shortage of vitamin B₂ causes poor growth in all species that need it, *curled toe paralysis* in chicks, *dermatitis* and *anemia* in hogs.—*See also* VITAMINS; ALFALFAS.

VITAMIN B₆, more correctly called *vitamin B₆ hydrochloride* or *pyridoxine hydrochloride,* is one of the water-soluble B-COMPLEX VITAMINS. It forms stable crystals and is widely distributed in grains (in CORN more than in grain SORGHUMS), grain by-products, ALFALFA, YEAST, liver, and milk. Rumen bacteria are able to synthesize this vitamin.

Most practical rations supply enough vitamin B₆ to meet the animals' need. Deficiencies, therefore, are practically never encountered under field conditions. —*See also* VITAMINS; LINSEED.

VITAMIN B₁₂ , or· *cyanocobalamin*, belongs to the B-COMPLEX VITAMINS. It is a major part of the *"animal protein factor,"* (a term now obsolete). It contains over 4 percent COBALT in its molecule—as well as some PHOSPHORUS—and forms dark-red, water-soluble crystals. Vitamin B₁₂ occurs in liver. It is produced by MOLDS (*Streptomyces* spp.) and BACTERIA (in rumen and in built-up poultry litter). It is important as a GROWTH factor in swine and poultry and should be added to all vegetable rations to insure rapid growth and satisfactory reproduction, unless the animals have access to cow MANURE or built-up litter. The table shows the vitamin B₁₂ content of various feedstuffs.—*See also* VITAMINS.; APF.

VITAMIN B₁₂ CONTENT OF FEEDS

	mcg./lb.
Fish solubles (dried)	65
Menhaden fish meal	28
Red fish meal	43
Sardine meal	70
Herring meal	118
Cod and haddock meal	43
Meat scrap	16-30
Liver meal	141
Crude casein	13-47
Wheat	0
Soybean meal	1
Yellow corn	0
Oats	1
Alfalfa leaf meal	1
Dried brewers' yeast	0
Milk powder	11
Egg yolk	12
Beef liver	230
Beef muscle	20
Pork muscle	13
Sheep muscle	16

VITAMIN B-COMPLEX.

See B-COMPLEX VITAMINS; ANTIBIOTICS.

VITAMIN B₁₂ SUPPLEMENT, a fermentation by-product, is a feed supplement containing at least 1.5 mg. VITAMIN B₁₂ activity per pound. Many *antibiotic feed supplements* also contain vitamin B₁₂. These combinations were once sold as APF supplements.—*See also* VITAMINS; APF; ANTIBIOTICS.

VITAMIN C, or *ascorbic acid*, forms white, water-soluble, stable crystals; it is not stable in aqueous solution. Vitamin C occurs in many vegetables and fruits (especially in citrus). It is not needed by farm animals. If added to feedstuffs, it acts as ANTIOXIDANT. Some people have used it as a drug in certain cases of sterility of cattle, but results are not satisfactory.— *See also* VITAMINS.

VITAMIN D, which is also called the *sunshine vitamin*, the *bone building vitamin*, or the *calcifying vitamin*, is a white, odorless crystal, soluble in vegetable oils or alcohol. It is in reality the activation product of several different sterols.

Only two of these sterol derivatives are of great importance. They are VITAMIN D₂ , (also called *calciferol* or *viosterol*), and VITAMIN D₃ , (known as the *antirachitic vitamin*). Poultry are able only to utilize the D₃ , but four-footed animals utilize both D₂ and D₃ as their source of the vitamin.

Vitamin D is quite stable, and ordinarily, no special precautions are required to prevent destruction. If, however, vitamin D is intimately mixed with minerals or with feeds that contain 20 to 30 percent minerals, it is destroyed, presumably by oxidation.

None of the common feeds are a highly potent source of vitamin D; LEGUMES rate best. Sun-cured roughages may have a high content, but so much variation is found that they are not always dependable. Neither is standard milk a dependable source of vitamin D, but the amount of the latter may be increased by feeding large amounts of irradiated yeast to the cows. A more efficient way to get vitamin-D milk is to add it at the rate of 400 I.U./qt. The action of ultraviolet sun rays on the skin produces sufficient vitamin D for normal dietary needs as long as the animals are daily exposed to direct sunshine. However, the response of animals to sunshine may be variable, and the effectiveness of the rays is diminished by clouds, haze, and fog, and destroyed by ordinary window glass.

NOTE: Vitamin D potency is always described in units, partly because there are several D vitamins, and partly because equal weights of these vitamins do not have the same potency for different animals.

1 U.S.P. unit or 1 I.U. is 0.025 mcg. crystalline vitamin D_3. Poultrymen use the *International chick unit* which is equal in biological activity for poultry to 1 U.S.P. unit vitamin D_3.

Vitamin D is essential for the proper utilization of calcium and phosphorus. Symptoms of its deficiency in animals are soft bones; leg bones, especially, become bent and deformed. Joints are stiff, swollen, and tender, and the back may become humped. Milk and egg production are lowered; eggs are thin-shelled and hatch poorly.

References: S.10; R.4; S.11; H.20.

See also VITAMINS; A.O.A.C.

VITAMIN D_2, which is also called *calciferol, viosterol,* or *D-activated plant sterol* (source of D_2), is prepared from ERGOSTEROL or YEAST by ultraviolet IRRADIATION or by electronic bombing of ergosterol.—*See also* STEROLS; VITAMIN D; YEAST; VITAMINS.

VITAMIN D_3—sometimes called the *antirachitic vitamin*—or *D-activated animal sterol* (source of D_3) is found in fish-liver oils, especially those of cod fish and sardines. It is also prepared by ultraviolet IRRADIATION of suitable precursors of the vitamin, e.g., 7-DEHYDROCHOLESTEROL, one of the CHOLESTEROL derivatives.—*See also* STEROL; VITAMIN D; VITAMINS; A.O.A.C.

VITAMIN D CONCENTRATES.

See VITAMINS.

VITAMIN DEFICIENCIES.

See table on page 539.

VITAMIN E, or *tocopherol,* has also been known as *antisterility, fertility,* or *reproductive vitamin.* It occurs in nature in the form of several active tocopherols; of them, *alpha-tocopherol* is the most important and the most potent. *Beta-, gamma-, delta-,* and other tocopherols have been isolated from some feeds.

Vitamin E is obtained largely from *wheat-germ oil* but is also found in the oils obtained from corn, cottonseed, soybean, etc. It is also contained in whole grains, alfalfa meal, and fresh greens.

This vitamin is considered important for the normal functioning of the reproductive organs of poultry and for normal fertility, but it is of far greater importance in preventing certain muscle disturbances (*white muscle disease* of calves, *yellow fat disease* of mink, *stiff lamb disease, enlarged hocks* in poultry, and *crazy chick disease*). It has been impossible to show that vitamin E is needed for reproduction in cattle. Vitamin E is the body's natural ANTIOXIDANT and protects vitamin A; it is also industrially used to retard rancidity of fats and oils.

NOTE: 1 I.U. vitamin E = 1 mg. standard *dl-alpha tocopheryl acetate* (which is an ester of vitamin E).

VITAMIN E SUPPLEMENT.

See VITAMINS.

VITAMIN F is a term sometimes used to indicate certain *unsaturated* FATTY ACIDS, e.g., *linoleic, linolenic,* or *arachidonic acid.* A lack of these acids may cause disturbances in skin, hair coat, and growth in pigs, calves, or lambs. Normal rations providing 1 or 2 percent of true FAT will carry enough of these essential acids to meet the animals' needs.

VITAMIN G = VITAMIN B_2.

VITAMIN K, commonly known as the *antihemorrhagic factor* or *coagulation vitamin,* is a yellow crystalline powder, insoluble in water but soluble in fat. It is present in green-leafed vegetables, fish meal, liver, etc. and necessary to maintain normal blood-clotting power; lack of vitamin K causes hemorrhages. In animals deficiency of vitamin K is rare, since normal rations supply the vitamin in sufficient amounts. In addition, ruminants are capable of synthesizing it with the help of rumen micro-organisms.—*See also* MENADIONE; SWEETCLOVER DISEASE; DICOUMAROL.

VITAMIN REQUIREMENTS.

See table on page 538.

VITAMINS are organic compounds needed in small amounts for normal body functions. Like the spark plugs of an engine, they make it possible for the body's fuel to be burned. Many vitamins play an important role in nutrition of farm animals and poultry. These NUTRIENTS are found in natural feeds, but it is sometimes more economical to buy vitamin supplements in small quantities than to use large amounts of a natural source. Many of the vitamins are somewhat unstable and lose potency when stored for a prolonged period of time, when mixed with certain minerals, or exposed to sunlight, air, or moisture; PELLETING is often very destructive. Thus it becomes necessary to insure the presence of sufficient amounts of needed vitamins in the rations at time of consumption. Many manufacturers now provide *stabilized* supplements by enclosing the vitamin in a protective coating.

The following officially accepted resolutions regarding vitamins are now in force:

1. The word *"vitamin"* or a contraction thereof, or any word suggesting a vitamin, may be used only in the brand name of a feed represented solely to be a vitamin supplement and which is labeled with the minimum vitamin content guaranteed.

2. No declaration of vitamin content of a feed or a feed supplement shall appear in the ingredient statement, or any other part of the label of a proprietary feed, excepting that such statement is a guarantee of minimum vitamin content of the entire product.

3. The common feed and/or mineral guaranties are not required for any product represented solely to be a vitamin supplement and which is labeled with a minimum vitamin guaranty.

4. The International chick unit of *vitamin D* is the activity produced by 1 unit vitamin D in the U.S. Pharmacopoeia "vitamin D reference standard," determined according to the method of the Association of Official Agricultural Chemists.

5. Guaranties of minimum vitamin content of feeds and feed supplements shall be stated in units of milligrams per pound: *vitamin E* activity in International units; *vitamin A*—other than precursors of vitamin A—in U.S.P. units; *vitamin D* in products offered for poultry feeding in International chick units; vitamin D for other uses, in U.S.P. units; all other vitamins as true vitamins, not compounds, excepting only *pyridoxine hydrochloride*

(VITAMIN B₆), *choline chloride*, and *thiamine hydrochloride* (VITAMIN B₁); oils and concentrates containing vitamin A, or vitamin D, or both, may be additionally labeled to show vitamin content in units per gram.

6. If a feeding material is represented to be a combined *vitamin B₁₂* and *antibiotic feed supplement*, it shall meet the potency standards and other label and informational requirements established for each component.

The officially recognized vitamins are defined as follows:

Cod liver oil—obtained from the livers of *Gadus morrhuae* or other species of the family Gadidae, either or both. It must contain not less than 385,900 U.S.P. units vitamin A per pound (or 850 units per gram), and not less than 29,510 International chick units per pound (65 units per gram).

Cod liver oil with added vitamin A and D concentrates consists of cod liver oil to which have been added small percentages of concentrates rich in vitamins A and D. The products shall contain not less than 136,200 International chick units vitamin D per pound (300 chick units per gram) and shall carry a minimum vitamin A guarantee.

Vitamin A feeding oil is either fish oil, or fish-liver oil, or a blend of two or more of the following: vitamin A concentrate, fish-liver oil, fish oil, marine animal oil, or edible vegetable oil. The vitamin potency shall be stated in U.S.P. units vitamin A per pound (or per gram).

Vitamin D feeding oil is either fish oil, or fish-liver oil, or a blend of two or more of the following: vitamin D concentrate, synthetic vitamin D, fish-liver oil, fish oil, marine animal oil, or edible vegetable oil. The vitamin potency shall be stated in International chick units vitamin D per pound (or per gram).

Vitamin A and D feeding oil is either fish oil, or fish-liver oil, or a blend of two or more of the following: vitamin A and/or D concentrate, synthetic vitamin D, fish-liver oil, fish oil, marine animal oil, or edible vegetable oil. The vitamin potency shall be stated in International chick units vitamin D and U.S.P. units vitamin A per pound (or per gram).

D-activated animal sterol is obtained by activation of a sterol fraction of animal origin with ultraviolet light or other means. For label identification it may be followed with this parenthetical phrase: (source of vitamin D₃).

D-activated plant sterol is obtained by activation of a sterol fraction of plant origin with ultraviolet light or other means. For label identification it may be followed with this parenthetical phrase: (source of vitamin D₂).

Riboflavin supplement is a feeding ma-

terial used chiefly for the riboflavin (VITAMIN B₂) content, and shall contain not less than 200 mg. riboflavin per pound. The label shall bear the statement of origin.

Dried . . . fermentation solubles is the product obtained by the concentration and dehydration of the liquid by-product resulting from the action of the ferment on the basic medium of grain, MOLASSES, WHEY, or other media; it shall contain not less than 18 mg. riboflavin (VITAMIN B₂) per pound on a moisture-free basis. For label identification the source shall be indicated as *"dried whey fermentation solubles."*

Dried fermentation solubles, which are

VITAMIN FUNCTIONS AND SOURCES

Vitamin	Physiological functions	Sources
A	Normal epithelia (cellular surface layer of skin, etc.), retinal (eye) pigments, growth of new cells	Green and yellow plants, high grade legume hay, yellow corn, good silage, fish oil, synthetics
D	Calcium and phosphorus absorption and metabolism	Exposure to sunshine, sun-cured hay, fish liver oil, irradiated feeds
E	Muscles, reproduction, antioxidation, cell maturation	Widely distributed in whole grains, germ oil, green forages, egg yolk
K	Normal blood clotting, synthesis of prothrombin in liver	Green feed and forages, alfalfa meal, liver meal, vegetable oils
B₁	Carbohydrate metabolism, decarboxylation (carbon dioxide removal)	Unheated grains, bran, liver, milk, yeast, alfalfa
B₂	Cellular oxidation, protein metabolism, retinal pigments, enzymes	Nonfat milk products, meat and fish, green feeds, fermentation solubles, yeast, synthetics
Ni (Niacin)	Carbohydrate and protein metabolism, coenzymes	Meat, liver, yeast, rice bran, wheat bran, middlings, peanut meal, synthetics
B₆	Amino acid metabolism, codecarboxylase (coenzyme)	Seeds, meat, liver, yeast, soybeans, peanuts, milk and fish products
Pa (Pantothenic acid)	Acetylation, coenzymes, carbohydrate and fat metabolism	Oats, yeast, milk, alfalfa, nuts, liver, molasses
Ch (Choline)	Fat metabolism, transmethylation (transfer of methyl groups)	Soybean meal, wheat germ meal, liver, fish meal, synthetics
Bi (Biotin)	Coenzyme for decarboxylation and deamination (removal of amino groups)	Green feeds, liver, yeast
Bc (Folic acid)*	Bone marrow, maturation of red blood cells	Green plants, yeast, liver
B₁₂	Maturation of red blood cells, formation of purines and pyrimidines (two organic nitrogen compounds), methylation	Fish meal, rumen contents, cow manure, fermentation residues, milk
C	Collagen deposition, adrenal (gland) function, amino acid metabolism	resh fruits and vegetables, especially citrus fruits, tomatoes, berries

* Folic acid = Pteroylglutamic acid

VITAMIN REQUIREMENTS

Requirements per lb. total feed or (if marked †) 100 lb. of body weight

Vitamin	Unit	Swine		Poultry			Horse	Cattle		Sheep
		Growth	Lact.	Chick	Hen hatching eggs	Poult	†	Beef	Dairy	†
A	I.U.	2500	8300	1200	2000	2400	16,600	12,000	13,300	16,600
D	I.U.	90	90	90	225	400	....	....	200	180
E	mg.	?	....	1.5	?	5	?	....	....	....
K	mg.	Usually met by intestinal synthesis		0.18	Usually met by intestinal synthesis		....			
B₁	mg.	0.5	0.5	0.8	0.9	....	?			
B₂	mg.	1.2	1.2	1.3	1.7	1.7	2			
Ni (Niacin)	mg.	5–8	5	12	....	35	Synthesis	Not required in rations of animals with a functioning rumen		
B₆	mg.	0.6	....	1.3	1.3	1.0	?			
Pa (Pantothenic acid)	mg.	5	4.5	4.2	4.2	5	?			
Ch (Choline)	mg.	500	....	600	?	750	?			
Bi (Biotin)	mg.	?	?	0.04	0.07	....	?			
Bc (Folic acid)	mg.	?	?	0.25	0.16	0.4	?			
B₁₂	mcg.	4–7	4	4	2	?	?			
C				not required						

ADDITIONAL VITAMINS AND DEFINITIONS

Recognized English name and English synonym, if any	Article and substance indicated by the name and synonym
Ascorbic acid (vitamin C)	Crystalline, relatively free of impurities.
Betaine hydrochloride	Crystalline, relatively free of impurities.
Biotin	Relatively free of impurities.
Calcium pantothenate	Crystalline, relatively free of impurities.
Carotene	The refined crystalline carotene fraction of plants.
Choline chloride	Relatively free of impurities.
Choline pantothenate	Crystalline, relatively free of impurities.
Folic acid	Crystalline, relatively free of impurities.
Herring oil	The oil extracted from whole, or parts of, herring.
Menhaden oil	The oil extracted from whole menhaden.
Niacin (nicotinic acid)	Crystalline, relatively free of impurities.
Niacinamide (nicotinamide)	Crystalline, relatively free of impurities.
Pyridoxine hydrochloride (vitamin B₆)	Crystalline, relatively free of impurities.
Riboflavin (vitamin B₂)	Crystalline, relatively free of impurities.
Salmon oil	The oil extracted from cannery refuse of salmon.
Salmon liver oil	The oil extracted from salmon livers.
Sardine oil (pilchard oil)	The oil extracted from whole, or parts of, pilchard.
Shark liver oil	The oil extracted from shark livers.
Stearin	The stearin fraction of animal or vegetable oils.
Thiamine hydrochloride or "thiamine" (vitamin B₁)	Crystalline hydrochloride of thiamine, relatively free of impurities.
alpha-Tocopherol (vitamin E)	Relatively free of impurities.
Tuna oil	The oil extracted from cannery refuse of tuna fish.
Vitamin A acetate	Relatively free of impurities.
Wheat germ oil	The oil extracted or expressed from wheat germ.

VITAMIN DEFICIENCIES

Vitamin	When is a deficiency likely under practical conditions?	Deficiency symptoms
A	Animals on low grade roughage, receiving no green feed or yellow corn	Low plasma vitamin A and carotene, sore eyes, weak young, urates in poultry, edema, night blindness
D	Calves and pigs indoors, poultry without added vitamin D in northern states during winter	Rickets, tetany, osteomalacia
E	In sheep fed field beans, in chicks on high corn rations	Stiff lambs, "crazy chick" disease, enlarged hearts, abortion, testes degeneration, abnormal estrus, creatinuria
K	Animals eating moldy sweetclover, hay, or silage	Loss of blood clotting power, hemorrhagic disease
B_1	Animals eating highly processed feeds containing no bran or whole grain	Anorexia, slow growth, paralysis, nervous conditions
B_2	Poultry and swine in dry lot unless supplemented	Poor growth, curled toe paralysis in chicks; photophobia (hypersensitivity to light), glossitis (inflammation of the tongue), dermatitis, neuritis (inflammation of a nerve), anemia
Ni (Niacin)	Poultry and swine in dry lot unless supplemented	Dermatitis, diarrhea, loss of hair
B_6	Not likely	Dermatitis, fits in pigs and dogs
Pa (Pantothenic acid)	Poultry and swine in dry lot unless supplemented	Dermatitis, diarrhea, "goose stepping" pigs, gray hair
Ch (Choline)	Poultry on some rations; young swine in dry lot on low quality feed	Fatty liver, hemorrhagic kidneys, perosis in chicks
Bi (Biotin)	Poultry if not fed high grade feeds, not likely in swine	Dermatitis, perosis in chicks
Bc (Folic acid)*	Not likely	Macrocytic anemia
B_{12}	Poultry on vegetable rations, swine on vegetable rations in dry lot	Pernicious anemia, poor growth, poor reproduction
C	Never in farm animals	Bleeding gums, delayed wound healing

*Folic acid = Pteroylglutamic acid

rich in vitamin B_2 and other B-complex vitamins, are used in poultry rations. However, if fed at a rate of 5 percent or more of the diet, they may have a laxative effect.

Condensed . . . fermentation solubles is the product resulting from the removal of a considerable portion of the liquid byproduct resulting from the action of the ferment on the basic medium of grain, MOLASSES, WHEY, or other media: it shall contain not less than 18 mg. riboflavin (VITAMIN B_2) per pound on a moisture-free basis. For label identification the source shall be indicated as *"condensed whey fermentation solubles."*

Vitamin B_{12} supplement is a feeding material used for its VITAMIN B_{12} activity. It shall contain a minimum vitamin B_{12} activity of 1.5 mg. per pound. The term should not be applied to products for which there are accepted names and definitions.

Antibiotic feed supplement is a feeding material used for its antibiotic activity. It shall contain a single ANTIBOTIC or a combination of antibiotics having growth-promoting properties. The name and amount of each antibiotic shall be declared on the label. It shall contain a minimum of 1 g. per pound of antibiotics. The label shall bear the legend "for feeding use only."

dl-Methionine must be of a minimum 95-percent purity.

NOTE: Generally, METHIONINE is classified as one of the essential AMINO ACIDS.

Vitamin E supplement is a feeding material used for its VITAMIN E activity. It shall contain a minimum vitamin E activity equal to 10,000 I.U. vitamin E per pound. The label shall bear a statement of vitamin E activity in terms of International units vitamin E per pound.

Additional vitamins, and definitions of some of the vitamins mentioned above, are compiled in the (official) table on page 538. The vitamin functions and sources are tabulated on page 537.

References: F.6; S.10.

See also FERMENTATION INDUSTRY BY-PRODUCTS; YEAST; MINERAL FEEDS; NUTRIENT REQUIREMENTS; FEEDSTUFF COMPOSITION; GROWTH; U.G.F; VITAMIN DEFICIENCIES; VITAMIN REQUIREMENTS.

VITAMIN SUPPLEMENTS.
See COMMERCIAL FEEDS.

VOLATILE describes a substance that evaporates rapidly.

VOLSTATE. *See* SOYBEAN (variety).

W

WABASH. *See* SOYBEAN (variety).

WACONIA AMBER SORGO, a *Black Amber*, belongs to the forage SORGHUMS.

WACONIA ORANGE SORGO is one of the ORANGE SORGO varieties which belong to the forage SORGHUMS.

WARD. *See* SIX-ROWED BARLEY (variety).

WART. *See* CROWN.

WASATCH is a COMMON WHEAT variety.

WASHED SULFUR. *See* SULFUR.

WASPS are four-winged, stinging INSECTS, which are closely related to the bees; they have yellow coloring. The so-called *social wasps* live in colonies and form papery nests. The workers are undeveloped females, the males are drones, and there is a queen wasp which lays eggs in the cells. Another group of wasps make nests of earth; these *solitary* and often insect-eating wasps are beneficial. Some of them are quite small and lay their eggs within the bodies of plant pests which are destroyed by the larvae feeding on them; e.g., the tiny wasp, *Eumicrosoma benefica*, is an enemy of the CHINCH BUG. Other species are parasites of the GARDEN WEBWORMS, the VELVETBEAN CATERPILLARS, and the

WHEAT STRAWWORMS. The small wasp, *Apanteles medicaginis*, is a natural enemy of the ALFALFA CATERPILLAR, and *A. militaris* attacks the ARMYWORM.

The WHEAT STEM SAWFLY is a slender wasp.—*See also* APANTELES; DITROPINOTUS; ERIDONTOMERUS; EUMICROSOMA; EUPELNUS; MERISOPORUS; MERISUS; TELENOMUS; TRICHOGRAMMA; "WEEVIL PARASITE."

WASTE PRODUCTS. *See* BODY WASTES.

WATER constitutes over 70 percent of most plants and of the animal body; it amounts to 70 to 80 percent of green, growing parts of crops, to over 90 percent of the BLOOD, 87 percent of cow's milk, etc. It is therefore not surprising that the daily consumption of water is greater than that of any other NUTRIENT—lack of water will retard GROWTH and reduce production.

Water is important because it softens the feed in the stomach, making it more digestible; it is absorbed into the blood stream all along the digestive tract and transports the dissolved nutrients to the cells, wherever required; it removes waste products from the body; it is the main constituent of the body liquids (e.g., blood); and it aids in regulating body temperature.

Water is needed for plant growth. On heavily grazed pastures, up to 70 percent of rainfall water is lost as runoff, while on lightly grazed pastures less than 10 percent is lost, and on ungrazed pastures and meadows no water is lost by runoff. Thus it becomes obvious that good PASTURE MANAGEMENT can help greatly to make more of the rainfalls available for use by grass and legume swards.

Drinking water must be free from PATHOGENS and should not contain more than traces of organic matter or heavy metal compounds and other inorganic salts. Its temperature should at no time be below 50° F.

> NOTE: When water is added in the preparation of canned foods for domestic animals, water shall be listed as an added ingredient.

See also GRAZING; RANGE MANAGEMENT; ABSORPTION; WATER REQUIREMENTS.

WATER CONSERVATION.
See DROUGHT.

WATER GRASSES.
See GRASSLIKE PLANTS.

WATERHEMLOCKS (*Cicuta* spp.)—especially the *spotted waterhemlock, western waterhemlock, tuber waterhemlock,* and *California waterhemlock*—are POISONOUS PLANTS containing ALKALOIDS and/or RESINOIDS in

Waterhemlock. Root and base of stem, showing broad basal leaf and characteristic partitions in interior of stem; upper left, flowering head; upper right, leaves. (D.19.)

the roots and rootstocks; eating even very small amounts of them causes violent spasms in cattle and sheep.

The waterhemlocks are stout, herbs, occasionally 10 ft. high, with large leaves. **WATER HYSSOP** (*Bacopa rotundifolia*) is an annual plant which looks like water cress and occurs as a WEED in RICE fields.

"WATERLILY" is a name sometimes applied to such WEEDS as ARROWHEAD and WATERPLANTAIN.

WATERPLANTAIN (*Alisma plantago-aquatica*), sometimes also called *"waterlily"*, is a serious WEED pest in RICE fields. It is a hardy perennial with small, white or rose-tinged flowers and large leaves.

Control is best accomplished by spraying with 2,4,D.

WATER REQUIREMENTS. A suitable water supply is needed for successful livestock production. Animals have many uses for water; it (1) serves as a solvent for digested NUTRIENTS; (2) transports food and wastes; (3) enters into structure, as the muscles are about three-fourths water; (4) protects the body from heat and shock; and (5) aids in cooling and temperature control.

The amount of drinking water needed will depend upon the class of animal, the level of production, the environmental temperature, the water content of the feed, and the type of RATION that is fed.

As a general guide, allow 5 parts of water for each part of air-dry feed. Sheep and poultry conserve water by excreting dry feces. These species can survive on less water than cattle. Horses also produce dry feces but their high losses through sweating may also increase requirements.

Under average conditions the species listed below may drink the following amounts of water: *beef cattle*, 5 to 8 lb. per 100 lb. of body weight, with greater amounts for young and lactating animals; *dairy cows*, 5 lb. for each pound of *milk* produced; *horses*, 5 to 8 lb. per 100 lb. of body weight; *poultry*, 2 to 3 lb. per pound of feed; *sheep*, 2 to 3 lb. per pound of feed; *swine*, 4 lb. per pound of feed.

When air temperatures are above 80° F. water requirements will rise rapidly.

WAXES are FATTY ACID esters and therefore related to FATS; they differ from them in being harder and less greasy. The waxes usually occur in small amounts as "bloom" on stems, fruits, and on other plant surfaces, protecting them from the weather.

WAXY CORN.

See CORN; AMYLOPECTIN; AMYLOSE.

WAXY KAFIR. *Combine waxy kafir* is one of the grain SORGHUMS.

WAYNE. *See* COMMON OAT (variety).

WEAK NECK is the result of overripeness accompanied by an inherent weakness of the tissues in the RACHIS (the center stem of the head) and the PEDUNCLE (the upper part of the stalk), especially of the main stalk. It affects certain dwarf varieties of SORGHUMS. The most objectionable feature of weak neck is the breaking over of the peduncles after the grain has matured, so that the heads fall to the ground and are missed by the combine at harvesting time. The boot, surrounding the base of the peduncle, generally contains a slimy liquid composed of water and honeydew in which decay-producing BACTERIA and FUNGI develop. These micro-organisms can invade and rot the broken stalk, thus bringing about the final stage of weak neck.

Poorly developed heads with lightweight, lusterless seeds sometimes are associated with the condition known as weak neck.

Some varieties of grain sorghums (especially Westland, Midland, and to a large extent the kafirs) and the sorgos are less subject to weak neck because in these varieties the upper stalks and the central stems of the heads remain green and solid long after the seeds are ripe and dry. Early Kalo, Sooner, and Colby are among the sorghums that are highly susceptible to weak neck.

Control: Weak neck is largely a varietal characteristic. The remedy, therefore, lies in growing combine types of grain sorghum, having stalks that remain green for a considerable period after the grain is ripe.

Reference: L.1.

See also STALK ROTS.

WEBWORMS. *See* GARDEN WEBWORM; BEET WEBWORM; SORGHUMS.

WEED can be defined as a plant which, in its location, is more harmful than beneficial.

Weed to the *farmer* is a plant out of place, especially an aggressive and pestiferous, often coarse, usually herbaceous plant species which takes possession of cultivated and fallow fields and pastures. Many kinds of weed can be used as SILAGE CROPS.

To the *stockman*, a weed is a *broadleaved herb;* i.e., a herbaceous, non-grasslike plant occurring on the range.

A considerable number of the farmer's weeds are good range forage plants for the western stockman, especially on those ranges where, under unfavorable growing conditions, the same quality that makes a species a pest in agricultural land, makes it useful for him, if it possesses palatability.

Weeds (as applied to herbaceous, non-grasslike plants)—with BROWSE, GRASSES, and GRASSLIKE PLANTS—constitute the four main groups into which western RANGE PLANTS are customarily subdivided.

Weeds are sometimes POISONOUS PLANTS; e.g., COPPERWEED, LOCOWEEDS, MILK-WEEDS, RUBBERWEEDS, TARWEED, and WESTERN SNEEZEWEED.

Other plants often considered as weeds are as follows: ARROWHEAD; BARNYARD-GRASS; BERMUDA-GRASS; BLUEGRASS; CANADA-THISTLE; CATTAIL; CHEATGRASS; CHESS; CHICKWEED; CHUFA SEDGE; COMMON WILD OAT; CORNCOCKLE; CRABGRASS; CURLED DOCK; DODDERS; ENCELIA; FOXTAIL; FOX-TAIL BARLEY; GLOBEMALLOWS; JERUSALEM ARTICHOKE; JOHNSONGRASS; KNOTGRASS; KNOTWEEDS; LAMBSQUARTERS; PENNY-CRESS; PENTSTEMONS; PEPPERGRASS; PIG-WEEDS; PLANTAIN; QUACKGRASS; RAGWEED; RED RICE; REDSTEM; ROUGH PIGWEED; RUSHES; RUSSIAN-THISTLE; RYEGRASS; SEDGES; SEEPWEEDS; SHEPHERD'S-PURSE; SHORT OAT; SHRUBBY CINQUEFOIL; SKELE-TONWEED; SLENDER OAT; SOW THISTLE; SPIKE RUSH; ST. JOHNSWORT; TARWEED; THISTLES; VELVETGRASS; WATERHYSSOP; "WATERLILY," WATERPLANTAIN; WHITETOP FLEABANE; WILD BARLEYS; WILD BUCK-WHEAT; WILD ONION; WILD RED OAT; YARROWS; YELLOW STAR-THISTLE.—*See also*

FLORIDA BEGGARWEED; FORBS; PASTURE MANAGEMENT; RANGE MANAGEMENT; RANGE PLANTS; GRAZING; SILAGE; HAY; ALFALFA; MUNGBEAN; RICE; CORN ROOT APHID; CUTWORMS; EUROPEAN CORN BORER; POTATO LEAFHOPPER; DUST STORMS; SCREENINGS.

WEEDER is an implement used in summer fallowing and seedbed preparation for pulling out the weeds.

WEED KILLER = HERBICIDE.

WEEPING LOVEGRASS (*Eragrostis curvula*) is sometimes erroneously called *African lovegrass;* there are several LOVEGRASSES from Africa now being grown in America, including LEHMANN'S LOVEGRASS and BOER'S LOVEGRASS, so the term African lovegrass is only confusing.

Particularly well adapted to the southern Great Plains, weeping lovegrass is a vigorous-growing, drought-resistant bunch-grass with an extensive root system and a large, densely tufted crown. The leaves are 10 to 20 in. long and curve in a very characteristic "weeping" manner, giving the grass its name. The stems, in mature growth, frequently branch at the nodes and in warm, wet weather may develop aerial roots as well. The seedstalks are 2 to 5 ft. tall and support the open, usually nodding, PANICLES. Plants in bloom have a distinctive odor that is pleasing to some, and offensive to others. This odor may be important in determining the reaction of livestock to the grass.

This lovegrass is adapted to a wide range of soil types; good results have been obtained on deep sands, eroded and leached clays, rocky outcroppings, and, indeed, practically all sites and situations. However, the forage is more nutritious and palatable when produced on a fertile soil.

Weeping lovegrass is used for grazing, soil and water conservation, and improving soils low in organic matter. It makes good winter roughage when supplemented by protein concentrates. Although a summer-growing, warm-season grass, it usually maintains some green growth in the crown during most of the winter, which undoubtedly improves its value for winter grazing. The first flush of growth in the spring is also taken readily, but as the plants mature and, especially as they go into bloom, the forage becomes quite unpalatable and is rejected by livestock whenever other forage is available. Cattle show an inclination to browse on the seedheads after seed is formed.

Weeping lovegrass. (H.1.)

Because young seedlings are sometimes killed by heavy frosts, late March and early April are preferred to earlier seeding dates. The seedbed must be firm; the seed should be drilled not over ½ in. deep. Packing of the seedbed may be desirable on sandy soils.

SERICEA (lespedeza) is a good companion species because it is a deep-rooted perennial that can withstand competition well. Other companion species that have been used successfully with weeping lovegrass are HOP CLOVER, SWEETCLOVER, WHITE CLOVER, HAIRY VETCH, and ALFALFA. It is usually desirable to establish the grass and the LEGUMES in separate rows.

Generally speaking, if the grass is to be used for summer grazing, it should be established on good, highly fertilized soil and kept short by grazing or mowing, or both. If high fertility levels are not practical, the grass could be used better as winter roughage and for early spring grazing. Weeping lovegrass responds readily to nitrogen fertilizers with both, increased production and improved palatability. Old forage should be removed by use of a mower or shredder before growth starts in the spring. Burning is very detrimental.

References: H.4; H.1.

See also GRASSES.

"WEEVIL PARASITE" is the common name of the WASP *Bathyplectes curculionis*. This parasite destroys the larvae of the ALFALFA WEEVIL.

WEEVILS are insects which belong to the BEETLE group; some of them attack stored seeds, others damage growing plants. Among the more important weevils are the GRAIN WEEVILS, CLOVER LEAF WEEVIL, CLOVER-HEAD WEEVIL, ALFALFA WEEVIL, COWPEA WEEVIL, PEA WEEVIL, and RICE WATER WEEVIL.—*See also* CORN EARWORM; VETCHES; FIELD PEA; COWPEA; WHEATS.

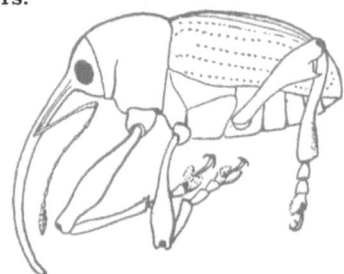

Weevil. (U.13.)

WEIGHT AND MEASURE. As yet, no international standards exist, but within nations, uniformity in these various standards is established. Therefore, it is necessary to designate clearly the scale used.

The avoirdupois weight—abbreviated *avd.* or *av.*—is the commercial standard in the United States.

1 dram	= 27.34375 grains (gr.)
1 ounce (oz.)	= 16 dr. or 437½ gr.
1 pound (lb.)	= 16 oz. or 7,000 gr.
1 hundredweight	
(cwt.)	= 100 lb.
1 ton (t.)	= 2,000 lb.

The U.S. fluid measures are as follows:

1 fluid dram (fl. dr.)	= 60 minims (min.)
1 fluid ounce (fl. oz.)	= 8 fl. dr. or 480 min.
1 pint (pt.)	= 16 fl. oz.
1 quart (qt.)	= 2 pt.
1 gallon (gal.)	= 8 pt. or 4 qt. or 231 cu. in.

Domestic measures are not accurate:

1 teaspoonful	= 1 fl. dr.
1 dessertspoonful	= 2 fl. dr.
1 tablespoonful	= ½ fl. oz.
1 wineglassful	= 2 fl. oz.
1 teacupful	= 4 to 5 fl. oz.
1 tumblerful	= 8 to 10 fl. oz.

The metric system is widely used, especially in foreign countries, and also among American scientists and progressive manufacturers. It is very simple and has good possibilities of developing into the internationally accepted weight and standard.

The unit of weights is the gram (gm.), equal to 15.432 gr.

1 dekagram (dgm.)	= 10 gm.
1 kilogram (kg.)	= 1,000 gm.
1 milligram (mg.)	= 1/1,000 gm. or 1,000 gamma (approximately 1/60 gr.)
1 microgram (mcg.) or gamma	= 1/1,000 mg. or 1/1,000,000 gm. (1/60,000 gr.)

The *liquid unit* is the cubic centimeter (cc.) or milliliter (ml.), i.e., approx. 16.23 minims.

1 liter (l.)	= 1,000 cc. (or ml.)
1 hectoliter (hl.)	= 100 l.

The *linear unit* is the millimeter (*mm*); 25.4mm = 1″.

1 centimeter (cm.)	= 10 mm.
1 meter (m.)	= 1,000 mm. = 100 cm.

Conversion table. To convert the weights and measures in ordinary use into metric weights and measures—or reversed—multiply the quantities by the corresponding equivalent:

To convert	Multiply by
gal. into l(iters)	3.785
pt. into l.	0.4731
fl. oz. into cc.	29.572
fl. dr. into cc.	3.697
gr. into gm.	0.0648
gr. into mg.	64.799
avd. oz. into gm.	28.3495
avd. lb. into kg.	0.4536
cc. into fl. oz.	0.0338
l. into gal.	0.2642
l. into pt.	2.113
mg. into gr.	0.01543
gm. into gr.	15.432

To convert	Multiply by
gm. into oz...................	0.03527
kg. into avd. lb..............	2.2046
mm. into inches..............	0.03937
cm. into inches..............	0.3937
m. into feet.................	3.2808
m. into yards...............	1.09361
inches into cm..............	2.54
feet into m.................	0.3048
yards into m................	0.9144

WESKAN is one of the intermediate-type grain SORGHUMS.

WESTAR. *See* COMMON WHEAT (variety).

WESTERN AZALEA (*Azalea occidentalis*) is a RANGE PLANT poisonous to sheep and cattle. Eating a few ounces of the RESINOID-containing leaves causes salivation, vomiting, and weakness in sheep.—*See also* POISONOUS PLANTS.

WESTERN BLACKHULL, a variety of the *Blackhull kafirs*, belongs to the grain SORGHUMS.

WESTERN CHOKECHERRY.

See WILD CHERRY; POISONOUS PLANTS.

WESTERN CUTWORM.

See PALE WESTERN CUTWORM.

WESTERN HARVESTER ANT (*Pogonomyrmex occidentalis*) may be distinguished from the RED HARVESTER ANT by its nesting habits. The western harvester ant builds a mound at the main opening in the center of the cleared area. The habits of the colony, methods of feeding, and means of control are practically the same for both species.

Reference: B.8.

See also ALFALFAS.

WESTERN HONEY MESQUITE.

See MESQUITE (variety).

WESTERN RYEGRASS is a widely used name for the COMMON RYEGRASS; however, in some areas this term is employed to designate SLENDER WHEATGRASS.

WESTERN SNEEZEWEED (*Helenium hoopesii*) is a perennial plant of the COMPOSITE family. When sheep or cattle feed on the leaves of this POISONOUS PLANT for two weeks or more, they develop weakness due to profuse vomiting.

WESTERN SPOTTED CUCUMBER BEETLE (*Diabrotica undecimpunctata*) is a pest of SUBCLOVER and other crops; it is particularly destructive to the seedlings of spring plantings.

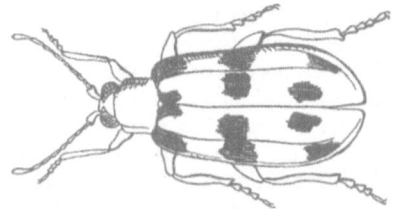

Western spotted cucumber beetle. (D.20.)

Control: This beetle is effectively controlled by the use of DDT dust.

Reference: R.3.

WESTERN WATERHEMLOCK.

See WATERHEMLOCKS; POISONOUS PLANTS.

WESTERN WHEATGRASS (*Agropyron smithii*), also called *bluestem wheatgrass*, or *Colorado bluestem*, is a native, perennial, sod-forming grass distributed generally throughout the United States, except for the humid southeastern states. Western wheatgrass is most abundant in the northern and central parts of the Great Plains. Plant growth is vigorous, with seed heads at a height of 2 to 3 ft. and with leaves 8 to 10 in. long and less than ¼ in. wide. The entire plant is usually covered with a grayish bloom.

Western wheatgrass. Plant and spikelet. (M.47; H.47.)

Although western wheatgrass is adapted to a wide range of soils (including alkaline soils), it seems to prefer the heavy soils characteristic of shallow lake beds, swales, and water courses that receive excess surface drainage water. It also occurs in nearly pure stands on abandoned cultivated fields where the original stand of wheatgrass was not entirely eliminated by cultivation. These "go back" fields are dependable for the production of hay or seed.

Western wheatgrass has several characteristics that make it exceedingly valuable for use in revegetation and erosion control. Its hardiness and drought resistance, and its capacity to spread rapidly, are outstanding values for conservation. It is excellent for terrace waterways and contour strip plantings for erosion control.

Growth starts fairly early in the spring and continues until limited by a shortage of moisture, or by continued hot summer periods. Abundant forage is produced which is relished by all classes of livestock until it becomes harsh and woody during late summer; sheep are fond of the heads of the plants. Western wheatgrass is very resistant to grazing pressure.

Mature plants cure well into a palatable, nutritious forage that provides excellent winter grazing. Leafy, high-quality hay also may be produced if proper precautions are taken to cut the grass while it is still succulent. Yields of hay depend upon moisture, particularly that available during the early part of the growing season. It is not uncommon to obtain yields of ¾ ton of hay per acre from "go back" fields.

Best results have been obtained when seedings were made on well-prepared, clean seedbeds; with full protection from grazing until the second growing season, the stand improves rapidly in vigor and density.

Western wheatgrass is one of the best GRASSES for revegetation and general farm use. Many acres of range and previously cropped farm lands have been seeded to this grass alone, or in combination with other adapted forage grasses.

References: H.1; B.1.

See also WHEATGRASSES; RANGE PLANTS.

WESTERN YARROW. *See* YARROWS.

WESTLAND is one of the intermediate-type grain SORGHUMS.

WET BREWERS' GRAINS contain very high amounts of water and spoil easily; therefore, they are fed not too far from the brewery and soon after arrival at the farm, except if used as SILAGE CROP. Approximately 4 lb. wet brewers' grains replace 1 lb. concentrates or 4 lb. silage; up to 30 lb. may be fed to dairy cows, up to 20 lb. to horses, and up to 2½ lb. per 100 lb. body weight to swine.

"WET MEADOW GRASSES."
See GRASSLIKE PLANTS.

WETTABLE POWDERS are finely ground materials, often mixed with suspending agents to increase their ability to make stable suspensions when mixed with water. Many insecticidal sprays are prepared from such wettable dusts.

WETTABLE SULFUR. *See* SULFUR.

WETTING AGENTS = SURFACTANTS.

WF 9 is one of the inbred lines of CORN hybrids.

WHALE MEAL. *See* ANIMAL PRODUCTS; FISHERY BY-PRODUCTS.

WHEAT *See* WHEATS.

WHEAT BRAN. *See* WHEATS.

"WHEAT BRAN AND CORN BRAN."
See LABELS.

WHEAT BROWN SHORTS.
See WHEATS.

WHEAT DISTILLERS' DRIED GRAINS. *See* DISTILLERS' PRODUCTS.

WHEAT DISTILLERS' DRIED GRAINS WITH SOLUBLES. *See* DISTILLERS' PRODUCTS.

WHEAT DISTILLERS' DRIED SOLUBLES. *See* DISTILLERS' PRODUCTS.

WHEAT FEED FLOUR. *See* WHEATS.

WHEAT FLOUR MIDDLINGS.
See WHEATS.

WHEAT GERM MEAL. *See* WHEATS.

WHEAT GERM OIL.
See VITAMIN E; VITAMINS.

WHEAT GERM OIL CAKE.
See WHEATS.

WHEAT GERM OIL MEAL.
See WHEATS.

WHEATGRASSES (*Agropyron* spp.) are hardy and drought-resistant. Usually the stems grow erect; the spikes resemble wheat (hence the name). These GRASSES are of great value in the northern Great Plains, the Intermountain region, and the higher altitudes of the Rocky Mountain states.

The wheatgrasses produce abundant forage that is acceptable to all classes of livestock. The production of lush forage occurs at the season when most needed by overwintering animals.

Wheatgrasses have also been used extensively for revegetating depleted range and abandoned farm lands. Many thousands of acres of previously cropped farm lands owe their present economic usefulness for grazing to these nutritious grasses. The sod-forming species are particularly valuable for erosion control.

Dangers: Wheatgrasses are attacked by many grass enemies, particularly ERGOT and the WHEAT STEM SAWFLY.

Species: The genus *Agropyron* contains approximately 150 species widely distributed in temperate regions of the world; about 30 species occur in North America. Most of them are perennials. The species probably best known is QUACKGRASS which invades the cultivated fields and gardens with such aggressive persistence that it has fully earned its place in the category of weeds.

It has been demonstrated that two species, TALL WHEATGRASS and INTERMEDIATE WHEATGRASS, will hybridize with common and durum wheats. Many promising possibilities are thus opened. Other important species are BLUEBUNCH WHEATGRASS, CRESTED WHEATGRASS, SLENDER WHEATGRASS, WESTERN WHEATGRASS, REE WHEATGRASS, and PUBESCENT WHEATGRASS, or *stiffhair wheatgrass*.

Reference: H.1.

See also STANDARD CRESTED WHEATGRASS; AGROPYRON; PRAIRIE HAY; SAGEBRUSHES; RANGE PLANTS; PASTURE PLANTS; FAIRWAY CRESTED WHEATGRASS.

Illustrations: See BLUEBUNCH WHEATGRASS; CRESTED WHEATGRASS; SLENDER WHEATGRASS; WESTERN WHEATGRASS.

WHEAT GRAY MIDDLINGS.
See WHEATS.

WHEAT GRAY SHORTS.
See WHEATS.

WHEAT HAY. *See* WHEATS.

WHEAT JOINTWORM (*Harmolita tritici*). In many eastern states it is one of the most consistently injurious insect enemies of soft WHEAT.

The wheat jointworm usually remains undetected, and losses caused by it are often attributed to unfavorable weather or cultural conditions. It robs the heads of nourishment by causing hard knots, or galls, in the stems. Where very abundant, it may cause lodging of the ripening grain, and thus call attention to its presence.

The adult jointworm looks like a small black ant with wings. It lays eggs early in the spring in the succulent plant stems. The larvae, small footless grubs, soon hatch and form cells in the wall of the stem, usually just above the second or third joint from the ground. At harvest time the larvae are yellowish in color and about ¼ in. long. Their cells, or "galls," have now become hard and woody. Sometimes they appear as wartlike swellings, and the wheat stems are badly twisted and bent. In winter the larvae change to pupae which are pale yellow at first but later turn black. In the spring the adults emerge through small circular holes which they gnaw through the walls of their cells. Mating soon takes place and the females leave the old stubble fields to find and infest green wheat fields in the vicinity. There is only one generation a year.

Control: The wheat jointworm may be controlled by plowing under the infested stubble, preferably late in the summer or early in the fall, to prevent the emergence of the adults during the following spring. When this is done the wheat should be cut as high as practicable, so that most of the jointworms will be left in the standing stubble to be plowed under. Objections to control by plowing, because of its interference with the growing of RED CLOVER and other crops useful in soil conservation, may be met by the temporary substitution of SOYBEANS, SWEETCLOVER, and other crops for forage and green manure.

Where infested stubble has not been plowed, the wheat should be sown, as far as practicable, from such stubble fields. This will make it more difficult for the jointworm adults, emerging from the old stubble, to reach the new crops.

In areas where severe losses from the jointworm have occurred during the preceding season, land sown to wheat should be top-dressed only with manure containing straw that has been well rotted or thoroughly trampled. This helps to insure the death of the jointworms that might otherwise emerge from the straw to reinfest the field.

If jointworm attacks are especially threatening, it may be advisable to substitute temporarily other crops, such as RYE, BARLEY, OATS, or BUCKWHEAT, for wheat. This can be done safely because the wheat jointworm attacks none of these crops.

References: B.16; U.3.

WHEATLAND and *Resistant wheatland-288* belong to the intermediate-type grain SORGHUMS.

WHEAT MALT SPROUTS.
　　　　　　　See BREWERS' PRODUCTS.
WHEAT MIDDLINGS.
　　　See WHEATS; PALMO MIDDLINGS.
"WHEAT MILL BY-PRODUCTS."
　　　　　　　　See LABELS.
WHEAT MILL FEEDS, or OFFAL, is a term used for various WHEAT by-products.
WHEAT MIXED FEED. *See* WHEATS.
WHEAT MOSAIC (*Marmor tritici*) is one of the MOSAICS; it attacks particularly WHEATS.
WHEAT RED DOG. *See* WHEATS.

WHEATS (*Triticum* spp.) belong to the annual or winter annual GRASSES and are true cereal crops. They are produced commercially in most parts of the United States under a wide range of environmental conditions. More than 200 distinct varieties of this small grain cereal are cultivated. Many of these are adapted only locally, whereas others are well adapted to varying conditions.

The identification and classification of varieties requires some knowledge of the appearance of plant and kernel. Confusion in names is frequent.

The principal parts of the (usually self-pollinated) plant are the roots, culms, leaves, and SPIKES. There are two sets of *roots*—the first (seminal or seed) roots, and the second (or coronal) roots which arise from the crown of the stem. The *culm* usually is a hollow, jointed cylinder comprising three to six nodes and internodes. The upper internode of the culm, which bears the *spike*, is called the peduncle. The spike is made up of the RACHIS and SPIKELETS, the latter in turn comprising the RACHILLAS, GLUMES, LEMMAS, PALEAS, and the sexual organs (the three STAMENS and the single ovary with its style and stigma). The *leaves* are composed of the sheath, blade, LIGULE, and AURICLE.

The following are the more important characteristics that distinguish the cultivated wheat varieties: (1) *Habit of growth:* winter habit, intermediate habit, or spring habit. (2) *Maturity:* early, midseason, or late. (3) *Height:* short, midtall, or tall; the height of the plants may vary from 12 to 60 in. (4) *Stem color:* white (including cream to golden-yellow) or purple (ranging from a pale violet to dark purple); the purple coloring may occur only in the short portion of the peduncle or in the sheath, and in some seasons it may not become apparent at all. (5) *Stem strength:* weak (varieties with such stems have a tendency to lodge), midstrong, or strong. (6) *Stem hollowness:* the internodes are hollow or (in some varieties) solid or nearly so. (7) *Spike awnedness:* awnless to awned. (8) *Spike shape:* varying in length from 2 to 6 in. and in width from $\frac{1}{2}$ to 1 in. (9) *Spike density:* lax, middense, or dense. (10) *Spike position* at maturity: erect, inclined, or nodding. (11) *Glume covering:* glabrous (without hair) or pubescent (covered with hair); the degree of pubescence varies. (12) *Glume color:* white, yellowish, brown, or black. (13) *Glume shape:* varying in length from 6 to 15 mm. and in width from 2 to 6 mm. (14) *Kernel color:* white (including cream and yellowish), or red (varying from light brown to dark red); the term "amber" is no longer in use but previously red as well as white wheat varieties were referred to as amber. (15) *Kernel length:* ranging from 4 to 10 mm. (16) *Kernel texture:* soft (with an endosperm entirely mealy or starchy), hard (with a horny or glassy endosperm), or semihard. (17) *Brush size* (i.e., size of the hairy area at the end opposite the germ): small, midsized, or large. (18) *Brush length* (average length of the hairs): varying from less than 0.5 mm. to more than 1 mm. (19) *Hardiness* (a) of winter wheats: the ability to survive low winter temperatures; (b) of spring wheats: the ability to resist injury from spring, summer, or fall frosts.

Wheats are adapted to a wide range of climate and soil. They are grown from the Equator to the Arctic Circle, and from sea levels to elevations above 10,000 ft. High rainfall, especially if accompanied by moderate or high temperature, is generally unfavorable for wheat production, because it intensifies disease and insect attacks and difficulties in harvesting and threshing unless there is a comparatively dry season at maturity. Areas with annual precipitation of 25 to 30 in. give high yields; cold winters and hot summers in areas with moderate precipitation produce high-quality wheats. When the weather is cool and there is enough soil moisture available after the blossom period, the kernels tend to be rich in starch and low in protein. On the other hand, soils rich in nitrogen and low in moisture at time of maturity tend to produce wheats of high protein content.

The wheats are best adapted to well-drained, fertile, medium-to-heavy textured soils, particularly to silt and clay loams; very sandy soils give poor crops, and rich bottom lands cause wheat plants to lodge.

Commercial *fertilizers* and manure may be used to good advantage in growing a wheat crop in most areas. The best fertilizer to use on wheat depends largely on the soil type, and the soil's state of productivity.

When the wheat plants have light green foliage it is usually an indication of a nitrogen shortage. The nitrogen deficiency of future crops may be corrected most economically by plowing under crops of LEGUMES, such as RED CLOVER, SWEET-CLOVER, COWPEAS, SOYBEANS, or LUPINES. However, it is often desirable to supplement this supply with commercial nitrogen, e.g., ammonium nitrate. A common practice is to apply nitrogen fertilizers broadcast on the surface of the ground

Wheat. Plant, awned spike (bearded wheat), and a nearly awnless spike (beardless wheat), spikelet, and floret. (H.26.)

early in the spring at the rate of about 20 lb. nitrogen per acre.

Phosphoric acid is used very widely in the growing of wheat; the crop on most soils responds well to application of such readily available form as superphosphate. Raw rock phosphate may be used for the immediate benefit of the sod-forming crops of the rotation; the wheat in turn benefits from the organic forms of phosphorus produced.

The potash content of soils differs greatly; on many soils potash requirements are very high. Potassium chloride is widely used as fertilizer for wheat.

A complete fertilizer is often applied at seeding time by an attachment to the grain drill. The grade of fertilizer frequently used is about 4-12-4 and the rate of application about 300 lb. per acre. This is often supplemented with a top dressing in the spring. The formula and rate will vary with locality, soil type, and cropping history.

When the soil is more than moderately acid, an application of 1 to 2 tons per acre of ground limestone is sometimes beneficial, but most of the benefit may be indirect by improving the legume crop. Wheat itself is not injured by a slightly acid soil, but most legumes grown in the rotation will be decidedly benefited by the proper use of lime.

Manure is a valuable fertilizer and soil conditioner for wheat, but it is used most economically when applied to the corn crop in the rotation. On both, light sandy soils and heavy clay soils, well-rotted manure plowed under supplies the humus necessary to improve the physical condition of the soil and also adds valuable nutrients, particularly nitrogen and potash. The manure may be applied before seeding, or as a top dressing to the wheat in late fall or early winter.

It is not advisable to grow wheat continuously on the same land, but it fits into many *crop rotations*. The crops to be used in a rotation, their sequence, management, and use are determined by the climate, length of growing season, soil type and fertility, and type of farming. The plowing under of green-manure crops is not practicable in growing wheat in the North, but is very beneficial in the South where the growing seasons are long and the winters mild. Crops like winter wheat, OATS, and BARLEY serve as good winter cover crops to reduce leaching and erosion during the winter. At the same time they may be of considerable value for pasture, especially in the South where they provide grazing during midwinter when it is not available from other crops.

A good winter wheat rotation contains at least one legume and one or more cultivated or row crops. In the Corn Belt these requirements are easily met with CORN, soybeans, and CLOVER. As a companion crop, wheat shades the ground less and is harvested earlier than oats and usually is more profitable than RYE. Inasmuch as cattle are raised on most farms, the growing of wheat is an advantage since wheat furnishes valuable pasture, feed, and bedding.

The fact that wheat is well suited as a companion crop for clover or grasses, whether alone or in combination, means that it should occupy a place in the rotation preceding them. Where corn can be grown successfully, it usually follows these crops.

Spring oats can be sown conveniently after corn, and the crop is harvested early enough to provide plenty of time to prepare the land for wheat. Thus, a natural and efficient rotation for most of the eastern United States consists of wheat one year, clover and TIMOTHY one or two years, corn one year, and oats one year. On very rich soil corn may be grown two years successively, thus increasing the length of the rotation and the acreage of the corn crop. Corn is likely to produce the greatest amount of feed or acre return in the rotation. Cowpeas may be substituted very profitably for oats, or for corn, in many areas. In tobacco-growing localities, tobacco may be grown in place of oats.

Wheat often is more profitable after soybeans than after corn. Hence, when these crops are grown, the rotation may well be corn, soybeans, wheat, and clover, each one year. Such a rotation is especially satisfactory if the soybeans are to be cut for hay.

LESPEDEZA is a popular crop in areas where an important objective is to improve the soil or prevent damage from erosion. As the lespedeza reseeds itself, the wheat-lespedeza rotation may be grown continuously or lespedeza may be left for two years.

Spring wheat is a satisfactory crop for grass and small legume seedings. In regions where POTATOES, BEETS, or PEAS are cultivated, spring wheat often follows these crops. In general, spring wheat occupies the same place in rotation as oats, namely after corn.

The time and method of *preparing the land* for winter wheat depends principally on the crop that precedes it. Unless rainfall is high and the land subject to erosion, it is desirable to have the land prepared considerably in advance of seeding to permit settling and the accumulation of moisture and available plant food, especially nitrates. When the land is to be plowed, at least a month should intervene between plowing and seeding. The soil can be plowed most easily soon after the harvest. If for any reason plowing must be delayed, disking the ground immediately after the harvest is beneficial.

The depth of plowing should be governed to a considerable extent by the quantity of stubble, weeds, and cover crop to be turned under, and should be sufficient to do this thoroughly. A common practice is to plow 6 to 7 in. deep in July and early August, gradually decreasing the depth to 4 to 5 in. in late August and September.

After plowing, the ground should be disked and cultipacked, as these operations are necessary to control weeds and volunteer grain and to get the soil in condition for seeding. A disk and a spike-tooth harrow are often used for the last operation of preparing a seedbed.

As early seeding is desirable for spring wheat, it is a good plan to plow in the fall, if the land is not subject to washing. However, a thorough disking and harrowing in the spring will prove reasonably satisfactory if fall plowing is not done. It is especially important that spring wheat be seeded as early as the ground can be prepared; there is a progressive decrease in the yield in later seedings.

Wheat is generally sown with a drill. Single-disk drills are used most commonly; hoe drills are satisfactory only on clean land, and have some advantages on stony land. If the land is subject to washing, the drilling should be crosswise, on the slope or on the contour to reduce leaching or erosion.

There is no advantage to seeding wheat deeper than necessary to insure sufficient moisture for good germination. In light soils the seed may be safely sown deeper than in heavy soils. Covering the seed from 1 to 1½ in. is usually sufficient.

Winter wheat should be sown early enough to become well established before winter, but not so early that it makes a rank growth or starts to shoot before winter, unless it is grown for pasture. To avoid fall infestation by the hessian fly it is advisable to delay fall seeding until the earliest day in the fall at which wheat can be seeded and still escape damage from this insect. These dates vary for different parts of the country.

An application of straw, not exceeding 2 tons per acre, applied as a top dressing in the fall or early winter, is sometimes recommended as a means of decreasing winter injury. Rolling wheat in the spring with a corrugated roller is beneficial only when the soil is badly cracked or the wheat plants have been heaved by alternate freezing and thawing. It must be done as soon as growth starts but not while the top soil is still wet.

Spring wheat may be seeded with a drill or sown broadcast, as with an endgate seeder. The former method produces larger yields and is to be preferred. However, the broadcast method is more rapid, less expensive, and sometimes allows earlier seeding.

Most wheat is now *harvested* with a combine. The binder is still common in some localities, and the cradle is occasionally used in harvesting small fields in the Southeast. The principal advantages of the combine are that the crop can be harvested and threshed more rapidly while the weather is favorable and with a saving in labor; an objection to it is that the grain cannot be harvested until it is fully ripe and the moisture down to 14 percent or less, thus increasing the risk from storm damage and the danger of spoilage in the bin, and of infestation of the grain by insects.

Where the wheat is to be harvested with a binder, the crop is ready when the straw is well-colored, either purple or yellow, and the grain is in the hard-dough stage. Immediately after being cut with a binder, wheat should be shocked. A well-built shock saves grain loss in wet weather. Storing wheat in the barn, or stacking it in the barnyard to await threshing, is occasionally practiced. Wheat should not be threshed when the straw is tough or the grain damp. It is much easier to dry the grain before threshing than after.

Wheat straw is commonly stacked in the open and used for roughage and bedding during the winter. Ultimately it should find its way back to the soil. Close to cities or poultry sections, wheat straw can often be baled and sold at a profit.

The wheat crop is used for seed, feed, and food (especially in the form of flour, breakfast foods, alcohol, or glucose). For milling, wheat is first cleaned, scoured, sometimes also washed and tempered, then ground and sifted; most flour is bleached. Wheat is usually a slightly better feed than corn which it exceeds in content of protein, minerals, and certain vitamins; corn, however, contains vitamin-A activity and more fat.

As *pasture* crop, wheat is highly nutritive. If grown primarily for grain, pasturage is of secondary importance; however, in some areas dairymen plant wheat for complete utilization as pasture crop. The young wheat plants contain nearly as much protein as does ALFALFA hay. The plants should be well established before pasturing begins; otherwise, they may be uprooted by GRAZING animals. The date for turning in livestock depends on the growing season. Livestock should be kept off when the ground is wet and soft. If grass and clover have been sown with the wheat, injury is likely to be especially severe if the wheat is pastured when the land is muddy. Spring pasturing may be practiced under favorable conditions, but a loss in grain yields should be expected.

In the South, wheat, rye, or winter oats provide pasture during the winter months when most other pasture crops are dormant. They are profitable crops when sown primarily for pasture during this period.

Wheat hay is best cut in the milk stage; however, care is necessary to cure it so that it will be sufficiently dry before it is stored. Hay has a satisfactory feeding value, especially if fed with a protein supplement.

Wheat silage must be prepared carefully; the crop should first be run through a silage cutter. The feeding value is fair.

Whole *wheat kernels* can be fed to poultry but require grinding if fed to livestock.

Offal is a term often used for the by-products of flour milling. A better term is *wheat mill feeds*. Their place in rations is well established. By-products with higher fiber contents are mostly used for cattle, sheep, and horses; those with lower fiber contents, for swine and poultry.

The officially recognized wheat products are as follows:

Wheat bran is the coarse outer covering of the wheat kernel as separated from cleaned and scoured wheat in the process of milling.

Bran is rich in niacin, vitamin B_1, phosphorus, and iron. A large part of the phosphorus is PHYTIN phosphorus. Bran is the coarsest of the wheat by-products, having the highest fiber content. Bran is especially prized as supplement in the rations of cows and ewes. It has been widely used in feeding horses because of its bulky nature and laxative effects; however, it is considered best to limit the proportion in the feed mixture for horses, or to feed it in large amounts only occasionally.

Wheat mixed feed consists of the coarse outer covering of the wheat kernel, fine particles of wheat bran, wheat germ, and wheat flour, and the offal from the "tail of the mill;" it must not contain more than 9.5 percent crude fiber.

Wheat feed flour consists principally of wheat flour together with fine particles of wheat bran, wheat germ, and the offal from the "tail of the mill;" it must not contain more than 1.5 percent crude fiber.

Wheat germ meal consists chiefly of wheat germ together with some bran and middlings or shorts; it must contain not less than 25 percent protein and 9 percent fat.

Germ meal is rich in vitamins B_1 and B_2, and in phosphorus and iron.

Wheat germ oil cake is the cake secured in the removal of part of the oil from wheat germ meal; it must contain not less than 29 percent protein.

The finer wheat products are known as *middlings, shorts,* and *red dog.* They exceed whole wheat in protein, niacin, and vitamins B_1 and B_2; except red dog and white middlings, which contain more fiber than whole wheat.

For swine, the various shorts and middlings, along with red dog, are used most efficiently with tankage, fish meal and milk by-products as supplements to corn and other cereal grains. Products like red dog, wheat flour middlings, and white middlings are useful in the diets of pigs.

Middlings, shorts, and red dog have also been widely used in poultry feed, but the increasing popularity of low-fiber diets, especially for growing birds, has limited their use. When they are omitted from starting and growing mashes, the addition of synthetic niacin may be necessary.

The list of official wheat products belonging to the group of middlings and shorts includes the following:

Wheat standard middlings consists of fine particles of wheat bran, wheat flour, and some of the offal from the "tail of the mill;" it must not contain more than 9.5 percent crude fiber.

Standard middlings is mostly derived from spring wheat.

Wheat brown shorts consists of fine particles of wheat bran, wheat germ, wheat flour, and some of the offal from the "tail of the mill;" it must not contain more than 7.5 percent crude fiber.

Brown shorts is generally derived from winter wheat.

Wheat gray shorts, wheat gray middlings, or *wheat flour middlings* consists of fine particles of wheat bran, wheat germ, wheat flour, and the offal from "the tail of the mill;" it must not contain more than 6 percent crude fiber.

Wheat red dog, wheat white shorts, or *wheat white middlings* consists of the offal from the "tail of the mill" together with some fine particles of wheat bran, wheat germ, and wheat flour; it must not contain more than 4 percent crude fiber. (It is rich in vitamin B_1).

Dangers: Among the most common diseases of wheat are ANTHRACNOSE; ERGOT; HELMINTHOSPORIUM LEAF SPOT; POWDERY MILDEW; RUSTS (STRIPE RUST, STEM RUST, LEAF RUST); SEPTORIA BLIGHT OF WHEAT; SMUTS (COMMON BUNT, DWARF BUNT, LOOSE SMUT OF WHEAT, FLY SMUT); TAKE-ALL; WHEAT SCAB; and MOSAIC. NEMATODES often interfere with the growth of the plants.

Among the insect pests attacking wheat are the ARMYWORM; BILLBUGS; the CHINCH BUG; the EUROPEAN CORN BORER; the FALL ARMYWORM; the GARDEN WEBWORM; GRASSHOPPERS; GREENBUGS; HARVESTER ANTS; HESSIAN FLY; the MORMON CRICKET; the PALE WESTERN CUTWORM; the SOUTHERN CORN ROOTWORM; STINK BUGS; the WHEAT JOINTWORM; the WHEAT STEM SAWFLY; the WHEAT STRAWWORM; WIREWORMS; and WEEVILS.

Weeds are often a problem in wheat fields; the most serious one is the FIELD BINDWEED and other BINDWEEDS; WILD ONION; CANADA-THISTLE; CHESS; CORNCOCKLE; CHICKWEED, and SKELETONWEED.

Species: The following wheat species have been cultivated in the United States: COMMON WHEAT; CLUB WHEAT; POULARD WHEAT; DURUM WHEAT; EMMER; SPELT; and POLISH WHEAT.

NOTE: Common and Durum wheat hybridize with TALL WHEAT GRASS.

Commercially, only the four species first mentioned are considered in the official *grain standards* of the United States; the latter divide wheat into six classes, namely: *Hard Red Spring* (wheat); *Hard Red Winter* (wheat); *Soft Red Winter* (wheat); *White wheat* (soft winter or soft spring, and hard spring); (white) *Durum;* and *Red Durum.*

Egyptian wheat, also called *California wheat* or *Mexican desert wheat corn,* is better known as *shallu;* it is a grain SORGHUM, not a wheat.

References: B.15; F.6; S.22; R.9; C.11; M.33; B.16; E.12.

See also WHEAT PRODUCTS; BUCKWHEATS;

STEM NEMATODE; RICE; VETCHES; SILAGE; SILAGE CROPS; DODDERS; PASTURES; PASTURE MANAGEMENT; DUST STORMS; SCREENINGS; DISTILLERS' PRODUCTS; BREWERS' PRODUCTS; DRIED CEREAL GRASSES; GRASS TETANY.

WHEAT SCAB, more correctly called *fusarium head blight*, is a FUNGUS DISEASE caused by several species of *Fusarium*. The most common species are *F. graminearum*, *F. culmorum*, and *F. avenaceum*. They attack WHEATS, BARLEY, RYE, and some other GRASSES, sometimes causing extensive losses; they also damage CORN and SORGHUM, especially in humid and subhumid areas.

The head blight develops during warm, humid weather. Infection occurs in the flowers and then spreads up and down the SPIKE. The diseased part of the head soon turns a straw color in wheat and rye, and light brown in barley. Pink mold growth frequently develops around the base of the flower and cements the chaff to the grain. The kernels of wheat and rye are shriveled, white or gray, and show a rough, scabby surface. In barley the kernels are grayish brown.

The *Fusarium* spp., which cause head blight, also incite such diseases as SEEDLING BLIGHT, ROOT ROT, and STALK ROT in the cereals.

Dangers: The fungus grows into the kernels and makes the starch floury and discolored, and changes it partly to sugar. The proteins are partly broken down into soluble nitrogen compounds. Some of the fats become rancid because of the formation of fatty acids. New compounds are produced that cause acute vomiting in dogs, pigs, and man, but do not affect sheep, cattle, or mature poultry. Grain containing 5 percent or more of badly scabbed kernels causes, when fed to pigs, vomiting, loss of appetite, and arrested growth. Scabbed kernels in grain, therefore, should not be processed into human food or used to feed swine.

Control: The *Fusarium* pathogens are carried over on crop residues. Spores are not produced or dispersed if all residues are covered by a thin layer of soil. All straw and stalks should therefore be plowed under and left below the surface to control the disease. (Surface mulching with straw and stalks invites head blight in small grains where wet, warm weather occurs after small grains are headed.)

No highly resistant varieties of wheat, rye, or barley have been produced but some are more susceptible than others.

SEED TREATMENT of well-cleaned grain with CERESAN M or other organic mercury compounds will control the seed-borne infection.

Reference: D.11.

See also COMMON WHEAT.

WHEAT SILAGE.
                    See WHEATS; SILAGE CROPS.

WHEAT STANDARD MIDDLINGS.
                    See WHEATS.

WHEAT STEM RUST (*Puccinia graminis tritici*) is characterized by pustules that develop and break through the surface of the stems, leaves, and sheaths and often the chaff and beards of the WHEAT plant. Myriads of brick-red spores escape from the pustules and are carried by the wind to other wheat plants.

The crop is damaged by the growth of the rust FUNGUS on the wheat stems and leaves and by the developing spores, both of which use up the water and nutrient materials needed for developing the wheat kernels. As a result, the kernels are badly shriveled, many of them being so light and chaffy that they are blown out with the chaff in threshing. The rusted straw turns brown, becomes dry and brittle, and soon breaks over.

Wheat stem rust also attacks BARLEY and occasionally RYE. It attacks many wild grasses. It does not attack oats.

The RUST lives over the summer on volunteer grains and wild grasses in the southern states and northern Mexico. These spores and those blown down from the north, in late summer and early fall, infect fall-sown wheat and barley. The rust lives over winter in the red rust stage in the southern part of the United States and northern Mexico but not in the northern states. If weather conditions are favorable in the spring, the rust multiplies, and the spores sweep northward with the advance of the crop season.

A new generation of rust spores may be produced every ten to fourteen days during the spring and summer, starting in Texas and advancing northward with the progressive development of the wheat crop at different latitudes. The rust can spread very rapidly.

An additional source of rust menaces the wheat in the northern half of the country—rust that develops on BARBERRY BUSHES. Usually the rust spreads from the barberries to the grainfields about two or three weeks before the general rust spread arrives from the south. Because of this early start rust may cause severe damage to grains growing near infected barberry bushes.

Reference: M.36.

WHEAT STEM SAWFLY (*Cephus cinctus*) is a slender but "broad-waisted" WASP, from ⅓ to ½ in. in length. It is black with several yellowish bands on the abdomen and remains in this stage—in which it is harmless—for about a month during the summer. The larva is a pale cream-colored, slender worm, about ½ in. in length when fully grown, and conforms to living inside GRASS stems. When removed from the stem, it assumes a characteristic "S"- or "?"-position, bending the head down and the tip of the abdomen up.

The adult wasps begin to emerge from last year's stubble about the middle of June and continue to appear until about the middle of July. They lay from 30 to 50 fertile eggs, whether the males are present or absent. The eggs are placed in the stems of the host plants; WHEAT is most attractive to egg-laying females when it is in the boot or just beginning to head. The eggs hatch in about a week, and the little larvae bore into the stem, where they spend the rest of their larval life. They travel up and down through the stem, boring through the nodes. When the plant begins to ripen, the larva migrates down to the crown, cutting a "V"-shaped notch around the inside of the stem at or near the ground line. It then plugs the stem just below this weakening and remains in the stub until the next spring when it pupates and emerges about the middle of June.

Wheat stem sawfly. (M. 36.)

Wheat stem sawflies are insects native to Montana and Canada. They may be found in the RYEGRASSES, WHEATGRASSES, NEEDLEGRASSES, and some of the less common species of the prairies; they will attack wheat, RUSSIAN WILD-RYE, and RYE extensively.

The most obvious damage to wheat this insect causes is the breaking-off and lodging of the crop before harvest. Additional damage involves some shriveling of the kernels and lightening of the yield.

Because the wheat stem sawfly is a poor flier and seldom goes farther than is necessary to find host plants, the margins of the field are the most highly infested.

Control of the wheat stem sawfly includes the following measures:

1. Early harvesting by swathing or binding while the grain is still a little on the green side.

2. Deep plowing—moldboard plowing 6 in. deep will reduce the number of sawflies; it may be desirable to burn tall stubble before plowing. Fall plowing is best, but plowing may be done any time before the first of June. Burning in itself is not a control measure.

3. Fallowing will drive the insects from the field, as will the planting of resistant crops, such as FLAX, CORN, BARLEY, or OATS.

4. Temporary trap crops—if a strip is planted completely around the block about ten days before the crop, it can catch many eggs. These trap strips should be approximately a rod wide and there should be a rod of well fallowed ground between the trap and the crop. Such traps must be cut as low as possible about the 20th of July,

while the larvae are still in the straws and have not descended to the crowns. The crop on the strips thus cut can be used for hay.

Reference: M.36.

See also COMMON WHEAT.

WHEAT STRAW. *See* WHEATS.

WHEAT STRAWWORM (*Harmolita grandis*) is an important pest in all WHEAT-growing areas. Even when it is very abundant and destructive, its presence may be readily overlooked, and the damage that it does may be ascribed to other causes.

The wheat strawworm apparently has only one food plant, wheat. It lays eggs in several other plants, but the larvae are unable to complete their development in any but wheat.

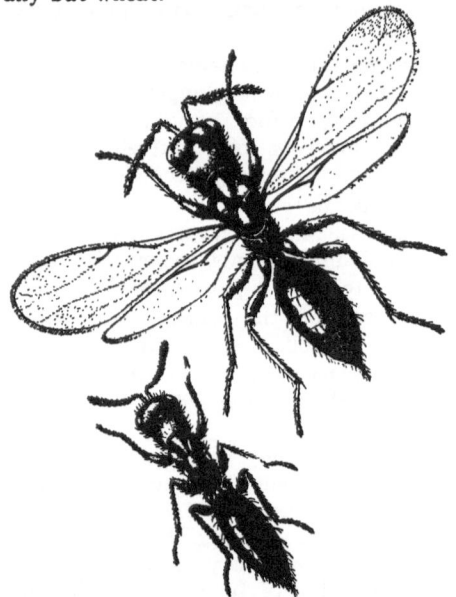

Wheat strawworm. Adults: bottom, spring form; top, summer form. (P.15.)

The wheat strawworm has two generations each year. Early in the spring the adult insect of the first generation deposits its eggs in or near the embryonic wheat head, when the young plants extend only a few inches above the ground. The larva, or grub, develops within and near the base of the plant. Thus, all tillers infested by the spring form of this pest are prevented from producing any grain and become a total loss.

The injury caused by the second generation is less severe, except where spring wheat is attacked. The adult deposits its eggs in winter wheat just above the youngest and succulent joints. The larva develops rapidly in the center of the stem or sometimes in the wall of the stem. The stems of winter wheat show no external evidence of injury; nevertheless, the yield of grain is reduced sometimes as much as 22 percent and it is of poorer quality.

The two generations of the wheat strawworm do not resemble each other very closely. They are designated as the *spring form* and the *summer form*. Each form passes through four stages of development —the egg, the larva (or grub), the pupa (or resting stage), and the adult insect.

The adults of the *spring form* are minute, shiny black insects closely resembling ants, usually without wings. If wings are present, they are rarely fully developed. The legs have light-yellowish bands at the knees. These adults, which are nearly all females, emerge in March and April in the eastern and central states; in the state of Washington emergence occurs in April; in Arizona it takes place during the latter part of January and continues through February. The eggs are white, nearly transparent, and pear-shaped. In about ten days the larva hatches from the egg; it is of a light straw color, has brown jaws, and when full-grown is 1/6 in. long. Full growth is reached in about twenty-seven days; the pupal stage begins thereafter. At first the pupa is the same color as the larva but later it changes to a shiny jet black. The pupal stage lasts about twelve days, after which the fully developed adult of the summer form gnaws a hole through the wall of the stem and comes out.

The adults of the *summer form* are much larger and more vigorous than those of the spring form. They have fully developed wings and can fly considerable distances. No males have been found among the adults of the summer form, and the females reproduce without mating. They emerge in May and June in most of the wheat-growing regions and deposit eggs singly in the growing wheat plant. They hatch in about five days. The larvae attain full growth before the plant tissues have hardened or else they perish. They remain within or near the joints to enter the pupal stage in the fall. No change occurs until early in the spring, when they develop into adults of the spring form to continue their life cycle.

Control: The wheat strawworm has a number of parasitic and predaceous enemies which help to keep down losses from this pest. A very small MITE, *Pediculoides ventricosus*, destroys the larvae in the stem. Important parasites of the wheat strawworm are *Eupelmus allynii, Ditropinotus aureoviridis, Merisus febriculosus, Eridontomerus isosomatis, Merisoporus chalcidiphagus*, and *Calosota metallica*. All these parasites are four-winged WASPS and are of about the same size as the adult strawworm.

An effective control measure is to avoid growing the wheat within 65 to 75 yd. of wheat straw or stubble' of the previous season. The spring form, being practically wingless, cannot travel great distances to infest young wheat and thus continue the life cycle.

In regions where spring wheat is grown to the exclusion of winter wheat, volunteer plants provide the only places in which the adults of the spring form can lay their eggs, since the egg-laying period of this form is over before any spring-sown wheat is up. From these volunteer plants, if not destroyed, emerge the adults of the summer form, which lay their eggs in the spring wheat and may cause considerable damage to the crop.

In localities where the wheat strawworm is injurious, wheat should not be top-dressed with manure containing unrotted straw infested with this insect.

Strawstacks are a greater source of infestation to growing wheat than is usually supposed. Volunteer wheat around the strawstacks should be destroyed early in the spring before the first generation develops.

Reference: P.15.

WHEAT WHITE MIDDLINGS.
                                    See WHEATS.
WHEAT WHITE SHORTS.
                                    See WHEATS.

WHEELER SUDANGRASS.
                        See SUDANGRASS (variety).
WHEY. *See* MILK PRODUCTS; VITAMINS.
WHEY FACTOR. *See* U.G.F.
WHIPPOORWILL is a COWPEA variety.
WHISK DWARF. *Japanese Whisk Dwarf* is a *broomcorn;* it belongs to the SORGHUMS.
WHITE AFRICAN SORGO is one of the forage SORGHUMS.
WHITE ARSENIC = ARSENIC TRIOXIDE.
WHITE BENT. *See* REDTOP.
WHITE CHINESE
                        = CHINESE VELVETBEAN.
WHITE CLAWSON = *Goldcoin. See* COMMON WHEAT.
WHITE CLOVER (*Trifolium repens*) is a widely distributed plant of pastures, lawns, roadsides, and thin woodlands—wherever there is sufficient moisture for it to survive. It is a perennial LEGUME, produces many seeds, and spreads by creeping stems that root at the nodes.

This TRUE CLOVER normally grows with GRASSES and, when seeded alone, creates an ideal condition for grass which soon appears and makes a good stand. When grown in mixtures with grass, it increases the carrying capacity of the pasture and provides a nutritious feed relished by livestock.

White clover is practically self-sterile; that is, the FLORETS have to be cross-pollinated before seed will form. Common white clover thrives best under cool, moist growing conditions in soils with plenty of lime, phosphate, and potash, but it will tolerate poor conditions better than other important true clovers. Clay and loam soils that have sufficient moisture to keep the plants growing seem better suited to white clover than sandy soils.

White clover is not regular in growth persistence. It occurs abundantly in some years but diminishes in others until it almost disappears. This spasmodic behavior may be due to (1) a deficiency of needed mineral nutrients, such as phosphate, potash, and lime; (2) crowding out by tall and thickly growing grasses; (3) insufficient rainfall; or (4) attacks by diseases and insect pests. The application of 200 to 400 lb. phosphate per acre and 50 to 100 lb. potash fertilizers has proved beneficial to its growth. Although it will grow on acid soil, it is never so thrifty as in the presence of lime. Occasionally, clover leaves become spotted with irregular light-brown areas, even in the absence of disease-producing· organisms, a condition that can be remedied by the application of potash. Repeated applications of fertilizers, containing a high percentage of nitrogen, stimulate the growth of grasses, and this frequently crowds out white clover. Close GRAZING in spring reduces the growth and density of the grass and gives the clover a better chance of surviving and spreading.

When new seedings are made in spring, the clover and grass seeds are mixed and the entire mixture sown at one time. Occasionally, grass seedings are made in the fall and those of white clover early in the following spring.

Regardless of the time of seeding, the seedbed should be well compacted. Inoculation is generally not necessary in the northern states.

White clover may be divided into three general types: (1) The *large type;* (2) the *intermediate type;* and (3) the *low-growing type.* Except under close, continual grazing, varieties of the low-growing type are not so productive as the intermediate or large types and are not recommended for forage purposes.

In each type there are usually plants of one or both of the others. Seeds of all types are similar in size and color and cannot be distinguished.

Dangers: More than 30 parasitic, bacterial, fungus, and virus diseases are known to attack white clover in the United States. Although usually of limited economic importance, the following occur commonly and may be destructive: CROWN ROT, PEPPER SPOT, BLACKPATCH, SOOTY BLOTCH, and CLOVER RUST. Insects infesting white clover are the LEAFHOPPERS, especially the POTATO LEAFHOPPER, GARDEN FLEA HOPPER, CLOVER ROOT CURCULIO, CLOVER LEAF WEEVIL, and the LESSER CLOVER LEAF WEEVIL.

Varieties: *Ladino clover,* or *Ladino white clover,* is a variety of the largest growing type of white clover. Except for size of leaves, flower heads, STOLONS, and length of internode, there is little difference between the Ladino clover and the common white clover as regards plant characteristics. Ladino clover can be successfully grown in many of the Lake and Corn Belt states and far enough south to include the Piedmont region. The high carrying capacity of Ladino clover for grazing and its adaptability for hay and silage enable farmers to grow most of the protein needed for livestock, dairy, and poultry production. Although Ladino clover will grow in medium to slightly acid soils, calcium is needed for plant growth; ordinarily calcium must be added by application of fertilizers. Large applications of phosphate fertilizers are generally necessary for continuous satisfactory growth. Potash is also essential.

The superior feeding qualities of Ladino are due to its high content of protein, minerals, and vitamins, and to its continuous palatability. It recovers rapidly following mowing and grazing, is relatively free of diseases, and is valuable as an aid in maintaining the nitrogen supply in soils and in controlling erosion. Excepting the year of seeding, mowing or clipping of Ladino fields after grazing should be as close to the ground as possible.

Common white clover, meaninglessly also called *White Dutch clover,* is truly common. It may be composed of either intermediate or low-growing types or a mixture of the two. Even plants of the large type may be present.

Louisiana white clover, a southern type of white clover, gives most productive clover PASTURES. It reseeds plentifully under proper management and may live from year to year if moisture and fertility conditions are favorable.

A long-lived, persistent, low-growing, rapidly spreading, and sparse blooming variety is *English wild white.* Compared with the intermediate type, the leaves, stems, and flower heads of English wild white are found to be much smaller, and blossoming is not so free.

New York wild white, too, is a variety of the low-growing type and similar to

English wild white, except that it usually has more bloom.

References: H.8; S.6; H.7; H.9; B.4.

See also POLLINATION; LEGUMES; INOCULATION; ANNUAL LESPEDEZAS; WEEPING LOVEGRASS; ALSIKE CLOVER; SUBCLOVER; PASTURE MANAGEMENT; PASTURE PLANTS.

WHITE COWPEAS is a group of COWPEA varieties.

WHITE DURA and *Dwarf white durra* belong to the grain SORGHUMS.

WHITE DURUM.
                            See WHEATS; DURUM WHEAT.

WHITE DUTCH CLOVER, correctly called *common white clover*, is a variety of WHITE CLOVER.

WHITE EGYPTIAN CORN, better known as *White durra*, is a variety of the durra group of the grain sorghums.

WHITE FEDERATION 38.
                            See COMMON WHEAT (variety).

WHITE FISH is often used in *fish meal.*— *See also* MARINE PRODUCTS.

WHITE FRENCH is a PROSO variety.

WHITE-FRINGED BEETLES, belonging to three species and several races, occur in some parts of the South; their grubs frequently feed on the roots of PEANUT plants. When feeding is severe, the plants turn yellow, wilt, and die.

These BEETLES (e.g., those of the widely distributed species *Graphognatus leucoloma*) emerge from the soil in the summer and attach their eggs in small masses to the plant stems, sticks, or pebbles near the soil surface. In warm, moist weather the eggs hatch in about two weeks, and the grubs immediately enter the soil where they remain until fully grown. There is usually one generation a year.

Control: White-fringed beetles can be suppressed by planting peanuts on the same land only once in three to four years, and a succession of OATS, CORN, and COTTON in the intervening years. Field infestation of the grubs can be controlled for several years with DDT.

Reference: B.10.

WHITE GRUBS are the young or immature stage of the common brown *May beetles*, of which there are more than 100 species. The grubs feed on the roots of CORN, SORGHUMS, SERICEA, and several

other crops. They sometimes ruin pastures in the Northeast and in the North Central States. The adult beetles eat the leaves of many trees. Most of the injurious species have a 3-year life cycle and cause serious outbreaks in certain years.

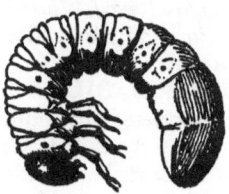

White grub. (U.13.)

The pearly-white eggs are deposited in the spring, 1 to 8 in. deep in the soil. They hatch three to four weeks later into young grubs which feed on decaying vegetation and even on living roots. The grubs do their greatest injury in their second year, but in their third year may sometimes damage early plantings.

Control: Populations of white grubs can be reduced by planting deep-rooted legumes, such as SWEETCLOVER and ALFALFA, in rotation with more susceptible crops.

Reference: U.3.

WHITE "GYP," commonly called *White durra*, is one of the grain SORGHUMS.

WHITEHEAD = TAKE-ALL.

WHITE KAFIR is one of the grain SORGHUMS.

WHITE LEAF-SPOT = WHITE SPOT.

WHITE LOCOWEED (*Oxytropis lambertii*), or *crazyweed*, is one of the poisonous LOCOWEEDS.—*See also* POISONOUS PLANTS.

WHITE LUPINE (*Lupinus albus*) is an annual species of the LUPINES. It is somewhat similar to the BLUE LUPINE in plant characteristics, except the leaves are larger and the flowers white. Seed of white lupine is large, wrinkled, flat, and creamy white. The plants grow best on alluvial soil in the lower Mississippi Delta. They need fertile, neutral soils for satisfactory growth.

Dangers: BROWN SPOT is one of the more common diseases of the white lupine.

Strains: *Sweet* and *bitter* strains of this species occur in this country; they can be distinguished by taste and by the fact that the sweet (nonalkaloid) plants make smaller growth.

Reference: S.7.

See also YELLOW LUPINE.

WHITE MAMMOTH

= WHITE AFRICAN SORGO.

WHITE MILO. *Dwarf White milo, Standard White milo,* and *Early White milo* are grain SORGHUMS.

WHITE MINERAL OIL, or *liquid petrolatum,* is a colorless, oily liquid obtained from PETROLEUM. A grade having a viscosity of 65 to 95 *Saybold* seconds at 100° F. is recommended for the preparation of DDT spray solutions.—*See also* NICOTINE; MINERAL OIL.

"WHITE MOLD" = SOUTHERN BLIGHT.

WHITE MUSCLE DISEASE is a vitamin-deficiency condition that affects calves. *See also* FISH LIVER OILS; VITAMIN E.

WHITE ORANGE is a term erroneously used for *Sourless sorgo,* one of the forage SORGHUMS.

WHITE PEA BEAN = NAVY BEAN.

WHITE POLISH.

See POLISH WHEAT (variety).

WHITESAGE = WINTERFAT.

WHITE-SEEDED SORGOS are those few forage SORGHUMS with seeds that are free from bitterness; e.g., *Atlas, White African,* and *Tricker.*

WHITE SMYRNA.

See TWO-ROWED BARLEY (variety).

WHITE SNAKEROOT (*Eupatorium rugosum*) belongs to the COMPOSITE family. It contains the toxic substance TREMETOL; when cattle or sheep feed on the leaves or stems for several days, they develop symptoms of poisoning, such as marked trembling and weakness, especially after exercise.—*See also* POISONOUS PLANTS.

WHITE SOONER. *Double Dwarf White Sooner* and the closely related *Miloca* are *milos* and belong to the grain SORGHUMS.

WHITE SPANISH is a Spanish-type PEANUT.

WHITE SPOT, or *white leaf-spot,* caused by the FUNGUS *Aristastoma oeconomicum,* often does severe damage to the leaves of the COWPEA.

No control is known.

Reference: W.15.

White snakeroot. Top of plant showing flower clusters, leaves with prominent veins, single flower, and fiberous roots. (D.16.)

WHITE SWEETCLOVER (*Melilotus alba*), formerly called *Bokhara clover,* is a species of the SWEETCLOVERS. It gives larger yields than the related YELLOW SWEETCLOVER and is later in maturing, thereby prolonging the grazing period of the second year.

Dangers: Among its enemies are the various diseases and insect pests which attack sweetclover in general, and also NORTHERN ANTHRACNOSE.

Varieties: There are two important varieties of this sweetclover species: (1) *Spanish sweetclover,* formerly called *Madrid White,* a valuable biennial crop and (2) *Hubam,* an annual variety; when a catch crop is wanted, especially when the field is to be fall-plowed, Hubam is very useful.—*See also* BLOAT; DICOUMAROL.

White sweetclover with flowers, stamens, pod, and taproot. (U.6.)

WHITE TARTAR.
                    See COMMON OAT (variety).

WHITE TIP, caused by the nematode *Aphelenchoides oryzae*, is characterized by the production of colorless leaf tips and occasionally other parts of the leaves, or PANICLES, of RICE plants.

The NEMATODES establish themselves inside the flowers, and after the seed is formed they remain there in a dormant condition until the seed is planted. Then the nematodes become active, feed on the developing seedling, and rapidly increase in numbers.

White tip symptoms appear when the rice plants are from two to three months old. The whitening of $\frac{1}{2}$ to $1\frac{1}{2}$ in. of the leaf tips is observed first, but this often spreads farther as the plants become older. The areas affected turn white, or greenish white, and are of a papery texture. This type of injury also produces sheath twisting, which results in imperfect emergence of the head and deformed grain. In severe cases the plants are dwarfed, and the panicle may not be formed.

Most commercial rice varieties appear to be susceptible, but most of the long-grain varieties are free from typical white tip symptoms.

Control of the seed-borne phase, without injury to germination, may be secured by fumigation with METHYL BROMIDE.
Reference: T.2.

WHITE TIP CLOVER (*Trifolium variegatum*), a productive native TRUE CLOVER species of the West, is of only minor importance as a forage crop.
Reference: M.3.

"WHITETOP" is a name sometimes used for REDTOP as well as for other plants.

WHITETOP FLEABANE (*Erigeron strigosus*) is an aster-like WEED.—*See also* MEADOW FESCUE.

WHITE WHEAT. *See* WHEATS.

WHITING = CALCIUM CARBONATE.

WHITTLE = *Taylor. See* COWPEA.

WHOLE CULL BEANS.
                    See FIELD BEANS.

WHOLE PRESSED COTTONSEED.
                    See COTTONSEED.

WHOLE PRESSED SAFFLOWER SEED —also known as *saffmeal, safflower oil feed, safflower meal,* or *safflower seed meal*—is an officially recognized feedstuff. It is the ground residue obtained after extracting the oil from whole SAFFLOWER seed. The name must include a term descriptive of the process of manufacture—hydraulic, expeller, or solvent.

The feed obtained from whole seed contains only 18 to 24 percent protein, but feed from decorticated seed approaches 40 percent protein, depending on the amount of hull removed; consequently, its fiber content will vary with the percentage of hull left in the feed. Good safflower feed is similar in composition and feeding value to COTTONSEED meal.

The high protein feed from decorticated safflower seed, when fed to growing calves and growing and fattening steers, is fully equal to SOYBEAN oil meal of approximately the same crude protein content. This feedstuff also is an economical protein source for fattening lambs and laying hens.

References: B.19; F.6.

See also OILSEED MEALS; MISCELLANEOUS PRODUCTS.

WHOLE STILLAGE.

See DISTILLERS' PRODUCTS.

WHORL is a cluster of several branches around the axis of an INFLORESCENCE.

WICHITA WHEAT.

See COMMON WHEAT (variety).

WILD ASPARAGUS = SKELETONWEED.

WILD BARLEYS are native species which often are considered detrimental WEEDS; e.g., FOXTAIL BARLEY.—*See also* ALFALFAS.

WILD BEES. *See* BEES.

WILD BUCKWHEAT (*Polygonum convolvulus*), also called *knot bindweed*, is one of the KNOTWEEDS which belong to the BUCKWHEAT family. This WEED abounds on poor soils, and occurs also in cultivated ground and on ranges.

The hairless plants are much-branched and erect, with alternate, stalkless leaves, small flowers without petals, and small, dark, three-angled fruits (ACHENES).

Reference: U.6.

See also SCREENINGS.

WILD CHERRY (*Prunus* spp.) is a term applied to the *western chokecherry*, the closely related *black chokecherry* of the western ranges, and the *common chokecherry* of the eastern United States. Wild cherry is an erect, leafy shrub or small tree. Sheep and cattle, when eating the leaves of this POISONOUS PLANT at the rate of 1 lb. per 100 lb. body weight within a few minutes, show soon afterwards characteristic effects of wild cherry poisoning, namely difficult breathing, spasms, and coma; the illness is mostly of short duration.—*See also* CYANOGENETIC PLANTS.

Wild cherry. Leaf, flowering shoot, and fruit cluster. (W.35.)

WILDFIRE is caused by *Pseudomonas tabaci*. This bacterial leaf spot is a serious disease on tobacco, but it appears also on SOYBEANS. Prominent yellow halos with well-marked margins develop around a central area of brown, dead leaf tissue. The brown area may increase to involve a large portion of the leaf. There is some reason to believe that wildfire is associated with BACTERIAL PUSTULE; a bacterial pustule can frequently be found in the center of a wildfire lesion.

Wildfire causes considerable damage in the southern states.

Among different soybean varieties, susceptibility appears to vary considerably.

No control measures for this disease are known.

Reference: C.9.

WILDGOOSE
    = TARTARIAN BUCKWHEAT.

WILD GRASSES are *native* or *naturally* introduced GRASSES; they are of varying importance as forage crops, depending on grass species available, animal species using them, and climatic and soil conditions. In general, the wild grasses of the upland prairies and mountain meadows make valuable pasturage and hay; this is particularly true of many WHEATGRASSES, GRAMA GRASSES, and BUFFALOGRASS.

Wild grasses, like all grasses, have many enemies; among them are ARMYWORMS, WHEAT STEM RUST, BROWN SPOT OF RICE, DODDERS, etc.—*See also* PRAIRIE HAY; RANGES; PASTURES; WEEDS.

WILD OATS.
See COMMON WILD OATS; HAY GRADING.

WILD ONION (*Allium canadense*) is a noxious WEED in pastures and fields of WHEAT or other small grains.

The wild onion is a perennial developing from a whitish bulb which is covered on the outside with a netlike, fibrous coating.

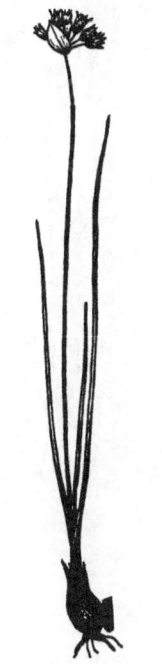

Wild onion. (D.19.)

Control: This weed is very difficult to control because it spreads from both, seed and bulbs. Intensive cultivation and turning up bulbs to kill young plants will keep infestation down. Land should be turned for two successive years, spring and fall, and clean-cultivated during the summer.
References: W.23; C.19.

WILD PASTURES
    = *native wild pastures. See* PASTURES.

WILDPEA = NARROWLEAF VETCH.

WILD RYE OAT (*Avena sterilis*) is a WEED; it is not as widely distributed as the COMMON WILD OAT.—*See also* OATS; AVENA.

WILDRICES (*Zizania* spp.) include annual and perennial species which grow from Canada to Florida. They are not true RICES but aquatic GRASSES occurring in marshes, lakes, and some tidal rivers. The plants grow in water at depths up to 6 ft. and reach a height of 3 to 11 ft. Wildrice is important as shelter and food for waterfowl, and is sometimes planted for this purpose in marshes and game preserves. The grain, expensive to gather and process, is considered a food delicacy.
Reference: H.26.
See also STEM ROT OF RICE.

WILD-RYE. *See* SALINA WILD-RYE.

WILD-RYE GRASSES (*Elymus* spp.), or *wild-ryes*, which are not to be confused with the RYEGRASSES, are well represented in the native grass species of the western states. Most wild-rye species are perennial and bunchgrasses; a few form sod.

These GRASSES have coarse and rough foliage and are relatively unpalatable, but they are most useful for revegetation because of their good seed habits, high forage production, wide adaption to a variety of soils, and relative ease of establishment.

Dangers: The wild-rye grasses are susceptible to ERGOT, a fungus disease that replaces the kernel of the seed head.

Species: The most important wild-rye grasses are BLUE WILD-RYE, RUSSIAN WILD-RYE, SALINA WILD-RYE, CANADA WILD-RYE, and GIANT WILD-RYE,
References: H.1; R.2.
See also MUTTON BLUEGRASS.
Illustration: See CANADA WILD-RYE.

WILD SUNFLOWER. *See* SUNFLOWER.

WILD TONGUE GRASS
= PEPPERGRASS.

WILD VETCH = NARROWLEAF VETCH.

WILD WHITE. *English wild white* and *New York wild white* are varieties of WHITE CLOVER.

WILD WINTER PEA = ROUGHPEA.

WILD WINTER PEA POISONING is caused by the ROUGHPEA, also called *wild winter pea*, when the animals graze during the bloom and early pod stage of this PASTURE plant. The affected animals will develop a wobbly gait, walk with their feet drawn under them, and appear to be sore on their feet. Deaths have not been observed.

There is no known treatment, but when animals are taken out of the pasture they make a spontaneous recovery in a few days. It is said that animals continuously grazed on the peas are less affected than those that have just been turned on them for a while.

Reference: S.28.

See also GRAZING.

WILLAMETTE VETCH.
See COMMON VETCH (variety).

WILLIAMSBURG ALFALFA is one of the common ALFALFAS.

WILMINGTON RUNNER is a runner-type PEANUT.

WILSON. *See* SOYBEAN (variety).

WILT. *See* BACTERIAL WILT; COWPEA WILT; FUSARIUM WILT OF ALFALFA; FUSARIUM WILT OF FIELD PEA; STEWART'S WILT; VIRUS DISEASES; ALFALFA CATERPILLAR.

WINDROW is a row of hay or of herbage that has been raked or dropped into a row for curing.

To windrow means: to cut or rake into windrows.—*See also* SWATH.

WINDROW BALING. *See* HAY.

WING is a relatively thin appendage, expansion, or part of a plant, especially (1) one of the two lateral (side) PETALS of a LEGUME blossom, or (2) a projection from certain fruits.

Winged means: provided with wings.

WINGED BUCKWHEAT (*Fagopyrum emarginatum*), or *notch-seeded buckwheat*, is very closely related to the Japanese variety of the COMMON BUCKWHEAT; however, the angles of its hulls are extended, forming wings and therefore making the seed very large.

Reference: Q.1.

See also BUCKWHEATS.

WINTER ANNUAL. *See* ANNUAL.

WINTER BARLEY. *See* BARLEYS.

WINTER BLUEGRASS
= MUTTON BLUEGRASS.

WINTER CLOVERS are frost-resistant; e.g., some varieties of WHITE CLOVER or RED CLOVER.—*See also* TRUE CLOVERS.

WINTER COVER CROPS are sown in the fall to prevent soil erosion during winter.

Winterfat with female and male flower and leaf. (U.6.)

WINTERFAT (*Eurotia lanata*), more correctly called *common winterfat* and known to many stockmen as *whitesage*, is a low, silvery-white shrub which grows 10 to 15 in. high. New stems are produced each year from the basal portion of the previous year's stems and from the woody crown of the plants. New growth averages 3 to 10 in. in height and is covered with a mat of silvery-white hair; the shrub is greenish-white in color but on maturity becomes characteristically silvery white.

Winterfat probably produces more forage on the *winter sheep range* than any other species because it is relatively abundant and is readily eaten. It is often injured or killed by heavy grazing and replaced by poor forage species. Winterfat commonly occurs in fairly dense, pure stands and covers large areas along the broad, flat drainages and lower valley slopes, where the soil is fine in texture and fairly alkaline.

Reference: H.41.

See also RANGE PLANTS.

WINTER GRAINS are often resistant to crop enemies, such as ROOT KNOT.—*See also* ITALIAN RYEGRASS; LESPEDEZAS; RED CLOVER.

"WINTERGRASS."
                    See MUTTON BLUEGRASS.
WINTER GRAZING. *See* GRAZING.
"WINTER GRAZING" = *Winter Turf*.
                    See COMMON OAT.
WINTERING RATIONS.
            See BEEF CATTLE RATIONS.
WINTER KING = *Nigger*.
                    See COMMON WHEAT.
WINTER LEGUMES. *See* ANNUAL LEGUMES; CRIMSON CLOVER; KUDZU; FIELD PEA; CROTALARIAS; VETCHES; BUR-CLOVERS.
WINTER OATS. *See* OATS; COWPEA.
WINTER PEAS.
        See PASTURES; CORN; LUPINES.
Wild winter pea = ROUGHPEA.
WINTER RANGES. *See* GRAZING; RANGE MANAGEMENT; RANGE PLANTS.
WINTER RYE. *See* RYE; GRAZING.
WINTER TURF.
                    See COMMON OAT (variety).
WINTER WHEAT. *See* WHEATS; COMMON WHEAT; TIMOTHY; OATS; ALFALFAS; HESSIAN FLY; FLAG SMUT OF WHEAT.

WINTEX.
            See SIX-ROWED BARLEY (variety).
WINTHEMIA. The gray, two-winged FLY *W. quadripustulata* is closely resembling, and slightly larger than, the housefly. It is one of the most effective insect foes of the ARMYWORM and CORN EARWORM. The fly fastens its eggs to the skin of the caterpillar, and the maggots, quickly hatching from the eggs, bore through the skin into the flesh where they soon devour the entire inside portions of the host's body. These flies multiply rapidly and often become numerous enough to control the worm pests completely in a given locality.

Reference: W.12.

WINTOK. *See* COMMON OAT (variety).
WIREGRASS, sometimes spelled *wire grass*, is a name used synonymously for at least two different genera of plants: (1) BERMUDA-GRASS and (2) SPIKE RUSH.
WIREWORMS (*Limonius* spp.) are the slender, about 1 in. long, tough, straw-colored or brown larvae of the CLICK BEETLES. Adults emerge from the ground very early in the spring. Eggs are laid in the soil, and the larvae feed on the roots of a wide variety of plants, especially POTATOES, and such cereal and forage crops as WHEAT, BUCKWHEAT, and CORN. Wireworms often require three or more years in the soil to complete their development.

Wireworm. (U.13.)

Control: The insecticides ALDRIN, DIELDRIN, and HEPTACHLOR are equally effective against the wireworms when used as a soil treatment. The fumigant ETHYLENE DIBROMIDE, in form of DOWFUME W-40, has also been recommended for the control of wireworms.

References: M.34; P.16.

WISCONSIN 416 is a CORN hybrid.
WISCONSIN BARBLESS is a winter variety of SIX-ROWED BARLEY.
WISCONSIN BLACK is a SOYBEAN variety.

WISCONSIN PEDIGREE 5, commonly called *Oderbruck*, and *Wisconsin Pedigree 38*, better known as *Wisconsin Barbless*, are SIX-ROWED BARLEY varieties.

WISCONSIN PEDIGREE NO. 2
= *Schlanstedt. See* RYE.

WITCHES'-BROOM is a VIRUS DISEASE of ALFALFA in the United States, in Canada, and in Australia. Except for a few localities where outbreaks were severe, it has been considered of minor importance.

Witches'-broom slowly changes the appearance of affected plants in several ways. Plants that show symptoms for the first time have many more stems, an erect habit of growth, and slight marginal CHLOROSIS of the younger leaves. In the advance stage of infection, the plants are severely dwarfed and bunchy because of excessive numbers of short, spindly shoots from the crown and axillary buds along the stems. Leaflets of affected plants are smaller. The plants usually have a yellowish cast. In advanced stages, the plants may develop a prostrate type of growth. Production of seed in diseased plants is almost prevented because flowers are sparse.

> NOTE: Witches'-broom can be distinguished from BACTERIAL WILT by the fact that the wood of the root of affected plants is not discolored.

The virus that causes witches'-broom in alfalfa is transmitted from diseased to healthy plants by LEAFHOPPERS.

Alfalfa shows some degree of tolerance to infection with witches'-broom virus. References: J.3; J.2.

WOLFF - LEHMANN FEEDING STANDARDS. *See* MORRISON FEEDING STANDARDS.

WONDER, sometimes called *Wonder kafir*, is not a true kafir but one of the intermediate-type grain SORGHUMS.

WONDERFUL is a "Clay" variety of COWPEA.

WONG. *See* SIX-ROWED BARLEY (variety).

WOOD ASHES contain CALCIUM, POTASSIUM, and other MINERALS and, therefore, may be used in mineral mixtures (especially for hogs) and other feeds.—*See also* MINERAL FEEDS; ASH.

WOODLAND CHAPARRAL. *See* CHAPARRALS; ALFILERIA; RANGE PLANTS.

WOODLAND PASTURES.
See PASTURES.

WOOD MOLASSES, or *wood-sugar molasses*, is a nonofficial feed MOLASSES obtained from wood wastes of the timber and lumber industries. The wood is converted to sugar by boiling it under pressure with diluted acid, neutralizing it with alkali, and evaporating it to wood molasses. Up to 200 gal. molasses may be obtained from 1 ton of dry wood waste.

Wood molasses is a satisfactory feed for feed-lot lambs and dairy cattle and may, depending on the ration, equal cane molasses when fed on an equal dry matter basis. In several tests it has not equaled cane-, beet-, or citrus molasses for beef steers. It is also consumed by swine and poultry. Wood molasses has a preservative value equal to that of cane molasses when used in alfalfa and alfalfa-broom SILAGE at a rate of 60 lb. per ton. Silage preserved with good quality wood molasses is palatable and readily consumed by cattle.— *See also* BY-PRODUCT FEEDSTUFFS.

WOODS PROLIFIC = *Leap.*
See COMMON WHEAT.

WOOD-SUGAR MOLASSES.
See WOOD MOLASSES.

WOODS YELLOW and *Early Woods Yellow* (better known as *Arksoy*) are SOYBEAN varieties.

WOODWARD = WARD.
See SIX-ROWED BARLEY.

WOOLLYFOOT = BLACK GRAMA.

WOOLLYPOD VETCH (*Vicia dasycarpa*), is one of the VETCHES and similar to smooth vetch (a variety of HAIRY VETCH); without the flowers and seed it can hardly be distinguished from it. The flowers are a little smaller than those of hairy vetch, and the seeds tend toward an oval shape. The seed scar has distinguished marks. Woollypod vetch is winter-hardy throughout the Cotton Belt and as far north as Washington, D.C.

This species makes a good winter growth in the South and is used for green manure and forage.

Reference: M.18.

See also LEGUMES; INOCULATION.

WORM is a term applied (1) to *true worms* and (2), popularly, to the wormlike larvae of many kinds of INSECTS.

The true worms, like the earthworms, are legless. Insects in the larval stages have leglike appendages.—*See also* CATER-PILLAR; GRUBS.

WORT. *See* BREWERS' PRODUCTS.

XYZ

XANTHIUM. *X. echinatum*

= COCKLEBUR.

XANTHOMONAS. The BACTERIUM *X. holcicola* is the cause of BACTERIAL STREAK; *X. phaseoli* var. *sojensis* causes BACTERIAL PUSTULE.

XEROPHTHALMIA is an eye condition. —*See also* VITAMIN A.

YARD-LONG BEAN

= ASPARAGUS BEAN. *See also* COWPEA.

YARROWS (*Achillea* spp.) are COMPOSITE perennials; some of them have limited forage value (e.g., Western yarrow), others (like the common yarrow) are considered WEEDS.

Western yarrow has silky-woolly hairs which cover the plant, giving it a somewhat grayish appearance. It is widely distributed in the western states of this country, Canada, and Mexico. The plant is drought-resistant, invades readily all areas, but avoids dense shade. On many ranges some classes of livestock graze this plant modestly throughout the season; in other areas, however, western yarrow is considered worthless or even classified as a weed, which, under certain conditions of overgrazing, has a tendency to dominate high summer ranges.

Common yarrow is a widely distributed weed found in the eastern part of the United States and in parts of the West. It is a smoother, greener, and taller plant than the very closely related western yarrow; its flowers are aromatic, and its leaves possess astringent properties.

Reference: U.6.

See also MUTTON BLUEGRASS.

Yarrow. Plant, flower, and leaf. (B.28.)

YEAST is a higher FUNGUS; the term is commonly used for the *Saccharomyces* spp. These vegetable organisms consist of single cells which reproduce vegetatively. Yeast ferments sugars to ALCOHOL and carbon dioxide by means of ENZYMES and, therefore, is used in breweries, distilleries, and bakeries.

All yeasts are vitamin-rich NUTRIENTS. They contain ERGOSTEROL, which on activation changes to VITAMIN D_2; the better grades of commercial yeast are good sources of the B-COMPLEX VITAMINS, especially of VITAMIN B_2 and PABA.

Officially recognized are the following yeast products:

Irradiated yeast or *irradiated . . . yeast* is yeast which has been subjected to ultraviolet rays in order to increase its antirachitic potency.

> NOTE: When irradiated yeast is used as an ingredient of proprietary feeds for four-footed animals, the name may be followed by this parenthetical phrase: (source of vitamin D_2).

Brewers' dried yeast—sometimes also called *dried brewers' yeast*—is the dried, nonfermentive, nonextracted yeast resulting as a by-product from the brewing of beer and ale; it shall contain not less than 40 percent crude protein on the moisture-free basis.

Brewers' dried yeast is valued primarily as a source of VITAMIN B_2, NIACIN, PANTHOTENIC ACID, and CHOLINE. It is also relatively high in protein.

In swine and poultry rations, brewers' yeast is used primarily as a vitamin-rich supplement; its high protein content is, however, another significant contribution to diets.

Although it does not furnish vitamin B_2 at as low a cost as the dried fermentation solubles—a valuable source of VITAMINS—brewers' dried yeast has some additional values as a source of choline and other B-complex vitamins for poultry.

Brewers' yeast also has been used successfully as a cattle feed.

Grain distillers' dried yeast—the properly dried yeast resulting from the fermentation of grains and yeast, separated from the mash, either before or after distillation.

Molasses distillers' dried yeast—the properly dried yeast resulting from the fermentation of molasses and yeast, separated from the medium, either before or after distillation.

Dried torula yeast is yeast of the botanical classification *Torulopsis* which has been separated from the media, in which it propagated, and dried. It shall contain not less than 40 percent crude protein on the moisture-free basis.

Active dry yeast is yeast which has been dried in such a manner as to preserve a large proportion of the fermenting power. It must contain no added cereal or filler

and not less than 15 billion live yeast cells per gram.

Yeast culture is composed of yeast in a suitable medium and capable of producing active fermentation. The medium shall be stated on the label.

> NOTE: Tentatively recognized are the following yeasts, all of which must contain not less than 40 percent crude protein on the moisture-free basis:

Dried candida yeast, i.e., yeast of the botanical classification *Candida* which has been separated from the medium, in which it propagated, and dried.

Dried saccharomyces yeast—yeast of the botanical classification *Saccharomyces*, which has been separated from the medium, in which it propagated, and dried.

Dried yeast is yeast of the botanical classifications *Saccharomyces*, *Torulopsis*, or *Candida*, which has been separated from the medium, in which it propagated, and dried. It shall contain not less than 12.2 mg. riboflavin (VITAMIN B_2) per pound. The term "dried yeast" may be used only as an ingredient name in mixed feeds. However, when sold as such, it must be specified as "brewers' dried yeast," "dried torula yeast," and so forth.

References: F.6; E.12.

See also FERMENTATION INDUSTRY BY-PRODUCTS; YEAST DRIED GRAINS OR VINEGAR DRIED GRAINS; SILAGE; INVERTASE; VITAMIN D.

YEAST CULTURE. *See* YEAST.

YEAST DRIED GRAINS OR VINEGAR DRIED GRAINS is the properly dried residue from the mixture of cereals, malt, and malt sprouts (or sometimes COTTONSEED MEAL), obtained in the manufacture of YEAST or VINEGAR. It consists of CORN, or corn and RYE, from which most of the starch has been extracted, plus malt added during the manufacturing process to change the starch to sugar, and malt sprouts (sometimes cottonseed meal), also added during the manufacturing process to aid in filtering the residue from the wort and to serve as a source of food supply for the yeast.

This is one of the officially recognized FERMENTATION INDUSTRY BY-PRODUCTS.

Reference: F.6.

YELLOW BEARDGRASS. *Texas yellow beardgrass*, better known as *King Ranch bluestem*, is a variety of TURKESTAN BLUESTEM.

YELLOW BERRY is a term used to indicate thick, humped, and very starchy, POULARD WHEAT kernels.

YELLOW BLUESTEM
= TURKESTAN BLUESTEM.

YELLOW CLOVER APHID (*Therioaphis ononidis*) has been found on CLOVERS, mostly RED CLOVERS, and on ALFALFA.

This insect is rather small and is pale yellow in color, with four rows of black spots on its back. The wings have conspicuously darkened veins. Like most other APHIDS, it has both winged and wingless adult females which produce young without mating. There are as many as 17 generations each summer.

Yellow clover aphids are found chiefly on the lower surfaces of the affected leaves, although individuals may be located on the upper leaf surfaces, in the buds, or even on the stems.

Control: PARATHION sprays are recommended.

Reference: D.8.

YELLOW CLOVERS = HOP CLOVERS.

YELLOW DWARF = RED LEAF.

YELLOW FAT DISEASE affects mink; it is due to VITAMIN E deficiency.

YELLOW-FLOWERED ALFALFAS include several *Medicago* spp., particularly *M. falcata*. All of them are characterized by yellow flowers and resistance to cold and drought, but they have little economic importance.—*See also* ALFALFAS.

YELLOW INDIANGRASS
= INDIAN GRASS.

YELLOWING. *See* YELLOWS; CHLOROSIS; POTATO LEAFHOPPER.

YELLOW LEAF BLOTCH, caused by the FUNGUS *Pseudopeziza jonesii*, occurs on ALFALFA. This disease is characterized by yellowish-orange blotches on the leaves, with the long diameters of the blotches parallel to the direction of the veins. At the time of the first appearance of the blotches, or shortly thereafter, small orange-colored points appear in the central portion of the blotches, primarily on the upper surface of the leaves; these points eventually turn dark brown to black. They are the spore-bearing bodies of the fungus. During the later stages of the disease, heavily infected leaves curl at the margins and eventually die.

Yellow leaf blotch occurs also on the stems. Stem lesions soon turn a dark chocolate-brown. These lesions are generally not numerous enough to cause death to the stems.

Control: There is no satisfactory method of controlling this disease. Losses may be reduced by mowing the crop before the disease becomes severe and causes defoliation.

Reference: G.7.

YELLOW LUPINE (*Lupinus luteus*), an annual species of the LUPINES, is a small plant 1 to 2 ft. tall with a number of stems on a short central stalk. This species does well on light sandy, moderately acid soils of low fertility. The seed is flat and usually speckled white and black; however, pure white and pure black seed strains exist in this species. Yellow lupine, because of some hard seed, may reseed slightly better than WHITE LUPINE or BLUE LUPINE, but is usually not to be depended on.

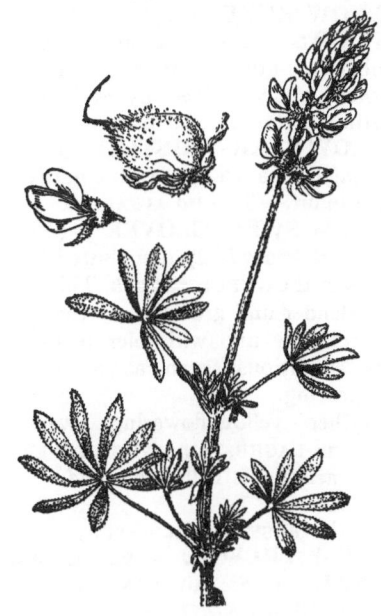

Yellow lupine. (M.3.)

Dangers: BROWN SPOT is a disease which often attacks yellow lupines.

Strains: There exist *sweet* as well as *bitter* strains of this species; they differ only in taste.

Reference: S.7.

YELLOW MANITOBA.

See PROSO (variety).

YELLOW MILO. *Dwarf Yellow milo, Standard Yellow milo,* and *Double Dwarf Yellow Sooner* are grain SORGHUMS.

YELLOW MOSAIC of SOYBEAN is caused by the *Phaseolus virus 2.* The leaves of infected soybean plants do not show the extreme distortion characteristic of other MOSAICS. The younger leaves show a yellow mottling. The mottling may be random spots over the leaf blade, an indefinite, yellow band along the major veins, or isolated yellow spots between the veins. Rusty spots of dead tissue appear later on the yellow portions of the leaf as the plants approach maturity. The plants are not usually noticeably stunted.

There is no evidence that the disease is seed-borne.

Reference: C.9.

YELLOW PEARL

= *Supergold.* See CORN.

YELLOW RUST = STRIPE RUST.

YELLOWS is a term used to indicate a number of FUNGUS DISEASES or VIRUS DISEASES characterized by *yellowing* and stunting of the affected plants.

YELLOW STAR-THISTLE (*Centaurea solstitialis*) is a WEED. It is a hardy composite plant.—*See also* ALFALFAS.

YELLOW SWEETCLOVER (*Melilotus officinalis*) is a widely distributed biennial species of the SWEETCLOVERS. This plant is more slender and grows lower than WHITE SWEETCLOVER and will tolerate more adverse conditions, such as drought and close grazing.

Another yellow-flowering sweetclover species is DAGHESTAN SWEETCLOVER; one of its varieties is called *Redfield.*

Reference: P.7.

See also RUSSIAN WILD-RYE.

YELLOW TREFOIL = BLACK TREFOIL.

YELNANDO. See SOYBEAN (variety).

YERBA-DE-PASMO. See BACCHARIS.

YIELDS. See HAY YIELDS.

YIELDS OF CROPS. See CROPS.

YOGO is a COMMON WHEAT variety.

YOKOHAMA VELVETBEAN (*Mucuna hassjoo*) produces a smaller vine growth than any other of the VELVETBEANS and is not a heavy yielder. It is an early type, maturing within 110 to 120 days. The purple flowers are borne in short RACEMES. The pods are 4 to 6 in. long, flat, quite pointed at each end, and covered with rather long, gray pubescence. The seeds are ash-colored, oblong, compressed, and about ⅔ in. long. Many of the pods form so close to the ground that they become water-soaked with each heavy rain and shatter in hot, dry weather. This plant is grown very little at the present time.

Reference: P.10.

See also LEGUMES; INOCULATION.

YOLREDO. See SOYBEAN (variety).

YORKWIN.

See COMMON WHEAT (variety).

YUCCAS (*Yucca* spp.) are southern desert shrubs with cup-shaped flowers and stiff, sword-shaped leaves. Finely chopped or shredded, these shrubs have supplied feed for cattle in emergencies.—*See also* RANGE PLANTS.

ZEA. *Z. mays* = CORN.

ZEELAND. See RYE (variety).

ZENITH. See RICE (variety).

ZIGZAG CLOVER (*Trifolium medium*), one of the TRUE CLOVERS, is of very little importance as a forage plant.

ZINC is a metal which occurs in nature in many ores. It is one of the TRACE ELEMENTS needed for normal GROWTH and hair development; zinc deficiency develops in swine that are fed rations containing too much calcium. The condition, called *parakeratosis,* may resemble mange; it causes severe skin disorders and stunts growth. The pigs seldom die and can be cured by adding zinc salts to their diet or by reducing their calcium intake.

ZINC CARBONATE (the commercial variety contains as much as 56 percent ZINC) is a white, water-insoluble powder. —*See also* MINERAL FEEDS.

ZINC SULFATE contains 22.7 percent ZINC (*dried zinc sulfate* contains about 50 percent more zinc). It forms a white powder or granules and is water-soluble. —*See also* MINERAL FEEDS.

ZINGIBER. *Zingiber* spp. are tropical perennial herbs. The dried rhizome of *Z. officianalle* is known as GINGER.

ZIZANIA. *Zizania* spp. = WILDRICES.

ZONAVE LEAF SPOT, caused by *Gleocercospora sorghi*, a FUNGUS, attacks forage and grain SORGHUMS, JOHNSONGRASS, and also SUGARCANE, CORN, and PEARLMILLET. It has been observed in Maryland, Virginia, North Carolina, South Carolina, Georgia, Florida, Mississippi, and Louisiana. This LEAF DISEASE is very conspicuous on sorghum leaves as reddish-purple bands alternate with tan or straw-colored areas, forming a zonate pattern. The spots often occur along the margins of the leaves, forming semicircular patterns, or they occur on other parts of the leaf where they are more nearly circular. These irregular spots vary greatly in size. At first they may be only a fraction of an inch in diameter, but, as they become older, they may reach several inches in length; when numerous they often unite to cover most of the leaf surface. When plants are so heavily infected that the leaves are killed prematurely, the forage value of the crop is reduced.

Control: The fungus has been found on the glumes and seed, which suggests that planting of disease-free or adequately treated seed would help to prevent the spread of zonate leaf spot. No highly resistant varieties are known.

Reference: L.1.

See also FUNGUS LEAF DISEASE; SEED TREATMENT.

ZOYSIA. *Z. japonica*, commonly known as *Korean lawngrass* or *Japanese lawngrass*, is tough, harsh, and unpalatable.

ZUMI = TURKEY. *See* COMMON WHEAT.

ZYGADEMUS. *Zyagademus* spp.
= DEATHCAMASSES.

AUTHORS OF PUBLICATIONS
Referred to in the Text*

The experts, whose publications were used as sources for this volume, are listed with the positions they held at the time their investigations were published. Following the dash are the keys to the publications that appear as References through out this volume: they are explained in the Bibliography beginning on page 581.

Åberg, E., Res. Asst. (U.Wisc. and U.S.D.A.) —A.6.

Ahlgren, G. H., Res. Spec. (Rutgers U.— N.J.AESt.)—A.1; A.3.

Ahlgren, H. L., Prof. Agr. (U.Wisc. and U.S.D.A.)—A.7.

Alexander, M. A., Prof. An. Husb. (Nebr. AESt.)—B.19.

Allison, J. L., Path. and Res. Prof. (N.C. StCol.)—W.15.

Anderson, K. L., Agr. (Kans.StCol.—AESt.) —F.9; S.22.

Anderson, L. D., Assoc. Ent. (U.Calif.)— —A.4.

Anderson, N. L., Asst. Ent. (Mont.StCol.— —AESt.)—A.2.

Arant, F. S., Ent. (Ala.Pol.Inst.—AESt.)— W.17.

Arnold, P. T. D., Assoc. Dair. Tech. (U.Fla. —AESt.)—M.5.

Arthaud, V. H., Asst. An. Husb. (U.Nebr.— AESt.)—B.25.

Atkins, I. M., Agr. (Tex.AESt. and U.S.D.A.) —A.5.

Bailey, R. Y., Chief, Agr. Div., Southeastern Reg. Soil Conserv. S.—B.3; B.5.

Baker, G. N., An. Husb. (U.Nebr.—AESt.) —H.30; B.19.

Baker, M. L., An. Husb. (U.Nebr.—AESt.) —H.30; B.19; B.25.

Baker, W. A., Sen. Ent. (B.E.P.Q.)—G.1.

Barnes, G. P., Chief Anal. (U.S.D.A.)—W.28.

Barnes, O. K., Asst. Agr. (U.Wyo.—AESt.) —B.24.

Barnes, O. L., Ent. (B.E.P.Q.)—B.8.

Barrentine, B. F., Prof. An. Husb. (Miss. StCol.)—B.17; H.37.

Bates, R. P., Asst. Agr. (Tex.AESt.)—W.4.

Battle, W. R., Asst. Res. Spec. (N.J.AESt.) —B.6.

Bayles, B. B., Princ. Agr. (U.S.D.A.)—B.15; B.16.

Bear, F. E., Res. Spec. (Rutgers U.—N.J. AESt.)—S.6.

Bearse, G. E., Poultry Sci. (StCol.Wash.— AESt.)—B.26.

Beattie, J. H., Sen. Hort. (U.S.D.A.)—B.10.

Beetle, A. A., Assoc. Agr. (U.Wyo.—AESt.) —B.1; B.24.

Bennion, N. L., Poultry Spec. (Oreg.StCol.— AExtS.)—B.22.

Benton, C., Ent. (B.E.P.Q.)—P.5.

Berg, L. R., Asst. Poultry Sci. (StCol.Wash. —AESt.)—B.26.

Bird, H. R., In Charge, Poultry Husb. (B.A.I.)—E.12; L.11.

Black, W. H., Sen. An. Husb. (B.A.I.)—W.25.

Blanchard, R. A., Ent. (B.E.P.Q.)—B.9.

Blank, L. M., In Charge, Cotton Disease Inv. (U.S.D.A.)—B.27.

Blinkensderfer, C. B., Nursery Mang. (U.S.D.A.)—K.1.

Bohmont, D. W., Asst. Agr. (U.Wyo.)— B.20; B.28; B.29.

Bolin, D. W., Asst. An. Nutr. (N.Dak.ACol. —AESt.)—R.10; D.13.

Booth, W. E., Bot. (Mont. StCol.—AESt.)— M.47.

Briggs, R. A., Ext. Assoc. (Rutgers U.— AExtS.)—B.7.

Brooks, J. S., Agr. (Okla.A&MCol.—AESt.) —B.12.

Brooks, O. L., Asst. Agr. (U.Ga.—AESt.)— B.23.

Brown, A. L., Collab. (U.Ariz.—AESt.)— H.47.

Brown, B. A., Agr. (U.Conn.—Storrs AESt.) —B.4.

Brown, D. C., Asst. Dair. Husb. (Wyo.AESt.) —B.11.

Brown, E. M., Prof., Field Crops (U.Mo. and U.S.D.A.)—B.21.

Brown, R. G., Vet. (U.Ky.—AExtS.)—H.48.

Brown, R. L., Asst. Nursery Spec. (U.S.D.A.) —H.22.

Brunson, A. M., Sen. Agr. (U.S.D.A.)—B.14.

Buchanan, M. L., Chief, Div. An. Ind. (N.Dak.ACol.—AESt.)—D.13; R.10.

Burkitt, W. H., Collab. (Mont.StCol.— AESt.)—B.18.

Burleson, C. A., Sup., Subst. No. 22 (Tex. AESt.)—W.4.

Burton, G. W., Sen. Genet. (U.S.D.A.)—B.2; H.24.

---

* For Key to Abbreviations, see page 591.

573

Campbell, L. E., Asst. Agr. Eng. (B.D.I.)—S.25.

Campbell, R. S., For. Ecol. (U.S.D.A.)—C.2.

Cardon, P. V., Asst. Chief (B.P.I.)—C.15.

Carther, J. L., Sen. Agr. (U.S.D.A.)—M.19; M.20; M.21.

Carver, J. S., Poultry Sci. (StCol.Wash.—AESt.)—B.26.

Case, A. A., Prof. Vet. Med. & Surgery (U.Mo.)—C.5.

Cave, H. W., Prof. Dair. Prod. (Okla.A&M Col.)—R.5.

Chaffin, W., Ext. Agr. (Okla.A&MCol.—AExtS.)—C.6; C.14.

Chamberlain, D. W., Pl. Path. (U.S.D.A.)—C.9; J.4.

Chamberlin, V. D., Collab. (Ohio AESt.)—K.10.

Chapline, W. R., Chief, Div. Range Res. (U.S.D.A.)—C.15; C.16.

Chapman, W. H., Asst. An. Husb. (U.Fla.—AESt.)—M.30.

Chase, A., Sen. Bot. (U.S.D.A.)—H.26.

Christenson, L. D., Ent. (B.E.P.Q.)—C.10.

Claassen, C. E., Asst. Agr. (Nebr.AESt.)—C.13.

Clark, A. J., Sen. Agr. (U.S.D.A.)—B.15; C.11.

Clemson Ext. Weed Committee, Clemson ACol., Clemson, S.C.—C.19.

Cline, R. A., Ext. Agr. (S.Dak.StCol.—AExtS.)—N.2.

Coffman, F. A., Sen. Agr. (U.S.D.A.)—S.16; S.17; S.18; S.19; S.20; S.21.

Cole, J. S., Sen. Agr. (B.P.I.)—M. 2

Conard, E. C., Assoc. Agr. (U.Nebr.—AESt.)—B.25.

Connin, R. V., Ent. (U.S.D.A.)—K.5.

Conrad, H. R., Inst. Dair. Sci. (Ohio St.U.—AESt.)—H.46.

Converse, H. T., Vitamin Spec. (B.D.I.)—M.15.

Cook, E. D., Agr. (Tex.A&MCol.—AESt.)—H.3; C.3; W.4.

Costello, D. F., Sen. For. Ecol. (U.S.D.A.)—C.17; S.30.

Couch, J. F., Sen. Chem. (U.S.D.A.)—H.43.

Cowsert, W. C., Assoc. Prof. Dair. Prod. (Miss.StCol.)—R.12.

Crafts, E. C., For. Ecol. (U.S.D.A.)—C.1; C.2.

Crane, P. L., Collab. (Purdue U.—AESt.)—M.32.

Craven, W. H., Ext. Agr. (Clemson ACol.—AExtS.)—W.2; W.23; W.27.

Creel, C. W., Ent. (B.E.P.Q.)—C.7.

Crider, F. J., In Charge, National Observational Nursery Project (U.S.D.A.)—C.8.

Crowder, L. V., Assoc. Agr. (U.Ga.—AESt.)—C.4.

Crowell, H. H., Ent. (Oreg.StCol.—AESt.)—M.34.

Cushman, C. G., Leader, Dair. Ext. Work (Clemson ACol.—AExtS.)—C.18.

Dale, T., Inf. Spec., Soil Conserv. S. (U.S.D.A.)—D.15; D.12.

Darnell, J. A., Collab. (Miss.StCol.—AESt.)—B.17.

Davies, F. F., Assoc. Agr. (Okla.AESt.)—S.8; D.4; D.5.

Davis, J. J., Asst. Ent. (U.S.D.A.)—D.10.

Davis, L. L., Ext. Spec. in Agr. (U.Calif.)—J.5; L.7; S.23.

Davis, R. E., Nutr. Spec. (B.A.I.)—L.11.

Davis, W. C., Sup., Rice Pasture ESt. No. 4 (Tex.AESt.)—J.6.

Davison, V. E., Chief, Biol. Div. Southeastern Reg. Soil Conserv. S. (U.S.D.A.)—D.7.

Dawson, J. R., Sen. Dair. Husb. (B.D.I.)—G.15.

Dayton, W. A., In Charge, Dendrology and Range Forage (U.S.D.A.)—H.1; D.1; D.9; D.16; D.17.

Deal, A. S., Asst. Farm Adv. (U.Calif.)—D.8.

Dearstyne, R. S., Collab. (U.N.C.—AESt.)—D.14.

Decker, L., Res. Tech. in Poultry Nutr. (U.Conn.—AESt.)—M.46.

Denman, C. E., Asst. in Grass Res. (Okla. A&MCol.—AESt.)—H.4; D.3.

Derscheid, L. A., Assoc. Agr. (S.Dak.StCol.—AESt.)—D.18.

Dickason, E. A., Asst. Ent. (Oreg.StCol.—AExtS.)—D.20.

Dickson, J. G., Prof. Pl. Path. (U.Wisc. and U.S.D.A.)—D.11.

Dickson, R. C., Assoc. Ent. (U.Calif.)—D.8.

Dinusson, W. E., Assoc. Prof. An. Husb. (N.Dak.ACol.)—R.10; D.13.

Dockins, J. O., Asst. Dir., Ark. Rice Branch ESt.—J.6.

Dotzenko, A. D., Asst. Agr. (N.Mex.A&M Col.—AESt.)—D.2; A.1.

Douglas, W. A., Ent. (B.E.P.Q.)—B.9.

Dreesen, J., Instr. Agr. (Okla.A&MCol.—AESt.)—E.15.

Dudley, D. I., Sup., Subst. No. 6. (Tex. A&MCol.—AESt.)—D.6.

Duncan, E. R., Asst. Prof. Soils (U.Minn.)—T.1.

Durrell, L. W., Prof. Bot. and Pl. Path. (Colo.A&MCol.)—D.19.

Earhart, R. W., Pl. Path. (U.Fla.—AESt.)—M.30.

Edwards, F. R., Collab. (Ga. AESt.)—E.10.

Elder, W. C., Asst. Agr. (Okla.A&MCol.—AESt.)—E.1; E.2; E.7; E.15; H.4; H.24.

Elliott, C., Path. (U.S.D.A.)—E.11.

Hodgson, R. E., Asst. Chief (B.D.I.)—S.24; E.13.

Hogan, A. G., Prof. A. Chem. (U.Mo.)— H.20; L.13.

Hogg, P. G., Agr. (Miss.StCol.—AESt.)— H.37.

Hollowell, E. A., Sen. Agr. (U.S.D.A.)—H.8; H.9; H.10; H.11; H.12; H.13; H.14; H.15; H.16; H.17; H.18; P.3; W.5.

Holton, C. S., Path. (U.S.D.A.)—H.28.

Hoover, M. M., Asst. Chief, Soil Conserv. S. (U.S.D.A.)—H.1.

Hormay, A. L., Assoc. For. Ecol. (U.S.D.A.) —H.45.

Hosterman, W. H., Marketing Spec. (U.S. D.A.)—W.19; H.33; H.34.

Huber, M. G., Ext. A. Eng. (Oreg. StCol.)— H.32.

Huffman, W. T., Vet. in Charge, Inv. of Poisoning by Plants (B.A.I.)—H.43.

Hull, A. C., Jr., Range Ecol. (U.S.D.A.)— P.18; P.22.

Humphrey, R. R., Collab. (U.Ariz.—AExtS.) —H.47.

Hurtt, L. C., In Charge, Range Res. for the Northern Rocky Mountain For. and Range ESt. (U.S.D.A.)—H.39.

Hutchings, S. S., Range Conserv. (U.S.D.A.) —H.35; H.41.

Hyatt, M. T., Asst. Bot. (U.Ky.—AExtS.) —H.48.

Hyland, H. L., Asst. Agr. (U.S.D.A.)—M.8.

Hyslop, G. R., Head, Pl. Ind. Div. (Oreg. StCol.—AExtS.)—H.29.

Ingebretsen, K. H., Farm Adv. (U.Calif.)— L.7.

Ingham, I. M., Ext. Agr. (Wash.StCol.— AExtS.)—L.12.

Jackman, E. R., Ext. Spec. (Oreg.StCol.)— H.32; B.22.

Jackson, W., An. Husb. (U.S.D.A.)—K.1.

Jensen, R., Prof. Vet. Path. (Col.A&MCol.) —D.19.

Johnson, H. W., Sen. Path. (U.S.D.A.)— —M.14; J.4.

Johnson, M. M., Home Econ. (N.Mex.A&M Col.—AESt.)—L.8.

Johnston, C. O., Path. (U.S.D.A.)—R.9.

Jones, D. W., Asst. Soil Tech. (Fla.AESt.)— H.7; H.38.

Jones, F. R., Assoc. in Res., Sen. Path. (U.Wisc. and U.S.D.A.)—J.2; J.3.

Jones, J. W., Princ. Agr. (U.S.D.A.)—J.5; J.6.

Jones, L. G., Assoc. Spec. Agr. (U.Calif.)— A.4.

Jones, M. P., Ext. Ent. (U.S.D.A.)—J.7.

Jordan, R. M., Asst. Husb. (S.Dak.StCol.— AESt.)—J.1; M.41.

Karper, R. E., Agr. in Charge, Sorghum Inv. (Tex.AESt.)—K.3; K.4.

Kendall, K. A., Assoc. Prof. Dair. Prod. Res. (U.Ill.)—N.3.

Kennard, D. C., Collab. (Ohio AESt.)—K.10.

Kernkamp, M. F., Assoc. Prof. Pl. Path. (U.Minn.)—T.1.

Kiesselbach, T. A., Agr. (Nebr.AESt.)—G.5; C.13.

Killinger, G. B., Agr. (Fla.AESt.)—K.1; W.13.

Kime, P. H., Agr. (U.N.C.—AExtS.)—D.14.

Kinman, M. L., Agr. (Tex.AESt. and U.S. D.A.)—K.7.

Kirk, D. E., Asst. Agr. Eng. (Oreg.StCol.— —AESt.)—K.9.

Kirk, W. G., Vice Dir. in Charge, Range Cattle St. (Fla.AESt.)—H.7; H.38.

Klinger, B., Assoc. Prof. Bot. and Pl. Path. (Colo.A&MCol.)—D.19.

Klosterman, E. W., Assoc. Prof. An. Sci. (Ohio StU.)—K.8.

Koehler, B., Chief, Crop Path. (Ill.AESt.)— C.9.

Kopland, D. V., Asst. Dair. Husb. (B.D.I.)— G.15.

Kozeff, A., Asst. Instr. Poultry Nutr. (U.Conn.)—M.46.

Kramer, N. W., Assoc. Agr. (Tex.AESt.)— K.2; K.3.

Kreitlow, K. W., Sen. Path. (U.S.D.A.)— H.21; K.11.

Kreizinger, E. J., Asst. Agr. (U.S.D.A.)— L.12.

Kuhlman, A. H., Collab. (Okla.A&MCol.— AESt.)—R.5; R.11.

Kuitert, L. C., Assoc. Ent. (U.Fla.—AESt.) —K.5.

Kulash, W. M., Assoc. Ent. (N.C.AESt.)— K.6.

Lang, R. L., Asst. Agr. (U.Wyo.—AESt.)— B.24.

Lange, W. H., Jr., Assoc. Prof. Ent. (U. Calif.)—L.7.

Lantz, E. M., Home Econ. (N.Mex.A&MCol. —AESt.)—L.8.

Larrimer, W. H., Ent. (B.E.P.Q.)—L.3.

Lasley, J. F., Prof. An. Husb. (U.Mo.)—L.13.

Laude, H. H., Agr. (Kans.StCol.)—S.3; S.4.

Law, A. G., Assoc. Prof. Farm Crops (StCol. Wash.)—L.12.

Lawritson, M. N., Assoc. Ext. Dair. (U.Nebr. —AExtS.)—H.25.

Lefebore, C. L., Path. (U.S.D.A.)—L.1.

Leighty, C. E., Agr. (U.S.D.A.)—M.33.

Leonard, W. H., Agr. (Colo.A&MCol.— AESt.)—L.5.

Leukel, R. W., Path. (U.S.D.A.)—L.1; L.2; L.10.

Lewis, J. K., Asst. Prof. An. Husb. (S.Dak. StCol.)—B.18.

Light, M. R., Asst. Prof. An. Husb. (N.Dak. A.Col.)—R.10.

Ligon, L. L., Assoc. Agr. (Okla.AESt.)—L.6.

Lindahl, I. L., Nutr. Spec. (B.A.I.)—L.11.

Love, R. M., Prof. Agr. (U.Calif.)—L.4.

Luginbill, P., Sr., Sen. Ent. (B.E.P.Q.)— P.5; L.9.

Lundquist, N. S., Assoc. Prof. Dair. Husb. (Purdue U.)—H.36.

Lusk, J. W., Instr. Dair. Prod. (Miss.StCol. —AESt.)—R.12.

Machlin, L. J., Nutr. Spec. (B.A.I.)—L.11.

Manis, H. C., Ent. (Idaho AESt.)—M.27.

Marshall, S. P., Assoc. Dair. Husb. (U.Fla.— AESt.)—M.5.

Martin, E., Collab. (U.Ark.)—M.29.

Martin, J. H., Sen. Agr. (U.S.D.A.)—L.1; M.1; M.2; M.31; M.33; M.36; M.39; V.2.

Martin, S. C., Range Conserv. (U.S.D.A.)— —P.11.

Massey, Z. A., Collab. (Ga.AESt.)—E.10.

Matlock, R. S., Agr. (U.S.D.A.)—M.22.

Matterson, L. D., Asst. Prof. Poultry Husb. (U.Conn.)—M.46.

Matthews, J. W., Asst. Ext. Poultryman (Clemson ACol.—AExtS.)—G.16.

Maynard, L. A., Prof. Nutr. (Cornell U.)— M.38.

McCalmont, J. R., A. Eng. (U.S.D.A.)— S.24; M.40.

McComas, E. W., An. Husb. (B.A.I.)—C.15.

McGinnis, J., Poultry Sci. (StCol.Wash.— AESt.)—B.26.

McKee, R., Sen. Agr. (U.S.D.A.)—M.3; M.6; M.7; M.8; M.9; M.10; M.11; M.12; M.13; M.14; M.17; M.18; M.24.

McNew, G. L., Mang. Dir., Boyce Thompson Inst. for Pl. Res.—M.37.

McSpadden, B. J., Poultry Husb. (U.Tenn. —AESt.)—P.21.

Meiners, J. P., Asst. Prof. Pl. Path. (StCol. Wash.)—M.26.

Milby, T. T., Collab. (Okla.A&MCol.— AESt.)—M.44.

Miles, J. T., Assoc. Prof. Dair. Prod. (Miss. StCol.)—R.12.

Miles, S. R., Collab. (Purdue U.—AESt.)— M.32.

Miller, E. A., Agr. (Tex.A&MCol.—AExtS.) —M.42.

Miller, M. R., Chem. (U.Nev.—AESt.)— F.10.

Milligan, J. L., Nutr. Spec. (B.A.I.)—L.11.

Mills, H. B., Ent. (Mont.AESt.)—M.35.

Mimms, O. L., A. Econ. (U.S.D.A.)—M.45.

Moncrief, J. B., Jun. Agr. (Tex.AESt. and U.S.D.A.)—M.28.

Moore, L. A., Head, Div. Nutr. (B.D.I.)— M.15.

Moore, W. J., Poultry Husb. (Tex.A&MCol. —AExtS.)—M.43.

Morey, D. D., Collab. (Fla.AESt.)—M.30.

Morris, H. E., Prof. Bot. (Mont.StCol.)— W.34; M.47.

Morrison, H. E., Assoc. Ent. (Oreg.St.Col.— AESt.)—M.34.

Morse, R. W., Ext. Dair. (Oreg.StCol.— AExtS.)—H.32; E.14.

Morse, W. J., Princ. Agr. (U.S.D.A.)—P.10; M.19; M.20; M.21; M.23; M.25.

Moseley, T. W., Assoc. Dair. Husb. (B.D.I.) —G.15.

Moxon, A. L., Chem. (Ohio AESt.)—M.41; K.8.

Mullen, L. A., Spec. Asst. (U.S.D.A.)—H.22.

Nerney, N. J., Biol. Aid (B.E.P.Q.)—B.8.

Nevens, W. B., Prof. Dair. Cattle Feeding Res. (U.Ill.)—N.3.

Newell, L. C., Asst. Agr. (U.Nebr.—AESt.) —B.25.

Newlander, J. A., Prof. Dair. Husb. (U.Vt.) —N.4.

Newman, J. E., Asst. Agr. (Purdue U.— Ind.AESt.)—M.32.

Noland, P. R., Asst. Prof. An. Nutr. (U.Ark.) —N.5.

Norgaard, U. J., Ext. Agr. (S.Dak.StCol.— AExtS.)—N.2.

Norris, J. J., Assoc. An. Husb. (N.Mex. A&MCol.)—N.1; N.6.

Packard, C. M., Princ. Ent. (U.S.D.A.)— P.5; P.16; W.12.

Parker, E. M., Asst. Agr. (U.Ga.—AESt.)— C.4.

Parker, J. E., Poultry Husb. (U.Tenn.— AESt.)—P.21.

Parker, J. R., In Charge, Res. on Grasshoppers (B.E.P.Q.)—P.6.

Parker, K. W., Range Conserv. (U.S.D.A.) —P.11; P.22.

Pass, H., Asst. Agr. (Okla.A&MCol.— AESt.)—B.12.

Payne, G. F., Collab. (Mont.StCol.—AESt.) —M.47.

Payne, L. F., In Charge, Dep. Poultry Husb. (Kans.StCol.)—P.20.

Pechanec, J. F., Assoc. For. Ecol. (U.S.D.A.) —P.18; P.19.

Peet, H. S., Collab. (U.Mo.—AESt.)—P.17.

Pepper, B. B., Assoc. Ent. (Rutgers U.— N.J.AESt.)—P.14.

Peterson, A. G., Collab. (Minn.AESt.)—T.1.

Pfeifer, R. P., Asst. Agr. (U.Wyo.—AESt.) —P.12.

Phillips, W. J., Ent. (U.S.D.A.)—P.15.

Phillips, W. M., Assoc. Agr. (U.S.D.A.)—P.23.

Pieters, A. J., Princ. Agr. (B.P.I.)—P.1; P.2; P.3; P.4; P.7.

Piper, C. V., Agr. (U.S.D.A.)—P.10.

Plummer, A. P., In Charge, For. and Range ESt. (U.S.D.A.)—P.18.

Poos, W. F., Ent. (B.E.P.Q.)—B.10; P.8; P.9; P.15.

Pope, L. S., Asst. Prof. An. Husb. (Okla. A&MCol.)—G.18.

Porter, L. B., Collab. (N.Mex.A&MCol.—AESt.)—N.6.

Portman, R. W., Ext. End. (Idaho ExtS.)—M.27.

Pounden, W. D., Prof. Vet. Sci. (Ohio StU.)—H.46.

Price, F. E., Dean and Dir. (Oreg.StCol.—AESt.)—K.9.

Price, R., Dir., Southwestern For. & Range ESt. (U.S.D.A.)—P.22.

Quinby, J. R., Sup., Subst. No. 8 (Tex. AESt.)—K.3; K.4.

Quinton, R. J., Collab. (U.Conn.—AESt.)—Q.2.

Quisenberry, K. S., Agr. (U.S.D.A.)—Q.1.

Ragsdale, A. C., Prof. Dair. Husb. (U.Mo.)—P.17.

Rampton, H. H., Assoc. Agr. (Oreg.StCol. and U.S.D.A.)—R.3.

Rea, H. E., Asst. Prof. Agr. (Tex.A&MCol.)—H.42.

Read, J., Dairy Herdsman (Wyo.AESt.)—B.11.

Reitz, L. P., Sen. Agr. (U.S.D.A.)—R.7; R.9.

Reynolds, H. G., Range Conserv. (U.S.D.A.)—R.1.

Reynolds, H. T., Asst. Ent. (U.Calif.)—A.4; D.8.

Richard, R. M., Asst. Prof. An. Husb. (N.Dak.ACol.)—R.10; D.13.

Richardson, D., An. Husb. (Kans.StCol.—AESt.)—F.9.

Richardson, W. L., Res. Asst. (Okla.A&M Col.—AESt.)—B.12.

Richey, F. D., Princ. Agr. (U.S.D.A.)—R.8.

Riggs, F. E., Asst. A. Econ. (Kans.StCol.—AESt.)—F.9.

Ritchey, G. E., Agr. (Fla.AESt.)—K.1; M.12; M.14.

Roark, D. B., Prof. Dair. Prod. (Miss.StCol.)—R.12.

Robertson, J. H., Chairman, Dep. Agr. (U.Nev. and U.S.D.A.)—P.18.

Robinson, R. R., Prof. Agr. (Pa.StCol.)—S.29.

Rockwood, L. P., Ent. (B.E.P.Q.)—C.7.

Rogler, G. A., Agr. (U.S.D.A.)—W.1; S.1; R.2.

Ronning, M., Asst. Prof. Dair. Prod. (Okla. A&MCol.)—R.5; R.11.

Rumery, M. G., Asst. Dair. Husb. (U.Nebr.—AESt.)—H.30.

Rusoff, L. L., Asst. An. Nutr. (U.Fla.—AESt.)—R.4.

Russell, E. E., Ent. (B.E.P.Q.)—R.6.

Salmon, S. C., Agr. (U.S.D.A.)—M.36.

Sanchez, A. B., Collab. (U.Fla.—AESt.)—M.5.

Sanderson, E. E., Ext. Agr. (S.Dak.StCol.—AExtS.)—N.2.

Satterthwait, A. F., Assoc. Ent. (U.S.D.A.)—S.31.

Savage, D. A., Sen. Agr. (U.S.D.A.)—S.30.

Scales, J. W., Head, Vet. Sci. Dep. (Miss. StCol.)—S.28.

Schmid, A. R., Assoc. Prof. Agr. (U.Minn.)—T.1.

Schoening, H. W., Chief, Path. Div. (B.A.I.)—S.33.

Schoth, H. A., Sen. Agr. (U.S.D.A.)—S.2; M.18; M.24; H.29.

Schroeder, W. T., Prof. Pl. Path. (Cornell U.)—S.13.

Scott, L. B., Chief, Nursery Div., Southeastern Reg. Soil Conserv. S. (U.S.D.A.)—B.3.

Seiden, R., Consultant—S.10; S.11; S.27.

Sell, O. E., Head, An. Ind. Dep. (U.Ga.—Ga.AESt.)—S.9.

Semple, A. T., Assoc. An. Husb. (B.A.I.)—M.2.

Shepherd, J. B., Dair. Husb. (B.D.I.)—S.24; S.25; W.26; W.30.

Shipley, M. A., Assoc. Range Man. (U.Nev.—AESt.)—F.10.

Sieglinger, J. B., Agr. in Charge, Soybean Res. (Okla.AESt. and U.S.D.A.)—S.8; D.4; D.5.

Singsen, E. P., Prof. Poultry Husb. (U.Conn.)—M.46.

Smith, F. L., Assoc. Prof. Agr. (U.Calif.)—S.23.

Smith, F. T., Sen. Ent. (B.E.P.Q.)—C.10.

Smith, G. M., Agr. (U.S.D.A.)—B.14.

Smith, J. W., Res. Asst. (Okla.A&MCol.—AESt.)—B.12.

Smith, L. H., Ext. Agr. (U.Vt.—AExtS.)—S.32.

Smith, O. F., Assoc. Path. (Nev.AESt. and U.S.D.A.)—J.3; S.12.

Smith, R. F., Asst. Prof. Ent. (U.Calif.)—A.4.

Smith, R. W., Assoc. Agr. (U.S.D.A.)—M.31.

Snell, R. S., Asst. Res. Spec. (Rutgers U.—N.J.AExtS.)—G.9.

Somers, H. L., Collab. (U.Fla.—AESt.)—M.5.

Spears, B. R., Ext. Agr. (Tex.A&MCol.—AExtS.)—S.5.

Spence, L. E., Ext. Spec. (U.Idaho—AExtS.)—S.26.

Spencer, D. A., Sen. An. Husb. (B.A.I.)—W.25.

Sprague, M. A., Assoc. Res. Spec. (Rutgers U.—N.J.AESt.)—A.3.

Sprague, V. G., Prof. Agr. (Pa.StCol.)—S.29.

Springfield, H. W., Range Conserv. (U.S.D.A.)—R.1.

Stanton, T. R., Sen. Agr. (U.S.D.A.)—S.16; S.17; S.18; S.19; S.20; S.21.

Staples, G. E., Asst. An. Husb. (S.Dak.ACol.—AESt.)—M.41.

Stefferud, A., Editor (U.S.D.A.)—U.13.

Stephens, J. C., Agr. (B.P.I.)—M.1; V.2.

Stephens, J. L., Agr. (Ga. Coastal Plain ESt. and U.S.D.A.)—M.9; M.14; S.7; S.15.

Stephenson, E. L., Asst. Prof. An. Ind. U.Ark)—N.5.

Stevenson, J. W., Nutr. Spec. (B.A.I.)—L.11.

Stewart, G., Sen. For. Ecol. (U.S.D.A.)—P.18; P.19; H.35.

Stewart, I., Res. Fellow (Rutgers U.—N.J.AESt.)—S.6.

Stitt, R. E., Bot. (Mont.StCol.—AESt. and U.S.D.A.)—M.47.

Sumner, D. C., Asst. Spec. Agr. (U.Calif.—AESt.)—L.4.

Swallen, J. R., Curator, Div. Grasses, U.S. Natl. Museum—S.1.

Swanson, A. F., Agr. (U.S.D.A.)—S.3; S.4; S.22.

Swift, J. E., Ext. Ent. (U.Calif.)—A.4.

Tapke, V. F., Sen. Path. (U.S.D.A.)—L.10; H.28.

Taylor, J. W., Sen. Agr. (U.S.D.A.)—Q.1; B.16.

Thatcher, L. E., Collab. (Ohio AESt.)—K.10.

Thomas, C. A., Path. (U.S.D.A.)—T.3.

Thomas, H. L., Assoc. Prof. Agr. (U.Minn.)—T.1.

Thompson, R. B., Asst. Prof. Hort. (Okla. A&MCol.)—M.44.

Timmons, F. L., Sen. Agr. (U.S.D.A.)—P.23.

Tribble, L. F., Instr. An. Husb. (U.Mo.—AESt.)—L.13.

Tucker, D. L., Collab. (U.Ark.—AESt.)—N.5.

Tullis, E. C., Pl. Path. (U.S.D.A.)—T.2.

Turner, E. C., Ext. Soil Conserv. (Clemson ACol.—AExtS.)—W.27; W.31.

Turner, G. T., Asst. Range Examiner (U.S. D.A.)—C.17.

Ullstrup, A. J., Path. (U.S.D.A.)—U.7.

Valentine, K. A., Assoc. An. Husb. (N.Mex. A&MCol.—AESt.)—N.6.

Van Horn, J. L., Prof. An. Ind. (Mont. StCol.)—B.18.

Vinall, H. N., Agr. (B.P.I.)—V.1; V.2.

Wagner, R. E., Assoc. Agr. (B.P.I.)—W.26.

Walker, E. D., Asst. Prof. Soil Ext. (U.Ill.)—W.32.

Walker, G. L., Nutr. Spec. (U.S.D.A.)—W.24.

Walker, R. K., Sup., Rice ESt. (LaAESt.)—J.6.

Wallace, A. T., Asst. Agr. (U.Fla.—AESt.)—W.13.

Wallace, K. E., Ext. Weed Spec. (S.Dak. StCol.—AESt.)—D.18.

Walter, E. V., Assoc. Ent. (B.E.P.Q.)—G.1.

Walton, W. R., Sen. Ent. (U.S.D.A.)—W.12; W.22.

Waters, R. E., Collab. (Miss.StCol.—AESt.)—R.12.

Watts, L. F., Chief, For. S. (U.S.D.A.)—P.19.

Weaver, L. A., Chairman, An. Husb. Dep. (U.Mo.)—W.29.

Webster, J. E., Pl. Chem. (Okla.AESt.)—S.8.

Webster, O. J., Agr. (Nebr.AESt. and U.S. D.A.)—W.21.

Weibel, R. O., Asst. Agr. (U.W.Va.—AESt.)—W.20.

Weihing, R. M., Agr. (Tex.AESt. and U.S. D.A.)—M.28.

Weimer, J. L., Sen. Path. (U.S.D.A.)—W.6; W.8; W.9; W.15.

Weintraub, F. C., Pl. Taxonomist (U.S.D.A.)—W.16.

Weir, W. C., Asst. Prof. An. Husb. (U.Calif.)—W.33.

Welch, H., Collab. (Mont.StCol.—AESt.)—W.34.

West, E., Bot. (U.Fla.—AESt.)—W.35.

Westover, H. L., Princ. Agr. (U.S.D.A.)—W.1; W.19.

Whitehair, C. K., Collab. (Okla.A&MCol.—AESt.)—G.18.

Whitfield, C. J., Project Sup. (Tex.A&MCol.—AESt.)—H.42.

Wiebe, G. A., Sen. Agr. (U.S.D.A.)—H.27; A.6.

Wildermuth, V. L., Sen. Ent. (U.S.D.A.)—W.11.

Willard, H. S., Dair. Husb. (Wyo.AESt.)—B.11.

Williams, A. H., Asst. Agr. (U.S.D.A.)—J.5.

Williams, J. O., Asst. An. Husb. (B.A.I.)—W.25.

Williams, L. F., Assoc. Agr. (U.S.D.A.)—M.19.

Willson, F. S., Head, An. Ind. Dep. (Mont. StCol.)—B.18.

Wilsie, C. P., Prof. Farm Crops (Iowa StCol.)—W.5; W.10.

Wilson, C., Asst. Dean, Pl. Path. (Ala.Pol. Inst.)—W.17; W.18.

Wise, L. N., Asst. Prof. Agr. (Miss.StCol.)—W.3.

Wood, J. R., Sup. Subst. No. 22 (Tex. AESt.)—W.4.

Woodle, H. A., Leader, Ext. Agr. (Clemson AESt. and U.S.D.A.)—W.2; W.7; W.14; W.23; W.27; W.31; W.36.

Woodruff, L. C., Jun. Ent. (B.E.P.Q.)—G.1.

Woodward, T. E., Sen. Dair. Husb. (B.D.I.)—C.15; W.25; W.30.

Wooten, H. H., Princ. A. Econ. (U.S.D.A.)—W.28.

Wright, J. C., Asst. Prof. Zool. and Ent. (Mont.StCol.)—A.2.

Zaumeyer, W. J., Path. (U.S.D.A.)—M.45.

BIBLIOGRAPHY*

A.1: Bromegrass Strain Performance Trials. *G. H. Ahlgren* and *A. Dotzenko*. Rutgers U. —N.J.AESt., Bu.753. (4 pp.)

A.2: Grasshopper Investigations on Montana Range Land. *N. L. Anderson* and *J. C. Wright*. Mont.St.Col.—AESt., Tech.Bu. 486. (46 pp., 6 tables.)

A.3: Bromegrass in New Jersey. *G. H. Ahlgren, H. D. Gross,* and *M. A. Sprague*. Rutgers U.—N.J.AESt., Bu.766. (11 pp.)

A.4: Lygus Bugs on Seed Alfalfa. *L. D. Anderson, L. G. Jones, H. T. Reynolds, R. F. Smith,* and *J. E. Swift*. U.Calif.— "Calif.A.", Nov. '52. (pp. 3-4.)

A.5: Cordova Barley. *I. M. Atkins*. Tex. A&MCol.—AESt., Bu.760. (9 pp.)

A.6: Classification of Barley Varieties Grown in the United States and Canada in 1945. *E. Åberg* and *G. A. Wiebe*. U.S.D.A., Tech. Bu.907. (190 pp.)

A.7: This Prosperous, Blessed Land. *H. L. Ahlgren*. U.S.D.A., Y.'48. (pp. 423-454.)

B.1: Wheatgrasses of Wyoming. *A. A. Beetle*. U.Wyo.—AESt., Bu.312. (26 pp.)

B.2: Coastal Bermuda Grass. *G. W. Burton*. U.Ga.—Coastal Plain AESt., Cr.10. (21 pp.)

B.3: Using Fall Fescue in Soil Conservation. *R. Y. Bailey* and *L. B. Scott*. U.S.D.A., Lf. 254. (8 pp.)

B.4: Research on the Management of Ladino Clover. *B. A. Brown*. U.Conn.—Storrs AESt., Inf.8. (4 pp.)

B.5: Kudzu for Erosion Control in the Southeast. *R. Y. Bailey*. U.S.D.A., F.Bu.1840. (30 pp.)

B.6: Atlantic Alfalfa. *W. R. Battle*. Rutgers U. —N.J.AESt., Bu.765: (4 pp.)

B.7: Growing Alfalfa in New Jersey. *R. A. Briggs*. Rutgers U.—AExtS., Lf.94. (8 pp.)

B.8: The Red Harvester Ant and How to Subdue It. *O. L. Barnes* and *N. J. Nerney*. U.S.D.A., F.Bu.1668. (11 pp.)

B.9: The Corn Earworm as an Enemy of Field Corn in the Eastern States. *R. A. Blanchard* and *W. A. Douglas*. U.S.D.A., F.Bu.1651. (16 pp.)

B.10: Growing Peanuts. *J. H. Beattie, F. W. Poos,* and *B. B. Higgins*. U.S.D.A., F.Bu. 2063. (54 pp.)

B.11: Alfalfa Pellets—Roughage or Concentrate? *D. C. Brown, H. S. Willard, J. W. Hamilton,* and *J. Read*. Wyo.AESt., Bu. 321. (8 pp.)

B.12: Oklahoma 301, A New Hybrid Corn for Oklahoma. *J. S. Brooks, H. Pass, W. L. Richardson,* and *J. W. Smith*. Okla.A&M Col.—AESt., Bu.B-390. (6 pp.)

B.13: Commercial Growing of Sweet Corn. *V. R. Boswell*. U.S.D.A., F.Bu.2042. (29 pp.)

B.14: Popcorn. *A. M. Brunson* and *G. M. Smith*. U.S.D.A., F.Bu.1679. (18 pp.)

B.15: Classification of Wheat Varieties Grown in the United States in 1949. *B. B. Bayles* and *A. J. Clark*. U.S.D.A., Tech. Bu.1083. (173 pp.)

B.16: Wheat Production in the Eastern United States. *B. B. Bayles* and *J. W. Taylor*. U.S.D.A., F.Bu.2006. (50 pp.)

B.17: Ammoniated Molasses Not Satisfactory as a Protein Supplement for Beef Calves. *B. F. Barrentine* and *J. A. Darnell*. Miss. StCol.—AESt., Inf. Sh. 498. (2 pp.)

B.18: Wood Molasses Compared with Cane Molasses for Lambs and Steers. *Wm. H. Burkitt, J. K. Lewis, J. L. Van Horn,* and *F. S. Willson*. Mont.StCol.—AESt., Bu. 498. (18 pp.)

B.19: Feeding Safflower Meal. *M. L. Baker, G. N. Baker, C. Ervin, L. C. Harris,* and *M. A. Alexander*. U.Nebr.—AESt., Bu.402. (11 pp.)

B.20: Chemical Sagebrush Control—Good and Bad. *D. W. Bohmont*. U.Wyo.—AESt,. Cr.54. (8 pp.)

B.21: The Management of Grazing. *E. M. Brown*. U.Mo. and U.S.D.A., Y.'48. (pp. 135-139.)

B.22: Green Feed, Sod and Pasture for Chickens and Turkeys. *N. L. Bennion, E. R. Jackman* and *O. S. Fletcher*, Oreg. StCol.—ExtS., Ext.Bu.659. (16 pp.)

B.23: The Production of Perennial Grazing and Forage Crops in North Georgia. *O. L. Brooks*. U.Ga.—AESt., Bu.270. (30 pp.)

* For Key to Abbreviations, see page 591.

B.24: Grass Establishment on Wyoming Dryland. *O. K. Barnes, R. L. Lang,* and *A. A. Beetle.* U.Wyo.—AESt., Bu.314. (24 pp.)

B.25: Effect of Time of Cutting on Yield and Feeding Value of Prairie Hay. *M. L. Baker, E. C. Conard, V. H. Arthaud,* and *L. C. Newell.* U.Nebr.—AESt., Bu.403. (19 pp.)

B.26: Antibiotics in the Nutrition of Laying Hens. *L. R. Berg, J. S. Carver, G. E. Bearse,* and *J. McGinnis.* StCol.Wash.—AESt., Bu.534. (12 pp.)

B.27: The Rot That Attacks 2,000 Species. *L. M. Blank.* U.S.D.A., Y.'53. (pp. 298-301.)

B.28: Weeds of Wyoming. *D. W. Bohmont.* U.Wyo., Bu.325. (160 pp.)

B.29: Chemical Control of Poisonous Range Plants. *D. W. Bohmont.* U.Wyo., Bu.313. (19 pp.)

C.1: How to Graze Blue Grama (on Southwestern Ranges). *E. C. Crafts* and *G. E. Glendening.* U.S.D.A., Lf.215. (8 pp.)

C.2: How to Keep and Increase Black Grama (on Southwestern Ranges). *R. S. Campbell* and *E. C. Crafts.* U.S.D.A., Lf.180. (8 pp.)

C.3: Chemical Control of Johnsongrass at the Blackland Station, 1952. *E. D. Cook.* Tex. A&M—AESt., Prog.Rep.1544. (1 p.)

C.4: Sudan Grass and Millet Selection for Grazing and Hay. *L. V. Crowder, E. M. Parker,* and *J. M. Elrod.* U.Ga.—AESt., P.Bu.644. (2 pp.)

C.5: Personal Communication. *A. A. Case.* U.Mo.

C.6: Sorghums for Grain and Forage. *W. Chaffin.* Okla.A&MCol.—AExtS., Cr.478. (20 pp.)

C.7: The Control of the Clover-Flower Midge. *C. W. Creel* and *L. P. Rockwood.* U.S.D.A., F.Bu.971. (9 pp.)

C.8: Natob—A New Bush Lespedeza for Soil Conservation. *F. J. Crider.* U.S.D.A., Cr. 900. (10 pp.)

C.9: Soybean Diseases in Illinois. *D. W. Chamberlain* and *B. Koehler.* U.Ill.—AExtS., Cr.676. (32 pp.)

C.10: Insects and the Plant Viruses. *L. D. Christenson* and *F. F. Smith.* U.S.D.A., Y.'52. (pp. 179-190.)

C.11: Varieties of Spring Wheat for the North Central States. *J. A. Clark.* U.S.D.A., F.Bu.1902. (22 pp.)

C.12: Facts About Cotton. Cotton Branch, Production & Marketing Adm., U.S.D.A., Lf.167. (8 pp.)

C.13: Experiments with Safflower in Western Nebraska. *C. E. Claassen* and *T. A. Kiesselbach.* U.Nebr.—AESt., Bu.376. (28 pp.)

C.14: Crops for Silage. *W. Chaffin.* Okla. A&MCol.—AExtS., Cr. 620. (8 pp.)

C.15: Pasture and Range in Livestock Feeding. *P. V. Cardon, W. R. Chapline, T. E. Woodward, E. W. McComas,* and *C. R. Enlow.* U.S.D.A., Y.'39. (pp. 925-955)

C.16: Grazing on Range Lands. *W. R. Chapline.* U.S.D.A., Y.'48. (pp. 212-216.)

C.17: Judging Condition and Utilization of Short-Grass Ranges on the Central Great Plains. *D. F. Costello* and *G. T. Turner.* U.S.D.A., F.Bu.1949. (21 pp.)

C.18: Trench and Box-Type Silos. *C. G. Cushman.* Clemson ACol.—AExtS., Cr.375. (23 pp.)

C.19: Weeds. Clemson Extension Weed Committee. Clemson ACol.—AExtS., Bu.113. (87 pp.)

D.1: Grass: Green, Grain, Grow. *W. A. Dayton.* Y.'48. (pp. 637-639.)

D.2: T.C.A. for Johnsongrass Control. *A. D. Dotzenko.* New Mex. A&M Col.—AESt., P.Bu.1058. (3 pp.)

D.3: Sudan Grass. *C. E. Denman.* Okla. A&MCol.—AESt., Forage Crops Lf.15. (2 pp.)

D.4: Darset, a Combine-Type Darso. *F. F. Davies* and *J. B. Sieglinger.* Okla.A&MCol.—AESt., Bu.B-391. (6 pp.)

D.5: Dwarf Kafir 44-14 and Redlan. *F. F. Davies* and *J. B. Sieglinger.* Okla.A&MCol.—AESt., Bu.B-384. (11 pp.)

D.6: Buffalograss. *D. I. Dudley.* Tex.A&M Col.—AESt., Prog.Rep. 1579. (3 pp.)

D.7: Bicolor Lespedeza for Quail and Soil Conservation in the Southeast. *V. E. Davison.* U.S.D.A., Lf.248. (8 pp.)

D.8: Yellow Clover Aphid in State. *A. S. Deal, R. C. Dickson,* and *H. T. Reynolds.* U.Calif.—"Calif.A." Sept. '54 (p. 5.)

D.9: Glossary of Botanical Terms Commonly Used in Range Research. *W. A. Dayton.* U.S.D.A., M.Pu.110. (41 pp.)

D.10: The Corn Root Aphid and Methods of Controlling It. *J. J. Davis.* U.S.D.A., F.Bu.891. (7 pp.)

D.11: Leaf and Head Blights of Cereals. *J. G. Dickson.* U.S.D.A., Y.'48. (pp. 344-349.)

D.12: For Insurance Against Drought—Soil and Water Conservation. *T. Dale.* U.S.D.A., F.Bu.2002. (22 pp.)

D.13: Comparative Value of Proso and Corn for Fattening Hogs. *W. E. Dinusson, D. W. Bolin, M. L. Buchanan,* and *R. M. Richard.* N.Dak.AESt., "Bimonthly Bu.", Nov.-Dec.'54. (pp. 74-76.)

D.14: Grazing Crops for Poultry. *R. S. Dearstyne* and *P. H. Kime.* U.N.C.—AExtS., Ext.Cr. 239. (12 pp.)

D.15: When Drought Returns to the Great Plains. *T. Dale*. U.S.D.A., F.Bu.1982. (14 pp.)

D.16: Stock Poisoning Plants of the Eastern Seaboard. *W. A. Dayton*. "Proceedings—Soc. of Am. Foresters", '48. (pp. 223-231.)

D.17: Historical Sketch of Barilla. *W. A. Dayton*. "Jrl. Range Management", Vol.4, No. 6. (pp. 375-381.)

D.18: Weed Control Research in South Dakota. *L. A. Derscheid* and *K. E. Wallace*. S.Dak.StCol.—AESt., Cr.122. (36 pp.)

D.19: Poisonous and Injurious Plants in Colorado. *L. W. Durrell, R. Jensen*, and *B. Klinger*. Colo.A&MCol.—AESt., Bu. 412-A. (88 pp.)

D.20: Legume Insects of Oregon. *E. A. Dickason, R. W. Every*, and *P. R. Hansen*, Oreg.StCol.—AExtSt., Ext.Bu.749.(38pp.)

E.1: Broomsedge. *W. C. Elder*. Okla.A&M Col.—AESt., Forage Crops Lf.13. (2 pp.)

E.2: Bermuda-Grass. *W. C. Elder*. Okla. A&MCol.—AESt., Forage Crops Lf.14. (4 pp.)

E.3: Sorghum Production in New Mexico. *I. M. Evans*. N.Mex.A&MCol.—AESt., Bu.371. (32 pp.)

E.4: Legume Inoculation. What It Is, What It Does. *L. W. Erdman*.—U.S.D.A., F.Bu. 2003. (20 pp.)

E.5: Diseases, Insects, and Other Factors in Relation to Red Clover Failure in West Virginia. *E. S. Elliott*. W.VA.U.—AESt., Bu.351T. (65 pp.)

E.6: Sericea Lespedeza. *J. M. Elrod*. U.Ga.—AESt., P.Bu.558. (2 pp.)

E.7: The Annual Lespedezas. *W. C. Elder*. U.S.D.A., Forage Crops Lf.9. (2 pp.)

E.8: Kudzu. *J. M. Elrod*. U.Ga.—AESt., P.Bu.575 Rev. (2 pp.)

E.9: Lupines for Green Manure. *J. M. Elrod*. U.Ga.—AESt., P.Bu.610 Rev. (2 pp.)

E.10: Peanut Meal in Livestock Production. *F. R. Edwards* and *Z. A. Massey*. U.Ga.—AESt., Bu.216. (20 pp.)

E.11: Bacterial Wilt of Corn. *C. Elliott*. U.S.D.A., F.Bu.1878. (21 pp.)

E.12: Byproducts as Feed for Livestock. *N R. Ellis* and *H. R. Bird*. U.S.D.A., Y,'50-51. (pp. 851-862.)

E.13: Feeding Cottonseed Products to Livestock. *N. R. Ellis* and *R. E. Hodgson*. U.S.D.A., F.Bu.1179. (13 pp.)

E.14: Feeding Dairy Cows on Pasture. *H. P. Ewalt* and *R. W. Morse*. Oreg.StCol.—AExtS., Ext.Bu.592. (4 pp.)

E.15: Chemical Control of Brush in Oklahoma. *W. C. Elder* and *J. Dreesen*. Okla.

A&MCol.—AESt., Mim.Cr.M-242. (7 pp.)

F.1: Rancher—A Low Hydrocyanic Acid Forage Sorghum. *C. J. Franzke*. S.Dak.St Col.—AESt., Cr.57. (8 pp.)

F.2: Reliance, an Early Grain Sorghum. *C. J. Franzke*. S.Dak.StCol.—AESt., Bu.426. (4 pp.)

F.3: Norghum Sorghum. *C. J. Franzke*. S.Dak.StCol.—AESt., Bu.397. (4 pp.)

F.4: Smuts That Parasitize Grasses. *G. W. Fisher*. U.S.D.A., Y.'53. (pp. 280-284.)

F.5: Winter Barley Offers Many Possibilities to Farming Programs. *V. C. Finkner* and *G. R. Gist*. "Ohio Farm and Home Res. Sept.-Oct.'53. (pp. 86 and 94.)

F.6: Official Publication: Association of American Feed Control Officials, Inc., 1955. (146 pp.)

F.7: 1953 Field Crop Recommendations for Oregon. Farm Crops Staff. Oreg.StCol.—AESt., St.Bu.533. (24 pp.)

F.8: The Digestibility of Perilla Meal, Hempseed Meal, and Babassu Meal, as Determined for Ruminants. *A. H. Folger*. U.Calif.—AESt., Bu.604. (8 pp.)

F.9: Making and Feeding Hay-Crop Silage. *F. C. Fountaine, D. Richardson, F. E. Riggs*, and *K. L. Anderson*. Kans.StCol.—AESt., Cr.282. (12 pp.)

F.10: Bronco Grass (*Bromus tectorum*) on Nevada Ranges. *Ch. E. Fleming, M. A. Shipley*, and *M. R. Miller*. U.Nev.—AESt., Bul.159. (21 pp.)

G.1: The Sorghum Midge. *C. H. Gable, W. A. Baker, L. C. Woodruff*, and *E. V. Walter*. U.S.D.A., F.Bu.1566. (9 pp.)

G.2: Cluster Clover (*Trifolium glomeratum*). *S. W. Greene*. U.S.D.A., Mim.Sh. '32. (2 pp.)

G.3: Better Strains of Red Clover. *C. S. Garrison*. Rutgers U.—N.J.AESt., Cr.419. (4 pp.)

G.4: Seeding Smooth Bromegrass. *C. S. Garrison*. Rutgers U.—AExtS., Lf. 26 (4 pp.)

G.5: Growing and Harvesting the Sweetclover Seed Crop. *S. Garver* and *T. A. Kiesselbach*. U.Nebr.—AESt., Bu. 387. (47 pp.)

G.6: Growing Alfalfa. *H. O. Graumann* and *C. H. Hanson*. U.S.D.A., F.Bu.1722. (38 pp.)

G.7: Alfalfa in Kansas. *C. O. Grandfield*. Kans. StCol.—AESt., Bu.346. (65 pp.)

G.8: Hairy Vetch for Nebraska. *T. H. Goodding*. U.Nebr.—AESt., Cr.89. (7 pp.)

G.9: Better Crops of Winter Barley. *C. S. Garrison* and *R. S. Snell*. Rutgers U.—Ext., AExt. Bu.240. (8 pp.)

G.10: Pierre Rye. *J. E. Grafius.* S.Dak.StCol. —AESt., Bu.406. (4 pp.)

G.11: Supplemented Cottonseed Hulls as a Roughage for Fattening Lambs in Drylot. *W. P. Garrigus.* U.Ky.—AESt., Bu.566. (23 pp.)

G.12: Distillers' Dried Solubles as a Protein Supplement for Steers in Drylot. *W. P. Garrigus.* U.Ky.—AESt., Bu.564. (15 pp.)

G.13: Alfalfa and Blue Grass Silages as Roughages for Fattening Steers in Drylot. *W. P. Garrigus.* U.Ky.—AESt., Bu.579. (18 pp.)

G.14: Digestible Nutrient Contents of Corn, Bluegrass, Alfalfa, Ladino, Fescue, and Soybean Silages for Steers. *W. P. Garrigus.* U.Ky.—AESt., Bu.573. (19 pp.)

G.15: Feeding Value for Milk Production of Pasture Grasses When Grazed, When Fed Green, and When Fed as Hay or Silage. *R. R. Graves, J. R. Dawson, D. V. Kopland,* and *T. W. Moseley.* U.S.D.A., Tech.Bu.381. (47 pp.)

G.16: Grazing Crops for Poultry. *P. H. Gooding* and *J. W. Matthews.* Clemson A Col.—AExtS., Cr.185. (11 pp.)

G.17: Grazing Crops for Poultry. *P. H. Gooding.* Clemson ACol.—AExtS., Cr.185 Revised. (11 pp.)

G.18: Urea as a Source of Protein in Livestock Rations. *W. D. Gallup, L. S. Pope,* and *C. K. Whitehair.* Okla.A&MCol.— AESt., Cr.C-137. (10 pp.)

G.19: Principal Poisonous Plants in Kansas. *F. C. Gates.* Kans.StCol., Tech.Bu.25. (67 pp.)

H.1: The Main Grasses for Farm and Home. *M. M. Hoover, M. A. Hein, W. A. Dayton,* and *C. O. Erlanson.* Y. '48. (pp. 639-700.)

H.2: King Ranch Bluestem. *J. R. Harlan.* Okla.A&MCol.—AESt., Forage Crops Lf. 11. (2 pp.)

H.3: Fescue Grass and Legumes for Soil Improvement and Forage Production at the Blackland Station. *R. J. Hervey* and *E. D. Cook.* Tex.A&MCol.—AESt., Prog. Rep. 1943. (3 pp.)

H.4: Weeping Lovegrass. *J. R. Harlan, C. E. Denman,* and *W. C. Elder.* Okla.A&MCol. -AESt., Forage Crops Lf.16. (4 pp.)

H.5: Manual of the Grasses of the United States. *A. S. Hitchcock.* U.S.D.A., M.Pu. 200. (1040 pp.)

H.6: Sweet Sorghum (Cane, Sorgo) for Silage and Forage. *C. A. Helm.* U.Mo.—AExtS., Cr.380. (4 pp.)

H.7: Winter Clovers in Central Florida. *E. M. Hodges, D. W. Jones,* and *W. G. Kirk.* U.Fla.—AESt., Bu.517. (23 pp.)

H.8: White Clover. *E. A. Hollowell.* U.S.D.A., Lf.119. (8 pp.)

H.9: Ladino White Clover for the Northeastern States. *E. A. Hollowell.* U.S.D.A., F.Bu.1910. (10 pp.)

H.10: Persian Clover. *E. A. Hollowell.* U.S.D.A., F.Bu.1929. (10 pp.)

H.11: Crimson Clover. *E. A. Hollowell.* U.S.D.A., Lf.160. (8 pp.)

H.12: Strawberry Clover. *E. A. Hollowell.* U.S.D.A., Lf.176. (8 pp.)

H.13: Lappa Clover (*Trifolium lappaceum*). *E. A. Hollowell.* Mim.Sh.1939. (2 pp.)

H.14: Why Red Clover Fails. *E. A. Hollowell.* U.S.D.A., Lf.110. (6 pp.)

H.15: Mammoth Red Clover. *E. A. Hollowell.* U.S.D.A., Mim.Sh.1932. (2 pp.)

H.16: Midland Red Clover. *E. A. Hollowell.* U.S.D.A., Mim.Sh.1941. (1 p.)

H.17: Registration of Varieties and Strains of Red Clover, II. *E. A. Hollowell.* "Agr. Jrl.", Vol. 43, No. 5. (p. 242.)

H.18: Registration of Varieties and Strains of Red Clover, III. *E. A. Hollowell.* "Agr. Jrl.", Vol. 45, No. 11, (p. 574.)

H.19: Dodder. *A. A. Hansen.* U.S.D.A., F.Bu. 1161. (21 pp.)

H.20: Vitamins for Livestock. *A. G. Hogan.* U.Mo.—AESt., Bu.453. (23 pp.)

H.21: The Many Ailments of Clover. *E. W. Hanson* and *K. W. Kreitlow.* U.S.D.A., Y.'53. (pp. 217-228.)

H.22: Grasses and Legumes for Soil Conservation in the Pacific Northwest. *A. L. Hafenrichter, L. A. Mullen,* and *R. L. Brown.* U.S.D.A., M.Pu.678. (56 pp.)

H.23: Side-Oats Grama. *J. R. Harlan.* Okla. A&MCol.—AESt., Forage Crops Lf.18. (2 pp.)

H.24: Midland Bermuda Grass. *J. R. Harlan, G. W. Burton,* and *W. C. Elder.* Okla.A&M Col.—AESt.,Bu.B-416. (10 pp.)

H.25: Pennycress and Peppergrass. *N. S. Hanson* and *M. N. Lawritson.* U.Nebr.— AExtS., Ext.Cr.162. (4 pp.)

H.26: Manual of the Grasses of the United States. *A. S. Hitchcock* and *A. Chase.* U.S.D.A., M.Pu.2000. (1051 pp.)

H.27: Growing Barley for Malt and Feed. *H. V. Harlan* and *G. A. Wiebe.* U.S.D.A., F.Bu.1732. (19 pp.)

H.28: The Smuts of Wheat, Oats, Barley. *C. S. Holton* and *V. F. Tapke.* U.S.D.A., Y.'48. (pp. 360-368.)

H.29: Rape. *G. R. Hyslop* and *H. A. Schoth.* Oreg.StCol.—AExtS., Ext.Bu.499. (4 pp.)

H.30: Distillers' Solubles in Market Pig Rations. *L. E. Hanson, M. L. Baker, G. N.*

Baker, and *M. G. A. Rumery*. U.Nebr.—AESt., Bu.415. (19 pp.)

H.31: Replacing Milk in Dairy Calf Feeding. *H. O. Henderson*. W.Va.U.—AExtS., Food for Freedom Pu. (6 pp.)

H.32: Making and Feeding Grass and Legume Silage in Western Oregon. *M. G. Huber, R. W. Morse*, and *E. R. Jackman*. Oreg.St Col.—AExtS., Ext.Bu.669. (28 pp.)

H.33: Measuring Hay in Stacks. *W. H. Hosterman*. U.S.D.A., Lf.72. (6 pp.)

H.34: High-Grade Timothy and Clover Hay. *W. H. Hosterman*. U.S.D.A. F.Bu.1770. (17 pp.)

H.35: Increasing Forage Yields and Sheep Production on Intermountain Winter Ranges. *S. S. Hutchings* and *G. Stewart*. U.S.D.A., Cr.925. (63 pp.)

H.36: Milk Production from Pasture. *D. L. Hill* and *N. S. Lundquist*. Purdue U.—AESt., St.Cr.386. (8 pp.)

H.37: Delta Bloat Studies Indicate Grass as the Answer. *P. G. Hogg* and *B. F. Barrentine*. Miss.StCol.—AESt.,Inf.Sh.420.(2pp.)

H.38: Grass Pastures in Central Florida. *E. M. Hodges, D. W. Jones*, and *W. G. Kirk*. U.Fla.—AESt., Bu.484. (32 pp.)

H.39: For a Better Range Management. *L. C. Hurtt*. U.S.D.A., Y.'48. (pp. 486-491.)

H.40: Seed Disorders of Forage Plants. *J. R. Hardison*. U.S.D.A., Y.'53. (pp. 272-276.)

H.41: Managing Winter Sheep Range for Greater Profit. *S. S. Hutchings*. U.S.D.A., F.Bu.2067. (46 pp.)

H.42: Control of Extensive Infestation of Bindweed in Northwest Texas. *R. D. Hamilton, C. J. Whitfield*, and *H. E. Rea*. Tex.A&MCol.—AESt., Prog. Rep. 1392. (5 pp.)

H.43: Plants Poisonous to Livestock. *W. T. Huffman* and *J. F. Couch*. U.S.D.A., Y.'42. (pp. 354-373.)

H.44: Leaf Diseases of Range Grasses. *J. R. Hardison*. U.S.D.A., Y.'53. (pp. 253-258.)

H.45: Moderate Grazing Pays on California Annual-Type Ranges. *A. L. Hormay*. U.S.D.A., Lf.239. (8 pp.)

H.46: Antibiotics Promote Calf Growth and Help Cure Diseases. *J. W. Hibbs, H. R. Conrad*, and *W. D. Pounden*. OhioStU.—AESt., "Ohio Farm and Home Res.", Vol. 40, No. 294. (pp. 38-39.)

H.47: Common Arizona Range Grasses. *R. R. Humphrey, A. L. Brown*, and *A. C. Everson*. U.Ariz.—AESt., Bu.243. (102 pp.)

H.48: Some Plants of Kentucky Poisonous to Livestock. *M. T. Hyatt, R. G. Brown*, and *J. W. Herron*. U.Ky.—AExtS., Cr.502. (57 pp.)

J.1: Sorghum . . . Its Feed Value for Lambs. *R. M. Jordan*. S.Dak.StCol.—AESt., "S. Dak. Farm and Home Res.", Vol. 1, No. 2. (pp. 40-42.)

J.2: Bacterial Wilt of Alfalfa and Its Control. *F. R. Jones*. U.S.D.A., Cr.573. (8 pp.)

J.3: Sources of Healthier Alfalfa. *F. R. Jones* and *O. F. Smith*. U.S.D.A., Y.'53. (pp. 228-237.)

J.4: Bacteria, Fungi, and Viruses on Soybeans. *H. W. Johnson* and *D. W. Chamberlain*. U.S.D.A., Y.'53. (pp. 238-247.)

J.5: Rice Culture in California. *J. W. Jones, L. L. Davis*, and *A. H. Williams*. U.S.D.A., F.Bu.2022. (30 pp.)

J.6: Rice Production in the Southern States, 1952. *J. W. Jones, J. O. Dockins, R. K. Walker*, and *W. C. Davis*. U.S.D.A., F.Bu. 2043. (36 pp.)

J.7: 4-H Club Insect Manual. *M. P. Jones*. U.S.D.A., A. Handbook 65. (64 pp.)

K.1: Argentine Bahia Grass. *G. B. Killinger, G. E. Ritchey, C. B. Blickensderfer*, and *W. Jackson*. U.Fla.—AESt., Cr.S-31. (4 pp.)

K.2: Personal Information. *N. W. Kramer*. Tex.AESt.

K.3: New Varieties of Sorghum. *R. E. Karper, J. R. Quinby*, and *N. W. Kramer*. Tex.A&MCol.—AESt., Prog. Rep. 1367. (6 pp.)

K.4: New Varieties of Sorghum. *R. E. Karper* and *J. R. Quinby*. Tex.A&MCol.—AESt., Prog. Rep. 1064. (7 pp.)

K.5: Grasshoppers and Their Control. *L. C. Kuitert* and *R. V. Connin*. U.Fla.—AESt., Bu.516. (30 pp.)

K.6: Insect Pests of Alfalfa in North Carolina. *W. M. Kulash* and *C. H. Hanson*. "Jrl. Economic Entomology". Vol. 42, No. 4. (p. 694.)

K.7: Sesame in Texas. *M. L. Kinman*. Tex. A&MCol.—AESt., M.Pu.98. (11 pp.)

K.8: Feeding Poor Hay. *E. W. Klosterman* and *A. L. Moxon*. OhioStU.—AESt., "Ohio Farm and Home Res." Vol. 37, No. 9-10. (p. 81.)

K.9: Hay Drying in Oregon. *D. E. Kirk* and *F. E. Price*. Oreg.StCol.—AESt., St.Bu. 506. (63 pp.)

K.10: Feed Supplements to Pasturage for Growth of Chickens. *D. C. Kennard, L. E. Thatcher*, and *V. D. Chamberlin*. Ohio StU. —AESt., "Ohio Farm and Home Res.", Vol. 34, No. 3-4. (pp. 64-66.)

K.11: The Northern Forage Grasses. *K. W. Kreitlow*. U.S.D.A., Y.'53. (pp. 262-267.)

L.1: Sorghum Diseases and Their Control. *R. W. Leukel, J. H. Martin*, and *C. L. Lefebore*. U.S.D.A., F.Bu.1959. (50 pp.)

L.2: *Periconia circinata* and Its Relation to Milo Disease. *R. W. Leukel.* U.S.D.A., "Jrl.A.Res.". Vol. 77, No. 7-8. (pp. 201-222.)

L.3: The Clover Leaf Weevil and Its Control. *W. H. Larrimer.* U.S.D.A., F.Bu.1484. (6 pp.)

L.4: Rose Clover, A New Winter Legume. *R. M. Love* and *D. C. Sumner.* U.Calif.—ExtS., Cr.407. (12 pp.)

L.5: Sorghums in Colorado. *W. H. Leonard.* Col.A&MCol.—AESt., Mim.8262-52.

L.6: Mungbeans. *L. L. Ligon.* Okla.A&MCol.—AESt., St.Bu.284. (12 pp.)

L.7: Rice Leaf Miner. *W. H. Lange, Jr., K. H. Ingebretsen,* and *L. L. Davis.* U.Calif., "Calif.A." Aug.'53. (pp. 8-9.)

L.8: The Niacin Content of Blue Corn. *E. Lantz, H. W. Gough,* and *M. M. Johnson.* N.Mex.A&MCol.—AESt., P.Bu.1066. (5 pp.)

L.9: The Southern Corn Rootworm and Farm Practices to Control It. *P. Luginbill.* U.S.D.A., F.Bu.950. (10 pp.)

L.10: Cereal Smuts and Their Control. *R. W. Leukel* and *V. F. Tapke.* U.S.D.A., F.Bu. 2069. (28 pp.)

L.11: Feeding Qualities for Livestock of Distillers' Corn By-Products from Fungal Amylase-Converted Mashes, Compared with Those from Malt Amylase-Converted Mashes. *I. L. Lindahl, R. E. Davis, J. L. Milligan, N. R. Ellis, H. R. Bird, J. W. Stevenson,* and *L. J. Machlin.* U.S.D.A., Tech.Bu.1053. (31 pp.)

L.12: Growing Green for Poultry. *A. G. Law, E. J. Kreizinger,* and *I. M. Ingham.* StCol. Wash.—AExtS., Ext.Bu.310. (16 pp.)

L.13: Value in Antibiotics in Swine Rations. *J. F. Lasley, L. F. Tribble,* and *A. G. Hogan.* U.Mo.—AESt., Res.Bu.543. (55 pp.)

M.1: The Culture and Use of Sorghums for Forage. *J. H. Martin* and *J. C. Stephens.* U.S.D.A., F.Bu.1844. (42 pp.)

M.2: Growing and Feeding Grain Sorghums. *J. H. Martin, J. S. Cole,* and *A. T. Semple.* U.S.D.A., F.Bu.1764. (46 pp.)

M.3: The Legumes of Many Uses. *R. McKee.* Y.'48. (pp. 701–726.)

M.4: Certified Kenland Red Clover. U.Md.—AExtS., Ext.Lf. L-11. (8 pp.)

M.5: Value of Pearl Millet Pasture for Dairy Cattle. *S. P. Marshall, A. B. Sanchez, H. L. Somers,* and *P. T. D. Arnold.* U.Fla.—AESt., Bu.527. (20 pp.)

M.6: Bur-Clover Cultivation and Utilization. *R. McKee.* U.S.D.A., F.Bu.1741. (12 pp.)

M.7: Lespedeza Culture and Utilization. *R. McKee.* U.S.D.A., F.Bu.1852. (14 pp.)

M.8: Kobe, A Superior Lespedeza. *R. McKee* and *H. L. Hyland.* U.S.D.A., Lf.240. (6 pp.)

M.9: Kudzu as a Farm Crop. *R. McKee* and *J. L. Stephens.* U.S.D.A., F.Bu.1923. (14 pp.)

M.10: The Other Pasture Legumes. *R. McKee.* Y.'48. (pp.363-366.)

M.11: Alyceclover. *R. McKee.* U.S.D.A., Mim.Sh.1940. (2 pp.)

M.12: Lupines: New Legumes for the South. *R. McKee* and *G. E. Ritchey.* U.S.D.A., F.Bu.1946. (10 pp.)

M.13: Florida Beggarweed. *R. McKee.* U.S.D.A., Mim. Sh. (4 pp.)

M.14: Crotalaria Culture and Utilization. *R. McKee, G. E. Ritchey, J. L. Stephens,* and *H. W. Johnson,* U.S.D.A., F.Bu.1980. (17 pp.)

M.15: Vitamin A for Dairy Cattle. *L. A. Moore, H. T. Converse,* and *S. R. Hall.* U.S.D.A., Y.'43-47. (pp. 133–142.)

M.16: Weather and Climate. *H. B. Mills.* U.S.D.A., Y,'52. (pp. 422–429.)

M.17: Alyceclover and Hairy Indigo Boost Southern Livestock Farming. *R. McKee.* U.S.D.A., Res. Achievement Sh. R.A.S.103. (2 pp.)

M.18: Vetch Culture and Uses. *R. McKee* and *H. A. Schoth.* U.S.D.A., F.Bu.1740. (23 pp.)

M.19: Soybeans: Culture and Varieties. *W. J. Morse, J. L. Cartter,* and *L. F. Williams.* U.S.D.A., F.Bu.1520. (38 pp.)

M.20: Soybeans for Feed, Food, and Industrial Products. *W. J. Morse* and *J. L. Cartter.* U.S.D.A., F.Bu.2038. (41 pp.)

M.21: Soybean Production for Hay and Beans. *W. J. Morse, J. L. Cartter,* and *E. E. Hartwig.* U.S.D.A., F.Bu.2024. (15 pp.)

M.22: Dorman Soybeans for Oklahoma. *R. S. Matlock.* Okla.A&MCol.—AESt., Bu.B-413. (16 pp.)

M.23: Cowpeas: Culture and Varieties. *W. J. Morse.* U.S.D.A., F.Bu.1148. (18 pp.)

M.24: Culture and Pests of Field Peas. *R, McKee* and *H. A. Schoth.* U.S.D.A. F.Bu.1803. (15 pp.)

M.25: Cowpeas: Utilization. *W. J. Morse.* U.S.D.A., F.Bu.1153. (24 pp.)

M.26: Seed Treatment to Control Smut in Grasses. *J. P. Meiners* and *G. W. Fischer.* St.Col.Wash.—AESt., St.Cr.139. (3 pp.)

M.27: Idaho Recommendations for Insect Control. *H. C. Manis* and *R. W. Portman.* U.Idaho.—AESt., Bu.279. (71 pp.)

M.28: Rapid, Low-Cost Conversion from Rice to Improved Pastures. *J. B. Moncrief* and *R. M. Weihing.* Tex.A&MCol.—AESt., Bu.729. (14 pp.)

M.29: A Study of Use of Rice By-Products for Feeding Swine and Effects on Quality of Pork. *E. Martin.* U.Ark.—AESt., Bu.303. (36 pp.)

M.30: Growing Oats in Florida. *D. D. Morey, W. H. Chapman,* and *R. W. Earhart.* U.Fla.—AESt., Bu.523. (36 pp.)

M.31: Growing Rye in the Western Half of the United States. *J. H. Martin* and *R. W. Smith.* U.S.D.A., F.Bu.1358. (17 pp.)

M.32: Performance of Dent Corn Hybrids in Indiana Through 1952. *S. R. Miles, J. E. Newman,* and *P. L. Crane.* PurdueU.—AESt., St.Bu.598. (44 pp.)

M.33: Emmer and Spelt. *J. H. Martin* and *C. E. Leighty.* U.S.D.A., F.Bu.1429. (12 pp.)

M.34: Control of Insect Pests of Potato Tubers. *H. E. Morrison* and *H. H. Crowell.* Oreg.St.Col.—AESt., Cr.Inf.538. (7 pp.)

M.35: The Wheat Stem Sawfly in Montana. *H. B. Mills.*—Mont.St.Col., AESt., WarCr.6. (6 pp.)

M.36: The Rusts of Wheat, Oats, Barley, Rye. *J. H. Martin* and *S. C. Salmon.* U.S.D.A., Y.'53. (pp. 329–343.)

M.37: The Effects of Soil Fertility. *G. L. McNew.* U.S.D.A., Y.'53. (pp.100–114.)

M.38: Meeting the Mineral Needs of Farm Animals. *L. A. Maynard.* CornellU.—AExtS., Bu.350. (30 pp.)

M.39: Proso or Hog Millet. *J. H. Martin.* U.S.D.A., F.Bu.1162. (12 pp.)

M.40: Silos: Types and Construction. *J. R. McCalmont.* U.S.D.A., F.Bu.1820. (66 pp.)

M.41: Grass Hay—at Its Best. *A. L. Moxon, G. Gastler, G. E. Staples,* and *R. M. Jordan.* S.Dak.St.Col.—AESt., Bu.405. (23 pp.)

M.42: Hints on Hay Making. *E. A. Miller.* Tex.A&MCol.—AExtS., P.Bu.52-22-175. (2 pp.)

M.43: Use Range for Growing Pullets. *W. J. Moore.* Tex.A&MCol.—AExtS., Release 52-18-139. (2 pp.)

M.44: Range Crops for Turkeys. *T. T. Milby* and *R. B. Thompson.* Okla.A&MCol.—AESt., Mim.Cr.M-134. (2 pp.)

M.45: Growing Dry Beans in the Western States. *O. L. Mimms* and *W. J. Zaumeyer.* U.S.D.A., F.Bu.1996. (42 pp.)

M.46: A Comparison of Several Antibiotics as Growth Stimulants in Practical Chick-Starting Rations. *L. D. Matterson, E. P. Singsen, L. Decker,* and *A. Kozeff.* U.Conn.—StorrsAESt., Bu.275. (20 pp.)

M.47: Important Grasses on Montana Ranges. *H. E. Morris, W. E. Booth, G. F.*

Payne, and *R. E. Stitt.* Mont.StCol.—AESt., Bu.470. (52 pp.)

N.1: Grasses of New Mexico. *J. J. Norris.* Mim. Pu.1947. (148 pp.)

N.2: Norghum Sorghum Culture. *U. J. Norgaard, E. E. Sanderson,* and *R. A. Cline.* S. Dak.StCol.—AExtS., Ext.Lf.127. (6 pp.)

N.3: Legumes and Grasses for Silage. *W. B. Nevens, K. E. Harshbarger,* and *K. A. Kendall.* U.Ill.—AESt., Bu.529. (27 pp.)

N.4: When Should the Hay Crop Be Cut? *J. A. Newlander.* U.Vt.—AESt., Pamphlet 7. (4 pp.)

N.5: Subcutaneous Implantation of Bacitracin in Pellet Form to Stimulate Growth of Suckling Pigs. *P. R. Noland, D. L. Tucker,* and *E. L. Stephenson.* U.Ark.—AESt., Rep.Ser.34. (4 pp.)

N.6: Principal Livestock-Poisoning Plants of New Mexico Ranges. *J. J. Norris, K. A. Valentine,* and *L. B. Porter.* N.Mex.A&M Col.—AESt., Bu.390. (78 pp.)

O.1: Indian Grass and Switch Grass. (No author.) Okla.A&MCol.—AESt., Forage Crops Lf.17. (2 pp.)

P.1: Legumes in Soil Conservation Practices. *A. J. Pieters.* U.S.D.A., Lf.163. (8 pp.)

P.2: Alsike Clover. *A. J. Pieters.* U.S.D.A., F.Bu.1151. (18 pp.)

P.3: Clover Improvement. *A. J. Pieters* and *E. A. Hollowell.* U.S.D.A., Y.'37. (pp. 1190–1214.)

P.4: Red Clover Culture. *A. J. Pieters.* U.S.D.A., F.Bu.1339. (30 pp.)

P.5: How to Fight the Chinch Bug. *C. M. Packard, P. Luginbill, Sr.,* and *C. Benton.* U.S.D.A., F.Bu.1780. (21 pp.)

P.6: Grasshoppers. *J. R. Parker.* U.S.D.A., Y.'52. (pp. 595–605.)

P.7: Sweetclover. *A. J. Pieters.* U.S.D.A., Lf.23. (8 pp.)

P.8: Control of the Garden Webworm in Alfalfa. *F. W. Poos.* U.S.D.A., Lf.304. (4 pp.)

P.9: The Potato Leafhopper. *F. W. Poos.* U.S.D.A., Lf.229. (4 pp.)

P.10: The Velvetbean. *C. V. Piper* and *W. J. Morse.* U.S.D.A., F.Bu.1276. (21 pp.)

P.11: The Mesquite Problem on Southern Arizona Ranges. *K. W. Parker* and *S. C. Martin.* U.S.D.A., Cr.908. (70 pp.)

P.12: Hiland—A Better Barley. *R. P. Pfeifer.* U.Wyo.—AESt., Bu.330. (8 pp.)

P.13: Winter Oats, Barley, Rye, and Spelt. (No author.) PurdueU.—AExtS., Mim. Pu. AY-86b. (4 pp.)

P.14: Control of European Corn Borer and Ear Smut on Sweet Corn with Dusts and

Sprays. *B. B. Pepper* and *C. M. Haenseler.* RutgersU.—AESt., Cr.486. (14 pp.)

P.15: The Wheat Strawworm and Its Control. *W. J. Phillips* and *F. W. Poos.* U.S.D.A., F.Bu.1323. (6 pp.)

P.16: Cereal and Forage Insects. *C. M. Packard.* U.S.D.A., Y.'52. (pp. 581–595.)

P.17: Cane *vs.* Wood Molasses Used as Preservatives for Grass Silage. *H. S. Peet* and *A. C. Ragsdale.* U.Mo.—AESt., Bu.605. (7 pp.)

P.18: Controlling Sagebrush on Range Lands. *J. F. Pechanec, G. Stewart, A. P. Plummer, J. H. Robertson,* and *A. C. Hull, Jr.* U.S.D.A., F.Bu.2072. (36 pp.)

P.19: Sagebrush Burning—Good and Bad. *J. F. Pechanec, G. Stewart,* and *L. F. Watts.* U.S.D.A., F.Bu.1948. (32 pp.)

P.20: Grass and Alfalfa as Silage Forage and Meal for Poultry. *L. F. Payne* and *C. L. Gish.* Kans.StCol.—AESt., Bu.320. (46 pp.)

P.21: Pastures for Growing Pullets. *J. E. Parker* and *B. J. McSpadden.* U.Tenn.—AESt., Bu.188. (14 pp.)

P.22: Grass as a Savings Account. *R. Price, K. W. Parker,* and *A. C. Hull, Jr.* U.S.D.A. Y.'48. (pp. 560–564.)

P.23: Bindweed—How to Control It. *W. M. Phillips* and *F. L. Timmons.* Kans.StCol.—AESt., Bu.366. (40 pp.)

Q.1: Growing Buckwheat. *K. S. Quisenberry* and *J. W. Taylor.* U.S.D.A., F.Bu.1835. (18 pp.)

Q.2: The Forage Insect Problem. *R. J. Quinton.* U.Conn.—AESt., Cr.197. (14 pp.)

R.1: Reseeding Southwestern Range Land with Crested Wheatgrass. *H. G. Reynolds* and *H. W. Springfield.* U.S.D.A., F.Bu. 2056. (20 pp.)

R.2: Russian Wild-Rye. *G. A. Rogler.* U.S.D.A., Lf.313. (8 pp.)

R.3: Growing Subclover in Oregon. *H. H. Rampton.* Oreg.StCol.—AESt., St.Bu.432. (12 pp.)

R.4: Vitamins in Poultry Feeding. *L. L. Rusoff,* U.Fla.—AESt., P.Bu.543. (4 pp.)

R.5: Feeding Tests with Mungbean Forage and Seed in Dairy Rations. *M. Ronning, A. H. Kuhlman, H. W. Cave,* and *W. D. Gallup.* Okla.A&MCol.—AESt., St.Bu.B-403. (8 pp.)

R.6: Stink Bugs on Seed Alfalfa in Southern Arizona. *E. E. Russell.* U.S.D.A., Cr.903. (19 pp.)

R.7: Oats in Nebraska. *L. P. Reitz.* U.Nebr.—AESt., Bu.408. (44 pp.)

R.8: The What and How of Hybrid Corn. *F. D. Richey.* U.S.D.A., F.Bu.1744. (13 pp.)

R.9: Varieties of Hard Red Winter Wheat in the United States. *L. P. Reitz* and *C. O. Johnston.* U.S.D.A., Cr.938. (24 pp.)

R.10: Urea and Molasses in Rations for Fattening Lambs. *R. M. Richard, M. R. Light, D. W. Bolin, W. E. Dinusson,* and *M. L. Buchanan.* N.Dak.U.—AESt., "Bymonthly Bu.," Vol. 17, No. 1. (pp. 40–45.)

R.11: The Value of Prairie Hay for Milk Production. *M. Ronning* and *A. H. Kuhlman.* Okla.A&MCol.—AESt., Bu.B-423. (4 pp.)

R.12: Pasture for Dairy Cattle. *D. B. Roark, J. W. Lusk, J. T. Miles, W. C. Cowsert,* and *R. E. Waters.* Miss.StCol.—AESt., Bu.507. (18 pp.)

S.1: The Status of Crested Wheatgrass. *J. R. Swallen* and *G. A. Rogler.* "Agr.Jrl.," Vol. 42, No. 11. (p. 571.)

S.2: The Ryegrasses. *H. A. Schoth* and *M. A. Hein.* U.S.D.A., Lf.196. (8 pp.)

S.3: Grain and Forage Sorghums for Kansas. *A. F. Swanson* and *H. H. Laude.* Kans. StCol.—AESt., Bu.349. (56 pp.)

S.4: Sorghums for Kansas. *A. F. Swanson* and *H. H. Laude.* Kans.StCol.—AESt., Bu.304. (63 pp.)

S.5: Give Legumes a Chance. *B. R. Spears.* Tex.A&MCol.—AExtS., P.Bu.53-36-251. (3 pp.)

S.6: Ladino Clover, Its Mineral Requirements and Chemical Composition. *I. Stewart* and *F. E. Bear.* RutgersU.—N.J.AESt., Bu.759. (32 pp.)

S.7: Winter Legume Cover Crops for the Coastal Plain of Georgia. *J. L. Stephens.* U.Ga.—AESt., Bu.23. (47 pp.)

S.8: Sirup Sorghum Varieties in Oklahoma Sorghums Performance Tests. *J. B. Sieglinger, F. F. Davies,* and *J. E. Webster.* Okla. A&MCol.—AESt., Bu.B-340. (19 pp.)

S.9: Italian Rye Grass for Pastures. *O. E. Sell.* U.Ga.—AESt., P.Bu.513 Rev. (2 pp.)

S.10: Livestock Health Encyclopedia (based on official publications). *R. Seiden.* Springer Publishing Co., New York. 1951. (614 pp.)

S.11: Poultry Handbook (based on official publications). *R. Seiden.* D. Van Nostrand, Princeton, N.J. 2nd ed. 1952. (444 pp.)

S.12: Diseases of Alfalfa in Nevada and Their Influence on Choice of Varieties. *O. F. Smith.* U.Nev.—AESt., Bu.182. (28 pp.)

S.13: Root Rots, Wilts, and Blights of Peas. *W. T. Schroeder.* U.S.D.A., Y.'53. (pp. 401–408.)

S.14: Feeding Peanut Meal and Hay. *W. J. Sheely, R. B. Becker, N. R. Mehrhof,* and *H. L. Brown.* U.Fla.—AExtS., Bu.115. (15 pp.)

S.15: Summer Legume Cover and Forage Crops for the Coastal Plains of Georgia. *J. L. Stephens.* U.Ga.—Coastal Plain AESt., Bu.50. (23 pp.)

S.16: Grow Disease-Resistant Oats. *T. R. Stanton* and *F. A. Coffman.* U.S.D.A., F.Bu.1941. (13 pp.)

S.17: Spring-Sown Red Oats. *T. R. Stanton* and *F. A. Coffman.* U.S.D.A., F.Bu.1583. (18 pp.)

S.18: Oats in the Western Half of the United States. *T. R. Stanton* and *F. A. Coffman.* U.S.D.A., F.Bu.1611. (22 pp.)

S.19: Oats in the Northeastern States. *T. R. Stanton* and *F. A. Coffman.* U.S.D.A., F.Bu.1659. (17 pp.)

S.20: Fall-Sown Oat Production. *T. R. Stanton* and *F. A. Coffman.* U.S.D.A., F.Bu.1640. (20 pp.)

S.21: Winter Oats for the South. *T. R. Stanton* and *F. A. Coffman.* U.S.D.A., F.Bu.2037. (19 pp.)

S.22: Winter Wheat for Pasture in Kansas. *A. F. Swanson* and *K. L. Anderson.* Kans. StCol.—AESt., Bu.345. (32 pp.)

S.23: Silage, Silage Crops, and Silos. *F. L. Smith* and *L. L. Davis.* U.Calif.—AESt., Cr.411. (27 pp.)

S.24: Ensiling Hay and Pasture Crops. *J. B. Shepherd, R. E. Hodgson, N. R. Ellis,* and *J. R. McCalmont.* U.S.D.A., Y.'48. (pp. 178–190.)

S.25: Developments and Problems in Making Grass Silage. *J. B. Shepherd, C. H. Gordon,* and *L. E. Campbell.* U.S.D.A., B.D.I.-Inf.149. (19 pp.)

S.26: Grassland Farming. *L. E. Spence.* U.Idaho.—AExtS., Ext.Bu.195. (8 pp.)

S.27: Vergleichende Untersuchungen ueber den Einfluss verschiedener aeusserer Faktoren insbesondere auf den Aschengehalt in den Pflanzen. *R. Seiden.* Paul Parey, Berlin. 1925. (50 pp.)

S.28: Livestock Ailments Related to Pasture. *J. W. Scales.* Miss.StCol.—AESt., Inf.Sh. 443. (2 pp.)

S.29: Management of Grasslands in the Northeastern States. *V. G. Sprague, R. R. Robinson,* and *R. J. Garber.* PaStCol.—AESt., Bu.554. (30 pp.)

S.30: Range Management. *D. A. Savage* and *D. F. Costello.* U.S.D.A., Y.'48. (pp. 522–537.)

S.31: How to Control Billbugs. *A. F. Satterthwait.* U.S.D.A., F.Bu.1003. (22 pp.)

S.32: New Hay-Pasture Crops. *L. H. Smith.* U.Vt.—AExtS., Cr.116. (19 pp.)

S.33: The Digestive Processes in Domestic Animals. *H. W. Schoening.* U.S.D.A., Y.'39. (pp. 418–430.)

T.1: Clovers for Minnesota. *H. L. Thomas, E. R. Duncan, M. F. Kernkamp, A. G. Peterson,* and *A. R. Schmid.* U.Minn.—AESt., Bu.415. (27 pp.)

T.2: Diseases of Rice. *E. C. Tullis.* U.S.D.A., F.Bu.1854. (16 pp.)

T.3: Herbs and Other Special Crops. *C. A. Thomas.* U.S.D.A., Y.'53. (pp. 863–868.)

U.1: Kenland Red Clover. (No author.) U.S.D.A., B.P.I., Mim.Sh.1951. (1 p.)

U.2: Control of the Alfalfa Caterpillar. (No author.) U.S.D.A., Div. Cereal and Forage Insect Investigation. B.E.P.Q., Pu.1952. (5 pp.)

U.3: Some Important Insects. (No author.) U.S.D.A., Y.'52, Plates LXII, LXIII, LXIV.

U.4: Cutworms in the Garden. (No author.) U.S.D.A., Div. Truck Crop and Garden Insect Investigations. B.E.P.Q., Home and Garden Bu.29. (4 pp.)

U.5: The Velvetbean Caterpillar. (No author.) U.S.D.A., Div. Cereal and Forage Insect Investigations, B.E.P.Q., Lf.348. (4 pp.)

U.6: Range Plant Handbook. (No author.) U.S.D.A., For.S., Pu.1937.

U.7: Leaf Blights of Corn. *A. J. Ullstrup.* PurdueU.—AESt., St.Bu.572. (24 pp.)

U.8: Corn and Its Uses as Food. (No author.) U.S.D.A., F.Bu.1236. (22 pp.)

U.9: How to Get More Milk from Your Pastures. (No author.) U.S.D.A., B.D.I., Pu.PA-47. (4 pp.)

U.10: Kill Barberry Bushes That Spread Stem Rust to Grains. (No author.) U.S.D.A., Div. Plant Disease Control, B.E.P.Q., F.Bu.2014. (11 pp.)

U.11: Handbook of Official Grain Standards of the United States. (No author.) U.S.D.A. Production & Marketing Adm., Grain Branch, Pu.1954. (102 pp.)

U.12: Barberry Bushes Spread Stem Rust. (No author.) U.S.D.A., B.E.P.Q., Lf.315. (6 pp.)

U.13: Insects. *A. Stefferud.* U.S.D.A., Y.'52. (780 pp.)

V.1: Sudan Grass. *H. N. Vinall.* U.S.D.A., F.Bu.1128. (23 pp.)

V.2: Identification, History, and Distribution of Common Sorghum Varieties. *H. N. Vinall, J. C. Stephens,* and *J. H. Martin.* U.S.D.A., Tech.Bu.506. (102 pp.)

W.1: Crested Wheatgrass. *H. L. Westover* and *G. A. Rogler.* U.S.D.A., Lf.104. (8 pp.)

W.2: Grain Sorghums for South Carolina. *H. A. Woodle* and *W. H. Craven.* Clemson A.Col.—AExtS., Cr.285. (8 pp.)

W.3: Legume Seed Yields Increased by Borax Fertilization. *L. N. Wise.* Miss.StCol.—AESt., Inf.Sh.475. (1 p.)

W.4: Clover Variety Tests at Kirbyville and Cleveland, 1950–52. *J. R. Wood, C. A. Burleson, E. D. Cook,* and *R. P. Bates.* Tex.A&MCol.—AESt., Prog.Rep.1497. (5 pp.)

W.5: Effect of Time of Cutting Red Clover on Forage Yields, Seed Setting, and Chemical Composition. *C. P. Wilsie* and *E. A. Hollowell.* Iowa StCol.—AESt., Res.Bu. 357. (32 pp.)

W.6: Lespedeza Anthracnose. *J. L. Weimer.* Ga.AESt., Jrl.Ser., Pap.143. (10 pp.)

W.7: Sericea and Other Perennial Lespedezas. *H. A. Woodle.* ClemsonA.Col.—AExtS., Cr.369. (8 pp.)

W.8: Diseases of Cultivated Lupines in the Southeast. *J. L. Weimer.* U.S.D.A., F.Bu. 2053. (18 pp.)

W.9: Lupines Anthracnose. *J. L. Weimer.* U.S.D.A., Cr.904. (17 pp.)

W.10: Alfalfa Varieties in Iowa. *C. P. Wilsie.* IowaStCol.—AESt., Bu.P.111. (20 pp.)

W.11: Chalcid Control in Alfalfa-Seed Production. *V. L. Wildermuth.* U.S.D.A., F.Bu.1642. (14 pp.)

W.12: The Armyworm and Its Control. *W. R. Walton* and *C. M. Packard.* U.S.D.A., F.Bu.1850. (10 pp.)

W.13: Big Trefoil—New Pasture Legume for Florida. *A. T. Wallace* and *G. B. Killinger.* U.Fla.—AESt., Cr.S-49. (6 pp.)

W.14: Soybeans. *H. A. Woodle.* Clemson A.Col.—AExtS., Cr.370. (8 pp.)

W.15: Legumes in the South. *J. L. Weimer* and *J. L. Allison.* U.S.D.A., Y.'53. (pp. 248–253.)

W.16: Grasses Introduced into the United States. *F. C. Weintraub.* U.S.D.A., A. Handbook 58. (79 pp.)

W.17: Control of Insects and Diseases of Peanuts. *C. Wilson* and *F. S. Arant.* Ala. Pol.Inst.—AESt., Lf.27. (4 pp.)

W.18: Preventing the Diseases of Peanuts. *C. Wilson.* U.S.D.A., Y.'53. (pp. 448–454.)

W.19: The Uses of Alfalfa. *H. L. Westover* and *W. H. Hosterman.* U.S.D.A., F.Bu. 1839. (36 pp.)

W.20: Winter Barley in West Virginia. *R. O. Weibel.* W.Va.U.—AESt., Bu.314. (27 pp.)

W.21: Barley in Nebraska. *O. J. Webster.* U.Nebr.—AESt., Bu.423. (35 pp.)

W.22: Cutworms and Their Control in Corn and Other Cereal Crops. *W. R. Walton.* U.S.D.A., F.Bu.739. (7 pp.)

W.23: Small Grains in South Carolina. *H. A. Woodle* and *W. H. Craven.* ClemsonA.Col. —AExtS., Cr.292. (22 pp.)

W.24: Feeding Molasses to Livestock. *G. L. Walker.* U.S.D.A., Lf.352. (8 pp.)

W.25: The Making and Feeding of Silage. *T. E. Woodward, W. H. Black, D. A. Spencer,* and *J. O. Williams.* U.S.D.A., F.Bu.578. (30 pp.)

W.26: Pastures on the Dairy Farm. *R. E. Wagner* and *J. B. Shepherd.* U.S.D.A., Y.'48. (pp. 127–134.)

W.27: Progress of Grassland Farming in South Carolina. *H. A. Woodle, E. C. Turner,* and *W. H. Craven.* ClemsonA.Col. —AExtS., Cr.373. (28 pp.)

W.28: A Billion Acres of Grassland. *H. H. Wooten* and *G. P. Barnes.* U.S.D.A., Y.'48. (pp. 25–34.)

W.29: Pastures for Hogs. *L. A. Weaver.* U.Mo. —AESt., Bu.247. (39 pp.)

W.30: The Hohenheim System in the Management of Permanent Pastures for Dairy Cattle. *T. E. Woodward, J. B. Shepherd,* and *M. A. Hein.* U.S.D.A., Tech.Bu.660. (33 pp.)

W.31: South Carolina Pastures. *H. A. Woodle* and *E. C. Turner.* ClemsonA.Col.—AExtS., Bu.115. (30 pp.)

W.32: Five Steps in Pasture Improvement. *E. D. Walker* and *J. C. Hackleman.* U.Ill.— AExtS., Cr.703. (16 pp.)

W.33: Almond Hulls as Feed. *W. C. Weir.* U.Calif. "Calif.A." Sept. 1951. (p. 31.)

W.34: Range Plants Poisonous to Livestock in Montana. *H. Welch* and *H. E. Morris.* Mont.StCol.—AESt., Cr.197. (35 pp.)

W.35: Poisonous Plants in Florida. *E. West* and *M. W. Emmel.* U.Fla.—AESt., Bu.510. (57 pp.)

W.36: Coastal Bermuda. *H. A. Woodle.* Clemson A.Col.—AExtS., Cr.374. (8 pp.)

KEY TO ABBREVIATIONS

These abbreviations appear in "Authors of Publications" and "Bibliography."

A.—*Agricultural, Agriculture*
ACol.—*Agricultural College*
Adv.—*Advisor*
AESt.—*Agricultural Experiment Station*
AExtS.—*Agricultural Extension Service*
Agr.—*Agronomist, Agronomy*
A&MCol.—*Agricultural and Mechanical College*
An.—*Animal*
Anal.—*Analyst*
Art.—*Article*
Assoc.—*Associate*
Asst.—*Assistant*
Bac.—*Bacteriologist*
B.A.I.—*Bureau of Animal Industry (U.S.D.A.)**
B.D.I.—*Bureau of Dairy Industry (U.S.D.A.)**
B.E.P.Q.—*Bureau of Entomology and Plant Quarantine (U.S.D.A.)**
Biol.—*Biologist, Biology*
Bot.—*Botanist, Botany*
B.P.I.—*Bureau of Plant Industry, Soils, and Agricultural Engineering (U.S.D.A.)**
Bu.—*Bulletin*
Chem.—*Chemist, Chemistry*
Col.—*College*
Collab.—*Collaborator*
Conserv.—*Conservator, Conservationist*
Conserv.S.—*Conservation Service*
Cr.—*Circular*
Dair.—*Dairy, Dairyman, Dairying*
Dep.—*Department (of); see also U.S.D.A.*
Dir.—*Director*
Div.—*Division (of)*
Ecol.—*Ecologist*
Econ.—*Economist, Economics*
Eng.—*Engineer, Engineering*
Ent.—*Entomology, Entomologist; see also B.E.P.Q.*
ESt.—*Experiment Station*
Ext.—*Extension*
F.Bu.—*Farmers' Bulletin*
F.D.A.—*Food and Drug Administration*
For.—*Forest, Forestry*
Genet.—*Geneticist*
Hort.—*Horticulturist, Horticulture*
Husb.—*Husbandry, Husbandman*
Inf.—*Information*

Inst.—*Institute (of)*
Instr.—*Instructor (in)*
Inv.—*Investigator, Investigation*
Jrl.—*Journal*
Jun.—*Junior*
Lab.—*Laboratory*
Lf.—*Leaflet*
M.—*Miscellaneous*
Mang.—*Management, Managing*
Mim.—*Mimeographed*
No.—*Number*
Nutr.—*Nutritionist, Nutrition*
Off.—*Official*
P.—*Press*
Pap.—*Paper*
Paras.—*Parasitologist, Parasitology*
Path.—*Pathologist, Pathological, Pathology*
Pl.—*Plant*
Pol.—*Polytechnic*
Princ.—*Principal*
Prod.—*Product(s), Production*
Prof.—*Professor*
Prog.—*Progress*
Pu.—*Publication*
Reg.—*Region, Regional*
Rep.—*Report*
Res.—*Research*
S.—*Service*
Sci.—*Science, Scientific*
Sen.—*Senior*
Ser.—*Series*
Sh.—*Sheet*
Spec.—*Specialist (in), Special*
St.—*Station, State; see also StCol., StU., AESt.*
StCol.—*State College (of)*
StU.—*State University (of)*
Subst.—*Substation*
Sup.—*Superintendent*
Tech.—*Technician, Technical*
U.—*University (of)*
U.S.D.A.—*U. S. Department of Agriculture*
Vet.—*Veterinary, Veterinarian*
Vol.—*Volume*
Y.—*Yearbook of Agriculture, U.S.D.A. (Y'48 = 1948 Yearbook)*
Zool.—*Zoologist, Zoology*

---

* These Bureaus no longer exist; they are now part of the *Agricultural Research Service* of the U.S.D.A.